T0206207

Graduate Texts in Mathematics 265

Graduate Texts in Mathematics

Graduate Texts in Mathematics bridge the gap between passive study and creative understanding, offering graduate-level introductions to advanced topics in mathematics. The volumes are carefully written as teaching aids and highlight characteristic features of the theory. Although these books are frequently used as textbooks in graduate courses, they are also suitable for individual study.

For further volumes:
www.springer.com/series/136

Konrad Schmüdgen

Unbounded
Self-adjoint Operators
on Hilbert Space

 Springer

Konrad Schmüdgen
Dept. of Mathematics
University of Leipzig
Leipzig, Germany

ISSN 0072-5285   Graduate Texts in Mathematics
ISBN 978-94-007-9741-3   ISBN 978-94-007-4753-1 (eBook)
DOI 10.1007/978-94-007-4753-1
Springer Dordrecht Heidelberg New York London

Mathematics Subject Classification: 47B25, 81Q10 (Primary), 47A, 47E05, 47F05, 35J (Secondary)

Printed on acid-free paper

Springer is part of Springer Science+Business Media (www.springer.com)

To ELISA

Preface and Overview

This book is designed as an advanced text on unbounded self-adjoint operators in Hilbert space and their spectral theory, with an emphasis on applications in mathematical physics and various fields of mathematics. Though in several sections other classes of unbounded operators (normal operators, symmetric operators, accretive operators, sectorial operators) appear in a natural way and are developed, the leitmotif is the class of *unbounded self-adjoint operators*.

Before we turn to the aims and the contents of this book, we briefly explain the two main notions occurring therein. Suppose that $\mathcal{H}$ is a Hilbert space with scalar product $\langle \cdot, \cdot \rangle$ and T is a linear operator on $\mathcal{H}$ defined on a dense linear subspace $\mathcal{D}(T)$. Then T is said to be *it symmetric* if

$$\langle Tx, y \rangle = \langle x, Ty \rangle \quad \text{for } x, y \in \mathcal{D}(T). \tag{0.1}$$

The operator T is called *self-adjoint* if it is symmetric and if the following property is satisfied: Suppose that $y \in \mathcal{H}$ and there exists a vector $u \in \mathcal{H}$ such that $\langle Tx, y \rangle = \langle x, u \rangle$ for all $x \in \mathcal{D}(T)$. Then y lies in $\mathcal{D}(T)$. (Since $\mathcal{D}(T)$ is assumed to be dense in $\mathcal{H}$, it follows then that $u = Ty$.)

Usually it is easy to verify that Eq. (0.1) holds, so that the corresponding operator is symmetric. For instance, if Ω is an open bounded subset of $\mathbb{R}^d$ and T is the Laplacian Δ on $\mathcal{D}(T) = C_0^\infty(\Omega)$ in the Hilbert space $L^2(\Omega)$, a simple integration-by-parts computation yields (0.1). If the symmetric operator is bounded, its continuous extension to $\mathcal{H}$ is self-adjoint. However, for unbounded operators, it is often difficult to prove or not true (as in the case $T = \Delta$) that a symmetric operator is self-adjoint. Differential operators and most operators occurring in mathematical physics are *not bounded*. Dealing with unbounded operators leads not only to many technical subtleties; it often requires the development of new methods or the invention of new concepts.

Self-adjoint operators are fundamental objects in mathematics and in quantum physics. The spectral theorem states that any self-adjoint operator T has an integral representation $T = \int \lambda \, dE(\lambda)$ with respect to some unique spectral measure E. This gives the possibility to define functions $f(T) = \int f(\lambda) \, dE(\lambda)$ of the operator and to develop a functional calculus as a powerful tool for applications. The spectrum of a self-adjoint operator is always a subset of the reals. In quantum physics it is postulated that each observable is given by a self-adjoint operator T. The spectrum of T is then the set of possible measured values of the observable, and for any unit vector $x \in \mathcal{H}$ and subset $M \subseteq \mathbb{R}$, the number $\langle E(M)x, x \rangle$ is the probability that the measured value in the state x lies in the set M. If T is the Hamilton operator, the one-parameter unitary group $t \to e^{itT}$ describes the quantum dynamics.

All this requires the operator T to be self-adjoint. For general symmetric operators T, the spectrum is no longer a subset of the reals, and it is impossible to get an integral representation $T = \int \lambda \, dE(\lambda)$ or to define the exponentiation e^{itT}. That is, the distinction between symmetric operators and self-adjoint operators is crucial! However, many symmetric operators that are not self-adjoint can be extended to a self-adjoint operator acting on the same Hilbert space.

The main aims of this book are the following:

- to provide a detailed study of unbounded self-adjoint operators and their properties,
- to develop methods for proving the self-adjointness of symmetric operators,
- to study and describe self-adjoint extensions of symmetric operators.

A particular focus and careful consideration is on the technical subtleties and difficulties that arise when dealing with unbounded operators.

Let us give an overview of the contents of the book. Part I is concerned with the basics of unbounded closed operators on a Hilbert space. These include fundamental general concepts such as regular points, defect numbers, spectrum and resolvent, and classes of operators such as symmetric and self-adjoint operators, accretive and sectorial operators, and normal operators.

Our first main goal is the theory of spectral integrals and the spectral decomposition of self-adjoint and normal operators, which is treated in detail in Part II. We use the bounded transform to reduce the case of unbounded operators to bounded ones and derive the spectral theorem in great generality for finitely many strongly commuting unbounded normal operators. The functional calculus for self-adjoint operators developed here will be will be essential for the remainder of the book.

Part III deals with generators of one-parameter groups and semigroups, as well as with a number of important and technical topics including the polar decomposition, quasi-analytic and analytic vectors, and tensor products of unbounded operators.

The second main theme of the book, addressed in Part IV, is perturbations of self-adjointness and of spectra of self-adjoint operators. The Kato–Rellich theorem, the invariance of the essential spectrum under compact perturbations, the Aronszajn–Donoghue theory of rank one perturbations and Krein's spectral shift and trace formula are treated therein. A guiding motivation for many results in the book, and in this part in particular, are applications to Schrödinger operators arising in quantum mechanics.

Part V contains a detailed and concise presentation of the theory of forms and their associated operators. This is the third main theme of the book. Here the central results are three representation theorems for closed forms, one for lower semibounded Hermitian forms and two others for bounded coercive forms and for sectorial forms. Other topics treated include the Friedrichs extension, the order relation of self-adjoint operators, and the min–max principle. The results on forms are applied to the study of differential operators. The Dirichlet and Neumann Laplacians on bounded open subsets of $\mathbb{R}^d$ and Weyl's asymptotic formula for the eigenvalues of the Dirichlet Laplacian are developed in detail.

The fourth major main theme of the book, featured in Part VI, is the self-adjoint extension theory of symmetric operators. First, von Neumann's theory of self-adjoint extensions, and Krein's theory and the Ando–Nishio theorem on positive self-adjoint extensions are investigated. The second chapter in Part VI gives an extensive presentation of the theory of boundary triplets. The Krein–Naimark resolvent formula and the Krein–Birman–Vishik theory on positive self-adjoint extensions are treated in this context. The two last chapters of Part VI are concerned with two important topics where self-adjointness and self-adjoint extensions play a crucial role. These are Sturm–Liouville operators and the Hamburger moment problem on the real line.

Throughout the book applications to Schrödinger operators and differential operators are our guiding motivation, and while a number of special operator-theoretic results on these operators are presented, it is worth stating that this is not a research monograph on such operators. Again, the emphasis is on the *general theory of unbounded self-adjoint Hilbert space operators*. Consequently, basic definitions and facts on such topics as Sobolev spaces are collected in an appendix; whenever they are needed for applications to differential operators, they are taken for granted.

This book is an outgrowth of courses on various topics around the theory of unbounded self-adjoint operators and their applications, given for graduate and Ph.D. students over the past several decades at the University of Leipzig. Some of these covered advanced topics, where the material was mainly to be found in research papers and monographs, with any suitable advanced text notably missing. Most chapters of this book are drawn from these lectures. I have tried to keep different parts of the book as independent as possible, with only one exception: The functional calculus for self-adjoint operators developed in Sect. 5.3 is used as an essential tool throughout.

The book contains a number of important subjects (Krein's spectral shift, boundary triplets, the theory of positive self-adjoint extensions, and others) and technical topics (the tensor product of unbounded operators, commutativity of self-adjoint operators, the bounded transform, Aronzajn–Donoghue theory) which are rarely if ever presented in text books. It is particularly hoped that the material presented will be found to be useful for graduate students and young researchers in mathematics and mathematical physics.

Advanced courses on unbounded self-adjoint operators can be built on this book. One should probably start with the general theory of closed operators by presenting the core material of Sects. 1.1, 1.2, 1.3, 2.1, 2.2, 2.3, 3.1, and 3.2. This could be followed by spectral integrals and the spectral theorem for unbounded self-adjoint operators based on selected material from Sects. 4.1, 4.2, 4.3 and 5.2, 5.3, avoiding technical subtleties. There are many possibilities to continue. One could choose relatively bounded perturbations and Schrödinger operators (Chap. 8), or positive form and their applications (Chap. 10), or unitary groups (Sect. 6.1), or von Neumann's extension theory (Sects. 13.1, 13.2), or linear relations (Sect. 14.1) and boundary triplets (Chap. 14). A large number of special topics treated in the book could be used as a part of an advanced course or a seminar.

The prerequisites for this book are the basics in functional analysis and of the theory of bounded Hilbert space operators as covered by a standard one semester course on functional analysis, together with a good working knowledge of measure theory. The applications on differential operators require some knowledge of ordinary and partial differential equations and of Sobolev spaces. In Chaps. 9 and 16 a few selected results from complex analysis are also needed. For the convenience of the reader, we have added six appendices; on bounded operators and classes of compact operators, on measure theory, on the Fourier transform, on Sobolev spaces, on absolutely continuous functions, and on Stieltjes transforms and Nevanlinna functions. These collect a number of special results that are used at various places in the text. For the results, here we have provided either precise references to standard works or complete proofs.

A few general notations that are repeatedly used are listed after the table of contents. A more detailed symbol index can be found at the end of the book. Occasionally, I have used either simplified or overlapping notations, and while this might at first sight seem careless, the meaning will be always clear from the context. Thus the symbol x denotes a Hilbert space vector in one section and a real variable or even the function $f(x) = x$ in others.

A special feature of the book is the inclusion of numerous examples which are developed in detail and which are accompanied by exercises at the ends of the chapters. A number of simple standard examples (for instance, multiplication operators or differential operators $-i\frac{d}{dx}$ or $-\frac{d^2}{dx^2}$ on intervals with different boundary conditions) are guiding examples and appear repeatedly throughout. They are used to illustrate various methods of constructing self-adjoint operators, as well as new notions even within the advanced chapters. The reader might also consider some examples as exercises with solutions and try to prove the statements therein first by himself, comparing the results with the given proofs. Often statements of exercises provide additional information concerning the theory. The reader who is interested in acquiring an ability to work with unbounded operators is of course encouraged to work through as many of the examples and exercises as possible. I have marked the somewhat more difficult exercises with a star. All stated exercises (with the possible exception of a few starred problems) are really solvable by students, as can be attested to by my experience in teaching this material. The hints added to exercises always contain key tricks or steps for the solutions.

In the course of teaching this subject and writing this book, I have benefited from many excellent sources. I should mention the volumes by Reed and Simon [RS1, RS2, RS4], Kato's monograph [Ka], and the texts (in alphabetic order) [AG, BSU, BEH, BS, D1, D2, DS, EE, RN, Yf, We]. I have made no attempt to give precise credit for a result, an idea or a proof, though the few names of standard theorems are stated in the body of the text. The notes at the end of each part contain some (certainly incomplete) information about a few sources, important papers and monographs in this area and hints for additional reading. Also I have listed a number of key pioneering papers. I felt it might be of interest for the reader to look at some of these papers and to observe, for instance, that H. Weyl's work around 1909–1911 contains fundamental and deep results about Hilbert space operators, while the corresponding general theory was only developed some 20 years later!

I am deeply indebted to Mr. René Gebhardt who read large parts of this book carefully and made many valuable suggestions. Finally, I would like to thank D. Dubray and K. Zimmermann for reading parts of the manuscript and Prof. M.A. Dritschel and Dr. Y. Savchuk for their help in preparing this book.

Leipzig   Konrad Schmüdgen
December 2011

Contents

General Notation

$\mathbb{N}_0$: set of nonnegative integers,
$\mathbb{N}$: set of positive integers,
$\mathbb{Z}$: set of integers,
$\mathbb{R}$: set of real numbers,
$\mathbb{C}$: set of complex numbers,
$\mathbb{T}$: set of complex numbers of modulus one,
i: complex unit,
χ_M: characteristic function of a set M.

For $\alpha = (\alpha_1, \ldots, \alpha_d) \in \mathbb{N}_0^d$, $k = 1, \ldots, d$, and $x = (x_1, \ldots, x_d) \in \mathbb{R}^d$, we set

$$x^\alpha := x_1^{\alpha_1} \cdots x_d^{\alpha_d}, \quad |\alpha| := \alpha_1 + \cdots + \alpha_d,$$

$$\partial_k := \frac{\partial}{\partial x_k}, \qquad D_k := -\mathrm{i}\frac{\partial}{\partial x_k},$$

$$\partial^\alpha := \partial_1^{\alpha_1} \cdots \partial_d^{\alpha_d} = \frac{\partial^{\alpha_1}}{\partial x_1^{\alpha_1}} \cdots \frac{\partial^{\alpha_d}}{\partial x_d^{\alpha_d}},$$

$$D^\alpha := D_1^{\alpha_1} \cdots D_d^{\alpha_d} = (-\mathrm{i})^{|\alpha|} \frac{\partial^{\alpha_1}}{\partial x_1^{\alpha_1}} \cdots \frac{\partial^{\alpha_d}}{\partial x_d^{\alpha_d}}$$

with the convention that terms with $\alpha_j = 0$ are set equal to one.

Sequences are written by round brackets such as $(x_n)_{\in \mathbb{N}}$ or (x_n), while sets are denoted by braces such as $\{x_i : i \in I\}$. Pairs of elements are written as (x, y).

For an open subset Ω of $\mathbb{R}^d$,

$C^n(\Omega)$ is the set of n times continuously differentiable complex functions on Ω,
$C_0^\infty(\Omega)$ is the set of functions in $C^\infty(\Omega)$ whose support is a compact subset of Ω,
$C^n(\overline{\Omega})$ is the set of functions $f \in C^n(\Omega)$ for which all functions $D^\alpha f$, $|\alpha| \leq n$, admit continuous extensions to the closure $\overline{\Omega}$ of the set Ω in $\mathbb{R}^d$,
$L^p(\Omega)$ is the L^p-space with respect to the Lebesgue measure on Ω.

We write $L^2(a, b)$ for $L^2((a, b))$ and $C^n(a, b)$ for $C^n((a, b))$.
"a.e." means "almost everywhere with respect to the Lebesgue measure."

The symbol $\mathcal{H}$ refers to a Hilbert space with scalar product $\langle\cdot,\cdot\rangle$ and norm $\|\cdot\|$. Scalar products are always denoted by angle brackets $\langle\cdot,\cdot\rangle$. Occasionally, indices or subscripts such as $\langle\cdot,\cdot\rangle_j$ or $\langle\cdot,\cdot\rangle_{H^2(\Omega)}$ are added.

The symbol $\oplus$ stands for the orthogonal sum of Hilbert spaces, while $\dotplus$ means the direct sum of vector spaces. By a projection we mean an orthogonal projection.

$\sigma(T)$: spectrum of T,

$\rho(T)$: resolvent set of T,

$\pi(T)$: regularity domain of T,

$\mathcal{D}(T)$: domain of T,

$\mathcal{N}(T)$: null space of T,

$\mathcal{R}(T)$: range of T,

$R_\lambda(T)$: resolvent $(T-\lambda I)^{-1}$ of T at λ,

E_T: spectral measure of T.

Part I
Basics of Closed Operators

Chapter 1
Closed and Adjoint Operators

Closed operators and closable operators are important classes of unbounded linear operators which are large enough to cover all interesting operators occurring in applications. In this chapter and the next, we develop basic concepts and results about general closed operators on Hilbert space. We define and study closed operators, closable operators, closures, and cores in Sect. 1.1 and adjoint operators in Sect. 1.2, while in Sect. 1.3 these concepts are discussed for differentiation operators on intervals and for linear partial differential operators. Section 1.4 deals with invariant subspaces and reducing subspaces of linear operators.

1.1 Closed and Closable Operators and Their Graphs

1.1.1 General Notions on Linear Operators

Definition 1.1 A *linear operator* from a Hilbert space $\mathcal{H}_1$ into a Hilbert space $\mathcal{H}_2$ is a linear mapping T of a linear subspace of $\mathcal{H}_1$, called the *domain* of T and denoted by $\mathcal{D}(T)$, into $\mathcal{H}_2$.

It should be emphasized that the domain is crucial for an unbounded operator. The same formal expression considered on different domains (for instance, by varying boundary conditions for the same differential expression) may lead to operators with completely different properties. We will see this by the examples in Sect. 1.3.

That T is a linear mapping means that $T(\alpha x + \beta y) = \alpha T(x) + \beta T(y)$ for all $x, y \in \mathcal{D}(T)$ and $\alpha, \beta \in \mathbb{C}$. The linear subspace

$$\mathcal{R}(T) := T\big(\mathcal{D}(T)\big) = \big\{T(x) : x \in \mathcal{D}(T)\big\}$$

of $\mathcal{H}_2$ is called the *range* or the *image* of T, and the linear subspace

$$\mathcal{N}(T) = \big\{x \in \mathcal{D}(T) : T(x) = 0\big\}$$

K. Schmüdgen, *Unbounded Self-adjoint Operators on Hilbert Space*,
Graduate Texts in Mathematics 265,
DOI 10.1007/978-94-007-4753-1_1, © Springer Science+Business Media Dordrecht 2012

of $\mathcal{H}_1$ is the *null space* or the *kernel* of T. We also write Tx for $T(x)$ if no confusion can arise.

By the *restriction* of T to a linear subspace $\mathcal{D}_0$ of $\mathcal{D}(T)$ we mean the linear operator $T\!\upharpoonright\!\mathcal{D}_0$ with domain $\mathcal{D}_0$ acting by $(T\!\upharpoonright\!\mathcal{D}_0)(x) = T(x)$ for $x \in \mathcal{D}_0$.

Let S and T be two linear operators from $\mathcal{H}_1$ into $\mathcal{H}_2$. By definition we have $S = T$ if and only if $\mathcal{D}(S) = \mathcal{D}(T)$ and $S(x) = T(x)$ for all $x \in \mathcal{D}(S) = \mathcal{D}(T)$. We shall say that T is an *extension* of S or that S is a *restriction* of T and write $S \subseteq T$, or equivalently $T \supseteq S$, when $\mathcal{D}(S) \subseteq \mathcal{D}(T)$ and $S(x) = T(x)$ for all $x \in \mathcal{D}(S)$. That is, we have $S \subseteq T$ if and only if $S = T\!\upharpoonright\!\mathcal{D}(S)$.

The *complex multiple* αT for $\alpha \in \mathbb{C}$, $\alpha \neq 0$, and the *sum* $S + T$ are the linear operators from $\mathcal{H}_1$ into $\mathcal{H}_2$ defined by

$$\mathcal{D}(\alpha T) = \mathcal{D}(T), \quad (\alpha T)(x) = \alpha T(x) \quad \text{for } x \in \mathcal{D}(\alpha T), \ \alpha \neq 0,$$

$$\mathcal{D}(S + T) = \mathcal{D}(S) \cap \mathcal{D}(T), \quad (S + T)(x) = S(x) + T(x) \quad \text{for } x \in \mathcal{D}(S + T).$$

Further, we define the multiple αT for $\alpha = 0$ to be the *null operator* 0 from $\mathcal{H}_1$ into $\mathcal{H}_2$; it has the domain $\mathcal{D}(0) = \mathcal{H}_1$ and acts by $0(x) = 0$ for $x \in \mathcal{H}_1$.

If R is a linear operator from $\mathcal{H}_2$ into a Hilbert space $\mathcal{H}_3$, then the *product* RT is the linear operator from $\mathcal{H}_1$ into $\mathcal{H}_3$ given by

$$\mathcal{D}(RT) = \big\{ x \in \mathcal{D}(T) : Tx \in \mathcal{D}(R) \big\}, \quad (RT)(x) = R(Tx) \quad \text{for } x \in \mathcal{D}(RT).$$

It is easily checked that the sum and product of linear operators are associative and that the two distributivity laws

$$(S + T)Q = SQ + TQ \quad \text{and} \quad R(S + T) \supseteq RS + RT \tag{1.1}$$

hold, where Q is a linear operator from a Hilbert space $\mathcal{H}_0$ into $\mathcal{H}_1$; see Exercises 1 and 2.

If $\mathcal{N}(T) = \{0\}$, then the *inverse operator* T^{-1} is the linear operator from $\mathcal{H}_2$ into $\mathcal{H}_1$ defined by $\mathcal{D}(T^{-1}) = \mathcal{R}(T)$ and $T^{-1}(T(x)) = x$ for $x \in \mathcal{D}(T)$. In this case we say that T is *invertible*. Note that $\mathcal{R}(T^{-1}) = \mathcal{D}(T)$. Clearly, if $\mathcal{N}(R) = \{0\}$ and $\mathcal{N}(T) = \{0\}$, then $\mathcal{N}(RT) = \{0\}$ and $(RT)^{-1} = T^{-1}R^{-1}$.

Definition 1.2 The *graph* of a linear operator T from $\mathcal{H}_1$ into $\mathcal{H}_2$ is the set

$$\mathcal{G}(T) = \big\{ (x, Tx) : x \in \mathcal{D}(T) \big\}.$$

The graph $\mathcal{G}(T)$ is a linear subspace of the Hilbert space $\mathcal{H}_1 \oplus \mathcal{H}_2$ which contains the full information about the operator T. Obviously, the relation $S \subseteq T$ is equivalent to the inclusion $\mathcal{G}(S) \subseteq \mathcal{G}(T)$. The next lemma contains an internal characterization of graphs. We omit the simple proof.

Lemma 1.1 *A linear subspace $\mathcal{E}$ of $\mathcal{H}_1 \oplus \mathcal{H}_2$ is the graph of a linear operator T from $\mathcal{H}_1$ into $\mathcal{H}_2$ if and only if $(0, y) \in \mathcal{E}$ for $y \in \mathcal{H}_2$ implies that $y = 0$.*

The operator T is then uniquely determined by $\mathcal{E}$; it acts by $Tx = y$ for $(x, y) \in \mathcal{E}$, and its domain is $\mathcal{D}(T) = \{x \in \mathcal{H}_1 : \text{There exists } y \in \mathcal{H}_2 \text{ such that } (x, y) \in \mathcal{E}\}$.

Often we are dealing with a linear operator that maps a Hilbert space $\mathcal{H}$ into itself. We usually express this by saying that we have an operator *on* $\mathcal{H}$. That is, whenever we speak about *an operator T on a Hilbert space $\mathcal{H}$*, we mean a linear mapping T of a linear subspace $\mathcal{D}(T)$ of $\mathcal{H}$ into $\mathcal{H}$.

Let $\mathcal{H}$ be a Hilbert space. The identity map of $\mathcal{H}$ is denoted by $I_{\mathcal{H}}$ or by I if no confusion is possible. Occasionally, we write λ instead of $\lambda \cdot I$ for $\lambda \in \mathbb{C}$.

Let T be a linear operator on $\mathcal{H}$. Then the powers of T are defined inductively by $T^n := T(T^{n-1})$ for $n \in \mathbb{N}$ and $T^0 := I$. If $p(x) = \alpha_n x^n + \cdots + \alpha_1 x + \alpha_0$ is a complex polynomial, $p(T)$ is defined by $p(T) = \alpha_n T^n + \cdots + \alpha_1 T + \alpha_0 I$. If p_1 and p_2 are complex polynomials, then $p_1(T)p_2(T) = (p_1 p_2)(T)$. In particular, $T^{n+k} = T^n T^k = T^k T^n$ for $k, n \in \mathbb{N}$ and $\mathcal{D}(p(T)) = \mathcal{D}(T^n)$ if p has degree n.

If T is an arbitrary linear operator on $\mathcal{H}$, we have the *polarization identity*

$$4\langle Tx, y \rangle = \langle T(x+y), x+y \rangle - \langle T(x-y), x-y \rangle$$
$$+ \, \mathrm{i}\langle T(x+\mathrm{i}y), x+\mathrm{i}y \rangle - \mathrm{i}\langle T(x-\mathrm{i}y), x-\mathrm{i}y \rangle \qquad (1.2)$$

for $x, y \in \mathcal{D}(T)$. It is proved by computing the right-hand side of (1.2).

Identity (1.2) is very useful and is often applied in this text. An immediate consequence of this identity is the following:

Lemma 1.2 *Let T be a linear operator on $\mathcal{H}$ such that $\mathcal{D}(T)$ is dense in $\mathcal{H}$. If $\langle Tx, x \rangle = 0$ for all $x \in \mathcal{D}(T)$, then $Tx = 0$ for all $x \in \mathcal{D}(T)$.*

The following simple fact will be used several times in this book.

Lemma 1.3 *Let S and T be linear operators such that $S \subseteq T$. If S is surjective and T is injective, then $S = T$.*

Proof Let $x \in \mathcal{D}(T)$. Since S is surjective, there is a $y \in \mathcal{D}(S)$ such that $Tx = Sy$. From $S \subseteq T$ we get $Tx = Ty$, so $x = y$, because T is injective. Hence, we have $x = y \in \mathcal{D}(S)$. Thus, $\mathcal{D}(T) \subseteq \mathcal{D}(S)$, whence $S = T$. $\qquad\square$

Except for the polarization identity and Lemma 1.2, all preceding notions and facts required only the vector space structure. In the next subsection we essentially use the norm and completeness of the underlying Hilbert spaces.

1.1.2 Closed and Closable Operators

Let T be a linear operator from a Hilbert space $\mathcal{H}_1$ into a Hilbert space $\mathcal{H}_2$. Let $\langle \cdot, \cdot \rangle_j$ and $\| \cdot \|_j$ denote the scalar product and norm of $\mathcal{H}_j$. It is easily seen that

$$\langle x, y \rangle_T = \langle x, y \rangle_1 + \langle Tx, Ty \rangle_2, \quad x, y \in \mathcal{D}(T), \qquad (1.3)$$

defines a scalar product on the domain $\mathcal{D}(T)$. The corresponding norm

$$\|x\|_T = \left(\|x\|_1^2 + \|Tx\|_2^2 \right)^{1/2}, \quad x \in \mathcal{D}(T), \qquad (1.4)$$

is called the *graph norm* of the operator T. It is equivalent to the norm

$$\|x\|'_T := \|x\|_1 + \|Tx\|_2, \quad x \in \mathcal{D}(T),$$

on $\mathcal{D}(T)$. Occasionally, the norm $\| \cdot \|'_T$ is called the graph norm of T as well.

Definition 1.3 An operator T is called *closed* if its graph $\mathcal{G}(T)$ is a closed subset of the Hilbert space $\mathcal{H}_1 \oplus \mathcal{H}_2$, and T is called *closable* (or *preclosed*) if there exists a closed linear operator S from $\mathcal{H}_1$ into $\mathcal{H}_2$ such that $T \subseteq S$.

The next propositions contain some slight reformulations of these definitions.

Proposition 1.4 *The following statements are equivalent:*

(i) *T is closed.*
(ii) *If $(x_n)_{n\in\mathbb{N}}$ is a sequence of vectors $x_n \in \mathcal{D}(T)$ such that $\lim_n x_n = x$ in $\mathcal{H}_1$ and $\lim_n T(x_n) = y$ in $\mathcal{H}_2$, then $x \in \mathcal{D}(T)$ and $Tx = y$.*
(iii) *$(\mathcal{D}(T), \| \cdot \|_T)$ is complete, or equivalently, $(\mathcal{D}(T), \langle \cdot, \cdot \rangle_T)$ is a Hilbert space.*

Proof (i) is equivalent to (ii), because (ii) is only a reformulation of the closedness of the graph $\mathcal{G}(T)$ in $\mathcal{H}_1 \oplus \mathcal{H}_2$.

(i) $\leftrightarrow$ (iii): By the definition (1.4) of the graph norm, the map $x \to (x, Tx)$ of $(\mathcal{D}(T), \| \cdot \|_T)$ onto the graph $\mathcal{G}(T)$ equipped with the norm of the direct sum Hilbert space $\mathcal{H}_1 \oplus \mathcal{H}_2$ is isometric. Therefore, $(\mathcal{D}(T), \| \cdot \|_T)$ is complete if and only if $\mathcal{G}(T)$ is complete, or equivalently, if $\mathcal{G}(T)$ is closed in $\mathcal{H}_1 \oplus \mathcal{H}_2$. $\qquad\square$

Proposition 1.5 *The following are equivalent:*

(i) *T is closable.*
(ii) *If $(x_n)_{n\in\mathbb{N}}$ is a sequence of vectors $x_n \in \mathcal{D}(T)$ such that $\lim_n x_n = 0$ in $\mathcal{H}_1$ and $\lim_n T(x_n) = y$ in $\mathcal{H}_2$, then $y = 0$.*
(iii) *The closure of the graph $\mathcal{G}(T)$ is the graph of a linear operator.*

Proof (i) $\to$ (iii): Let S be a closed operator such that $T \subseteq S$. Then $\mathcal{G}(T) \subseteq \mathcal{G}(S)$, and so $\overline{\mathcal{G}(T)} \subseteq \mathcal{G}(S)$. From Lemma 1.1 it follows that $\overline{\mathcal{G}(T)}$ is the graph of an operator.

(iii) $\to$ (i): If $\overline{\mathcal{G}(T)} = \mathcal{G}(S)$ for some operator S, then $T \subseteq S$, and S is closed, because $\mathcal{G}(S)$ is closed.

(ii) $\leftrightarrow$ (iii): Condition (ii) means that $(0, y) \in \overline{\mathcal{G}(T)}$ implies that $y = 0$. Hence, (ii) and (iii) are equivalent by Lemma 1.1. $\qquad\square$

Suppose that T is closable. Let $\overline{T}$ denote the closed linear operator such that

$$\overline{\mathcal{G}(T)} = \mathcal{G}(\overline{T})$$

by Proposition 1.5(iii). By this definition, $\mathcal{D}(\overline{T})$ is the set of vectors $x \in \mathcal{H}_1$ for which there exists a sequence $(x_n)_{n\in\mathbb{N}}$ from $\mathcal{D}(T)$ such that $x = \lim x_n$ in $\mathcal{H}_1$ and $(T(x_n))_{n\in\mathbb{N}}$ converges in $\mathcal{H}_2$. For such a sequence (x_n), we then set

$\overline{T}(x) := \lim_n T(x_n)$. Since $\overline{\mathcal{G}(T)}$ is the graph of an operator, this definition of $\overline{T}(x)$ is independent of the sequence (x_n). Clearly, $\overline{T}$ is the smallest (with respect to the operator inclusion $\subseteq$) closed extension of the closable operator T.

Definition 1.4 The operator $\overline{T}$ is called the *closure* of the closable operator T.

Let us briefly discuss these concepts for continuous operators. Recall that a linear operator T is *continuous* (that is, if $\lim_n x_n = x$ in $\mathcal{H}_1$ for $x_n, x \in \mathcal{D}(T)$, $n \in \mathbb{N}$, then $\lim_n T(x_n) = T(x)$ in $\mathcal{H}_2$) if and only if T is *bounded* (that is, there is a constant $c > 0$ such that $\|T(x)\|_2 \leq c\|x\|_1$ for all $x \in \mathcal{D}(T)$).

If an operator T is continuous, then it is clearly closable by Proposition 1.5(ii). In fact, closability can be considered as a weakening of continuity for linear operators. Let $(x_n)_{n \in \mathbb{N}}$ be a sequence from $\mathcal{D}(T)$ such that $\lim_n x_n = 0$ in $\mathcal{H}_1$. If the operator T is continuous, then $\lim_n T(x_n) = 0$ in $\mathcal{H}_2$. For T being closable, we require only that *if* (!) the sequence $(T(x_n))_{n \in \mathbb{N}}$ converges in $\mathcal{H}_2$, then it has to converge to the "*correct limit*", that is, $\lim_n T(x_n) = 0$.

If T is bounded, the graph norm $\|\cdot\|_T$ on $\mathcal{D}(T)$ is obviously equivalent to the Hilbert space norm $\|\cdot\|_1$. Therefore, a bounded linear operator T is closed if and only if its domain $\mathcal{D}(T)$ is closed in $\mathcal{H}_1$. Conversely, if T is a closed linear operator whose domain $\mathcal{D}(T)$ is *closed* in $\mathcal{H}_1$, then the closed graph theorem implies that T is bounded. In particular, it follows that a closed linear operator which is not bounded cannot be defined on the whole Hilbert space.

Another useful notion is that of a core of an operator.

Definition 1.5 A linear subspace $\mathcal{D}$ of $\mathcal{D}(T)$ is called a *core* for T if $\mathcal{D}$ is dense in $(\mathcal{D}(T), \|\cdot\|_T)$, that is, for each $x \in \mathcal{D}(T)$, there exists a sequence $(x_n)_{n \in \mathbb{N}}$ of vectors $x_n \in \mathcal{D}$ such that $x = \lim_n x_n$ in $\mathcal{H}_1$ and $Tx = \lim_n T(x_n)$ in $\mathcal{H}_2$.

If T is closed, a linear subspace $\mathcal{D}$ of $\mathcal{D}(T)$ is a core for T if and only if T is the closure of its restriction $T \lceil \mathcal{D}$. That is, a closed operator can be restored from its restriction to any core. The advantage of a core is that closed operators are often easier to handle on appropriate cores rather than on full domains.

Example 1.1 (*Nonclosable operators*) Let $\mathcal{D}$ be a linear subspace of a Hilbert space $\mathcal{H}$, and let $e \neq 0$ be a vector of $\mathcal{H}$. Let F be a linear functional on $\mathcal{D}$ which is not continuous in the Hilbert space norm. Define the operator T by $\mathcal{D}(T) = \mathcal{D}$ and $T(x) = F(x)e$ for $x \in \mathcal{D}$.

Statement T *is not closable*.

Proof Since F is not continuous, there exists a sequence $(x_n)_{n \in \mathbb{N}}$ from $\mathcal{D}$ such that $\lim_n x_n = 0$ in $\mathcal{H}$ and $(F(x_n))$ does not converge to zero. By passing to a subsequence if necessary we can assume that there is a constant $c > 0$ such that $|F(x_n)| \geq c$ for all $n \in \mathbb{N}$. Putting $x'_n = F(x_n)^{-1} x_n$, we have $\lim_n x'_n = 0$ and $T(x'_n) = F(x'_n)e = e \neq 0$. Hence, T is not closable by Proposition 1.5(ii). $\qquad\square$

The preceding proof has also shown that $(0, e) \in \overline{\mathcal{G}(T)}$, so $\overline{\mathcal{G}(T)}$ is not the graph of a linear operator.

Explicit examples of discontinuous linear functionals are easily obtained as follows: If $\mathcal{D}$ is the linear span of an orthonormal sequence $(e_n)_{n \in \mathbb{N}}$ of a Hilbert space, define F on $\mathcal{D}$ by $F(e_n) = 1, n \in \mathbb{N}$. If $\mathcal{H} = L^2(\mathbb{R})$ and $\mathcal{D} = C_0^\infty(\mathbb{R})$, define $F(f) = f(0)$ for $f \in \mathcal{D}$.   ○

1.2 Adjoint Operators

In this section the scalar products of the underlying Hilbert spaces are essentially used to define adjoints of densely defined linear operators.

Let $(\mathcal{H}_1, \langle \cdot, \cdot \rangle_1)$ and $(\mathcal{H}_2, \langle \cdot, \cdot \rangle_2)$ be Hilbert spaces. Let T be a linear operator from $\mathcal{H}_1$ into $\mathcal{H}_2$ such that *the domain $\mathcal{D}(T)$ is dense in $\mathcal{H}_1$*. Set

$$\mathcal{D}(T^*) = \{ y \in \mathcal{H}_2 : \text{There exists } u \in \mathcal{H}_1 \text{ such that } \langle Tx, y \rangle_2 = \langle x, u \rangle_1$$
$$\text{for } x \in \mathcal{D}(T) \}.$$

By Riesz' theorem, a vector $y \in \mathcal{H}_2$ belongs to $\mathcal{D}(T^*)$ if and only if the map $x \to \langle Tx, y \rangle_2$ is a continuous linear functional on $(\mathcal{D}(T), \| \cdot \|_1)$, or equivalently, there is a constant $c_y > 0$ such that $|\langle Tx, y \rangle_2| \leq c_y \|x\|_1$ for all $x \in \mathcal{D}(T)$. An explicit description of the set $\mathcal{D}(T^*)$ is in general a very difficult matter.

Since $\mathcal{D}(T)$ is dense in $\mathcal{H}_1$, the vector $u \in \mathcal{H}_1$ satisfying $\langle Tx, y \rangle_2 = \langle x, u \rangle_1$ for all $x \in \mathcal{D}(T)$ is uniquely determined by y. Therefore, setting $T^* y = u$, we obtain a well-defined mapping T^* from $\mathcal{H}_2$ into $\mathcal{H}_1$. It is easily seen that T^* is linear.

Definition 1.6 The linear operator T^* is called the *adjoint operator* of T.

By the preceding definition we have

$$\langle Tx, y \rangle_2 = \langle x, T^* y \rangle_1 \quad \text{for all } x \in \mathcal{D}(T), \; y \in \mathcal{D}(T^*). \tag{1.5}$$

Let T be a densely defined linear operator on $\mathcal{H}$. Then T is called *symmetric* if $T \subseteq T^*$. Further, we say that T is *self-adjoint* if $T = T^*$ and that T is *essentially self-adjoint* if its closure $\overline{T}$ is self-adjoint. These are fundamental notions studied extensively in this book.

The domain $\mathcal{D}(T^*)$ of T^* may be not dense in $\mathcal{H}_2$ as the next example shows. There are even operators T such that $\mathcal{D}(T^*)$ consists of the null vector only.

Example 1.2 (*Example* 1.1 *continued*) Suppose that $\mathcal{D}(T)$ is dense in $\mathcal{H}$. Since the functional F is discontinuous, the map $x \to \langle Tx, y \rangle = F(x) \langle e, y \rangle$ is continuous if and only if $y \perp e$. Hence, $\mathcal{D}(T^*) = e^\perp$ in $\mathcal{H}$ and $T^* y = 0$ for $y \in \mathcal{D}(T^*)$.   ○

Example 1.3 (*Multiplication operators by continuous functions*) Let $\mathcal{J}$ be an interval. For a continuous function φ on $\mathcal{J}$, we define the multiplication operator M_φ on the Hilbert space $L^2(\mathcal{J})$ by

$$(M_\varphi f)(x) = \varphi(x) f(x) \quad \text{for } f \in \mathcal{D}(M_\varphi) := \{ f \in L^2(\mathcal{J}) : \varphi \cdot f \in L^2(\mathcal{J}) \}.$$

Since $\mathcal{D}(M_\varphi)$ contains all continuous functions with compact support, $\mathcal{D}(M_\varphi)$ is dense, so the adjoint operator $(M_\varphi)^*$ exists. We will prove the following:

Statement $(M_\varphi)^* = M_{\overline{\varphi}}$.

Proof From the relation

$$\langle M_\varphi f, g \rangle = \int_{\mathcal{J}} \varphi f \overline{g} \, dx = \int_{\mathcal{J}} f \overline{\overline{\varphi} g} \, dx = \langle f, M_{\overline{\varphi}} g \rangle$$

for $f, g \in \mathcal{D}(M_\varphi) = \mathcal{D}(M_{\overline{\varphi}})$ we conclude that $M_{\overline{\varphi}} \subseteq (M_\varphi)^*$.

To prove the converse inclusion, let $g \in \mathcal{D}((M_\varphi)^*)$ and set $h = (M_\varphi)^* g$. Let χ_K be the characteristic function of a compact subset K of $\mathcal{J}$. For $f \in \mathcal{D}(M_\varphi)$, we have $f \cdot \chi_K \in \mathcal{D}(M_\varphi)$ and $\langle M_\varphi(f \chi_K), g \rangle = \langle \varphi f \chi_K, g \rangle = \langle f \chi_K, h \rangle$, so that

$$\int_{\mathcal{J}} f \chi_K (\varphi \overline{g} - \overline{h}) \, dx = 0.$$

Since $\mathcal{D}(M_\varphi)$ is dense, the element $\chi_K(\varphi \overline{g} - \overline{h})$ of $L^2(\mathcal{J})$ must be zero, so that $\overline{\varphi} g = h$ on K and hence on the whole interval $\mathcal{J}$. That is, we have $g \in \mathcal{D}(M_{\overline{\varphi}})$ and $h = (M_\varphi)^* g = M_{\overline{\varphi}} g$. This completes the proof of the equality $(M_\varphi)^* = M_{\overline{\varphi}}$.   □

The special case where $\varphi(x) = x$ and $\mathcal{J} = \mathbb{R}$ is of particular importance. The operator $Q := M_\varphi = M_x$ is then the position operator of quantum mechanics. By the preceding statement we have $Q = Q^*$, so Q is a self-adjoint operator on $L^2(\mathbb{R})$.

Multiplication operators by measurable functions on general measure spaces will be studied in Sect. 3.4.   ○

We now begin to develop basic properties of adjoint operators.

Proposition 1.6 *Let S and T be linear operators from $\mathcal{H}_1$ into $\mathcal{H}_2$ such that $\mathcal{D}(T)$ is dense in $\mathcal{H}_1$. Then:*

(i)   *T^* is a closed linear operator from $\mathcal{H}_2$ into $\mathcal{H}_1$.*
(ii) *$\mathcal{R}(T)^\perp = \mathcal{N}(T^*)$.*
(iii) *If $\mathcal{D}(T^*)$ is dense in $\mathcal{H}_2$, then $T \subseteq T^{**}$, where $T^{**} := (T^*)^*$.*
(iv) *If $T \subseteq S$, then $S^* \subseteq T^*$.*
(v)   *$(\lambda T)^* = \overline{\lambda} T^*$ for $\lambda \in \mathbb{C}$.*
(vi) *If $\mathcal{D}(T + S)$ is dense in $\mathcal{H}_1$, then $(T + S)^* \supseteq T^* + S^*$.*
(vii) *If S is bounded and $\mathcal{D}(S) = \mathcal{H}_1$, then $(T + S)^* = T^* + S^*$.*

Proof (i): Let $(y_n)_{n \in \mathbb{N}}$ be a sequence from $\mathcal{D}(T^*)$ such that $\lim_n y_n = y$ in $\mathcal{H}_2$ and $\lim_n T^* y_n = v$ in $\mathcal{H}_1$. For $x \in \mathcal{D}(T)$, we then have

$$\langle Tx, y_n \rangle = \langle x, T^* y_n \rangle \to \langle Tx, y \rangle = \langle x, v \rangle,$$

so that $y \in \mathcal{D}(T^*)$ and $v = T^* y$. This proves that T^* is closed.

(ii): By (1.5) it is obvious that $y \in \mathcal{N}(T^*)$ if and only if $\langle Tx, y \rangle_2 = 0$ for all $x \in \mathcal{D}(T)$, or equivalently, $y \in \mathcal{R}(T)^\perp$. That is, $\mathcal{R}(T)^\perp = \mathcal{N}(T^*)$.

(iii)–(vi) follow from the definition of the adjoint operator. We omit the details.

(vii): By (vi) it suffices to prove that $\mathcal{D}((T+S)^*) \subseteq \mathcal{D}(T^*+S^*)$. Since $\mathcal{D}(S) = \mathcal{H}_1$ by assumption and $\mathcal{D}(S^*) = \mathcal{H}_1$ because S is bounded, we have $\mathcal{D}(T + S) = \mathcal{D}(T)$ and $\mathcal{D}(T^*+S^*) = \mathcal{D}(T^*)$. Let $y \in \mathcal{D}((T+S)^*)$. For $x \in \mathcal{D}(T+S)$,

$$\langle Tx, y \rangle = \langle (T + S)x, y \rangle - \langle Sx, y \rangle = \langle x, (T + S)^*y - S^*y \rangle.$$

This implies that $y \in \mathcal{D}(T^*) = \mathcal{D}(T^* + S^*)$. $\square$

Let T be a densely defined closable linear operator from $\mathcal{H}_1$ into $\mathcal{H}_2$. Then $\overline{T} = (T^*)^*$ as shown in Theorem 1.8(ii) below. Taking this for granted, Proposition 1.6(ii), applied to T and T^*, yields

$$\mathcal{N}(T^*) = \mathcal{R}(T)^\perp, \qquad \mathcal{N}(\overline{T}) = \mathcal{R}(T^*)^\perp,$$
$$\mathcal{H}_2 = \mathcal{N}(T^*) \oplus \overline{\mathcal{R}(T)}, \qquad \mathcal{H}_1 = \mathcal{N}(\overline{T}) \oplus \overline{\mathcal{R}(T^*)}.$$
(1.6)

Note that the ranges $\mathcal{R}(T)$ and $\mathcal{R}(T^*)$ are not closed in general. However, the closed range theorem (see, e.g., [K2, IV, Theorem 5.13]) states that for a closed linear operator T, the range $\mathcal{R}(T)$ is closed in $\mathcal{H}_2$ if and only if $\mathcal{R}(T^*)$ is closed in $\mathcal{H}_1$.

Replacing T by $T - \lambda I$ in (1.6), one obtains the useful decomposition

$$\mathcal{H} = \mathcal{N}(T^* - \overline{\lambda}I) \oplus \overline{\mathcal{R}(T - \lambda I)} = \mathcal{N}(\overline{T} - \lambda I) \oplus \overline{\mathcal{R}(T^* - \overline{\lambda}I)} \qquad (1.7)$$

for any densely defined closable operator T on a Hilbert space $\mathcal{H}$ and $\lambda \in \mathbb{C}$.

Assertions (vi) and (vii) of Proposition 1.6 have counterparts for the product.

Proposition 1.7 *Let* $T : \mathcal{H}_1 \to \mathcal{H}_2$ *and* $S : \mathcal{H}_2 \to \mathcal{H}_3$ *be linear operators such that* $\mathcal{D}(ST)$ *is dense in* $\mathcal{H}_1$.

(i) *If* $\mathcal{D}(S)$ *is dense in* $\mathcal{H}_2$, *then* $(ST)^* \supseteq T^*S^*$.
(ii) *If* S *is bounded and* $\mathcal{D}(S) = \mathcal{H}_2$, *then* $(ST)^* = T^*S^*$.

Proof (i): Note that $\mathcal{D}(T) \supseteq \mathcal{D}(ST)$; hence, $\mathcal{D}(T)$ is dense in $\mathcal{H}_2$. Suppose that $y \in \mathcal{D}(T^*S^*)$. Let $x \in \mathcal{D}(ST)$. Then we have $Tx \in \mathcal{D}(S)$, $y \in \mathcal{D}(S^*)$, and

$$\langle STx, y \rangle = \langle Tx, S^*y \rangle = \langle x, T^*S^*y \rangle.$$

Therefore, $y \in \mathcal{D}((ST)^*)$ and $(ST)^*y = T^*S^*y$.

(ii): Because of (i), it is sufficient to show that $\mathcal{D}((ST)^*) \subseteq \mathcal{D}(T^*S^*)$. Suppose that $y \in \mathcal{D}((ST)^*)$. Let $x \in \mathcal{D}(T)$. Since S is bounded and $\mathcal{D}(S) = \mathcal{H}_1$, we have $x \in \mathcal{D}(ST)$ and $y \in \mathcal{D}(S^*) = \mathcal{H}_2$, so that

$$\langle Tx, S^*y \rangle = \langle STx, y \rangle = \langle x, (ST)^*y \rangle.$$

Consequently, $S^*y \in \mathcal{D}(T^*)$, and so $y \in \mathcal{D}(T^*S^*)$. $\square$

A number of important and useful properties of closable and adjoint operators are derived by the so-called *graph method*. They are collected in the following theorem. Clearly, assertions (v) and (vi) are still valid if $\mathcal{D}(T)$ is not dense in $\mathcal{H}_1$.

Theorem 1.8 *Let T be a densely defined linear operator from $\mathcal{H}_1$ into $\mathcal{H}_2$.*

(i) *T is closable if and only if $\mathcal{D}(T^*)$ is dense in $\mathcal{H}_2$.*

(ii) *If T is closable, then $(\overline{T})^* = T^*$, and setting $T^{**} := (T^*)^*$, we have*

$$\overline{T} = T^{**}.$$

(iii) *T is closed if and only if $T = T^{**}$.*

(iv) *Suppose that $\mathcal{N}(T) = \{0\}$ and $\mathcal{R}(T)$ is dense in $\mathcal{H}_2$. Then T^* is invertible and*

$$(T^*)^{-1} = (T^{-1})^*.$$

(v) *Suppose that T is closable and $\mathcal{N}(T) = \{0\}$. Then the inverse T^{-1} of T is closable if and only if $\mathcal{N}(\overline{T}) = \{0\}$. If this holds, then*

$$(\overline{T})^{-1} = \overline{(T^{-1})}.$$

(vi) *If T is invertible, then T is closed if and only if T^{-1} is closed.*

Before we turn to the proof we state an important consequence separately as the following:

Corollary 1.9 *If T is a self-adjoint operator such that $\mathcal{N}(T) = \{0\}$, then T^{-1} is also a self-adjoint operator.*

Proof Since $T = T^*$, we have $\mathcal{R}(T)^{\perp} = \mathcal{N}(T) = \{0\}$. Hence, $\mathcal{R}(T)$ is dense, and the assertion follows from Theorem 1.8(iv). $\qquad\square$

The main technical tool for the graph method are the two unitary operators U, V of $\mathcal{H}_1 \oplus \mathcal{H}_2$ onto $\mathcal{H}_2 \oplus \mathcal{H}_1$ defined by

$$U(x, y) = (y, x) \quad \text{and} \quad V(x, y) = (-y, x) \quad \text{for } x \in \mathcal{H}_1, \ y \in \mathcal{H}_2. \quad (1.8)$$

Clearly, $V^{-1}(y, x) = (x, -y)$ for $y \in \mathcal{H}_2$ and $x \in \mathcal{H}_1$, and $U^{-1}V = V^{-1}U$, since

$$U^{-1}V(x, y) = U^{-1}(-y, x) = (x, -y) = V^{-1}(y, x) = V^{-1}U(x, y).$$

Let T be an invertible operator from $\mathcal{H}_1$ into $\mathcal{H}_2$. If $x \in \mathcal{D}(T)$ and $y = Tx$, then $U(x, Tx) = (y, T^{-1}y)$. Therefore, U maps the graph of T onto the graph of T^{-1}, that is,

$$U\big(\mathcal{G}(T)\big) = \mathcal{G}(T^{-1}). \quad (1.9)$$

The next lemma describes the graph of T^* by means of the unitary V. In particular, it implies that $\mathcal{G}(T^*)$ is closed. This gives a second proof of the fact (see Proposition 1.6(i)) that the operator T^* is closed.

Lemma 1.10 *For any densely defined linear operator T of $\mathcal{H}_1$ into $\mathcal{H}_2$, we have*

$$\mathcal{G}(T^*) = V\big(\mathcal{G}(T)\big)^{\perp} = V\big(\mathcal{G}(T)^{\perp}\big).$$

Proof Let $x \in \mathcal{D}(T)$ and $y \in \mathcal{D}(T^*)$. By (1.5),

$$\langle V(x, Tx), (y, T^*y)\rangle = \langle(-Tx, x), (y, T^*y)\rangle = \langle -Tx, y\rangle_2 + \langle x, T^*y\rangle_1 = 0.$$

Thus, $\mathcal{G}(T^*) \subseteq V(\mathcal{G}(T))^\perp$.

Conversely, suppose that $(y, u) \in V(\mathcal{G}(T))^\perp$. For $x \in \mathcal{D}(T)$, we obtain

$$\langle V(x, Tx), (y, u)\rangle = \langle -Tx, y\rangle_2 + \langle x, u\rangle_1 = 0,$$

that is, $\langle Tx, y\rangle_2 = \langle x, u\rangle_1$. By the definition of the adjoint operator this implies that $y \in \mathcal{D}(T^*)$ and $u = T^*y$, so $(y, u) \in \mathcal{G}(T^*)$. Hence $V(\mathcal{G}(T))^\perp \subseteq \mathcal{G}(T^*)$.

The second equality follows at once from the fact that V is unitary. $\qquad\square$

Proof of Theorem 1.8

(i): Assume that T is closable. Let $u \in \mathcal{D}(T^*)^\perp$. Then $(-u, 0) \in \mathcal{G}(T^*)^\perp$, so

$$(0, u) = V^{-1}(-u, 0) \in V^{-1}(\mathcal{G}(T^*)^\perp) = V^{-1}(V(\mathcal{G}(T))^{\perp\perp}) = \mathcal{G}(T)^{\perp\perp} = \overline{\mathcal{G}(T)},$$

where the second equality holds by Lemma 1.10 and the equality before last used the fact that V is unitary. Since T is closable, $\overline{\mathcal{G}(T)}$ is the graph of an operator by Proposition 1.5(ii). Therefore, $u = 0$. This proves that $\mathcal{D}(T^*)$ is dense in $\mathcal{H}_2$.

Conversely, suppose that $\mathcal{D}(T^*)$ is dense in $\mathcal{H}_2$. Then $T^{**} \equiv (T^*)^*$ exists. Since $T \subseteq T^{**}$ by Proposition 1.6(iii) and T^{**} is closed, T has a closed extension. That is, T is closable by Definition 1.3.

(ii): Using Lemma 1.10, we derive

$$\mathcal{G}((\overline{T})^*) = V(\mathcal{G}(\overline{T})^\perp) = V((\overline{\mathcal{G}(T)})^\perp) = V(\mathcal{G}(T)^\perp) = \mathcal{G}(T^*),$$

which implies that $(\overline{T})^* = T^*$.

Since T is closable, $\mathcal{D}(T^*)$ is dense in $\mathcal{H}_2$ by (i), and hence T^{**} is defined. Applying Lemma 1.10 twice, first to T^* with $-V^{-1}$ being the corresponding unitary and then to T with V, we deduce

$$\mathcal{G}(T^{**}) = ((-V^{-1})(\mathcal{G}(T^*))^\perp) = V^{-1}(V(\mathcal{G}(T))^{\perp\perp}) = \mathcal{G}(T)^{\perp\perp} = \overline{\mathcal{G}(T)} = \mathcal{G}(\overline{T}).$$

Therefore, $T^{**} = \overline{T}$.

(iii) is an immediate consequence of (ii).

(iv): In this proof we use both unitaries U and V. Since $\mathcal{R}(T)$ is assumed to be dense in $\mathcal{H}_2$, we have $\mathcal{N}(T^*) = \{0\}$ by Proposition 1.6(ii). Hence, the inverse $(T^*)^{-1}$ exists. The adjoint of T^{-1} exists, because $\mathcal{D}(T^{-1}) = \mathcal{R}(T)$ is dense in $\mathcal{H}_2$. Note that $-V^{-1}$ is the corresponding unitary for the adjoint of T^{-1}, that is, we have $\mathcal{G}((T^{-1})^*) = (-V^{-1})(\mathcal{G}(T^{-1})^\perp)$ by Lemma 1.10. Using formula (1.9) and Lemma 1.10 twice and the fact that U is unitary, we derive

$$\mathcal{G}((T^*)^{-1}) = U^{-1}(\mathcal{G}(T^*)) = U^{-1}(V(\mathcal{G}(T)^\perp)) = U^{-1}V(\mathcal{G}(T)^\perp)$$
$$= V^{-1}U(\mathcal{G}(T)^\perp) = V^{-1}(U(\mathcal{G}(T))^\perp) = (-V^{-1})(\mathcal{G}(T^{-1})^\perp)$$
$$= \mathcal{G}((T^{-1})^*).$$

This proves that $(T^*)^{-1} = (T^{-1})^*$.

(v): By (1.9) we have $U(\mathcal{G}(\overline{T})) = U(\overline{\mathcal{G}(T)}) = \overline{U(\mathcal{G}(T))} = \overline{\mathcal{G}(T^{-1})}$. Hence, $(0, x) \in \overline{\mathcal{G}(T^{-1})}$ if and only if $(x, 0) \in \mathcal{G}(\overline{T})$, or equivalently, if $x \in \mathcal{N}(\overline{T})$. Thus, by Lemma 1.1, $\overline{\mathcal{G}(T^{-1})}$ is the graph of an operator if and only if $\mathcal{N}(\overline{T}) = \{0\}$. Therefore, by Proposition 1.5, (i) $\leftrightarrow$ (iii), T^{-1} is closable if and only if $\mathcal{N}(\overline{T}) = \{0\}$.

Further, if T^{-1} is closable, then $\mathcal{G}(\overline{T^{-1}}) = \overline{\mathcal{G}(T^{-1})} = U(\mathcal{G}(\overline{T})) = \mathcal{G}((\overline{T})^{-1})$ again by formula (1.9). Hence, $\overline{T^{-1}} = (\overline{T})^{-1}$.

(vi) follows from (v) or directly from formula (1.9), which implies that $\mathcal{G}(T)$ is closed if and only if $\mathcal{G}(T^{-1})$ is. $\square$

1.3   Examples: Differential Operators

In the first subsection we discuss the notions from the preceding sections for the differentiation operator $-i\frac{d}{dx}$ and its square $-\frac{d^2}{dx^2}$ on various intervals. We develop these examples and their continuations in later sections in great detail and as elementarily as possible by using absolutely continuous functions. In the second subsection distributions are used to define maximal and minimal operators for linear partial differential expressions.

1.3.1   Differentiation Operators on Intervals I

We begin with two examples on a *bounded* interval (a, b). First, let us repeat some definitions and facts on absolutely continuous functions (see Appendix E).

A function f on an interval $[a, b]$, $a, b \in \mathbb{R}$, $a < b$, is absolutely continuous if and only if there exists a function $h \in L^1(a, b)$ such that for all $x \in [a, b]$,

$$f(x) = f(a) + \int_a^x h(t)\, dt. \tag{1.10}$$

The set of absolutely continuous functions on $[a, b]$ is denoted by $AC[a, b]$.

If $f \in AC[a, b]$, then $f \in C([a, b])$, and f is a.e. differentiable with $f'(x) = h(x)$ a.e. on $[a, b]$. We call h the derivative of f and write $f' = h$. Set

$$H^1(a, b) = \{f \in AC[a, b] : f' \in L^2(a, b)\}, \tag{1.11}$$

$$H^2(a, b) = \{f \in C^1([a, b]) : f' \in H^1(a, b)\}. \tag{1.12}$$

If $\langle \cdot, \cdot \rangle$ denotes the scalar product of the Hilbert space $L^2(a, b)$, then the formula for integration by parts (E.2) yields

$$\langle f', g \rangle + \langle f, g' \rangle = f\overline{g}|_a^b \equiv f(b)\overline{g(b)} - f(a)\overline{g(a)} \tag{1.13}$$

for $f, g \in H^1(a, b)$ and

$$\langle f'', g \rangle - \langle f, g'' \rangle = (f'\overline{g} - f\overline{g'})|_a^b$$

$$\equiv f'(b)\overline{g(b)} - f(b)\overline{g'(b)} - f'(a)\overline{g(a)} + f(a)\overline{g'(a)} \tag{1.14}$$

for $f, g \in H^2(a, b)$, where we have set $h|_a^b := h(b) - h(a)$.

Example 1.4 (*Bounded interval*) Suppose that $a, b \in \mathbb{R}$, $a < b$. Let T be the linear operator on the Hilbert space $L^2(a, b)$ defined by $Tf = -if'$ for f in

$$\mathcal{D}(T) = H_0^1(a, b) := \{ f \in H^1(a, b) : f(a) = f(b) = 0 \}.$$

Clearly, its square T^2 acts as $T^2 f = -f''$ for f in

$$\mathcal{D}(T^2) = H_0^2(a, b) := \{ f \in H^2(a, b) : f(a) = f(b) = f'(a) = f'(b) = 0 \}.$$

Obviously, $\mathcal{D}(T)$ and $\mathcal{D}(T^2)$ are dense in $L^2(a, b)$. Our aim in this example is to describe the adjoints of both operators T and T^2 and a core for T.

Statement 1 $\mathcal{R}(T)^\perp \subseteq \mathbb{C} \cdot 1$ and $\mathcal{R}(T^2)^\perp \subseteq \mathbb{C} \cdot 1 + \mathbb{C} \cdot x$.

(Here 1 and x denote the functions $f_1(x) = 1$ and $f_2(x) = x$ for $x \in [a, b]$, respectively. Moreover, we have even equality in both cases.)

Proof It suffices to prove that $(\mathbb{C} \cdot 1)^\perp \subseteq \mathcal{R}(T)$ and $(\mathbb{C} \cdot 1 + \mathbb{C} \cdot x)^\perp \subseteq \mathcal{R}(T^2)$. Suppose that $h_1 \in (\mathbb{C} \cdot 1)^\perp$ and $h_2 \in (\mathbb{C} \cdot 1 + \mathbb{C} \cdot x)^\perp$. We define functions on $[a, b]$ by

$$k_1(x) = \int_a^x h_1(t) \, dt, \qquad k_2(x) = \int_a^x \left(\int_a^t h_2(s) \, ds \right) dt. \tag{1.15}$$

Since $h_1, h_2 \in L^2(a, b) \subseteq L^1(a, b)$, we conclude that $k_1 \in H^1(a, b)$, $k_1' = h_1$, and $k_2 \in H^2(a, b)$, $k_2'' = h_2$. Obviously, $k_1(a) = k_2(a) = k_2'(a) = 0$. Moreover, $k_1(b) = \langle h_1, 1 \rangle = 0$ and $k_2'(b) = \langle h_2, 1 \rangle = 0$. Hence, $k_2'(x) x |_a^b = 0$, and using formula (1.13), we derive

$$k_2(b) = \int_a^b k_2'(t) \, dt = \langle k_2', 1 \rangle = -\langle k_2'', x \rangle = -\langle h_2, x \rangle = 0.$$

Thus, we have shown that $k_1 \in \mathcal{D}(T)$ and $k_2 \in \mathcal{D}(T^2)$, so $T(ik_1) = h_1 \in \mathcal{R}(T)$ and $T^2(-k_2) = h_2 \in \mathcal{R}(T^2)$. $\qquad\square$

Statement 2 $\mathcal{D}(T^*) = H^1(a, b)$ and $T^* g = -ig'$ for $g \in \mathcal{D}(T^*)$, $\mathcal{D}((T^2)^*) = \mathcal{D}((T^*)^2) = H^2(a, b)$ and $(T^2)^* g = -g''$ for $g \in \mathcal{D}((T^2)^*)$.

Proof First suppose that $g \in H^1(a, b)$. Let $f \in \mathcal{D}(T)$. Since $f(a) = f(b) = 0$, the right-hand of formula (1.13) vanishes, so we obtain

$$\langle Tf, g \rangle = -i \langle f', g \rangle = i \langle f, g' \rangle = \langle f, -ig' \rangle.$$

By the definition of T^* it follows that $g \in \mathcal{D}(T^*)$ and $T^* g = -ig'$.

Now let $g \in H^2(a, b)$. Then, by the definition of $H^2(a, b)$, g and g' are in $H^1(a, b)$. Applying the result of the preceding paragraph twice, first to g and then to $T^* g = -ig'$, we conclude that $g \in \mathcal{D}((T^*)^2)$ and $(T^*)^2 g = -g''$.

Conversely, suppose that $g_1 \in \mathcal{D}(T^*)$ and $g_2 \in \mathcal{D}((T^2)^*)$. We set $h_1 := T^* g_1$ and $h_2 := (T^2)^* g_2$ and define the functions k_1 and k_2 by (1.15). As noted above, we then have $k_1 \in H^1(a, b)$, $k_2 \in H^2(a, b)$, $k_1' = h_1$, and $k_2'' = h_2$.

Let $f_1 \in \mathcal{D}(T)$ and $f_2 \in \mathcal{D}(T^2)$. Since the boundary values $f_1(a)$, $f_1(b)$, $f_2(a)$, $f_2(b)$, $f_2'(a)$, and $f_2'(b)$ vanish, it follows from (1.13) and (1.14) that

$$-\langle f_1', k_1 \rangle = \langle f_1, k_1' \rangle = \langle f_1, h_1 \rangle = \langle f_1, T^* g_1 \rangle = \langle T f_1, g_1 \rangle = \langle -i f_1', g_1 \rangle,$$
$$\langle f_2'', k_2 \rangle = \langle f_2, k_2'' \rangle = \langle f_2, h_2 \rangle = \langle f_2, (T^2)^* g_2 \rangle = \langle T^2 f_2, g_2 \rangle = \langle -f_2'', g_2 \rangle.$$

Hence, $\langle -i f_1', g_1 - i k_1 \rangle = 0$ and $\langle f_2'', g_2 + k_2 \rangle = 0$, so that $g_1 - i k_1 \in \mathcal{R}(T)^\perp \subseteq \mathbb{C} \cdot 1$ and $g_2 + k_2 \in \mathcal{R}(T^2)^\perp \subseteq \mathbb{C} \cdot 1 + \mathbb{C} \cdot x$ by Statement 1. Since the functions k_1 and 1 are in $H^1(a, b)$ and k_2, 1, and x are in $H^2(a, b)$, we conclude that $g_1 \in H^1(a, b)$ and $g_2 \in H^2(a, b)$.

By the preceding, we have proved the assertions about T^* and the relations $H^2(a, b) \subseteq \mathcal{D}((T^*)^2)$, $(T^*)^2 g = -g''$ for $g \in H^2(a, b)$, and $\mathcal{D}((T^2)^*) \subseteq H^2(a, b)$. Since $(T^*)^2 \subseteq (T^2)^*$ by Proposition 1.7(i), this implies the assertions concerning $(T^2)^*$ and $(T^*)^2$.   □

By Statement 2 we have $T \subseteq T^*$, so the operator T is closable. But we even have the following:

Statement 3 $T = T^{**}$, *that is, the operator T is closed.*

Proof Since always $T \subseteq T^{**}$, it suffices to prove that $T^{**} \subseteq T$. Let $g \in \mathcal{D}(T^{**})$. Since $T \subseteq T^*$ by Statement 2 and hence $T^{**} \subseteq T^*$, $g \in \mathcal{D}(T^*) = H^1(a, b)$ and $T^{**} g = -i g'$ again by Statement 2. For $f \in \mathcal{D}(T^*)$, we therefore obtain

$$0 = \langle T^* f, g \rangle - \langle f, T^{**} g \rangle = -i \langle f', g \rangle - i \langle f, g' \rangle = -i \big(f(b) \overline{g(b)} - f(a) \overline{g(a)} \big),$$

where the last equality follows from (1.13). Since the values $f(b)$ and $f(a)$ of $f \in \mathcal{D}(T^*)$ are arbitrary, it follows that $g(b) = g(a) = 0$. Thus, $g \in \mathcal{D}(T)$.   □

The reason for defining the operator T on absolutely continuous functions was to get a *closed* operator. Often it is more convenient to work with C^∞-functions with zero boundary values. Let T_0 denote the restriction of T to the dense domain

$$\mathcal{D}_0 = \big\{ f \in C^\infty([a, b]) : f(a) = f(b) = 0 \big\}.$$

Statement 4 $\mathcal{D}_0$ *is a core for T, so $\overline{T_0} = T$.*

Proof We first show that $\mathcal{R}(T_0)^\perp \subseteq \mathbb{C} \cdot 1$. Indeed, upon applying the linear transformation $x \to \pi(b - a)^{-1}(x - a)$, we may assume that $a = 0$ and $b = \pi$. Then, $\sin nx \in \mathcal{D}(T_0)$, and so $\cos nx = i n^{-1} T_0(\sin nx) \in \mathcal{R}(T_0)$ for $n \in \mathbb{N}$. From the theory of Fourier series it is known that the linear span of functions $\cos nx$, $n \in \mathbb{N}_0$, is dense in $L^2(0, \pi)$. Hence, $\mathcal{R}(T_0)^\perp \subseteq \{\cos nx : n \in \mathbb{N}\}^\perp = \mathbb{C} \cdot 1$.

Since $\mathcal{R}(T_0)^\perp \subseteq \mathbb{C} \cdot 1$, the reasoning used in the proof of Statement 2 yields $\mathcal{D}(T_0^*) \subseteq H^1(0, \pi) = \mathcal{D}(T^*)$ and $T_0^* \subseteq T^*$. Hence, $T = T^{**} \subseteq T_0^{**} = \overline{T_0}$. Since $T_0 \subseteq T$ and therefore $\overline{T_0} \subseteq T$, we get $\overline{T_0} = T$.

We give a second proof of Statement 4 by a direct approximation of $f \in \mathcal{D}(T)$. Since $f' \in L^2(a, b)$, there exists a sequence $(f_n)_{n \in \mathbb{N}}$ from $C^{\infty}([a, b])$ such that $g_n \to f'$ in $L^2(a, b)$. We define the functions $f_n \in \mathcal{D}_0$ by

$$f_n(x) := \int_a^x g_n(t) \, dt - (x - a)(b - a)^{-1} \int_a^b g_n(t) \, dt.$$

By the Hölder inequality we also have $g_n \to f'$ in $L^1(a, b)$, which implies that

$$\int_a^b g_n(t) \, dt \to \int_a^b f'(t) \, dt = f(b) - f(a) = 0. \tag{1.16}$$

Hence, $T f_n = -i g_n + i (b - a)^{-1} \int_a^b g_n(t) \, dt \to -i f' = T f$ in $L^2(a, b)$. Since $f(a) = 0$, from the definition of f_n we obtain for $x \in [a, b]$,

$$\left| f_n(x) - f(x) \right| \le \int_a^b \left| g_n(t) - f'(t) \right| dt + \left| \int_a^b g_n(t) \, dt \right|.$$

As $g_n \to f'$ in $L^1(a, b)$, by (1.16) the right-hand side tends to zero as $n \to \infty$. Therefore, $f_n(x) \to f(x)$ uniformly on $[a, b]$, and hence $f_n \to f$ in $L^2(a, b)$. Thus, $f_n \to f$ and $T_0 f_n = T f_n \to T f$ in $L^2(a, b)$. □ ∘

Example 1.5 (*Bounded interval continued*) Let $a, b \in \mathbb{R}$, $a < b$. For any z in $\overline{\mathbb{C}} := \mathbb{C} \cup \{\infty\}$, we define a linear operator S_z on the Hilbert space $L^2(a, b)$ by $S_z f = -i f'$, $f \in \mathcal{D}(S_z)$, where

$$\mathcal{D}(S_z) = \left\{ f \in H^1(a, b) : f(b) = z f(a) \right\}, \quad z \in \mathbb{C},$$
$$\mathcal{D}(S_{\infty}) = \left\{ f \in H^1(a, b) : f(a) = 0 \right\}.$$

The following result describes the adjoint of the operator S_z.

Statement *For each $z \in \overline{\mathbb{C}}$, we have $(S_z)^* = S_{\bar{z}^{-1}}$, where we set $0^{-1} := \infty$ and $\overline{\infty}^{-1} := 0$. In particular, the operator S_z is self-adjoint if and only if $|z| = 1$.*

Proof We carry out the proof for $z \in \mathbb{C}$, $z \ne 0$; the cases $z = 0$ and $z = \infty$ are treated similarly. For $f \in \mathcal{D}(S_z)$ and $g \in H^1(a, b)$, we use (1.13) to compute

$$\langle S_z f, g \rangle - \langle f, -i g' \rangle = -i \left(\langle f', g \rangle + \langle f, g' \rangle \right) = -i \left(f(b) \overline{g(b)} - f(a) \overline{g(a)} \right)$$
$$= -i f(a) \left(z \overline{g(b)} - \overline{g(a)} \right). \tag{1.17}$$

If $g \in \mathcal{D}(S_{\bar{z}^{-1}})$, then $\bar{z} g(b) = g(a)$, so the right-hand side of (1.17) vanishes. Hence, $g \in \mathcal{D}((S_z)^*)$ and $(S_z)^* g = -i g'$.

Conversely, let $g \in \mathcal{D}((S_z)^*)$. We choose $\lambda \in \mathbb{C}$ such that $e^{\lambda(b-a)} = z$. Then $f(x) := e^{\lambda x}$ is in $\mathcal{D}(S_z)$. Since $T \subseteq S_z$ and hence $(S_z)^* \subseteq T^*$, we have $g \in H^1(a, b)$ and $(S_z)^* g = -i g'$ by the preceding example, so the left-hand side of (1.17) is zero. Since $f(a) \ne 0$, this yields $\bar{z} g(b) = g(a)$, so $g \in \mathcal{D}(S_{\bar{z}^{-1}})$. □ ∘

Before we continue we derive two useful technical lemmas.

Lemma 1.11

(i) *If* $f \in H^1(0, +\infty)$, *then* $\lim_{b \to +\infty} f(b) = 0$.
(ii) *If* $f \in H^1(\mathbb{R})$, *then* $\lim_{b \to +\infty} f(b) = \lim_{a \to -\infty} f(a) = 0$.

Proof We prove (i). The proof of (ii) is similar. Formula (1.13) yields

$$\int_0^b \left(f(t)\overline{f'(t)} + f'(t)\overline{f(t)} \right) dt = |f(b)|^2 - |f(0)|^2$$

for any $b > 0$. By the definition of $H^1(0, +\infty)$, f and f' are in $L^2(0, +\infty)$, so $f'\overline{f}$ belongs to $L^1(0, +\infty)$. Therefore, the left-hand side of the preceding equality converges to the integral over $[0, +\infty)$ as $b \to +\infty$. Hence, $\lim_{b \to +\infty} |f(b)|^2$ exists. Since $f \in L^2(0, +\infty)$, this limit has to be zero.   □

Lemma 1.12 *Let* $\mathcal{J}$ *be an open interval, and let* c *be in the closure of* $\mathcal{J}$. *For any* $\varepsilon > 0$, *there is a constant* $b_\varepsilon > 0$ *such that*

$$|f(c)| \leq \varepsilon \|f'\| + b_\varepsilon \|f\| \quad \text{for } f \in H^1(\mathcal{J}). \tag{1.18}$$

Proof For notational simplicity, assume that c is not the left end point of $\mathcal{J}$. Take $a \in \mathcal{J}$, $a < c$, and a function $\eta \in C_0^\infty(\mathbb{R})$ such that $\eta(c) = 1$, $\eta(a) = 0$, and $|\eta(t)| \leq 1$ on $\mathbb{R}$. Let M be the supremum of $|\eta'(t)|$ over $\mathbb{R}$. Then

$$|f(c)|^2 = |(\eta f)^2(c) - (\eta f)^2(a)| = \left| \int_a^c ((\eta f)^2)'(t)\, dt \right| = \left| \int_a^c 2\eta f (\eta f' + \eta' f)\, dt \right|$$

$$\leq 2\|f'\|\|f\| + 2M\|f\|^2 = 2\varepsilon\|f'\|\varepsilon^{-1}\|f\| + 2M\|f\|^2$$

$$\leq \varepsilon^2\|f'\|^2 + (\varepsilon^{-2} + 2M)\|f\|^2 \leq (\varepsilon\|f'\| + (\varepsilon^{-1} + 1 + M)\|f\|)^2. \qquad \square$$

Next we investigate the differentiation operator $-i\frac{d}{dx}$ on the *half-axis* and on the *whole axis*. For $\mathcal{J} = (0, \infty)$ or $\mathcal{J} = \mathbb{R}$, we define

$$H^1(\mathcal{J}) = \left\{ f \in L^2(\mathcal{J}) : f \in AC[a, b] \text{ for all } [a, b] \subseteq \overline{\mathcal{J}} \text{ and } f' \in L^2(\mathcal{J}) \right\}.$$

Let $\mathcal{J} = (0, +\infty)$. Taking the limit $b \to +\infty$ in formula (1.13) applied with $a = 0$ and using Lemma 1.11(i), we obtain for $f, g \in H^1(0, +\infty)$,

$$\langle f', g \rangle + \langle f, g' \rangle = -f(0)\overline{g(0)}. \tag{1.19}$$

Example 1.6 (*Half-axis*) Let T be the operator on $L^2(0, +\infty)$ defined by $Tf = -if'$ for f in the domain $\mathcal{D}(T) = H_0^1(0, \infty) := \{ f \in H^1(0, +\infty) : f(0) = 0 \}$.

Statement $\mathcal{D}(T^*) = H^1(0, +\infty)$ *and* $T^*g = -ig'$ *for* $g \in \mathcal{D}(T^*)$.

Proof Suppose that $g \in H^1(0, +\infty)$ and let $f \in \mathcal{D}(T)$. Since $f(0) = 0$, formula (1.19) yields $\langle f', g \rangle + \langle f, g' \rangle = 0$. Hence, $\langle Tf, g \rangle = \langle f, -ig' \rangle$ for all $f \in \mathcal{D}(T)$. This implies that $g \in \mathcal{D}(T^*)$ and $T^*g = -ig'$.

Now let $g \in \mathcal{D}(T^*)$. We denote by T^b the operator on $L^2(0, b)$ obtained from T by restricting functions of $\mathcal{D}(T)$ to $[0, b]$ and by T_b the operator T on $L^2(0, b)$ from Example 1.4. From Statement 2 therein, $g \in H^1(0, b)$ and $(T_b)^*g = -ig'$ on $[0, b]$. Since $T_b \subseteq T^b$ and hence $(T^b)^* \subseteq (T_b)^*$, this yields $(T^b)^*g = -ig'$ on $[0, b]$. Obviously, $((T^b)^*g)(x) = (T^*g)(x)$ for $x \in [0, b]$ and any $b > 0$. Hence, $T^*g = -ig'$ on $[0, +\infty)$. Since $g, g' \in L^2(0, +\infty)$, we have $g \in H^1(0, +\infty)$. $\square$

Repeating almost verbatim the proof of Statement 3 from Example 1.4, it can be shown that the operator T is closed. This follows also from the facts that the graph norm of T is the norm of the Hilbert space $H^1(0, +\infty)$ and that the functional $f \to f(0)$ is continuous on $H^1(0, +\infty)$ by (1.18).   ○

Example 1.7 (*Real line*) Let T be the operator $-i\frac{d}{dx}$ with domain $\mathcal{D}(T) = H^1(\mathbb{R})$ on the Hilbert space $L^2(\mathbb{R})$. Proceeding as in Example 1.6, by using Lemma 1.11(ii) we derive $T = T^*$, that is, the operator T is self-adjoint.   ○

Finally, we mention the relations to distributive derivatives (see Appendix D).
Let $\mathcal{J}$ be an open interval, and let $f, g \in L^1_{\mathrm{loc}}(\mathcal{J})$. Recall that the function g is called the *weak derivative* (or derivative in the sense of distributions) of f if

$$\int_{\mathcal{J}} f(t)\varphi'(t)\, dt = -\int_{\mathcal{J}} g(t)\varphi(t)\, dt \quad \text{for all } \varphi \in C_0^\infty(\mathcal{J}).$$

Then g is uniquely determined a.e. by f. For $f \in H^1(\mathcal{J})$, the weak derivative g is a.e. equal to the derivative f' of f as an absolutely continuous function. In particular, this implies that T^* is the adjoint of $T \upharpoonright C_0^\infty(\mathcal{J})$ and that $C_0^\infty(\mathcal{J})$ is a core for T.

1.3.2 Linear Partial Differential Operators

Throughout this subsection we assume that Ω is an open subset of $\mathbb{R}^d$ and that for each $\alpha \in \mathbb{N}_0^d$, $|\alpha| \leq n$, a function $a_\alpha \in C^\infty(\Omega)$ is given.
Let us consider the formal *linear partial differential expression*

$$\mathcal{L} = \sum_{|\alpha| \leq n} a_\alpha(x) D^\alpha \tag{1.20}$$

and its *formal adjoint* $\mathcal{L}^+$ defined by

$$\mathcal{L}^+ := \sum_{|\alpha| \leq n} D^\alpha \overline{a_\alpha(x)} \equiv \sum_{|\alpha| \leq n} a_\alpha^+(x) D^\alpha, \quad \text{where } a_\alpha^+(x) := \sum_{\alpha \leq \beta} \binom{\beta}{\alpha} D^{\beta - \alpha} \overline{a_\beta(x)}.$$

The function $a_\alpha^+(x)$ is chosen such that by the Leibniz rule

$$\mathcal{L}^+ f \equiv \sum_{|\alpha| \leq n} D^\alpha \big(\overline{a_\alpha(x)} f \big) = \sum_{|\alpha| \leq n} a_\alpha^+(x) D^\alpha f, \quad f \in C^\infty(\Omega).$$

For $\alpha = (\alpha_1, \ldots, \alpha_d)$ and $\beta = (\beta_1, \ldots, \beta_d) \in \mathbb{N}_0^d$, we used the multi-index notation

$$D^\alpha := (-i)^{|\alpha|} \partial^\alpha, \qquad \partial^\alpha := \frac{\partial^{\alpha_1}}{\partial x_1^{\alpha_1}} \cdots \frac{\partial^{\alpha_d}}{\partial x_d^{\alpha_d}}, \qquad |\alpha| := \alpha_1 + \cdots + \alpha_d,$$

$$\binom{\alpha}{\beta} := \binom{\alpha_1}{\beta_1} \cdots \binom{\alpha_d}{\beta_d}, \quad \text{and} \quad \beta \leq \alpha \quad \text{if and only if} \quad \beta_1 \leq \alpha_1, \ldots, \beta_d \leq \alpha_d.$$

Note that $\mathcal{L}^+ = \mathcal{L}$ if all functions $a_\alpha(x)$ are real constants.

Our aim is to define two closed linear operators on the Hilbert space $\mathcal{H} = L^2(\Omega)$ associated with $\mathcal{L}$ and $\mathcal{L}^+$, the maximal operator $L_{\max}$ and the minimal operator $(L^+)_{\min}$, and to prove that they are adjoints of each other.

The next lemma follows essentially by integration by parts.

Lemma 1.13 $\langle \mathcal{L}f, g \rangle_{L^2(\Omega)} = \langle f, \mathcal{L}^+ g \rangle_{L^2(\Omega)}$ *for* $f, g \in C_0^\infty(\Omega)$.

Proof Since $f, g \in C_0^\infty(\Omega)$, we can choose a bounded open subset $\widetilde{\Omega}$ of Ω with C^∞-boundary such that $\operatorname{supp} f \subseteq \widetilde{\Omega}$ and $\operatorname{supp} g \subseteq \widetilde{\Omega}$. Then we compute

$$\langle \mathcal{L}f, g \rangle = \sum_\alpha \langle a_\alpha D^\alpha f, g \rangle = \sum_\alpha \int_\Omega a_\alpha (-i)^{|\alpha|} (\partial^\alpha f) \overline{g} \, dx$$

$$= \sum_\alpha \int_{\widetilde{\Omega}} (-i)^{|\alpha|} (\partial^\alpha f) \overline{a_\alpha} g \, dx = \sum_\alpha \int_{\widetilde{\Omega}} i^{|\alpha|} f \partial^\alpha (\overline{a_\alpha g}) \, dx$$

$$= \sum_\alpha \int_\Omega i^{|\alpha|} f \partial^\alpha (\overline{a_\alpha g}) \, dx = \sum_\alpha \langle f, D^\alpha \overline{a_\alpha} g \rangle = \langle f, \mathcal{L}^+ g \rangle.$$

The fourth equality follows from the Gauss formula (D.4) by using that all terms of f and $\overline{a_\alpha} g$ on the boundary $\partial \widetilde{\Omega}$ vanish, since $\operatorname{supp} f \subseteq \widetilde{\Omega}$ and $\operatorname{supp} g \subseteq \widetilde{\Omega}$. □

Clearly, the formal adjoint expression $\mathcal{L}^+$ is uniquely determined by the equation in Lemma 1.13. From this characterization it follows at once that $(\mathcal{L}^+)^+ = \mathcal{L}$.

First we define linear operators L_0 and $(L^+)_0$ on the Hilbert space $L^2(\Omega)$ by $L_0 = \mathcal{L}f$ and $(L^+)_0 f = \mathcal{L}^+ f$ for $f \in \mathcal{D}(L_0) = \mathcal{D}((L^+)_0) := C_0^\infty(\Omega)$.

By Lemma 1.13 we have $L_0 \subseteq ((L^+)_0)^*$. Hence, $(L^+)_0$ has a densely defined adjoint, so $(L^+)_0$ is closable by Theorem 1.8(i). The closure of $(L^+)_0$ is denoted by $(L^+)_{\min}$ and called the *minimal operator* associated with the expression $\mathcal{L}^+$.

To define $L_{\max}$, we need some notions on distributions, see Appendix D. Let $\mathcal{D}'(\Omega)$ be the space of distributions on Ω. Let $f \in \mathcal{D}'(\Omega)$ and $\varphi \in C_0^\infty(\Omega)$. Since $a_\alpha \in C^\infty(\Omega)$, $\mathcal{L}f$ is again a distribution of $\mathcal{D}'(\Omega)$ which acts on $\overline{\varphi} \in C_0^\infty(\Omega)$ by

$$\mathcal{L}f(\overline{\varphi}) = \sum_\alpha (a_\alpha D^\alpha f)(\overline{\varphi}) = \sum_\alpha D^\alpha f(a_\alpha \overline{\varphi}) = \sum_\alpha (-1)^{|\alpha|} f(D^\alpha (a_\alpha \overline{\varphi}))$$

$$= \sum_\alpha f(\overline{D^\alpha (\overline{a_\alpha} \varphi)}) = f(\overline{\mathcal{L}^+ \varphi}), \tag{1.21}$$

where the bars refer to conjugate complex functions.

If both distributions f and $g = \mathcal{L}f$ are given by functions $f, g \in L^2(\Omega)$, we say that the equation $\mathcal{L}f = g$ holds *weakly* in $L^2(\Omega)$. This means that the expressions D^α in (1.20) act on the L^2-function f in the sense of distributions.

Let $\mathcal{D}(L_{\max})$ denote the set of all $f \in L^2(\Omega)$ for which the distribution $g := \mathcal{L}f$ is given by a function $g \in L^2(\Omega)$ and define $L_{\max}f := g \equiv \mathcal{L}f$. The linear operator $L_{\max}$ is called the *maximal operator* associated with the differential expression $\mathcal{L}$.

Proposition 1.14 *The operators* $(L^+)_0$, $(L^+)_{\min}$, *and* $L_{\max}$ *satisfy the relations*

$$\left((L^+)_0\right)^* = \left((L^+)_{\min}\right)^* = L_{\max} \quad and \quad (L_{\max})^* = \left(L^+\right)_{\min}. \quad (1.22)$$

Proof Let $f \in \mathcal{D}(L_{\max})$. Then Eq. (1.21) can be written as $\langle L_{\max}f, \varphi \rangle = \langle \mathcal{L}f, \varphi \rangle = \langle f, (L^+)_0\varphi \rangle$ for $\varphi \in \mathcal{D}((L^+)_0)$. Hence, $L_{\max} \subseteq ((L^+)_0)^*$ by the definition of the adjoint operator. Now suppose that $f \in \mathcal{D}(((L^+)_0)^*)$ and set $g = ((L^+)_0)^* f$. Then $\langle g, \varphi \rangle = \langle f, (L^+)_0\varphi \rangle = \langle f, \mathcal{L}^+\varphi \rangle$ for all $\varphi \in \mathcal{D}((L^+)_0)$. On the other hand, we have $\mathcal{L}f(\overline{\varphi}) = f(\overline{\mathcal{L}^+\varphi}) = \langle f, \mathcal{L}^+\varphi \rangle$ by (1.21). Comparing both formulas, we conclude that the distribution $\mathcal{L}f$ is given by the function $g \in L^2(\Omega)$, that is, $f \in \mathcal{D}(L_{\max})$. Thus, we have shown that $L_{\max} = ((L^+)_0)^*$.

Since $(L^+)_{\min}$ is the closure of $(L^+)_0$, Theorem 1.8(ii) yields $((L^+)_{\min})^* = L_{\max}$ and $(L_{\max})^* = ((L^+)_{\min})^{**} = (L^+)_{\min}$, which completes the proof. $\qquad\square$

1.4 Invariant Subspaces and Reducing Subspaces

Let T_0 and T_1 be linear operators on Hilbert spaces $\mathcal{H}_0$ and $\mathcal{H}_1$, respectively. We denote by $T_0 \oplus T_1$ the operator with domain $\mathcal{D}(T_0 \oplus T_1) = \mathcal{D}(T_0) \oplus \mathcal{D}(T_1)$ on the Hilbert space $\mathcal{H}_0 \oplus \mathcal{H}_1$ defined by

$$(T_0 \oplus T_1)(x_0, x_1) = (T_0x_0, T_1x_1), \quad \text{where } x_0 \in \mathcal{D}(T_0),\; x_1 \in \mathcal{D}(T_1).$$

Let T be a linear operator on a Hilbert space $\mathcal{H}$, and let $\mathcal{H}_0$ be a closed linear subspace of $\mathcal{H}$. The basic notions of this section are defined as follows.

Definition 1.7 If T maps $\mathcal{D}(T_0) := \mathcal{D}(T) \cap \mathcal{H}_0$ into $\mathcal{H}_0$, then $\mathcal{H}_0$ is called an *invariant subspace* for T, and the operator $T_0 := T \!\restriction\! \mathcal{D}(T_0)$ is called the *part of the operator T on* $\mathcal{H}_0$.

Definition 1.8 We say that $\mathcal{H}_0$ is a *reducing subspace* for T if there exist linear operators T_0 on $\mathcal{H}_0$ and T_1 on $\mathcal{H}_0^\perp$ such that $T = T_0 \oplus T_1$. The operator T is called *irreducible* if $\{0\}$ and $\mathcal{H}$ are the only reducing subspaces for T.

Further, the *commutant* of the operator T is defined by

$$\{T\}' := \left\{B \in \mathbf{B}(\mathcal{H}) : BT \subseteq TB\right\}. \quad (1.23)$$

That is, an operator $B \in \mathbf{B}(\mathcal{H})$ is in the commutant $\{T\}'$ if and only if B maps the domain $\mathcal{D}(T)$ into itself and $BTx = TBx$ for all $x \in \mathcal{D}(T)$.

Proposition 1.15 *Let P_0 be the projection of $\mathcal{H}$ on the closed linear subspace $\mathcal{H}_0$. The following statements are equivalent:*

(i) *$\mathcal{H}_0$ is a reducing subspace for T.*
(ii) *$\mathcal{H}_0^\perp$ is a reducing subspace for T.*
(iii) *$\mathcal{H}_0$ and $\mathcal{H}_0^\perp$ are invariant subspaces for T, and P_0 maps $\mathcal{D}(T)$ into itself.*
(iv) *$P_0 \in \{T\}'$.*

The parts of T on $\mathcal{H}_0$ and $\mathcal{H}_0^\perp$ are then $T_0 = T \restriction P_0 \mathcal{D}(T)$ and $T_1 = T \restriction (I - P_0)\mathcal{D}(T)$.

Proof All assertions are easily derived from the corresponding definitions. As a sample, we prove that (iv) implies (iii). By (1.23) we have $P_0 \mathcal{D}(T) \subseteq \mathcal{D}(T)$. Let $x_0 \in \mathcal{D}(T) \cap \mathcal{H}_0$. Then $T x_0 = T P_0 x_0 = P_0 T x_0$ by (1.23), so that $T x_0 \in \mathcal{H}_0$. Thus, $\mathcal{H}_0$ is an invariant subspace for T. Replacing P_0 by $I - P_0$, it follows that $\mathcal{H}_0^\perp$ is also an invariant subspace for T. $\square$

Corollary 1.16 *An operator T is irreducible if and only if 0 and I are the only projections contained in the commutant $\{T\}'$.*

If T is a *bounded* self-adjoint operator, an invariant subspace $\mathcal{H}_0$ for T is always reducing. For *unbounded* self-adjoint operators, this is no longer true, but by Proposition 1.17 below, $\mathcal{H}_0$ is reducing if the restriction $T \restriction (\mathcal{D}(T) \cap \mathcal{H}_0)$ is self-adjoint.

Example 1.8 By Example 1.7 the operator $T = -\mathrm{i}\frac{d}{dx}$ on $H^1(\mathbb{R})$ is self-adjoint. Clearly, $\mathcal{H}_0 := L^2(0, 1)$ is an invariant subspace for T. If $f \in H^1(\mathbb{R})$ and $f(0) \neq 0$, then $P_0 f \notin \mathcal{D}(T)$. Hence, $\mathcal{H}_0$ is not reducing for T.   ○

Proposition 1.17 *Let T be a closed symmetric operator on $\mathcal{H}$. Let $\mathcal{D}_0$ be a dense linear subspace of a closed subspace $\mathcal{H}_0$ of $\mathcal{H}$ such that $\mathcal{D}_0 \subseteq \mathcal{D}(T)$ and $T\mathcal{D}_0 \subseteq \mathcal{H}_0$, and let P_0 denote the projection onto $\mathcal{H}_0$. Suppose that $T_0 := T \restriction \mathcal{D}_0$ is essentially self-adjoint on $\mathcal{H}_0$. Then $\mathcal{H}_0$ is a reducing subspace for T, and $\overline{T_0}$ is the part of T on $\mathcal{H}_0$.*

Proof Let $u \in \mathcal{D}(T)$. Using the facts that T is symmetric and $T_0 v \in \mathcal{H}_0$, we derive

$$\langle T_0 v, P_0 u\rangle = \langle P_0 T_0 v, u\rangle = \langle T_0 v, u\rangle = \langle v, Tu\rangle = \langle P_0 v, Tu\rangle = \langle v, P_0 Tu\rangle$$

for arbitrary $v \in \mathcal{D}_0$. Therefore, $P_0 u \in \mathcal{D}((T_0)^*)$ and $(T_0)^* P_0 u = P_0 Tu$. Since T_0 is essentially self-adjoint, we have $(T_0)^* = \overline{T_0}$. Because T is closed, $\overline{T_0} \subseteq T$. From these facts it follows that $P_0 u \in \mathcal{D}(\overline{T_0}) \subseteq \mathcal{D}(T)$ and $T P_0 u = \overline{T_0} P_0 u = P_0 Tu$. This proves that $P_0 \in \{T\}'$. Hence, $\mathcal{H}_0$ is reducing by Proposition 1.15.

Clearly, $\mathcal{D}(\overline{T_0}) = P_0 \mathcal{D}(\overline{T_0}) \subseteq P_0 \mathcal{D}(T)$. Since $P_0 u \in \mathcal{D}(\overline{T_0})$ for $u \in \mathcal{D}(T)$ as shown in the preceding paragraph, we have $P_0 \mathcal{D}(T) = \mathcal{D}(\overline{T_0})$. Hence, $\overline{T_0}$ is the part of T on $\mathcal{H}_0$ by the last assertion of Proposition 1.15. $\square$

1.5 Exercises

1. Prove the two distributivity laws (1.1), that is,
$$(S+T)Q = SQ + TQ \quad \text{and} \quad R(S+T) \supseteq RS + RT.$$

2. Find linear operators R, S, T on $\mathcal{H}$ such that $R(S+T) \neq RS + RT$.
   Hint: Look for $S = -T$.

3. Let A and B be linear operators on $\mathcal{H}$, and let $C = AB$. Suppose that B is injective. Show that $CB^{-1} \subseteq A$. Give an example of operators for which $CB^{-1} \neq A$.

4. Let T and S be linear operators on $\mathcal{H}$. Suppose that $\mathcal{D}$ is a dense linear subspace of $\mathcal{H}$ such that $\mathcal{D} \subseteq \mathcal{D}(T) \cap \mathcal{D}(S)$ and $\langle Tx, x \rangle = \langle Sx, x \rangle$ for all $x \in \mathcal{D}$. Prove that $Tx = Sx$ for all $x \in \mathcal{D}$.

5. Show that a continuous linear operator $T : \mathcal{H}_1 \to \mathcal{H}_2$ is closed if and only if its domain $\mathcal{D}(T)$ is closed in $\mathcal{H}_1$.

6. Let T be a linear operator on $\mathcal{H}$, and let $F \not\equiv 0$ be a linear functional on $\mathcal{D}(T)$.
   a. Show that $\mathcal{N}(F)$ is a core for T if and only if F is not continuous on the normed space $(\mathcal{D}(T), \|\cdot\|_T)$.
   b. Suppose that T is closed and F is continuous on $(\mathcal{D}(T), \|\cdot\|_T)$. Show that $T{\restriction}\mathcal{N}(F)$ is also closed.

7. Let T be a closed operator and $\lambda \in \mathbb{C}$. Show that $\mathcal{N}(T - \lambda I)$ is closed.

8. Find examples of densely defined closed operators T and S on a Hilbert space for which $(T+S)^* \neq T^* + S^*$.

9. Let T be a closed operator on $\mathcal{H}$, and let $B \in \mathbf{B}(\mathcal{H})$. Show that the operators TB and $T+B$ are closed.

10. Find a closed operator T on $\mathcal{H}$ and an operator $B \in \mathbf{B}(\mathcal{H})$ such that BT is not closed.
    Hint: Take $B = \langle \cdot, x \rangle y$, where $x \notin \mathcal{D}(T^*)$ and $y \neq 0$.

11. Let T denote the multiplication operator by $\varphi \in C(\mathbb{R})$ on $L^2(\mathbb{R})$ with domain $\mathcal{D}(T) = \{f \in L^2(\mathbb{R}) : \varphi \cdot f \in L^2(\mathbb{R})\}$. Show that $C_0^\infty(\mathbb{R})$ is a core for T.

12. Let $\alpha = (\alpha_n)_{n \in \mathbb{N}}$ be a complex sequence. Define the linear operators L_α, R_α, and T_α on $l^2(\mathbb{N})$ by $T_\alpha(\varphi_n) = (\alpha_n \varphi_n)$, $L_\alpha(\varphi_1, \varphi_2, \ldots) = (\alpha_1 \varphi_2, \alpha_2 \varphi_3, \ldots)$, $R_\alpha(\varphi_1, \varphi_2, \ldots) = (0, \alpha_1 \varphi_1, \alpha_2 \varphi_2, \ldots)$ with domains
    $\mathcal{D}(T_\alpha) = \{(\varphi_n) \in l^2(\mathbb{N}) : (\alpha_n \varphi_n) \in l^2(\mathbb{N})\}$,
    $\mathcal{D}(L_\alpha) = \{(\varphi_n) \in l^2(\mathbb{N}) : (\alpha_1 \varphi_2, \alpha_2 \varphi_3, \ldots) \in l^2(\mathbb{N})\}$,
    $\mathcal{D}(R_\alpha) = \{(\varphi_n) \in l^2(\mathbb{N}) : (0, \alpha_1 \varphi_1, \alpha_2 \varphi_2, \ldots) \in l^2(\mathbb{N})\}$.
    a. Prove that these three operators are closed.
    b. Determine the adjoints of these operators.
    c. Prove that $\mathcal{D}_0 = \{(\varphi_1, \ldots, \varphi_n, 0, \ldots) : \varphi_k \in \mathbb{C}, n \in \mathbb{N}\}$ is a core for each of these operators.

13. Consider the linear subspace $\mathcal{D} = \{(f_1, f_2) : f_1 \in H^1(0, 1), f_2 \in H^1(2, 3)\}$ of the Hilbert space $\mathcal{H} = L^2(0, 1) \oplus L^2(2, 3)$. Let $z, w \in \mathbb{C}$. Define the linear operators T_k, $k = 1, 2, 3$, by $T_k(f_1, f_2) = (-if_1', -if_2')$ with domains
    $\mathcal{D}(T_1) = \{(f_1, f_2) \in \mathcal{D} : f_1(0) = zf_2(3)\}$,
    $\mathcal{D}(T_2) = \{(f_1, f_2) \in \mathcal{D} : f_1(0) = zf_2(3), f_1(1) = wf_2(2)\}$,
    $\mathcal{D}(T_3) = \{(f_1, f_2) \in \mathcal{D} : f_1(0) = f_2(3) = 0, f_1(1) = zf_2(2)\}$.
    Determine the adjoint operators T_1^*, T_2^*, T_3^*.

14. Define the linear operator T on the Hilbert space $L^2(0, 1)$ by $Tf = -if'$ for
    $f \in \mathcal{D}(T) := \{f \in C^1([0, 1]) : f(0) = f(1) = 0, f'(0) = f'(1)\}$.
    a. Determine the operators T^*, $(T^2)^*$, $(T^*)^2$. Compare $(T^2)^*$ and $(T^*)^2$.
    b. Is $C_0^\infty(0, 1)$ a core for T or T^2?

15. Let $\mathcal{J}$ be an open interval, and $T = -i\frac{d}{dx}$ on $\mathcal{D}(T) = H_0^1(\mathcal{J})$. For $c \in \mathcal{J}$, define
    $\mathcal{D}_0 := \{f \in H_0^1(\mathcal{J}) : f(c) = 0\}$, $\mathcal{D}_1 = \{f \in H_0^2(\mathcal{J}) : f(c) = f'(c) = 0\}$, and
    $\mathcal{D}_2 = \{f \in H_0^2(\mathcal{J}); f(c) = 0\}$.
    a. Are $\mathcal{D}_0$ or $\mathcal{D}_2$ cores for T?
    b. Are $\mathcal{D}_1$ or $\mathcal{D}_2$ cores for T^2?
    c. Which of the operators $T \restriction \mathcal{D}_0$, $T \restriction \mathcal{D}_1$, $T^2 \restriction \mathcal{D}_1$, and $T^2 \restriction \mathcal{D}_2$ are closed?

16. Let T be a closed operator on a Hilbert space $\mathcal{H}$.
    a. Show that the commutant $\{T\}'$ is a subalgebra of $\mathbf{B}(\mathcal{H})$.
    b. Show that $\{T\}'$ is closed in $\mathbf{B}(\mathcal{H})$ with respect to the strong convergence.
    c. Give an example for which $B \in \{T\}'$ but $B^* \notin \{T\}'$.

17. Let T be a linear operator on $\mathcal{H}$ and $B \in \mathbf{B}(\mathcal{H})$.
    a. Suppose that T is closable. Show that $B \in \{T\}'$ implies $B \in \{\overline{T}\}'$.
    b. Suppose that T is densely defined. Show that $B \in \{T\}'$ implies $B^* \in \{T^*\}'$.
    c. Suppose that T is self-adjoint and BT is symmetric. Show that $B \in \{T\}'$.

Chapter 2
The Spectrum of a Closed Operator

The main themes of this chapter are the most important concepts concerning general closed operators, spectrum and resolvent. Section 2.2 is devoted to basic properties of these notions for arbitrary closed operators. In Sect. 2.3 we treat again differentiation operators as illustrating examples. First however, in Sect. 2.1 we introduce regular points and defect numbers and derive some technical results that are useful for the study of spectra (Sect. 2.2) and for self-adjointness criteria (Sect. 13.2).

2.1 Regular Points and Defect Numbers of Operators

Let T be a linear operator on a Hilbert space $\mathcal{H}$.

Definition 2.1 A complex number λ is called a *regular point* for T if there exists a number $c_\lambda > 0$ such that

$$\|(T - \lambda I)x\| \geq c_\lambda \|x\| \quad \text{for all } x \in \mathcal{D}(T). \tag{2.1}$$

The set of regular points of T is the *regularity domain* of T and denoted by $\pi(T)$.

Remark There is no unique symbol for the regularity domain of an operator in the literature. It is denoted by $\hat{\rho}(T)$ in [BS], by $\Pi(T)$ in [EE], and by $\Gamma(T)$ in [We]. Many books have no special symbol for this set.

Recall that the dimension of a Hilbert space $\mathcal{H}$, denoted by $\dim \mathcal{H}$, is defined by the cardinality of an orthonormal basis of $\mathcal{H}$.

Definition 2.2 For $\lambda \in \pi(T)$, we call the linear subspace $\mathcal{R}(T - \lambda I)^\perp$ of $\mathcal{H}$ the *deficiency subspace* of T at λ and its dimension $d_\lambda(T) := \dim \mathcal{R}(T - \lambda I)^\perp$ the *defect number* of T at λ.

Deficiency spaces and defect numbers will play a crucial role in the theory of self-adjoint extensions of symmetric operators developed in Chap. 13.

A number of properties of these notions are collected in the next proposition.

K. Schmüdgen, *Unbounded Self-adjoint Operators on Hilbert Space*,
Graduate Texts in Mathematics 265,
DOI 10.1007/978-94-007-4753-1_2, © Springer Science+Business Media Dordrecht 2012

Proposition 2.1 *Let T be a linear operator on $\mathcal{H}$, and $\lambda \in \mathbb{C}$.*

(i) *$\lambda \in \pi(T)$ if and only if $T - \lambda I$ has a bounded inverse $(T - \lambda I)^{-1}$ defined on $\mathcal{R}(T - \lambda I)$. In this case inequality (2.1) holds with $c_\lambda = \|(T - \lambda I)^{-1}\|^{-1}$.*

(ii) *$\pi(T)$ is an open subset of $\mathbb{C}$. More precisely, if $\lambda_0 \in \pi(T)$, $\lambda \in \mathbb{C}$, and $|\lambda - \lambda_0| < c_{\lambda_0}$, where c_{λ_0} is a constant satisfying (2.1) for λ_0, then $\lambda \in \pi(T)$.*

(iii) *If T is closable, then $\pi(\overline{T}) = \pi(T)$, $d_\lambda(\overline{T}) = d_\lambda(T)$, and $\mathcal{R}(\overline{T} - \lambda I)$ is the closure of $\mathcal{R}(T - \lambda I)$ in $\mathcal{H}$ for each $\lambda \in \pi(T)$.*

(iv) *If T is closed and $\lambda \in \pi(T)$, then $\mathcal{R}(T - \lambda I)$ is a closed linear subspace of $\mathcal{H}$.*

Proof (i): First suppose that $\lambda \in \pi(T)$. Then $\mathcal{N}(T - \lambda I) = \{0\}$ by (2.1), so the inverse $(T - \lambda I)^{-1}$ exists. Let $y \in \mathcal{D}((T - \lambda I)^{-1}) = \mathcal{R}(T - \lambda I)$. Then we have $y = (T - \lambda I)x$ for some $x \in \mathcal{D}(T)$, and hence,

$$\left\|(T - \lambda I)^{-1} y\right\| = \|x\| \le c_\lambda^{-1} \left\|(T - \lambda I)x\right\| = c_\lambda^{-1} \|y\|$$

by (2.1). That is, $(T - \lambda I)^{-1}$ is bounded, and $\|(T - \lambda I)^{-1}\| \le c_\lambda^{-1}$.

Assume now that $(T - \lambda I)^{-1}$ has a bounded inverse. Then, with x and y as above,

$$\|x\| = \left\|(T - \lambda I)^{-1} y\right\| \le \left\|(T - \lambda I)^{-1}\right\| \|y\| = \left\|(T - \lambda I)^{-1}\right\| \left\|(T - \lambda I)x\right\|.$$

Hence, (2.1) holds with $c_\lambda = \|(T - \lambda I)^{-1}\|^{-1}$.

(ii): Let $\lambda_0 \in \pi(T)$ and $\lambda \in \mathbb{C}$. Suppose that $|\lambda - \lambda_0| < c_{\lambda_0}$, where c_{λ_0} is a constant such that (2.1) holds. Then for $x \in \mathcal{D}(T)$, we have

$$\left\|(T - \lambda I)x\right\| = \left\|(T - \lambda_0 I)x - (\lambda - \lambda_0)x\right\| \ge \left\|(T - \lambda_0 I)x\right\| - |\lambda - \lambda_0| \|x\|$$
$$\ge \left(c_{\lambda_0} - |\lambda - \lambda_0|\right) \|x\|.$$

Thus, $\lambda \in \pi(T)$, since $|\lambda - \lambda_0| < c_{\lambda_0}$. This shows that the set $\pi(T)$ is open.

(iii): Let y be in the closure of $\mathcal{R}(T - \lambda I)$. Then there is a sequence $(x_n)_{n \in \mathbb{N}}$ of vectors $x_n \in \mathcal{D}(T)$ such that $y_n := (T - \lambda I)x_n \to y$ in $\mathcal{H}$. By (2.1) we have

$$\|x_n - x_k\| \le c_\lambda^{-1} \left\|(T - \lambda I)(x_n - x_k)\right\| = c_\lambda^{-1} \|y_n - y_k\|.$$

Hence, (x_n) is a Cauchy sequence in $\mathcal{H}$, because (y_n) is a Cauchy sequence. Let $x := \lim_n x_n$. Then $\lim_n T x_n = \lim_n (y_n + \lambda x_n) = y + \lambda x$. Since T is closable, $x \in \mathcal{D}(\overline{T})$ and $\overline{T}x = y + \lambda x$, so that $y = (\overline{T} - \lambda I)x \in \mathcal{R}(\overline{T} - \lambda I)$. This proves that $\overline{\mathcal{R}(T - \lambda I)} \subseteq \mathcal{R}(\overline{T} - \lambda I)$. The converse inclusion follows immediately from the definition of the closure $\overline{T}$. Thus, $\overline{\mathcal{R}(T - \lambda I)} = \mathcal{R}(\overline{T} - \lambda I)$.

Clearly, $\pi(\overline{T}) = \pi(T)$ by (2.1). Since $\mathcal{R}(\overline{T} - \lambda I)$ is the closure of $\mathcal{R}(T - \lambda I)$, both have the same orthogonal complements, so $d_\lambda(\overline{T}) = d_\lambda(T)$ for $\lambda \in \pi(T)$.

(iv) follows at once from (iii). □

Combining Proposition 2.1(iii) and formula (1.7), we obtain

Corollary 2.2 *If T is a closable densely defined linear operator, and $\lambda \in \pi(T)$, then $\mathcal{H} = \mathcal{R}(\overline{T} - \lambda I) \oplus \mathcal{N}(T^* - \overline{\lambda} I)$.*

The following technical lemma is needed in the proof of the next proposition.

Lemma 2.3 *If $\mathcal{F}$ and $\mathcal{G}$ are closed linear subspaces of a Hilbert space $\mathcal{H}$ such that $\dim \mathcal{F} < \dim \mathcal{G}$, then there exists a nonzero vector $y \in \mathcal{G} \cap \mathcal{F}^{\perp}$.*

Proof In this proof we denote by $|M|$ the cardinality of a set M. First, we suppose that $k = \dim \mathcal{F}$ is finite. We take a $(k+1)$-dimensional subspace $\mathcal{G}_0$ of $\mathcal{G}$ and define the mapping $\Phi : \mathcal{G}_0 \to \mathcal{F}$ by $\Phi(x) = Px$, where P is the projection of $\mathcal{H}$ onto $\mathcal{F}$. If Φ would be injective, then $k + 1 = \dim \mathcal{G}_0 = \dim \Phi(\mathcal{G}_0) \le \dim \mathcal{F} = k$, which is a contradiction. Hence, there is a nonzero vector $y \in \mathcal{N}(\Phi)$. Clearly, $y \in \mathcal{G} \cap \mathcal{F}^{\perp}$.

Now suppose that $\dim \mathcal{F}$ is infinite. Let $\{ f_k : k \in K \}$ and $\{ g_l : l \in L \}$ be orthonormal bases of $\mathcal{F}$ and $\mathcal{G}$, respectively. Set $L_k := \{ l \in L : \langle f_k, g_l \rangle \ne 0 \}$ for $k \in K$ and $L' = \bigcup_{k \in K} L_k$. Since each set L_k is at most countable and $\dim \mathcal{F} = |K|$ is infinite, we have $|L'| \le |K||\mathbb{N}| = |K|$. Since $|K| = \dim \mathcal{F} < \dim \mathcal{G} = |L|$ by assumption, we deduce that $L' \ne L$. Each vector g_l with $l \in L \setminus L'$ is orthogonal to all f_k, $k \in K$, and hence, it belongs to $\mathcal{G} \cap \mathcal{F}^{\perp}$. $\qquad\square$

The next proposition is a classical result of *M.A. Krasnosel'skii* and *M.G. Krein*.

Proposition 2.4 *Suppose that T is a closable linear operator on $\mathcal{H}$. Then the defect number $d_\lambda(T)$ is constant on each connected component of the open set $\pi(T)$.*

Proof By Proposition 2.1(iii), we can assume without loss of generality that T is closed. Then $\mathcal{R}(T - \mu I)$ is closed for all $\mu \in \pi(T)$ by Proposition 2.1(iv). Therefore, setting $\mathcal{K}_\mu := \mathcal{R}(T - \mu I)^{\perp}$, we have

$$(\mathcal{K}_\mu)^{\perp} = \mathcal{R}(T - \mu I) \quad \text{for } \mu \in \pi(T). \tag{2.2}$$

Suppose that $\lambda_0 \in \pi(T)$ and $\lambda \in \mathbb{C}$ are such that $|\lambda - \lambda_0| < c_{\lambda_0}$. Then $\lambda \in \pi(T)$ by Proposition 2.1(ii). The crucial step is to prove that $d_\lambda(T) = d_{\lambda_0}(T)$.

Assume to the contrary that $d_\lambda(T) \ne d_{\lambda_0}(T)$. First suppose that $d_\lambda(T) < d_{\lambda_0}(T)$. By Lemma 2.3 there exists a nonzero vector $y \in \mathcal{K}_{\lambda_0}$ such that $y \in (\mathcal{K}_\lambda)^{\perp}$. Then $y \in \mathcal{R}(T - \lambda I)$ by (2.2), say $y = (T - \lambda I)x$ for some nonzero $x \in \mathcal{D}(T)$. Since $y = (T - \lambda I)x \in \mathcal{K}_{\lambda_0}$, we have

$$\langle (T - \lambda I)x, (T - \lambda_0 I)x \rangle = 0. \tag{2.3}$$

Equation (2.3) is symmetric in λ and λ_0, so it holds also when $d_{\lambda_0}(T) < d_\lambda(T)$. Using (2.3), we derive

$$\| (T - \lambda_0 I)x \|^2 = \langle (T - \lambda I)x + (\lambda - \lambda_0)x, (T - \lambda_0 I)x \rangle$$
$$\le |\lambda - \lambda_0| \|x\| \|(T - \lambda_0 I)x\|.$$

Thus, $\|(T - \lambda_0 I)x\| \le |\lambda - \lambda_0| \|x\|$. Since $x \ne 0$ and $|\lambda - \lambda_0| < c_{\lambda_0}$, we obtain

$$|\lambda - \lambda_0| \|x\| < c_{\lambda_0} \|x\| \le \| (T - \lambda_0 I)x \| \le |\lambda - \lambda_0| \|x\|$$

by (2.1), which is a contradiction. Thus, we have proved that $d_\lambda(T) = d_{\lambda_0}(T)$.

The proof will be now completed by using a well-known argument from elementary topology. Let α and β be points of the same connected component $\mathcal{U}$ of the open set $\pi(T)$ in the complex plane. Then there exists a polygonal path $\mathcal{P}$ contained in $\mathcal{U}$ from α to β. For $\lambda \in \mathcal{P}$, let $\mathcal{U}_\lambda = \{\lambda' \in \mathbb{C} : |\lambda' - \lambda| < c_\lambda\}$. Then $\{\mathcal{U}_\lambda : \lambda \in \mathcal{P}\}$ is an open cover of the compact set $\mathcal{P}$, so there exists a finite subcover $\{\mathcal{U}_{\lambda_1}, \ldots, \mathcal{U}_{\lambda_s}\}$ of $\mathcal{P}$. Since $d_\lambda(T)$ is constant on each open set $\mathcal{U}_{\lambda_k}$ as shown in the preceding paragraph, we conclude that $d_\alpha(T) = d_\beta(T)$. $\square$

The *numerical range* of a linear operator T in $\mathcal{H}$ is defined by

$$\Theta(T) = \{\langle Tx, x \rangle : x \in \mathcal{D}(T),\ \|x\| = 1\}.$$

A classical result of F. Hausdorff (see, e.g., [K2, V, Theorem 3.1]) says that $\Theta(T)$ is a convex set. In general, the set $\Theta(T)$ is neither closed nor open for a bounded or closed operator. However, we have the following simple but useful fact.

Lemma 2.5 *Let T be a linear operator on $\mathcal{H}$. If $\lambda \in \mathbb{C}$ is not in the closure of $\Theta(T)$, then $\lambda \in \pi(T)$.*

Proof Set $\gamma_\lambda := \mathrm{dist}(\lambda, \Theta(T)) > 0$. For $x \in \mathcal{D}(T)$, $\|x\| = 1$, we have

$$\|(T - \lambda I)x\| \geq |\langle (T - \lambda I)x, x \rangle| = |\langle Tx, x \rangle - \lambda| \geq \gamma_\lambda,$$

so that $\|(T - \lambda I)y\| \geq \gamma_\lambda \|y\|$ for arbitrary $y \in \mathcal{D}(T)$. Hence, $\lambda \in \pi(T)$. $\square$

2.2 Spectrum and Resolvent of a Closed Operator

In this section we assume that T is a *closed* linear operator on a Hilbert space $\mathcal{H}$.

Definition 2.3 A complex number λ belongs to the *resolvent set* $\rho(T)$ of T if the operator $T - \lambda I$ has a bounded everywhere on $\mathcal{H}$ defined inverse $(T - \lambda I)^{-1}$, called the *resolvent* of T at λ and denoted by $R_\lambda(T)$.

The set $\sigma(T) := \mathbb{C} \setminus \rho(T)$ is called the *spectrum* of the operator T.

Remarks 1. Formally, the preceding definition could be also used to define the spectrum for a not necessarily closed operator T. But if $\lambda \in \rho(T)$, then the bounded everywhere defined operator $(T - \lambda I)^{-1}$ is closed, so is its inverse $T - \lambda I$ by Theorem 1.8(vi) and hence T. Therefore, if T is *not closed*, we would always have that $\rho(T) = \emptyset$ and $\sigma(T) = \mathbb{C}$ according to Definition 2.3, so the notion of spectrum becomes trivial. For this reason, we assumed above that the operator T is closed.

2. The reader should notice that in the literature the resolvent $R_\lambda(T)$ is often defined by $(\lambda I - T)^{-1}$ rather than $(T - \lambda I)^{-1}$ as we do.

By Definition 2.3, a complex number λ is in $\rho(T)$ if and only if there is an operator $B \in \mathbf{B}(\mathcal{H})$ such that

$$B(T - \lambda I) \subseteq I \quad \text{and} \quad (T - \lambda I)B = I.$$

The operator B is then uniquely determined and equal to the resolvent $R_\lambda(T)$.

Proposition 2.6

(i) $\rho(T) = \{\lambda \in \pi(T) : d_\lambda(T) = 0\}$.
(ii) $\rho(T)$ is an open subset, and $\sigma(T)$ is a closed subset of $\mathbb{C}$.

Proof (i) follows at once from Proposition 2.1, (i) and (iv). Since $\pi(T)$ is open and $d_\lambda(T)$ is locally constant on $\pi(T)$ by Proposition 2.4, the assertion of (i) implies that $\rho(T)$ is open. Hence, $\sigma(T) = \mathbb{C} \setminus \rho(T)$ is closed. □

The requirement that the inverse $(T - \lambda I)^{-1}$ is *bounded* can be omitted in Definition 2.3. This is the first assertion of the next proposition.

Proposition 2.7 *Let T be a closed operator on $\mathcal{H}$.*

(i) $\rho(T)$ *is the set of all numbers $\lambda \in \mathbb{C}$ such that $T - \lambda I$ is a bijective mapping of $\mathcal{D}(T)$ on $\mathcal{H}$ (or equivalently, $\mathcal{N}(T - \lambda I) = \{0\}$ and $\mathcal{R}(T - \lambda I) = \mathcal{H}$).*
(ii) *Suppose that $\mathcal{D}(T)$ is dense in $\mathcal{H}$ and let $\lambda \in \mathbb{C}$. Then $\lambda \in \sigma(T)$ if and only if $\bar{\lambda} \in \sigma(T^*)$. Moreover, $R_\lambda(T)^* = R_{\bar{\lambda}}(T^*)$ for $\lambda \in \rho(T)$.*

Proof (i): Clearly, $T - \lambda I$ is bijective if and only if the inverse $(T - \lambda I)^{-1}$ exists and is everywhere defined on $\mathcal{H}$. It remains to prove that $(T - \lambda I)^{-1}$ is bounded if $T - \lambda I$ is bijective. Since T is closed, $T - \lambda I$ is closed, and so is its inverse $(T - \lambda I)^{-1}$ by Theorem 1.8(vi). That is, $(T - \lambda I)^{-1}$ is a closed linear operator defined on the whole Hilbert space $\mathcal{H}$. Hence, $(T - \lambda I)^{-1}$ is bounded by the closed graph theorem.

(ii): It suffices to prove the corresponding assertion for the resolvent sets.

Let $\lambda \in \rho(T)$. Then, by Theorem 1.8(iv), $(T - \lambda I)^* = T^* - \bar{\lambda} I$ is invertible, and $(T^* - \bar{\lambda} I)^{-1} = ((T - \lambda I)^{-1})^*$. Since $(T - \lambda I)^{-1} \in \mathbf{B}(\mathcal{H})$ by $\lambda \in \rho(T)$, we have $((T - \lambda I)^{-1})^* \in \mathbf{B}(\mathcal{H})$, and hence $(T^* - \bar{\lambda} I)^{-1} \in \mathbf{B}(\mathcal{H})$, that is, $\bar{\lambda} \in \rho(T^*)$.

Replacing T by T^* and λ by $\bar{\lambda}$ and using the fact that $T = T^{**}$, it follows that $\bar{\lambda} \in \rho(T^*)$ implies $\lambda \in \rho(T)$. Thus, $\lambda \in \rho(T)$ if and only if $\bar{\lambda} \in \rho(T^*)$. □

Proposition 2.8 *Let T be a closed operator on $\mathcal{H}$. Let $\mathcal{U}$ be a connected open subset of $\mathbb{C} \setminus \overline{\Theta(T)}$. If there exists a number $\lambda_0 \in \mathcal{U}$ which is contained in $\rho(T)$, then $\mathcal{U} \subseteq \rho(T)$. Moreover, $\|(T - \lambda I)^{-1}\| \le (\mathrm{dist}(\lambda, \Theta(T)))^{-1}$ for $\lambda \in \mathcal{U}$.*

Proof By Lemma 2.5 we have $\mathcal{U} \subseteq \pi(T)$. Therefore, since T is closed, it follows from Proposition 2.1(iv) that $\mathcal{R}(T - \lambda I)$ is closed in $\mathcal{H}$ for all $\lambda \in \mathcal{U}$. By Proposition 2.4, the defect number $d_\lambda(T)$ is constant on the connected open set $\mathcal{U}$. But $d_{\lambda_0}(T) = 0$ for $\lambda_0 \in \mathcal{U}$, since $\lambda_0 \in \rho(T)$. Hence, $d_\lambda(T) = 0$ on the whole set $\mathcal{U}$. Consequently, $\mathcal{U} \subseteq \rho(T)$ by Proposition 2.6(i).

From the inequality $\|(T - \lambda I)y\| \ge \gamma_\lambda \|y\|$ for $y \in \mathcal{D}(T)$ shown in the proof of Lemma 2.5 we get $\|(T - \lambda I)^{-1}\| \le \gamma_\lambda^{-1}$ for $\lambda \in \mathcal{U}$, where $\gamma_\lambda = \mathrm{dist}(\lambda, \Theta(T))$. □

Next we define an important subset of the spectrum.

Definition 2.4 $\sigma_p(T) := \{\lambda \in \mathbb{C} : \mathcal{N}(T - \lambda I) \neq \{0\}\}$ is the *point spectrum* of T. We call $\lambda \in \sigma_p(T)$ an *eigenvalue* of T, the dimension of $\mathcal{N}(T - \lambda I)$ its *multiplicity*, and any nonzero element of $\mathcal{N}(T - \lambda I)$ an *eigenvector* of T at λ.

Let λ be a point of the spectrum $\sigma(T)$. Then, by Proposition 2.7(i), the operator $(T - \lambda I) : \mathcal{D}(T) \to \mathcal{H}$ is *not bijective*. This means that $T - \lambda I$ is not injective *or* $T - \lambda I$ is not surjective. Clearly, the point spectrum $\sigma_p(T)$ is precisely the set of all $\lambda \in \sigma(T)$ for which $T - \lambda I$ is not injective. Let us look now at the numbers where the surjectivity of the operator $T - \lambda I$ fails.

The set of all $\lambda \in \mathbb{C}$ for which $T - \lambda I$ has a bounded inverse which is not defined on the whole Hilbert space $\mathcal{H}$ is called the *residual spectrum* of T and denoted by $\sigma_r(T)$. Note that $\sigma_r(T) = \{\lambda \in \pi(T) : d_\lambda(T) \neq 0\}$. By Proposition 2.4 this description implies that $\sigma_r(T)$ is an open set. It follows from Proposition 3.10 below that for self-adjoint operators T, the residual spectrum $\sigma_r(T)$ is empty.

Further, the set of $\lambda \in \mathbb{C}$ for which the range of $T - \lambda I$ is not closed, that is, $\mathcal{R}(T - \lambda I) \neq \overline{\mathcal{R}(T - \lambda I)}$, is called the *continuous spectrum* $\sigma_c(T)$ of T. Then $\sigma(T) = \sigma_p(T) \cup \sigma_r(T) \cup \sigma_c(T)$, but the sets $\sigma_c(T)$ and $\sigma_p(T)$ are in general *not disjoint*, see Exercise 5.

Remark The reader should be cautioned that some authors (for instance, [RN, BEH]) define $\sigma_c(T)$ as the complement of $\sigma_p(T) \cup \sigma_r(T)$ in $\sigma(T)$; then $\sigma(T)$ becomes the *disjoint* union of the three parts.

Example 2.1 (*Example* 1.3 *continued*) Let φ be a continuous function on an interval $\mathcal{J}$. Recall that the operator M_φ was defined by $M_\varphi f = \varphi \cdot f$ for f in the domain $\mathcal{D}(M_\varphi) = \{f \in L^2(\mathcal{J}) : \varphi \cdot f \in L^2(\mathcal{J})\}$.

Statement $\sigma(M_\varphi)$ *is the closure of the set* $\varphi(\mathcal{J})$.

Proof Let $\lambda \in \varphi(\mathcal{J})$, say $\lambda = \varphi(t_0)$ for $t_0 \in \mathcal{J}$. Given $\varepsilon > 0$, by the continuity of φ there exists an interval $K \subseteq \mathcal{J}$ of positive length such that $|\varphi(t) - \varphi(t_0)| \leq \varepsilon$ for all $t \in K$. Then $\|(M_\varphi - \lambda I)\chi_K\| \leq \varepsilon \|\chi_K\|$. If λ would be in $\rho(M_\varphi)$, then

$$\|\chi_K\| = \|R_\lambda(M_\varphi)(M_\varphi - \lambda I)\chi_K\| \leq \|R_\lambda(M_\varphi)\| \varepsilon \|\chi_K\|,$$

which is impossible if $\varepsilon \|R_\lambda(M_\varphi)\| < 1$. Thus, $\lambda \in \sigma(M_\varphi)$ and $\varphi(\mathcal{J}) \subseteq \sigma(M_\varphi)$. Hence, $\overline{\varphi(\mathcal{J})} \subseteq \sigma(M_\varphi)$.

Suppose that $\lambda \notin \overline{\varphi(\mathcal{J})}$. Then there is a $c > 0$ such that $|\lambda - \varphi(t)| \geq c$ for all $t \in \mathcal{J}$. Hence, $\psi(t) := (\varphi(t) - \lambda)^{-1}$ is a bounded function on $\mathcal{J}$, so M_ψ is bounded, $\mathcal{D}(M_\psi) = L^2(\mathcal{J})$, and $M_\psi = (M_\varphi - \lambda I)^{-1}$. Therefore, $\lambda \in \rho(M_\varphi)$.   □ ∘

Now we turn to the resolvents. Suppose that T and S are closed operators on $\mathcal{H}$ such that $\mathcal{D}(S) \subseteq \mathcal{D}(T)$. Then the following *resolvent identities* hold:

$$R_\lambda(T) - R_\lambda(S) = R_\lambda(T)(S - T)R_\lambda(S) \quad \text{for } \lambda \in \rho(S) \cap \rho(T), \quad (2.4)$$

$$R_\lambda(T) - R_{\lambda_0}(T) = (\lambda - \lambda_0)R_\lambda(T)R_{\lambda_0}(T) \quad \text{for } \lambda, \lambda_0 \in \rho(T). \quad (2.5)$$

Indeed, if $\lambda \in \rho(S) \cap \rho(T)$ and $x \in \mathcal{H}$, we have $R_\lambda(S)x \in \mathcal{D}(S) \subseteq \mathcal{D}(T)$ and

$$R_\lambda(T)(S - T)R_\lambda(S)x = R_\lambda(T)\big((S - \lambda I) - (T - \lambda I)\big)R_\lambda(S)x$$
$$= R_\lambda(T)x - R_\lambda(S)x,$$

which proves (2.4). The second formula (2.5) follows at once from the first (2.4) by setting $S = T + (\lambda - \lambda_0)I$ and using the relation $R_\lambda(S) = R_{\lambda_0}(T)$.

Both identities (2.4) and (2.5) are very useful for the study of operator equations. In particular, (2.5) implies that $R_\lambda(T)$ and $R_{\lambda_0}(T)$ commute.

The next proposition shows that the resolvent $R_\lambda(T)$ is an analytic function on the resolvent set $\rho(T)$ with values in the Banach space $(\mathbf{B}(\mathcal{H}), \|\cdot\|)$.

Proposition 2.9 *Suppose that* $\lambda_0 \in \rho(T)$, $\lambda \in \mathbb{C}$, *and* $|\lambda - \lambda_0| < \|R_{\lambda_0}(T)\|^{-1}$. *Then we have* $\lambda \in \rho(T)$ *and*

$$R_\lambda(T) = \sum_{n=0}^{\infty} (\lambda - \lambda_0)^n R_{\lambda_0}(T)^{n+1}, \quad (2.6)$$

where the series converges in the operator norm. In particular,

$$\lim_{\lambda \to \lambda_0} \|R_\lambda(T) - R_{\lambda_0}(T)\| = 0 \quad \text{for } \lambda_0 \in \rho(T). \quad (2.7)$$

Proof As stated in Proposition 2.1(i), (2.1) holds with $c_{\lambda_0} = \|R_{\lambda_0}(T)\|^{-1}$, so that $|\lambda - \lambda_0| < c_{\lambda_0}$ by our assumption. Therefore, $\lambda \in \pi(T)$ and $d_\lambda(T) = d_{\lambda_0}(T) = 0$, and hence, $\lambda \in \rho(T)$ by Propositions 2.4 and 2.6.

Since $\|(\lambda - \lambda_0)R_{\lambda_0}(T)\| < 1$ by assumption, the operator $I - (\lambda - \lambda_0)R_{\lambda_0}(T)$ has a bounded inverse on $\mathcal{H}$ which is given by the Neumann series

$$\big(I - (\lambda - \lambda_0)R_{\lambda_0}(T)\big)^{-1} = \sum_{n=0}^{\infty} (\lambda - \lambda_0)^n R_{\lambda_0}(T)^n. \quad (2.8)$$

On the other hand, we have $R_\lambda(T)(I - (\lambda - \lambda_0)R_{\lambda_0}(T)) = R_{\lambda_0}(T)$ by (2.5), and hence, $R_\lambda(T) = R_{\lambda_0}(T)(I - (\lambda - \lambda_0)R_{\lambda_0}(T))^{-1}$. Multiplying (2.8) by $R_{\lambda_0}(T)$ from the left and using the latter identity, we obtain $R_\lambda(T)$. This proves (2.6).

Since analytic operator-valued functions are continuous, (2.6) implies (2.7). $\square$

From formula (2.6) it follows in particular that for arbitrary vectors $x, y \in \mathcal{H}$, the complex function $\lambda \to \langle R_\lambda(T)x, y \rangle$ is analytic on the resolvent set $\rho(T)$.

For $T \in \mathbf{B}(\mathcal{H})$, it is well known (see, e.g., [RS1, Theorem VI.6]) that the spectrum $\sigma(T)$ is not empty and contained in a circle centered at the origin with radius

$$r(T) := \lim_{n \to \infty} \|T^n\|^{1/n}.$$

This number $r(T)$ is called the *spectral radius* of the operator T. Clearly, we have $r(T) \le \|T\|$. If $T \in \mathbf{B}(\mathcal{H})$ is self-adjoint, then $r(T) = \|T\|$.

By Proposition 2.6 the spectrum of a closed operator is a closed subset of $\mathbb{C}$. Let us emphasize that *any closed subset* (!) of the complex plane arises in this manner. Example 2.2 shows that each *nonempty* closed subset is spectrum of some closed operator. A closed operator with *empty* spectrum is given in Example 2.4 below.

Example 2.2 Suppose that M is a nonempty closed subset of $\mathbb{C}$. Since $\mathbb{C}$ is separable, so is M, that is, there exists a countable subset $\{r_n : n \in \mathbb{N}\}$ of M which is dense in M. Define the operator T on $l^2(\mathbb{N})$ by $\mathcal{D}(T) = \{(x_n) \in l^2(\mathbb{N}) : (r_n x_n) \in l^2(\mathbb{N})\}$ and $T(x_n) = (r_n x_n)$ for $(x_n) \in \mathcal{D}(T)$. It is easily seen that $\mathcal{D}(T) = \mathcal{D}(T^*)$ and $T^*(x_n) = (\overline{r_n} x_n)$ for $(x_n) \in \mathcal{D}(T^*)$. Hence, $T = T^{**}$, so T is closed. Each number r_n is an eigenvalue of T, and we have $\sigma(T) = \overline{\{r_n : n \in \mathbb{N}\}} = M$.   ○

The next propositions relate the spectrum of the resolvent to the spectrum of the operator. Closed operators with compact resolvents will play an important role in several later chapters of this book.

Proposition 2.10 *Let λ_0 be a fixed number of $\rho(T)$, and let $\lambda \in \mathbb{C}$, $\lambda \ne \lambda_0$.*

(i) *$\lambda \in \rho(T)$ if and only if $(\lambda - \lambda_0)^{-1} \in \rho(R_{\lambda_0}(T))$.*
(ii) *λ is an eigenvalue of T if and only if $(\lambda - \lambda_0)^{-1}$ is an eigenvalue of $R_{\lambda_0}(T)$. In this case both eigenvalues have the same multiplicities.*

Proof Both assertions are easy consequences of the following identity:

$$T - \lambda I = \big(R_{\lambda_0}(T) - (\lambda - \lambda_0)^{-1} I\big)(T - \lambda_0 I)(\lambda_0 - \lambda). \tag{2.9}$$

(i): Since $(T - \lambda_0 I)(\lambda_0 - \lambda)$ is a bijection from $\mathcal{D}(T)$ to $\mathcal{H}$, it follows from (2.9) that $T - \lambda I$ is a bijection from $\mathcal{D}(T)$ to $\mathcal{H}$ if and only if $R_{\lambda_0}(T) - (\lambda - \lambda_0)^{-1} I$ is a bijection of $\mathcal{H}$. By Proposition 2.7(i) this gives the assertion.

(ii): From (2.9) we conclude that $(T - \lambda_0 I)(\lambda_0 - \lambda)$ is a bijection of $\mathcal{N}(T - \lambda I)$ on $\mathcal{N}(R_{\lambda_0}(T) - (\lambda - \lambda_0)^{-1} I)$.   □

We shall say that a closed operator T has a *purely discrete spectrum* if $\sigma(T)$ consists only of eigenvalues of finite multiplicities which have no finite accumulation point.

Proposition 2.11 *Suppose that there exists a $\lambda_0 \in \rho(T)$ such that $R_{\lambda_0}(T)$ is compact. Then $R_\lambda(T)$ is compact for all $\lambda \in \rho(T)$, and T has a purely discrete spectrum.*

Proof The compactness of $R_\lambda(T)$ follows at once from the resolvent identity (2.5). By Theorem A.3 all nonzero numbers in the spectrum of the compact operator $R_{\lambda_0}(T)$ are eigenvalues of finite multiplicities which have no nonzero accumulation point. By Proposition 2.10 this implies that the operator T has a purely discrete spectrum.   □

2.3 Examples: Differentiation Operators II

In this section we determine spectra and resolvents of the differentiation operators $-i\frac{d}{dx}$ on intervals from Sect. 1.3.1.

Example 2.3 (*Example* 1.4 *continued: bounded interval* (a, b)) Recall that $\mathcal{D}(T^*) = H^1(a, b)$ and $T^*f = -if'$ for $f \in \mathcal{D}(T^*)$. For each $\lambda \in \mathbb{C}$, $f_\lambda(x) := e^{i\lambda x}$ is in $\mathcal{D}(T^*)$, and $T^*f_\lambda = \lambda f_\lambda$, so $\lambda \in \sigma_p(T^*)$. Thus, $\sigma(T^*) = \mathbb{C}$. Since $T = (T^*)^*$, Proposition 2.7(ii) implies that $\sigma(T) = \mathbb{C}$.   ∘

Example 2.4 (*Example* 1.5 *continued*)

Statement $\sigma(S_z) = \{\lambda \in \mathbb{C} : e^{i\lambda(a-b)}z = 1\}$ *for* $z \in \mathbb{C}$ *and* $\sigma(S_\infty) = \emptyset$.

Proof Let $\lambda \in \mathbb{C}$ and $g \in L^2(a, b)$. In order to "guess" the formula for the resolvent of S_z, we try to find an element $f \in \mathcal{D}(S_z)$ such that $(S_z - \lambda I)f \equiv -if' - \lambda f = g$. The general solution of the differential equation $-if' - \lambda f = g$ is

$$f(x) = ie^{i\lambda x}\left(\int_a^x e^{-i\lambda t}g(t)\, dt + c_{\lambda,g}\right), \quad \text{where } c_{\lambda,g} \in \mathbb{C}. \tag{2.10}$$

Clearly, $f \in H^1(a, b)$, since $g \in L^2(a, b)$ and hence $e^{-i\lambda t}g(t) \in L^1(a, b)$. Hence, f is in $\mathcal{D}(S_z)$ if and only if f satisfies the boundary condition $f(b) = zf(a)$ for $z \in \mathbb{C}$ resp. $f(a) = 0$ for $z = \infty$.

First suppose that $z \in \mathbb{C}$ and $e^{i\lambda(a-b)}z \neq 1$. Then $f \in \mathcal{D}(S_z)$ if and only if

$$c_{\lambda,g} = \left(e^{i\lambda(a-b)}z - 1\right)^{-1} \int_a^b e^{-i\lambda t}g(t)\, dt. \tag{2.11}$$

We therefore define

$$\left(R_\lambda(S_z)g\right)(x) = ie^{i\lambda x}\left(\int_a^x e^{-i\lambda t}g(t)\, dt + \left(e^{i\lambda(a-b)}z - 1\right)^{-1} \int_a^b e^{-i\lambda t}g(t)\, dt\right).$$

Next suppose that $z = \infty$. Then $f \in \mathcal{D}(S_\infty)$ if and only if $c_{\lambda,g} = 0$, so we define

$$\left(R_\lambda(S_\infty)g\right)(x) = ie^{i\lambda x}\int_a^x e^{-i\lambda t}g(t)\, dt.$$

We prove that $R_\lambda(S_z)$, $z \in \mathbb{C} \cup \{\infty\}$, is the resolvent of S_z. Let $g \in L^2(a, b)$ and set $f := R_\lambda(S_z)g$. By the preceding considerations, we have $f \in \mathcal{D}(S_z)$ and $(S_z - \lambda I)f = (S_z - \lambda I)R_\lambda(S_z)g = g$. Hence, $S_z - \lambda I$ is surjective. From (2.10) and (2.11) we conclude that $g = 0$ implies that $f = 0$, so $S_z - \lambda I$ is injective. Therefore, by Proposition 2.7(i), $\lambda \in \rho(S_z)$ and $(S_z - \lambda I)^{-1} = R_\lambda(S_z)$. Thus, we have shown that $\{\lambda : e^{i\lambda(a-b)}z \neq 1\} \subseteq \rho(S_z)$ for $z \in \mathbb{C}$ and $\rho(S_\infty) = \mathbb{C}$.

Suppose that $z \in \mathbb{C}$ and $e^{i\lambda(a-b)}z = 1$. Then $f_\lambda(x) := e^{i\lambda x}$ belongs to $\mathcal{D}(S_z)$, and $S_z f_\lambda = \lambda f_\lambda$. Hence, $\lambda \in \sigma(S_z)$. This completes the proof of the statement.   □

Let us consider the special case where $|z| = 1$, say $z = e^{i\mu(b-a)}$ with $\mu \in \mathbb{R}$. Then the operator S_z is self-adjoint (by Example 1.5) and the above statement yields

$$\sigma(S_z) = \{\mu + (b-a)^{-1}2\pi k : k \in \mathbb{Z}\}.$$

○

Example 2.5 (*Example 1.6 continued: half-axis*) Recall that $\mathcal{D}(T) = H_0^1(0, +\infty)$. We prove that $\sigma(T) = \{\lambda \in \mathbb{C} : \operatorname{Im}\lambda \leq 0\}$.

Assume that $\operatorname{Im}\lambda < 0$. Then $f_\lambda(x) := e^{i\bar{\lambda}x} \in \mathcal{D}(T^*)$ and $T^* f_\lambda = \bar{\lambda} f_\lambda$, so that $\bar{\lambda} \in \sigma_p(T^*)$ and $\lambda \in \sigma(T)$ by Proposition 2.7(ii). Hence, $\{\lambda : \operatorname{Im}\lambda \leq 0\} \subseteq \sigma(T)$.

Suppose now that $\operatorname{Im}\lambda > 0$ and define

$$(R_\lambda(T)g)(x) = i\int_0^x e^{i\lambda(x-t)}g(t)\,dt, \quad g \in L^2(0, +\infty).$$

That is, $R_\lambda(T)$ is the convolution operator with the function $h(t) := ie^{i\lambda t}$ on the half-axis $[0, +\infty)$. Since $\operatorname{Im}\lambda > 0$ and hence $h \in L^1(0, +\infty)$, $R_\lambda(T)$ is a bounded operator on $L^2(0, +\infty)$. Indeed, using the Cauchy–Schwarz inequality, we derive

$$
\begin{aligned}
\left\| (R_\lambda(T)g) \right\|^2 &= \int_0^\infty \left| \int_0^x h(x-t)g(t)\,dt \right|^2 dx \\
&\leq \int_0^\infty \left(\int_0^x |h(x-t)|\,dt \right) \left(\int_0^x |h(x-t)||g(t)|^2\,dt \right) dx \\
&\leq \|h\|_{L^1(0,+\infty)} \int_0^\infty \int_0^x |h(x-t)||g(t)|^2\,dt\,dx \\
&\leq \|h\|_{L^1(0,+\infty)} \int_0^\infty \int_0^\infty |h(x')||g(t)|^2\,dt\,dx' = \|h\|_{L^1(0,+\infty)}^2 \|g\|^2.
\end{aligned}
$$

Set $f := R_\lambda(T)g$. Clearly, $f \in AC[a,b]$ for all intervals $[a,b] \subseteq (0, +\infty)$. Since $f \in L^2(0, +\infty)$, $f' = i(\lambda f + g) \in L^2(0, +\infty)$ and $f(0) = 0$, we have $f \in H_0^1(0, +\infty) = \mathcal{D}(T)$ and $(T - \lambda I)f = (T - \lambda I)R_\lambda(T)g = g$. This shows that $T - \lambda I$ is surjective. Since $\operatorname{Im}\lambda > 0$, $\mathcal{N}(T - \lambda I) = \{0\}$. Thus, $T - \lambda I$ is bijective, and hence $\lambda \in \rho(T)$ by Proposition 2.7(i). From the equality $(T - \lambda I)R_\lambda(T)g = g$ for $g \in L^2(0, +\infty)$ it follows that $R_\lambda(T) = (T - \lambda I)^{-1}$ is the resolvent of T.

By the preceding we have proved that $\sigma(T) = \{\lambda : \operatorname{Im}\lambda \leq 0\}$.

○

Example 2.6 (*Example 1.7 continued: real line*) Then the operator $T = -i\frac{d}{dx}$ on $H^1(\mathbb{R})$ is self-adjoint. We show that $\sigma(T) = \mathbb{R}$.

Suppose that $\lambda \in \mathbb{R}$. Let us choose a function $\omega \in C_0^\infty(\mathbb{R})$, $\omega \neq 0$, and put $h_\epsilon(x) := \epsilon^{1/2}e^{i\lambda x}\omega(\epsilon x)$ for $\epsilon > 0$. Since $\|h_\epsilon\| = \|\omega\|$ and $\|(T - \lambda I)h_\epsilon\| = \epsilon\|\omega'\|$, it follows that λ is not in $\pi(T)$ and so not in $\rho(T)$. Hence, $\lambda \in \sigma(T)$. Since T is self-adjoint, $\sigma(T) \subseteq \mathbb{R}$ by Corollary 3.14 below. Thus, $\sigma(T) = \mathbb{R}$.

The resolvents of T for $\lambda \in \mathbb{C}\backslash\mathbb{R}$ are given by the formulas

$$(R_\lambda(T)g)(x) = i\int_{-\infty}^x e^{i\lambda(x-t)}g(t)\,dt, \quad \operatorname{Im}\lambda > 0, \tag{2.12}$$

$$(R_\lambda(T)g)(x) = -i\int_x^{+\infty} e^{i\lambda(x-t)}g(t)\,dt, \quad \operatorname{Im}\lambda < 0. \tag{2.13}$$

○

2.4 Exercises

1. Find a bounded operator T such that $\Theta(T)$ is not the convex hull of $\sigma(T)$.
   Hint: Look for some lower triangular 2×2-matrix.
2. Let $\alpha = (\alpha_n)_{n \in \mathbb{N}}$ be a complex sequence. Define an operator T_α on $l^2(\mathbb{N})$ with domain $\mathcal{D}(T) = \{(\varphi_n) \in l^2(\mathbb{N}) : (\alpha_n \varphi_n) \in l^2(\mathbb{N})\}$ by $T_\alpha(\varphi_n) = (\alpha_n \varphi_n)$.
   a. Determine the spectrum $\sigma(T_\alpha)$ and the point spectrum $\sigma_p(T_\alpha)$.
   b. When has T_α a discrete spectrum?
3. Let M_φ be the multiplication operator from Example 2.1. Find necessary and/or sufficient conditions for a number belonging to the point spectrum $\sigma_p(M_\varphi)$.
4. Let T_1 and T_2 be closed operators on $\mathcal{H}_1$ and $\mathcal{H}_2$, respectively.
   a. Show that $T_1 \oplus T_2$ is a closed operator on $\mathcal{H}_1 \oplus \mathcal{H}_2$.
   b. Show that $\sigma(T_1 \oplus T_2) = \sigma(T_1) \cup \sigma(T_2)$.
5. Find a bounded operator T and a $\lambda \in \sigma_p(T)$ such that $\mathcal{R}(T - \lambda I) \neq \overline{\mathcal{R}(T - \lambda I)}$.
   Hint: Look for some operator $T = T_1 \oplus T_2$.
6. Let $T = -\mathrm{i}\frac{d}{dx}$ on $\mathcal{D}(T) = \{f \in H^1(0, 1) : f(0) = 0\}$ in $\mathcal{H} = L^2(0, 1)$.
   a. Show that T is a closed operator.
   b. Determine the adjoint operator T^*.
   c. Show that $\rho(T) = \mathbb{C}$ and determine the operator $R_\lambda(T)$ for $\lambda \in \mathbb{C}$.
7. Prove the two resolvent formulas (2.12) and (2.13) in Example 2.6. Show that none of these operators is compact.
8. Let q be a real-valued continuous function on $[a, b]$, $a, b \in \mathbb{R}$, $a < b$. For $z \in \mathbb{T}$, define an operator T_z on $L^2(a, b)$ by $(T_z f)(x) = -\mathrm{i}f'(x) + q(x)f(x)$ with domain $\mathcal{D}(T_z) = \{f \in H^1(a, b) : f(b) = zf(a)\}$.
   a. Show that T_z is a self-adjoint operator on $L^2(a, b)$.
   b. Determine the spectrum and the resolvent $R_\lambda(T_z)$ for $\lambda \in \rho(T_z)$.
   Hint: Find a unitary operator U on $L^2(a, b)$ such that $T_z = UTU^*$, where T is the operator from Example 2.3.
9. Find a densely defined closed operator T such that each complex number is an eigenvalue of T^*, but T has no eigenvalue.
10. Let T be a closed operator on $\mathcal{H}$. Use formula (2.6) to prove that
$$\frac{dR_\lambda(T)}{d\lambda} := \lim_{h \to 0} \frac{R_{\lambda+h}(T) - R_\lambda(T)}{h} = R_\lambda(T)^2, \quad \lambda \in \rho(T),$$
    in the operator norm on $\mathcal{H}$.
11. Prove that $\sigma(TS) \cup \{0\} = \sigma(ST) \cup \{0\}$ for $T \in \mathbf{B}(\mathcal{H}_1, \mathcal{H}_2)$ and $S \in \mathbf{B}(\mathcal{H}_2, \mathcal{H}_1)$.
    Hint: Verify that $(ST - \lambda I)^{-1} = \lambda^{-1}[S(TS - \lambda I)^{-1}T - I]$ for $\lambda \neq 0$.
*12. (*Volterra integral operator*)
    Let K be a bounded measurable function on $\{(x, y) \in \mathbb{R}^2 : 0 \leq y \leq x \leq 1\}$. Prove that the spectrum of the Volterra operator V_K is equal to $\{0\}$, where
$$(V_K f)(x) = \int_0^x K(x, t)f(t)\,dt, \quad f \in L^2(0, 1).$$

Hints: Show that $(V_K)^n$ is an integral operator with kernel K_n satisfying

$$\left|K_n(x, y)\right| \leq M^n |x - y|^{n-1}/(n - 1)!, \quad \text{where } M := \|K\|_{L^\infty(0,1)}.$$

Then deduce that $\|(V_K)^n\| \leq M^n/(n - 1)!$ and hence $\lim_n \|(V_K)^n\|^{1/n} = 0$.

Chapter 3
Some Classes of Unbounded Operators

The notion of a closed operator is too general to develop a deeper theory. This chapter is devoted to some important classes of closed or closable operators. Symmetric operators are investigated in Sect. 3.1, and self-adjoint operators in Sect. 3.2. The interplay between symmetric and self-adjoint operators, or more precisely, the problem of when a symmetric operator is self-adjoint is one of the main themes in this book. Section 3.2 contains some basic self-adjointness criteria. Section 3.3 is concerned with classes of operators (sectorial operators, accretive operators, dissipative operators) that will be met later as operators associated with forms or as generators of contraction semigroups. Section 3.4 deals with unbounded normal operators.

3.1 Symmetric Operators

Throughout this section, T is a linear operator on a Hilbert space $\mathcal{H}$.

Definition 3.1 The operator T is called *symmetric* (or *Hermitian*) if

$$\langle Tx, y \rangle = \langle x, Ty \rangle \quad \text{for all } x, y \in \mathcal{D}(T). \tag{3.1}$$

Note that we do not assume that a symmetric operator is densely defined.

Lemma 3.1 *T is symmetric if and only if $\langle Tx, x \rangle$ is real for all $x \in \mathcal{D}(T)$.*

Proof If T is symmetric, then $\langle Tx, x \rangle = \langle x, Tx \rangle = \overline{\langle Tx, x \rangle}$, so $\langle Tx, x \rangle \in \mathbb{R}$. Conversely, if $\langle Tx, x \rangle$ is real for all $x \in \mathcal{D}(T)$, it follows immediately from the polarization formula (1.2) that (3.1) holds, that is, T is symmetric. $\qquad\square$

Definition 3.2 Let T be a symmetric operator. T is said to be *lower semibounded* (resp. *upper semibounded*) if there exists a real number m such that

$$\langle Tx, x \rangle \geq m\|x\|^2 \quad \left(\text{resp. } \langle Tx, x \rangle \leq m\|x\|^2\right) \quad \text{for all } x \in \mathcal{D}(T).$$

K. Schmüdgen, *Unbounded Self-adjoint Operators on Hilbert Space*,
Graduate Texts in Mathematics 265,
DOI 10.1007/978-94-007-4753-1_3, © Springer Science+Business Media Dordrecht 2012

Any such number m is called a *lower bound* (resp. an *upper bound*) for T. If T is lower semibounded or upper semibounded, T is called *semibounded*. We say that T is *positive* and write $T \geq 0$ if $\langle Tx, x \rangle \geq 0$ for all $x \in \mathcal{D}(T)$. If $\langle Tx, x \rangle > 0$ for all nonzero $x \in \mathcal{D}(T)$, then T is called *strictly positive*.

Clearly, each lower semibounded operator T has a *greatest lower bound* given by

$$m_T := \inf\{\langle Tx, x \rangle \|x\|^{-2}; \ x \in \mathcal{D}(T), \ x \neq 0\}.$$

Example 3.1 (*A nonclosable symmetric operator*) Let S be a nonclosable operator on a Hilbert space $\mathcal{H}_1$ (see Example 1.1). Then the operator T on the Hilbert space $\mathcal{H} := \mathcal{H}_1 \oplus \mathcal{H}_1$ defined by $T(x_1, 0) = (0, Sx_1)$ for $(x_1, 0) \in \mathcal{D}(T) := \mathcal{D}(S) \oplus \{0\}$ is not closable. But T is symmetric, since for $x_1, y_1 \in \mathcal{D}(S)$, we have

$$\langle T(x_1, 0), (y_1, 0) \rangle = \langle 0, y_1 \rangle_1 + \langle Sx_1, 0 \rangle_1 = 0 = \langle (x_1, 0), T(y_1, 0) \rangle. \qquad \circ$$

Suppose that T is densely defined. Comparing formulas (3.1) and (1.5), we conclude that T is symmetric if and only if $T \subseteq T^*$. Since T^* is closed and $(\overline{T})^* = T^*$, it follows that each *densely defined* symmetric operator T is closable, its closure $\overline{T}$ is again symmetric, and we have

$$T \subseteq \overline{T} = T^{**} \subseteq T^*.$$

If the domain $\mathcal{D}(T)$ of a symmetric operator T on $\mathcal{H}$ is the whole (!) Hilbert space $\mathcal{H}$, then T is a closed operator with domain $\mathcal{H}$. Hence, T is bounded by the closed graph theorem. This is a classical result called the *Hellinger–Toeplitz theorem*.

The next proposition deals with the regularity domain of a symmetric operator.

Proposition 3.2 *If T is a symmetric operator on $\mathcal{H}$, then*:

(i) $\mathbb{C} \setminus \mathbb{R} \subseteq \pi(T)$.

(ii) *If T is lower semibounded (resp. upper semibounded) and m is a lower bound (resp. an upper bound) for T, then $(-\infty, m) \subseteq \pi(T)$ (resp. $(m, +\infty) \subseteq \pi(T)$).*

(iii) *If T is densely defined and $\lambda \in \pi(T)$, then $\mathcal{R}(T^* - \overline{\lambda}I) = \mathcal{H}$.*

Proof (i): Let $\lambda = \alpha + i\beta$, where $\alpha, \beta \in \mathbb{R}$, and $x \in \mathcal{D}(T)$. Then

$$\|(T - \lambda I)x\|^2 = \langle (T - \alpha I)x - i\beta x, (T - \alpha I)x - i\beta x \rangle$$

$$= \|(T - \alpha I)x\|^2 + \|\beta x\|^2 - i\beta\{\langle x, (T - \alpha I)x \rangle - \langle (T - \alpha I)x, x \rangle\}$$

$$= \|(T - \alpha I)x\|^2 + |\beta|^2 \|x\|^2. \tag{3.2}$$

Here the expression in braces vanishes, because T is symmetric and α is real. Equation (3.2) implies that

$$\|(T - \lambda I)x\| \geq |\operatorname{Im}\lambda| \|x\| \quad \text{for } x \in \mathcal{D}(T), \ \lambda \in \mathbb{C}. \tag{3.3}$$

Therefore, $\lambda \in \pi(T)$ if $\operatorname{Im}\lambda \neq 0$.

(ii): We prove the assertion for lower semibounded T. For $\lambda \in (-\infty, m)$,

$$(m - \lambda)\|x\|^2 \leq \langle (T - \lambda I)x, x \rangle \leq \|(T - \lambda I)x\| \|x\|,$$

and hence $(m - \lambda)\|x\| \leq \|(T - \lambda I)x\|$ for $x \in \mathcal{D}(T)$. Since $m > \lambda$, $\lambda \in \pi(T)$.

(iii): Let $y \in \mathcal{H}$. We define a linear functional F_y on $\mathcal{R}(T - \lambda I)$ by $F_y((T - \lambda I)x) = \langle x, y \rangle$, $x \in \mathcal{D}(T)$. Since $\lambda \in \pi(T)$, it follows from (2.1) that the functional F_y is well defined and bounded. By Riesz' theorem, applied to the closure of $\mathcal{R}(T - \lambda I)$ in $\mathcal{H}$, there is a vector $u \in \mathcal{H}$ such that $F_y((T - \lambda I)x) = \langle (T - \lambda I)x, u \rangle$ for $x \in \mathcal{D}(T)$. Since $\langle (T - \lambda I)x, u \rangle = \langle x, y \rangle$ for all $x \in \mathcal{D}(T)$, we conclude that $u \in \mathcal{D}((T - \lambda I)^*) = \mathcal{D}(T^* - \bar{\lambda}I)$ and $(T - \lambda I)^* u = (T^* - \bar{\lambda}I)u = y$. $\qquad\square$

Recall from Proposition 2.4 that for a closable operator T, the defect number $d_\lambda(T)$ is constant on connected subsets of the regularity domain $\pi(T)$. Therefore, if T is closable and symmetric, by Proposition 3.2 the number $d_\lambda(T)$ is constant on the upper half-plane and on the lower half-plane. If, in addition, T is semibounded, then $\pi(T)$ is connected, and hence $d_\lambda(T)$ is constant on the whole set $\pi(T)$.

Definition 3.3 The *deficiency indices* (or the *defect numbers*) of a closable symmetric operator T are the cardinal numbers

$$d_+(T) := d_\lambda(T) = \dim \mathcal{R}(T - \bar{\lambda}I)^\perp, \quad \text{Im}\,\lambda > 0, \tag{3.4}$$

$$d_-(T) := d_\lambda(T) = \dim \mathcal{R}(T - \bar{\lambda}I)^\perp, \quad \text{Im}\,\lambda < 0. \tag{3.5}$$

If T is densely defined and symmetric, then T is closable, and by formula (1.7),

$$d_+(T) = \dim \mathcal{N}(T^* - iI) = \dim \mathcal{N}(T^* - \lambda I), \quad \text{Im}\,\lambda > 0, \tag{3.6}$$

$$d_-(T) = \dim \mathcal{N}(T^* + iI) = \dim \mathcal{N}(T^* - \lambda I), \quad \text{Im}\,\lambda < 0. \tag{3.7}$$

Remark By our definition, $d_\pm(T) = \dim \mathcal{N}(T^* \mp iI)$. Some authors define $d_\pm(T)$ to be $\dim \mathcal{N}(T^* \pm iI)$.

Proposition 3.3 *Suppose that T is a densely defined symmetric operator. If T is semibounded or if $\pi(T)$ contains a real number, then $d_+(T) = d_-(T)$.*

Proof In both cases, $\pi(T)$ is connected by Proposition 3.2, so the assertion follows from Proposition 2.4. $\qquad\square$

Example 3.2 (*Examples* 1.4, 1.6, *and* 1.7 *continued*) Recall that in all three examples we have $Tf = -\mathrm{i}f'$ for $f \in \mathcal{D}(T) = H_0^1(\mathcal{J})$ and $T^*g = -\mathrm{i}g'$ for $g \in \mathcal{D}(T^*) = H^1(\mathcal{J})$, where $\mathcal{J} = (a, b)$ with $a, b \in \mathbb{R}$, $\mathcal{J} = (0, \infty)$ or $\mathcal{J} = \mathbb{R}$, respectively. Since $T \subseteq T^*$, T and hence T^2 are symmetric. Moreover, $g \in \mathcal{N}(T^* - \lambda I)$ if and only if $g \in \mathcal{D}(T^*)$ and $-\mathrm{i}g'(x) = \lambda g(x)$ on $\mathcal{J}$.

First suppose that $\mathcal{J}$ is a bounded interval (a, b). Then $\mathcal{N}(T^* - \lambda I) = \mathbb{C} \cdot e^{i\lambda x}$ and $d_+(T) = d_-(T) = 1$. Likewise we have $\mathcal{N}((T^2)^* - \lambda I) = \mathbb{C} \cdot e^{i\sqrt{\lambda}x} + \mathbb{C} \cdot e^{-i\sqrt{\lambda}x}$. Thus, $d_\pm(T) = 1$ and $d_\pm(T^2) = 2$.

Next let $\mathcal{J} = (0, \infty)$. Since $e^{i\lambda x} \in L^2(0, \infty)$ if and only if $\operatorname{Im} \lambda > 0$, we have $\mathcal{N}(T^* - \lambda I) = \mathbb{C} \cdot e^{i\lambda x}$ if $\operatorname{Im} \lambda > 0$ and $\mathcal{N}(T^* - \lambda I) = \{0\}$ if $\operatorname{Im} \lambda < 0$. That is, $d_+(T) = 1$ and $d_-(T) = 0$.

Finally, let $\mathcal{J} = \mathbb{R}$. Then $e^{i\lambda x} \notin L^2(\mathbb{R})$ for all $\lambda \in \mathbb{C}$, and hence $d_\pm(T) = 0$.   ○

Example 3.3 Let $S = -\frac{d^2}{dx^2}$ with domain $\mathcal{D}(S) = \{f \in H^2(\mathbb{R}) : f(0) = f'(0) = 0\}$ on $L^2(\mathbb{R})$. Then $S^* f = -f''$ for $f \in \mathcal{D}(S^*) = H^2(-\infty, 0) \oplus H^2(0, +\infty)$. Clearly, S is symmetric and densely defined.

Fix $\lambda \in \mathbb{C} \setminus [0, \infty)$. Let $\sqrt{\lambda}$ denote the square root of λ with $\operatorname{Im} \sqrt{\lambda} > 0$, and let χ_+ and χ_- be the characteristic functions of $[0, \infty)$ and $(-\infty, 0)$, respectively. Then

$$\mathcal{N}(S^* - \lambda I) = \mathbb{C} \cdot \chi_+(x)e^{i\sqrt{\lambda}x} + \mathbb{C} \cdot \chi_-(x)e^{-i\sqrt{\lambda}x},$$

and S has deficiency indices $(2, 2)$.

For $f, g \in H^2(-\infty, 0) \oplus H^2(0, +\infty)$, we have the integration-by-parts formula

$$\langle f'', g \rangle_{L^2(\mathbb{R})} - \langle f, g'' \rangle_{L^2(\mathbb{R})}$$
$$= f'(-0)\overline{g(-0)} - f(-0)\overline{g'(-0)} - f'(+0)\overline{g(+0)} + f(+0)\overline{g'(+0)}. \quad (3.8)$$

To derive this formula, we apply (1.14) to the intervals $(0, b)$ and $(a, 0)$, add both formulas, and pass to the limits $a \to -\infty$ and $b \to +\infty$. Since $f(b), f'(b), g(b),$ $g'(b), f(a), f'(a), g(a), g'(a)$ tend to zero by Lemma 1.11, we then obtain (3.8).

Now let $T = -\frac{d^2}{dx^2}$ on $\mathcal{D}(T) = \{f \in H^2(\mathbb{R}) : f'(0) = 0\}$ and $\lambda \in \mathbb{C} \setminus [0, \infty)$. Since $S \subseteq T$, we have $\mathcal{N}(T^* - \lambda I) \subseteq \mathcal{N}(S^* - \lambda I)$. A function $g \in \mathcal{N}(S^* - \lambda I)$ is in $\mathcal{N}(T^* - \lambda I)$ if and only if $\langle Tf, g \rangle \equiv \langle -f'', g \rangle = \langle f, \lambda g \rangle$ for all $f \in \mathcal{D}(T)$. Choosing $f \in \mathcal{D}(T)$ such that $f(0) \neq 0$ and applying (3.8), we conclude that $g'(+0) = g'(-0)$. This implies that T has deficiency indices $(1, 1)$ and

$$\mathcal{N}(T^* - \lambda I) = \mathbb{C} \cdot \left(\chi_+(x)e^{i\sqrt{\lambda}x} - \chi_-(x)e^{-i\sqrt{\lambda}x} \right).$$ ○

Lemma 3.4 *Let T be a symmetric operator.*

(i) *Any eigenvalue of T is real.*
(ii) *Eigenvectors belonging to different eigenvalues of T are mutually orthogonal.*
(iii) *Suppose that T is densely defined and $T \geq 0$. If $\langle Tx, x \rangle = 0$ for some $x \in \mathcal{D}(T)$, then $Tx = 0$.*

Proof (i): Let x be a nonzero vector from $\mathcal{N}(T - \lambda I)$. Since $\langle Tx, x \rangle = \lambda \|x\|^2$ and $\langle Tx, x \rangle$ is real, λ is real.

(ii): Let $x \in \mathcal{N}(T - \lambda I)$ and $y \in \mathcal{N}(T - \mu I)$, where $\lambda \neq \mu$. Since μ is real and T is symmetric, $\mu \langle x, y \rangle = \langle x, Ty \rangle = \langle Tx, y \rangle = \lambda \langle x, y \rangle$, and hence $\langle x, y \rangle = 0$.

(iii): Since $T \geq 0$, $\langle T \cdot, \cdot \rangle$ is a positive semidefinite sesquilinear form. Hence, the Cauchy–Schwarz inequality applies and yields

$$\left| \langle Tx, y \rangle \right|^2 \leq \langle Tx, x \rangle \langle Ty, y \rangle = 0$$

for all $y \in \mathcal{D}(T)$. Since $\mathcal{D}(T)$ is dense, the latter implies that $Tx = 0$. □

The following notion will play a crucial role for the study of boundary triplets in Chap. 14.

Definition 3.4 Let T be a densely defined symmetric operator on $\mathcal{H}$. The *boundary form* of T^* is the sequilinear form $[\cdot, \cdot]_{T^*}$ on $\mathcal{D}(T^*)$ defined by

$$[x, y]_{T^*} = \langle T^*x, y \rangle - \langle x, T^*y \rangle, \quad x, y \in \mathcal{D}(T^*). \tag{3.9}$$

A linear subspace $\mathcal{D}$ of $\mathcal{D}(T^*)$ is called *symmetric* if $[x, y]_{T^*} = 0$ for all $x, y \in \mathcal{D}$.

The reason for the latter terminology stems from the following simple fact which follows immediately from the corresponding definitions.

Lemma 3.5 *The symmetric extensions of a densely defined symmetric operator T are the restrictions of T^* to symmetric subspaces $\mathcal{D}$ of $\mathcal{D}(T^*)$ which contain $\mathcal{D}(T)$.*

Proposition 3.6 *Let T be a densely defined closed symmetric operator, and let $\mathcal{E}$ be a finite-dimensional linear subspace of $\mathcal{D}(T^*)$. Define*

$$\mathcal{D}(T_\mathcal{E}) := \mathcal{D}(T) + \mathcal{E} \quad \text{and} \quad T_\mathcal{E} := T^* \!\restriction\! \mathcal{D}(T_\mathcal{E}).$$

Suppose that the linear subspace $\mathcal{D}(T_\mathcal{E})$ of $\mathcal{D}(T^*)$ is symmetric. Then $T_\mathcal{E}$ is a closed symmetric operator. If $\mathcal{E}$ has dimension k modulo $\mathcal{D}(T)$, then $T_\mathcal{E}$ has the deficiency indices $(d_+(T) - k, d_-(T) - k)$.

Proof Since $\mathcal{G}(T_\mathcal{E})$ is the sum of the closed subspace $\mathcal{G}(T)$ and the finite-dimensional vector space $\{(x, T^*x) : x \in \mathcal{E}\}$, $\mathcal{G}(T_\mathcal{E})$ is closed, and so is the operator $T_\mathcal{E}$.

By Lemma 3.5, $T_\mathcal{E}$ is symmetric, because $\mathcal{D}(T_\mathcal{E})$ is a symmetric subspace.

It remains to prove the assertion concerning the deficiency indices of $T_\mathcal{E}$. Let $\lambda \in \mathbb{C} \setminus \mathbb{R}$. Without loss of generality we can assume that $\mathcal{D}(T) + \mathcal{E}$ is a direct sum and $\dim \mathcal{E} = k$. First we show that $\mathcal{R}(T - \lambda I) + (T^* - \lambda I)\mathcal{E}$ is also a direct sum and $\dim(T^* - \lambda I)\mathcal{E} = k$. Indeed, assume that $(T - \lambda I)x + (T^* - \lambda I)u = 0$ for $x \in \mathcal{D}(T)$ and $u \in \mathcal{E}$. Then $x + u \in \mathcal{N}(T^* - \lambda I)$. But $x + u \in \mathcal{D}(T_\mathcal{E})$, so that $x + u \in \mathcal{N}(T_\mathcal{E} - \lambda I)$. Therefore, since $T_\mathcal{E}$ is a symmetric operator and λ is not real, $x + u = 0$ by Lemma 3.4(i). Since the sum $\mathcal{D}(T) + \mathcal{E}$ is direct, $x = 0$ and $u = 0$. This proves also that $(T^* - \lambda I)\!\restriction\!\mathcal{E}$ is injective. Hence, $\dim(T^* - \lambda I)\mathcal{E} = \dim \mathcal{E} = k$.

Obviously, $\mathcal{R}(T_\mathcal{E} - \lambda I)^\perp \subseteq \mathcal{R}(T - \lambda I)^\perp$. Since $\mathcal{R}(T_\mathcal{E} - \lambda I)$ is the direct sum of $\mathcal{R}(T - \lambda I)$ and the k-dimensional space $(T^* - \lambda I)\mathcal{E}$, $\mathcal{R}(T_\mathcal{E} - \lambda I)^\perp$ has the codimension k in $\mathcal{R}(T - \lambda I)^\perp$. By Definition 3.3 this means that $d_{\overline{\lambda}}(T_\mathcal{E}) = d_{\overline{\lambda}}(T) - k$. □

3.2 Self-adjoint Operators

Self-adjointness is the most important notion on unbounded operators in this book. The main results about self-adjoint operators are the spectral theorem proved in Sect. 5.2 and the corresponding functional calculus based on it. A large effort is made in this book to prove that certain symmetric operators are self-adjoint or to extend symmetric operators to self-adjoint ones.

Definition 3.5 A densely defined symmetric operator T on a Hilbert space $\mathcal{H}$ is called *self-adjoint* if $T = T^*$ and *essentially self-adjoint*, briefly *e.s.a.*, if $\overline{T}$ is self-adjoint, or equivalently, if $\overline{T} = T^*$.

Let us state some simple consequences that will be often used without mention.
A self-adjoint operator T is symmetric and closed, since T^* is always closed.

Let T be a densely defined symmetric operator. Since then $T \subseteq T^*$, T is self-adjoint if and only if $\mathcal{D}(T) = \mathcal{D}(T^*)$. Likewise, T is essentially self-adjoint if and only if $\mathcal{D}(\overline{T}) = \mathcal{D}(T^*)$.

Any self-adjoint operator T on $\mathcal{H}$ is *maximal symmetric*, that is, if S is a symmetric operator on $\mathcal{H}$ such that $T \subseteq S$, then $T = S$. Indeed, $T \subseteq S$ implies that $S^* \subseteq T^*$. Combined with $S \subseteq S^*$ and $T^* = T$, this yields $S \subseteq T$, so that $T = S$.

Some self-adjointness criteria follow easily from the next result. The nice direct sum decomposition (3.10) of the domain $\mathcal{D}(T^*)$ is called *von Neumann's formula*.

Proposition 3.7 *Let T be a densely defined symmetric operator. Then*

$$\mathcal{D}(T^*) = \mathcal{D}(\overline{T}) \dotplus \mathcal{N}(T^* - \lambda I) \dotplus \mathcal{N}(T^* - \overline{\lambda} I) \quad \text{for } \lambda \in \mathbb{C} \setminus \mathbb{R}, \quad (3.10)$$

$$\dim \mathcal{D}(T^*)/\mathcal{D}(\overline{T}) = d_+(T) + d_-(T). \quad (3.11)$$

Proof Let us abbreviate $\mathcal{N}_\lambda := \mathcal{N}(T^* - \lambda I)$ and $\mathcal{N}_{\overline{\lambda}} := \mathcal{N}(T^* - \overline{\lambda} I)$. The inclusion $\mathcal{D}(\overline{T}) + \mathcal{N}_\lambda + \mathcal{N}_{\overline{\lambda}} \subseteq \mathcal{D}(T^*)$ is obvious. We prove that $\mathcal{D}(T^*) \subseteq \mathcal{D}(\overline{T}) + \mathcal{N}_\lambda + \mathcal{N}_{\overline{\lambda}}$.

Let $x \in \mathcal{D}(T^*)$. By Corollary 2.2 we have

$$\mathcal{H} = \mathcal{R}(\overline{T} - \lambda I) \oplus \mathcal{N}_{\overline{\lambda}}. \quad (3.12)$$

We apply (3.12) to the vector $(T^* - \lambda I)x \in \mathcal{H}$. Then there exist $x_0 \in \mathcal{D}(\overline{T})$ and $x'_- \in \mathcal{N}_{\overline{\lambda}}$ such that $(T^* - \lambda I)x = (\overline{T} - \lambda I)x_0 + x'_-$. Since $\lambda \notin \mathbb{R}$, we can set $x_- := (\overline{\lambda} - \lambda)^{-1} x'_-$. The preceding can be rewritten as $(T^* - \lambda I)(x - x_0 - x_-) = 0$, that is, $x_+ := x - x_0 - x_-$ is in $\mathcal{N}_\lambda$. Hence, $x = x_0 + x_+ + x_- \in \mathcal{D}(\overline{T}) + \mathcal{N}_\lambda + \mathcal{N}_{\overline{\lambda}}$.

To prove that the sum in (3.10) is a direct sum, we assume that $x_0 + x_+ + x_- = 0$ for $x_0 \in \mathcal{D}(\overline{T})$, $x_+ \in \mathcal{N}_\lambda$, and $x_- \in \mathcal{N}_{\overline{\lambda}}$. Then

$$(T^* - \lambda I)(x_0 + x_+ + x_-) = (\overline{T} - \lambda I)x_0 + (\overline{\lambda} - \lambda)x_- = 0.$$

The vector $(\lambda - \overline{\lambda})x_- = (\overline{T} - \lambda I)x_0$ is in $\mathcal{R}(\overline{T} - \lambda I) \cap \mathcal{N}_{\overline{\lambda}}$. Therefore, it follows from (3.12) that $x_- = 0$ and $(\overline{T} - \lambda I)x_0 = 0$. Since $\overline{T}$ is symmetric and λ is not real, the latter yields $x_0 = 0$ by Lemma 3.4. Since $x_0 = x_- = 0$, we get $x_+ = 0$.

The preceding proves (3.10). (3.11) is an immediate consequence of (3.10). $\square$

Recall that a densely defined symmetric operator T is essentially self-adjoint if and only if $\mathcal{D}(\overline{T}) = \mathcal{D}(T^*)$. By (3.10) (or (3.11)) the latter is satisfied if and only if $d_+(T) = d_-(T) = 0$. That is, using formulas (3.4)–(3.7), we obtain the following:

Proposition 3.8 *Let T be a densely defined symmetric operator on $\mathcal{H}$, and let λ_+ and λ_- be complex numbers such that $\operatorname{Im}\lambda_+ > 0$ and $\operatorname{Im}\lambda_- < 0$.*
Then the operator T is essentially self-adjoint if and only $d_+(T) = 0$ (equivalently, $\mathcal{N}(T^* - \lambda_+ I) = \{0\}$, or equivalently, $\overline{\mathcal{R}(T - \overline{\lambda}_+ I)} = \mathcal{H}$) and $d_-(T) = 0$ (equivalently, $\mathcal{N}(T^* - \lambda_- I) = \{0\}$, or equivalently, $\overline{\mathcal{R}(T - \overline{\lambda}_- I)} = \mathcal{H}$).

For lower semibounded operators, we have the following stronger result.

Proposition 3.9 *Let T be a densely defined symmetric operator such that there is a real number in $\pi(T)$; in particular, if T is lower semibounded, the latter is fulfilled, and $\mathbb{C} \setminus [m_T, \infty) \subseteq \pi(T)$. Then T is essentially self-adjoint if and only if $d_\lambda(T) = 0$ (equivalently, $\mathcal{N}(T^* - \lambda I) = \{0\}$, or equivalently, $\overline{\mathcal{R}(T - \lambda I)} = \mathcal{H}$) for one, hence all, $\lambda \in \pi(T)$.*

Proof Since $d_\lambda(T) = d_\pm(T)$ by Proposition 3.3 and $\mathbb{C} \setminus [m_T, \infty) \subseteq \pi(T)$ by Proposition 3.2, the assertions follow at once from Proposition 3.8. □

The next proposition characterizes the numbers of the resolvent set of a self-adjoint operator. Condition (ii) therein is often useful to detect the spectrum of the operator.

Proposition 3.10 *Let T be a self-adjoint operator on a Hilbert space $\mathcal{H}$. For any complex number λ, the following conditions are equivalent:*

(i) $\lambda \in \rho(T)$.
(ii) $\lambda \in \pi(T)$, *that is, there exists a constant $c_\lambda > 0$ such that $\|(T - \lambda I)x\| \geq c_\lambda \|x\|$ for all $x \in \mathcal{D}(T)$.*
(iii) $\mathcal{R}(T - \lambda I) = \mathcal{H}$.

Moreover, if $\lambda \in \mathbb{C} \setminus \mathbb{R}$, then $\lambda \in \rho(T)$ and $\|R_\lambda(T)\| \leq |\operatorname{Im}\lambda|^{-1}$.

Proof Since T is self-adjoint, by Corollary 2.2 we have

$$\mathcal{R}(T - \lambda I)^\perp = \mathcal{N}(T - \overline{\lambda}I). \tag{3.13}$$

First suppose that $\lambda \in \mathbb{C} \setminus \mathbb{R}$. Then $\lambda \in \pi(T)$, and hence $\mathcal{R}(T - \lambda I)$ is closed by Proposition 2.1(iv). Further, $\mathcal{N}(T - \lambda I) = \{0\}$ and $\mathcal{N}(T - \overline{\lambda}I) = \{0\}$ by Lemma 3.4(i). Therefore, it follows from (3.13) that $\mathcal{R}(T - \lambda I) = \mathcal{H}$. Hence, $\lambda \in \rho(T)$ by Proposition 2.7(i), and all three conditions (i)–(iii) are satisfied.

Let $y \in \mathcal{H}$. Then $x := R_\lambda(T)y \in \mathcal{D}(T)$ and $y = (T - \lambda I)x$. Inserting x into (3.3) yields $\|y\| \geq |\operatorname{Im}\lambda| \|R_\lambda(T)y\|$. That is, $\|R_\lambda(T)\| \leq |\operatorname{Im}\lambda|^{-1}$.

From now on assume that $\lambda \in \mathbb{R}$.
(i) $\rightarrow$ (ii) is obvious, since $\rho(T) \subseteq \pi(T)$.

(ii) → (iii): Since $\lambda \in \pi(T)$ by (ii), $\mathcal{N}(T - \lambda I) = \{0\}$, so $\mathcal{R}(T - \lambda I)$ is dense in $\mathcal{H}$ by (3.13). Because $\mathcal{R}(T - \lambda I)$ is closed by Proposition 2.1(iv), $\mathcal{R}(T - \lambda I) = \mathcal{H}$.

(iii) → (i): Since $\mathcal{R}(T - \lambda I) = \mathcal{H}$ by (iii), we have $\mathcal{N}(T - \lambda I) = \{0\}$ again by (3.13). Therefore, it follows from Proposition 2.7(i) that $\lambda \in \rho(T)$.   □

Now we give some self-adjointness criteria that *do not assume that the symmetric operator is densely defined*. They are essentially based on the following proposition.

Proposition 3.11 *Let T be a symmetric operator on $\mathcal{H}$. If there exists a complex number λ such that $\mathcal{R}(T - \lambda I) = \mathcal{H}$ and $\mathcal{R}(T - \bar{\lambda}I)$ is dense in $\mathcal{H}$, then T is self-adjoint, and λ and $\bar{\lambda}$ are in $\rho(T)$.*

Proof We first show that $\mathcal{D}(T)$ is dense in $\mathcal{H}$. Let $y \in \mathcal{D}(T)^{\perp}$. Since $\mathcal{R}(T - \lambda I) = \mathcal{H}$, there exists a vector $u \in \mathcal{D}(T)$ such that $y = (T - \lambda I)u$. Therefore, we have $0 = \langle y, x \rangle = \langle (T - \lambda I)u, x \rangle = \langle u, (T - \bar{\lambda}I)x \rangle$ for $x \in \mathcal{D}(T)$, so $u \in \mathcal{R}(T - \bar{\lambda}I)^{\perp}$. Since $\mathcal{R}(T - \bar{\lambda}I)$ is dense, $u = 0$ and hence $y = 0$. Thus, $\mathcal{D}(T)$ is dense.

Hence, T^* is well defined. Let $w \in \mathcal{D}(T^*)$. Applying once more the assumption $\mathcal{R}(T - \lambda I) = \mathcal{H}$, there is a $v \in \mathcal{D}(T)$ such that $(T^* - \lambda I)w = (T - \lambda I)v$. Then

$$\langle (T - \bar{\lambda}I)x, w \rangle = \langle x, (T^* - \lambda I)w \rangle = \langle x, (T - \lambda I)v \rangle = \langle (T - \bar{\lambda}I)x, v \rangle$$

for $x \in \mathcal{D}(T)$. Because $\mathcal{R}(T - \bar{\lambda}I)$ is dense, $w = v$ and so $w \in \mathcal{D}(T)$, that is, $\mathcal{D}(T^*) \subseteq \mathcal{D}(T)$. Since T is symmetric, this implies that T is self-adjoint.

Since T is self-adjoint and $\mathcal{R}(T - \lambda I) = \mathcal{H}$, $\lambda \in \rho(T)$ by Proposition 3.10.   □

In particular, the preceding result implies again Proposition 3.8.

Note that the assumption $\mathcal{R}(T - \lambda I) = \mathcal{H}$ in Proposition 3.11 is crucial. It cannot be replaced by the density of $\mathcal{R}(T - \lambda I)$ in $\mathcal{H}$; see Exercise 12.

The case $\lambda = 0$ in Proposition 3.11 and Corollary 1.9 yields the following.

Corollary 3.12 *If T is a symmetric operator on $\mathcal{H}$ such that $\mathcal{R}(T) = \mathcal{H}$, then T is self-adjoint, and its inverse T^{-1} is a bounded self-adjoint operator on $\mathcal{H}$.*

Proposition 3.13 *Let T be a closed symmetric operator on $\mathcal{H}$, and let $\lambda_+, \lambda_- \in \mathbb{C}$, where $\operatorname{Im}\lambda_+ > 0$ and $\operatorname{Im}\lambda_- < 0$. The operator T is self-adjoint if and only if $d_+(T) = 0$ (equivalently, $\lambda_- \in \rho(T)$, or equivalently, $\mathcal{R}(T - \lambda_-I) = \mathcal{H}$) and $d_-(T) = 0$ (equivalently, $\lambda_+ \in \rho(T)$, or equivalently, $\mathcal{R}(T - \lambda_+I) = \mathcal{H}$).*

Proof From Proposition 3.10, a self-adjoint operator satisfies all these conditions.

To prove the converse, we first note that $\mathcal{R}(T - \lambda I)$ is closed for any $\lambda \in \mathbb{C} \setminus \mathbb{R}$ by Proposition 2.1(iv), because $\lambda \in \pi(T)$ and T is closed. Therefore, $d_{\pm}(T) = 0$ if and only if $\mathcal{R}(T - \lambda_{\pm}I) = \mathcal{H}$, or equivalently, $\lambda_{\pm} \in \rho(T)$ by Proposition 3.10. Thus, it suffices to assume that $d_+(T) = d_-(T) = 0$. But then $\mathcal{R}(T \pm iI) = \mathcal{H}$, so T is self-adjoint by Proposition 3.11.   □

By the preceding proposition a closed symmetric operator T is self-adjoint if and only if $\mathbb{C} \setminus \mathbb{R} \subseteq \rho(T)$. We restate this fact by the following corollary.

Corollary 3.14 *A closed symmetric linear operator T on $\mathcal{H}$ is self-adjoint if and only if $\sigma(T) \subseteq \mathbb{R}$.*

Since the deficiency indices are constant on connected subsets of $\pi(T)$, Proposition 3.13 implies the following result.

Proposition 3.15 *Let T be a closed symmetric operator. Suppose that $\pi(T)$ contains a real number. Then T is self-adjoint if and only if $d_\lambda(T) = 0$ (equivalently, $\lambda \in \rho(T)$, or equivalently, $\mathcal{R}(T - \lambda I) = \mathcal{H}$) for one, hence all, $\lambda \in \pi(T)$.*

In the following examples we construct self-adjoint extensions of symmetric operators by adding appropriate elements to the domains and using Proposition 3.6.

Example 3.4 (*Examples* 1.4, 1.5, *and* 3.2 *continued*) Let $a, b \in \mathbb{R}$, $a < b$. Recall from Examples 1.4 and 3.2 that $T = -i\frac{d}{dx}$ on $\mathcal{D}(T) = H_0^1(a, b)$ is a closed symmetric operator with deficiency indices $(1, 1)$. In Example 1.5 it was shown that for any $z = e^{i\varphi} \in \mathbb{T}$, the operator $S_z = -i\frac{d}{dx}$ with boundary condition $f(b) = zf(a)$ is self-adjoint and hence a self-adjoint extension of T.

We rederive this result by applying Proposition 3.6 to the subspace $\mathcal{E}_z := \mathbb{C} \cdot u_\varphi$, where $u_\varphi(x) := \exp(i\varphi(b-a)^{-1}x)$. Using formula (1.13), one verifies that $\mathcal{D}(T_{\mathcal{E}_z})$ is a symmetric subspace. Thus, $T_{\mathcal{E}_z}$ is a closed symmetric operator with deficiency indices $(0, 0)$ by Proposition 3.6 and hence self-adjoint by Proposition 3.13. Since $u_\varphi(b) = zu_\varphi(a)$, we have $\mathcal{D}(T_{\mathcal{E}_z}) \subseteq \mathcal{D}(S_z)$ and $T_{\mathcal{E}_z} \subseteq S_z$. Hence, $T_{\mathcal{E}_z} = S_z$.   ∘

Example 3.5 (*Example* 3.3 *continued*) As in Example 3.3, we let $T = -\frac{d^2}{dx^2}$ on $\mathcal{D}(T) = \{f \in H^2(\mathbb{R}) : f'(0) = 0\}$. By Lemma 1.12, applied to f', the functional $f \to f'(0)$ is continuous on $H^2(\mathbb{R})$, so $\mathcal{D}(T)$ is a closed subspace of $H^2(\mathbb{R})$. Since the graph norm of T is equivalent to the norm of $H^2(\mathbb{R})$, $\mathcal{D}(T)$ is complete in the graph norm of T. Therefore, T is closed.

For $B \in \mathbb{R}$, we define the operator $T_B = -\frac{d^2}{dx^2}$ on the domain

$$\mathcal{D}(T_B) = \big\{f \in H^2(-\infty, 0) \oplus H^2(0, +\infty):$$
$$f'(+0) = f'(-0), \, f(+0) - f(-0) = Bf'(0)\big\}.$$

Statement *T_B is a self-adjoint extension of T.*

Proof Using the integration-by-parts formula (3.8), one easily verifies that T_B is symmetric. From Example 3.3 we know that the symmetric operator T has deficiency indices $(1, 1)$. Clearly, $\mathcal{D}(T) \subseteq \mathcal{D}(T_B)$, and T_B is an extension of T.

Choose numbers $a, b > 0$ and $\alpha, \beta \in \mathbb{C}$ such that $\alpha(2 + aB) + \beta(2 + bB) = 0$ and $a\alpha + b\beta \neq 0$. Define the function f on $\mathbb{R}$ by $f(x) = \alpha e^{-ax} + \beta e^{-bx}$ and $f(-x) = -f(-x)$ if $x > 0$. Then $f \in \mathcal{D}(T_B)$ and $f \notin \mathcal{D}(T)$. Put $\mathcal{E} := \mathbb{C} \cdot f$. Since $T_{\mathcal{E}} \subseteq T_B$ and T_B is symmetric, $\mathcal{D}(T_{\mathcal{E}})$ is a symmetric subspace of $\mathcal{D}(T^*)$. Therefore, by Proposition 3.6, $T_{\mathcal{E}}$ is a closed symmetric operator with deficiency indices $(0, 0)$. Hence, $T_{\mathcal{E}}$ is self-adjoint by Proposition 3.13 and $T_{\mathcal{E}} = T_B$.   □

If $B < 0$, then $-4B^{-2}$ is an eigenvalue of T_B. An eigenvector is the function f with $a = -2B^{-1}$, $\beta = 0$, that is, $f(x) = e^{2B^{-1}x}$ and $f(-x) = -f(x)$, $x > 0$.   ∘

The problem of describing self-adjoint extensions (if there are any) of a general symmetric operator will be studied in detail in Chaps. 13 and 14. In general, there is no self-adjoint extension in the same Hilbert space. Here we prove only two simple results. The second shows by an easy construction that a densely defined symmetric operator has always a self-adjoint extension on a *larger* Hilbert space.

Proposition 3.16 *If T is a densely defined closed symmetric operator on $\mathcal{H}$ and μ is a real number in $\pi(T)$, then there exists a self-adjoint extension A of T on $\mathcal{H}$ such that $\mu \in \rho(A)$.*

Proof Upon replacing T by $T + \mu I$, we can assume without loss of generality that $\mu = 0$. Let P denote the projection of $\mathcal{H}$ onto $\mathcal{N}(T^*)$. Since $\mu = 0 \in \pi(T)$, we have $\mathcal{R}(T^*) = \mathcal{H}$ by Proposition 3.2(iii). Hence, there exists a linear subspace $\mathcal{D}_0$ of $\mathcal{D}(T^*)$ such that $T^*\mathcal{D}_0 = \mathcal{N}(T^*)$. Define $\mathcal{D}(A) := \mathcal{D}(T) + (I - P)\mathcal{D}_0$ and $A := T^* \upharpoonright \mathcal{D}(A)$. Obviously, A is an extension of T.

We show that A is symmetric. Let $u_k = x_k + y_k$, where $x_k \in \mathcal{D}(T)$ and $y_k \in (I - P)\mathcal{D}_0, k = 1, 2$. From the relation $T^*(I - P)\mathcal{D}_0 = T^*\mathcal{D}_0 = \mathcal{N}(T^*) \perp (I - P)\mathcal{D}_0$ we obtain $\langle T^*y_j, y_k \rangle = 0$ for $j, k = 1, 2$. Therefore,

$$
\begin{aligned}
\langle Au_1, u_2 \rangle &= \langle Tx_1 + T^*y_1, x_2 + y_2 \rangle \\
&= \langle Tx_1, x_2 \rangle + \langle T^*y_1, y_2 \rangle + \langle Tx_1, y_2 \rangle + \langle T^*y_1, x_2 \rangle \\
&= \langle x_1, Tx_2 \rangle + \langle x_1, T^*y_2 \rangle + \langle y_1, Tx_2 \rangle \\
&= \langle x_1, Tx_2 \rangle + \langle y_1, T^*y_2 \rangle + \langle x_1, T^*y_2 \rangle + \langle y_1, Tx_2 \rangle \\
&= \langle x_1 + y_1, Tx_2 + T^*y_2 \rangle = \langle u_1, Au_2 \rangle,
\end{aligned}
$$

that is, A is symmetric.

Since $0 \in \pi(T)$ and T is closed, we have $\mathcal{H} = \mathcal{R}(T) \oplus \mathcal{N}(T^*)$ by Corollary 2.2. By construction, $A(I - P)\mathcal{D}_0 = T^*(I - P)\mathcal{D}_0 = \mathcal{N}(T^*)$, and so $\mathcal{R}(A) = \mathcal{H}$. Therefore, by Corollary 3.12, A is self-adjoint, and $0 \in \rho(A)$.   □

Proposition 3.17 *Let T be a densely defined symmetric operator on a Hilbert space $\mathcal{H}$. Then there exists a self-adjoint operator A on a Hilbert space $\mathcal{G}$ which contains $\mathcal{H}$ as a subspace such that $T \subseteq A$ and $\mathcal{D}(\overline{T}) = \mathcal{D}(A) \cap \mathcal{H}$.*

Proof Let T_d be the operator on the "diagonal" subspace $\mathcal{H}_d = \{(x, x) : x \in \mathcal{H}\}$ of the Hilbert space $\mathcal{G} := \mathcal{H} \oplus \mathcal{H}$ defined by $T_d(x, x) = (Tx, Tx)$, $x \in \mathcal{D}(T)$, with domain $\mathcal{D}(T_d) := \{(x, x) : x \in \mathcal{D}(T)\}$. The map U given by $U(x) := \frac{1}{\sqrt{2}}(x, x)$ is an isometry of $\mathcal{H}$ onto $\mathcal{H}_d$ such that $UTU^{-1} = T_d$. Hence, the operators T and T_d are unitarily equivalent, so it suffices to prove the assertion for T_d.

Consider the following block matrix on the Hilbert space $\mathcal{G} = \mathcal{H} \oplus \mathcal{H}$:

$$A = \begin{pmatrix} 0 & \overline{T} \\ T^* & 0 \end{pmatrix}.$$

That is, A is defined by $A(x, y) = (\overline{T}y, T^*x)$ for $(x, y) \in \mathcal{D}(A) = \mathcal{D}(T^*) \oplus \mathcal{D}(\overline{T})$. A straightforward verification shows that A is self-adjoint operator on $\mathcal{G}$. Obviously, $T_d \subseteq A$ and $\mathcal{D}(A) \cap \mathcal{H}_d = \mathcal{D}(\overline{T_d})$. $\qquad\qquad\square$

We close this section by proving two simple technical results. They will be used in Sects. 3.4, 5.2, 7.1, and 7.3.

Proposition 3.18 *Let T be a densely defined closed linear operator from a Hilbert space $\mathcal{H}_1$ into a Hilbert space $\mathcal{H}_2$. Then:*

(i) *$I + T^*T$ is a bijective mapping of $\mathcal{H}_1$. Its inverse $C := (I + T^*T)^{-1}$ is a bounded self-adjoint operator on $\mathcal{H}_1$ such that $0 \le C \le I$.*
(ii) *T^*T is a positive self-adjoint operator on $\mathcal{H}_1$, and $\mathcal{D}(T^*T)$ is a core for T.*

Proof (i): Recall that $\mathcal{G}(T^*) = V(\mathcal{G}(T))^{\perp}$ by Lemma 1.10, where $V(x, y) = (-y, x)$, $x \in \mathcal{H}_1$, $y \in \mathcal{H}_2$. Hence, $\mathcal{H}_2 \oplus \mathcal{H}_1 = \mathcal{G}(T^*) \oplus V(\mathcal{G}(T))$. Therefore, for each $u \in \mathcal{H}_1$, there exist vectors $x \in \mathcal{D}(T)$ and $y \in \mathcal{D}(T^*)$ such that

$$(0, u) = (y, T^*y) + V(x, Tx) = (y - Tx, T^*y + x),$$

so $y = Tx$ and $u = x + T^*y = x + T^*Tx = (I + T^*T)x$. That is, $I + T^*T$ is surjective. The operator $I + T^*T$ is injective, because

$$\begin{aligned} \left\| (I + T^*T)x \right\|^2 &= \langle x + T^*Tx, x + T^*Tx \rangle \\ &= \|x\|^2 + \left\| T^*Tx \right\|^2 + 2\|Tx\|^2 \end{aligned} \qquad (3.14)$$

for $x \in \mathcal{D}(T^*T)$. Thus, $I + T^*T$ is bijective.

Let $u = (I + T^*T)x$, where $x \in \mathcal{D}(T^*T)$. Then $x = Cu$, and by (3.14),

$$\|Cu\| = \|x\| \le \left\| (I + T^*T)x \right\| = \|u\|.$$

Therefore, C is a bounded operator on $\mathcal{H}_1$. From (3.14) we also derive

$$\langle Cu, u \rangle = \langle x, u \rangle = \|x\|^2 + \|Tx\|^2 \le \left\| (I + T^*T)x \right\|^2 = \|u\|^2.$$

Therefore, C is symmetric and hence self-adjoint, since C is bounded. Further, $0 \le C \le I$.

(ii): Since $C = (I + T^*T)^{-1}$ is self-adjoint as just proved, so are its inverse $I + T^*T$ by Theorem 1.8(iv) and hence T^*T. We have $\langle T^*Tx, x \rangle = \|Tx\|^2 \ge 0$ for $x \in \mathcal{D}(T^*T)$, that is, the operator T^*T is positive.

That $\mathcal{D}(T^*T)$ is a core for T means that $\mathcal{D}(T^*T)$ is dense in the Hilbert space $(\mathcal{D}(T), \langle \cdot, \cdot \rangle_T)$, where $\langle \cdot, \cdot \rangle_T$ is defined by (1.3). If $y \in \mathcal{D}(T)$ is orthogonal to $\mathcal{D}(T^*T)$ in $(\mathcal{D}(T), \langle \cdot, \cdot \rangle_T)$, then

$$0 = \langle y, x \rangle_T = \langle y, x \rangle + \langle Ty, Tx \rangle = \langle y, (I + T^*T)x \rangle$$

for $x \in \mathcal{D}(T^*T)$. Hence, $y = 0$, since $\mathcal{R}(I + T^*T) = \mathcal{H}_1$. This proves that $\mathcal{D}(T^*T)$ is a core for T. $\qquad\qquad\square$

3.3 Accretive and Sectorial Operators

In this section we study classes of operators which can be considered as generalizations of lower semibounded or positive symmetric (resp. self-adjoint) operators.

Throughout, T denotes a linear operator on a Hilbert space $\mathcal{H}$.

Definition 3.6 We shall say that the operator T is *accretive* if $\mathrm{Re}\langle Tx, x\rangle \geq 0$ for all $x \in \mathcal{D}(T)$ and that T is *m-accretive* if T is closed, accretive, and $\mathcal{R}(T - \lambda_0 I)$ is dense in $\mathcal{H}$ for some $\lambda_0 \in \mathbb{C}$, $\mathrm{Re}\,\lambda_0 < 0$.

T is called *dissipative* if $-T$ is accretive and *m-dissipative* if $-T$ is *m*-accretive.

Definition 3.7 The operator T is said to be *sectorial* if its numerical range $\Theta(T)$ is contained in a sector

$$S_{c,\theta} := \left\{\lambda \in \mathbb{C} : |\mathrm{Im}\,\lambda| \leq \tan\theta\,(\mathrm{Re}\,\lambda - c)\right\} = \left\{\lambda \in \mathbb{C} : \left|\arg(\lambda - c)\right| \leq \theta\right\}$$

$$(3.15)$$

for some $c \in \mathbb{R}$, called *vertex* of $S_{c,\theta}$, and $\theta \in [0, \pi/2)$, called *semi-angle* of $S_{c,\theta}$. We say that T is *m-sectorial* if T is closed, $\Theta(T) \subseteq S_{c,\theta}$ for some $c \in \mathbb{R}$ and $\theta \in [0, \pi/2)$, and $\mathcal{R}(T - \lambda_0 I)$ is dense in $\mathcal{H}$ for some $\lambda_0 \in \mathbb{C} \backslash S_{c,\theta}$.

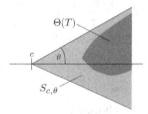

In Theorem 6.12 we will prove that *m*-dissipative operators are the generators of contraction semigroups, while *m*-sectorial operators will appear as operators associated with sectorial or elliptic forms (see Theorem 11.8).

Clearly, T is accretive if and only if $\Theta(T)$ is a subset of the closed right half-plane $S_{0,\pi/2}$. For a sectorial operator, it is required that $\Theta(T)$ is contained in a sector $S_{c,\theta}$ with opening angle 2θ *strictly less* than π, so accretive operators are not necessarily sectorial. Because of their resemblance to sectorial operators, we will develop all results for accretive operators rather than for dissipative operators.

Example 3.6 Let A and B be symmetric operators on $\mathcal{H}$ and set $T := A + iB$. Then T is *accretive* if and only if $A \geq 0$, and T is *sectorial* if and only if there are numbers $c \in \mathbb{R}$ and $m \geq 0$ such that $B \leq m(A - cI)$ and $-B \leq m(A - cI)$. In the latter case, $\Theta(T) \subseteq S_{c,\theta}$, where $\theta \in [0, \pi/2)$ is determined by $\tan\theta = m$.   ○

More examples can be found in the Exercises.

The next proposition describes *m*-sectorial operators among sectorial operators.

Proposition 3.19 *Let T be a closed operator on $\mathcal{H}$, and $\Theta(T) \subseteq S_{c,\theta}$, where $c \in \mathbb{R}$ and $\theta \in [0, \pi/2)$. Then the following assertions are equivalent:*

(i) *T is m-sectorial.*
(ii) *$\lambda \in \rho(T)$ for some, hence all, $\lambda \in \mathbb{C} \backslash S_{c,\theta}$.*
(iii) *$\mathcal{D}(T)$ is dense in $\mathcal{H}$, and T^* is m-sectorial.*

If T is m-sectorial and $\Theta(T) \subseteq S_{c,\theta}$, we have $\Theta(T^*) \subseteq S_{c,\theta}$, $\sigma(T) \subseteq S_{c,\theta}$,

$$\left\| (T - \lambda I)^{-1} \right\| \le \left(\mathrm{dist}(\lambda, \Theta(T)) \right)^{-1} \quad \text{for } \lambda \in \mathbb{C} \backslash S_{c,\theta}. \tag{3.16}$$

Proof Upon replacing T by $T + cI$, we can assume throughout this proof that $c = 0$.

(i) $\rightarrow$ (ii): Let λ_0 be as in Definition 3.7. Then, by Lemma 2.5, $\lambda_0 \in \pi(T)$. Therefore, since T is closed, $\mathcal{R}(T - \lambda_0 I)$ is closed by Proposition 2.1(iv), so $\mathcal{R}(T - \lambda_0 I) = \mathcal{H}$ by Definition 3.7. Hence, $\lambda_0 \in \rho(T)$ by Proposition 2.6(i). Applying Proposition 2.8 with $\mathcal{U} = \mathbb{C} \backslash S_{c,\theta}$, we obtain $\mathbb{C} \backslash S_{c,\theta} \subseteq \rho(T)$, and so $\sigma(T) \subseteq S_{c,\theta}$.

(ii) $\rightarrow$ (iii): First we show that $\mathcal{D}(T)$ is dense in $\mathcal{H}$. Assuming the contrary, there exists a nonzero vector $u \in \mathcal{H}$ which is orthogonal to $\mathcal{D}(T)$. Since $-1 \in \mathbb{C} \backslash S_{0,\theta}$, it follows from (ii) that $-1 \in \rho(T)$. Then $x := (T + I)^{-1} u \in \mathcal{D}(T)$, and hence $0 = \langle u, x \rangle = \langle (T + I)x, x \rangle$. From the latter we deduce that $-1 = \|x\|^{-2} \langle Tx, x \rangle$ belongs to $\Theta(T) \subseteq S_{0,\theta}$, which is the desired contradiction. Thus, $\mathcal{D}(T)$ is dense, and the operator T^* is defined.

Next we prove that $\Theta(T^*) \subseteq S_{0,\theta}$. Let x be a unit vector of $\mathcal{D}(T^*)$, and let $\delta > 0$. Since $-\delta \in \mathbb{C} \backslash S_{0,\theta}$, we have $-\delta \in \rho(T)$ by (ii). In particular, $\mathcal{R}(T + \delta I) = \mathcal{H}$, so there exists $y \in \mathcal{D}(T)$ such that $(T^* + \delta)x = (T + \delta)y$. Now we compute

$$\langle T^*x, x \rangle + \delta = \langle (T^* + \delta)x, x \rangle = \langle (T + \delta)y, x \rangle = \langle y, (T^* + \delta)x \rangle = \langle y, (T + \delta)y \rangle.$$

The right-hand side belongs to $\mathbb{R}_+ \cdot \Theta(T + \delta) \subseteq \mathbb{R}_+ \cdot S_{\delta,\theta} \subseteq S_{0,\theta}$. Hence, we have $\langle T^*x, x \rangle + \delta \in S_{0,\theta}$ for all $\delta > 0$ which implies that $\langle T^*x, x \rangle \in S_{0,\theta}$. Therefore, $\Theta(T^*) \subseteq S_{0,\theta}$. This proves that T^* is sectorial.

Let $\lambda \in \mathbb{C} \backslash S_{c,\theta}$. Since $\bar{\lambda}$ also belongs to $\mathbb{C} \backslash S_{c,\theta}$ and hence to $\rho(T)$, we have $\mathcal{N}(T - \bar{\lambda}I) = \{0\}$. Therefore, by (1.7), $\mathcal{R}(T^* - \lambda I)$ is dense. Since T^* is closed, we have proved that T^* is m-sectorial according to Definition 3.7.

(iii) $\rightarrow$ (i): Applying the implication (i) $\rightarrow$ (iii) already proved with T replaced by T^* and using that T is closed, we conclude that $T = (T^*)^*$ is m-sectorial.

Formula (3.16) follows from Proposition 2.8. $\qquad \square$

The preceding proofs carry over almost verbatim to the sector $S_{0,\pi/2}$ and yield the following characterization of m-accretive operators among accretive operators.

Proposition 3.20 *If T is a closed accretive operator, the following are equivalent:*

(i) *T is m-accretive.*
(ii) *$\lambda \in \rho(T)$ for some, hence all, $\lambda \in \mathbb{C}$, $\mathrm{Re}\,\lambda < 0$.*
(iii) *$\mathcal{D}(T)$ is dense in $\mathcal{H}$, and T^* is m-accretive.*

If T is m-accretive, then $\sigma(T) \subseteq \{\lambda \in \mathbb{C} : \operatorname{Re}\lambda \geq 0\}$, and

$$\|(T - \lambda I)^{-1}\| \leq |\operatorname{Re}\lambda|^{-1} \quad \text{for } \operatorname{Re}\lambda < 0.$$

In particular, if T is closed and m-accretive, so is T^* by Proposition 3.20.

Example 3.7 (*Examples* 1.6 *and* 2.5 *continued*) Recall from Example 2.5 that the operator T on $L^2(0, +\infty)$ defined by $Tf = -if'$, $f \in \mathcal{D}(T) = H_0^1(0, +\infty)$ has the resolvent set $\rho(T) = \{\lambda \in \mathbb{C} : \operatorname{Im}\lambda > 0\}$. Set $D := iT$, that is, $Df = f'$ for $f \in \mathcal{D}(D) = H_0^1(0, +\infty)$. Because T is symmetric, $\operatorname{Re}\langle Df, f \rangle = 0$ for $f \in \mathcal{D}(D)$, so D is accretive. Since $\rho(D) = i\rho(T) = \{\lambda \in \mathbb{C} : \operatorname{Re}\lambda < 0\}$, the operators D and D^* are m-accretive by Proposition 3.20.   ∘

Next we give some characterizations of accretive resp. m-accretive operators.

Lemma 3.21 *A linear operator T is accretive if and only if $\|(T + \lambda I)x\| \geq \lambda \|x\|$ for all $\lambda > 0$ and $x \in \mathcal{D}(T)$.*

Proof Clearly, $\|(T + \lambda I)x\| \geq \lambda \|x\|$ means that $\|Tx\|^2 + 2\lambda \operatorname{Re}\langle Tx, x \rangle \geq 0$. Since $\lambda > 0$ is arbitrary, the latter is obviously equivalent to $\operatorname{Re}\langle Tx, x \rangle \geq 0$.   □

Proposition 3.22 *A linear operator T on $\mathcal{H}$ is m-accretive if and only if $\mathcal{D}(T)$ is dense in $\mathcal{H}$, T is closed, $(-\infty, 0) \subseteq \rho(T)$, and $\|(T + \lambda I)^{-1}\| \leq \lambda^{-1}$ for $\lambda > 0$.*

Proof The necessity of these conditions follows at once from Proposition 3.20. The sufficiency will follow from Proposition 3.20, (ii) → (i), once we have shown that T is accretive. Indeed, if $\lambda > 0$, then $\|(T + \lambda I)^{-1}y\| \leq \lambda^{-1}\|y\|$ for $y \in \mathcal{H}$. Setting $y = (T + \lambda I)x$, the latter yields that $\|(T + \lambda I)x\| \geq \lambda \|x\|$. Therefore, T is accretive by Lemma 3.21.   □

The numerical range of a symmetric operator is obviously a subset of $\mathbb{R}$. Hence, a symmetric operator is sectorial (resp. accretive) if and only if it is lower semibounded (resp. positive). Combining Proposition 3.19 (resp. 3.22) and Corollary 3.14, we therefore obtain the following:

Corollary 3.23 *A symmetric operator is m-sectorial (resp. m-accretive) if and only if it is lower semibounded (resp. positive) and self-adjoint.*

Proposition 3.24 *An m-sectorial (resp. m-accretive) operator T is maximal sectorial (resp. maximal accretive), that is, it has no proper sectorial (resp. accretive) extension acting on the same Hilbert space.*

Proof Let us prove this for sectorial operators. Suppose that T_1 is a sectorial extension of T, say $\Theta(T) \subseteq S_{c,\theta}$ and $\Theta(T_1) \subseteq S_{c_1,\theta_1}$. Take λ from the complement of both sectors. Then $T - \lambda I$ is surjective, because $\lambda \in \rho(T)$ by Proposition 3.19(ii), and $T_1 - \lambda I$ is injective, since $\lambda \in \pi(T_1)$ by Lemma 2.5. Therefore, by Lemma 1.3, $T - \lambda I = T_1 - \lambda I$, and hence $T = T_1$.   □

3.4 Normal Operators

Normal operators are the largest class of single operators for which we prove the spectral theorem (see Sect. 5.5).

In this section, T denotes a densely defined linear operator on a Hilbert space $\mathcal{H}$.

Definition 3.8 We say that T is *formally normal* if $\mathcal{D}(T) \subseteq \mathcal{D}(T^*)$ and

$$\|Tx\| = \|T^*x\| \quad \text{for all } x \in \mathcal{D}(T) \tag{3.17}$$

and that T is *normal* if T is formally normal and $\mathcal{D}(T) = \mathcal{D}(T^*)$.

From Proposition 3.25 it is clear that T is normal if and only if T^* is normal. Moreover, by (3.17) we have $\mathcal{N}(T) = \mathcal{N}(T^*)$ for any normal operator T.

Obviously, if T is a densely defined symmetric operator, then T is formally normal, and T is normal if and only if T is self-adjoint.

However, there is a striking difference to the relations between symmetric and self-adjoint operators: While each densely defined symmetric operator has a self-adjoint extension on a larger Hilbert space (Proposition 3.17), there exists a formally normal operator which has no normal extension on some larger Hilbert space (see Example 5.5).

Proposition 3.25 *T is normal if and only if T is closed and $T^*T = TT^*$.*

Proof First suppose that T is normal. By (3.17), the graph norms $\|\cdot\|_T$ and $\|\cdot\|_{T^*}$ coincide on $\mathcal{D}(T) = \mathcal{D}(T^*)$. Since T^* is closed, $(\mathcal{D}(T^*), \|\cdot\|_{T^*})$ is complete by Proposition 1.4(iii), and so is $(\mathcal{D}(T), \|\cdot\|_T)$. Therefore, T is closed.

For $x, y \in \mathcal{D}(T)$, we have $\|T(x + \tau y)\| = \|T^*(x + \tau y)\|$ for $\tau = 1, -1, i, -i$ by (3.17), and hence $\langle Tx, Ty \rangle = \langle T^*x, T^*y \rangle$ by the polarization formula (1.2).

Let $y \in \mathcal{D}(T^*T)$. Then $y \in \mathcal{D}(T)$, and so $\langle x, T^*Ty \rangle = \langle Tx, Ty \rangle = \langle T^*x, T^*y \rangle$ for all $x \in \mathcal{D}(T)$. Therefore, $T^*y \in \mathcal{D}(T^{**}) = \mathcal{D}(T)$ and $TT^*y = T^*Ty$. This proves that $T^*T \subseteq TT^*$. Since $T = T^{**}$ by Theorem 1.8(iii), T^* is also normal, so we can interchange T and T^*, which yields $TT^* \subseteq T^*T$. Thus, $T^*T = TT^*$.

Conversely, suppose that T is closed and $T^*T = TT^*$. For $x \in \mathcal{D} := \mathcal{D}(T^*T) = \mathcal{D}(TT^*)$, we have

$$\|Tx\|^2 = \langle T^*Tx, x \rangle = \langle TT^*x, x \rangle = \|T^*x\|^2,$$

and hence $\|x\|_T = \|x\|_{T^*}$. By Proposition 3.18(ii), $\mathcal{D}(T^*T)$ is a core for T, and $\mathcal{D}(TT^*)$ is a core for T^*. That is, both $\mathcal{D}(T)$ and $\mathcal{D}(T^*)$ are completions of $(\mathcal{D}, \|\cdot\|_T) = (\mathcal{D}, \|\cdot\|_{T^*})$. This implies $\|Tx\| = \|T^*x\|$ for $x \in \mathcal{D}(T) = \mathcal{D}(T^*)$. $\square$

Another characterization of normal operators will be given in Proposition 5.30. The next proposition collects a number of basic properties of normal operators.

Proposition 3.26 *Let T be a normal operator on $\mathcal{H}$. Then we have:*

(i) *For any $\lambda \in \mathbb{C}$, the operator $T - \lambda I$ is normal, and $\mathcal{N}(T - \lambda I) = \mathcal{N}(T^* - \overline{\lambda}I)$.*
In particular, λ is an eigenvalue of T and $x \in \mathcal{H}$ is an eigenvector of T for λ if and only if $\overline{\lambda}$ is an eigenvalue of T^* and x is an eigenvector of T^* for $\overline{\lambda}$.
(ii) *Eigenvectors of T belonging to different eigenvalues are orthogonal.*
(iii) $\mathcal{H} = \overline{\mathcal{R}(T - \lambda I)} \oplus \mathcal{N}(T - \lambda I)$.
(iv) *If S is an operator such that $T \subseteq S$ and $\mathcal{D}(S) \subseteq \mathcal{D}(S^*)$, then $T = S$.*
In particular, each normal operator is maximal formally normal, that is, if T is normal, S is formally normal and $T \subseteq S$, then $T = S$.
(v) *If T is injective, then its inverse T^{-1} is also normal. In particular, the resolvents $R_\lambda(T)$, $\lambda \in \rho(T)$, of a normal operator T are normal.*

Proof (i): For $x \in \mathcal{D}(T - \lambda I) = \mathcal{D}(T) = \mathcal{D}(T^*) = \mathcal{D}((T - \lambda I)^*)$, we have

$$\left\| (T - \lambda I)x \right\|^2 = \|Tx\|^2 + |\lambda|^2 \|x\|^2 + \langle \lambda x, Tx \rangle + \langle Tx, \lambda x \rangle$$
$$= \left\| T^*x \right\|^2 + |\lambda|^2 \|x\|^2 + \langle T^*x, \overline{\lambda}x \rangle + \langle \overline{\lambda}x, T^*x \rangle = \left\| (T - \lambda I)^*x \right\|^2,$$

that is, $T - \lambda I$ is normal. Hence, $\mathcal{N}(T - \lambda I) = \mathcal{N}((T - \lambda I)^*) = \mathcal{N}(T^* - \overline{\lambda}I)$.

(ii): Suppose that $Tx_1 = \lambda_1 x_1$, $Tx_2 = \lambda_2 x_2$, and $\lambda_1 \neq \lambda_2$. Then $T^*x_2 = \overline{\lambda}_2 x_2$ by (i), and hence

$$\lambda_1 \langle x_1, x_2 \rangle = \langle Tx_1, x_2 \rangle = \langle x_1, T^*x_2 \rangle = \lambda_2 \langle x_1, x_2 \rangle,$$

so that $\langle x_1, x_2 \rangle = 0$.

(iii) follows at once from $\mathcal{N}((T - \lambda I)^*) = \mathcal{N}(T^* - \overline{\lambda}I) = \mathcal{N}(T - \lambda I)$ by (i) and the first equality of (1.7).

(iv): Since $T \subseteq S$, $S^* \subseteq T^*$ and $\mathcal{D}(T) \subseteq \mathcal{D}(S) \subseteq \mathcal{D}(S^*) \subseteq \mathcal{D}(T^*) = \mathcal{D}(T)$, that is, $\mathcal{D}(T) = \mathcal{D}(S)$, and hence $T = S$.

(v): Since T is injective, $\mathcal{R}(T)^\perp = \mathcal{N}(T^*) = \mathcal{N}(T) = \{0\}$, so $\mathcal{R}(T)$ is dense. Hence, Theorem 1.8(iv) applies and yields $(T^{-1})^* = (T^*)^{-1}$. Therefore,

$$\left(T^{-1}\right)^* T^{-1} = \left(T^*\right)^{-1} T^{-1} = \left(TT^*\right)^{-1} = \left(T^*T\right)^{-1} = T^{-1}\left(T^*\right)^{-1} = T^{-1}\left(T^{-1}\right)^*,$$

which proves that T^{-1} is normal. $\qquad\square$

The next proposition is *Fuglede's theorem*. First we recall a result from elementary functional analysis.

Lemma 3.27 *For $B \in \mathbf{B}(\mathcal{H})$, there is an operator $e^B \in \mathbf{B}(\mathcal{H})$ defined by the series*

$$e^B := \sum_{n=0}^{\infty} \frac{B^n}{n!} \tag{3.18}$$

which converges in the operator norm. If $B, C \in \mathbf{B}(\mathcal{H})$ and $BC = CB$, then we have $e^{B+C} = e^B e^C$. Moreover, $e^{-B} = (e^B)^{-1}$.

Proof From the convergence of the series $e^{\|B\|} = \sum_{n=0}^{\infty} \frac{1}{n!} \|B\|^n$ it follows that $(S_k := \sum_{n=0}^{k} \frac{1}{n!} B^n)_{k \in \mathbb{N}}$ is a Cauchy sequence in the operator norm. Hence, the series in (3.18) converges in the operator norm.

If B and C commute, then $(B+C)^n = \sum_{k=0}^{n} \binom{n}{k} B^k C^{n-k}$, and therefore

$$
\begin{aligned}
e^{B+C} &= \sum_{n=0}^{\infty} \frac{(B+C)^n}{n!} = \sum_{n=0}^{\infty} \sum_{k=0}^{n} \binom{n}{k} \frac{B^k C^{n-k}}{n!} \\
&= \sum_{n=0}^{\infty} \sum_{k=0}^{n} \frac{B^k}{k!} \frac{C^{n-k}}{(n-k)!} = \sum_{k=0}^{\infty} \sum_{l=0}^{\infty} \frac{B^k}{k!} \frac{C^l}{l!} = e^B e^C.
\end{aligned}
$$

Since $e^B e^{-B} = e^{-B} e^B = e^0 = I$, we conclude that $e^{-B} = (e^B)^{-1}$. $\qquad\square$

Proposition 3.28 (*Fuglede's theorem*) Let $T, S \in \mathbf{B}(\mathcal{H})$. If the operator T is normal and $TS = ST$, then $T^*S = ST^*$.

Proof Fix $x, y \in \mathcal{H}$. We define the function f on the complex plane by

$$
f(z) = \langle e^{zT^*} S e^{-zT^*} x, y \rangle, \quad z \in \mathbb{C}. \tag{3.19}
$$

Inserting the power series expansions of e^{zT^*} and e^{-zT^*} (by (3.18)), it follows that f is an entire function.

Set $A_z := -\mathrm{i}(zT^* - \bar{z}T)$. Then $zT^* = \mathrm{i}A_z + \bar{z}T$. Since T is normal, $\mathrm{i}A_z$ and $\bar{z}T$ commute, so $e^{\mathrm{i}A_z} e^{\bar{z}T} = e^{zT^*}$ by Lemma 3.18. Similarly, $e^{-\bar{z}T} e^{-\mathrm{i}A_z} = e^{-zT^*}$. The assumption $ST = TS$ implies that $Se^{\bar{z}T} = e^{\bar{z}T} S$. Using these facts, we derive

$$
e^{zT^*} S e^{-zT^*} = e^{\mathrm{i}A_z} e^{\bar{z}T} S e^{-\bar{z}T} e^{-\mathrm{i}A_z} = e^{\mathrm{i}A_z} S e^{\bar{z}T} e^{-\bar{z}T} e^{-\mathrm{i}A_z} = e^{\mathrm{i}A_z} S e^{-\mathrm{i}A_z}.
$$

$$\tag{3.20}$$

Since $A_z = A_z^*$ (by definition) and $e^{-\mathrm{i}A_z} = (e^{\mathrm{i}A_z})^{-1}$ (by Lemma 3.18), we obtain

$$
(e^{\mathrm{i}A_z})^* = \left(\sum_n \frac{1}{n!} (\mathrm{i}A_z)^n \right)^* = \sum_n \frac{1}{n!} (-\mathrm{i}A_z)^n = e^{-\mathrm{i}A_z} = (e^{\mathrm{i}A_z})^{-1},
$$

that is, the operator $e^{\mathrm{i}A_z}$ is unitary. Hence, $|f(z)| \leq \|S\| \|x\| \|y\|$ by (3.19) and (3.20). By Liouville's theorem the bounded entire function f is constant, that is, $f(z) = f(0)$. Since $x, y \in \mathcal{H}$ were arbitrary, the latter implies $e^{zT^*} S e^{-zT^*} = S$. Comparing the coefficients of z, we obtain $T^*S - ST^* = 0$. $\qquad\square$

Example 3.8 (*Multiplication operators on general measure spaces*) Let $(X, \mathfrak{A}, \mu)$ be a σ-finite measure space. Suppose that $\varphi : X \to \mathbb{C} \cup \{\infty\}$ is an $\mathcal{A}$-measurable function on X which is μ-a.e. finite (i.e., $K_\infty := \{t \in X : \varphi(t) = \infty\}$ has measure zero). The multiplication operator M_φ is defined as in Example 1.3 by $(M_\varphi f)(t) = \varphi(t) f(t)$ for f in $\mathcal{D}(M_\varphi) := \{f \in L^2(X, \mu) : \varphi \cdot f \in L^2(X, \mu)\}$.

Such a function φ can be much more complicated than in the continuous case. We illustrate this by giving an example of an a.e. finite function φ_0 on $\mathbb{R}$ which is ∞ at all rationals. Let $(r_n)_{n\in\mathbb{N}}$ be an enumeration of the rationals, and let $\omega \in C_0^\infty(\mathbb{R})$ be a nonnegative function which is 1 in some neighborhood of 0. Fix $\alpha \in (0, 1)$ and define

$$
\varphi_0(t) = \sum_{n=1}^{\infty} 2^{-n} \frac{\omega(t - r_n)}{|t - r_n|^\alpha}, \quad t \in \mathbb{R}.
$$

Since the integral of $|\omega(t - r_n)||t - r_n|^{-\alpha}$ over $\mathbb{R}$ does not depend on n, it follows from Fatou's Lemma B.3 that $\varphi_0 \in L^1(\mathbb{R})$. Hence, φ_0 is a.e. finite on $\mathbb{R}$.

Now we continue the study of the operator M_φ in the general case.

First we prove that the domain $\mathcal{D}(M_\varphi)$ is dense in $L^2(X, \mu)$. Let $f \in L^2(X, \mu)$. Let χ_n denote the characteristic function of $K_n := \{t : |\varphi(t)| \le n\}$, $n \in \mathbb{N}$. Clearly, $\chi_n f \in \mathcal{D}(M_\varphi)$. Since $\mu(K_\infty) = 0$, we have $\lim_{n \to \infty}(\chi_n f - f)(t) = 0$ μ-a.e. on X. Hence, it follows from Lebesgue's dominated convergence theorem (Theorem B.1) that $\lim_{n \to \infty} \|\chi_n f - f\| = 0$ in $L^2(X, \mu)$. This shows that $\mathcal{D}(M_\varphi)$ is dense.

By the same reasoning as in Example 1.3 we derive $(M_\varphi)^* = M_{\overline{\varphi}}$.

Since $(M_\varphi)^* = M_{\overline{\varphi}}$, we have $\mathcal{D}((M_\varphi)^*) = \mathcal{D}(M_\varphi)$ and $\|M_\varphi f\| = \|(M_\varphi)^* f\|$ for $f \in \mathcal{D}(M_\varphi)$. Therefore, by Definition 3.8, the operator M_φ is *normal*.

Finally, we determine the spectrum of M_φ. Recall that the case of a continuous function was treated in Example 2.1. The set of all complex numbers λ for which $\mu(\{t : |\varphi(t) - \lambda| < \varepsilon\}) > 0$ for each $\varepsilon > 0$ is called the *essential range* of the function φ with respect to the measure μ and is denoted by $\mathrm{sp}(\varphi)$.

Statement $\sigma(M_\varphi) = \mathrm{sp}(\varphi)$.

Proof First let $\lambda \in \mathrm{sp}(\varphi)$. Then the measure of the set

$$N_n := \left\{t \in X : \left|\varphi(t) - \lambda\right| < n^{-1}\right\}$$

is positive for each $n \in \mathbb{N}$. Because μ is σ-finite, upon replacing N_n by a subset, we can assume that $0 < \mu(N_n) < \infty$. Since φ is bounded on N_n, $\chi_{N_n} \in \mathcal{D}(M_\varphi)$ and

$$\left\|(M_\varphi - \lambda I)\chi_{N_n}\right\|^2 = \int_{N_n} \left|\varphi(t) - \lambda\right|^2 d\mu(t) \le n^{-2}\|\chi_{N_n}\|^2.$$

Hence, λ is not a regular point. Therefore, $\lambda \notin \rho(M_\varphi)$, and so $\lambda \in \sigma(M_\varphi)$.

Now suppose that $\lambda \notin \mathrm{sp}(\varphi)$. Then there exists $n \in \mathbb{N}$ such that $\mu(N_n) = 0$. The function $\psi := (\varphi - \lambda)^{-1}$ is μ-a.e. finite and essentially bounded on X with essential bound n, so the operator M_ψ is bounded. Since $M_\psi = (M_\varphi - \lambda I)^{-1}$, we obtain $\lambda \in \rho(M_\varphi)$, that is, $\lambda \notin \sigma(M_\varphi)$.   □ o

Example 3.9 (*Example 2.2 continued*) The operator T in Example 2.2 is normal. Hence, *any nonempty (!) closed* subset of $\mathbb{C}$ is spectrum of some normal operator. From the spectral theorem proved in Sect. 5.5 it follows that the spectrum of a normal operator is nonempty.   o

3.5 Exercises

*1. Suppose that T is a densely defined closed linear operator of $\mathcal{H}_1$ into $\mathcal{H}_2$. Show that the orthogonal projection of $\mathcal{H}_1 \oplus \mathcal{H}_2$ onto the graph $\mathcal{G}(T)$ of T is given by the operator block matrix

$$\begin{pmatrix} (I + T^*T)^{-1} & T^*(I + TT^*)^{-1} \\ T(I + T^*T)^{-1} & TT^*(I + TT^*)^{-1} \end{pmatrix}. \tag{3.21}$$

2. Let T and S be symmetric operators on $\mathcal{H}$.
   a. Show that $\alpha T + \beta S$ is a symmetric operator for any $\alpha, \beta \in \mathbb{R}$.
   b. Suppose that $\mathcal{N}(T) = \{0\}$. Show that T^{-1} is a symmetric operator.
3. Let T_1 and T_2 be symmetric operators on $\mathcal{H}_1$ and $\mathcal{H}_2$, respectively. Show that $T_1 \oplus T_2$ is a symmetric operator on $\mathcal{H}_1 \oplus \mathcal{H}_2$ with deficiency indices $d_{\pm}(T_1 \oplus T_2) = d_{\pm}(T_1) + d_{\pm}(T_2)$.
4. Let T be a symmetric operator, and let $p, q \in \mathbb{R}[x]$ be polynomials.
   a. Show that if $p(x) \geq 0$ for all $x \in \mathbb{R}$, then $p(T) \geq 0$.
   b. Suppose that $T \geq 0$ and $q(x) \geq 0$ for all $x \in [0, +\infty)$. Show that $q(T) \geq 0$.
   Hint: Use the fundamental theorem of algebra to prove that $p(x) = \overline{r(x)}r(x)$ and $q(x) = \overline{r_1(x)}r_1(x) + x\overline{r_2(x)}r_2(x)$, where $r, r_1, r_2 \in \mathbb{C}[x]$.
*5. Let T be a symmetric operator on $\mathcal{H}$. Prove by induction on n that $\|T^m x\| \leq \|T^n x\|^{m/n} \|x\|^{1-m/n}$ for $x \in \mathcal{D}(T^n)$ and $m, n \in \mathbb{N}$, $m \leq n$.
6. Let T be a densely defined symmetric operator on $\mathcal{H}$. Prove that

$$\langle x_0 + x_+ + x_-, y_0 + y_+ + y_- \rangle_{T^*} = \langle x_0, y_0 \rangle_{\overline{T}} + 2\langle x_+, y_+ \rangle + 2\langle x_-, y_- \rangle$$

for $x_0, y_0 \in \mathcal{D}(\overline{T})$, $x_+, y_+ \in \mathcal{N}(T^* - iI)$, $x_-, y_- \in \mathcal{N}(T^* + iI)$.
(Recall that $\langle \cdot, \cdot \rangle_{T^*}$ is the scalar product on the domain $\mathcal{D}(T^*)$ defined by (1.3).)

*7. (Uncertainty principle)
   Let A and B be symmetric operators on $\mathcal{H}$ and define its commutator by $[A, B]x := (AB - BA)x$ for $x \in \mathcal{D}(AB) \cap \mathcal{D}(BA)$. Let $\lambda, \mu \in \mathbb{R}$.
   a. Show that for any $x \in \mathcal{D}(AB) \cap \mathcal{D}(BA)$, we have

$$|\langle [A, B]x, x \rangle| \leq 2\|(A - \lambda I)x\| \|(B - \mu I)x\|. \tag{3.22}$$

   b. Show that equality holds in (3.22) if and only if there is a $\varphi \in [0, \pi)$ such that $(\sin\varphi)(A - \lambda I)x = i(\cos\varphi)(B - \mu I)x$.
   Hint: First verify that $\langle [A, B]x, x \rangle = 2i\,\mathrm{Im}(\langle (A - \lambda I)x, (B - \mu I)x \rangle)$.
   Apply the Cauchy–Schwarz inequality and recall that equality occurs if and only if the vectors $(A - \lambda I)x$ and $(B - \mu I)x$ are linearly dependent.

*8. (Uncertainty principle continued)
   Let $Q = x$ be the position operator, and $P = -i\hbar\frac{d}{dx}$ the momentum operator on $\mathcal{D}(Q) = \{f \in L^2(\mathbb{R}) : xf(x) \in L^2(\mathbb{R})\}$ and $\mathcal{D}(P) = H^1(\mathbb{R})$ on $L^2(\mathbb{R})$.
   a. Suppose that $\lambda, \mu \in \mathbb{R}$ and $f \in \mathcal{D}(QP) \cap \mathcal{D}(PQ)$. Prove that

$$\hbar\|f\|^2 \leq 2\|(Q - \lambda I)f\| \|(P - \mu I)f\|. \tag{3.23}$$

   b. Show that equality occurs in (3.23) if $f(x) = c\exp(i\mu x/\hbar - \alpha(x - \lambda)^2/\hbar)$ for some $\alpha > 0$ and $c \in \mathbb{C}$.
   Hint: For a., use Exercise 7.a. For b., apply Exercise 7.b, set $\alpha = \tan\varphi$, and solve the corresponding first-order differential equation.

9. Use Proposition 3.6 to show that in Example 3.4 the self-adjoint operators S_z, $z \in \mathbb{T}$, exhaust all self-adjoint extensions on $L^2(a, b)$ of the operator $T = -i\frac{d}{dx}$.

10. Let X be a bounded self-adjoint operator on $\mathcal{H}$ such that $\mathcal{N}(X) = \{0\}$, and let P be a projection on $\mathcal{H}$. Define $\mathcal{D}(T) = X(I - P)\mathcal{H}$ and $T := X^{-1}{\restriction}\mathcal{D}(T)$.

a. Show that T is a closed symmetric operator and $d_{\pm}(T) = \dim P\mathcal{H}$.

b. Show that $\mathcal{D}(T)$ is dense in $\mathcal{H}$ if and only if $\mathcal{R}(X) \cap P\mathcal{H} = \{0\}$.

11. Define the operator T by $(Tf)(x) = -if(x) - 2ixf'(x)$ for $f \in \mathcal{D}(T) := C_0^{\infty}(0, \infty)$ on the Hilbert space $L^2(0, \infty)$. Show that T is essentially self-adjoint.

12. Find a symmetric linear operator T on $\mathcal{H}$ such that $\mathcal{R}(T + iI)$ and $\mathcal{R}(T - iI)$ are dense in $\mathcal{H}$, but T is *not* essentially self-adjoint.

    (By Proposition 3.8, $\mathcal{D}(T)$ cannot be dense in $\mathcal{H}$; however, see Proposition 3.11.)

    Hints: Define T as in Example 1.1 with domain $\mathcal{D}(T) = e^{\perp}$. Show that then $\mathcal{R}(T - \lambda I)$ is dense for all $\lambda \in \mathbb{C}$, $\lambda \neq 0$.

13. Let T be a self-adjoint operator on $\mathcal{H}$. Show that a (real) number λ is an eigenvalue of T if and only if $\mathcal{R}(T - \lambda I)$ is *not* dense in $\mathcal{H}$.

14. Let T be a densely defined closed operator on $\mathcal{H}$. Show that the linear operator $S := (I + T^*T)^{-1}T^*$ with dense domain $\mathcal{D}(S) = \mathcal{D}(T^*)$ is bounded and $\|S\| \leq \frac{1}{2}$.

15. Let T be a densely defined closed linear operator on $\mathcal{H}$. Show that if both T and T^* are accretive, then T and T^* are m-accretive.

16. Let T be a symmetric operator which is not essentially self-adjoint. Show that iT is accretive, but $(iT)^*$ is not.

17. Let T be an accretive operator such that $\mathcal{N}(T) = \{0\}$.

    a. Show that T^{-1} is also accretive.

    b. Show that if T is m-accretive, so is T^{-1}.

18. Let T be an m-accretive operator.

    a. Show that $(T + \lambda I)^{-1}$ is m-accretive for $\lambda \in \mathbb{C}$, $\operatorname{Re}\lambda > 0$.

    b. Show that s-$\lim_{n\to\infty}(I + \frac{1}{n}T)^{-1} = I$.

19. Let p be a real-valued function from $C^1([0, 1])$, and let T be the operator on $L^2(0, 1)$ defined by $Tf = -pf'$ for $f \in \mathcal{D}(T) := H_0^1(0, 1)$.

    Show that T is accretive if and only if $p'(x) \geq 0$ for all $x \in [0, 1]$.

20. Let $p \in C^1([0, 1])$ and $q \in C([0, 1])$ be real-valued functions. Define the operator T on $L^2(0, 1)$ by $Tf = -pf'' + qf$ for $f \in \mathcal{D}(T) := H_0^2(0, 1)$.

    a. Suppose that $p(x) > 0$ for all $x \in [0, 1]$. Show that T is a sectorial operator.

    b. Define $m := \inf\{p(x) : x \in [0, 1]\}$, $M := \sup\{p'(x) : x \in [0, 1]\}$, $L := \inf\{q(x) : x \in [0, 1]\}$, and suppose that $m > 0$ and $4mL \geq M^2$.

       Prove that the operator T is accretive.

21. (*Putnam's corollary to Fuglede's theorem*)

    Let $T_1, T_2, S \in \mathbf{B}(\mathcal{H})$. Suppose that T_1 and T_2 are normal and $T_1 S = S T_2$. Prove that $T_1^* S = S T_2^*$.

    Hint: Apply Fuglede's theorem (Proposition 3.28) to the operator matrices

$$\begin{pmatrix} T_1 & 0 \\ 0 & T_2 \end{pmatrix} \quad \text{and} \quad \begin{pmatrix} 0 & S \\ 0 & 0 \end{pmatrix}.$$

*22. (*q-normal operators; see* [CSS])

    Let $q \in (1, +\infty)$. For $\alpha > 0$, we define the operator T_α on $\mathcal{H} = l^2(\mathbb{Z})$ by $(T_\alpha(\varphi_n))_{k+1} = \alpha q^{-k/2}\varphi_k$ on $\mathcal{D}(T_\alpha) = \{(\varphi_n) \in l^2(\mathbb{Z}) : (q^{-n/2}\varphi_n) \in l^2(\mathbb{Z})\}$.

  a. Prove that $\|T^*\varphi\| = \sqrt{q}\|T\varphi\|$ for $\varphi \in \mathcal{D}(T) = \mathcal{D}(T^*)$.
  b. Prove that $TT^* = qT^*T$.
  *c. Prove that $\|((T + T^*) - \lambda I)\varphi\| \geq |\lambda|(q - 1)(q + 1)^{-1}\|\varphi\|$ for $\lambda \in \mathbb{R}$, $\varphi \in \mathcal{D}(T)$.
  Hint: Show that the following identity holds for vectors of $\mathcal{D}(T^2)$:

$$(T + T^*)^2 - 2\lambda(T + T^*) + 4\lambda^2 q(q + 1)^{-2}I$$
$$= (T + T^* - \lambda I)^2 - \lambda^2(q - 1)^2(q + 1)^{-2}I.$$

  d. Deduce that the symmetric operator $T + T^*$ is not essentially self-adjoint.
  e. When are two operators T_α and $T_{\alpha'}$ unitarily equivalent?

3.6 Notes to Part I

Important pioneering works on unbounded symmetric and self-adjoint operators include Carleman [Cl], von Neumann [vN1, vN2, vN3, vN4], and Stone [St1, St2]. The graph method and Theorem 1.8 have been developed in [vN4].

Early papers on accretive or dissipative operators are [Lv, F2, Ph].

Fuglede's theorem was discovered in [Fu]; the ingenious proof given in the text is due to Rosenblum [Ro].

Unbounded normal operators have been investigated by Stochel and Szafraniec [SS1, SS2].

Part II
Spectral Theory

Chapter 4
Spectral Measures and Spectral Integrals

This chapter is devoted to a detailed treatment of spectral measures and operator-valued integrals of measurable functions. These spectral integrals are unbounded normal operators on Hilbert space. They will be the main tools for the functional calculus of self-adjoint operators in the next chapter. The theory of spectral measures is presented in Sect. 4.2. Spectral integrals are studied in Sect. 4.3.

A large part of this chapter is occupied by measure-theoretic considerations, and the reader may wish to skip some of these technical parts or even avoid them. For this reason, we develop in Sect. 4.1 operator-valued integrals of continuous functions at an elementary level as operator Riemann–Stieltjes integrals.

4.1 Resolutions of the Identity and Operator Stieltjes Integrals

In order to define operator Stieltjes integrals of continuous functions, we need the notion of a resolution of the identity.

Definition 4.1 A *resolution of the identity* on a Hilbert space $\mathcal{H}$ is a one-parameter family $\{E(\lambda) : \lambda \in \mathbb{R}\}$ of orthogonal projections on $\mathcal{H}$ such that

(i) $E(\lambda_1) \leq E(\lambda_2)$ if $\lambda_1 \leq \lambda_2$ (monotonicity),
(ii) $\lim_{\lambda \to \lambda_0 + 0} E(\lambda)x = E(\lambda_0)x$ for $x \in \mathcal{H}$ and $\lambda_0 \in \mathbb{R}$ (strong right continuity),
(iii) $\lim_{\lambda \to -\infty} E(\lambda)x = 0$ and $\lim_{\lambda \to +\infty} E(\lambda)x = x$ for $x \in \mathcal{H}$.

Before we discuss this notion, let us give three examples. They will be our guiding examples in this and in the next chapter.

The reader who encounters this subject for the first time might look at special cases of these examples, for instance, if we have only finitely many nonzero projections in Example 4.1, if μ is the Lebesgue measure on $\mathcal{J} = [0, 1]$ in Example 4.2, or if μ is the Lebesgue measure on $\Omega = [0, 1]$, and $h(t) = \sqrt{t}$ or $h(t) = t^2$ in Example 4.3.

K. Schmüdgen, *Unbounded Self-adjoint Operators on Hilbert Space*,
Graduate Texts in Mathematics 265,
DOI 10.1007/978-94-007-4753-1_4, © Springer Science+Business Media Dordrecht 2012

Example 4.1 Let $(\lambda_n)_{n \in \mathbb{N}}$ be a real sequence, and let $(P_n)_{n \in \mathbb{N}}$ be a sequence of orthogonal projections on $\mathcal{H}$ such that $P_k P_n = 0$ for $k \neq n$ and $\sum_{n=1}^{\infty} P_n = I$. (Recall that any infinite sum of pairwise orthogonal projections converges strongly.) Then

$$E(\lambda) := \sum_{\lambda_n \leq \lambda} P_n, \quad \lambda \in \mathbb{R}, \tag{4.1}$$

defines a resolution of the identity on $\mathcal{H}$.

Since the projections P_n are pairwise orthogonal, $E(\lambda)$ is indeed a projection.

Let $x \in \mathcal{H}$ and $\varepsilon > 0$. Since $\sum_k P_k = I$ and hence $\sum_k \| P_k x \|^2 = \| x \|^2 < \infty$, there exists $k(\varepsilon) \in \mathbb{N}$ such that

$$\sum_{k \geq k(\varepsilon)} \| P_k x \|^2 < \varepsilon^2. \tag{4.2}$$

We show that axioms (i)–(iii) are fulfilled. Axiom (i) is obvious.

(ii): Let $\lambda_0 \in \mathbb{R}$. We choose $\lambda > \lambda_0$ such that the numbers $\lambda_1, \dots, \lambda_{k(\varepsilon)}$ are not in the left-open interval $(\lambda_0, \lambda]$. By (4.1) and (4.2) we then have

$$\left\| E(\lambda)x - E(\lambda_0)x \right\|^2 = \sum_{\lambda_0 < \lambda_n \leq \lambda} \| P_n x \|^2 < \varepsilon^2.$$

This in turn implies that axiom (ii) is satisfied.

(iii): If $\lambda, \lambda' \in \mathbb{R}$ are chosen such that $\lambda_1, \dots, \lambda_{k(\varepsilon)} < \lambda$ and $\lambda' < \lambda_1, \dots, \lambda_{k(\varepsilon)}$, it follows from (4.1) and (4.2) that

$$\left\| x - E(\lambda)x \right\|^2 = \sum_{\lambda_n > \lambda} \| P_n x \|^2 < \varepsilon^2 \quad \text{and} \quad \left\| E(\lambda')x \right\|^2 = \sum_{\lambda_n \leq \lambda'} \| P_n x \|^2 < \varepsilon^2.$$

Hence, axiom (iii) holds.   ∘

Example 4.2 Let μ be a positive regular Borel measure on an interval $\mathcal{J}$ and $\mathcal{H} = L^2(\mathcal{J}, \mu)$. Define

$$\bigl(E(\lambda)f\bigr)(t) = \chi_{(-\infty,\lambda]}(t) \cdot f(t) \quad \text{for } t \in \mathcal{J}, \ \lambda \in \mathbb{R}, \ f \in \mathcal{H},$$

that is, $E(\lambda)$ is the multiplication operator by the characteristic function of the interval $(-\infty, \lambda]$. Then $\{E(\lambda) : \lambda \in \mathbb{R}\}$ is a resolution of the identity.

We verify axiom (ii). Let $f \in \mathcal{H}$, and let $(\lambda_n)_{n \in \mathbb{N}}$ be a real sequence such that $\lambda_n > \lambda_0$ and $\lim_{n \to \infty} \lambda_n = \lambda_0$. Put $f_n = \chi_{(\lambda_0, \lambda_n]} |f|^2$. Then $f_n(t) \to 0$ as $n \to \infty$ and $|f_n(t)| \leq |f(t)|^2$ for $n \in \mathbb{N}$ and $t \in \mathcal{J}$. Since $f \in L^2(\mathcal{J}, \mu)$, Lebesgue's dominated convergence theorem (Theorem B.1) applies and yields $\lim_{n \to \infty} \int_{\mathcal{J}} f_n \, d\mu = 0$. By the definition of $E(\lambda)$ we therefore have

$$\left\| E(\lambda_n)f - E(\lambda_0)f \right\|^2 = \int_{\mathcal{J}} \chi_{(\lambda_0, \lambda_n]} \bigl| f(t) \bigr|^2 \, d\mu(t) = \int_{\mathcal{J}} f_n(t) \, d\mu(t) \to 0,$$

so that $\lim_{n \to \infty} E(\lambda_n)f = E(\lambda_0)f$. This proves that axiom (ii) holds.

Axiom (iii) follows by a similar reasoning. Axiom (i) is obvious.   ∘

Example 4.2 is just the special case $\Omega = \mathcal{J}, h(t) = t$, of the following more general example.

Example 4.3 Let $(\Omega, \mathfrak{A}, \mu)$ be a measure space, $\mathcal{H} = L^2(\Omega, \mu)$, and $h \colon \Omega \to \mathbb{R}$ an $\mathfrak{A}$-measurable function. Define $\Omega(\lambda) = \{t \in \Omega : h(t) \leq \lambda\}$ and

$$\big(E(\lambda)f\big)(t) = \chi_{\Omega(\lambda)}(t) \cdot f(t) \quad \text{for } t \in \Omega, \ \lambda \in \mathbb{R}.$$

Then $\{E(\lambda) : \lambda \in \mathbb{R}\}$ is a resolution of the identity on the Hilbert space $\mathcal{H}$. Axioms (ii) and (iii) are proved as in Example 4.2 using Lebesgue's theorem. ○

Let us discuss the axioms of Definition 4.1 and some simple consequences.

The essential requirement in Definition 1 is that $\{E(\lambda) : \lambda \in \mathbb{R}\}$ is a monotone increasing one-parameter family of orthogonal projection on $\mathcal{H}$.

Axiom (ii) is only a normalization condition. Also, we could have required strong left continuity instead of strong right continuity as some authors do. If (i) holds, then the family of projections $\{\widetilde{E}(\lambda) := \text{s-lim}_{\mu \to \lambda+0} E(\mu) : \lambda \in \mathbb{R}\}$ satisfies (i) and (ii). Note that the strong limit $\text{s-lim}_{\mu \to \lambda+0} E(\mu)$ always exists because of (i).

Axiom (iii) is a completeness condition. If (iii) is not fulfilled, then we have a resolution of the identity on the Hilbert space $(E(+\infty) - E(-\infty))\mathcal{H}$, where $E(\pm\infty) := \text{s-lim}_{\lambda \to \pm\infty} E(\lambda)$.

Since the operators $E(\lambda)$ are orthogonal projections, axiom (i) is equivalent to

$$(\text{i})' \quad E(\lambda_1)E(\lambda_2) = E\big(\min(\lambda_1, \lambda_2)\big) \quad \text{for } \lambda_1, \lambda_2 \in \mathbb{R}.$$

From (i)$'$ it follows that two projections $E(\lambda_1)$ and $E(\lambda_2)$ for $\lambda_1, \lambda_2 \in \mathbb{R}$ commute.

The monotonicity axiom (i) implies that for all $\lambda_0 \in \mathbb{R}$, the strong limit $E(\lambda_0 - 0) := \text{s-lim}_{\lambda \to \lambda_0 - 0} E(\lambda)$ exists and is again an orthogonal projection satisfying

$$E(\lambda_1) \leq E(\lambda_0 - 0) \leq E(\lambda_2) \quad \text{if } \lambda_1 < \lambda_0 \leq \lambda_2. \tag{4.3}$$

Now we define spectral projections for intervals by

$$E\big([a, b]\big) := E(b) - E(a - 0), \qquad E\big([a, b)\big) := E(b - 0) - E(a - 0), \tag{4.4}$$
$$E\big((a, b]\big) := E(b) - E(a), \qquad E\big((c, d)\big) := E(d - 0) - E(c) \tag{4.5}$$

for $a, b, c, d \in \mathbb{R} \cup \{-\infty\} \cup \{+\infty\}$, $a \leq b, c < d$, where we set $E(-\infty) = 0$ and $E(+\infty) = I$. By axiom (i) and (4.3), the preceding definitions give indeed an orthogonal projection $E(\mathcal{J})$ for each of the intervals $\mathcal{J} = [a, b], [a, b), (a, b], (c, d)$. In particular, $E(\{\lambda\}) = E(\lambda) - E(\lambda - 0)$ and $E(\lambda) = E((-\infty, \lambda])$ for $\lambda \in \mathbb{R}$.

In the case of Example 4.2 the projection $E(\mathcal{J})$ is just the multiplication operator by the characteristic function $\chi_{\mathcal{J}}$ of the interval $\mathcal{J}$.

Now we suppose that $\{E(\lambda) : \lambda \in \mathbb{R}\}$ is a resolution of the identity. Our next aim is to define operator-valued Stieltjes integrals $\int_{\mathcal{J}} f \, dE$ and $\int_{\mathbb{R}} f \, dE$ for a continuous function f on $\mathcal{J} = [a, b]$ resp. $\mathbb{R}$.

Let $f \in C(\mathcal{J})$, where $\mathcal{J} := [a, b], a, b \in \mathbb{R}, a < b$. We extend f to a continuous function on $[a - 1, b]$. By a partition $\mathcal{Z}$ we mean a sequence $\{\lambda_0, \ldots, \lambda_n\}$ such that

$$a - 1 < \lambda_0 < a < \lambda_1 < \cdots < \lambda_n = b.$$

Set $|Z| := \max_k |\lambda_k - \lambda_{k-1}|$. We choose $\zeta_k \in [\lambda_{k-1}, \lambda_k]$ and define the Riemann sum

$$S(f, Z) = \sum_{k=1}^{n} f(\zeta_k)\big(E(\lambda_k) - E(\lambda_{k-1})\big).$$

Proposition 4.1 *For any function $f \in C(\mathcal{J})$, there exists a bounded operator $\int_{\mathcal{J}} f\, dE$ on $\mathcal{H}$ which is uniquely defined by the following property:*

For each $\varepsilon > 0$, there exists $\delta(\varepsilon) > 0$ such that $\| \int_{\mathcal{J}} f\, dE - S(f, Z)\| \leq \varepsilon$ for each partition Z satisfying $|Z| < \delta(\varepsilon)$.

Moreover, for any $x \in \mathcal{H}$, we have

$$\left\langle \left(\int_{\mathcal{J}} f\, dE\right)x, x\right\rangle = \int_{\mathcal{J}} f(\lambda)\, d\langle E(\lambda)x, x\rangle, \tag{4.6}$$

$$\left\|\left(\int_{\mathcal{J}} f\, dE\right)x\right\|^2 = \int_{\mathcal{J}} |f(\lambda)|^2\, d\langle E(\lambda)x, x\rangle. \tag{4.7}$$

Proof Fix $\varepsilon > 0$. Suppose that $Z = \{\lambda_0, \ldots, \lambda_n\}$ and Z' are partitions such that Z' is a refinement of Z and

$$|f(t) - f(t')| < \varepsilon/2 \quad \text{when } |t - t'| \leq |Z|, \ t, t' \in [a - 1, b]. \tag{4.8}$$

We write the points of Z' as $\{\lambda'_{kl} : l = 0, \ldots, r_k\}$ with $\lambda_k = \lambda'_{k0} < \cdots < \lambda'_{k, r_k} = \lambda'_{k+1, 0} = \lambda_{k+1}$. Let $x \in \mathcal{H}$. Using the mutual orthogonality of the projections $E(\lambda'_{kl}) - E(\lambda'_{k, l-1})$ (by Exercise 2) and (4.8), we obtain

$$\|2\big(S(f, Z) - S(f, Z')\big)x\|^2$$

$$= \left\|\sum_k \sum_l 2\big(f(\zeta_k) - f(\zeta'_{kl})\big)\big(E(\lambda'_{kl}) - E(\lambda'_{k, l-1})\big)x\right\|^2$$

$$= \sum_k \sum_l 4\big|f(\zeta_k) - f(\zeta'_{kl})\big|^2 \big\|\big(E(\lambda'_{kl}) - E(\lambda'_{k, l-1})\big)x\big\|^2$$

$$\leq \sum_k \sum_l \varepsilon^2 \big\|\big(E(\lambda'_{kl}) - E(\lambda'_{k, l-1})\big)x\big\|^2 = \varepsilon^2 \big\|\big(E(b) - E(\lambda_0)\big)x\big\|^2 \leq \varepsilon^2 \|x\|^2,$$

so that

$$\|S(f, Z) - S(f, Z')\| \leq \varepsilon/2. \tag{4.9}$$

By the uniform continuity of the function f on $[a - 1, b]$, for each $\varepsilon > 0$ there exists a $\delta(\varepsilon) > 0$ such that $|f(t) - f(t')| < \varepsilon/2$ whenever $|t - t'| < \delta(\varepsilon)$. Suppose that Z_1 and Z_2 are partitions such that $|Z_1|, |Z_2| < \delta(\varepsilon)$. Let Z' be a common refinement of Z_1 and Z_2. Since $|Z'| < \delta(\varepsilon)$, we have by (4.9),

$$\|S(f, Z_1) - S(f, Z_2)\| \leq \|S(f, Z_1) - S(f, Z')\| + \|S(f, Z') - S(f, Z_2)\| \leq \varepsilon. \tag{4.10}$$

Therefore, if $(\mathcal{Z}_n)_{n \in \mathbb{N}}$ is a sequence of partitions such that $\lim_{n \to \infty} |\mathcal{Z}_n| = 0$, then $(S(f, \mathcal{Z}_n))_{n \in \mathbb{N}}$ is a Cauchy sequence in the operator norm, so it converges to a bounded operator on $\mathcal{H}$, denoted by $\int_{\mathcal{J}} f \, dE$.

If $\mathcal{Z}$ is a partition satisfying $|\mathcal{Z}| < \delta(\varepsilon)$, then we have $\|S(f, \mathcal{Z}_n) - S(f, \mathcal{Z})\| \leq \varepsilon$ if $|\mathcal{Z}_n| < \delta(\varepsilon)$ by (4.10). Letting $n \to \infty$, we obtain $\|\int_{\mathcal{J}} f \, dE - S(f, \mathcal{Z})\| \leq \varepsilon$.

The sums $\langle S(f, \mathcal{Z}_n)x, x \rangle$ and $\|S(f, \mathcal{Z}_n)x\|^2$ are Riemann sums for the scalar Stieltjes integrals on the right-hand sides of (4.6) and (4.7), respectively. Therefore, formulas (4.6) and (4.7) follow as $n \to \infty$. □

Proposition 4.1 says that $\int_{\mathcal{J}} f \, dE$ is an operator-valued Stieltjes integral with approximating Riemann sums converging in the *operator norm*. Clearly, this operator does not depend on the particular continuous extension of f to $[a - 1, b]$. The reason for requiring that $\lambda_0 < a < \lambda_1$ for the partition points λ_0, λ_1 is that we wanted to include a possible jump $E(a) - E(a - 0)$ at a into the integral over the closed interval $\mathcal{J} = [a, b]$. A possible jump at b is already included because of the right continuity of $\{E(\lambda)\}$.

A Stieltjes integral over the open interval (a, b) can be defined by

$$\int_{(a,b)} f \, dE = \int_{[a,b]} f \, dE - f(a)\big(E(a) - E(a - 0)\big) - f(b)\big(E(b) - E(b - 0)\big).$$

Proposition 4.2 *For $f \in C(\mathbb{R})$, there exists a linear operator $\int_{\mathbb{R}} f \, dE$ such that*

$$\mathcal{D} := \mathcal{D}\left(\int_{\mathbb{R}} f \, dE \right) = \left\{ x \in \mathcal{H} : \int_{\mathbb{R}} |f(\lambda)|^2 \, d\langle E(\lambda)x, x \rangle < \infty \right\}, \quad (4.11)$$

$$\left(\int_{\mathbb{R}} f \, dE \right) x = \lim_{a \to -\infty} \lim_{b \to +\infty} \left(\int_{[a,b]} f \, dE \right) x \quad \text{for } x \in \mathcal{D}, \quad (4.12)$$

$$\left\langle \left(\int_{\mathbb{R}} f \, dE \right) x, x \right\rangle = \int_{\mathbb{R}} f(\lambda) \, d\langle E(\lambda)x, x \rangle \quad \text{for } x \in \mathcal{D}. \quad (4.13)$$

Proof For $x \in \mathcal{H}$ and $a, b, c, d \in \mathbb{R}$, $a < b$, we abbreviate $x_{a,b} := (\int_{[a,b]} f \, dE)x$, $\{c, d\} = [c, d]$ if $c \leq d$ and $\{c, d\} = [d, c]$ if $d < c$. Suppose that $x \in \mathcal{D}$. By considering approximating Riemann sums one easily verifies that

$$\|x_{a,b} - x_{a',b'}\|^2 \leq \left(\int_{\{a,a'\}} + \int_{\{b,b'\}} \right) |f(\lambda)|^2 \, d\langle E(\lambda)x, x \rangle \quad (4.14)$$

for $a, b, a', b' \in \mathbb{R}$, $a < b$, $a' < b'$. Since the Stieltjes integral $\int_{\mathbb{R}} |f|^2 \, d\langle E(\lambda)x, x \rangle$ is finite by (4.11), if follows from (4.14) that the strong limit in (4.12) exists.

Conversely, if for a vector $x \in \mathcal{H}$, the limit in (4.12) exists, then we conclude from (4.7) that the Stieltjes integral in (4.11) is finite, that is, x is in $\mathcal{D}$. In particular, since $\mathcal{D}$ is the set of vectors $x \in \mathcal{H}$ for which the limit in (4.12) exists, $\mathcal{D}$ is a linear subspace of $\mathcal{H}$. Formula (4.13) follows at once from (4.6) and (4.12). □

The operator integrals from Propositions 4.1 and 4.2 are special cases of spectral integrals developed in Section 4.3 below. By comparing formulas (4.11) and (4.6),

(4.13) with formulas (4.26) and (4.32) below it follows easily that the Stieltjes integrals $\int_{\mathcal{J}} f \, dE$ and $\int_{\mathbb{R}} f \, dE$ coincide with the spectral integrals $\mathbb{I}(f \chi_{\mathcal{J}})$ and $\mathbb{I}(f)$, respectively, with respect to the spectral measure E which is associated with the resolution of the identity $\{E(\lambda) : \lambda \in \mathbb{R}\}$ by Theorem 4.6 below. That is, for *continuous* functions, one may use these very elementary and easily feasible definitions as operator-valued Stieltjes integrals given in Propositions 4.1 and 4.2 rather than the technically more involved approach of spectral integrals in Section 4.3.

4.2 Spectral Measures

4.2.1 Definitions and Basic Properties

Let $\mathfrak{A}$ be an algebra of subsets of a set Ω, and let $\mathcal{H}$ be a Hilbert space.

Definition 4.2 A *spectral premeasure* on $\mathfrak{A}$ is a mapping E of $\mathfrak{A}$ into the orthogonal projections on $\mathcal{H}$ such that

(i) $E(\Omega) = I$,
(ii) E is countably additive, that is, $E(\bigcup_{n=1}^{\infty} M_n) = \sum_{n=1}^{\infty} E(M_n)$ for any sequence $(M_n)_{n \in \mathbb{N}}$ of pairwise disjoint sets from $\mathfrak{A}$ whose union is also in $\mathfrak{A}$.

If $\mathfrak{A}$ is a σ-algebra, then a spectral premeasure on $\mathfrak{A}$ is called a *spectral measure*.

Infinite sums as above are always meant in the strong convergence, that is, the equation in (ii) says that $E(\bigcup_{n=1}^{\infty} M_n)x = \lim_{k \to \infty} \sum_{n=1}^{k} E(M_n)x$ for $x \in \mathcal{H}$.

Let us begin with the standard example of a spectral measure.

Example 4.4 Let $(\Omega, \mathfrak{A}, \mu)$ be a measure space, and $\mathcal{H} = L^2(\Omega, \mu)$. For $M \in \mathfrak{A}$, let $E(M)$ be the multiplication operator by the characteristic function χ_M, that is,

$$(E(M)f)(t) = \chi_M(t) \cdot f(t), \quad f \in \mathcal{H}. \tag{4.15}$$

Since $\chi_M^2 = \chi_M = \overline{\chi}_M$, we have $E(M)^2 = E(M) = E(M)^*$, so $E(M)$ is an orthogonal projection. Obviously, $E(\Omega) = I$.

We verify axiom (ii). Let $(M_n)_{n \in \mathbb{N}}$ be a sequence of disjoint sets of $\mathfrak{A}$ and set $M := \bigcup_n M_n$. For $f \in \mathcal{H}$, set $f_k := \sum_{n=1}^{k} \chi_{M_n} f$. Since $|\chi_M f - f_k|^2 \to 0$ as $k \to \infty$ and $|\chi_M f - f_k|^2 \leq 4|f|^2$ on Ω, it follows from Lebesgue's theorem (Theorem B.1) that $\lim_k \|\chi_M f - f_k\|^2 = 0$, and so $\chi_M f = \lim_k \sum_{n=1}^{k} \chi_{M_n} f$. The latter means that $E(M) = \sum_{n=1}^{\infty} E(M_n)$. Hence, E is a spectral measure on $\mathfrak{A}$.   ∘

Now we discuss some simple consequences of Definition 4.2 and suppose that E is a spectral premeasure on an algebra $\mathfrak{A}$.

The case $M_n = \emptyset$ for all n in (ii) yields $E(\emptyset) = 0$.

Setting $M_n = \emptyset$ if $n \geq k+1$ in (ii), we obtain the *finite additivity* of E, that is, for pairwise disjoint sets $M_1, \ldots, M_k \in \mathfrak{A}$, we have

$$E(M_1 \cup \cdots \cup M_k) = E(M_1) + \cdots + E(M_k).$$

Lemma 4.3 *If E is a finitely additive map of an algebra $\mathfrak{A}$ into the orthogonal projections on a Hilbert space $\mathcal{H}$, then we have*

$$E(M)E(N) = E(M \cap N) \quad \text{for } M, N \in \mathfrak{A}. \tag{4.16}$$

In particular, $E(M)E(N) = 0$ if $M, N \in \mathfrak{A}$ are disjoint.

Proof First we note that $E(M_1)E(M_2) = 0$ if $M_1, M_2 \in \mathfrak{A}$ are disjoint. Indeed, by the finite additivity of E, the sum of the two projections $E(M_1)$ and $E(M_2)$ is again a projection. Therefore, $E(M_1)E(M_2) = 0$.

Now we put $M_0 := M \cap N$, $M_1 := M \backslash M_0$, $M_2 := N \backslash M_0$. Since $M_1 \cap M_2 = M_0 \cap M_2 = M_1 \cap M_0 = \emptyset$, by the preceding we have

$$E(M_1)E(M_2) = E(M_0)E(M_2) = E(M_1)E(M_0) = 0. \tag{4.17}$$

Since $M = M_1 \cup M_0$ and $N = M_2 \cup M_0$, from the finite additivity of E and from formula (4.17) we derive

$$\begin{aligned}
E(M)E(N) &= \big(E(M_1) + E(M_0)\big)\big(E(M_2) + E(M_0)\big) = E(M_0)^2 \\
&= E(M \cap N).
\end{aligned}$$
$\square$

Note that for scalar measures, equality (4.16) does not holds in general.

By (4.16), two arbitrary projections $E(M)$ and $E(N)$ for $M, N \in \mathfrak{A}$ *commute*. Moreover, if $M \supseteq N$ for $M, N \in \mathfrak{A}$, then $E(M) = E(N) + E(M \backslash N) \geq E(N)$.

The next lemma characterizes a spectral measure in terms of scalar measures.

Lemma 4.4 *A map E of an algebra (resp. σ-algebra) $\mathfrak{A}$ on a set Ω into the orthogonal projections on $\mathcal{H}$ is a spectral premeasure (resp. spectral measure) if and only if $E(\Omega) = I$ and for each vector $x \in \mathcal{H}$, the set function $E_x(\cdot) := \langle E(\cdot)x, x \rangle$ on $\mathfrak{A}$ is countably additive (resp. is a measure).*

Proof The *only if* assertion follows at once from Definition 4.2.

We prove the *if* direction. Let $(M_n)_{n \in \mathbb{N}}$ be a sequence of disjoint sets in $\mathfrak{A}$ such that $M := \bigcup_n M_n$ is also in $\mathfrak{A}$. Since E_x is finitely additive for each element $x \in \mathcal{H}$, E is finitely additive as well. Therefore, by Lemma 4.3, $(E(M_n))$ is a sequence of pairwise orthogonal projections. Hence, the series $\sum_n E(M_n)$ converges strongly. Because E_x is countably additive by assumption, we have

$$\big\langle E(M)x, x \big\rangle = E_x(M) = \sum_{n=1}^{\infty} E_x(M_n) = \sum_{n=1}^{\infty} \big\langle E(M_n)x, x \big\rangle = \bigg\langle \sum_{n=1}^{\infty} E(M_n)x, x \bigg\rangle$$

for each $x \in \mathcal{H}$, and hence $E(M) = \sum_n E(M_n)$ by the polarization formula (1.2). Thus, E is a spectral premeasure on $\mathfrak{A}$. $\square$

Let E be a spectral measure on the σ-algebra $\mathfrak{A}$ in $\mathcal{H}$. As noted in Lemma 4.4, each vector $x \in \mathcal{H}$ gives rise to a scalar positive measure E_x on $\mathfrak{A}$ by

$$E_x(M) := \big\| E(M)x \big\|^2 = \big\langle E(M)x, x \big\rangle, \quad M \in \mathfrak{A}.$$

The measure E_x is finite, since $E_x(\Omega) = \|E(\Omega)x\|^2 = \|x\|^2$. The family of these measures E_x plays a crucial role for the study of spectral integrals in Section 4.3.

Let $x, y \in \mathcal{H}$. Then there is a complex measure $E_{x,y}$ on $\mathfrak{A}$ given by $E_{x,y}(M) = \langle E(M)x, y \rangle$. The complex measure $E_{x,y}$ is a linear combination of four positive measures $E_z, z \in \mathcal{H}$. Indeed, by the polarization formula (1.2),

$$E_{x,y} = \frac{1}{4}(E_{x+y} - E_{x-y} + iE_{x-iy} - iE_{x-iy}).$$

The following lemma is very similar to the scalar case. We omit the simple proof.

Lemma 4.5 *Let E be a spectral premeasure on $\mathfrak{A}$. Suppose that $(M_n)_{n\in\mathbb{N}}$ is a decreasing sequence and $(N_n)_{n\in\mathbb{N}}$ is an increasing sequence of sets in $\mathfrak{A}$ such that $M := \bigcap_n M_n$ and $N := \bigcup_n N_n$ are in $\mathfrak{A}$.*
Then we have $E(M) = \text{s-}\lim_{n\to\infty} E(M_n)$ and $E(N) = \text{s-}\lim_{n\to\infty} E(N_n)$.

The following theorem states a one-to-one correspondence between resolutions of the identity and spectral measures on the Borel σ-algebra $\mathfrak{B}(\mathbb{R})$. Its proof uses Lemma 4.9 below which deals with the main technical difficulty showing that the corresponding operators $E(M)$, $M \in \mathfrak{B}(\mathbb{R})$, are indeed projections.

Theorem 4.6 *If E is a spectral measure on the Borel σ-algebra $\mathfrak{B}(\mathbb{R})$ in $\mathcal{H}$, then*

$$E(\lambda) := E\big((-\infty, \lambda]\big), \quad \lambda \in \mathbb{R}, \tag{4.18}$$

defines a resolution of the identity. Conversely, for each resolution of the identity $\{E(\lambda) : \lambda \in \mathbb{R}\}$, there is a unique spectral measure E on $\mathfrak{B}(\mathbb{R})$ such that (4.18) holds.

Proof Let E be a spectral measure on $\mathfrak{B}(\mathbb{R})$. We verify the axioms in Definition 4.1. Axiom (i) is obvious. We prove axiom (ii). Let $(\lambda_n)_{n\in\mathbb{N}}$ be a decreasing real sequence tending to λ_0. Then $((-\infty, \lambda_n])_{n\in\mathbb{N}}$ is a decreasing sequence of sets such that $\bigcap_n(-\infty, \lambda_n] = (-\infty, \lambda_0]$. Hence, $\lim_{n\to\infty} E(\lambda_n)x = E(\lambda_0)x$ for $x \in \mathcal{H}$ by Lemma 4.5. The proof of axiom (iii) is similar.

The main part of the theorem is the opposite direction. Let $\{E(\lambda) : \lambda \in \mathbb{R}\}$ be a resolution of the identity. Equations (4.4) and (4.5) define a projection $E(\mathcal{J})$ for each interval $\mathcal{J}$. One easily verifies that $E(\mathcal{J} \cap \mathcal{I}) = 0$ if $\mathcal{J}$ and $\mathcal{I}$ are disjoint (Exercise 2). Let $\mathfrak{A}_0$ be the algebra of all finite unions of arbitrary intervals. Each $N \in \mathfrak{A}_0$ is a finite union $\bigcup_{l=1}^s \mathcal{J}_s$ of disjoint intervals. Then $E(\mathcal{J}_l)E(\mathcal{J}_k) = 0$ for $l \neq k$, so $E(N) := \sum_{l=1}^s E(\mathcal{J}_l)$ is a projection. Since $\langle E(\cdot)x, x \rangle$ is an nondecreasing right-continuous function such that $\langle E(-\infty)x, x \rangle = 0$, by a standard result from measure theory ([Cn], Proposition 1.3.8) there is a unique measure μ_x on $\mathfrak{B}(\mathbb{R})$ such that $\mu_x((-\infty, \lambda]) = \langle E(\lambda)x, x \rangle$ for $\lambda \in \mathbb{R}$. By (4.4) and (4.5) we then have $\mu_x(\mathcal{J}) = \langle E(\mathcal{J})x, x \rangle$ for each interval $\mathcal{J}$, and hence

$$E_x(N) = \langle E(N)x, x \rangle = \sum_l \langle E(\mathcal{J}_l)x, x \rangle = \sum_l \mu_x(\mathcal{J}_l) = \mu_x(N), \quad N \in \mathfrak{A}_0.$$

Thus, E_x is countably additive on $\mathfrak{A}_0$, because μ_x is, so E is a spectral premeasure by Lemma 4.4. By Lemma 4.9 below, E extends to a spectral measure, denoted also by E, on $\mathfrak{B}(\mathbb{R})$. By construction, $E((-\infty, \lambda]) = E(\lambda)$. The uniqueness assertion follows from the fact that the intervals $(-\infty, \lambda]$ generate the σ-algebra $\mathfrak{B}(\mathbb{R})$. $\qquad\square$

Proposition 4.7 *For* $j = 1, 2$, *let* E_j *be a spectral measure on* $\mathfrak{B}(\mathbb{R})$ *in a Hilbert space* $\mathcal{H}_j$. *For any operator* $S \in \mathbf{B}(\mathcal{H}_1, \mathcal{H}_2)$, *we have* $SE_1(\lambda) = E_2(\lambda)S$ *for all* $\lambda \in \mathbb{R}$ *if and only if* $SE_1(M) = E_2(M)S$ *for all* $M \in \mathfrak{B}(\mathbb{R})$.

Proof The *if* part follows at once from (4.18). We prove the *only if* direction.

Let $\mathfrak{A}$ denote the family of sets $M \in \mathfrak{B}(\mathbb{R})$ for which $SE_1(M) = E_2(M)S$. It is easy to check that $\mathfrak{A}$ is a σ-algebra. Since $SE_1(\lambda) = E_2(\lambda)S$, it follows from (4.18) that $\mathfrak{A}$ contains all intervals $(-\infty, \lambda]$. Therefore, $\mathfrak{A} = \mathfrak{B}(\mathbb{R})$. $\qquad\square$

It is not difficult to "guess" the corresponding spectral measures for our three guiding examples of resolutions of the identity.

Example 4.5 (*Example* 4.1 *continued*) $E(M) = \sum_{\lambda_n \in M} P_n$ for $f \in \mathcal{H}$. $\qquad\circ$

Example 4.6 (*Example* 4.2 *continued*) $E(M)f = \chi_M \cdot f$ for $f \in \mathcal{H}$. $\qquad\circ$

Example 4.7 (*Example* 4.3 *continued*) $E(M)f = \chi_{h^{-1}(M)} \cdot f$ for $f \in \mathcal{H}$, where $h^{-1}(M) := \{t \in \Omega : h(t) \in M\}$. $\qquad\circ$

Just as in the case of scalar measures, we can say that a property of points of $\mathfrak{A}$ holds E-*almost everywhere*, abbreviated E-a.e., if there is a set $N \in \mathfrak{A}$ such that $E(N) = 0$ and the property holds on $\Omega \setminus N$. Note that $E(N) = 0$ if and only if $E_x(N) \equiv \langle E(N)x, x \rangle = 0$ for all $x \in \mathcal{H}$. If the Hilbert space $\mathcal{H}$ is separable, it can be shown (see, e.g., [AG], No. 76) that there exists a vector $x_{\max} \in \mathcal{H}$ of "maximal type," that is, $E(N) = 0$ if and only if $E_{x_{\max}}(N) = 0$ for $N \in \mathfrak{A}$.

Next, let us define and discuss the *support of a spectral measure*. Suppose that Ω is a topological Hausdorff space which has a countable base of open sets. For instance, each subspace of $\mathbb{R}^n$ or $\mathbb{C}^n$ has this property. Recall that the Borel algebra $\mathfrak{B}(\Omega)$ is the smallest σ-algebra in Ω which contains all open sets in Ω.

Definition 4.3 The *support* of a spectral measure E on $\mathfrak{B}(\Omega)$ is the complement in Ω of the union of all open sets N such that $E(N) = 0$. It is denoted by $\operatorname{supp} E$.

By this definition, a point $t \in \Omega$ is in $\operatorname{supp} E$ if and only if $E(U) \neq 0$ for each open set U containing t. Since Ω has a countable base of open sets, $\Omega \setminus \operatorname{supp} E$ is a union of at most countably many open sets N_n with $E(N_n) = 0$. By the countable additivity of E the latter implies that $E(\Omega \setminus \operatorname{supp} E) = 0$ and so $E(\operatorname{supp} E) = I$. Being the complement of a union of open sets, $\operatorname{supp} E$ is closed. Summarizing, we have shown that $\operatorname{supp} E$ *is the smallest closed subset* M *of* Ω *such that* $E(M) = I$.

4.2.2 Three Technical Results on Spectral Measures

The aim of this subsection is to prove three technical facts. The third is needed in
the proof of the multidimensional spectral theorem (see Theorem 5.21 below), while
the second is used in the proofs of Theorem 4.6 above and of Theorem 4.10 below.
The first result will be used in the proof of formula (4.32) in Sect. 4.3.2.

Recall that the *total variation* $|\mu|$ of a complex measure μ on $\mathfrak{A}$ is a finite
positive measure (see Appendix B) and $|\mu|(M)$ is defined as the supremum of
$\sum_{k=1}^{n} |\mu(M_n)|$ over all disjoint unions $M = \bigcup_{k=1}^{n} M_n$, where $M_k \in \mathfrak{A}$ and $n \in \mathbb{N}$.
Clearly, $|\mu(M)| \le |\mu|(M)$.

Lemma 4.8 *Let E be a spectral measure on $(\Omega, \mathfrak{A})$ in a Hilbert space $\mathcal{H}$.*

(i) $|E_{x,y}|(M) \le E_x(M)^{1/2} E_y(M)^{1/2}$ *for $x, y \in \mathcal{H}$ and $M \in \mathfrak{A}$.*
(ii) *If $f \in L^2(\Omega, E_x)$, $g \in L^2(\Omega, E_y)$, then*

$$\left| \int_{\Omega} fg \, dE_{x,y} \right| \le \int_{\Omega} |fg| \, d|E_{x,y}| \le \|f\|_{L^2(\Omega, E_x)} \|g\|_{L^2(\Omega, E_y)}. \quad (4.19)$$

Proof (i): Let M be a disjoint union of sets $M_1, \dots, M_n \in \mathfrak{A}$, $n \in \mathbb{N}$. Since

$$\begin{aligned}
|E_{x,y}(M_k)| &= |\langle E(M_k)x, E(M_k)y \rangle| \\
&\le \|E(M_k)x\| \|E(M_k)y\| = E_x(M_k)^{1/2} E_y(M_k)^{1/2},
\end{aligned}$$

using the Cauchy–Schwarz inequality and the σ-additivity of the scalar measures
E_x and E_y, we derive

$$\begin{aligned}
\sum_k |E_{x,y}(M_k)| &\le \sum_k E_x(M_k)^{1/2} E_y(M_k)^{1/2} \\
&\le \left(\sum_k E_x(M_k) \right)^{1/2} \left(\sum_k E_y(M_k) \right)^{1/2} = E_x(M)^{1/2} E_y(M)^{1/2}.
\end{aligned}$$

Taking the supremum over all such disjoint unions of M, we get the assertion.

(ii): It suffices to prove (4.19) for simple functions, because they are dense in the
L^2-spaces. Let $f = \sum_k a_k \chi_{M_k}$ and $g = \sum_k b_k \chi_{M_k}$ be finite sums of characteristic
functions χ_{M_k} of sets $M_k \in \mathfrak{A}$. By (i) and the Cauchy–Schwarz inequality we get

$$\begin{aligned}
\left| \sum_k a_k b_k E_{x,y}(M_k) \right| &\le \sum_k |a_k b_k| |E_{x,y}|(M_k) \\
&\le \left(\sum_k |a_k|^2 E_x(M_k) \right)^{1/2} \left(\sum_k |b_k|^2 E_y(M_k) \right)^{1/2},
\end{aligned}$$

which gives inequalities (4.19) for the simple functions f and g. $\square$

Lemma 4.9 *Let E_0 be a spectral premeasure on an algebra $\mathfrak{A}_0$ of subsets of
a set Ω. Then there is a spectral measure E on the σ-algebra $\mathfrak{A}$ generated by $\mathfrak{A}_0$
such that $E(M) = E_0(M)$ for all $M \in \mathfrak{A}_0$.*

Proof In this proof we use the following simple fact from measure theory ([Ha, p. 27]): Let $\mathfrak{B}$ be a class of sets which contains an algebra $\mathfrak{A}_0$. If $\mathfrak{B}$ is a monotone class (that is, $\bigcup_n M_n \in \mathfrak{B}$ for each sequence $(M_n)_{n\in\mathbb{N}}$ of sets $M_n \in \mathfrak{B}$ such that $M_n \subseteq M_{n+1}$ for $n \in \mathbb{N}$), then $\mathfrak{B}$ contains the σ-algebra $\mathfrak{A}$ generated by $\mathfrak{A}_0$.

Let $x \in \mathcal{H}$. By Lemma 4.4, $\mu'_x(\cdot) := \langle E_0(\cdot)x, x\rangle$ is a finite scalar premeasure on $\mathfrak{A}_0$. Hence, it has a unique extension μ_x to a measure on the σ-algebra $\mathfrak{A}$ (see, e.g., [Ha, p. 54]). Fix $M \in \mathfrak{A}$. For $x, y \in \mathcal{H}$, we define

$$\mu_{x,y}(M) = \frac{1}{4}\big(\mu_{x+y}(M) - \mu_{x-y}(M) + i\mu_{x+iy}(M) - i\mu_{x-iy}(M)\big). \quad (4.20)$$

Let $\mathfrak{B}$ be the set of all $M \in \mathfrak{A}$ for which the map $\mathcal{H} \ni x \to \mu_{x,y}(M)$ is linear for each $y \in \mathcal{H}$. Inserting (4.20), the linearity of this map is equivalent to a linear equation in certain terms $\mu_z(M)$, $z \in \mathcal{H}$. Since $\mu_z(\bigcup_n M_n) = \lim_{n\to\infty} \mu_z(M_n)$ for each increasing sequence $(M_n)_{n\in\mathbb{N}}$ in $\mathfrak{A}$, this equation holds for the set $\bigcup_n M_n$ if it holds for each M_n. That is, $\mathfrak{B}$ is a monotone class. By construction, $\mathfrak{B}$ contains $\mathfrak{A}_0$. Therefore, by the result stated above, $\mathfrak{B} \supseteq \mathfrak{A}$, and so $\mathfrak{B} = \mathfrak{A}$.

In a similar manner one proves that the map $\mathcal{H} \ni y \to \mu_{x,y}(M)$ is antilinear for all $x \in \mathcal{H}$ and $M \in \mathfrak{A}$. Clearly, $\mu'_{x+ix} = \mu'_{x-ix}$. Hence, $\mu_{x+ix} = \mu_{x-ix}$ by the uniqueness of the extension. Inserting the latter into (4.20), we get $\mu_{x,x}(M) = \mu_x(M) \geq 0$. That is, the map $(x, y) \to \mu_{x,y}(M)$ is a positive sesquilinear form on $\mathcal{H}$. Since

$$\big|\mu_{x,y}(M)\big|^2 \leq \mu_x(M)\mu_y(M) \leq \mu'_x(\Omega)\mu'_y(\Omega) = \|x\|^2\|y\|^2$$

by the Cauchy–Schwarz inequality, this sesquilinear form is bounded, so there exists a bounded self-adjoint operator $E(M)$ on $\mathcal{H}$ such that $\mu_{x,y}(M) = \langle E(M)x, y\rangle$ for all $x, y \in \mathcal{H}$. For $M \in \mathfrak{A}_0$, we have

$$\langle E_0(M)x, x\rangle = \mu'_x(M) = \mu_x(M) = \langle E(M)x, x\rangle \quad \text{for } x \in \mathcal{H}.$$

Hence, we obtain $E_0(M) = E(M)$ by polarization.

We denote by $\mathcal{C}$ the set of $M \in \mathfrak{A}$ for which $E(M)$ is a projection. Since the weak limit of an increasing sequence of projections is again a projection, $\mathcal{C}$ is a monotone class. Since $E(M)$ is a projection for $M \in \mathfrak{A}_0$, $\mathcal{C}$ contains $\mathfrak{A}_0$ and hence the σ-algebra $\mathfrak{A}$. That is, $E(M)$ is a projection for $M \in \mathfrak{A}$. Since $\mu_x(\cdot) = \langle E(\cdot)x, x\rangle$ is a measure for all $x \in \mathcal{H}$, E is a spectral measure on $\mathfrak{A}$ by Lemma 4.4. $\qquad\square$

The following theorem is about the *product of commuting spectral measures*. In the special case where $\Omega_j = \mathbb{R}$ it will be used in Sect. 5.5.

Theorem 4.10 *For $j = 1, \ldots, k$, let Ω_j be a locally compact Hausdorff space which has a countable base of open sets, and let E_j be a spectral measure on the Borel algebra $\mathfrak{B}(\Omega_j)$. Suppose that these spectral measures act on the same Hilbert space $\mathcal{H}$ and pairwise commute, that is, $E_j(M)E_l(N) = E_l(N)E_j(M)$ for $M \in \mathfrak{B}(\Omega_j)$, $N \in \mathfrak{B}(\Omega_l)$, $j, l = 1, \ldots, k$. Then there exists a unique spectral measure E on the Borel algebra $\mathfrak{B}(\Omega)$ of the product space $\Omega = \Omega_1 \times \cdots \times \Omega_k$ such that*

$$E(M_1 \times \cdots \times M_k) = E_1(M_1) \cdots E(M_k) \quad \text{for } M_j \in \mathfrak{B}(\Omega_j), \ j = 1, \ldots, k. \quad (4.21)$$

The unique spectral measure E satisfying Eq. (4.21) is called the *product of the spectral measures* $E_1, \ldots, E_k$ and denoted by $E_1 \times \cdots \times E_k$.

Proof For simplicity, we carry out the proof in the case $k = 2$. Let $\mathfrak{A}_0$ be the algebra generated by the set $\mathfrak{A}_{00}$ of "measurable rectangles" $M_1 \times M_2$, where $M_1 \in \mathfrak{B}(\Omega_1)$ and $M_2 \in \mathfrak{B}(\Omega_2)$. Each set $N \in \mathfrak{A}_0$ is a disjoint union of finitely many sets

$$N_1 = M_{11} \times M_{21}, \quad \ldots, \quad N_s = M_{1s} \times M_{2s} \in \mathfrak{A}_{00}.$$

We define $E(N) := \sum_{l=1}^{s} E(N_l)$, where $E(N_l) := E_1(M_{1l}) E_2(M_{2l})$.

Using essentially that the spectral measures E_1 and E_2 commute, we show that $E(N)$ is a projection. If $j \neq l$, then $N_j \cap N_l = \emptyset$, and hence $M_{1j} \cap M_{1l} = \emptyset$ or $M_{2j} \cap M_{2l} = \emptyset$. By Lemma 4.3 we obtain in either case

$$0 = E_1(M_{1j} \cap M_{1l}) E_2(M_{2j} \cap M_{2l}) = E_1(M_{1j}) E_1(M_{1l}) E_2(M_{2j}) E_2(M_{2l})$$

$$= E_1(M_{1l}) E_2(M_{2j}) E_1(M_{1l}) E_2(M_{2l}) = E(N_j) E(N_l).$$

Being the product of two commuting projections, $E(N_l)$ is a projection. As a sum of mutually orthogonal projections, $E(N) = \sum_{l=1}^{s} E(N_l)$ is also a projection.

Just as in the case of scalar measures, one can show that the definition of E is independent of the particular representation $N = \bigcup_l N_l$ (see, e.g., [Ha, p. 35] or [Be, pp. 89–96] for details). From its definition it is clear that E is finitely additive on $\mathfrak{A}_0$. The crucial part of this proof is to show that E is *countably additive* on the algebra $\mathfrak{A}_0$. This is where the assumption on the spaces Ω_j is used.

First we prove the following technical fact. Let $N \in \mathfrak{A}_0$, $x \in \mathcal{H}$, and $\varepsilon > 0$. Then there exist a compact set C and an open set U in $\mathfrak{A}_0$ such that $C \subseteq N \subseteq U$ and

$$\langle E(U)x, x \rangle - \varepsilon \leq \langle E(N)x, x \rangle \leq \langle E(C)x, x \rangle + \varepsilon. \tag{4.22}$$

Because E is finitely additive on the algebra $\mathfrak{A}_0$, it suffices to prove the latter for $N = M_1 \times M_2 \in \mathfrak{A}_{00}$. Since E_j is a spectral measure, $\langle E_j(\cdot)x, x \rangle$ is a finite Borel measure on Ω_j. Each positive finite Borel measure on a locally compact Hausdorff space that has a countable basis is regular [Cn, Proposition 7.2.3], that is, we have

$$\mu(M) = \sup\{\mu(C) : C \subseteq M, C \text{ compact}\} = \inf\{\mu(U) : U \supseteq M, U \text{ open}\}$$

for any Borel set M. Hence, there exist compact sets $C_j \subseteq M_j$ such that

$$\|x\| \left(\langle E_j(M_j)x, x \rangle - \langle E_j(C_j)x, x \rangle \right)^{1/2} = \|x\| \| E_j(M_j \backslash C_j)x \| < \varepsilon/2.$$

Since $E(M_1 \times M_2) - E(C_1 \times C_2) = E_1(M_1 \backslash C_1) E_2(C_2) + E_1(M_1) E_2(M_2 \backslash C_2)$, we obtain

$$\langle E(M_1 \times M_2)x, x \rangle - \langle E(C_1 \times C_2)x, x \rangle$$
$$= \langle E_2(C_2)x, E_1(M_1 \backslash C_1)x \rangle + \langle E_2(M_2 \backslash C_2)x, E_1(M_1)x \rangle$$
$$\leq \|x\| \| E_1(M_1 \backslash C_1)x \| + \| E_2(M_2 \backslash C_2)x \| \|x\| < \varepsilon.$$

This proves the second inequality of (4.22). The proof of the first one is similar.

Now let $N_n \in \mathfrak{A}_0$, $n \in \mathbb{N}$, be pairwise disjoint sets such that $N := \bigcup_{n=1}^{\infty} N_n \in \mathfrak{A}_0$. Let $\varepsilon > 0$. By the result proved in the preceding paragraph we can find a compact set C and open sets U_n in $\mathfrak{A}_0$ such that $C \subseteq N$, $N_n \subseteq U_n$, and

$$\langle E(N)x, x \rangle \leq \langle E(C)x, x \rangle + \varepsilon, \qquad \langle E(U_n)x, x \rangle \leq \langle E(N_n)x, x \rangle + \varepsilon 2^{-n}.$$

Since $\{U_n : n \in \mathbb{N}\}$ is an open cover of the compact set C, there exists $s \in \mathbb{N}$ such that $U_1 \cup \cdots \cup U_s \supseteq C$. Using the finite additivity of E on $\mathfrak{A}_0$, we derive

$$\langle E(N)x, x \rangle \leq \langle E(C)x, x \rangle + \varepsilon \leq \left\langle E\left(\bigcup_{n=1}^{s} U_n \right)x, x \right\rangle + \varepsilon$$

$$\leq \sum_{n=1}^{s} \langle E(U_n)x, x \rangle + \varepsilon \leq \sum_{n=1}^{\infty} \langle E(N_n)x, x \rangle + 2\varepsilon.$$

Thus, $E(N) \leq \sum_{n=1}^{\infty} E(N_n)$, because $\varepsilon > 0$ and $x \in \mathcal{H}$ are arbitrary.

Conversely, the finite additivity of E implies that

$$\sum_{n=1}^{s} E(N_n) = E\left(\bigcup_{n=1}^{s} N_n \right) \leq E(N),$$

and hence $\sum_{n=1}^{\infty} E(N_n) \leq E(N)$. Putting now both inequalities together, we conclude that $E(N) = \sum_{n=1}^{\infty} E(N_n)$. This proves that E is a spectral premeasure on $\mathfrak{A}_0$.

The σ-algebra generated by $\mathfrak{A}_0$ is the Borel σ-algebra $\mathfrak{B}(\Omega)$. Therefore, by Lemma 4.9 the spectral premeasure E on $\mathfrak{A}_0$ has an extension to a spectral measure, denoted again by E, on $\mathfrak{B}(\Omega)$. By the definition of E, (4.21) is satisfied. Since $\mathfrak{A}_{00}$ generates the σ-algebra $\mathfrak{B}(\Omega)$ as well, the spectral measure E is uniquely determined by (4.21). $\qquad \square$

Remark Without additional assumptions as in Theorem 4.10 the product spectral measure of two commuting spectral measures does not exist in general. That is, if we define E by (4.21), then the countable additivity of E may fail. Such a counterexample has been given in [BVS].

4.3 Spectral Integrals

Throughout this section, $\mathfrak{A}$ is a σ-algebra of subsets of a set Ω, and E is a spectral measure on $(\Omega, \mathfrak{A})$. Our aim is to investigate spectral integrals. These are the integrals

$$\mathbb{I}(f) = \int_{\Omega} f(t) \, dE(t) = \int_{\Omega} f \, dE$$

of E-a.e. finite $\mathfrak{A}$-measurable functions $f : \Omega \to \mathbb{C} \cup \{\infty\}$ with respect to the spectral measure E. Roughly speaking, the idea of our construction is to define

$$\mathbb{I}(\chi_M) \equiv \int_{\Omega} \chi_M(t) \, dE(t) := E(M)$$

for characteristic functions χ_M of sets $M \in \mathfrak{A}$ and to extend this definition by linearity and by taking limits to general measurable functions.

In the first subsection we shall carry out this for bounded functions and obtain *bounded* operators. In the second subsection we treat the more subtle case of E-a.e. finite unbounded functions which leads to *unbounded* operators.

4.3.1 Spectral Integrals of Bounded Measurable Functions

Let $\mathcal{B} = \mathcal{B}(\Omega, \mathfrak{A})$ be the Banach space of all bounded $\mathfrak{A}$-measurable functions on Ω equipped with the norm

$$\|f\|_\Omega = \sup\{|f(t)| : t \in \Omega\}.$$

Let $\mathcal{B}_s$ denote the subspace of simple functions in $\mathcal{B}$, that is, of all functions which have only a finite number of values. Each $f \in \mathcal{B}_s$ can be written in the form

$$f = \sum_{r=1}^{n} c_r \chi_{M_r}, \tag{4.23}$$

where $c_1, \ldots, c_n \in \mathbb{C}$, and $M_1, \ldots, M_n$ are pairwise disjoint sets from $\mathfrak{A}$. For such a function f, we define

$$\mathbb{I}(f) = \sum_{r=1}^{n} c_r E(M_r). \tag{4.24}$$

From the finite additivity of E it follows as in "ordinary" integration theory that $\mathbb{I}(f)$ is independent of the particular representation (4.23) of f. In order to extend the definition of $\mathbb{I}(f)$ to arbitrary functions in $\mathcal{B}$, we need the following lemma.

Lemma 4.11 $\|\mathbb{I}(f)\| \leq \|f\|_\Omega$ *for* $f \in \mathcal{B}_s$.

Proof Since the sets $M_1, \ldots, M_n \in \mathfrak{A}$ in (4.23) are disjoint, $E(M_k)\mathcal{H}$ and $E(M_l)\mathcal{H}$ are orthogonal for $k \neq l$ by Lemma 4.3. Using this fact, we get for $x \in \mathcal{H}$,

$$\|\mathbb{I}(f)x\|^2 = \left\|\sum_r c_r E(M_r)x\right\|^2 = \sum_r |c_r|^2 \|E(M_r)x\|^2$$

$$\leq \sum_r \|f\|_\Omega^2 \|E(M_r)x\|^2 = \|f\|_\Omega^2 \left\|\sum_r E(M_r)x\right\|^2 \leq \|f\|_\Omega^2 \|x\|^2. \quad \square$$

Let $f \in \mathcal{B}$. Since the subspace $\mathcal{B}_s$ is dense in the Banach space $(\mathcal{B}, \|\cdot\|_\Omega)$, there is a sequence $(f_n)_{n \in \mathbb{N}}$ from $\mathcal{B}_s$ such that $\lim_n \|f - f_n\|_\Omega = 0$. Since $(f_n)_{n \in \mathbb{N}}$ is a $\|\cdot\|_\Omega$-Cauchy sequence, Lemma 4.11 implies that $(\mathbb{I}(f_n))_{n \in \mathbb{N}}$ is a Cauchy sequence in the operator norm. Hence, there exists a bounded operator $\mathbb{I}(f)$ on $\mathcal{H}$ such that $\lim_n \|\mathbb{I}(f) - \mathbb{I}(f_n)\| = 0$. From Lemma 4.11 it follows also that the operator $\mathbb{I}(f)$ does not depend on the particular sequence $(f_n)_{n \in \mathbb{N}}$ for which $\lim_n \|f - f_n\|_\Omega = 0$.

Basic properties of the operators $\mathbb{I}(f)$ are summarized in the next proposition.

Proposition 4.12 *For $f, g \in \mathcal{B}(\Omega, \mathfrak{A})$, $\alpha, \beta \in \mathbb{C}$, and $x, y \in \mathcal{H}$, we have:*

(i) $\mathbb{I}(\overline{f}) = \mathbb{I}(f)^*$, $\mathbb{I}(\alpha f + \beta g) = \alpha \mathbb{I}(f) + \beta \mathbb{I}(g)$, $\mathbb{I}(fg) = \mathbb{I}(f)\mathbb{I}(g)$,
(ii) $\langle \mathbb{I}(f)x, y \rangle = \int_\Omega f(t) \, d\langle E(t)x, y \rangle$,
(iii) $\|\mathbb{I}(f)x\|^2 = \int_\Omega |f(t)|^2 \, d\langle E(t)x, x \rangle$,
(iv) $\|\mathbb{I}(f)\| \leq \|f\|_\Omega$.
(v) *Let $f_n \in \mathcal{B}$ for $n \in \mathbb{N}$. If $f_n(t) \to f(t)$ E-a.e. on Ω and if there is a constant c such that $|f_n(t)| \leq c$ for $n \in \mathbb{N}$ and $t \in \Omega$, then s-$\lim_{n \to \infty} \mathbb{I}(f_n) = \mathbb{I}(f)$.*

Proof It suffices to prove assertions (i)–(iv) for simple functions f and g, because by continuity all relations extend then to arbitrary functions from $\mathcal{B}$. For simple functions, the proofs of (i)–(iv) are straightforward verifications.

We carry out the proof of the equality $\mathbb{I}(fg) = \mathbb{I}(f)\mathbb{I}(g)$. Let $f = \sum_r a_r \chi_{M_r}$ and $g = \sum_s b_s \chi_{N_s}$, where the sets $M_r \in \mathfrak{A}$ resp. $N_s \in \mathfrak{A}$ are pairwise disjoint. Then $fg = \sum_{r,s} a_r b_s \chi_{M_r \cap N_s}$. Using (4.24) and (4.16), we obtain

$$\mathbb{I}(fg) = \sum_{r,s} a_r b_s E(M_r \cap N_s) = \sum_{r,s} a_r b_s E(M_r) E(N_s)$$
$$= \left(\sum_r a_r E(M_r) \right) \left(\sum_s b_s E(N_s) \right) = \mathbb{I}(f)\mathbb{I}(g).$$

(v) follows from Lebesgue's dominated convergence theorem (Theorem B.1), since by (iii),

$$\left\| \left(\mathbb{I}(f_n) - \mathbb{I}(f) \right) x \right\|^2 = \int_\Omega |f_n - f|^2 \, d\langle E(t)x, x \rangle \quad \text{for } x \in \mathcal{H}. \qquad \square$$

4.3.2 *Integrals of Unbounded Measurable Functions*

Let $\mathcal{S} = \mathcal{S}(\Omega, \mathfrak{A}, E)$ denote the set of all $\mathfrak{A}$-measurable functions $f : \Omega \to \mathbb{C} \cup \{\infty\}$ which are E-a.e. finite, that is, $E(\{t \in \Omega : f(t) = \infty\}) = 0$.

In Theorem 4.13 below we define an (unbounded) linear operator $\mathbb{I}(f)$ for each $f \in \mathcal{S}$. Our main technical tool for this is the notion of a bounding sequence.

Definition 4.4 A sequence $(M_n)_{n \in \mathbb{N}}$ of sets $M_n \in \mathfrak{A}$ is a *bounding sequence* for a subset $\mathcal{F}$ of $\mathcal{S}$ if each function $f \in \mathcal{F}$ is bounded on M_n and $M_n \subseteq M_{n+1}$ for $n \in \mathbb{N}$, and $E(\bigcup_{n=1}^\infty M_n) = I$.

If (M_n) is a bounding sequence, then by the properties of the spectral measure,

$$E(M_n) \leq E(M_{n+1}) \quad \text{for } n \in \mathbb{N}, \qquad \lim_{n \to \infty} E(M_n)x = x \quad \text{for } x \in \mathcal{H}, \quad (4.25)$$

and the set $\bigcup_{n=1}^\infty E(M_n)\mathcal{H}$ is dense in $\mathcal{H}$.

Each *finite* set of elements $f_1, \ldots, f_r \in \mathcal{S}$ has a bounding sequence. Indeed, set

$$M_n = \left\{ t \in \Omega : |f_j(t)| \leq n \text{ for } j = 1, \ldots, r \right\}$$

and $M = \bigcup_{n=1}^{\infty} M_n$. Then we have $\Omega \setminus M \subseteq \bigcup_{j=1}^{r} \{t : f_j(t) = \infty\}$, and hence $E(\Omega \setminus M) = 0$, so that $E(M) = E(\Omega) = I$. Therefore, $(M_n)_{n \in \mathbb{N}}$ is a bounding sequence for $\{f_1, \ldots, f_r\}$ and also for the $*$-subalgebra of S generated by $f_1, \ldots, f_r$.

Theorem 4.13 *Suppose that $f \in S$ and define*

$$\mathcal{D}(\mathbb{I}(f)) = \left\{ x \in \mathcal{H} : \int_{\Omega} |f(t)|^2 \, d\langle E(t)x, x \rangle < \infty \right\}. \qquad (4.26)$$

Let $(M_n)_{n \in \mathbb{N}}$ be a bounding sequence for f. Then we have:

(i) *A vector $x \in \mathcal{H}$ is in $\mathcal{D}(\mathbb{I}(f))$ if and only if the sequence $(\mathbb{I}(f \chi_{M_n})x)_{n \in \mathbb{N}}$ converges in $\mathcal{H}$, or equivalently, if $\sup_{n \in \mathbb{N}} \|\mathbb{I}(f \chi_{M_n})x\| < \infty$.*

(ii) *For $x \in \mathcal{D}(\mathbb{I}(f))$, the limit of the sequence $(\mathbb{I}(f \chi_{M_n})x)$ does not depend on the bounding sequence (M_n). There is a linear operator $\mathbb{I}(f)$ on $\mathcal{D}(\mathbb{I}(f))$ defined by*

$$\mathbb{I}(f)x = \lim_{n \to \infty} \mathbb{I}(f \chi_{M_n})x \quad \text{for } x \in \mathcal{D}(\mathbb{I}(f)). \qquad (4.27)$$

(iii) $\bigcup_{n=1}^{\infty} E(M_n)\mathcal{H}$ *is contained in $\mathcal{D}(\mathbb{I}(f))$ and is a core for $\mathbb{I}(f)$. Further,*

$$E(M_n)\mathbb{I}(f) \subseteq \mathbb{I}(f)E(M_n) = \mathbb{I}(f \chi_{M_n}) \quad \text{for } n \in \mathbb{N}. \qquad (4.28)$$

Proof (i): First suppose that $x \in \mathcal{D}(\mathbb{I}(f))$. Since f is bounded on M_n and hence $f \chi_{M_n} \in \mathcal{B}$, the bounded operator $\mathbb{I}(f \chi_{M_n})$ is defined by the preceding subsection. Using Proposition 4.12(iii), we obtain

$$
\begin{aligned}
\left\| \mathbb{I}(f \chi_{M_k})x - \mathbb{I}(f \chi_{M_n})x \right\|^2 &= \left\| \mathbb{I}(f \chi_{M_k} - f \chi_{M_n})x \right\|^2 \\
&= \int_{\Omega} |f \chi_{M_k} - f \chi_{M_n}|^2 \, d\langle E(t)x, x \rangle \\
&= \| f \chi_{M_k} - f \chi_{M_n} \|_{L^2(\Omega, E_x)}^2 \qquad (4.29)
\end{aligned}
$$

for $k, n \in \mathbb{N}$. Since $f \in L^2(\Omega, E_x)$ by (4.26) and hence $f \chi_{M_n} \to f$ in $L^2(\Omega, E_x)$ by Lebesgue's dominated convergence theorem (Theorem B.1), $(f \chi_{M_n})_{n \in \mathbb{N}}$ is a Cauchy sequence in $L^2(\Omega, E_x)$, and so is $(\mathbb{I}(f \chi_{M_n})x))_{n \in \mathbb{N}}$ in $\mathcal{H}$ by (4.29). Therefore, the sequence $(\mathbb{I}(f \chi_{M_n})x)_{n \in \mathbb{N}}$ converges in $\mathcal{H}$.

If the sequence $(\mathbb{I}(f \chi_{M_n})x)_{n \in \mathbb{N}}$ converges, the set $\{\|\mathbb{I}(f \chi_{M_n})x\| : n \in \mathbb{N}\}$ is obviously bounded.

Now we suppose that the set $\{\|\mathbb{I}(f \chi_{M_n})x\| : n \in \mathbb{N}\}$ is bounded, that is, $c := \sup_n \|\mathbb{I}(f \chi_{M_n})x\| < \infty$. Since $(|f \chi_{M_n}(t)|^2)_{n \in \mathbb{N}}$ converges monotonically to $|f(t)|^2$ E_x-a.e. on Ω, Lebesgue's monotone convergence theorem (Theorem B.2) yields

$$
\begin{aligned}
\int_{\Omega} |f|^2 \, dE_x &= \lim_{n \to \infty} \int_{\Omega} |f \chi_{M_n}|^2 \, dE_x \\
&= \lim_{n \to \infty} \left\| \mathbb{I}(f \chi_{M_n})x \right\|^2 \leq c^2 < \infty, \qquad (4.30)
\end{aligned}
$$

where the second equality follows from Proposition 4.12(iii). Hence, we obtain $f \in L^2(\Omega, E_x)$, and so $x \in \mathcal{D}(\mathbb{I}(f))$ by (4.26).

(ii): Let $(M_n')_{n\in\mathbb{N}}$ be another bounding sequence for f. From Proposition 4.12, (i) and (iii), we obtain

$$\left\|\mathbb{I}(f\chi_{M_n})x - \mathbb{I}(f\chi_{M_k'})x\right\| = \|f\chi_{M_n} - f\chi_{M_k'}\|_{L^2(\Omega,E_x)}$$
$$\leq \|f\chi_{M_n} - f\|_{L^2(\Omega,E_x)} + \|f - f\chi_{M_k'}\|_{L^2(\Omega,E_x)} \to 0$$

as $k, n \to \infty$, since $f\chi_{M_n} \to f$ and $f\chi_{M_k'} \to f$ in $L^2(\Omega, E_x)$ as noted above. Therefore, $\lim_n \mathbb{I}(f\chi_{M_n})x = \lim_k \mathbb{I}(f\chi_{M_k'})x$.

By the characterization given in (i), $\mathcal{D}(\mathbb{I}(f))$ is a linear subspace of $\mathcal{H}$, and (4.27) defines a linear operator $\mathbb{I}(f)$ with domain $\mathcal{D}(\mathbb{I}(f))$.

(iii): Let $x \in \mathcal{H}$. Since $E(M_k) = \mathbb{I}(\chi_{M_k})$, using Proposition 4.12, we compute

$$\mathbb{I}(f\chi_{M_k})x = \mathbb{I}(f\chi_{M_n}\chi_{M_k})x = \mathbb{I}(f\chi_{M_n})E(M_k)x = E(M_k)\mathbb{I}(f\chi_{M_n})x \quad (4.31)$$

for $n \geq k$. Hence, $\sup_n \|\mathbb{I}(f\chi_{M_n})E(M_k)x\| < \infty$, and so $E(M_k)x \in \mathcal{D}(\mathbb{I}(f))$. That is, $\bigcup_k E(M_k)\mathcal{H} \subseteq \mathcal{D}(\mathbb{I}(f))$.

Letting $n \to \infty$ in (4.31) and using (4.27), we get $\mathbb{I}(f)E(M_k)x = \mathbb{I}(f\chi_{M_k})x$ for $x \in \mathcal{H}$. Now suppose that $x \in \mathcal{D}(\mathbb{I}(f))$. Letting again $n \to \infty$ in (4.31), we obtain $E(M_k)\mathbb{I}(f)x = \mathbb{I}(f)E(M_k)x$. This proves (4.28).

Since $E(M_n)x \to x$ by (4.25) and $\mathbb{I}(f)E(M_n)x = E(M_n)\mathbb{I}(f)x \to \mathbb{I}(f)x$ for $x \in \mathcal{D}(\mathbb{I}(f))$, the linear subspace $\bigcup_n E(M_n)\mathcal{H}$ of $\mathcal{H}$ is a core for $\mathbb{I}(f)$. □

Corollary 4.14 *Let $n \in \mathbb{N}$ and $f_1, \ldots, f_n \in S$. Then $\bigcap_{k=1}^n \mathcal{D}(\mathbb{I}(f_k))$ is a core for each operator $\mathbb{I}(f_k)$.*

Proof Choose a bounding sequence for all f_j and apply Theorem 4.13(iii). □

Proposition 4.15 *Let $f, g \in S$ and $x \in \mathcal{D}(\mathbb{I}(f))$, $y \in \mathcal{D}(\mathbb{I}(g))$. Then we have*

$$\langle \mathbb{I}(f)x, \mathbb{I}(g)y \rangle = \int_\Omega f(t)\overline{g(t)}\, d\langle E(t)x, y \rangle, \tag{4.32}$$

$$\|\mathbb{I}(f)x\|^2 = \int_\Omega |f(t)|^2\, d\langle E(t)x, x \rangle. \tag{4.33}$$

Proof Equation (4.33) follows from (4.32) by setting $f = g$, $x = y$.

We prove formula (4.32). Applying Proposition 4.12(ii) to the bounded function $f\overline{g}\chi_{M_n}$ and Proposition 4.12(i), we obtain

$$\int_\Omega f\overline{g}\chi_{M_n}\, dE_{x,y} = \langle \mathbb{I}(f\overline{g}\chi_{M_n})x, y \rangle = \langle \mathbb{I}(f\chi_{M_n})x, \mathbb{I}(g\chi_{M_n})y \rangle. \tag{4.34}$$

Since $x \in \mathcal{D}(\mathbb{I}(f))$ and $y \in \mathcal{D}(\mathbb{I}(g))$, we have $f \in L^2(\Omega, E_x)$ and $g \in L^2(\Omega, E_y)$ by (4.26). Hence, by Lemma 4.8 and (4.19), the integral $\int f\overline{g}\, dE_{x,y}$ exists, so that

$$\left| \int_\Omega f\overline{g}\chi_{M_n}\, dE_{x,y} - \int_\Omega f\overline{g}\, dE_{x,y} \right|$$
$$= \left| \int_\Omega (f\chi_{M_n} - f)\overline{g}\, dE_{x,y} \right|$$
$$\leq \|f\chi_{M_n} - f\|_{L^2(\Omega,E_x)}\|g\|_{L^2(\Omega,E_y)} \to 0 \quad \text{as } n \to \infty,$$

because $f \chi_{M_n} \to f$ in $L^2(\Omega, E_x)$ as shown in the proof of Theorem 4.13(i). Therefore, since $\mathbb{I}(f)x = \lim_n \mathbb{I}(f \chi_{M_n})x$ and $\mathbb{I}(g)y = \lim_n \mathbb{I}(g \chi_{M_n})y$ by (4.27), letting $n \to \infty$ in (4.34), we obtain (4.32).   □

If $f \in \mathcal{B}$, then $(M_n = \Omega)_{n \in \mathbb{N}}$ is a bounding sequence for f, and hence the operator $\mathbb{I}(f)$ given by (4.27) coincides with the operator $\mathbb{I}(f)$ defined in Theorem 4.13.

The main algebraic properties of the map $f \to \mathbb{I}(f)$ are collected in the following theorem. Here $\overline{f}$ denotes the complex conjugate of a function f.

Theorem 4.16 *For $f, g \in \mathcal{S}(\Omega, \mathfrak{A}, E)$ and $\alpha, \beta \in \mathbb{C}$, we have:*

(i) $\mathbb{I}(\overline{f}) = \mathbb{I}(f)^*$,
(ii) $\mathbb{I}(\alpha f + \beta g) = \alpha \mathbb{I}(f) + \beta \mathbb{I}(g)$,
(iii) $\mathbb{I}(fg) = \overline{\mathbb{I}(f)\mathbb{I}(g)}$,
(iv) $\mathbb{I}(f)$ *is a closed normal operator on* $\mathcal{H}$, *and* $\mathbb{I}(f)^*\mathbb{I}(f) = \mathbb{I}(\overline{f}f) = \mathbb{I}(f)\mathbb{I}(f)^*$,
(v) $\mathcal{D}(\mathbb{I}(f)\mathbb{I}(g)) = \mathcal{D}(\mathbb{I}(g)) \cap \mathcal{D}(\mathbb{I}(fg))$.

Proof We choose a bounding sequence $(M_n)_{n \in \mathbb{N}}$ for f and g and abbreviate $E_n := E(M_n)$ and $h_n := h \chi_{M_n}$ for $h \in \mathcal{S}$. Recall that $\lim_n E_n x = x$ for $x \in \mathcal{H}$ by (4.25). Clearly, (M_n) is also a bounding sequence for $\overline{f}$, $f + g$ and fg. Hence, by Theorem 4.13(iii), $\mathcal{D}_0 := \bigcup_{n=1}^{\infty} E_n \mathcal{H}$ is a core for the operators $\mathbb{I}(f + g)$ and $\mathbb{I}(fg)$.

(i): Let $x \in \mathcal{D}(\mathbb{I}(f))$ and $y \in \mathcal{D}(\mathbb{I}(\overline{f}))$. Applying (4.28) twice and Proposition 4.12(i), we derive

$$\langle E_n \mathbb{I}(f)x, y \rangle = \langle \mathbb{I}(f_n)x, y \rangle = \langle x, \mathbb{I}(\overline{f_n})y \rangle = \langle x, \mathbb{I}((\overline{f})_n)y \rangle = \langle x, E_n \mathbb{I}(\overline{f})y \rangle.$$

As $n \to \infty$, we obtain $\langle \mathbb{I}(f)x, y \rangle = \langle x, \mathbb{I}(\overline{f})y \rangle$. This shows that $\mathbb{I}(\overline{f}) \subseteq \mathbb{I}(f)^*$.

Now suppose that $y \in \mathcal{D}(\mathbb{I}(f)^*)$. Again by (4.28) and Proposition 4.12(i) we get

$$\langle x, E_n \mathbb{I}(f)^* y \rangle = \langle \mathbb{I}(f) E_n x, y \rangle = \langle \mathbb{I}(f_n)x, y \rangle = \langle x, \mathbb{I}(\overline{f_n})y \rangle$$

for all $x \in \mathcal{H}$. Hence, $E_n \mathbb{I}(f)^* y = \mathbb{I}(\overline{f_n})y$, and so $\sup_n \|\mathbb{I}(\overline{f_n})y\| \leq \|\mathbb{I}(f)^* y\|$. Therefore, $y \in \mathcal{D}(\mathbb{I}(\overline{f}))$ by Proposition 4.13(i). Thus, $\mathcal{D}(\mathbb{I}(f)^*) \subseteq \mathcal{D}(\mathbb{I}(\overline{f}))$.

Putting the preceding two paragraphs together, we have proved that $\mathbb{I}(\overline{f}) = \mathbb{I}(f)^*$.

In particular, each operator $\mathbb{I}(f)$ is closed, since $\mathbb{I}(f) = \mathbb{I}(\overline{f})^*$.

(ii): Obviously, it suffices to prove that $\mathbb{I}(f + g) = \overline{\mathbb{I}(f) + \mathbb{I}(g)}$. First, we note that the operator $\mathbb{I}(f) + \mathbb{I}(g)$ is closable, because

$$\mathcal{D}\big((\mathbb{I}(f) + \mathbb{I}(g))^*\big) \supseteq \mathcal{D}\big(\mathbb{I}(f)^* + \mathbb{I}(g)^*\big) = \mathcal{D}\big(\mathbb{I}(\overline{f}) + \mathbb{I}(\overline{g})\big) \supseteq \mathcal{D}_0$$

and $\mathcal{D}_0$ is dense in $\mathcal{H}$. Since (M_n) is a bounding sequence for $f, g, f + g$, we have

$$\big(\mathbb{I}(f) + \mathbb{I}(g)\big) E_n = \mathbb{I}(f_n) + \mathbb{I}(g_n) = \mathbb{I}(f_n + g_n) = \mathbb{I}(f + g) E_n, \quad (4.35)$$

$$E_n\big(\mathbb{I}(f) + \mathbb{I}(g)\big) = E_n \mathbb{I}(f) + E_n \mathbb{I}(g) \subseteq \big(\mathbb{I}(f) + \mathbb{I}(g)\big) E_n$$

$$= \mathbb{I}(f + g) E_n \quad (4.36)$$

by (4.28) and Proposition 4.12(i). Letting $n \to \infty$ in (4.35) and using that $\mathcal{D}_0 = \bigcup_n E_n \mathcal{H}$ is a core for the closed operator $\mathbb{I}(f + g)$, we get $\mathbb{I}(f + g) \subseteq \overline{\mathbb{I}(f) + \mathbb{I}(g)}$.

Recall that $\mathbb{I}(f+g)E_n$ is a bounded operator defined on $\mathcal{H}$. From (4.36) it follows therefore that $\overline{E_n\mathbb{I}(f)+\mathbb{I}(g)} \subseteq \mathbb{I}(f+g)E_n$. Letting again $n \to \infty$ and using (4.27), we obtain $\overline{\mathbb{I}(f)+\mathbb{I}(g)} \subseteq \mathbb{I}(f+g)$. Thus, $\mathbb{I}(f+g) = \overline{\mathbb{I}(f)+\mathbb{I}(g)}$.

(iii): Formula (4.28) implies that the dense linear subspace $\mathcal{D}_0 = \bigcup_n E_n\mathcal{H}$ is contained in the domain of $\mathbb{I}(\bar{g})\mathbb{I}(\bar{f})$. Hence, the operator $\mathbb{I}(f)\mathbb{I}(g)$ is closable, since $(\mathbb{I}(f)\mathbb{I}(g))^* \supseteq \mathbb{I}(\bar{g})\mathbb{I}(\bar{f})$ by (i). Again by (4.28) and Proposition 4.12(i),

$$\mathbb{I}(fg)E_n = \mathbb{I}\big((fg)_n\big) = \mathbb{I}(f_n)\mathbb{I}(g_n) = \mathbb{I}(f)E_n\mathbb{I}(g_n) = \mathbb{I}(f)\mathbb{I}(g_n)E_n = \mathbb{I}(f)\mathbb{I}(g)E_n,$$
(4.37)

$$E_n\mathbb{I}(f)\mathbb{I}(g) \subseteq \mathbb{I}(f)\mathbb{I}(g)E_n = \mathbb{I}(fg)E_n. \tag{4.38}$$

Letting $n \to \infty$ in (4.37) and using that $\mathcal{D}_0$ is a core for the closed operator $\mathbb{I}(fg)$, we conclude that $\mathbb{I}(fg) \subseteq \overline{\mathbb{I}(f)\mathbb{I}(g)}$.

Since $\mathbb{I}(fg)E_n$ is bounded and defined on $\mathcal{H}$, we obtain $\overline{E_n\mathbb{I}(f)\mathbb{I}(g)} \subseteq \mathbb{I}(fg)E_n$ from (4.38), and hence $\overline{\mathbb{I}(f)\mathbb{I}(g)} \subseteq \mathbb{I}(fg)$ as $n \to \infty$ by (4.27).

Thus, we have shown that $\mathbb{I}(fg) = \overline{\mathbb{I}(f)\mathbb{I}(g)}$.

(iv): By (i) and (iii) we have $\mathbb{I}(f)^*\mathbb{I}(f) = \overline{\mathbb{I}(\bar{f})\mathbb{I}(f)} \subseteq \mathbb{I}(\bar{f}f)$. Since $\mathbb{I}(f)$ is closed, $\mathbb{I}(f)^*\mathbb{I}(f)$ is self-adjoint by Proposition 3.18(ii). By (i), $\mathbb{I}(\bar{f}f)$ is self-adjoint. Hence, $\mathbb{I}(f)^*\mathbb{I}(f) = \mathbb{I}(\bar{f}f)$. A similar reasoning shows that $\mathbb{I}(f)\mathbb{I}(f)^* = \mathbb{I}(\bar{f}f)$. Therefore, $\mathbb{I}(f)^*\mathbb{I}(f) = \mathbb{I}(\bar{f}f) = \mathbb{I}(f)\mathbb{I}(f)^*$, that is, $\mathbb{I}(f)$ is normal.

(v): Since $\mathbb{I}(f)\mathbb{I}(g) \subseteq \mathbb{I}(fg)$ by (iii), $\mathcal{D}(\mathbb{I}(f)\mathbb{I}(g)) \subseteq \mathcal{D}(\mathbb{I}(g)) \cap \mathcal{D}(\mathbb{I}(fg))$. To prove the converse inclusion, let $x \in \mathcal{D}(\mathbb{I}(g)) \cap \mathcal{D}(\mathbb{I}(fg))$. Recall that by (4.37),

$$\mathbb{I}(fg)E_n x = \mathbb{I}(f)\mathbb{I}(g)E_n x. \tag{4.39}$$

As $n \to \infty$, we get $\mathbb{I}(g)E_n x \to \mathbb{I}(g)x$ (by $x \in \mathcal{D}(\mathbb{I}(g))$) and $\mathbb{I}(fg)E_n x \to \mathbb{I}(fg)x$ (by $x \in \mathcal{D}(\mathbb{I}(fg))$). Therefore, since the operator $\mathbb{I}(f)$ is closed, it follows from (4.39) that $\mathbb{I}(g)x \in \mathcal{D}(\mathbb{I}(f))$, that is, $x \in \mathcal{D}(\mathbb{I}(f)\mathbb{I}(g))$. $\quad\square$

4.3.3 Properties of Spectral Integrals

In this subsection we develop a number of further properties of spectral integrals which will be used later. Throughout, f and g are functions from $\mathcal{S} = \mathcal{S}(\Omega, \mathfrak{A}, E)$.

Proposition 4.17

(i) *If $f(t) = g(t)$ E-a.e. on Ω, then $\mathbb{I}(f) = \mathbb{I}(g)$.*

(ii) *If $f(t)$ is real-valued E-a.e. on Ω, then $\mathbb{I}(f)$ is self-adjoint.*

(iii) *If $f(t) \geq 0$ E-a.e. on Ω, then $\mathbb{I}(f)$ is positive and self-adjoint.*

(iv) *If $g(t) \geq 0$ E-a.e. on Ω, then $\mathbb{I}(\sqrt{g})$ is a positive self-adjoint operator such that $\mathbb{I}(\sqrt{g})^2 = \mathbb{I}(g)$.*

Proof (i) and (iii) follow at once from (4.32). (ii) is an immediate consequence of Theorem 4.16(i). (iv) follows from Theorem 4.16, (i) and (iv), applied with $\sqrt{f} = g$, and from (iii). $\quad\square$

The assertion of (iv) means that $\mathbb{I}(\sqrt{g})$ is a "positive square root" of the positive self-adjoint operator $\mathbb{I}(g)$. We shall see below (Proposition 5.13) that *each* positive self-adjoint operator has a *unique* positive square root.

Recall that a function $f \in S$ is in $L^\infty(\Omega, E)$ if and only if f is bounded E-a.e. on Ω and that $\|f\|_\infty$ is the E-essential supremum of f, that is, $\|f\|_\infty$ is the infimum of $\sup\{|f(t)| : t \in \Omega \setminus N\}$ over all sets $N \in \mathfrak{A}$ such that $E(N) = 0$. By this definition, $\|f\|_\infty < \infty$ if and only if $f \in L^\infty(\Omega, E)$.

Proposition 4.18 *The operator $\mathbb{I}(f)$ is bounded if and only if $f \in L^\infty(\Omega, E)$. In this case, $\|\mathbb{I}(f)\| = \|f\|_\infty$.*

Proof Obviously, if $f \in L^\infty(\Omega, E)$, then $\mathbb{I}(f)$ is bounded and $\|\mathbb{I}(f)\| \leq \|f\|_\infty$. Conversely, suppose $\mathbb{I}(f)$ is bounded. Set $M_n := \{t : |f(t)| \geq \|\mathbb{I}(f)\| + 2^{-n}\}$ for $n \in \mathbb{N}$. By Theorem 4.16(iii) and by (4.33), we have for $x \in \mathcal{H}$,

$$\|\mathbb{I}(f)\|^2 \|E(M_n)x\|^2 \geq \|\mathbb{I}(f)E(M_n)x\|^2 = \|\mathbb{I}(f\chi_{M_n})x\|^2$$

$$= \int_\Omega |f\chi_{M_n}|^2 \, dE_x = \int_{M_n} |f|^2 \, dE_x$$

$$\geq (\|\mathbb{I}(f)\| + 2^{-n})^2 \|E(M_n)x\|^2.$$

This is only possible if $E(M_n)x = 0$. Thus, $E(M_n) = 0$, and hence $E(M) = 0$, where $M := \bigcup_{n=1}^\infty M_n$. Since obviously $M = \{t \in \Omega : |f(t)| > \|\mathbb{I}(f)\|\}$, it follows that $|f(t)| \leq \|\mathbb{I}(f)\|$ E-a.e., and hence $\|f\|_\infty \leq \|\mathbb{I}(f)\|$.   □

Proposition 4.19 *The operator $\mathbb{I}(f)$ is invertible if and only if $f(t) \neq 0$ E-a.e. on Ω. In this case we have $\mathbb{I}(f)^{-1} = \mathbb{I}(f^{-1})$. Here f^{-1} denotes the function from S which is defined by $f^{-1}(t) := \frac{1}{f(t)}$, where we set $\frac{1}{0} = \infty$ and $\frac{1}{\infty} = 0$.*

Proof Set $\mathcal{N}(f) := \{t \in \Omega : f(t) = 0\}$. Then $\mathbb{I}(f)E(\mathcal{N}(f)) = \mathbb{I}(f\chi_{\mathcal{N}(f)}) = \mathbb{I}(0) = 0$ by Theorem 4.16. Hence, $\mathbb{I}(f)$ is not invertible if $E(\mathcal{N}(f)) \neq 0$.

Suppose now that $E(\mathcal{N}(f)) = 0$. Clearly, then $f^{-1} \in S$. Since $f^{-1}f = 1$ E-a.e., we have $\mathcal{D}(\mathbb{I}(f^{-1}f)) = \mathcal{D}(\mathbb{I}(1)) = \mathcal{H}$, and hence $\mathcal{D}(\mathbb{I}(f^{-1})\mathbb{I}(f)) = \mathcal{D}(\mathbb{I}(f))$ by Theorem 4.16(v). Further, $\mathbb{I}(f^{-1})\mathbb{I}(f) \subseteq \mathbb{I}(f^{-1}f) = I$ by Theorem 4.16(iii). From these two relations we conclude that $\mathbb{I}(f)$ is invertible and that $\mathbb{I}(f)^{-1} \subseteq \mathbb{I}(f^{-1})$.

Replacing f by f^{-1} in the preceding, we obtain $\mathbb{I}(f^{-1})^{-1} \subseteq \mathbb{I}(f)$ which in turn implies that $(\mathbb{I}(f^{-1})^{-1})^{-1} = \mathbb{I}(f^{-1}) \subseteq \mathbb{I}(f)^{-1}$. Thus, $\mathbb{I}(f)^{-1} = \mathbb{I}(f^{-1})$.   □

Proposition 4.20

(i) *The spectrum of $\mathbb{I}(f)$ is the* essential range *of f, that is,*

$$\sigma(\mathbb{I}(f)) = \{\lambda \in \mathbb{C} : E(\{t \in \Omega : |f(t) - \lambda| < \varepsilon\}) \neq 0 \text{ for all } \varepsilon > 0\}.$$

$$(4.40)$$

If $\lambda \in \rho(\mathbb{I}(f))$, then we have $R_\lambda(\mathbb{I}(f)) = \mathbb{I}((f - \lambda)^{-1})$.

(ii) $\lambda \in \mathbb{C}$ *is an eigenvalue of* $\mathbb{I}(f)$ *if and only if* $E(\{t \in \Omega : f(t) = \lambda\}) \neq 0$. *In this case,* $E(\{t \in \Omega : f(t){=}\lambda\})$ *is the projection on the eigenspace of* $\mathbb{I}(f)$ *at* λ.

Proof Upon replacing f by $f - \lambda$, we can assume throughout that $\lambda = 0$.

(i): Clearly, $0 \in \rho(\mathbb{I}(f))$ if and only if $\mathbb{I}(f)$ has a bounded everywhere defined inverse. Hence, it follows from Propositions 4.18 and 4.19 that $0 \in \rho(\mathbb{I}(f))$ if and only if $f \neq 0$ E-a.e. and $f^{-1} \in L^\infty(\Omega, E)$, or equivalently, if there is a constant $c > 0$ such that $E(\{t : |f(t)| \geq c\}) = 0$. Therefore, $0 \in \sigma(\mathbb{I}(f))$ if and only if $E(\{t : |f(t)| < \varepsilon\}) \neq 0$ for all $\varepsilon > 0$. By Proposition 4.19, we then have $R_0(\mathbb{I}(f)) = \mathbb{I}(f)^{-1} = \mathbb{I}(f^{-1})$.

(ii): Let $x \in \mathcal{H}$. Put $\mathcal{N}(f) := \{t \in \Omega : f(t) = 0\}$. From Eq. (4.33) it follows that $\mathbb{I}(f)x = 0$ if and only if $f(t) = 0$ E_x-a.e. on Ω, that is, if $\langle E(\Omega \backslash \mathcal{N}(f))x, x \rangle = 0$, or equivalently, if $E(\Omega \backslash \mathcal{N}(f))x \equiv x - E(\mathcal{N}(f))x = 0$. That is, $x \in \mathcal{N}(\mathbb{I}(f))$ if and only if $x = E(\mathcal{N}(f))x$. Hence, $E(\mathcal{N}(f))$ is the projection onto $\mathcal{N}(\mathbb{I}(f))$. In particular, $\mathcal{N}(\mathbb{I}(f)) \neq \{0\}$ if only and only if $E(\mathcal{N}(f)) \neq 0$. $\square$

Simple examples show that for arbitrary functions $f, g \in \mathcal{S}$, the bars in Theorem 4.16, (ii) and (iii), cannot be avoided (see Exercise 7), that is, in general, we only have $\mathbb{I}(f) + \mathbb{I}(g) \subseteq \mathbb{I}(f + g)$ and $\mathbb{I}(f)\mathbb{I}(g) \subseteq \mathbb{I}(fg)$. However, if $\mathcal{D}(\mathbb{I}(fg)) \subseteq \mathcal{D}(\mathbb{I}(g))$ (for instance, if the operator $\mathbb{I}(g)$ is bounded), then we have $\mathbb{I}(fg) = \mathbb{I}(f)\mathbb{I}(g)$ by Theorem 4.16, (iii) and (v). We illustrate this by a simple example.

Example 4.8 Let $\lambda \in \rho(\mathbb{I}(f))$. By Theorem 4.16(v) and Proposition 4.40(i) we then have $\mathbb{I}(f(f - \lambda)^{-1}) = \mathbb{I}(f)\mathbb{I}((f - \lambda)^{-1}) = \mathbb{I}(f)(\mathbb{I}(f) - \lambda I)^{-1}$; this operator is bounded, and its domain is $\mathcal{H}$.

Also, $\mathbb{I}(f(f - \lambda)^{-1}) \supseteq \mathbb{I}((f - \lambda)^{-1})\mathbb{I}(f)$, but the domain of $\mathbb{I}((f - \lambda)^{-1})\mathbb{I}(f)$ is only $\mathcal{D}(\mathbb{I}(f))$. However, we have $\mathbb{I}(f(f - \lambda)^{-1}) = \overline{\mathbb{I}((f - \lambda)^{-1})\mathbb{I}(f)}$. $\circ$

The following two propositions describe some cases where the bars can be omitted.

Proposition 4.21

(i) *If there exists a* $c > 0$ *such that* $|f(t) + g(t)| \geq c|f(t)|$ *E-a.e., then we have* $\mathbb{I}(f) + \mathbb{I}(g) = \mathbb{I}(f + g)$.
(ii) *If* $f(t) \geq 0$ *and* $g(t) \geq 0$ *E-a.e., then* $\mathbb{I}(f) + \mathbb{I}(g) = \mathbb{I}(f + g)$.
(iii) *If there is a constant* $c > 0$ *such that* $|f(t)| \geq c$ *E-a.e., then* $\mathbb{I}(fg) = \mathbb{I}(f)\mathbb{I}(g)$.

Proof (i): By Theorem 4.16(ii), $\mathbb{I}(f) + \mathbb{I}(g) \subseteq \mathbb{I}(f + g)$. Thus, it suffices to show that $\mathcal{D}(\mathbb{I}(f + g)) \subseteq \mathcal{D}(\mathbb{I}(f) + \mathbb{I}(g))$. Since $c|f| \leq |f + g|$, we have $|g| \leq (1 + c^{-1})|f + g|$, so (4.26) implies that $\mathcal{D}(\mathbb{I}(f + g)) \subseteq \mathcal{D}(\mathbb{I}(f)) \cap \mathcal{D}(\mathbb{I}(g)) = \mathcal{D}(\mathbb{I}(f) + \mathbb{I}(g))$.

(ii) follows immediately from (i).

(iii): Since $\mathbb{I}(f)\mathbb{I}(g) \subseteq \mathbb{I}(fg)$ by Theorem 4.16(iii), it is enough to verify that $\mathcal{D}(\mathbb{I}(fg)) \subseteq \mathcal{D}(\mathbb{I}(f)\mathbb{I}(g))$. But $\mathcal{D}(\mathbb{I}(f)\mathbb{I}(g)) = \mathcal{D}(\mathbb{I}(g)) \cap \mathcal{D}(\mathbb{I}(fg))$ by Theorem 4.16(v), so it suffices to check that $\mathcal{D}(\mathbb{I}(fg)) \subseteq \mathcal{D}(\mathbb{I}(g))$. The latter follows at once from (4.26) using the assumption $|f| \geq c > 0$. $\qquad\square$

Proposition 4.22 $\mathbb{I}(p(f)) = p(\mathbb{I}(f))$ *for any polynomial* $p \in \mathbb{C}[x]$.

Proof We proceed by induction on the degree of the polynomial p.

Suppose that the assertion $\mathbb{I}(q(f)) = q(\mathbb{I}(f))$ is true for all $q \in \mathbb{C}[x]$ such that degree $q < n$. Let $p(x) \in \mathbb{C}[x]$ be of degree n. We write p as $p(x) = a_n x^n + q(x)$, where $a_n \neq 0$ and degree $q < n$. We first prove that

$$\mathbb{I}(f^n) = \mathbb{I}(f)^n. \tag{4.41}$$

Since $|f|^2 \leq 1 + |f^n|^2$, we have $\mathcal{D}(\mathbb{I}(f^n)) \subseteq \mathcal{D}(\mathbb{I}(f))$ by (4.26). Using the induction hypothesis $\mathbb{I}(f^{n-1}) = \mathbb{I}(f)^{n-1}$, Theorem 4.16(v), and the latter inclusion, we derive

$$\mathcal{D}(\mathbb{I}(f)^n) = \mathcal{D}(\mathbb{I}(f)^{n-1}\mathbb{I}(f)) = \mathcal{D}(\mathbb{I}(f^{n-1})\mathbb{I}(f))$$
$$= \mathcal{D}(\mathbb{I}(f)) \cap \mathcal{D}(\mathbb{I}(f^n)) = \mathcal{D}(\mathbb{I}(f^n)).$$

Therefore, since

$$\mathbb{I}(f^n) \supseteq \mathbb{I}(f^{n-1})\mathbb{I}(f) = \mathbb{I}(f)^{n-1}\mathbb{I}(f) = \mathbb{I}(f)^n$$

by Theorem 4.16(iii), Eq. (4.41) follows.

Using (4.41), the hypothesis $\mathbb{I}(q(f)) = q(\mathbb{I}(f))$, and Theorem 4.16(ii), we obtain

$$p(\mathbb{I}(f)) = a_n\mathbb{I}(f)^n + q(\mathbb{I}(f)) = a_n\mathbb{I}(f^n) + \mathbb{I}(q(f)) \subseteq \mathbb{I}(p(f)). \tag{4.42}$$

Since degree $p = n$, we have $f^n \in L^2(\Omega, E_x)$ if and only if $p(f) \in L^2(\Omega, E_x)$, and so $\mathcal{D}(\mathbb{I}(f^n)) = \mathcal{D}(\mathbb{I}(p(f)))$ by (4.26). On the other hand, $\mathcal{D}(p(\mathbb{I}(f))) = \mathcal{D}(\mathbb{I}(f)^n)$ by the definition of the operator $p(\mathbb{I}(f))$. Therefore, $\mathcal{D}(\mathbb{I}(p(f))) = \mathcal{D}(p(\mathbb{I}(f)))$ by (4.41), and hence $\mathbb{I}(p(f)) = p(\mathbb{I}(f))$ by (4.42). $\qquad\square$

Proposition 4.23 *An operator* $T \in \mathbf{B}(\mathcal{H})$ *commutes with a spectral measure* E *on* $\mathcal{H}$ *(that is,* $TE(M) = E(M)T$ *for all* $M \in \mathfrak{A}$*) if and only if* $T\mathbb{I}(f) \subseteq \mathbb{I}(f)T$ *for all* $f \in \mathcal{S}$.

Proof The *if* part is clear, because $\mathbb{I}(\chi_M) = E(M)$. We prove the *only if* assertion. For $x \in \mathcal{H}$ and $M \in \mathfrak{A}$, we have

$$E_{Tx}(M) = \|E(M)Tx\|^2 = \|TE(M)x\|^2 \leq \|T\|^2 E_x(M).$$

Hence, $Tx \in \mathcal{D}(\mathbb{I}(f))$ if $x \in \mathcal{D}(\mathbb{I}(f))$ by (4.26). Since T commutes with E, T commutes with $\mathbb{I}(g)$ for simple functions g and by taking limits with $\mathbb{I}(g)$ for all $g \in \mathcal{B}$. In particular, T commutes with all operators $\mathbb{I}(f\chi_{M_k})$. Therefore, since $T\mathcal{D}(\mathbb{I}(f)) \subseteq \mathcal{D}(\mathbb{I}(f))$, it follows from (4.27) that $T\mathbb{I}(f) \subseteq \mathbb{I}(f)T$. $\qquad\square$

Let $N \in \mathfrak{A}$. Since $E(N)$ commutes with E, $E(N)\mathbb{I}(f) \subseteq \mathbb{I}(f)E(N)$ by Proposition 4.23, so the closed subspace $E(N)\mathcal{H}$ *reduces* $\mathbb{I}(f)$ by Proposition 1.15. That is, setting $\mathbb{I}(f)_0 = \mathbb{I}(f){\restriction}E(N)\mathcal{H}$ and $\mathbb{I}(f)_1 = \mathbb{I}(f){\restriction}(I - E(N))\mathcal{H}$, we have

$$\mathbb{I}(f) = \mathbb{I}(f)_0 \oplus \mathbb{I}(f)_1 \quad \text{on } \mathcal{H} = E(N)\mathcal{H} \oplus \big(I - E(N)\big)\mathcal{H}.$$

The last result in this section is about *transformations of spectral measures*.

Let φ be a mapping of Ω onto another set Ω_0, and let $\mathfrak{A}_0$ be the σ-algebra of all subsets M of Ω_0 whose inverse image $\varphi^{-1}(M)$ is in $\mathfrak{A}$. It is easily checked that

$$F(M) := E\big(\varphi^{-1}(M)\big), \quad M \in \mathfrak{A}_0, \tag{4.43}$$

defines a spectral measure on $(\Omega_0, \mathfrak{A}_0)$ acting on the same Hilbert space $\mathcal{H}$.

Proposition 4.24 *If* $h \in \mathcal{S}(\Omega_0, \mathfrak{A}_0, F)$, *then* $h \circ \varphi \in \mathcal{S}(\Omega, \mathfrak{A}, E)$, *and*

$$\int_{\Omega_0} h(s)\, dF(s) = \int_{\Omega} h\big(\varphi(t)\big)\, dE(t). \tag{4.44}$$

Proof If follows at once from definition (4.43) that $h \circ \varphi$ is in $\mathcal{S}(\Omega, \mathfrak{A}, E)$. From the transformation formula for scalar measures (see (B.6)) we derive

$$\int_{\Omega_0} |h(s)|^2\, d\langle F(s)x, x\rangle = \int_{\Omega} |h(\varphi(t))|^2\, d\langle E(t)x, x\rangle, \tag{4.45}$$

$$\int_{\Omega_0} h(s)\, d\langle F(s)y, y\rangle = \int_{\Omega} h\big(\varphi(t)\big)\, d\langle E(t)y, y\rangle \tag{4.46}$$

for all $x \in \mathcal{H}$ and those $y \in \mathcal{H}$ for which h is $\langle F(\cdot)y, y\rangle$-integrable on Ω_0. Combining (4.45) and (4.26), we obtain $\mathcal{D}(\mathbb{I}_F(h)) = \mathcal{D}(\mathbb{I}_E(h \circ \varphi))$. Here $\mathbb{I}_F$ and $\mathbb{I}_E$ denote the spectral integrals with respect to the spectral measures F and E. From formulas (4.46) and (4.32), applied with $g(t) = 1$, we conclude that $\langle \mathbb{I}_F(h)y, y\rangle = \langle \mathbb{I}_E(h \circ \varphi)y, y\rangle$ for $y \in \mathcal{D}(\mathbb{I}_F(h)) = \mathcal{D}(\mathbb{I}_E(h \circ \varphi))$. Therefore, $\mathbb{I}_F(h) = \mathbb{I}_E(h \circ \varphi)$ by the polarization formula (1.2). This proves (4.44). $\qquad \square$

We illustrate the preceding considerations by describing the spectral integrals $\mathbb{I}(f) = \int_{\Omega} f\, dE(t)$ for our three guiding examples of spectral measures.

Example 4.9 (*Example* 4.5 *continued*)

$$\mathcal{D}\big(\mathbb{I}(f)\big) = \Big\{ x \in \mathcal{H} : \sum_n |f(\lambda_n)|^2 \|P_n x\|^2 < \infty \Big\},$$

$$\mathbb{I}(f)x = \sum_n f(\lambda_n) P_n x \quad \text{for } x \in \mathcal{D}\big(\mathbb{I}(f)\big). \qquad \circ$$

Example 4.10 (*Example* 4.6 *continued*)

$$\mathcal{D}\big(\mathbb{I}(f)\big) = \Big\{ x \in L^2(\mathcal{J}, \mu) : \int_{\mathcal{J}} |f(t)|^2 |x(t)|^2\, d\mu(t) < \infty \Big\},$$

$$\big(\mathbb{I}(f)x\big)(t) = f(t)x(t), \quad t \in \mathcal{J}, \text{ for } x \in \mathcal{D}\big(\mathbb{I}(f)\big). \qquad \circ$$

Example 4.11 (*Example* 4.7 *continued*)

$$\mathcal{D}\big(\mathbb{I}(f)\big) = \left\{ x \in L^2(\Omega, \mu) : \int_\Omega \big|f\big(h(t)\big)\big|^2 \big|x(t)\big|^2 \, d\mu(t) < \infty \right\},$$

$$\big(\mathbb{I}(f)x\big)(t) = f\big(h(t)\big)x(t), \quad t \in \Omega, \ \text{for } x \in \mathcal{D}\big(\mathbb{I}(f)\big). \qquad \circ$$

For characteristic functions $f = \chi_M$, these formulas are just the definitions of the spectral measures given in Examples 4.5, 4.6, and 4.7. Hence, the formulas hold for simple functions by linearity and for arbitrary functions $f \in \mathcal{S}$ by taking limits.

It might be instructive for the reader to check the properties and constructions developed above in the case of these examples.

4.4 Exercises

1. Prove that the family $\{E(\lambda)\}$ in Example 4.3 is a resolution of the identity.
2. Use the definitions (4.4) and (4.5) to prove that $E(\mathcal{J}_1)E(\mathcal{J}_2) = 0$ if $\mathcal{J}_1$ and $\mathcal{J}_2$ are disjoint intervals.
3. Let $\{E(\lambda) : \lambda \in \mathbb{R}\}$ be a resolution of the identity. Prove Eq. (4.3), that is,

$$E(\lambda_1) \le E(\lambda_0 - 0) \le E(\lambda_2) \quad \text{if } \lambda_1 < \lambda_0 \le \lambda_2.$$

4. Let $P_n, n \in \mathbb{N}$, and P be projections on a Hilbert space $\mathcal{H}$. Show that $\lim_{n\to\infty}\langle P_n x, y\rangle = \langle Px, y\rangle$ for $x, y \in \mathcal{H}$ if and only if $\lim_{n\to\infty} P_n x = Px$ for $x \in \mathcal{H}$.
5. Let E be a spectral measure on $(\mathfrak{A}, \Omega)$ on a Hilbert space $\mathcal{H}$.
   a. Show that $E_{x+y}(M) + E_{y-x}(M) = 2E_x(M) + 2E_y(M)$.
   b. Show that $E(M \cup N) + E(M \cap N) = E(M) + E(N)$ for $M, N \in \mathfrak{A}$.
   c. Show that $E(M_1 \cup \cdots \cup M_n) \le E(M_1) + \cdots + E(M_n)$ for $M, N \in \mathfrak{A}$.
   d. Show that $\langle E(\bigcup_{n=1}^\infty M_n)x, x\rangle \le \sum_{n=1}^\infty \langle E(M_n)x, x\rangle$ for each sequence $(M_n)_{n\in\mathbb{N}}$ of sets in $\mathfrak{A}$ for $x \in \mathcal{H}$.
6. Let $\mathfrak{A}$ be a σ-algebra on Ω, and let E be a mapping of $\mathfrak{A}$ into the projections of a Hilbert space $\mathcal{H}$ such that $E(\Omega) = I$. Show that E is a spectral measure if and only if one of the following conditions is satisfied:
   a. $E(\bigcup_{n=1}^\infty M_n) = \text{s-}\lim_{n\to\infty} E(M_n)$ for each sequence $(M_n)_{n\in\mathbb{N}}$ of sets $M_n \in \mathfrak{A}$ such that $M_n \subseteq M_{n+1}$ for $n \in \mathbb{N}$.
   b. $E(\bigcap_{n=1}^\infty N_n) = \text{s-}\lim_{n\to\infty} E(N_n)$ for each sequence $(N_n)_{n\in\mathbb{N}}$ of sets $N_n \in \mathfrak{A}$ such that $N_{n+1} \subseteq N_n$ for $n \in \mathbb{N}$.
7. Prove the formula in Proposition 4.12(iii) for functions $f \in \mathcal{B}_s$.
8. Find functions $f_1, f_2, g_1, g_2 \in \mathcal{S}$ for which $\mathbb{I}(f_1 + g_1) \ne \mathbb{I}(f_1) + \mathbb{I}(g_1)$ and $\mathbb{I}(f_2 g_2) \ne \mathbb{I}(f_2)\mathbb{I}(g_2)$.
   Hint: Take $g_1 = -f_1$ and $g_2 = \frac{1}{f_2}$ for some unbounded functions.
9. Let $f, g \in \mathcal{S}$. Prove or disprove the equalities $\mathbb{I}(e^{f+g}) = \mathbb{I}(e^f)\mathbb{I}(e^g)$ and $\mathbb{I}(\sin(f + g)) = \mathbb{I}(\sin f)\mathbb{I}(\cos g) + \mathbb{I}(\cos f)\mathbb{I}(\sin g)$.
10. Carry out the proofs of some formulas stated in Examples 4.9–4.11.

Further exercises on spectral integrals are given after Chap. 5 in terms of the functional calculus for self-adjoint or normal operators.

Chapter 5
Spectral Decompositions of Self-adjoint and Normal Operators

If E is a spectral measure on the real line, the spectral integral $\int_{\mathbb{R}} \lambda \, dE(\lambda)$ is a self-adjoint operator. The spectral theorem states that *any* self-adjoint operator A is of this form and that the spectral measure E is *uniquely* determined by A. The proof given in Sect. 5.2 reduces this theorem to the case of bounded self-adjoint operators by using the bounded transform. Section 5.1 contains a proof of the spectral theorem in the bounded case. It should be emphasized that this proof establishes the existence of a spectral measure rather than only a resolution of the identity.

The spectral theorem can be considered as a structure theorem for self-adjoint operators. It allows us to define functions $f(A) = \int_{\mathbb{R}} f(\lambda) \, dE(\lambda)$ of the operator A. In Sect. 5.3 this functional calculus is developed, and a number of applications are derived. Section 5.4 is about self-adjoint operators with simple spectra.

In Sect. 5.5 the spectral theorem for n-tuples of strongly commuting unbounded normal operators is proved, and the joint spectrum of such n-tuples is investigated. Permutability problems involving unbounded operators are a difficult matter. The final Sect. 5.6 deals with the strong commutativity of unbounded normals and contains a number of equivalent characterizations.

5.1 Spectral Theorem for Bounded Self-adjoint Operators

The main result proved in this section is the following theorem.

Theorem 5.1 *Let A be a bounded self-adjoint operator on a Hilbert space $\mathcal{H}$. Let $\mathcal{J} = [a, b]$ be a compact interval on $\mathbb{R}$ such that $\sigma(A) \subseteq \mathcal{J}$. Then there exists a unique spectral measure E on the Borel σ-algebra $\mathfrak{B}(\mathcal{J})$ such that*

$$A = \int_{\mathcal{J}} \lambda \, dE(\lambda).$$

If F is another spectral measure on $\mathfrak{B}(\mathbb{R})$ such that $A = \int_{\mathbb{R}} \lambda \, dF(\lambda)$, then we have $E(M \cap \mathcal{J}) = F(M)$ for all $M \in \mathfrak{B}(\mathbb{R})$.

K. Schmüdgen, *Unbounded Self-adjoint Operators on Hilbert Space*,
Graduate Texts in Mathematics 265,
DOI 10.1007/978-94-007-4753-1_5, © Springer Science+Business Media Dordrecht 2012

A bounded operator T on $\mathcal{H}$ commutes with A if and only if it commutes with all projections $E(M)$, $M \in \mathfrak{B}(\mathcal{J})$.

Before we begin the proof of this theorem, we develop some preliminaries.

Let $\mathbb{C}[t]$ be the $*$-algebra of complex polynomials $p(t) = \sum_{k=0}^{n} a_k t^k$ with involution given by $\overline{p}(t) = \sum_{k=0}^{n} \overline{a}_k t^k$. Recall that $p(A) = \sum_{k=0}^{n} a_k A^k$, where $A^0 = I$. Clearly, the map $p \to p(A)$ is a $*$-homomorphism of $\mathbb{C}[t]$ onto the $*$-algebra $\mathbb{C}[A]$ of operators $p(A)$ in $\mathbf{B}(\mathcal{H})$.

For $f \in C(\mathcal{J})$, we define $\|f\|_{\mathcal{J}} = \sup\{|f(t)| : t \in \mathcal{J}\}$.

Lemma 5.2 $\sigma(p(A)) \subseteq p(\sigma(A))$ *for* $p \in \mathbb{C}[t]$.

Proof Let $\gamma \in \sigma(p(A))$. Clearly, we can assume that $n := \deg p > 0$. By the fundamental theorem of algebra, there are complex numbers $\alpha_1, \ldots, \alpha_n$ such that $p(t) - \gamma = a_n(t - \alpha_1) \cdots (t - \alpha_n)$. Then

$$p(A) - \gamma I = a_n(A - \alpha_1 I) \cdots (A - \alpha_n I). \tag{5.1}$$

If all α_j are in $\rho(A)$, then the right-hand side of (5.1) has a bounded inverse, so has $p(A) - \gamma I$, and hence $\gamma \in \rho(p(A))$. This is a contradiction, since $\gamma \in \sigma(p(A))$. Hence, at least one α_j is in $\sigma(A)$. Since $p(\alpha_j) - \gamma = 0$, we get $\gamma \in p(\sigma(A))$. $\qquad \square$

Lemma 5.3 *For* $p \in \mathbb{C}[t]$, *we have* $\|p(A)\| \leq \|p\|_{\mathcal{J}}$.

Proof Using Lemma 5.2, applied to $\overline{p}p$, and the relation $\sigma(A) \subseteq \mathcal{J}$, we obtain

$$\|p(A)\|^2 = \|p(A)^* p(A)\| = \|(\overline{p}p)(A)\| = \sup\{|\gamma| : \gamma \in \sigma((\overline{p}p)(A))\}$$
$$\leq \sup\{\overline{p}p(\lambda) : \lambda \in \sigma(A)\} \leq \|\overline{p}p\|_{\mathcal{J}} = \|p\|_{\mathcal{J}}^2. \qquad \square$$

The main technical ingredient of our proof is the Riesz representation theorem. More precisely, we shall use the following lemma.

Lemma 5.4 *For each continuous linear functional F on the normed linear space $(\mathbb{C}[t], \| \cdot \|_{\mathcal{J}})$, there exists a unique complex regular Borel measure μ on $\mathcal{J}$ such that*

$$F(p) = \int_{\mathcal{J}} p(\lambda) \, d\mu(\lambda), \quad p \in \mathbb{C}[t].$$

Moreover, $|\mu(M)| \leq \|F\|$ *for all* $M \in \mathfrak{B}(\mathcal{J})$.

Proof By the Weierstrass theorem, the polynomials are dense in $(C(\mathcal{J}), \| \cdot \|_{\mathcal{J}})$, so each continuous linear functional on $(\mathbb{C}[t], \| \cdot \|_{\mathcal{J}})$ has a unique extension to a continuous linear functional on $C(\mathcal{J})$. The existence and uniqueness of μ follow then from the Riesz representation theorem (Theorem B.7). $\qquad \square$

Proof of Theorem 5.1

Existence: For arbitrary $x, y \in \mathcal{H}$, we define the linear functional $F_{x,y}$ on $\mathbb{C}[t]$ by $F_{x,y}(p) = \langle p(A)x, y \rangle$, $p \in \mathbb{C}[t]$. Since

$$\left| F_{x,y}(p) \right| \le \| p(A) \| \, \| x \| \, \| y \| \le \| p \|_{\mathcal{J}} \| x \| \, \| y \|$$

by Lemma 5.3, $F_{x,y}$ is continuous on $(\mathbb{C}[t], \| \cdot \|_{\mathcal{J}})$, and $\| F_{x,y} \| \le \| x \| \, \| y \|$. By Lemma 5.4, there is a unique regular complex Borel measure $\mu_{x,y}$ on $\mathcal{J}$ such that

$$F_{x,y}(p) = \langle p(A)x, y \rangle = \int_{\mathcal{J}} p(\lambda) \, d\mu_{x,y}(\lambda), \quad p \in \mathbb{C}[t]. \tag{5.2}$$

Our aim is to show that there exists a spectral measure E on the Borel σ-algebra $\mathfrak{B}(\mathcal{J})$ such that $\mu_{x,y}(M) = \langle E(M)x, y \rangle$ for $x, y \in \mathcal{H}$ and $M \in \mathfrak{B}(\mathcal{J})$.

Let $\alpha_1, \alpha_2 \in \mathbb{C}$ and $x_1, x_2 \in \mathcal{H}$. Using the linearity of the scalar product in the first variable and (5.2), we obtain

$$\int p \, d\mu_{\alpha_1 x_1 + \alpha_2 x_2, y} = \alpha_1 \int p \, d\mu_{x_1, y} + \alpha_2 \int p \, d\mu_{x_2, y} = \int p \, d(\alpha_1 \mu_{x_1, y} + \alpha_2 \mu_{x_2, y})$$

for $p \in \mathbb{C}[t]$. Therefore, by the uniqueness of the representing measure of the functional $F_{\alpha_a x_1 + \alpha_2 x_2, y}$ (see Lemma 5.4) we obtain

$$\mu_{\alpha_1 x_1 + \alpha_2 x_2, y}(M) = \alpha_1 \mu_{x_1, y}(M) + \alpha_2 \mu_{x_2, y}(M)$$

for $M \in \mathfrak{B}(\mathcal{J})$. Similarly, for $y_1, y_2 \in \mathcal{H}$,

$$\mu_{x, \alpha_1 y_1 + \alpha_2 y_2}(M) = \overline{\alpha}_1 \mu_{x, y_1}(M) + \overline{\alpha}_2 \mu_{x, y_2}(M).$$

Moreover, $|\mu_{x,y}(M)| \le \| F_{x,y} \| \le \| x \| \, \| y \|$ by the last assertion of Lemma 5.4. That is, for each $M \in \mathfrak{B}(\mathcal{J})$, the map $(x, y) \rightarrow \mu_{x,y}(M)$ is a continuous sesquilinear form on $\mathcal{H}$. Hence, there exists a bounded operator $E(M)$ on $\mathcal{H}$ such that

$$\mu_{x,y}(M) = \langle E(M)x, y \rangle, \quad x, y \in \mathcal{H}. \tag{5.3}$$

Setting $p(t) = 1$ in (5.2), we obtain $\langle x, y \rangle = \mu_{x,y}(\mathcal{J}) = \langle E(\mathcal{J})x, y \rangle$ for $x, y \in \mathcal{H}$, and so $E(\mathcal{J}) = I$.

The crucial step of this proof is to show that $E(M)$ is a projection. Once this is established, E is a spectral measure by Lemma 4.4. Let $x, y \in \mathcal{H}$ and $p \in \mathbb{C}[t]$. By (5.2) we have

$$\int p \, d\mu_{x,y} = \langle p(A)x, y \rangle = \overline{\langle \overline{p}(A)y, x \rangle} = \overline{\int \overline{p} \, d\mu_{y,x}} = \int p \, d\overline{\mu}_{y,x},$$

and hence $\mu_{x,y}(M) = \overline{\mu_{y,x}(M)}$ for $M \in \mathfrak{B}(\mathcal{J})$ again by the uniqueness of the representing measure (Lemma 5.4). By (5.3), the latter yields $E(M) = E(M)^*$.

To prove that $E(M)^2 = E(M)$, we essentially use the multiplicativity property $(pq)(A) = p(A)q(A)$ for $p, q \in \mathbb{C}[t]$. Since

$$\int p \, d\mu_{q(A)x, y} = \langle p(A)q(A)x, y \rangle = \langle (pq)(A)y, x \rangle = \int pq \, d\mu_{y,x},$$

the uniqueness assertion of Lemma 5.4 implies that $d\mu_{q(A)x,y} = q\,d\mu_{x,y}$, and so $\langle E(M)q(A)x, y\rangle = \int_M q\,d\mu_{x,y}$ for $M \in B(\mathcal{J})$ by (5.3). Hence, since $E(M) = E(M)^*$, we have

$$\int q\,d\mu_{x,E(M)y} = \langle q(A)x, E(M)y\rangle = \langle E(M)q(A)x, y\rangle = \int q\chi_M\,d\mu_{x,y}$$

for $q \in \mathbb{C}[t]$. Using once more the uniqueness of the representing measure, we obtain $d\mu_{x,E(M)y} = \chi_M\,d\mu_{x,y}$, so $\mu_{x,E(M)y}(N) = \int_N \chi_M\,d\mu_{x,y} = \mu_{x,y}(M \cap N)$ for $N \in \mathfrak{B}(\mathcal{J})$, and hence $\langle E(N)x, E(M)y\rangle = \langle E(M \cap N)x, y\rangle$ by (5.3). Therefore, $E(M)E(N) = E(M \cap N)$. Setting $M = N$, we get $E(M)^2 = E(M)$, so $E(M)$ is indeed a projection for $M \in \mathfrak{B}(\mathcal{J})$. Hence, E is a spectral measure on $\mathfrak{B}(\mathcal{J})$.

Let $x, y \in \mathcal{H}$. From (5.2) and (5.3) we obtain

$$\langle p(A)x, y\rangle = \int_{\mathcal{J}} p(\lambda)\,d\langle E(\lambda)x, y\rangle \tag{5.4}$$

for $p \in \mathbb{C}[t]$. In particular, $\langle Ax, y\rangle = \int_{\mathcal{J}} \lambda\,d\langle E(\lambda)x, y\rangle$. For the operator $\mathbb{I}(f_0) = \int_{\mathcal{J}} \lambda\,dE(\lambda)$ with $f_0(\lambda) = \lambda$, $\lambda \in \mathcal{J}$, we also have $\langle \mathbb{I}(f_0)x, y\rangle = \int_{\mathcal{J}} \lambda\,d\langle E(\lambda)x, y\rangle$ by (4.32). Thus, $A = \mathbb{I}(f_0) = \int_{\mathcal{J}} \lambda\,dE(\lambda)$.

Uniqueness: Let $\mathbb{I}(f)$ denote the spectral integral $\int_{\mathbb{R}} f\,dF$ with respect to a second spectral measure F. By assumption, we have $A = \mathbb{I}(f_0)$, where $f_0(\lambda) = \lambda$. Let $\mathcal{J} = [a, b]$ and $N_n := [b + n^{-1}, +\infty)$. By the properties stated in Theorem 4.16,

$$\langle AF(N_n)x, F(N_n)x\rangle = \langle \mathbb{I}(f_0\chi_{N_n})x, x\rangle$$
$$= \int_{N_n} \lambda\,d\langle F(\lambda)x, x\rangle \geq (b + n^{-1})\int_{N_n} d\langle F(\lambda)x, x\rangle = (b + n^{-1})\|F(N_n)x\|^2.$$

Since $\sigma(A) \subseteq [a, b]$ and hence $\langle Ay, y\rangle \leq b\|y\|^2$ for $y \in \mathcal{H}$, we get $F(N_n)x = 0$ for all $x \in \mathcal{H}$. Thus, s-$\lim_{n\to\infty} F(N_n) = F((b, +\infty)) = 0$ by Lemma 4.5.

A similar reasoning shows that $F((-\infty, a)) = 0$. Thus, we have

$$\langle p(A)x, x\rangle = \int_{\mathcal{J}} p(\lambda)\,d\langle F(\lambda)x, x\rangle$$

for $p \in \mathbb{C}[t]$. Comparing the latter with (5.4), we obtain $\langle E(N)x, x\rangle = \langle F(N)x, x\rangle$ for all $N \in \mathfrak{B}(\mathcal{J})$ and $x \in \mathcal{H}$ by the uniqueness assertion of Lemma 5.4. Consequently, $E(N) = F(N)$ for $N \in \mathfrak{B}(\mathcal{J})$. Since $F((b, +\infty)) = F((-\infty, a)) = 0$ and hence $F(\mathbb{R}\setminus\mathcal{J}) = 0$, we get $E(M \cap \mathcal{J}) = F(M)$ for $M \in \mathfrak{B}(\mathbb{R})$.

Finally, let $T \in \mathbf{B}(\mathcal{H})$ and suppose that $TA = AT$. Since $A = \int_{\mathcal{J}} \lambda\,dE(\lambda)$ and hence $p(A) = \int_{\mathcal{J}} p(\lambda)\,dE(\lambda)$ for each polynomial p, we obtain

$$\int_{\mathcal{J}} p(\lambda)\,d\langle E(\lambda)Tx, y\rangle = \langle p(A)Tx, y\rangle = \langle p(A)x, T^*y\rangle = \int_{\mathcal{J}} p(\lambda)\,d\langle E(\lambda)x, T^*y\rangle$$

for $x, y \in \mathcal{H}$. Therefore, by the uniqueness assertion in Lemma 5.4 the measures $\langle E(\cdot)Tx, y\rangle$ and $\langle E(\cdot)x, T^*y\rangle$ coincide. That is,

$$\langle E(M)Tx, y\rangle = \langle E(M)x, T^*y\rangle = \langle TE(M)x, y\rangle,$$

and so $E(M)T = TE(M)$ for $M \in \mathfrak{B}(\mathcal{J})$. Conversely, if $TE(\cdot) = E(\cdot)T$, then we have $TA = AT$ by Proposition 4.23. $\square$

Corollary 5.5 *Let A be a positive bounded self-adjoint operator on $\mathcal{H}$. There exists a unique positive self-adjoint operator B on $\mathcal{H}$, denoted by $A^{1/2}$, such that $B^2 = A$. If $(p_n)_{n \in \mathbb{N}}$ is a sequence of polynomials such that $p_n(\lambda) \to \lambda^{1/2}$ uniformly on $[0, b]$ and $\sigma(A) \subseteq [0, b]$, then we have $\lim_{n \to \infty} \| p_n(A) - A^{1/2} \| = 0$.*

Proof Since A is positive and bounded, $\sigma(A) \subseteq [0, b]$ for some $b > 0$. By Theorem 5.1, there is a spectral measure E such that $A = \int_{[0,b]} \lambda \, dE(\lambda)$. Setting $B := \int_{[0,b]} \lambda^{1/2} \, dE(\lambda)$, B is a positive self-adjoint operator and $B^2 = A$. By Proposition 4.12(iv) we have $\lim_n \| p_n(A) - B \| = 0$ if $p_n(\lambda) \to \lambda^{1/2}$ uniformly on $[0, b]$.

We prove the uniqueness assertion. Let C be an arbitrary positive self-adjoint operator such that $C^2 = A$. Then C commutes with A, so with polynomials of A, and hence with $B = \lim_n p_n(A)$. Let $x \in \mathcal{H}$ and put $y := (B - C)x$. Using that $BC = CB$ and $B^2 = C^2 (= A)$, we derive

$$\langle By, y \rangle + \langle Cy, y \rangle = \langle (B + C)(B - C)x, y \rangle = \langle (B^2 - C^2)x, y \rangle = 0.$$

Therefore, $\langle By, y \rangle = \langle Cy, y \rangle = 0$, since $B \geq 0$ and $C \geq 0$. Hence, $By = Cy = 0$ by Lemma 3.4(iii). Thus, $\| (B - C)x \|^2 = \langle (B - C)^2 x, x \rangle = \langle (B - C)y, x \rangle = 0$, which yields $Bx = Cx$. This proves that $B = C$. $\qquad\square$

Applying the last assertion of Theorem 5.1 twice, first with $T = B$ and then with $T = E_A(M)$, we obtain the following corollary.

Corollary 5.6 *Let $A = \int_{\mathbb{R}} \lambda \, dE_A(\lambda)$ and $B = \int_{\mathbb{R}} \lambda \, dE_B(\lambda)$ be bounded self-adjoint operators on $\mathcal{H}$ with spectral measures E_A and E_B on $\mathfrak{B}(\mathbb{R})$. Then we have $AB = BA$ if and only if $E_A(M)E_B(N) = E_B(N)E_A(M)$ for $M, N \in \mathfrak{B}(\mathbb{R})$.*

5.2 Spectral Theorem for Unbounded Self-adjoint Operators

The spectral theorem of unbounded self-adjoint operators is the following result.

Theorem 5.7 *Let A be a self-adjoint operator on a Hilbert space $\mathcal{H}$. Then there exists a unique spectral measure $E = E_A$ on the Borel σ-algebra $\mathfrak{B}(\mathbb{R})$ such that*

$$A = \int_{\mathbb{R}} \lambda \, dE_A(\lambda). \tag{5.5}$$

First, we explain the idea of our proof of Theorem 5.7. Clearly, the mapping $t \to z_t := t(1 + t^2)^{-1/2}$ is a homeomorphism of $\mathbb{R}$ on the interval $(-1, 1)$. We define an operator analog of this mapping to transform the unbounded self-adjoint operator A into a bounded self-adjoint operator Z_A. Then we apply Theorem 5.1 to Z_A and transform the spectral measure of Z_A into a spectral measure of A.

A similar idea will be used for the spectral theorem for unbounded normals in Section 5.5. To develop the preliminaries for this result as well, we begin slightly more generally and suppose that T is a *densely defined closed operator* on a Hilbert space $\mathcal{H}$. By Proposition 3.18(i), $C_T := (I + T^*T)^{-1}$ is a positive bounded self-adjoint operator, so it has a positive square root $C_T^{1/2}$ by Corollary 5.5. The operator

$$Z_T := T C_T^{1/2} \tag{5.6}$$

is called the *bounded transform* of T. It will be studied in more detail in Sect. 7.3. Here we prove only the following lemma.

Lemma 5.8 *If T is a densely defined closed operator on $\mathcal{H}$, then*:

(i) *Z_T is a bounded operator defined on $\mathcal{H}$ such that*

$$\|Z_T\| \leq 1 \quad and \quad C_T = (I + T^*T)^{-1} = I - Z_T^* Z_T. \tag{5.7}$$

(ii) *$(Z_T)^* = Z_{T^*}$. In particular, Z_T is self-adjoint if T is self-adjoint.*
(iii) *If T is normal, so is Z_T.*

Proof In this proof we abbreviate $C = C_T$, $Z = Z_T$, $C_* = C_{T^*}$, and $Z_* = Z_{T^*}$.

(i): Clearly, $C\mathcal{H} = \mathcal{D}(I + T^*T) = \mathcal{D}(T^*T)$. Hence, for $x \in \mathcal{H}$, we have

$$\|T C^{1/2} C^{1/2} x\|^2 = \langle T^*TCx, Cx \rangle \leq \langle (I + T^*T)Cx, Cx \rangle = \langle x, Cx \rangle = \|C^{1/2}x\|^2,$$

that is, $\|Zy\| = \|TC^{1/2}y\| \leq \|y\|$ for $y \in C^{1/2}\mathcal{H}$. Since $\mathcal{N}(C) = \{0\}$ and hence $\mathcal{N}(C^{1/2}) = \{0\}$, $C^{1/2}\mathcal{H}$ is dense in $\mathcal{H}$. The operator $Z = TC^{1/2}$ is closed, since T is closed and $C^{1/2}$ is bounded. Therefore, the preceding implies that $C^{1/2}\mathcal{H} \subseteq \mathcal{D}(T)$, $\mathcal{D}(Z) = \mathcal{H}$, and $\|Z\| \leq 1$.

Using the relations $C\mathcal{H} = \mathcal{D}(T^*T)$ and $Z^* \supseteq C^{1/2}T^*$, we obtain

$$Z^*ZC^{1/2} \supseteq C^{1/2}T^*TC^{1/2}C^{1/2} = C^{1/2}(I + T^*T)C - C^{1/2}C = (I - C)C^{1/2},$$

so $Z^*ZC^{1/2} = (I - C)C^{1/2}$. Since $C^{1/2}\mathcal{H}$ is dense, this yields $I - Z^*Z = C$.

(ii): Recall that $C = (I + T^*T)^{-1}$ and $C_* = (I + TT^*)^{-1}$. Let $x \in \mathcal{D}(T^*)$. Setting $y = C_*x$, we obtain $x = (I + TT^*)y$ and $T^*x = T^*(I + TT^*)y = (I + T^*T)T^*y$, so that $C_*x \in \mathcal{D}(T^*)$ and $CT^*x = T^*y = T^*C_*x$. This implies that $p(C_*)x \in \mathcal{D}(T^*)$ and $p(C)T^*x = T^*p(C_*)x$ for each polynomial p.

By the Weierstrass theorem, there is a sequence $(p_n)_{n \in \mathbb{N}}$ of polynomials such that $p_n(t) \to t^{1/2}$ uniformly on $[0, 1]$. From Corollary 5.5 it follows that

$$\lim_{n \to \infty} \|p_n(C) - C^{1/2}\| = \lim_{n \to \infty} \|p_n(C_*) - (C_*)^{1/2}\| = 0.$$

Taking the limit in the equation $p_n(C)T^*x = T^*p_n(C_*)x$, by using that T^* is closed, we get $C^{1/2}T^*x = T^*(C_*)^{1/2}x$ for $x \in \mathcal{D}(T^*)$. Since $Z^* = (TC^{1/2})^* \supseteq C^{1/2}T^*$, we obtain $Z^*x = C^{1/2}T^*x = T^*(C_*)^{1/2}x = Z_*x$ for $x \in \mathcal{D}(T^*)$. Since $\mathcal{D}(T^*)$ is dense in $\mathcal{H}$, we conclude that $Z^* = Z_*$ on $\mathcal{H}$.

(iii): Suppose that T is normal, that is, $T^*T = TT^*$. Using the last equality in (5.7) twice, first for T and then for T^*, and the relation $Z^* = Z_*$, we derive

$$I - Z^*Z = (I + T^*T)^{-1} = (I + TT^*)^{-1} = I - (Z_*)^*Z_* = I - ZZ^*.$$

Hence, Z is normal. $\square$

Proof of Theorem 5.7 Let us abbreviate $C = C_A$ and $Z = Z_A$. By Lemma 5.8, $Z = AC^{1/2}$ is a bounded self-adjoint operator on $\mathcal{H}$ such that $\sigma(Z) \subseteq [-1, 1]$ and $I - Z^2 = C = (I + A^2)^{-1}$. By the spectral Theorem 5.1, applied to the operator Z, there is a unique spectral measure F on $\mathfrak{B}([-1, 1])$ such that

$$Z = \int_{[-1,1]} z \, dF(z). \tag{5.8}$$

Since $F(\{1\})\mathcal{H} + F(\{-1\})\mathcal{H} \subseteq \mathcal{N}(I - Z^2) = \mathcal{N}((I + A^2)^{-1}) = \{0\}$, it follows that $F(\{1\}) = F(\{-1\}) = 0$. Hence, $\varphi(z) := z(1 - z^2)^{-1/2}$ is an F-a.e. finite Borel function on $[-1, 1]$.

We prove that $A = \mathbb{I}(\varphi)$. For this, we use some properties of spectral integrals. Since $\mathcal{D}(\mathbb{I}(\varphi)) = \mathcal{D}(\mathbb{I}((1 - z^2)^{-1/2}))$ by (4.26), we have $\mathbb{I}(\varphi) = \mathbb{I}(z)\mathbb{I}((1 - z^2)^{-1/2})$ by Theorem 4.16, (iii) and (v). From Proposition 4.19 we obtain $\mathbb{I}((1 - z^2)^{-1/2}) = \mathbb{I}((1 - z^2)^{1/2})^{-1}$. By the uniqueness of the positive square root (Corollary 5.5) of the operator $C = I - Z^2 \geq 0$ we get $C^{1/2} = (I - Z^2)^{1/2} = \mathbb{I}((1 - z^2)^{1/2})$. Recall that $\mathbb{I}(z) = Z$ by (5.8). Summarizing the preceding, we have shown that $\mathbb{I}(\varphi) = Z(C^{1/2})^{-1}$.

Since $Z = AC^{1/2}$ is everywhere defined, $C^{1/2}\mathcal{H} \subseteq \mathcal{D}(A)$, and hence $\mathbb{I}(\varphi) = Z(C^{1/2})^{-1} \subseteq A$. Because $\mathbb{I}(\varphi)$ and A are self-adjoint operators, $\mathbb{I}(\varphi) = A$.

Now we apply Proposition 4.24 to the map $(-1, 1) \ni z \to \lambda = \varphi(z) \in \mathbb{R}$. Then $E(M) := F(\varphi^{-1}(M))$, $M \in \mathfrak{B}(\mathbb{R})$, defines a spectral measure, and we have

$$\int_{\mathbb{R}} \lambda \, dE(\lambda) = \int_{\mathcal{J}} \varphi(z) \, dF(z) = \mathbb{I}(\varphi) = A$$

by (4.44). This proves the existence assertion of Theorem 5.7.

The uniqueness assertion of Theorem 5.7 will be reduced to the uniqueness in the bounded case. The inverse φ^{-1} of the mapping φ is $\varphi^{-1}(\lambda) = \lambda(1 + \lambda^2)^{-1/2}$. If E' is another spectral measure on $\mathfrak{B}(\mathbb{R})$ such that $A = \int_{\mathbb{R}} \lambda \, dE'(\lambda)$ holds, then $F'(\varphi(M)) := E'(M)$, $M \in \mathfrak{B}((-1, 1))$, defines a spectral measure, and

$$Z_A = A((1 + A^2)^{-1})^{1/2} = \int_{\mathbb{R}} \varphi^{-1}(\lambda) \, dE'(\lambda) = \int_{(-1,1)} z \, dF(z).$$

Extending F' to a spectral measure on $[-1, 1]$ by $F'(M) := F'(M \cap (-1, 1))$, $M \in \mathfrak{B}([-1, 1])$, and using the uniqueness assertion in Theorem 5.1, we conclude that $F' = F$ and hence $E' = E$. $\qquad\square$

5.3 Functional Calculus and Various Applications

Throughout this section A is a *self-adjoint operator* on a Hilbert space $\mathcal{H}$, and E_A denotes the spectral measure of A. By Theorem 5.7 we have

$$A = \int_{\mathbb{R}} \lambda \, dE_A(\lambda). \tag{5.9}$$

We will omit the set $\mathbb{R}$ under the integral sign. Recall that $S \equiv S(\mathbb{R}, \mathfrak{B}(\mathbb{R}), E_A)$ is the set of all E_A-a.e. finite Borel functions $f: \mathbb{R} \to \mathbb{C} \cup \{\infty\}$. For a function $f \in S$, we write $f(A)$ for the spectral integral $\mathbb{I}(f)$ from Sect. 4.3. Then

$$f(A) = \int f(\lambda)\, dE_A(\lambda) \tag{5.10}$$

is a *normal operator* on $\mathcal{H}$ with dense domain

$$\mathcal{D}(f(A)) = \left\{ x \in \mathcal{H} : \int |f(\lambda)|^2 \, d\langle E_A(\lambda)x, x \rangle < \infty \right\}. \tag{5.11}$$

If f is a polynomial $p(\lambda) = \sum_n \alpha_n \lambda^n$, it follows from (5.9) and Proposition 4.22 that the operator $p(A)$ given by (5.10) coincides with the "ordinary" operator polynomial $p(A) = \sum_n \alpha_n A^n$. A similar result is true for inverses by Proposition 4.19. If A is the multiplication operator by a real function h, then $f(A)$ is the multiplication operator with $f \circ h$ (see Example 5.3 below). All these and more facts justify using the notation $f(A)$ and call the operator $f(A)$ a *function* of A.

The assignment $f \to f(A)$ is called the *functional calculus* of the self-adjoint operator A. It is a very powerful tool for the study of self-adjoint operators.

In the case of our three guiding examples, the operators $f(A)$ can be described explicitly. We collect the formulas for the spectral measures from Examples 4.5, 4.6, 4.7 and rewrite the corresponding spectral integrals $\mathbb{I}(f)$ from Examples 4.9, 4.10, 4.11.

Example 5.1 (*Examples* 4.5 *and* 4.9 *continued*) Let $(\lambda_n)_{n\in\mathbb{N}}$ be a real sequence, and $(P_n)_{n\in\mathbb{N}}$ be a sequence of pairwise orthogonal projections on $\mathcal{H}$ such that $\sum_n P_n = I$. For the self-adjoint operator A defined by

$$Ax = \sum_{n=1}^{\infty} \lambda_n P_n x, \qquad \mathcal{D}(A) = \left\{ x \in \mathcal{H} : \sum_{n=1}^{\infty} |\lambda_n|^2 \|P_n x\|^2 < \infty \right\},$$

the spectral measure and functions act by

$$E_A(M) = \sum_{\lambda_n \in M} P_n, \qquad M \in \mathfrak{B}(\mathbb{R}),$$

$$f(A)x = \sum_{n=1}^{\infty} f(\lambda_n) P_n x, \qquad \mathcal{D}(f(A)) = \left\{ x \in \mathcal{H} : \sum_{n=1}^{\infty} |f(\lambda_n)|^2 \|P_n x\|^2 < \infty \right\}. \circ$$

Example 5.2 (*Examples* 4.6 *and* 4.10 *continued*) Let μ be a positive regular Borel measure on an interval $\mathcal{J}$. For the self-adjoint operator A on $\mathcal{H} = L^2(\mathcal{J}, \mu)$ defined by

$$(Ag)(t) = tg(t), \qquad \mathcal{D}(A) = \left\{ g \in L^2(\mathcal{J}, \mu) : \int_{\mathcal{J}} t^2 |g(t)|^2 \, d\mu(t) < \infty \right\},$$

the spectral measure and functions are given by

$$(E_A(M)g)(t) = \chi_M(t) \cdot g(t), \quad g \in \mathcal{H}, \ M \in \mathfrak{B}(\mathcal{J}),$$

$$(f(A)g)(t) = f(t) \cdot g(t), \quad g \in \mathcal{D}(f(A)),$$

$$\mathcal{D}(f(A)) = \left\{ g \in L^2(\mathcal{J}, \mu) : \int_{\mathcal{J}} |f(t)|^2 |g(t)|^2 \, d\mu(t) < \infty \right\}. \qquad \circ$$

Example 5.3 (*Examples* 4.7 *and* 4.11 *continued*) Let $(\Omega, \mathfrak{A}, \mu)$ be a measure space, and $h : \Omega \to \mathbb{R}$ be a measurable function on Ω. The multiplication operator A by the function h on $\mathcal{H} = L^2(\Omega, \mu)$, that is,

$$(Ag)(t) = h(t) \cdot g(t), \qquad \mathcal{D}(A) = \left\{ g \in L^2(\Omega, \mu) : \int_{\Omega} |h(t)|^2 |g(t)|^2 \, d\mu(t) < \infty \right\},$$

is self-adjoint. Setting $h^{-1}(M) := \{t \in \Omega : h(t) \in M\}$, we have

$$(E_A(M)g)(t) = \chi_{h^{-1}(M)}(t) \cdot g(t), \quad g \in \mathcal{H}, \ M \in \mathfrak{B}(\mathbb{R}),$$

$$(f(A)g)(t) = f(h(t)) \cdot g(t), \quad g \in \mathcal{D}(f(A)),$$

$$\mathcal{D}(f(A)) = \left\{ g \in L^2(\Omega, \mu) : \int_{\Omega} |f(h(t))|^2 |g(t)|^2 \, d\mu(t) < \infty \right\}. \qquad \circ$$

The basic properties of the functional calculus $f \to f(A) = \mathbb{I}(f)$ have been established in Sect. 4.3. We restate them in the following theorem.

Theorem 5.9 (*Properties of the functional calculus*) Let $f, g \in \mathcal{S}$, $\alpha, \beta \in \mathbb{C}$, $x, y \in \mathcal{D}(f(A))$, and $B \in \mathbf{B}(\mathcal{H})$. Then we have:

1) $\langle f(A)x, y \rangle = \int f(\lambda) d\langle E_A(\lambda)x, y \rangle$.
2) $\|f(A)x\|^2 = \int |f(\lambda)|^2 d\langle E_A(\lambda)x, x \rangle$.
   In particular, if $f(t) = g(t)$ E_A-a.e. on $\mathbb{R}$, then $f(A) = g(A)$.
3) $f(A)$ is bounded if and only if $f \in L^\infty(\mathbb{R}, E_A)$. In this case, $\|f(A)\| = \|f\|_\infty$.
4) $\overline{f}(A) = f(A)^*$. In particular, $f(A)$ is self-adjoint if f is real E_A-a.e. on $\mathbb{R}$.
5) $(\alpha f + \beta g)(A) = \alpha f(A) + \beta g(A)$.
6) $(fg)(A) = f(A)g(A)$.
7) $p(A) = \sum_n \alpha_n A^n$ for any polynomial $p(t) = \sum_n \alpha_n t^n \in \mathbb{C}[t]$.
8) $\chi_M(A) = E_A(M)$ for $M \in \mathfrak{B}(\mathbb{R})$.
9) If $f(t) \neq 0$ E_A-a.e. on $\mathbb{R}$, then $f(A)$ is invertible, and $f(A)^{-1} = (1/f)(A)$.
10) If $f(t) \geq 0$ E_A-a.e. on $\mathbb{R}$, then $f(A) \geq 0$.

The next proposition describes the spectrum and resolvent in terms of the spectral measure.

Proposition 5.10

(i) *The support* $\operatorname{supp} E_A$ *of the spectral measure* E_A *is equal to the spectrum* $\sigma(A)$ *of the operator* A. *That is, a real number* λ_0 *is in* $\sigma(A)$ *if and only if* $E_A(\lambda_0 + \epsilon) \neq E_A(\lambda_0 - \epsilon)$ *for all* $\varepsilon > 0$. *For* $\lambda_0 \in \rho(A)$, *we have*

$$R_{\lambda_0}(A) \equiv (A - \lambda_0 I)^{-1} = \int (\lambda - \lambda_0)^{-1} dE_A(\lambda). \tag{5.12}$$

(ii) *A real number λ_0 is an eigenvalue of A if and only if $E_A(\{\lambda_0\}) \neq 0$. In this case, $E_A(\{\lambda_0\})$ is the projection of $\mathcal{H}$ on the eigenspace of A at λ_0.*

Proof By (5.9), A is the spectral integral $\mathbb{I}(f)$, where $f(\lambda) = \lambda$ on $\mathbb{R}$. Therefore, except for the statement about $\mathrm{supp}\, E_A$, all above assertions are already contained in Proposition 4.20. By Definition 4.3, we have $\lambda_0 \in \mathrm{supp}\, E_A$ if and only if $E_A((\lambda_0 - \varepsilon, \lambda_0 + \varepsilon)) \neq 0$ for all $\varepsilon > 0$, or equivalently, $E_A(\lambda_0 + \varepsilon) \neq E_A(\lambda_0 - \varepsilon)$ for all $\varepsilon > 0$. By Proposition 4.20(i) the latter is equivalent to $\lambda_0 \in \sigma(A)$. $\square$

A simple consequence of this proposition is the following:

Corollary 5.11 *Each isolated point of the spectrum $\sigma(A)$ of a self-adjoint operator A is an eigenvalue for A.*

The next proposition characterizes self-adjoint operators A with purely discrete spectra. Among others, it shows that the underlying Hilbert space is then separable and has an orthonormal basis of eigenvectors of A.

Proposition 5.12 *For a self-adjoint operator A on an infinite dimensional Hilbert space $\mathcal{H}$, the following are equivalent:*

(i) *There exist a real sequence $(\lambda_n)_{n \in \mathbb{N}}$ and an orthonormal basis $\{e_n : n \in \mathbb{N}\}$ of $\mathcal{H}$ such that $\lim_{n \to \infty} |\lambda_n| = +\infty$ and $A e_n = \lambda_n e_n$ for $n \in \mathbb{N}$.*
(ii) *A has a purely discrete spectrum.*
(iii) *The resolvent $R_\lambda(A)$ is compact for one, hence for all, $\lambda \in \rho(T)$.*
(iv) *The embedding map $J_A : (\mathcal{D}(A), \|\cdot\|_A) \to \mathcal{H}$ is compact.*

Proof (i) $\to$ (ii): From the properties stated in (i) one easily verifies that A has a purely discrete spectrum $\sigma(A) = \{\lambda_n : n \in \mathbb{N}\}$.

(ii) $\to$ (iii) and (ii) $\to$ (i): We choose a maximal orthonormal set $E = \{e_i : i \in J\}$ of eigenvectors of A. (A priori this set E might be empty.) Let $A e_i = \lambda_i e_i$ for $i \in J$. By Lemma 3.4(i), each λ_i is real. Then we have $\sigma(A) = \{\lambda_i : i \in J\}$. (Indeed, if there were a $\lambda \in \sigma(A)$ such that $\lambda \neq \lambda_i$ for all i, then λ should be an eigenvalue of A, say $A e = \lambda e$ for some unit vector e, because A has a purely discrete spectrum. But, since $\lambda \neq \lambda_i$, we have $e \perp e_i$ by Lemma 3.4(ii), which contradicts the maximality of the set E.) Therefore, it follows from Proposition 2.10(i) that

$$\sigma\big(R_\lambda(A)\big) \setminus \{0\} = \big\{(\lambda_i - \lambda)^{-1} : i \in J\big\}. \tag{5.13}$$

Let $\mathcal{G}$ denote the closed linear span of e_i, $i \in J$. Since $\sigma(A)$ has no finite accumulation point, there exists a real number $\lambda \in \rho(A)$. Then $R_\lambda(A) = (A - \lambda I)^{-1}$ is a bounded self-adjoint operator on $\mathcal{H}$ which leaves $\mathcal{G}$, hence $\mathcal{G}^\perp$, invariant.

Next we prove that $\mathcal{G} = \mathcal{H}$. Assume the contrary. Then $B := R_\lambda(A) {\restriction} \mathcal{G}^\perp$ is a self-adjoint operator on $\mathcal{G}^\perp \neq \{0\}$. Since $\mathcal{N}(B) = \{0\}$, there exists a nonzero μ in $\sigma(B)$. Clearly, $\sigma(B) \subseteq \sigma(R_\lambda(A))$. Thus, $\mu = (\lambda_{i_0} - \lambda)^{-1}$ for some $i_0 \in J$ by (5.13).

Since $\sigma(A)$ has no finite accumulation point, $\sigma(R_\lambda(A))$, hence its subset $\sigma(B)$, has no nonzero accumulation point by (5.13), so μ is an isolated point of $\sigma(B)$. Therefore, by Corollary 5.11, μ is an eigenvalue of B, say $Bu = \mu u$ for some unit vector $u \in \mathcal{G}^\perp$. Since $B \subseteq R_\lambda(A)$, $R_\lambda(A)u = (\lambda_{i_0} - \lambda)^{-1}u$, so that $Au = \lambda_{i_0}u$. This contradicts the maximality of the set E, because $u \in \mathcal{G}^\perp \perp E$. This proves that $\mathcal{G} = \mathcal{H}$.

Hence, E is an orthonormal basis of $\mathcal{H}$. Since $\mathcal{H}$ is infinite-dimensional and $\sigma(A)$ has no finite accumulation point, we can take $J = \mathbb{N}$ and have $\lim_n |\lambda_n| = \infty$. This proves (i).

The operator $R_\lambda(A)$ has an orthonormal basis E of eigenvectors, and the sequence $((\lambda_n - \lambda)^{-1})_{n \in \mathbb{N}}$ of the corresponding eigenvalues is a null sequence, since $\lim_n |\lambda_n| = \infty$. This implies that $R_\lambda(A)$ is the limit of a sequence of finite rank operators in the operator norm. Therefore, $R_\lambda(A)$ is compact. This proves (iii).

(iii) $\rightarrow$ (ii) follows from Proposition 2.11.

(iii) $\rightarrow$ (iv): Let M be a bounded subset of $(\mathcal{D}(A), \|\cdot\|_A)$. Then $N := (A - \lambda I)M$ is bounded in $\mathcal{H}$. By (iii), $(A - \lambda I)^{-1}N = M$ is relatively compact in $\mathcal{H}$. This proves that J_A is compact.

(iv) $\rightarrow$ (iii): Suppose that M is a bounded subset of $\mathcal{H}$. From the inequality

$$\left\| A(A - \lambda I)^{-1}x \right\| = \left\| x + \lambda(A - \lambda I)^{-1}x \right\| \le \left(1 + |\lambda| \left\| (A - \lambda I)^{-1} \right\| \right) \|x\|, \quad x \in \mathcal{H},$$

it follows that $(A - \lambda I)^{-1}M$ is bounded in $(\mathcal{D}(A), \|\cdot\|_A)$. Since J_A is compact, $(A - \lambda I)^{-1}M$ is relatively compact in $\mathcal{H}$. That is, $(A - \lambda I)^{-1}$ is compact. $\square$

Corollary 5.5 dealt with the square root of a *bounded* positive operator. The next proposition contains the corresponding result in the unbounded case.

Proposition 5.13 *If A is a positive self-adjoint operator on $\mathcal{H}$, then there is unique positive self-adjoint operator B on $\mathcal{H}$ such that $B^2 = A$.*

The operator B is called the *positive square root* of A and denoted by $A^{1/2}$.

Proof Since $A \ge 0$, we have $\operatorname{supp} E_A = \sigma(A) \subseteq [0, +\infty)$ by Proposition 5.10. Hence, the function $f(\lambda) := \sqrt{\lambda}$ of S is nonnegative E_A-a.e. on $\mathbb{R}$. Therefore, the operator $B := f(A)$ is positive, and $B^2 = f(A)^2 \subseteq (f^2)(A) = A$ by Theorem 5.9, 10) and 6). Since B^2 is self-adjoint, $B^2 = A$.

Suppose that $\widetilde{B}$ is another positive self-adjoint operator such that $\widetilde{B}^2 = A$. Let $\widetilde{B} = \int_{[0,+\infty)} t\, dF(t)$ be the spectral decomposition of $\widetilde{B}$. We define a spectral measure F' by $F'(M) = F(M^2)$, $M \in \mathfrak{B}([0, +\infty))$, where $M^2 = \{t^2 : t \in M\}$. Using Proposition 4.24, we then obtain

$$\int_{[0,+\infty)} \lambda\, dF'(\lambda) = \int_{[0,+\infty)} t^2\, dF(t) = \widetilde{B}^2 = A = \int_{[0,+\infty)} \lambda\, dE_A(\lambda).$$

From the uniqueness of the spectral measure of A (Theorem 5.7) it follows that $F'(M) \equiv F(M^2) = E_A(M)$ for $M \in \mathfrak{B}([0, +\infty))$. Hence, by formula (4.44) and the definition of B,

$$\widetilde{B} = \int_{[0,+\infty)} t \, dF(t) = \int_{[0,+\infty)} \sqrt{\lambda} \, dF'(\lambda) = \int_{[0,+\infty)} \sqrt{\lambda} \, dE_A(\lambda) = B. \qquad \square$$

By formula (5.5) the resolvent $R_{\lambda_0}(A) = (A - \lambda_0 I)^{-1}$ was written as an integral with respect to the spectral measure of A. We now proceed in reverse direction and express the spectral projections of A in terms of the resolvent.

Proposition 5.14 (*Stone's formulas*)   *Suppose that* $a, b \in \mathbb{R} \cup \{-\infty\} \cup \{+\infty\}$, $a < b$, *and* $c \in \mathbb{R}$. *Then we have*

$$E_A([a, b]) + E_A((a, b))$$

$$= \text{s-lim}_{\varepsilon \to 0+} \frac{1}{\pi i} \int_a^b \left((A - (t + i\varepsilon)I)^{-1} - (A - (t - i\varepsilon)I)^{-1} \right) dt, \qquad (5.14)$$

$$E_A(c) + E_A(c - 0)$$

$$= \text{s-lim}_{\varepsilon \to 0+} \frac{1}{\pi i} \int_{-\infty}^c \left((A - (t + i\varepsilon)I)^{-1} - (A - (t - i\varepsilon)I)^{-1} \right) dt. \qquad (5.15)$$

For $a, b \in \mathbb{R}$, the integral in (5.14) is meant as a Riemann integral in the *operator norm* convergence. Improper integrals are always understood as *strong* limits of the corresponding integrals $\int_\alpha^\beta$ as $\alpha \to -\infty$ resp. $\beta \to +\infty$.

Proof For $a, b \in \overline{\mathbb{R}} := \mathbb{R} \cup \{-\infty\} \cup \{+\infty\}$ and $\varepsilon > 0$, we define the functions

$$f_\varepsilon(\lambda, t) = \frac{1}{\pi i} \left((\lambda - (t + i\varepsilon))^{-1} - (\lambda - (t - i\varepsilon))^{-1} \right) \quad \text{and}$$

$$f_{\varepsilon,a,b}(\lambda) = \int_a^b f_\varepsilon(\lambda, t) \, dt,$$

where $\lambda, t \in \mathbb{R}$.

First suppose that $a, b \in \mathbb{R}$, $a < b$. Let $(g_n(\lambda, t))_{n \in \mathbb{N}}$ be a sequence of Riemann sums for the integral $f_{\varepsilon,a,b}(\lambda) = \int_a^b f_\varepsilon(\lambda, t) \, dt$. Since $|f_\varepsilon(\lambda, t)| \leq 2/\varepsilon\pi$ and hence $|g_n(\lambda, t)| \leq 2(b - a)/\varepsilon\pi$, we have $f_{\varepsilon,a,b}(A) = \text{s-lim}_{n \to \infty} g_n(A, t)$ by Proposition 4.12(v). The resolvent formula (5.12) implies that

$$f_\varepsilon(A, t) = \frac{1}{\pi i} \left((A - (t + i\varepsilon)I)^{-1} - (A - (t - i\varepsilon)I)^{-1} \right). \qquad (5.16)$$

Hence, $(g_n(A, t))_{n \in \mathbb{N}}$ is a sequence of Riemann sums for the operator-valued Riemann integral $\int_a^b f_\varepsilon(A, t) \, dt$. Thus, $f_{\varepsilon,a,b}(A) = \int_a^b f_\varepsilon(A, t) \, dt$ for $a, b \in \mathbb{R}$.

Let $a = -\infty$ and $b \in \mathbb{R}$. Since $|f_{\varepsilon,\alpha,b}(\lambda)| \leq 2$ on $\mathbb{R}$ and $f_{\varepsilon,\alpha,b}(\lambda) \to f_{\varepsilon,-\infty,b}(\lambda)$ as $\alpha \to -\infty$, we obtain $f_{\varepsilon,-\infty,b}(A) = \text{s-lim}_{\alpha \to -\infty} f_{\varepsilon,\alpha,b}(A) = \int_{-\infty}^b f_\varepsilon(A, t) \, dt$ again by Proposition 4.12(v). The case $b = +\infty$ is treated similarly.

Putting the results of the preceding two paragraphs together, we have shown that

$$f_{\varepsilon,a,b}(A) = \int_a^b f_\varepsilon(A,t)\,dt \quad \text{for } a,b \in \mathbb{R}, a < b. \tag{5.17}$$

An explicit computation of the integral $\int_a^b f_\varepsilon(\lambda,t)\,dt$ yields

$$f_{\varepsilon,a,b}(\lambda) = \frac{2}{\pi}\left(\arctan\frac{b-\lambda}{\varepsilon} - \arctan\frac{a-\lambda}{\varepsilon}\right) \quad \text{for } \lambda \in \mathbb{R}, a,b \in \overline{\mathbb{R}},\ a < b, \tag{5.18}$$

where we have set $\arctan(-\infty) := -\frac{\pi}{2}$ and $\arctan(+\infty) := \frac{\pi}{2}$.

From (5.18) we easily obtain that $f_{\varepsilon,a,b}(\lambda) \to 0$ if $\lambda \notin [a,b]$, $f_{\varepsilon,a,b}(\lambda) \to 1$ if $\lambda = a, b$ and $f_{\varepsilon,a,b}(\lambda) \to 2$ if $\lambda \in (a,b)$ as $\varepsilon \to 0+$. That is, we have

$$\lim_{\varepsilon\to0+} f_{\varepsilon,a,b}(\lambda) = \chi_{[a,b]}(\lambda) + \chi_{(a,b)}(\lambda).$$

Since $|f_{\varepsilon,a,b}(\lambda)| \le 2$ on $\mathbb{R}$, we can apply Proposition 4.12(v) once again and derive

$$\text{s-}\lim_{\varepsilon\to0+} f_{\varepsilon,a,b}(A) = E_A\big([a,b]\big) + E_A\big((a,b)\big). \tag{5.19}$$

Inserting (5.16) into (5.17) and then (5.17) into (5.19), we obtain (5.14).

Formula (5.15) follows from (5.14) by setting $b = c, a = -\infty$ and recalling that $E_A(c) = E_A((-\infty, c])$ and $E_A(c - 0) = E_A((-\infty, c))$. $\qquad\square$

For $c \in \mathbb{R}$, it follows immediately from (5.15) that

$$E_A(c) = \text{s-}\lim_{\delta\to0+}\text{s-}\lim_{\varepsilon\to0+}\frac{1}{2\pi i}\int_{-\infty}^{c+\delta}\left((A-(t+i\varepsilon)I)^{-1} - (A-(t-i\varepsilon)I)^{-1}\right)dt. \tag{5.20}$$

By combining (5.20) and (5.14) one easily derives formulas for the spectral projections of arbitrary intervals. We shall use (5.20) in the proof of the next proposition.

Proposition 5.15 *Let A_1 and A_2 be self-adjoint operators on Hilbert spaces $\mathcal{H}_1$ and $\mathcal{H}_2$, respectively, and let $S \in \mathbf{B}(\mathcal{H}_1, \mathcal{H}_2)$. Then the following are equivalent:*

(i) $SA_1 \subseteq A_2 S$.
(ii) $SR_\lambda(A_1) = R_\lambda(A_2)S$ *for one (hence for all)* $\lambda \in \rho(A_1) \cap \rho(A_2)$.
(iii) $SE_{A_1}(\lambda) = E_{A_2}(\lambda)S$ *for all* $\lambda \in \mathbb{R}$.
(iv) $SE_{A_1}(M) = E_{A_2}(M)S$ *for all* $M \in \mathfrak{B}(\mathbb{R})$.

Proof (i) $\leftrightarrow$ (ii): Fix $\lambda \in \rho(A_1) \cap \rho(A_2)$. Clearly, (i) is equivalent to $S(A_1 - \lambda I) \subseteq (A_2 - \lambda I)S$ and so to $R_\lambda(A_2)S = SR_\lambda(A_1)$, because $\mathcal{R}(A_1 - \lambda I) = \mathcal{H}$.

The equivalence of (iii) and (iv) follows from Proposition 4.7. Formulas (5.20) and (5.12) yield the implications (ii) $\to$ (iii) and (iv) $\to$ (ii), respectively. $\qquad\square$

The fractional power A^α of a positive self-adjoint operator A is defined by

$$A^\alpha = \int_0^\infty \lambda^\alpha\,dE_A(\lambda).$$

Now we give another integral representation of A^α, where $\alpha \in (0, 1)$, by means of the resolvent of A, that is, in terms of the operator $A(A + tI)^{-1} = I - tR_{-t}(A)$. It can be considered as an operator-theoretic counterpart of the classical formula

$$\lambda^\alpha = \pi^{-1} \sin \pi \alpha \int_0^\infty t^{\alpha-1} \lambda(\lambda + t)^{-1} dt, \quad \lambda > 0, \ \alpha \in (0, 1). \quad (5.21)$$

Proposition 5.16 *Let A be a positive self-adjoint operator on $\mathcal{H}$ and $\alpha \in (0, 1)$.*

(i) *A vector $x \in \mathcal{H}$ belongs to the domain $\mathcal{D}(A^{\alpha/2})$ if and only if the improper integral $\int_0^\infty t^{\alpha-1} \langle A(A + tI)^{-1} x, x \rangle dt$ is finite. We then have*

$$\|A^{\alpha/2} x\|^2 = \pi^{-1} \sin \pi \alpha \int_0^\infty t^{\alpha-1} \langle A(A + tI)^{-1} x, x \rangle dt \quad \text{for } x \in \mathcal{D}(A^{\alpha/2}),$$

(ii) $$A^\alpha x = \pi^{-1} \sin \pi \alpha \lim_{\substack{\varepsilon \to +0 \\ a \to +\infty}} \int_\varepsilon^a t^{\alpha-1} A(A + tI)^{-1} x \, dt \quad \text{for } x \in \mathcal{D}(A).$$

Proof All formulas in Proposition 5.16 remain unchanged when x is replaced by $x - E_A(\{0\})x$. Hence, we can assume without loss of generality that $E_A(\{0\})x = 0$. Further, we abbreviate $b_\alpha := \pi^{-1} \sin \pi \alpha$, $P_\varepsilon := E_A([\varepsilon, \infty))$, and $P_{\varepsilon,a} := E_A([\varepsilon, a])$.

(i): Let $x \in \mathcal{H}$ and $\varepsilon > 0$. Employing formula (5.21), we compute

$$\int_\varepsilon^\infty \lambda^\alpha d\langle E_A(\lambda)x, x \rangle = \int_\varepsilon^\infty \left(b_\alpha \int_0^\infty t^{\alpha-1}(\lambda + t)^{-1} dt \right) d\langle E_A(\lambda)x, x \rangle$$

$$= b_\alpha \int_0^\infty \left(\int_\varepsilon^\infty \lambda(\lambda + t)^{-1} d\langle E_A(\lambda)x, x \rangle \right) t^{\alpha-1} dt$$

$$= b_\alpha \int_0^\infty \langle P_\varepsilon A(A + tI)^{-1} x, x \rangle t^{\alpha-1} dt. \quad (5.22)$$

The two integrals in the first line can be interchanged by Fubini's theorem (Theorem B.5). From Theorem 4.16, (iii) and (v), it follows that the interior integral in the second line is equal to $\langle P_\varepsilon A(A + tI)^{-1} x, x \rangle$.

Now we consider the limit $\varepsilon \to +0$ in (5.22). Using the monotone convergence theorem (Theorem B.2) and the fact that $\langle P_\varepsilon A(A + tI)^{-1}x, x \rangle \to \langle A(A + tI)^{-1}x, x \rangle$, we get

$$\int_0^\infty \lambda^\alpha d\langle E_A(\lambda)x, x \rangle = b_\alpha \int_0^\infty t^{\alpha-1} \langle A(A + tI)^{-1}x, x \rangle dt.$$

By formula (4.26) the first integral is finite if and only if $x \in \mathcal{D}(A^{\alpha/2})$. In this case this integral is equal to $\|A^{\alpha/2} x\|^2$ by (4.33).

(ii): Let $x \in \mathcal{D}(A)$. Fix $\varepsilon > 0$ and $a > \varepsilon$. Using again formula (5.21), we derive

$$x_{\varepsilon,a} := \int_\varepsilon^a \lambda^\alpha dE_A(\lambda)x = \int_\varepsilon^a \left(b_\alpha \int_0^\infty t^{\alpha-1} \lambda(\lambda + t)^{-1} dt \right) dE_A(\lambda)x$$

$$= b_\alpha \int_0^\infty t^{\alpha-1} \int_\varepsilon^a \lambda(\lambda + t)^{-1} dE_A(\lambda)x \, dt$$

$$= b_\alpha \int_0^\infty t^{\alpha-1} P_{\varepsilon,a} A(A + tI)^{-1} x \, dt.$$

Now we pass to the limits $\varepsilon \to +0$ and $a \to \infty$ in this equality. Clearly,

$$\left\| (I - P_{\varepsilon,a}) A (A + tI)^{-1} x \right\| \leq \left\| (I - P_{\varepsilon,a}) x \right\|, \tag{5.23}$$

$$\left\| (I - P_{\varepsilon,a}) A (A + tI)^{-1} x \right\| \leq t^{-1} \left\| (I - P_{\varepsilon,a}) Ax \right\|, \tag{5.24}$$

$(I - P_{\varepsilon,a})x \to E_A(\{0\})x = 0$, and $(I - P_{\varepsilon,a})Ax \to E_A(\{0\})Ax = 0$. Therefore, using inequalities (5.23) at zero and (5.24) at $+\infty$, we conclude that

$$b_\alpha \int_0^\infty t^{\alpha-1} P_{\varepsilon,a} A (A + tI)^{-1} x \, dt \to b_\alpha \int_0^\infty t^{\alpha-1} A (A + tI)^{-1} x \, dt.$$

Since $x_{\varepsilon,a}$ tends to $A^\alpha x$, we obtain the desired equality. $\qquad\square$

5.4 Self-adjoint Operators with Simple Spectra

In this section A is a *self-adjoint operator* on $\mathcal{H}$ with spectral measure E_A.

Lemma 5.17 *Let $\mathcal{N}$ be a subset of $\mathcal{H}$, and let $\mathcal{H}_\mathcal{N}$ be the closed linear span of vectors $E_A(M)x$, where $x \in \mathcal{N}$ and $M \in \mathfrak{B}(\mathbb{R})$. Then $\mathcal{H}_\mathcal{N}$ is smallest reducing subspace for A which contains $\mathcal{N}$. Further, $\mathcal{H}_\mathcal{N}$ is the closed subspace generated by $R_z(A)x$, where $x \in \mathcal{N}$ and $z \in \mathbb{C} \backslash \mathbb{R}$.*

Proof Let $\mathcal{H}_0$ be a closed subspace of $\mathcal{H}$, and let P_0 be the projection onto $\mathcal{H}_0$. By Proposition 1.15, $\mathcal{H}_0$ is reducing for A if and only if $P_0 A \subseteq A P_0$, or equivalently by Proposition 5.15, if $P_0 E_A(\lambda) = E_A(\lambda) P_0$ for all $\lambda \in \mathbb{R}$. Therefore, if $\mathcal{H}_0$ is reducing for A and $\mathcal{N} \subseteq \mathcal{H}_0$, then $\mathcal{H}_0 \subseteq \mathcal{H}_\mathcal{N}$. Obviously, each projection $E_A(\lambda)$ leaves $\mathcal{H}_\mathcal{N}$ invariant. Hence, the projection onto $\mathcal{H}_\mathcal{N}$ commutes with $E_A(\lambda)$, so that $\mathcal{H}_\mathcal{N}$ is reducing for A. The last assertion follows from formulas (5.12) and (5.20) which express resolvents by means of spectral projections and vice versa. $\qquad\square$

Definition 5.1 A vector $x \in \mathcal{H}$ is called a *generating vector* or a *cyclic vector* for A if the linear span of vectors $E_A(M)x$, $M \in \mathfrak{B}(\mathbb{R})$, is dense in $\mathcal{H}$. We say that A has a *simple spectrum* if A has a generating vector.

That is, by Lemma 5.17, a vector $x \in \mathcal{H}$ is generating for A if and only if $\mathcal{H}$ is the smallest reducing subspace for A containing x.

Example 5.4 (*Multiplication operators on* $\mathbb{R}$) Suppose that μ is a positive regular Borel measure on $\mathbb{R}$. Let A_t be the operator on $\mathcal{H} = L^2(\mathbb{R}, \mu)$ defined by $(A_t f)(t) = t f(t)$ for $f \in \mathcal{D}(A_t) = \{ f \in \mathcal{H} : t f(t) \in \mathcal{H} \}$.

By Example 5.2, the spectral projection $E_{A_t}(M)$ acts as multiplication operator by the characteristic functions χ_M. Hence, $\mathrm{supp}\, \mu = \mathrm{supp}\, E_{A_t}$. Since $\mathrm{supp}\, E_{A_t} = \sigma(A_t)$ by Proposition 5.10(i), we have

$$\mathrm{supp}\, \mu = \sigma(A_t). \tag{5.25}$$

Statement 1 $\lambda \in \mathbb{R}$ is an eigenvalue of A_t if and only if $\mu(\{\lambda\}) \neq 0$. Each eigenvalue of A_t has multiplicity one.

Proof Both assertions follow immediately from that fact that $\mathcal{N}(A_t - \lambda I)$ consists of complex multiples of the characteristic function of the point λ.   □

Statement 2 A_t has a simple spectrum.

Proof If the measure μ is finite, then the constant function 1 is in $\mathcal{H}$ and a generating vector for A_t. In the general case we set

$$x(t) := \sum_{k=-\infty}^{\infty} 2^{-|k|} \mu\big([k, k+1)\big)^{-1/2} \chi_{[k,k+1)}(t).$$

Then $x \in L^2(\mathbb{R}, \mu)$. Clearly, the linear span of functions $\chi_M \cdot x$, $M \in \mathcal{B}(\mathbb{R})$, contains the characteristic functions χ_N of all bounded Borel sets N. Since the span of such functions χ_N is dense in $L^2(\mathbb{R}, \mu)$, $x(t)$ is a generating vector for A_t.   □ ∘

The next proposition shows that up to unitary equivalence each self-adjoint operator with simple spectrum is of this form for some *finite* Borel measure μ.

For $x \in \mathcal{H}$, we denote by $\mathcal{F}_x$ the set of all E_A-a.e. finite measurable functions $f : \mathbb{R} \to \mathbb{C} \cup \{\infty\}$ for which $x \in \mathcal{D}(f(A))$.

Proposition 5.18 *Let x be a generating vector for the self-adjoint operator A on $\mathcal{H}$. Set $\mu(\cdot) := \langle E_A(\cdot)x, x \rangle$. Then the map $(U(f(A)x))(t) = f(t)$, $f \in \mathcal{F}_x$, is a unitary operator of $\mathcal{H}$ onto $L^2(\mathbb{R}, \mu)$ such that $A = U^{-1}A_t U$ and $(Ux)(t) = 1$ on $\mathbb{R}$, where A_t is the multiplication operator on $L^2(\mathbb{R}, \mu)$ from Example 5.4.*

Proof For $f \in \mathcal{F}_x$ it follows from Theorem 5.9, 2) that

$$\big\| f(A)x \big\|^2 = \int_{\mathbb{R}} \big| f(t) \big|^2 d\langle E_A(t)x, x \rangle = \| f \|^2_{L^2(\mathbb{R}, \mu)}.$$

If $f \in L^2(\mathbb{R}, \mu)$, then $x \in \mathcal{D}(f(A))$ by (5.11). Therefore, the preceding formula shows that U is a well-defined isometric linear map of $\{f(A)x : f \in \mathcal{F}_x\}$ onto $L^2(\mathbb{R}, \mu)$. Letting $f(\lambda) \equiv 1$, we get $(Ux)(t) = 1$ on $\mathbb{R}$.

Let $g \in \mathcal{F}_x$. By (5.11), applied to the function $f(\lambda) = \lambda$ on $\mathbb{R}$, we conclude that $g(A)x \in \mathcal{D}(A)$ if and only if $Ug(A)x \in \mathcal{D}(A_t)$. From the functional calculus we obtain $U A g(A)x = A_t U g(A)x$. This proves that $A = U^{-1}A_t U$.   □

Corollary 5.19 *If the self-adjoint operator A has a simple spectrum, each eigenvalue of A has multiplicity one.*

Proof Combine Proposition 5.18 with Statement 1 of Example 5.4.   □

Proposition 5.20 *A self-adjoint operator A on $\mathcal{H}$ has a simple spectrum if and only if there exists a vector $x \in \bigcap_{n=0}^{\infty} \mathcal{D}(A^n)$ such that $\mathrm{Lin}\{A^n x : n \in \mathbb{N}_0\}$ is dense in $\mathcal{H}$.*

Proof Suppose that the condition is fulfilled. By the definition of spectral integrals, $A^n x$ is a limit of linear combinations of vectors $E_A(M)x$. Therefore, the density of $\mathrm{Lin}\{A^n x : n \in \mathbb{N}_0\}$ in $\mathcal{H}$ implies that the span of vectors $E_A(M)x$, $M \in \mathfrak{B}(\mathbb{R})$, is dense in $\mathcal{H}$. This means that x is a generating vector for A.

The converse direction will be proved at the end of Sect. 7.4. $\square$

Finally, we state without proof another criterion (see, e.g., [RS1, Theorem VII.5]): *A self-adjoint operator A has a simple spectrum if and only if its commutant*

$$\{A\}' := \big\{ B \in \mathbf{B}(\mathcal{H}) : BA \subseteq AB \big\}$$

is commutative, or equivalently, if each operator $B \in \{A\}'$ is a (bounded) function $f(A)$ of A.

5.5 Spectral Theorem for Finitely Many Strongly Commuting Normal Operators

In this section we develop basic concepts and results on the multidimensional spectral theory for strongly commuting unbounded normal operators.

5.5.1 The Spectral Theorem for n-Tuples

Definition 5.2 We say that two unbounded normal operators S and T acting on the same Hilbert space *strongly commute* if their bounded transforms Z_T and Z_S (see (5.6)) commute.

This definition is justified by Proposition 5.27 below which collects various equivalent conditions. Among others this proposition contains the simple fact that two *bounded* normals T and S strongly commute if and only if they commute.

The following theorem is the most general spectral theorem in this book.

Theorem 5.21 *Let $T = \{T_1, \dots, T_n\}$, $n \in \mathbb{N}$, be an n-tuple of unbounded normal operators acting on the same Hilbert space $\mathcal{H}$ such that T_k and T_l strongly commute for all $k, l = 1, \dots, n$, $k \neq l$. Then there exists a unique spectral measure $E = E_T$ on the Borel σ-algebra $\mathfrak{B}(\mathbb{C}^n)$ on the Hilbert space $\mathcal{H}$ such that*

$$T_k = \int_{\mathbb{C}^n} t_k \, dE_T(t_1, \dots, t_n), \quad k = 1, \dots, n. \tag{5.26}$$

In the proof of Theorem 5.21 we use the following lemma.

Lemma 5.22 *Let $E = E_1 \times \cdots \times E_m$ be the product measure (by Theorem 4.10) of spectral measures $E_1, \dots, E_m$ on $\mathfrak{B}(\mathbb{R})$. If f is a Borel function on $\mathbb{R}$, then*

$$\int_{\mathbb{R}} f(\lambda_k) \, dE_k(\lambda_k) = \int_{\mathbb{R}^m} f(\lambda_k) \, dE(\lambda_1, \dots, \lambda_m). \tag{5.27}$$

Proof Formula (5.27) holds for any characteristic function χ_M, $M \in \mathfrak{B}(\mathbb{R})$, since

$$\int_{\mathbb{R}} \chi_M(\lambda_k) \, dE_k(\lambda_k) = E_k(M) = E(\mathbb{R} \times \cdots \times M \times \cdots \times \mathbb{R})$$

$$= \int_{\mathbb{R}^m} \chi_M(\lambda_k) \, dE(\lambda_1, \ldots, \lambda_m)$$

by (4.21) and Theorem 5.9, 8.). Hence, formula (5.27) is valid for simple functions by linearity and for arbitrary Borel functions by passing to limits. $\qquad\square$

Proof of Theorem 5.21 By Lemma 5.8, (i) and (iii), the bounded transform $Z_k := Z_{T_k}$ of the normal operator T_k is a bounded normal operator. The operators $Z_1, \ldots, Z_n$ pairwise commute, because $T_1, \ldots, T_n$ pairwise strongly commute by assumption. We first prove the existence of a spectral measure for $Z_1, \ldots, Z_n$, and from this we then derive a spectral measure for $T_1, \ldots, T_n$.

To be precise, we first show that if $Z_1, \ldots, Z_n$ are pairwise commuting *bounded* normal operators on $\mathcal{H}$, there exists a spectral measure F on $\mathfrak{B}(\mathbb{C}^n)$ such that

$$Z_k = \int_{\mathbb{C}^n} z_k \, dF(z_1, \ldots, z_n), \quad k = 1, \ldots, n. \qquad (5.28)$$

Since Z_k is normal, there are commuting bounded self-adjoint operators A_{k1} and A_{k2} such that $Z_k = A_{k1} + iA_{k2}$. By Theorem 5.1, there exists a spectral measure F_{kr} on $\mathfrak{B}(\mathbb{R})$ such that $A_{kr} = \int_{\mathbb{R}} \lambda \, dF_{kr}(\lambda)$, $k = 1, \ldots, n$, $r = 1, 2$.

Let $k, l \in \{1, \ldots, n\}$ and $r, s \in \{1, 2\}$. Since $Z_k Z_l = Z_l Z_k$, it follows from Fuglede's theorem (Proposition 3.28) that Z_k and Z_l commute with Z_l^* and Z_k^*, respectively. This implies that $A_{kr} A_{ls} = A_{ls} A_{kr}$. Therefore, the spectral measures F_{kr} and F_{ls} commute by Corollary 5.6. Hence, by Theorem 4.10, the product measure $F = F_{11} \times F_{12} \times \cdots \times F_{n1} \times F_{n2}$ on $\mathfrak{B}(\mathbb{C}^n)$ exists, where we identify $\mathbb{R}^{2n}$ and $\mathbb{C}^n$ in the obvious way. By Lemma 5.22, we have

$$A_{kr} = \int_{\mathbb{R}} \lambda \, dF_{kr}(\lambda) = \int_{\mathbb{C}^n} \lambda_{kr} \, dF(\lambda_{11}, \lambda_{12}, \ldots, \lambda_{n1}, \lambda_{n2}),$$

which in turn implies Eq. (5.28).

Now we begin the proof of the *existence* of a spectral measure for the *unbounded* n-tuple $\{T_1, \ldots, T_n\}$. Consider the sets

$$D := \{(z_1, \ldots, z_n) \in \mathbb{C}^n : |z_k| < 1 \text{ for all } k = 1, \ldots, n\},$$

$$S_k := \{(z_1, \ldots, z_n) \in \overline{D} : |z_k| = 1\}, \quad S := \bigcup_{k=1}^{n} S_k.$$

Let $x \in \mathcal{H}$ and $M \in \mathfrak{B}(\mathbb{C}^n)$. Using (5.28) and Theorem 4.16, we obtain

$$\langle (I - Z_k^* Z_k) F(M)x, F(M)x \rangle = \int_M (1 - |z_k|^2) \, d\langle F(z_1, \ldots, z_n)x, x \rangle. \qquad (5.29)$$

We have $I - Z_k^* Z_k = (I + T_k^* T_k)^{-1} \geq 0$ by Lemma 5.8(i) and $\mathcal{N}(I - Z_k^* Z_k) = \{0\}$. Therefore, it follows from (5.29) that supp $F \subseteq \overline{D}$ and that $F(S_k) = 0$ for each $k = 1, \ldots, n$. Hence, $F(S) = 0$, and so $F(D) = F(\overline{D} \backslash S) = I$.

Set $\varphi_k(z_1, \ldots, z_n) := (1 - \bar{z}_k z_k)^{-1/2}$. Since $F(D) = I$, φ_k is an F-a.e. fi-
nite Borel function on $\mathbb{C}^n$. Repeating verbatim the proof of the equality $\mathbb{I}(\varphi) = Z(C^{1/2})^{-1}$ from the proof of Theorem 5.7, we get $\mathbb{I}(\varphi_k) = Z_k((I - Z_k^* Z_k)^{1/2})^{-1}$.
On the other hand, we have $Z_k = T_k C_{T_k}^{1/2}$ by (5.6), and hence $Z_k(C_{T_k}^{1/2})^{-1} \subseteq T_k$.
Since $C_{T_k}^{1/2} = (I - Z_k^* Z_k)^{1/2}$ by (5.7), we obtain $\mathbb{I}(\varphi_k) \subseteq T_k$. Since both operators
$\mathbb{I}(\varphi_k)$ and T_k are normal, it follows from Proposition 3.26(iv) that $\mathbb{I}(\varphi_k) = T_k$.

Let $\varphi : D \to \mathbb{C}^n$ denote the mapping given by $\varphi = (\varphi_1, \ldots, \varphi_n)$. Then $E(\cdot) := F(\varphi^{-1}(\cdot))$ defines a spectral measure on $\mathfrak{B}(\mathbb{C}^n)$. Applying Proposition 4.24 to the
mapping φ and the function $h(t) = t_k$, $t = (t_1, \ldots, t_n) \in \mathbb{C}^n$, we derive

$$\int_{\mathbb{C}^n} t_k \, dE(t) = \int_D h(\varphi(z)) \, dF(z) = \int_D \varphi_k(z) \, dF(z) = \mathbb{I}(\varphi_k) = T_k,$$

which proves formula (5.26).

To prove the *uniqueness* assertion, we argue again as in the proof of Theorem 5.7.
Let E' be another spectral measure satisfying (5.26). Then $F'(\cdot) := E'(\varphi(\cdot))$ defines a spectral measure on $\mathfrak{B}(D)$ such that $Z_k = \int_D z_k \, dF'(z)$, $k = 1, \ldots, n$.
On the other hand, since $F(D) = I$, we have $Z_k = \int_D z_k \, dF(z)$. We consider
the Cartesian decompositions $Z_k = A_{k1} + iA_{k2}$ and $z_k = \lambda_{k1} + i\lambda_{k2}$ and identify
$z = (z_1, \ldots, z_n) \in \mathbb{C}^n$ with $\lambda = (\lambda_{11}, \lambda_{12}, \ldots, \lambda_{n2}) \in \mathbb{R}^{2n}$ as in the first paragraph
of this proof. Using the properties of spectral integrals from Theorem 4.16, we de-
rive

$$\langle p(A_{11}, A_{12}, \ldots, A_{n2})x, x \rangle = \int_D p(\lambda) \, d\langle F(\lambda)x, x \rangle = \int_D p(\lambda) \, d\langle F'(\lambda)x, x \rangle$$

for any polynomial $p \in \mathbb{C}[\lambda_{11}, \lambda_{12}, \ldots, \lambda_{n1}, \lambda_{n2}]$ and $x \in \mathcal{H}$. Since the polynomials
are dense in $C(\overline{D})$, the measures $\langle F(\cdot)x, x \rangle$ and $\langle F'(\cdot)x, x \rangle$ coincide for $x \in \mathcal{H}$.
This implies that $F = F'$ on $\mathfrak{B}(D)$, and so $E = E'$ on $\mathfrak{B}(\mathbb{C}^n)$. $\square$

An important special case of Theorem 5.21 is when all operators T_j are *self-
adjoint*. Since the bounded transform Z_j of the self-adjoint operator T_j is self-
adjoint by Lemma 5.8(ii), the spectral measure E in the above proof is supported
by $\mathbb{R}^n$. Thus we obtain the following:

Theorem 5.23 *For each n-tuple $T = \{T_1, \ldots, T_n\}$ of pairwise strongly commuting
self-adjoint operators on a Hilbert space $\mathcal{H}$, there exists a unique spectral measure
$E = E_T$ on the σ-algebra $\mathfrak{B}(\mathbb{R}^n)$ such that*

$$T_k = \int_{\mathbb{R}^n} t_k \, dE_T(t_1, \ldots, t_n), \quad k = 1, \ldots, n.$$

Next, let us consider the case of a *single normal operator* T. Just as for a self-
adjoint operator (Proposition 5.10), it follows from Proposition 4.20(i) that the spectrum $\sigma(T)$ is equal to the support of the spectral measure E_T. Let K be a Borel set
in $\mathbb{C}$ containing $\sigma(T)$. Since $\text{supp } E_T = \sigma(T)$, Eq. (5.26) yields

$$T = \int_K t \, dE_T(t). \tag{5.30}$$

Formula (5.30) covers spectral theorems for several important classes of operators: If T is self-adjoint, we have $\sigma(T) \subseteq \mathbb{R}$, so we can set $K = \mathbb{R}$ in (5.30), and we rediscover the spectral theorem for a self-adjoint operator (Theorem 5.7). If T is bounded and normal, we can choose the disk in $\mathbb{C}$ centered at the origin with radius $\|T\|$ as K. Finally, if T is unitary, we can take K to be the unit circle.

5.5.2 Joint Spectrum for n-Tuples

In this subsection, $T = \{T_1, \ldots, T_n\}$ is an n-tuple of pairwise strongly commuting normal operators on a Hilbert space $\mathcal{H}$, and E_T denotes its spectral measure.

As noted above, the spectrum of a single normal operator is equal to the support of its spectral measure. This suggests the following:

Definition 5.3 The support of the spectral measure E_T is called the *joint spectrum* of the n-tuple T and denoted by $\sigma(T) = \sigma(T_1, \ldots, T_n)$.

The next proposition describes the set $\sigma(T)$ in terms of the operators $T_1, \ldots, T_n$.

Proposition 5.24 *Let $s = (s_1, \ldots, s_n) \in \mathbb{C}^n$. Then:*

(i) *s belongs to $\sigma(T)$ if and only if there exists a sequence $(x_k)_{k \in \mathbb{N}}$ of unit vectors $x_k \in \mathcal{D}(T_j)$ such that $\lim_{k \to \infty}(T_j x_k - s_j x_k) = 0$ for $j = 1, \ldots, n$.*

(ii) *$\sigma(T) \subseteq \sigma(T_1) \times \cdots \times \sigma(T_n)$.*

(iii) *$\bigcap_{j=1}^{n} \mathcal{N}(T_j - s_j I) = E_T(\{s\})\mathcal{H}$. That is, there exists a nonzero vector $x \in \mathcal{H}$ such that $T_j x = s_j x$ for all $j = 1, \ldots, n$ if and only if $E_T(\{s\}) \neq 0$.*

In case (i) we say that s is a *joint approximate eigenvalue* of $T = \{T_1, \ldots, T_n\}$ and in case (iii) that s is a *joint eigenvalue* of $T = \{T_1, \ldots, T_n\}$ if $E_T(\{s\}) \neq 0$.

Proof Using properties of spectral integrals, we compute for $M \in \mathfrak{B}(\mathbb{C}^n)$,

$$\sum_{j=1}^{n} \|(T_j - s_j I)E_T(M)x\|^2 = \sum_{j=1}^{n} \int_M |t_j - s_j|^2 d\langle E_T(t)x, x\rangle$$

$$= \int_M \|t - s\|^2 d\langle E_T(t)x, x\rangle. \qquad (5.31)$$

(i): Let $M_k = \{t \in \mathbb{C}^n : \|t - s\| \leq 1/k\}$, $k \in \mathbb{N}$. If $s \in \sigma(T) = \operatorname{supp} E_T$, then $E_T(M_k) \neq 0$, and we can choose unit vectors $x_k = E_T(M_k)x_k$. Then we have $\|(T_j - s_j I)x_k\| \leq 1/k$ by (5.31), so that $\lim_k (T_j x_k - s_j x_k) = 0$.

Conversely, if $s \notin \sigma(T)$, then $E_T(M_k) = 0$ for some $k \in \mathbb{N}$. By (5.31), applied with $M = \mathbb{C}^n$, we therefore obtain

$$\sum_{j=1}^{n} \|(T_j - s_j I)x\|^2 = \int_{\mathbb{C}^n \setminus M_k} \|t - s\|^2 d\langle E_T(t)x, x\rangle \geq k^{-2}\|x\|^2.$$

Hence, the condition in (i) cannot hold.

(ii): If s satisfies the condition in (i) for the n-tuple $T = \{T_1, \ldots, T_n\}$, then s_j obviously does for the single operator T_j. Hence, $\sigma(T) \subseteq \sigma(T_1) \times \cdots \times \sigma(T_n)$.

(iii): We apply formula (5.31) with $M = \mathbb{C}^n$. Then $x \in \bigcap_{j=1}^n \mathcal{N}(T_j - s_j I)$ if and only if the left-hand side and so the right-hand side of (5.31) vanishes. The latter holds if and only if $\operatorname{supp}\langle E_T(\cdot)x, x\rangle = \{s\}$, or equivalently, $x \in E_T(\{s\})\mathcal{H}$. □

As for a single self-adjoint operator (Theorem 5.9), the spectral theorem allows us to develop a *functional calculus* for the n-tuple T by assigning the normal operator

$$f(T) \equiv f(T_1, \ldots, T_n) = \mathbb{I}(f) = \int_{\mathbb{C}^n} f(t)\, dE_T(t) \tag{5.32}$$

to each E_T-a.e. finite Borel function $f: \mathbb{C}^n \to \mathbb{C} \cup \{\infty\}$.

Since $\sigma(T) = \operatorname{supp} E_T$, it clearly suffices to integrate over $\sigma(T)$ in (5.32), and $f(T)$ is well defined for any E_T-a.e. finite Borel function $f: \sigma(T) \to \mathbb{C} \cup \{\infty\}$. The basic rules of this calculus are easily obtained by rewriting the corresponding properties of spectral integrals $\mathbb{I}(f)$. We do not restate these rules.

The next result follows essentially from Proposition 4.20.

Proposition 5.25 *If $f : \sigma(T) \to \mathbb{C}$ is a continuous function, then $\sigma(f(T))$ is the closure of the set $f(\sigma(T))$, that is,*

$$\sigma\big(f(T)\big) = \overline{f\big(\sigma(T)\big)}. \tag{5.33}$$

Proof Let λ_0 be in the closure of $f(\sigma(T))$. Let $\varepsilon > 0$. Then there exists $t_0 \in \sigma(T)$ such that $|\lambda_0 - f(t_0)| < \varepsilon/2$. By the continuity of f there is a $\delta > 0$ such that

$$\mathcal{U}_\delta := \{t : |t - t_0| < \delta\} \subseteq \{t : |f(t) - f(t_0)| < \varepsilon/2\} \subseteq \mathcal{V}_\varepsilon := \{t : |f(t) - \lambda_0| < \varepsilon\}.$$

Since $t_0 \in \sigma(T) = \operatorname{supp} E_T$, we have $E_T(\mathcal{U}_\delta) \neq 0$, and so $E_T(\mathcal{V}_\varepsilon) \neq 0$. Hence, $\lambda_0 \in \sigma(f(T))$ by Proposition 4.20(i).

Conversely, if λ_0 is not in the closure of $f(\sigma(T))$, then $\{t : |f(t) - \lambda| < \varepsilon\}$ is empty for small $\varepsilon > 0$, so $\lambda_0 \notin \sigma(f(T))$ again by Proposition 4.20(i). □

Formula (5.33) is usually called the *spectral mapping theorem*.

Note that the range $f(\sigma(T))$ is not closed in general. For instance, if $f(t_1, t_2) = t_1^2 + (t_1 t_2 - 1)^2$, then 0 is not in $f(\mathbb{R}^2)$, but it is in the closure of $f(\mathbb{R}^2)$. However, if $\sigma(T)$ is *compact* or f has a *bounded support*, then $f(\sigma(T))$ is closed, and

$$\sigma\big(f(T)\big) = f\big(\sigma(T)\big).$$

5.6 Strong Commutativity of Unbounded Normal Operators

Our first proposition collects a number of equivalent conditions describing when a *bounded* operator commutes with (possibly unbounded) normal operators $T_1, \ldots, T_n$.

Proposition 5.26 *Let $T = \{T_1, \ldots, T_n\}$ and $E = E_T$ be as in Theorem 5.21. Recall that $Z_k = Z_{T_k}$ is the bounded transform of T_k, $k = 1, \ldots, n$, defined by (5.6). For each bounded operator $S \in \mathbf{B}(\mathcal{H})$, the following statements are equivalent:*

(i) $ST_k \subseteq T_k S$ *for* $k = 1, \ldots, n$.

(ii) $ST_k \subseteq T_k S$ *and* $ST_k^* \subseteq T_k^* S$ *for* $k = 1, \ldots, n$.

(iii) $SZ_{T_k} = Z_{T_k} S$ *for* $k = 1, \ldots, n$.

(iv) $SE_T(M) = E_T(M)S$ *for all* $M \in \mathfrak{B}(\mathbb{C}^n)$.

If all resolvent sets $\rho(T_1), \ldots, \rho(T_n)$ are nonempty, these are also equivalent to

(v) $S(T_k - s_k I)^{-1} = (T_k - s_k I)^{-1} S$ *for one (hence for all) $s_k \in \rho(T_k)$, $k = 1, \ldots, n$.*

Proof (i) $\to$ (iv): First we prove this in the case where all operators $T_1, \ldots, T_n$ are bounded. Then supp E is compact, since supp $E = \sigma(T) \subseteq \sigma(T_1) \times \cdots \times \sigma(T_n)$ by Proposition 5.24(ii). We proceed as in the proof of Theorem 5.21 and write $T_k = A_{k1} + iA_{k2}$ with A_{k1} and A_{k2} self-adjoint. Since the normal operators $T_1, \ldots, T_n$ pairwise commute, $A_{11}, A_{12}, \ldots, A_{n2}$ are pairwise commuting bounded self-adjoint operators. Using properties of spectral integrals, we compute

$$\langle p(A_{11}, A_{12}, \ldots, A_{n1}, A_{n2})x, y \rangle = \int p(\lambda) \, d\langle E(\lambda)x, y \rangle \qquad (5.34)$$

for any polynomial $p \in \mathbb{C}[\lambda_{11}, \ldots, \lambda_{n2}]$ and $x, y \in \mathcal{H}$. Now we argue as in the proof of the last assertion of Theorem 5.1. Since S commutes with T_k and so with T_k^* by Proposition 3.28, S commutes with $A_{11}, A_{12}, \ldots, A_{n2}$. Let R be a bounded rectangle in $\mathbb{R}^{2n}$ which contains the compact set supp E. Therefore, by (5.34),

$$\int_R p(\lambda) \, d\langle E(\lambda)Sx, y \rangle = \langle p(A_{11}, \ldots, A_{n2})Sx, y \rangle$$

$$= \langle p(A_{11}, \ldots, A_{n2})x, S^*y \rangle$$

$$= \int_R p(\lambda) \, d\langle E(\lambda)x, S^*y \rangle \qquad (5.35)$$

for each polynomial p. Since the polynomials are dense in $C(R)$, (5.35) implies that the complex measures $\langle E(\cdot)Sx, y \rangle$ and $\langle E(\cdot)x, S^*y \rangle$ coincide. The latter implies that $E(\cdot)S = SE(\cdot)$. This completes the proof of the assertion in the bounded case.

Now we turn to the *general case*. We fix $r > 0$ and denote by $C = C_r$ the set $\{t \in \mathbb{C}^n : |t_1| < r, \ldots, |t_n| < r\}$. Let $\varphi : \mathbb{C}^n \to \mathbb{C}^n$ be given by $\varphi(t) = t\chi_C(t)$. By Proposition 4.24, $F(\cdot) := E(\varphi^{-1}(\cdot))$ is a spectral measure F on $\mathfrak{B}(\mathbb{C}^n)$ such that

$$\int_{\mathbb{C}^n} t_k \, dF(t) = \int_{\mathbb{C}^n} t_k \chi_C(t) \, dE(t) \equiv \mathbb{I}(t_k \chi_C), \quad k = 1, \ldots, n.$$

The spectral integrals $\mathbb{I}(t_1 \chi_C), \ldots, \mathbb{I}(t_n \chi_C)$ are pairwise commuting bounded normals, and F is their spectral measure by the uniqueness assertion of Theorem 5.21.

Note that $\mathbb{I}(t_k) = T_k$ and $\mathbb{I}(\chi_C) = E(C)$. Using standard properties of spectral integrals (especially formula (4.28)) we obtain

$$E(C)\mathbb{I}(t_k\chi_C) = \mathbb{I}(t_k\chi_C) = T_k E(C),$$

$$E(C)T_k \subseteq T_k E(C) = E(C)\mathbb{I}(t_k\chi_C) = \mathbb{I}(t_k\chi_C)E(C).$$

Inserting these two equalities and the assumption $ST_k \subseteq T_k S$, we derive

$$E(C)SE(C)\mathbb{I}(t_k\chi_C) = E(C)ST_k E(C) \subseteq E(C)T_k SE(C) \subseteq \mathbb{I}(t_k\chi_C)E(C)SE(C).$$

Hence, $E(C)SE(C)\mathbb{I}(t_k\chi_C) = \mathbb{I}(t_k\chi_C)E(C)SE(C)$, since $E(C)SE(C)\mathbb{I}(t_k\chi_C)$ is everywhere defined on $\mathcal{H}$. Therefore, by the corresponding result in the bounded case proved above, $E(C)SE(C)$ commutes with the spectral measure F of the operators $\mathbb{I}(t_1\chi_C), \dots, \mathbb{I}(t_n\chi_C)$.

Let $M \in \mathfrak{B}(\mathbb{C}^n)$ be bounded. We choose $r > 0$ such that $M \subseteq C_r$. Then we have $E(C_r)F(M) = F(M)E(C_r) = E(M)$ by the definition of F, and hence

$$E(C_r)SE(M) = E(C_r)SE(C_r)F(M) = F(M)E(C_r)SE(C_r) = E(M)SE(C_r).$$

Since $I = \text{s-lim}_{r\to\infty} E(C_r)$, we conclude that $SE(M) = E(M)S$. Because each $M \in \mathfrak{B}(\mathbb{C}^n)$ is a union of an increasing sequence of bounded sets of $\mathfrak{B}(\mathbb{C}^n)$, we obtain $SE(M) = E(M)S$. This completes the proof of (i) $\to$ (iv).

(iv) $\to$ (ii) follows from Proposition 4.23, since $T_k = \int t_k \, dE(t)$ and $T_k^* = \int \overline{t_k} \, dE(t)$.

(ii) $\to$ (i) is trivial.

(iii) $\leftrightarrow$ (iv): Recall that Z_k is normal by Lemma 5.8 and that Z_k and Z_l commute, since T_k and T_l strongly commute. From the proof of Theorem 5.21 we know that the set $\{E_T(M) : M \in \mathfrak{B}(\mathbb{C}^n)\}$ of spectral projections of the n-tuple T coincides with the set $\{E_Z(N) : N \in \mathfrak{B}(\mathbb{C}^n)\}$ of spectral projections of the n-tuple $Z = (Z_1, \dots, Z_n)$. Applying the equivalence of (i) and (iv) proved above to Z, S commutes with $Z_1, \dots, Z_n$ if and only if it does with E_Z and so with E_T.

Finally, assume that the sets $\rho(T_1), \dots, \rho(T_n)$ are not empty. Let $s_k \in \rho(T_k)$. The relation $ST_k \subseteq T_k S$ is equivalent to $S(T_k - s_k I) \subseteq (T_k - s_k I)S$ and therefore to $(T_k - s_k I)^{-1}S = S(T_k - s_k I)^{-1}$. Hence, (i) and (v) are equivalent. □

The next proposition collects various characterizations of the strong commutativity of two unbounded normal operators. The second condition is often taken as the definition of strong commutativity in the literature, see also Proposition 6.15 below.

Proposition 5.27 *Let T_1 and T_2 be normal operators on a Hilbert space $\mathcal{H}$, and let E_{T_1} and E_{T_2} be their spectral measures. Consider the following statements:*

(i) *T_1 and T_2 strongly commute, that is, $Z_{T_1}Z_{T_2} = Z_{T_2}Z_{T_1}$.*

(ii) *E_{T_1} and E_{T_2} commute, that is, $E_{T_1}(M)E_{T_2}(N) = E_{T_2}(N)E_{T_1}(M)$ for all $M, N \in \mathfrak{B}(\mathbb{C})$.*

(iii) *$(T_1 - s_1 I)^{-1}T_2 \subseteq T_2(T_1 - s_1 I)^{-1}$ for one (hence for all) $s_1 \in \rho(T_1)$.*

(iv) *$(T_1 - s_1 I)^{-1}(T_2 - s_2 I)^{-1} = (T_2 - s_2 I)^{-1}(T_1 - s_1 I)^{-1}$ for one (hence for all) $s_1 \in \rho(T_1)$ and for one (hence for all) $s_2 \in \rho(T_2)$.*

Then (i) $\leftrightarrow$ (ii). *Further, we have* (i) $\leftrightarrow$ (iii) *if* $\rho(T_1)$ *is not empty and* (i) $\leftrightarrow$ (iv) *if* $\rho(T_1)$ *and* $\rho(T_2)$ *are not empty. If* T_1 *is bounded, then* (i) *is equivalent to*

(v) $T_1 T_2 \subseteq T_2 T_1$.

Proof All assertions are easily derived from Proposition 5.26. As a sample, we verify the equivalence of (i) and (iv). Indeed, $Z_{T_1} Z_{T_2} = Z_{T_2} Z_{T_1}$ is equivalent to $Z_{T_1}(T_2 - s_2 I)^{-1} = (T_2 - s_2 I)^{-1} Z_{T_1}$ by Proposition 5.26, (iii) $\leftrightarrow$ (v), applied to T_2 and $S = Z_{T_1}$, and hence to $(T_1 - s_1 I)^{-1}(T_2 - s_2 I)^{-1} = (T_2 - s_2 I)^{-1}(T_1 - s_1 I)^{-1}$ again by Proposition 5.26, (iii) $\leftrightarrow$ (v), now applied to T_1 and $S = (T_2 - s_2 I)^{-1}$. $\square$

Corollary 5.28 *Suppose that* T_1 *and* T_2 *are strongly commuting normal operators on* $\mathcal{H}$. *Then there exists a dense linear subspace* $\mathcal{D}$ *of* $\mathcal{H}$ *such that*

(i) $T_k \mathcal{D} \subseteq \mathcal{D}$ *and* $T_k^* \mathcal{D} \subseteq \mathcal{D}$ *for* $k = 1, 2$.
(ii) $T_1 T_2 x = T_2 T_1 x$ *for all* $x \in \mathcal{D}$.
(iii) $\mathcal{D}$ *is a core for* T_k *and* T_k^*.

Proof Let E_T be the spectral measure of the strongly commuting pair $T = (T_1, T_2)$ and put $\mathcal{D} := \bigcup_{n=1}^{\infty} E_T(M_n)\mathcal{H}$, where $M_n = \{(t_1, t_2) \in \mathbb{C}^2 : |t_1| \leq n, |t_2| \leq n\}$. It is easy to verify that $\mathcal{D}$ has the desired properties. $\square$

Proposition 5.29 *Let* E *be a spectral measure on* $(\Omega, \mathfrak{A})$, *and* $f, g \in \mathcal{S}(\Omega, \mathfrak{A}, E)$.

(i) *The spectral measure* $E_{\mathbb{I}(f)}$ *of the normal operator* $\mathbb{I}(f)$ *is given by*

$$E_{\mathbb{I}(f)}(M) = E\big(f^{-1}(M \cap f(\Omega))\big) \quad \text{for } M \in \mathfrak{B}(\mathbb{C}). \tag{5.36}$$

(ii) $\mathbb{I}(f)$ *and* $\mathbb{I}(g)$ *are strongly commuting normal operators.*

Proof (i): Let F denote the spectral measure on $\mathfrak{B}(\mathbb{C})$ defined by (5.36). From Proposition 4.24, applied with $\Omega_0 = f(\Omega)\backslash\{\infty\}$, $\varphi(t) = f(t)$, $h(s) = s$, we obtain

$$\int_{\mathbb{C}} s \, dF(s) = \int_{\Omega_0} s \, dF(s) = \int_{\Omega} f(t) \, dE(t) = \mathbb{I}(f).$$

Therefore, from the uniqueness of the spectral measure (Theorem 5.21) of the normal operator $\mathbb{I}(f)$ it follows that F is the spectral measure $E_{\mathbb{I}(f)}$.

(ii): By (5.36) the spectral projections of $\mathbb{I}(f)$ and $\mathbb{I}(g)$ are contained in the set $\{E(N) : N \in \mathfrak{A}\}$. Since these projections pairwise commute, $\mathbb{I}(f)$ and $\mathbb{I}(g)$ strongly commute by Proposition 5.27(ii). $\square$

The next proposition shows that unbounded normal operators are in one-to-one correspondence to pairs of strongly commuting self-adjoint operators.

Proposition 5.30 *If* A *and* B *are strongly commuting self-adjoint operators on a Hilbert space* $\mathcal{H}$, *then* $N = A + iB$ *is a normal operator, and* $N^* = A - iB$.

Each normal operator is of this form, that is, if N is a normal operator, then

$$A := \overline{(N + N^*)/2} \quad and \quad B := \overline{(N - N^*)/2i} \tag{5.37}$$

are strongly commuting self-adjoint operators such that $N = A + iB$.

Proof Let E_T be the spectral measure of the pair $T = (A, B)$. Let f be the function on $\mathbb{R}^2$ defined by $f = f_1 + if_2$, where $f_j(\lambda_1, \lambda_2) = \lambda_j$. The spectral integral $N := \mathbb{I}(f)$ is normal. Since $A = \mathbb{I}(f_1)$, $B = \mathbb{I}(f_2)$, and $\mathcal{D}(\mathbb{I}(f_1)) \cap \mathcal{D}(\mathbb{I}(f_2)) = \mathcal{D}(\mathbb{I}(f))$ by (4.26), we have $N = A + iB$ and $N^* = A - iB$ by Theorem 4.16.

If N is a normal operator and E_N is its spectral measure, we proceed in reversed order. Letting $f(\lambda) = \lambda$, $f_1(\lambda) = \text{Re}\,\lambda$ and $f_1(\lambda) = \text{Im}\,\lambda$ for $\lambda \in \mathbb{C}$, we then obtain $N = \mathbb{I}(f)$. Then the self-adjoint operators $A := \mathbb{I}(f_1)$ and $B := \mathbb{I}(f_2)$ strongly commute (by Proposition 5.29(ii)) and satisfy (5.37). $\quad\square$

By Corollary 5.28, strongly commuting normal operators commute pointwise on a common core. But the converse of this statement is not true! Example 5.5 below shows that the *pointwise* commutativity on a common core does *not* imply the *strong* commutativity of self-adjoint operators. The existence of such examples was discovered by Nelson [Ne1]. Another example of this kind is sketched in Exercise 6.16. These examples show that commutativity for *unbounded* self-adjoint or normal operators is indeed a delicate matter.

Lemma 5.31 *Let X and Y be bounded self-adjoint operators on a Hilbert space $\mathcal{H}$ such that $\mathcal{N}(X) = \mathcal{N}(Y) = \{0\}$, and let Q be the projection onto the closure of $[X, Y]\mathcal{H}$. Set $\mathcal{D} = XY(I - Q)\mathcal{H}$. Then $A := X^{-1}$ and $B := Y^{-1}$ are self-adjoint operators such that:*

(i) $\mathcal{D} \subseteq \mathcal{D}(AB) \cap \mathcal{D}(BA)$ and $ABx = BAx$ for $x \in \mathcal{D}$.
(ii) If $Q\mathcal{H} \cap X\mathcal{H} = Q\mathcal{H} \cap Y\mathcal{H} = \{0\}$, then $\mathcal{D}$ is a core for A and B.

Proof (i): Let $y \in \mathcal{H}$. For arbitrary $u \in \mathcal{H}$, we have by the definition of Q,

$$\langle (YX - XY)(I - Q)y, u \rangle = \langle (I - Q)y, [X, Y]u \rangle = 0.$$

Hence, $x := XY(I - Q)y = YX(I - Q)y \in \mathcal{D}(AB) \cap \mathcal{D}(BA)$, and we have $ABx = (I - Q)y = BAx$.

(ii): We first prove that $B\mathcal{D}$ is dense in $\mathcal{H}$. Suppose that $u \in \mathcal{H}$ and $u \perp B\mathcal{D}$. Since $\mathcal{D} = YX(I - Q)\mathcal{H}$ as shown in the preceding proof of (i), we have

$$0 = \langle u, BYX(I - Q)y \rangle = \langle u, X(I - Q)y \rangle = \langle Xu, (I - Q)y \rangle$$

for all $y \in \mathcal{H}$. Thus, $Xu \in Q\mathcal{H}$. By the assumption, $Xu = 0$, and hence $u = 0$.

Let $x \in \mathcal{D}(B)$. By the density of $B\mathcal{D}$, there is a sequence (x_n) of vectors $x_n \in \mathcal{D}$ such that $Bx_n \to Bx$. Since $Y = B^{-1}$ is bounded, $x_n = YBx_n \to YBx = x$. This shows that $\mathcal{D}$ is a core for B. The proof for A is similar. $\quad\square$

Example 5.5 (*Pointwise commuting self-adjoints that do not commute strongly*) Let S be the unilateral shift operator on $\mathcal{H} = l^2(\mathbb{N}_0)$, that is,

$$S(\varphi_0, \varphi_1, \varphi_2, \ldots) = (0, \varphi_0, \varphi_1, \ldots) \quad \text{for } (\varphi_n) \in l^2(\mathbb{N}_0).$$

Then $X := S + S^*$ and $Y := -\mathrm{i}(S - S^*)$ are bounded self-adjoint operators on $\mathcal{H}$, and $[X, Y]\mathcal{H} = \mathbb{C} \cdot e_0$, where $e_0 := (1, 0, , 0, \ldots)$. Then the assumptions of Lemma 5.31 are fulfilled, that is, $\mathcal{N}(X) = \mathcal{N}(Y) = \{0\}$ and $Q\mathcal{H} \cap X\mathcal{H} = Q\mathcal{H} \cap Y\mathcal{H} = \{0\}$.

As a sample, we verify that $Q\mathcal{H} \cap X\mathcal{H} = \{0\}$. Suppose that $X(\varphi_n) = \alpha e_0$ for some $(\varphi_n) \in \mathcal{H}$ and $\alpha \in \mathbb{C}$. It follows that $\varphi_{k+2} = -\varphi_k$ for $k \in \mathbb{N}_0$ and $\varphi_1 = \alpha$. Since $(\varphi_n) \in l^2(\mathbb{N}_0)$, the latter implies that $\varphi_k = \alpha = 0$ for all k.

Statement 1 $A := X^{-1}$ *and* $B := Y^{-1}$ *are self-adjoint operators on* $\mathcal{H}$*. They commute pointwise on the common core* $\mathcal{D} := XY(I - Q)\mathcal{H}$ *for* A *and* B*, but they do not commute strongly.*

Proof Since $\mathcal{N}(X) = \mathcal{N}(Y) = \{0\}$, A and B are well defined. They are self-adjoint by Corollary 1.9 and commute on the common core $\mathcal{D}$ by Lemma 5.31. Since their resolvents $R_0(A) = X$ and $R_0(B) = Y$ do not commute, A and B do not commute strongly by Proposition 5.27(iv). □

Now we define a linear operator T by $Tx = Ax + \mathrm{i}Bx$ for $x \in \mathcal{D}(T) := \mathcal{D}$.

Statement 2 T *is a formally normal operator on* $\mathcal{H}$ *which has no normal extension in a possible larger Hilbert space (that is, there is no normal operator* N *acting on a Hilbert space* $\mathcal{G}$ *which contains* $\mathcal{H}$ *as a subspace such that* $T \subseteq N$*).*

Proof Let $x \in \mathcal{D}$. From the equality $\langle Ty, x \rangle = \langle Ay + \mathrm{i}By, x \rangle = \langle y, Ax - \mathrm{i}Bx \rangle$ for $y \in \mathcal{D}$ it follows that $x \in \mathcal{D}(T^*)$ and $T^*x = Ax - \mathrm{i}Bx$. Using Lemma 5.31(i), we compute

$$\|A \pm \mathrm{i}Bx\|^2 = \langle Ax \pm \mathrm{i}Bx, Ax \pm \mathrm{i}Bx \rangle$$
$$= \|Ax\|^2 + \|Bx\|^2 \mp \mathrm{i}\langle Ax, Bx \rangle \pm \mathrm{i}\langle Bx, Ax \rangle$$
$$= \|Ax\|^2 + \|Bx\|^2 \mp \mathrm{i}\langle BAx, x \rangle \pm \mathrm{i}\langle ABx, x \rangle = \|Ax\|^2 + \|Bx\|^2.$$

Therefore, $\|Tx\| = \|T^*x\|$ for $x \in \mathcal{D}(T) \subseteq \mathcal{D}(T^*)$, that is, T is formally normal.

Assume that T has a normal extension N on a larger Hilbert space $\mathcal{G}$. We first show that $T^* \restriction \mathcal{D} \subseteq N^*$. Let $y \in \mathcal{D}(\subseteq \mathcal{D}(T) \subseteq \mathcal{D}(N) = \mathcal{D}(N^*))$. Let P denote the projection of $\mathcal{G}$ onto $\mathcal{H}$. Since $\langle x, T^*y \rangle = \langle Tx, y \rangle = \langle Nx, y \rangle = \langle x, PN^*y \rangle$ for all $x \in \mathcal{D}(T)$, we conclude that $T^*y = PN^*y$. Using that T and N are formally normal, we obtain

$$\|Ty\| = \|T^*y\| = \|PN^*y\| \leq \|N^*y\| = \|Ny\| = \|Ty\|.$$

Therefore, $\|PN^*y\| = \|N^*y\|$. This implies that $N^*y \in \mathcal{H}$. Hence, $T^*y = N^*y$, which completes the proof of the inclusion $T^* \restriction \mathcal{D} \subseteq N^*$.

Let $\tilde{A}$ and $\tilde{B}$ be the closures of $(N + N^*)/2$ and $(N - N^*)/2i$, respectively. From the inclusions $T \subseteq N$ and $T^* \lceil \mathcal{D} \subseteq N^*$ it follows that $A \lceil \mathcal{D} \subseteq \tilde{A}$. Since $\mathcal{D}$ is a core for A by Lemma 5.31(ii), $A \lceil \mathcal{D}$ is essentially self-adjoint. Therefore, by Proposition 1.17, $\mathcal{H}$ is reducing for $\tilde{A}$, and $A = \overline{A \lceil \mathcal{D}}$ is the part of $\tilde{A}$ on $\mathcal{H}$. That is, we have $\tilde{A} = A \oplus A_1$ on $\mathcal{G} = \mathcal{H} \oplus \mathcal{H}^{\perp}$, where A_1 is a certain operator on $\mathcal{H}^{\perp}$. Similarly, $\tilde{B} = B \oplus B_1$. By Proposition 5.30, $\tilde{A}$ and $\tilde{B}$ are strongly commuting self-adjoint operators. This implies that A and B strongly commute, which contradicts Statement 1. $\qquad \square$ $\circ$

5.7 Exercises

1. Let A be a bounded self-adjoint operator, and $p \in \mathbb{C}[t]$. Show without using the spectral theorem that $p(\sigma(A)) \subseteq \sigma(p(A))$.
   (Note that the opposite inclusion was proved in Lemma 5.2.)
2. Let T be a self-adjoint operator on $\mathcal{H}$ such that $0 \leq T \leq I$. Use the spectral theorem to prove that there is a sequence of pairwise commuting projections on $\mathcal{H}$ such that $T = \sum_{n=1}^{\infty} 2^{-n} P_n$.
   Hint: Take $P_1 = E((\frac{1}{2}, 1])$, $P_2 = E((\frac{1}{4}, \frac{1}{2}] \cup (\frac{3}{4}, 1])$, etc.
3. Let T be a self-adjoint operator on $\mathcal{H}$ with spectral measure E_T. Show that T is compact if and only if $\dim E_T(\mathcal{J})\mathcal{H} < \infty$ for each compact interval $\mathcal{J}$ which does not contain 0.
4. (*Spectral theorem for compact self-adjoint operators*)
   Let T be a compact self-adjoint operator on an infinite-dimensional separable Hilbert space $\mathcal{H}$. Show that there are an orthonormal basis $\{x_n : n \in \mathbb{N}\}$ of $\mathcal{H}$ and a real sequence $(\lambda_n)_{n \in \mathbb{N}}$ such that $\lim_{n \to \infty} \lambda_n = 0$ and

$$Tx = \sum_{n=1}^{\infty} \lambda_n \langle x, x_n \rangle x_n, \quad x \in \mathcal{H}.$$

5. Let A be a bounded self-adjoint operator, and let $f(z) = \sum_{n=0}^{\infty} a_n z^n$ be a power series with radius of convergence $r > \|A\|$. Show that the series $\sum_{n=0}^{\infty} a_n A^n$ converges in the operator norm to the operator $f(A)$ defined by (5.10).
6. Let A be a bounded self-adjoint operator.
   a. Show that for any $z \in \mathbb{C}$, the series $\sum_{n=0}^{\infty} \frac{1}{n!} z^n A^n$ converges in the operator norm to e^{zA}, where e^{zA} is defined by (5.10).
   b. Show that $\lim_{z \to z_0} \|e^{zA} - e^{z_0 A}\| = 0$ for $z_0 \in \mathbb{C}$.
7. Let A be a self-adjoint operator on $\mathcal{H}$. Suppose that $Ax = \lambda x$, where $\lambda \in \mathbb{R}$ and $x \in \mathcal{H}$. Prove that $f(A)x = f(\lambda)x$ for each function $f \in C(\mathbb{R})$.
8. Let A be a self-adjoint operator, and f and g be real-valued Borel functions on $\mathbb{R}$. Prove that $f(g(A)) = (f \circ g)(A)$.
   Hint: Use Proposition 4.24.
9. Let A be a positive self-adjoint operator such that $\mathcal{N}(A) = \{0\}$.
   a. Show that $(A^{-1})^{1/2} = (A^{1/2})^{-1} = A^{-1/2}$.
   b. Prove that $\lim_{\varepsilon \to +0} \frac{A^{\varepsilon} - 1}{\varepsilon} x = (\log A)x$ for $x \in \mathcal{D}(\log A)$.

10. Let A be a positive self-adjoint operator.
    a. Prove that $A^\alpha x = \lim_{\beta \to \alpha - 0} A^\beta x$ for any $\alpha > 0$ and $x \in \mathcal{D}(A^\alpha)$.
    b. Prove that $\alpha A + (1 - \alpha)I - A^\alpha \geq 0$ for any $\alpha \in (0, 1]$.
    c. Prove $A^\alpha - \alpha A - (1 - \alpha)I \geq 0$ for any $\alpha > 1$.
    d. Suppose that $\mathcal{N}(A) = \{0\}$. Prove that $A^\alpha - \alpha A - (1 - \alpha)I \geq 0$ for $\alpha < 0$.

11. Let A be a self-adjoint operator which is *not* bounded. Show that there exists a vector $x \in \mathcal{D}(A)$ such that $x \notin \mathcal{D}(A^2)$.

12. (*Characterizations of a spectral gap*)
    Let A be a self-adjoint operator with spectral measure E_A, and $a, b \in \mathbb{R}$, $a < b$. Show that the following three statements are equivalent:
    a. $(a, b) \subseteq \rho(A)$.
    b. $\|2Ax - (a + b)x\| \geq (b - a)\|x\|$ for all $x \in \mathcal{D}(A)$.
    c. $(A - bI)(A - aI) \geq 0$.

13. (*McCarthy inequalities* [Mc])
    Let A be a positive self-adjoint operator on $\mathcal{H}$, and $x \in \mathcal{D}(A)$.
    a. Prove that $\langle A^\alpha x, x \rangle \leq \langle Ax, x \rangle^\alpha \|x\|^{2 - 2\alpha}$ for $\alpha \in (0, 1]$.
    b. Prove that $\langle A^\alpha x, x \rangle \geq \langle Ax, x \rangle^\alpha \|x\|^{2 - 2\alpha}$ for $\alpha > 1$.
    c. Suppose that $\mathcal{N}(A) = \{0\}$ and $\alpha < 0$. Prove that

    $$\langle A^\alpha x, x \rangle \geq \langle Ax, x \rangle^\alpha \|x\|^{2 - 2\alpha}.$$

    d. Suppose that $\alpha \neq 1$ and $x \neq 0$. Assume that equality holds in one of the preceding inequalities. Show that x is an eigenvector of A.
    Hints: For (b), apply the Hölder inequality to $\langle Ax, x \rangle = \int_0^\infty \lambda \, d\langle E_A(\lambda)x, x \rangle$. For (a), apply (b) to A^α and α^{-1}.

14. Show that each self-adjoint operator with simple spectrum acts on a separable Hilbert space.

15. Let A be a self-adjoint operator on a separable Hilbert space $\mathcal{H}$.
    a. Show that there is an orthogonal direct sum decomposition $A = \bigoplus_{n=0}^{N} A_n$ on $\mathcal{H} = \bigoplus_{n=0}^{N} \mathcal{H}_n$, where $N \in \mathbb{N} \cup \{\infty\}$, such that each A_n is a self-adjoint operator with simple spectrum on the Hilbert space $\mathcal{H}_n$.
    Hint: Use Lemma 5.17.
    b. Show that there are a finite measure space (Λ, ν), a real-valued measurable function F on Λ, and a unitary operator U of $\mathcal{H}$ onto $L^2(\Lambda, \nu)$ such that $(U A U^{-1} f)(\lambda) = F(\lambda) f(\lambda)$ for $f \in L^2(\Lambda, \nu)$.
    Hint: Use (a) and Proposition 5.18.

16. Let Ω be an open subset of $\mathbb{R}^n$, and $\mathcal{H} = L^2(\Omega)$. Define $(T_k f)(t) = t_k f(t)$ for $f \in \mathcal{H}$ and $k = 1, \ldots, n$. Determine the joint spectrum $\sigma(T)$ of the n-tuple $T = \{T_1, \ldots, T_n\}$ of self-adjoint operators T_k and the spectra $\sigma(T_k)$.

17. Let T be a bounded normal operator. Use the spectral theorem to prove that for each $\varepsilon > 0$, there are numbers $\lambda_1, \ldots, \lambda_n \in \mathbb{C}$ and pairwise orthogonal projections $P_1, \ldots, P_n$ such that $P_1 + \cdots + P_n = I$ and $\|T - \sum_{k=1}^{n} \lambda_n P_n\| < \varepsilon$.

18. A linear operator T on $\mathcal{H}$ is called *diagonalizable* if there are an orthonormal basis $\{e_j : j \in I\}$ of $\mathcal{H}$ and a set $\{\lambda_j : j \in I\}$ of complex numbers such that

$$\mathcal{D}(T) = \left\{ x \in \mathcal{H} : \sum_j |\lambda_j|^2 |\langle x, e_j \rangle|^2 < \infty \right\} \quad \text{and} \quad Tx = \sum_j \lambda_j \langle x, e_j \rangle e_j$$

for $x \in \mathcal{D}(T)$. Suppose that T is diagonalizable.
a. Determine the adjoint operator T^* and show that T^* is diagonalizable.
b. Show that T is normal and determine the spectral measure of T.
c. When is T self-adjoint? When is $T \geq 0$?

19. Show that for each normal operator $T \in \mathbf{B}(\mathcal{H})$, there is a sequence $(T_k)_{k \in \mathbb{N}}$ of diagonalizable operators $T_k \in \mathbf{B}(\mathcal{H})$ such that $\lim_{k \to \infty} \|T - T_k\| = 0$.

20. Let T be a normal operator.
a. Prove that $p(T)^* = p(T^*)$ for any polynomial $p \in \mathbb{R}[t]$.
b. Formulate and prove the corresponding result for $p \in \mathbb{C}[t]$.

21. Let A be a normal operator.
a. Show that A is a projection if and only if $\sigma(A) \subseteq \{0, 1\}$.
b. Show that A is a self-adjoint unitary if and only if $\sigma(A) \subseteq \{-1, 1\}$.
c. Show that $e^{iA} = I$ if and only if A is self-adjoint and $\sigma(A) \subseteq 2\pi\mathbb{Z}$.
d. Let $n \in \mathbb{N}$. Show that $A^n = I$ if and only if $\lambda^n = 1$ for each $\lambda \in \sigma(A)$.

22. Let T be a self-adjoint operator, and let E_T be its spectral measure.
a. Determine the spectral measure and resolution of the identity of the operators T^2, $E_T(\mathcal{J})T$ for an interval $\mathcal{J}$, and $|T|$.
b. Determine the spectral measure of T^3 and $T^{1/2}$ if $T \geq 0$.
c. Determine the spectral measure of $(T - \lambda I)^{-1}$ for $\lambda \in \rho(T)$.

23. Let A and B be strongly commuting positive self-adjoint operators on $\mathcal{H}$. Prove that $\overline{AB}$ is a positive self-adjoint operator.

24. Let T and S be strongly commuting normal operators on $\mathcal{H}$. Suppose that there exists a dense subset $\mathcal{M} \subseteq \mathcal{D}(T) \cap \mathcal{D}(S)$ of $\mathcal{H}$ such that $Tx = Sx$ for $x \in \mathcal{M}$. Prove that $T = S$.
Hint: Show that $T E_T(C_r) E_S(C_s) x = S E_T(C_r) E_S(C_s) x$ for $x \in \mathcal{M}$, where $C_r := \{z \in \mathbb{C} : |z| < r\}$.

25. Prove that a self-adjoint operator A on $\mathcal{H}$ has a purely discrete spectrum if and only if $\mathcal{D}(A)$ does not contain an *infinite-dimensional* closed subspace of $\mathcal{H}$.

5.8 Notes to Part II

The spectral theorem for bounded self-adjoint operators was discovered by Hilbert [Hi, 4. Mitteilung, 1906], who expressed the result in terms of quadratic forms. The present formulation of this theorem including the functional calculus is due to Riesz [Ri1]. Details on the history of the spectral theorem can be found in [HT, RN], and [Sn]. The spectral theorem for unbounded self-adjoint operators was proved by von Neumann [vN1], Stone [St2], Riesz [Ri2], and others.

   The spectral theorem is developed in various versions in most books on Hilbert space operators, see, e.g., [AG, BS, BSU, DS, RN, RS1, Ru3]. Now there exist numerous proofs of the spectral theorem. The most prominent approach is probably the one based on Gelfand's theory of abelian C^*-algebras, see, e.g., [Cw], IX, § 2. A careful treatment of spectral measures can be found in the notes [Bn].

   Example 5.5 is taken from [Sch3]. The existence of such pairs of commuting self-adjoint operators and of a formally normal operator without normal extension was first shown by Nelson [Ne1] and Coddington [Cd1], respectively.

Part III
Special Topics

Chapter 6
One-Parameter Groups and Semigroups of Operators

One-parameter groups or semigroups of operators and their generators have wide ranges of applications in various parts of mathematics and physics. In this chapter we develop basic results of this theory. In Sect. 6.1 one-parameter unitary groups are investigated, and two fundamental theorems, Stone's theorem and Trotter's formula, are proved. In Sect. 6.2 semigroups are used to solve the Cauchy problems for some abstract differential equations on Hilbert space. In Sect. 6.3 the generators of semigroups of contractions on Banach spaces are studied, and the Hille–Yosida theorem is derived. In the short Sect. 6.4 the generators of contraction semigroups on Hilbert spaces are characterized as m-dissipative operators.

6.1 Groups of Unitaries

Let us begin by giving a precise definition of a one-parameter unitary group.

Definition 6.1 A *strongly continuous one-parameter unitary group*, briefly a *unitary group*, is a family $U = \{U(t) : t \in \mathbb{R}\}$ of unitaries $U(t)$ on a Hilbert space $\mathcal{H}$ such that

(i) $U(t)U(s) = U(t + s)$ for $t, s \in \mathbb{R}$,
(ii) $\lim_{h \to 0} U(t + h)x = U(t)x$ for $x \in \mathcal{H}$ and $t \in \mathbb{R}$.

Axiom (i) means that U is a group homomorphism of the additive group $\mathbb{R}$ into the group of unitary operators on $\mathcal{H}$. In particular, this implies that $U(0) = I$ and

$$U(-t) = U(t)^{-1} = U(t)^* \quad \text{for } t \in \mathbb{R}.$$

Axiom (ii) is the strong continuity of U. It clearly suffices to require (ii) for $t = 0$ and for x from a dense subset of $\mathcal{H}$ (see Exercise 1.a). Since the operators $U(t)$ are unitaries, it is even enough to assume that $\lim_{t \to 0} \langle U(t)x, x \rangle = \langle x, x \rangle$ for x from a dense set (see Exercise 1.b). If the Hilbert space $\mathcal{H}$ is separable, axiom

K. Schmüdgen, *Unbounded Self-adjoint Operators on Hilbert Space*,   117
Graduate Texts in Mathematics 265,
DOI 10.1007/978-94-007-4753-1_6, © Springer Science+Business Media Dordrecht 2012

(ii) can be further weakened by requiring only that for all $x, y \in \mathcal{H}$, the function $t \to \langle U(t)x, y \rangle$ on $\mathbb{R}$ is Lebesgue measurable (see, e.g., [RS1, Theorem VIII.9]).

Let A be a self-adjoint operator with spectral measure E_A. Recall that by the functional calculus of self-adjoint operators the operator e^{itA} is defined by

$$e^{itA} := \int_{\mathbb{R}} e^{it\lambda} dE(\lambda), \quad t \in \mathbb{R}. \tag{6.1}$$

Proposition 6.1 *Let A be a self-adjoint operator on a Hilbert space $\mathcal{H}$. Then $U = \{U(t) := e^{itA} : t \in \mathbb{R}\}$ is a strongly continuous one-parameter unitary group on $\mathcal{H}$. The operator A is uniquely determined by U, that is,*

$$\mathcal{D}(A) = \left\{ x \in \mathcal{H} : \frac{d}{dt}\bigg|_{t=0} U(t)x := \lim_{h \to 0} h^{-1}(U(h) - I)x \text{ exists} \right\}, \tag{6.2}$$

$$\mathcal{D}(A) = \left\{ x \in \mathcal{H} : \frac{d^+}{dt}\bigg|_{t=0} U(t)x := \lim_{h \to +0} h^{-1}(U(h) - I)x \text{ exists} \right\}, \tag{6.3}$$

$$iAx = \frac{d}{dt}\bigg|_{t=0} U(t)x = \frac{d^+}{dt}\bigg|_{t=0} U(t)x \quad \text{for } x \in \mathcal{D}(A). \tag{6.4}$$

Further, for $x \in \mathcal{D}(A)$ and $t \in \mathbb{R}$, we have $U(t)x \in \mathcal{D}(A)$ and

$$\frac{d}{dt} U(t)x = iAU(t)x = iU(t)Ax. \tag{6.5}$$

Proof From the functional calculus of A (Theorem 5.9) it follows at once that axiom (i) of Definition 6.1 holds and that $U(t)$ is unitary.

Let $x \in \mathcal{H}$. Put $f_h(\lambda) := e^{ih\lambda} - 1$. Clearly, $f_h(\lambda) \to 0$ as $h \to 0$, and $|f_h(\lambda)| \leq 2$ on $\mathbb{R}$. Therefore, by Lebesgue's dominated convergence theorem (Theorem B.1),

$$\|U(h)x - x\|^2 = \int_{\mathbb{R}} |f_h(\lambda)|^2 d\langle E_A(\lambda)x, x \rangle \to 0 \quad \text{as } h \to 0,$$

that is, $\lim_{h \to 0} U(h)x = x$. Hence, axiom (ii) of Definition 6.1 is fulfilled, so U is a unitary group.

We define the linear operator T by $Tx := -i \lim_{h \to 0} h^{-1}(U(h) - I)x$ with domain

$$\mathcal{D}(T) = \left\{ x \in \mathcal{H} : \lim_{h \to 0} h^{-1}(U(h) - I)x \text{ exists} \right\}.$$

The operator T is symmetric. Indeed, using that $U(h)^* = U(-h)$, we derive

$$\langle Tx, y \rangle = \lim_{h \to 0} \langle -ih^{-1}(U(h) - I)x, y \rangle = \lim_{h \to 0} \langle x, -i(-h)^{-1}(U(-h) - I)y \rangle$$
$$= \langle x, Ty \rangle.$$

Let $x \in \mathcal{D}(A)$. Set $g_h(\lambda) := h^{-1}(e^{ih\lambda} - 1) - i\lambda$ for $\lambda \in \mathbb{R}$. Then $g_h(\lambda) \to 0$ as $h \to 0$. Since $|g_h'(\lambda)| = |i(e^{ih\lambda} - 1)| \leq 2$, $|g_h(\lambda)|^2 = |g_h(\lambda) - g_h(0)|^2 \leq 4\lambda^2$ by the mean value theorem. Hence, since $\int 4\lambda^2 d\langle E_A(\lambda)x, x \rangle = \|2Ax\|^2 < \infty$, the dominated convergence theorem applies and yields

$$\|h^{-1}(U(h) - I)x - iAx\|^2 = \int_{\mathbb{R}} |g_h(\lambda)|^2 d\langle E_A(\lambda)x, x \rangle \to 0 \quad \text{as } h \to 0,$$

that is, $iAx = \lim_{h\to 0} h^{-1}(U(h) - I)x$. Therefore, $A \subseteq T$. Since A is self-adjoint and T is symmetric, $A = T$. This proves (6.2) and the first equality of (6.4).

Now suppose that $y := \frac{d^+}{dt}|_{t=0} U(t)x = \lim_{h\to +0} h^{-1}(U(h) - I)x$ exists. Then

$$\lim_{h\to +0} (-h)^{-1}(U(-h) - I)x = \lim_{h\to +0} U(-h) h^{-1}(U(h) - I)x = Iy = y.$$

Thus, $\lim_{h\to 0} h^{-1}(U(h) - I)x = y$. This means that $x \in \mathcal{D}(T)$ and $y = iTx$. Since $T = A$ as just shown, this proves (6.3) and the second equality of (6.4).

Finally, we prove (6.5). Suppose that $x \in \mathcal{D}(A)$ and $t \in \mathbb{R}$. From

$$\lim_{h\to 0} h^{-1}(U(h) - I)U(t)x = \lim_{h\to 0} h^{-1}(U(t+h) - U(t))x$$
$$= U(t) \lim_{h\to 0} h^{-1}(U(h) - I)x = U(t)iAx$$

and from (6.2) it follows that $U(t)x \in \mathcal{D}(A)$ and $iAU(t)x = \frac{d}{dt} U(t)x = U(t)iAx$. $\square$

In general, it is difficult to obtain "explicit" formulas for the unitary groups of self-adjoint operators (see, however, Examples 6.1 and 8.1).

The following theorem states that *each* unitary group is of the form described in Proposition 6.1.

Theorem 6.2 (*Stone's theorem*) *If U is a strongly continuous one-parameter unitary group on $\mathcal{H}$, then there is a unique self-adjoint operator A on $\mathcal{H}$ such that $U(t) = e^{itA}$ for $t \in \mathbb{R}$.*

The operator iA is called the *infinitesimal generator* of the unitary group U.

Proof The uniqueness assertion is already contained in Proposition 6.1. It remains to prove the existence.

For $x \in \mathcal{H}$ and $f \in C_0^\infty(\mathbb{R})$, we define a "smoothed" element $x_f \in \mathcal{H}$ by

$$x_f = \int_\mathbb{R} f(t)U(t)x\, dt. \tag{6.6}$$

The integral in (6.6) exists as an $\mathcal{H}$-valued Riemann integral, since $f(t)U(t)x$ is a continuous $\mathcal{H}$-valued function with compact support.

First, we show that the linear span $\mathcal{D}_G$ of vectors x_f is dense in $\mathcal{H}$. Take a "δ-sequence" $(f_n)_{n\in\mathbb{N}}$, that is, a sequence of functions $f_n \in C_0^\infty(\mathbb{R})$ such that $f_n(t) \geq 0$ on $\mathbb{R}$, $\int_\mathbb{R} f_n(t)\, dt = 1$ and supp $f_n \subseteq [-\frac{1}{n}, \frac{1}{n}]$. For $x \in \mathcal{H}$, we have

$$\|x_{f_n} - x\| = \left\| \int_\mathbb{R} f_n(t)(U(t) - I)x\, dt \right\| \leq \int_\mathbb{R} f_n(t) \| (U(t) - I)x \|\, dt$$
$$\leq \sup\left\{ \|(U(t) - I)x\| : t \in \left[-\frac{1}{n}, \frac{1}{n} \right] \right\} \to 0 \quad \text{as } n \to \infty$$

by the strong continuity of U. Hence, $x = \lim_{n\to\infty} x_{f_n}$. Thus, $\mathcal{D}_G$ is dense in $\mathcal{H}$.

We define an operator T with domain $\mathcal{D}(T)$ as in the proof of Proposition 6.1. By the same reasoning as given therein it follows that T is symmetric. For $x \in \mathcal{H}$, $h > 0$, and $f \in C_0^\infty(\mathbb{R})$, we compute

$$h^{-1}\big(U(h) - I\big)x_f = \int_{\mathbb{R}} h^{-1} f(t)\big(U(t+h) - U(t)\big)x \, dt$$

$$= \int_{\mathbb{R}} h^{-1}\big(f(s-h) - f(s)\big)U(s)x \, ds.$$

Since $f \in C_0^\infty(\mathbb{R})$ and hence $h^{-1}(f(s-h) - f(s)) \to -f'(s)$ uniformly on $\mathbb{R}$ as $h \to 0$, we conclude that $\lim_{h\to0} h^{-1}(U(h) - I)x_f = x_{-f'}$ in $\mathcal{H}$. Thus, $x_f \in \mathcal{D}(T)$ and $iTx_f = x_{-f'}$. Therefore, $\mathcal{D}_G \subseteq \mathcal{D}(T)$, so that T is densely defined.

Next, we prove that $\overline{T}$ is self-adjoint. Let $y \in \mathcal{N}(T^* - iI)$. From the definition of T it follows immediately that $U(t)\mathcal{D}(T) \subseteq \mathcal{D}(T)$ and $U(t)Tx = TU(t)x$ for $x \in \mathcal{D}(T)$ and $t \in \mathbb{R}$. Using the latter, we derive for $t_0 \in \mathbb{R}$,

$$\frac{d}{dt}\bigg|_{t=t_0} \langle U(t)x, y \rangle = \frac{d}{ds}\bigg|_{s=0} \langle U(s)U(t_0)x, y \rangle = \langle iTU(t_0)x, y \rangle$$

$$= \langle iU(t_0)x, T^*y \rangle = \langle U(t_0)x, y \rangle.$$

That is, the function $g(t) := \langle U(t)x, y \rangle$ satisfies the differential equation $g' = g$ on $\mathbb{R}$. Therefore, $g(t) = g(0)e^t$ for $t \in \mathbb{R}$. Since $|g(t)| \le \|x\|\|y\|$, we conclude that $g(0) = 0$ and hence $\langle x, y \rangle = 0$ for all $x \in \mathcal{D}(T)$. Since $\mathcal{D}(T)$ is dense in $\mathcal{H}$, $y = 0$. Similarly, $\mathcal{N}(T^* + iI) = \{0\}$. By Proposition 3.8, $\overline{T}$ is self-adjoint.

Put $V(t) := e^{it\overline{T}}$ for $t \in \mathbb{R}$. We prove that $U(t) = V(t)$. Let $x \in \mathcal{D}(T)$. First, we note that the definition of T implies that

$$\frac{d}{dt}\bigg|_{t=t_0} U(t)x = \frac{d}{ds}\bigg|_{s=0} U(s)U(t_0)x = iTU(t_0)x.$$

On the other hand, applying formula (6.5) to the group V, we obtain

$$\frac{d}{dt}\bigg|_{t=t_0} V(t)x = i\overline{T}V(t_0)x.$$

That is, setting $y(t) = (U(t) - V(t))x$, we have $\frac{d}{dt}y(t) = i\overline{T}y(t)$, and hence,

$$\frac{d}{dt}\|y(t)\|^2 = \frac{d}{dt}\langle y(t), y(t) \rangle = \langle i\overline{T}y(t), y(t) \rangle + \langle y(t), i\overline{T}y(t) \rangle = 0,$$

because $y(t) \in \mathcal{D}(\overline{T})$ and $\overline{T}$ is symmetric. Hence, $\|y(t)\|$ is constant on $\mathbb{R}$. Since $y(0) = (U(0) - V(0))x = 0$, we have $y(t) = 0$ on $\mathbb{R}$, that is, $U(t)x = V(t)x$ for $t \in \mathbb{R}$ and $x \in \mathcal{D}(T)$. Because $\mathcal{D}(T)$ is dense in $\mathcal{H}$, $U(t) = V(t)$ on $\mathcal{H}$ for all $t \in \mathbb{R}$. Setting $A := \overline{T}$, we complete the proof of Theorem 6.2. (By applying (6.2) and (6.4) to $U = V$ we could even infer that $T = \overline{T} \equiv A$.) $\qquad\square$

Stone's theorem was a cornerstone in the development of functional analysis. Among its important mathematical applications is the representation theory of Lie groups. A strongly continuous one-parameter unitary group is nothing but a unitary representation of the Lie group $\mathbb{R}$. The main part of the above proof generalizes to unitary representations of Lie groups to show that they give rise to $*$-representations of the corresponding Lie algebra, where the domain $\mathcal{D}_G$ becomes the Gårding domain of the representation. Detailed treatments can be found in [Sch1] and [Wa]. Another proof of Stone's theorem will be given in Sect. 6.4.

The next proposition shows that the unitary group can be useful for proving that some domain is a core for a self-adjoint operator.

Proposition 6.3 *Let A be a self-adjoint operator on a Hilbert space $\mathcal{H}$, and let $\mathcal{D}_1$ and $\mathcal{D}_2$ be linear subspaces of $\mathcal{H}$ such that $\mathcal{D}_2 \subseteq \mathcal{D}(A)$ and $\mathcal{D}_1$ is dense in $\mathcal{H}$. If there exists $c > 0$ such that $e^{itA}\mathcal{D}_1 \subseteq \mathcal{D}_2$ for all $t \in (-c, c)$, then $\mathcal{D}_2$ is a core for A, that is, $A \upharpoonright \mathcal{D}_2$ is essentially self-adjoint.*

Proof Suppose that $\tau \in \{1, -1\}$ and $y \in \mathcal{N}((A \upharpoonright \mathcal{D}_2)^* - \tau iI)$. Let $x \in \mathcal{D}_1$. By assumption, $e^{itA}x \in \mathcal{D}_2 \subseteq \mathcal{D}(A)$ for $|t| < c$. Using (6.4), we compute

$$\frac{d}{dt}\langle e^{itA}x, y\rangle = \langle iAe^{itA}x, y\rangle = \langle ie^{itA}x, \tau iy\rangle = \tau\langle e^{itA}x, y\rangle, \quad t \in (-c, c).$$

Thus, the function $g(t) := \langle e^{itA}x, y\rangle$ satisfies the differential equation $g' = \tau g$ on $(-c, c)$. Therefore, $g(t) = g(0)e^{\tau t}$, and so $\langle x, e^{-itA}y\rangle = \langle x, e^{\tau t}y\rangle$ on $(-c, c)$. Since $\mathcal{D}_1$ is dense in $\mathcal{H}$, we get $e^{-itA}y = e^{\tau t}y$ on $(-c, c)$. Hence, $t \to e^{-itA}y$ is differentiable at $t = 0$ and $\frac{d}{dt}|_{t=0}e^{-itA}y = \tau y = -iAy$ by (6.2) and (6.4). Since A is self-adjoint, $y = 0$. By Proposition 3.8, $A \upharpoonright \mathcal{D}_2$ is essentially self-adjoint. $\square$

Example 6.1 (*Translation group on $\mathbb{R}$*)

Statement *The translation group $(U(t)f)(x) = f(x + t)$, $t \in \mathbb{R}$, is a unitary group on $\mathcal{H} = L^2(\mathbb{R})$ with generator iA given by $iAf = f'$ for $f \in \mathcal{D}(A) = H^1(\mathbb{R})$. Moreover, $C_0^\infty(\mathbb{R})$ is a core for the self-adjoint operator A.*

Proof Clearly, $U(t)$ is unitary and axiom (i) of Definition 6.1 is satisfied.

Let $f \in C_0^\infty(\mathbb{R})$. For all $h \in \mathbb{R}$, $0 < |h| < 1$, the functions $U(h)f$ and $h^{-1}((U(h) - I)f)(x) = h^{-1}(f(x + h) - f(x))$ are supported by some bounded interval J. As $h \to 0$, they converge to f and f', respectively, uniformly on J and hence in $L^2(\mathbb{R})$. Therefore,

$$\lim_{h \to 0} U(h)f = f \quad \text{and} \quad \lim_{h \to 0} h^{-1}(U(h) - I)f = f'. \qquad (6.7)$$

Because $f \in C_0^\infty(\mathbb{R})$ is dense in $L^2(\mathbb{R})$, the first equality of (6.7) implies the strong continuity of U (see, e.g., Exercise 1), so U is indeed a unitary group.

By (6.4), the second equality of (6.7) yields $f \in \mathcal{D}(A)$ and $iAf = f'$. Thus, the operators A and $-i\frac{d}{dx}$ (on $H^1(\mathbb{R})$) coincide on $C_0^\infty(\mathbb{R})$. Since $C_0^\infty(\mathbb{R})$ is invariant under U, it is a core for A by Proposition 6.3. Hence, $A \subseteq -i\frac{d}{dx}$. Since both operators are self-adjoint, it follows that $A = -i\frac{d}{dx}$ and $\mathcal{D}(A) = H^1(\mathbb{R})$. $\square$ $\circ$

The next theorem generalizes a classical result that was obtained by S. Lie for matrices. The formulas (6.8) and (6.9) therein are called *Trotter product formulas*.

Formula (6.8) expresses the unitary group $e^{it(A+B)}$ as a strong limit of terms built from the unitary groups e^{isA} and e^{isB}.

Theorem 6.4 *Let A and B be self-adjoint operators on $\mathcal{H}$. Suppose that the sum $A + B$ is self-adjoint on the domain $\mathcal{D}(A + B) = \mathcal{D}(A) \cap \mathcal{D}(B)$. Then*

$$e^{it(A+B)}x = \lim_{n \to \infty} \left(e^{itA/n} e^{itB/n}\right)^n x \quad \text{for } x \in \mathcal{H}, \quad t \in \mathbb{R}. \tag{6.8}$$

If, in addition, both operators A and B are semibounded from below, then

$$e^{-t(A+B)}x = \lim_{n \to \infty} \left(e^{-tA/n} e^{-tB/n}\right)^n x \quad \text{for } x \in \mathcal{H}, \, t \geq 0. \tag{6.9}$$

Proof Throughout this proof we abbreviate $C := A + B$ and

$$x(s) := e^{isC}x, \quad T(t) := t^{-1}\left(e^{itA} e^{itB} - e^{itC}\right), \quad x_t(s) := T(t)x(s) \tag{6.10}$$

for $x \in \mathcal{D}(C)$ and $t, s \in \mathbb{R}$, $t \neq 0$.

Recall that the domain $\mathcal{D}(C)$ is a Hilbert space, denoted by $\mathcal{D}_C$, when it is equipped with the graph scalar product (1.3) for the operator C. Obviously, $T(t)$ maps $\mathcal{D}_C$ continuously into $\mathcal{H}$. Using (6.4), we obtain

$$T(t)x = t^{-1}\left(e^{itA} - I\right)x + e^{itA}t^{-1}\left(e^{itB} - I\right)x - t^{-1}\left(e^{itC} - I\right)x$$
$$\to iAx + iBx - iCx = 0 \quad \text{as } t \to 0. \tag{6.11}$$

This implies that $\sup_{t \in \mathbb{R}} \|T(t)x\| < \infty$ for $x \in \mathcal{D}(C)$. Therefore, by the principle of uniform boundedness the family of continuous linear mappings $T(t) : \mathcal{D}_C \to \mathcal{H}$ is uniformly bounded, that is, there is a constant $\gamma > 0$ such that

$$\|T(t)x\| \leq \gamma \|x\|_C \quad \text{for } x \in \mathcal{D}(C), \, t \in \mathbb{R}. \tag{6.12}$$

Fix $x \in \mathcal{D}(C)$. We prove that $\lim_{t \to 0} x_t(s) = 0$ *uniformly* on each bounded interval J. Let $\varepsilon > 0$. We choose subintervals J_k, $k = 1, \ldots, l$, of length $|J_k|$ such that $\gamma |J_k| \|Cx\|_C \leq \varepsilon$ and fix $s_k \in J_k$. Since $x(s) \in \mathcal{D}(C)$, (6.11) applies with x replaced by $x(s_k)$ and yields $\lim_{t \to 0} x_t(s_k) = 0$. Hence, there exists $\delta > 0$ such that $\|x_t(s_k)\| < \varepsilon$ for $|t| < \delta$ and all k. If $s \in J$, then $s \in J_k$ for some k. Let E be the spectral measure of the self-adjoint operator C. Inserting (6.10) and (6.12) and using the functional calculus for C (see, e.g., Theorem 4.16 and (4.33)), we derive

$$\|x_t(s) - x_t(s_k)\|^2 = \|T(t)\left(x(s) - x(s_k)\right)\|^2 \leq \gamma^2 \|x(s) - x(s_k)\|_C^2$$
$$= \gamma^2 \left(\|\left(e^{isC} - e^{is_kC}\right)x\|^2 + \|C\left(e^{isC} - e^{is_kC}\right)x\|^2\right)$$
$$= \gamma^2 \int_{\mathbb{R}} \left(1 + \lambda^2\right)\left|e^{is\lambda} - e^{is_k\lambda}\right|^2 d\langle E(\lambda)x, x\rangle$$
$$\leq \gamma^2 \int_{\mathbb{R}} \left(1 + \lambda^2\right)|s - s_k|^2 \lambda^2 d\langle E(\lambda)x, x\rangle$$
$$= \gamma^2 |s - s_k|^2 \|Cx\|_C^2 \leq \varepsilon^2$$

for $t \in \mathbb{R}$. Hence, if $|t| < \delta$, then $\|x_t(s)\| \le \|x_t(s) - x_t(s_k)\| + \|x_t(s_k)\| < 2\varepsilon$. Thus, we have shown that $\lim_{t \to 0} x_t(s) = 0$ uniformly on J.

Now we are ready to prove the Trotter formula (6.8). Let $t \in \mathbb{R}$. First suppose that $x \in \mathcal{D}(C)$. Put $J = [-|t|, |t|]$. Then we compute

$$\left(e^{itA/n} e^{itB/n}\right)^n x - e^{it(A+B)} x$$

$$= \left(\left(e^{itA/n} e^{itB/n}\right)^n - \left(e^{itC/n}\right)^n\right) x$$

$$= \sum_{k=0}^{n-1} \left(e^{itA/n} e^{itB/n}\right)^k \left[e^{itA/n} e^{itB/n} - e^{itC/n}\right]\left(e^{itC/n}\right)^{n-k-1} x$$

$$= \frac{t}{n} \sum_{k=0}^{n-1} \left(e^{itA/n} e^{itB/n}\right)^k x_{t/n}\left(t(n-k-1)/n\right).$$

Clearly, the norm of the last vector is less than or equal to

$$|t| \sup_{s \in J} \|x_{t/n}(s)\|.$$

By the result of the preceding paragraph this converges to zero as $n \to \infty$. Therefore, $(e^{itA/n} e^{itB/n})^n x \to e^{it(A+B)} x$ for $x \in \mathcal{D}(C)$. Since $\|(e^{itA/n} e^{itB/n})^n\| = 1$ and $\mathcal{D}(C)$ is dense in $\mathcal{H}$, it follows that $(e^{itA/n} e^{itB/n})^n x \to e^{it(A+B)} x$ for all $x \in \mathcal{H}$.

The second Trotter formula (6.9) is proved by almost the same reasoning. $\quad\square$

The preceding proof shows that for any fixed $x \in \mathcal{H}$, in (6.8) we have uniform convergence on bounded intervals.

With more refined technique it can be shown [RS1, Theorem VIII.31] that both formulas (6.8) and (6.9) remain valid (with $A + B$ replaced by $\overline{A + B}$) if the self-adjointness assumption of $A + B$ is weakened by the *essential self-adjointness* of $A + B$ on $\mathcal{D}(A) \cap \mathcal{D}(B)$.

6.2 Differential Equations on Hilbert Spaces

Throughout this section, A is a self-adjoint operator on a Hilbert space $\mathcal{H}$.

For an open interval $\mathcal{J} \subseteq \mathbb{R}$, let $C^1(\mathcal{J}, \mathcal{H})$ denote the set of all $\mathcal{H}$-valued functions $u : \mathcal{J} \to \mathcal{H}$ for which the derivative

$$u'(t) := \lim_{h \to 0} h^{-1}\left(u(t+h) - u(t)\right)$$

exists for all $t \in \mathcal{J}$ and is continuous on $\mathcal{J}$. Further, $C^2(\mathcal{J}, \mathcal{H})$ is the set of all $u \in C^1(\mathcal{J}, \mathcal{H})$ such that $u' \in C^1(\mathcal{J}, \mathcal{H})$.

Let us begin with the abstract *Schrödinger equation*

$$u'(t) = -iAu(t), \quad t \in \mathbb{R}. \tag{6.13}$$

Given $u_0 \in \mathcal{H}$, the *Cauchy problem* for Eq. (6.13) consists of finding a $u \in C^1(\mathbb{R}, \mathcal{H})$ such that $u(t) \in \mathcal{D}(A)$ and (6.13) holds for all $t \in \mathbb{R}$ and $u(0) = u_0$. (Note that these conditions imply that $u_0 \in \mathcal{D}(A)$.)

Proposition 6.5 *For any $u_0 \in \mathcal{D}(A)$, the Cauchy problem of the Schrödinger equation* (6.13) *has a unique solution, which is given by*

$$u(t) = e^{-itA}u_0, \quad t \in \mathbb{R}.$$

Proof That $u(t)$ satisfies Eq. (6.13) follows at once from Eq. (6.4). Since obviously $u(0) = u_0$, $u(t)$ solves the Cauchy problem.

To prove the uniqueness, let u_1 and u_2 be two solutions and set $v := u_1 - u_2$. Then we have $v'(t) = -iAv(t)$ and $v(0) = u_1(0) - u_2(0) = u_0 - u_0 = 0$. Hence,

$$\frac{d}{dt}\langle v(t), v(t)\rangle = \langle v'(t), v(t)\rangle + \langle v(t), v'(t)\rangle = \langle -iAv(t), v(t)\rangle + \langle v(t), -iAv(t)\rangle = 0.$$

Here the last equality holds because the operator A is symmetric. Therefore, the function $\|v(t)\|$ is constant on $\mathbb{R}$. Since $v(0) = 0$, it follows that $v(t) = 0$ and hence $u_1(t) = u_2(t)$ on $\mathbb{R}$.   □

Now we assume that $A \geq 0$ and let $u_0 \in \mathcal{H}$. We consider the *Cauchy problem* of the abstract *heat equation*

$$u'(t) = -Au(t), \quad t \in (0, \infty),$$
$$u(0) = u_0,$$

where $u \in C^1((0, \infty), \mathcal{H})$, $u(t) \in \mathcal{D}(A)$ for $t > 0$, and $u(0) := \lim_{t \to +0} u(t)$.

Proposition 6.6 *The Cauchy problem for the heat equation has a unique solution*

$$u(t) = e^{-tA}u_0, \quad t > 0.$$

Proof Fix $t > 0$. Put $f_h(\lambda) := h^{-1}(e^{-(t+h)\lambda} - e^{-t\lambda}) + \lambda e^{-t\lambda}$ for $0 < |h| < t/2$ and $g_s(\lambda) := e^{-s\lambda} - 1$ for $s > 0$. Then $f_h(\lambda) \to 0$ as $h \to 0$ and $g_s(\lambda) \to 0$ as $s \to 0$. Clearly, $|g_s(\lambda)| \leq 2$ on $[0, +\infty)$. From the mean value theorem it follows that $|e^{-x} - 1 + x| \leq e^{|x|}|x|$ for $x \in \mathbb{R}$. Using this fact, we obtain for $\lambda \in [0, +\infty)$,

$$\left|f_h(\lambda)\right| = |h|^{-1}e^{-t\lambda}\left|e^{-h\lambda} - 1 + h\lambda\right| \leq |h|^{-1}e^{-t\lambda}e^{|h|\lambda}|h|\lambda \leq e^{-t\lambda/2}\lambda \leq 2t^{-1}.$$

Therefore, Lebesgue's dominated convergence theorem (Theorem B.1) applies and yields

$$\left\|h^{-1}\big(u(t+h) - u(t)\big) + Au(t)\right\|^2 = \int_0^\infty \left|f_h(\lambda)\right|^2 d\langle E_A(\lambda)u_0, u_0\rangle \to 0 \quad \text{as } h \to 0,$$

$$\left\|u(s) - u_0\right\|^2 = \int_0^\infty \left|g_s(\lambda)\right|^2 d\langle E_A(\lambda)u_0, u_0\rangle \to 0 \quad \text{as } s \to 0.$$

(In fact, this proves that $\{e^{-tA} : t \geq 0\}$ is a contraction semigroup with generator $-A$, see, e.g., Proposition 6.14 below.) Thus, we have shown that $u'(t) = -Au(t)$ and $u(0) = u_0$, so u is a solution of the Cauchy problem. The uniqueness is proved similarly as in case of the Schrödinger equation (6.13).   □

Next, we suppose that $A \geq 0$ and $\mathcal{N}(A) = \{0\}$. Let $u_0 \in \mathcal{D}(A)$ and $u_1 \in \mathcal{H}$ be given. We investigate the *Cauchy problem* for the abstract *wave equation*

$$u''(t) = -Au(t), \quad t \in \mathbb{R},$$
$$u(0) = u_0, \quad u'(0) = u_1,$$

where $u \in C^2(\mathbb{R}, \mathcal{H})$ and $u(t) \in \mathcal{D}(A)$ for all $t \in \mathbb{R}$.

Proposition 6.7 *Suppose that $u_0 \in \mathcal{D}(A)$ and $u_1 \in \mathcal{D}(A^{1/2})$. Then*

$$u(t) := \left(\cos A^{1/2}t\right)u_0 + \left(A^{-1/2}\sin A^{1/2}t\right)u_1, \quad t \in \mathbb{R},$$

is the unique solution of the Cauchy problem for the wave equation, and

$$u'(t) = \left(\cos A^{1/2}t\right)u_1 - \left(A^{1/2}\sin A^{1/2}t\right)u_0, \quad t \in \mathbb{R}.$$

Proof First we note that $u(t)$ is well defined, since $\mathcal{N}(A^{1/2}) = \{0\}$ by the assumption $\mathcal{N}(A) = \{0\}$. Proceeding as in the proof of Proposition 6.6, one shows that $u(t)$ is in $C^2(\mathbb{R}, \mathcal{H})$, $u'(t)$ is given by the above formula, and that $u''(t) = -Au(t)$ for all $t \in \mathbb{R}$. This is where the assumptions $u_0 \in \mathcal{D}(A)$ and $u_1 \in \mathcal{D}(A^{1/2})$ are used. Since obviously $u(0) = u_0$ and $u'(0) = u_1$, $u(t)$ solves the Cauchy problem.

We prove the uniqueness assertion. Let w be another solution of the Cauchy problem and put $v := u - w$. Then $v(t) \in \mathcal{D}(A)$ for all $t > 0$. From the identity

$$\langle Av(t+h), v(t+h)\rangle - \langle Av(t), v(t)\rangle$$
$$= \langle v(t+h) - v(t), Av(t+h)\rangle + \langle Av(t), v(t+h) - v(t)\rangle$$

we derive $\frac{d}{dt}\langle Av, v\rangle = \langle v', Av\rangle + \langle Av, v'\rangle$ on $\mathbb{R}$. Since $v'' = -Av$, we compute

$$\frac{d}{dt}\left(\langle v', v'\rangle + \langle Av, v\rangle\right) = \langle v'', v'\rangle + \langle v', v''\rangle + \frac{d}{dt}\langle Av, v\rangle$$
$$= \langle -Av, v'\rangle + \langle v', -Av\rangle + \langle v', Av\rangle + \langle Av, v'\rangle = 0,$$

so $\|v'(t)\|^2 + \|A^{1/2}v(t)\|^2$ is constant on $\mathbb{R}$. Since $v(0) = v'(0) = 0$, this constant is zero. Therefore, $A^{1/2}v(t) = 0$. Hence, $v(t) = 0$ and $u(t) = w(t)$ on $\mathbb{R}$. $\square$

There is a nice "matrix trick" that allows one to reduce the wave equation for A to the Schrödinger equation for the self-adjoint operator B on the Hilbert space $\mathcal{H} \oplus \mathcal{H}$ given by the block matrix

$$B = \begin{pmatrix} 0 & -iA^{1/2} \\ iA^{1/2} & 0 \end{pmatrix}, \tag{6.14}$$

that is, $B(u, v) = (-iA^{1/2}v, iA^{1/2}u)$ for $u, v \in \mathcal{D}(A^{1/2})$.

Indeed, let $u_0 \in \mathcal{D}(A)$ and $u_1 \in \mathcal{D}(A^{1/2}) \cap \mathcal{D}(A^{-1/2})$ be given. Then $x_0 := (u_0, -A^{-1/2}u_1) \in \mathcal{D}(B)$, so by Proposition 6.5 there is a solution $x(t) = (u(t), v(t))$ of the Cauchy problem $x'(t) = -iBx(t)$, $x(0) = x_0$.

Since $x_0 \in \mathcal{D}(B^2)$, we have $x \in C^2(\mathbb{R}, \mathcal{H})$ and $x''(t) = -B^2x(t)$ (see Exercise 8). The latter implies that $u''(t) = -Au(t)$ on $\mathbb{R}$. From $x'(t) = -iBx(t)$ we obtain $u'(t) = -A^{1/2}v(t)$. Therefore, the equality $x(0) = x_0$ yields $u(0) = u_0$ and $u'(0) = -A^{1/2}v(0) = u_1$. That is, we have shown that $u(t)$ *is a solution of the Cauchy problem $u(0) = u_0$, $u'(0) = u_1$ for the wave equation $u''(t) = -Au(t)$.*

6.3 Semigroups of Contractions on Banach Spaces

In this section we develop some basics of generators of operator semigroups on Banach spaces. All notions and facts about closed operators we will use remain valid verbatim in this case. Throughout the symbol E will denote a *Banach space*.

Definition 6.2 A *strongly continuous one-parameter semigroup* on a Banach space E is a family $T = \{T(t) : t \geq 0\}$ of bounded linear operators $T(t)$ on E satisfying

(i) $T(0) = I$ and $T(t)T(s) = T(t + s)$ for all $t, s \in (0, \infty)$,
(ii) $\lim_{t \to +0} T(t)x = x$ and $\lim_{t \to t_0} T(t)x = T(t_0)x$ for $t_0 \in (0, \infty)$, $x \in E$.

If, in addition, $\|T(t)\| \leq 1$ for all $t > 0$, then T is called a *strongly continuous one-parameter semigroup of contractions* (or briefly, a *contraction semigroup*).

Definition 6.3 The *(infinitesimal) generator* of a strongly continuous one-parameter semigroup is the linear operator B on E defined by

$$Bx = \lim_{h \to +0} h^{-1}(T(h) - I)x,$$

$$\mathcal{D}(B) = \left\{ x \in E : \lim_{h \to +0} h^{-1}(T(h) - I)x \text{ exists} \right\}.$$

Remark Note that in [RS2] the operator $-B$ is called the generator of T.

Example 6.2 Suppose that A is a self-adjoint operator on a Hilbert space $\mathcal{H}$. Then, by Proposition 6.2, $\{e^{itA} : t \geq 0\}$ is a contraction semigroup. Equations (6.3) and (6.4) say that the operator iA is the generator of this contraction semigroup. ○

Example 6.3 (*Translations on intervals*) Let $\mathcal{J}$ be an interval, and $E = L^p(\mathcal{J})$, $p \in [1, +\infty)$. For $f \in E$ and $t \in [0, +\infty)$, we define the *right translation* $T(t)f$ by

$$(T(t)f)(x) = f(x - t) \quad \text{if } x - t \in \mathcal{J}, \qquad (T(t)f)(x) = 0 \quad \text{if } x - t \notin \mathcal{J}, \ x \in \mathcal{J}.$$

It is not difficult to verify that $\{T(t) : t \geq 0\}$ is a contraction semigroup on E. ○

More examples of contraction semigroups can be found in the Exercises.

The following two propositions contain some basic properties of generators.

Proposition 6.8 *The generator B is a densely defined closed operator on E which determines the strongly continuous one-parameter semigroup T uniquely. We have*

$$\frac{d}{dt}T(t)x = BT(t)x = T(t)Bx \quad \text{for } x \in \mathcal{D}(B) \text{ and } t > 0. \tag{6.15}$$

Let $a \geq 0$, $t > 0$, and $x \in E$. The main technical tool in the proof of Proposition 6.8 are "smoothed" elements defined by

$$x_{a,t} := \int_a^{a+t} T(s)x \, ds \quad \text{and} \quad x_t := x_{0,t}.$$

Since the map $[a, a+t] \ni s \to T(s)x \in E$ is continuous by Definition 6.2(ii), the integral exists as a limit of Riemann sums.

Lemma 6.9 $\lim_{t \to +0} t^{-1} x_{a,t} = T(a)x$. In particular, $\lim_{t \to +0} t^{-1} x_t = x$.

Proof By axiom (ii) in Definition 6.2, given $\varepsilon > 0$, there exists $\delta > 0$ such that $\|(T(s) - T(a))x\| \leq \varepsilon$ when $s \in [a, a+\delta]$. For $0 < t \leq \delta$, we then obtain

$$\|t^{-1} x_{a,t} - T(a)x\| = \left\| t^{-1} \int_a^{a+t} (T(s) - T(a))x \, ds \right\|$$

$$\leq t^{-1} \int_a^{a+t} \|(T(s) - T(a))x\| \, ds \leq \varepsilon. \qquad \square$$

Proof of Proposition 6.8 Suppose that $x \in E$. Using Lemma 6.9, we compute

$$\lim_{h \to +0} h^{-1}(T(h) - I)x_t = \lim_{h \to +0} h^{-1} \left(\int_h^{t+h} T(s)x \, ds - \int_0^t T(s)x \, ds \right)$$

$$= \lim_{h \to +0} h^{-1} \left(\int_t^{t+h} T(s)x \, ds - \int_0^h T(s)x \, ds \right)$$

$$= \lim_{h \to +0} h^{-1}(x_{t,h} - x_h) = T(t)x - x.$$

Therefore, from the definition of the operator B we conclude that $x_t \in \mathcal{D}(B)$ and

$$Bx_t \equiv B \int_0^t T(s)x \, ds = T(t)x - x, \quad x \in E. \tag{6.16}$$

Since $x_t \in \mathcal{D}(B)$ and $t^{-1} x_t \to x$ as $t \to +0$ by Lemma 6.9, $\mathcal{D}(B)$ is dense in E.

Now suppose that $x \in \mathcal{D}(B)$ and $t > 0$. Then we have

$$\lim_{h \to +0} h^{-1}(T(h) - I)T(t)x = \lim_{h \to +0} T(t)h^{-1}(T(h) - I)x = T(t)Bx,$$

$$\lim_{h \to +0} (-h)^{-1}(T(-h) - I)T(t)x = \lim_{h \to +0} T(t - h)h^{-1}(T(h) - I)x = T(t)Bx.$$

By Definition 6.3 it follows from these two equations that $T(t)x \in \mathcal{D}(B)$ and $\frac{d}{dt}T(t)x = T(t)Bx = BT(t)x$. This proves (6.15) and that $[0, \infty) \ni t \to T(t)x \in E$ is a C^1-map. Therefore,

$$T(t)x - x = \int_0^t \frac{d}{ds} T(s)x \, ds = \int_0^t T(s)Bx \, ds, \quad x \in \mathcal{D}(B). \tag{6.17}$$

Next, we show that B is closed. Let $x_n \in \mathcal{D}(B)$, $x_n \to x$, and $Bx_n \to y$ in E as $n \to \infty$. Using formula (6.17), we derive

$$T(h)x - x = \lim_{n \to \infty} (T(h)x_n - x_n) = \lim_{n \to \infty} \int_0^h T(s)Bx_n \, ds = \int_0^h T(s)y \, ds,$$

$$\lim_{h \to +0} h^{-1}(T(h) - I)x = \lim_{h \to +0} h^{-1} \int_0^h T(s)y \, ds = y.$$

(Considering the integrals as limits of Riemann sums, the change of order of limits and integrals is justified.) Thus, $x \in \mathcal{D}(B)$ and $Bx = y$. This proves that B is closed.

Finally, we prove that B determines T uniquely. Let S be another such semigroup which has the same generator B. Let $x \in \mathcal{D}(B)$ and $s > 0$. Set $z(t) = T(t)S(s - t)x$ for $t \in [0, s]$. Using (6.15) for T and S and the Leibniz rule, we differentiate

$$\frac{d}{dt}z(t) = T(t)(-B)S(s - t)x + T(t)BS(s - t)x = 0, \quad t \in (0, s).$$

Hence, $z(t)$ is constant on $[0, s]$, so that

$$T(s)x = T(s)S(0)x = z(s) = z(0) = T(0)S(s)x = S(s)x$$

for $x \in \mathcal{D}(B)$. Since $\mathcal{D}(B)$ is dense in E, it follows that $T(s) = S(s)$ on E. □

Proposition 6.10 *Let B be the generator of a contraction semigroup T on E. Each number $\lambda \in \mathbb{C}$, $\operatorname{Re}\lambda > 0$, is in $\rho(B)$ and $\|(B - \lambda I)^{-1}\| \leq (\operatorname{Re}\lambda)^{-1}$. Moreover,*

$$(B - \lambda I)^{-1}x = -\int_0^\infty e^{-\lambda s}T(s)x\,ds \quad \text{for } x \in E, \ \operatorname{Re}\lambda > 0. \tag{6.18}$$

Proof Fix a number $\lambda \in \mathbb{C}$, $\operatorname{Re}\lambda > 0$, and let S_λ denote the operator defined by the right-hand side of Eq. (6.18). Since $\|T(s)\| \leq 1$ and hence

$$\|S_\lambda x\| \leq \int_0^\infty e^{-(\operatorname{Re}\lambda)s}\|x\|\,ds = (\operatorname{Re}\lambda)^{-1}\|x\|,$$

S_λ is a bounded linear operator defined on E and $\|S_\lambda\| \leq (\operatorname{Re}\lambda)^{-1}$.

Clearly, $\{e^{-\lambda t}T(t) : t \geq 0\}$ is also a contraction semigroup, and its infinitesimal generator is $B - \lambda I$. Applying Eqs. (6.16) and (6.17) to this semigroup, we get

$$e^{-\lambda t}T(t)x - x = (B - \lambda I)\int_0^t e^{-\lambda s}T(s)x\,ds, \quad x \in E,$$

$$e^{-\lambda t}T(t)y - y = \int_0^t e^{-\lambda s}T(s)(B - \lambda I)y\,ds, \quad y \in \mathcal{D}(B).$$

Now we let $t \to \infty$. Then $e^{-\lambda t}T(t)x \to 0$ and $\int_0^t e^{-\lambda s}T(s)x\,ds \to S_\lambda(-x)$, since $\operatorname{Re}\lambda > 0$. Therefore, because $B - \lambda I$ is a closed operator, the first of the preceding equalities yields $-x = (B - \lambda I)S_\lambda(-x)$ for $x \in E$. Likewise, we obtain $-y = S_\lambda(B - \lambda I)(-y)$ for $y \in \mathcal{D}(B)$ by the second equality. These two relations imply that $\lambda \in \rho(B)$ and S_λ is the resolvent $(B - \lambda I)^{-1}$. □

Example 6.4 (*Bounded generators*)

Statement *Let B be a bounded operator on E. Then $T = \{T(t) = e^{tB} : t \geq 0\}$, where*

$$e^{tB} := \sum_{n=0}^\infty \frac{t^n B^n}{n!}, \tag{6.19}$$

is a strongly continuous one-parameter semigroup with generator B satisfying

$$\lim_{t \to +0}\|T(t) - I\| = 0. \tag{6.20}$$

Proof In the proof of Lemma 3.27 it was shown that the series (6.19) converges in the operator norm and that $e^{tB}e^{sB} = e^{(t+s)B}$ for $t, s \in \mathbb{R}$.

Estimating the power series, we easily derive

$$\|T(t) - I\| = \left\| \sum_{n \geq 1} t^n B^n / n! \right\| \leq t \|B\| e^{t\|B\|}, \quad t > 0, \tag{6.21}$$

$$\|h^{-1}(T(h) - I) - B\| = \left\| \sum_{n \geq 2} h^{n-1} B^n / n! \right\| \leq h \|B\|^2 e^{h\|B\|}, \quad h > 0. \tag{6.22}$$

Clearly, (6.21) implies (6.20), so axiom (ii) in Definition 6.2 is satisfied. From (6.22) we conclude that B is the generator of T. $\qquad\square$

An operator semigroup T is called *uniformly continuous* if condition (6.20) holds. As shown above, each bounded operator on E is generator of a one-parameter uniformly continuous operator semigroup. Conversely, the generator of a uniformly continuous semigroup is always a *bounded* operator (see Exercise 13). $\qquad\circ$

The following result is the famous *Hille–Yosida theorem*. It gives a complete characterization of generators of *contraction semigroups* on Banach spaces.

Theorem 6.11 *A linear operator B on a Banach space E is generator of a strongly continuous one-parameter contraction semigroup if and only if B is densely defined, closed, $(0, \infty) \subseteq \rho(B)$, and*

$$\|(B - \lambda I)^{-1}\| \leq \lambda^{-1} \quad \text{for } \lambda > 0. \tag{6.23}$$

Proof The necessity of these conditions was already shown by Propositions 6.8 and 6.10. It remains to prove the sufficiency.

Let $n \in \mathbb{N}$. Since $n \in \rho(B)$, we can define a bounded operator B_n on E by

$$B_n := nB(nI - B)^{-1} = n^2(nI - B)^{-1} - nI. \tag{6.24}$$

From the assumption $\|(nI - B)^{-1}\| \leq n^{-1}$ (by (6.23)) it follows that e^{tB_n} is a contraction for $t \geq 0$, because

$$\|e^{tB_n}\| = \|e^{-tn}e^{tn^2(nI-B)^{-1}}\| \leq e^{-tn}e^{tn^2\|(nI-B)^{-1}\|} \leq e^{-tn}e^{tn} = 1.$$

Thus, by Example 6.4, $T_n := \{T_n(t) = e^{tB_n} : t \geq 0\}$ is a contraction semigroup.

Now we prove that $\lim_{n\to\infty} B_n x = Bx$ for $x \in \mathcal{D}(B)$. For $y \in \mathcal{D}(B)$, we have

$$\|n(nI - B)^{-1}y - y\| = \|(nI - B)^{-1}(ny - (nI - B))y\|$$
$$= \|(nI - B)^{-1}By\| \leq n^{-1}\|By\|$$

by (6.23), and hence $\lim_{n\to\infty} n(nI - B)^{-1}y = y$. Since $\|n(nI - B)^{-1}\| \leq 1$ (again by (6.23)) and $\mathcal{D}(B)$ is dense in E, it follows that $\lim_{n\to\infty} n(nI - B)^{-1}y = y$ for all $y \in E$. Therefore, for $x \in \mathcal{D}(B)$,

$$\lim_{n\to\infty} B_n x = \lim_{n\to\infty} nB(nI - B)^{-1}x = \lim_{n\to\infty} n(nI - B)^{-1}Bx = Bx.$$

The main step of the proof is to show that the sequence $(T_n)_{n \in \mathbb{N}}$ converges strongly to a contraction semigroup T. Fix $x \in \mathcal{D}(B)$ and $t > 0$. For $k, n \in \mathbb{N}$ and $s \in (0, t]$, we set $z_{kn}(s) = T_k(s)T_n(t - s)x$. Since $B_k B_n = B_n B_k$ by (6.24) and hence $B_k T_n(t - s) = T_n(t - s)B_k$, using the product rule and Eq. (6.15), we compute

$$\frac{d}{ds} z_{kn}(s) = T_k(s)B_k T_n(t - s)x + T_k(s)T_n(t - s)(-B_n)x = z_{kn}(s)(B_k x - B_n x).$$

Therefore,

$$T_k(t)x - T_n(t)x = z_{kn}(t) - z_{kn}(0) = \int_0^t \frac{d}{ds} z_{kn}(s)\, ds = \int_0^t z_{kn}(s)(B_k x - B_n x)\, ds.$$

Because $T_k(s)$ and $T_n(t - s)$ are contractions and hence $\|z_{kn}(s)\| \leq 1$, this yields

$$\left\| T_k(t)x - T_n(t)x \right\| \leq t \| B_k x - B_n x \|. \tag{6.25}$$

Since $(B_n x)$ is a Cauchy sequence (because of $B_n x \to Bx$), so is $(T_n(t)x)$ by (6.25). Let $T(t)x := \lim_{n \to \infty} T_n(t)x$. Passing to the limit $n \to \infty$ in (6.25), we obtain

$$\left\| T_k(t)x - T(t)x \right\| \leq t \| B_k x - Bx \| \quad \text{for } x \in \mathcal{D}(B). \tag{6.26}$$

Since $\mathcal{D}(B)$ is dense in E by assumption and $\|T_n(t)\| \leq 1$, it follows that the sequence $(T_n(t))_{n \in \mathbb{N}}$ converges strongly to some contraction $T(t)$ on E. Obviously, since $T_k(0) = I$, this is also true for $t = 0$, and we have $T(0) = I$. Axiom (i) of Definition 6.2 holds for T_n, so it holds also for $T := \{T(t) : t \geq 0\}$.

Let $x \in \mathcal{D}(B)$. From (6.26) it follows that $T_k(t)x \to T(t)x$ as $k \to \infty$ uniformly on each bounded interval. Being a uniform limit of continuous maps on bounded intervals, the map $t \to T(t)x$ of $[0, \infty)$ into E is continuous. Since $\mathcal{D}(B)$ is dense, this implies that $t \to T(t)x$ is continuous for all $x \in E$ (see Exercise 1). This proves that T is a contraction semigroup.

Let A denote the generator of T. To complete the proof, it suffices to show that $A = B$. Let $x \in \mathcal{D}(B)$ and $h > 0$. By formula (6.17), applied to T_n, we obtain

$$T_n(h)x - x = \int_0^h T_n(s)Bx\, ds + \int_0^h T_n(s)(B_n x - Bx)\, ds.$$

Passing to the limit $n \to \infty$ by using that $B_n x \to Bx$ and $T_n(s)x \to T(s)x$ uniformly on $[0, h]$, we get $T(h)x - x = \int_0^h T(s)Bx\, ds$. Using Lemma 6.9 and Definition 6.3, applied to the generator A of T, the latter equation implies that $x \in \mathcal{D}(A)$ and $Ax = \lim_{h \to +0} h^{-1}(T(h) - I)x = Bx$. Thus, we have shown that $B \subseteq A$.

Fix $\lambda > 0$. Then $\lambda \in \rho(A)$ by Proposition 6.10 and $\lambda \in \rho(B)$ by assumption. Since $B - \lambda I \subseteq A - \lambda I$, Lemma 1.3 applies and yields $B - \lambda I = A - \lambda I$, so that $A = B$. $\qquad \square$

Remark The above proof is K. Yosida's approach [Yo]. Because of Example 6.4, one may think of an operator semigroup T with generator B as "$T(t) = e^{tB}$", so given B, one has to define the "exponential function" e^{tB}. E. Hille's proof [Hl] used the definition

$$e^{tB}x = \lim_{n \to \infty} (I - tB/n)^{-n}x, \quad t > 0. \tag{6.27}$$

Note that the operator $(I - tB/n)^{-n}$ exists, since $(0, \infty) \subseteq \rho(B)$. This approach is presented in [K2, Chap. IX]. The reader should be cautioned that $T(t) = e^{-tB}$ in [K2].

6.4 Semigroups of Contractions on Hilbert Spaces

In this section we assume again that the underlying space is a *Hilbert space* $\mathcal{H}$.

The following result is the Hilbert space version of the *Lumer–Phillips theorem*.

Theorem 6.12 *A linear operator B on a Hilbert space $\mathcal{H}$ is the generator of a strongly continuous one-parameter contraction semigroup if and only if B is m-dissipative, or equivalently, $-B$ is m-accretive.*

Proof Combine Theorem 6.11 and Proposition 3.22, applied to $-B$. $\qquad\square$

If B is the generator of a contraction semigroup T, then in particular B is dissipative, and hence $\mathrm{Re}\langle Bx, x \rangle \leq 0$ for $x \in \mathcal{D}(B)$ by Definition 3.6. The latter fact can be easily derived directly. Indeed, using that $T(h)$ is a contraction, we obtain

$$\mathrm{Re}\langle (T(h) - I)x, x \rangle = \mathrm{Re}\langle T(h)x, x \rangle - \|x\|^2 \leq \|T(h)x\|\|x\| - \|x\|^2 \leq 0.$$

Dividing by $h > 0$ and passing to the limit $h \to +0$, we get $\mathrm{Re}\langle Bx, x \rangle \leq 0$.

Proposition 6.13 *If $T = \{T(t)\}$ is a contraction semigroup on $\mathcal{H}$ with generator B, then $T^* := \{T(t)^*\}$ is also a contraction semigroup whose generator is B^*.*

Proof Obviously, $T(t)^*$ is also a contraction, and axiom (i) of Definition 6.2 holds.

To prove axiom (ii), it suffices to show that $\lim_{t \to +0} T^*(t)x = x$ for $x \in \mathcal{H}$. For $t > 0$, we have

$$\begin{aligned}
\|(T(t)^* - I)x\|^2 &= -\langle (T(t)^* - I)x, x \rangle + \|T(t)^*x\|^2 - \langle x, T(t)^*x \rangle \\
&\leq -\langle x, (T(t) - I)x \rangle + \|x\|^2 - \langle T(t)x, x \rangle \\
&= -\langle x, (T(t) - I)x \rangle - \langle (T(t) - I)x, x \rangle \to 0
\end{aligned}$$

as $t \to +0$ by the strong continuity of T. Hence, $T(t)^*x \to x$ as $t \to +0$. This proves that T^* is a contraction semigroup.

Let A denote the generator of T^*. For $x \in \mathcal{D}(B)$ and $y \in \mathcal{D}(A)$, we derive

$$\langle Bx, y \rangle = \lim_{h \to +0} \langle h^{-1}(T(h) - I)x, y \rangle = \lim_{h \to +0} \langle x, h^{-1}(T(h)^* - I)y \rangle = \langle x, Ay \rangle.$$

Therefore, $A \subseteq B^*$ and hence $A - \lambda I \subseteq B^* - \lambda I$ for $\lambda > 0$. But $\lambda \in \rho(A) \cap \rho(B)$ by Proposition 6.10, so $A - \lambda I$ and $B - \lambda I$ are surjective. The latter implies that $B^* - \lambda I = (B - \lambda I)^*$ is injective. Thus, Lemma 1.3 applies and yields $A = B^*$. $\qquad\square$

Proposition 6.14 *If A is a positive self-adjoint operator on a Hilbert space $\mathcal{H}$, then $T := \{T(t):=e^{-tA} : t \geq 0\}$ is a contraction semigroup of self-adjoint operators on $\mathcal{H}$ whose generator is $-A$. Each contraction semigroup of self-adjoint operators is of this form.*

Proof That T is a contraction semigroup and $\lim_{h \to +0} h^{-1}(T(h) - I)x = -Ax$ for $x \in \mathcal{D}(A)$ can be shown in a similar manner as in the case of unitary groups (see the proof of Proposition 6.6). Thus, if B is the generator of T, we have $-A \subseteq B$. Since $T = T^*$ and hence $B = B^*$ by Proposition 6.13, it follows that $-A = B$.

Conversely, let T be a contraction semigroup of self-adjoint operators. From Proposition 6.13 it follows that its generator B is self-adjoint. Because B is dissipative, $A := -B \geq 0$. Since the two contraction semigroups T and $\{e^{-tA} : t \geq 0\}$ have the same generator $B = -A$, they coincide by Proposition 6.8. □

Explicit formulas for the unitary group and the contraction semigroup associated with the operator $-A = \Delta$ on $\mathbb{R}^d$ can be found in Example 8.1 below.

The next result characterizes the strong commutativity of self-adjoint operators in terms of their unitary groups.

Proposition 6.15 *Two self-adjoint operators A_1 and A_2 acting on the Hilbert space $\mathcal{H}$ commute strongly if and only if $e^{itA_1}e^{isA_2} = e^{isA_2}e^{itA_1}$ for all $t, s \in \mathbb{R}$.*

Proof First, suppose that A_1 and A_2 strongly commute. Let E_A be the spectral measure of the pair $A = (A_1, A_2)$ (see Theorem 5.23). Then e^{itA_j} is the corresponding spectral integral $\mathbb{I}(f_{j,t})$ of the function $f_{j,t}(\lambda_1, \lambda_2) = e^{it\lambda_j}$ on $\mathbb{R}^2$, where $t \in \mathbb{R}$, $j = 1, 2$. Hence, e^{itA_1} and e^{isA_2} commute by Proposition 5.29(ii) or 4.12(i).

Now suppose that the unitary groups e^{itA_1} and e^{isA_2} commute. From formula (6.18) it follows then that the resolvents $R_{\lambda_1}(A_1)$ and $R_{\lambda_2}(A_2)$ commute for $\operatorname{Re}\lambda_j > 0$, so A_1 and A_2 strongly commute by Proposition 5.27. □

We close this section by using Propositions 6.8 and 6.13 (but not the Hille–Yosida Theorem 6.11!) to give a very short proof of Stone's theorem.

Second proof of Stone's Theorem 6.2 Suppose that U is a unitary group on $\mathcal{H}$. Let B be the generator of the contraction semigroup $U_+ = \{U(t) : t \geq 0\}$. Obviously, $-B$ is then the generator of $U_- = \{U(-t) : t \geq 0\}$. Since $(U_+)^* = U_-$, we have $B^* = -B$ by Proposition 6.13. Hence, $A := -iB$ is self-adjoint.

It remains to check that $U(t)$ is equal to the operator e^{itA} defined by the functional calculus of A, that is, by formula (6.1). The operator iA is the generator of the semigroups $U_+ = \{U(t) : t \geq 0\}$ (by definition) and $\{e^{itA} : t \geq 0\}$ (by Example 6.2 or Proposition 6.1). Therefore, by Proposition 6.8, $U(t) = e^{itA}$ for all $t \geq 0$ and by applying the adjoint also for $t \leq 0$. □

6.5 Exercises

1. Let $T(t)$, $t \in (-a, a)$, where $a > 0$, be bounded operators on a Hilbert space $\mathcal{H}$, and let $\mathcal{D}$ be a dense subset of $\mathcal{H}$.
    a. Suppose that $\sup_{t \in (-a,a)} \|T(t)\| < \infty$ and $\lim_{t \to 0} T(t)x = x$ for $x \in \mathcal{D}$. Prove that $\lim_{t \to 0} T(t)x = x$ for all $x \in \mathcal{H}$.
    b. Suppose that $T(t)$, $t \in (-a, a)$, is unitary and $\lim_{t \to 0} \langle T(t)x, x \rangle = \langle x, x \rangle$ for $x \in \mathcal{D}$. Prove that $\lim_{t \to 0} T(t)x = x$ for all $x \in \mathcal{H}$.
2. Let A be a self-adjoint operator, and $U(t) = e^{itA}$, $t \in \mathbb{R}$. Show that A is bounded if and only if $\lim_{h \to 0} \|U(h) - I\| = 0$.
*3. Let A be a self-adjoint operator, and $U(t) = e^{itA}$, $t \in \mathbb{R}$. Let $\mathcal{D}$ be a linear subspace of $\mathcal{D}(A)$ such that $A\mathcal{D} \subseteq \mathcal{D}$ and $U(t)\mathcal{D} \subseteq \mathcal{D}$ for all $t \in \mathbb{R}$. Let p be a real polynomial. Prove that $\mathcal{D}$ is a core for the self-adjoint operator $p(A)$.
    Hint: Prove this for $p(A) = A^n$ by modifying the proof of Proposition 6.3.
4. (*Multiplication operators*)
    Let μ be a positive Borel measure on a Borel subset Ω of $\mathbb{R}^d$, and φ a real-valued measurable function on Ω. Define $U(t)f = e^{it\varphi} \cdot f$ for $t \in \mathbb{R}$ and $f \in L^2(\Omega, \mu)$. Show that U is a unitary group on $L^2(\Omega, \mu)$ and determine its generator.
5. Let S_z be the self-adjoint operator from Example 1.5, where $|z| = 1$. Determine the action of the unitary group $\{U(t) = e^{itS_z} : t \in \mathbb{R}\}$ on $L^2(a, b)$.
6. Let e be a unit vector of $\mathbb{R}^2$ and define $(U(t)f)(x) = f(x + te)$, $t \in \mathbb{R}$, for $f \in L^2(\mathbb{R}^2)$.
    a. Show that U is a unitary group on $L^2(\mathbb{R}^2)$ and determine its generator iA.
    b. If N is a finite subset of $\mathbb{R}^2$, prove that for each real polynomial, the subspace $C_0^\infty(\mathbb{R}^2 \backslash N)$ is a core for $p(A)$.
    Hint: Use Exercise 3.
7. (*Dilations*)
    Define $(U(t)f)(x) = e^t f(e^{2t}x)$ for $t \in \mathbb{R}$ and $f \in L^2(0, \infty)$. Show that U is a unitary group on $L^2(0, \infty)$ and determine its generator.
    Hint: See Exercise 11 of Chap. 3 and the reasoning used in Example 6.1.
8. Let A be a self-adjoint operator on $\mathcal{H}$, and let $\mathcal{J}$ be an open interval. Define $C^k(\mathcal{J}, \mathcal{H})$ and $u^{(k)}(t)$ inductively, that is, $C^k(\mathcal{J}, \mathcal{H})$ is the set of all $u \in C^{k-1}(\mathcal{J}, \mathcal{H})$ such that $u^{(k-1)} \in C^1(\mathcal{J}, \mathcal{H})$ and $u^{(k)} := (u^{(k-1)})'$.
    Suppose that $u_0 \in \mathcal{D}(A^n)$ and set $u(t) = e^{-itA}u_0$. Show that $u(t) \in C^n(\mathbb{R}, \mathcal{H})$, $u^{(k)}(t) = (-iA)^k u(t)$, and $u^{(k)}(0) = (-iA)^k u_0$ for $k = 1, \ldots, n$.
9. Let A be a positive self-adjoint operator on $\mathcal{H}$ and $u_0 \in \mathcal{H}$. Set $u(t) = e^{-tA}u_0$ for $t > 0$.
    a. Show that $u(t) \in \mathcal{D}(A^n)$ for $t > 0$ and $u \in C^n((0, +\infty), \mathcal{H})$ for all $n \in \mathbb{N}$.
    b. Show that for any $n \in \mathbb{N}$, there exists a positive constant c_n such that $\|A^k u^{(n-k)}(t)\| \le c_n t^{-n} \|u_0\|$ for all $k = 0, \ldots, n$, $t > 0$, and $u_0 \in \mathcal{H}$.
10. Let A be a positive self-adjoint operator on $\mathcal{H}$, and $U(t) = e^{-tA}$, $t \ge 0$. Show that 0 is not an eigenvalue of A if and only if $\lim_{t \to \infty} U(t)x = 0$ for all $x \in \mathcal{H}$. (Semigroups having this property are called *strongly stable*.)

11. (*Left translations on* $(0, \infty)$)
    Define $(T(t)f)(x) = f(x + t)$, $t \geq 0$. Show that $T := \{T(t) : t \geq 0\}$ is a contraction semigroup on $L^2(0, \infty)$ and determine its generator.

12. Let $a > 0$. Find a contraction semigroup such that $T(t) = 0$ if and only if $t \geq a$.
    Hint: See Example 6.3.

*13. Prove that the generator of a uniformly continuous semigroup T on a Banach space E is bounded.
    Idea of the proof: Show that $S(t) := t^{-1} \int_0^t T(s) \, ds \to I$ in the operator norm as $t \to +0$ and conclude that $S(t)$ has a bounded inverse for small t.

14. Let A be a positive self-adjoint operator on $\mathcal{H}$. Suppose that there exists an orthonormal basis $\{e_i : i \in I\}$ of $\mathcal{H}$ consisting of eigenvectors of A. Determine the action of the contraction semigroup $\{e^{-tA} : t \geq 0\}$ in terms of this basis.

15. (*One-dimensional diffusion semigroup*)
    Let T be the operator with domain

    $$\mathcal{D}(T) = \{f \in H^2(0, 1) : f'(0) = f'(1) = 0\}$$

    on $\mathcal{H} = L^2(0, 1)$ defined by $Tf = -f''$.
    a. Show that T is a positive self-adjoint operator.
    b. Let $e_0(x) = 1$ and $e_n(x) = \sqrt{2}\cos(\pi nx)$, $n \in \mathbb{N}$. Show that $\{e_n : n \in \mathbb{N}_0\}$ is an orthonormal basis of $\mathcal{H}$ such that $Te_n = \pi^2 n^2 e_n$ for $n \in \mathbb{N}_0$.
    c. Show that $(e^{-tA} f)(x) = \int_0^1 K_t(x, y) f(y) dy$, $t \geq 0$, where K_t is the kernel

    $$K_t(x, y) = \sum_{n \in \mathbb{Z}} e^{-\pi^2 n^2 t} \cos(\pi nx) \cdot \cos(\pi ny).$$

    Hint: Use Exercise 14.

*16. (*An example of Nelson type*)
    Let $z \in \mathbb{C}$, $|z| = 1$. Consider the unitary groups U and V on the Hilbert space $\mathcal{H} = L^2(\mathbb{R}^2)$ which are defined by $(U(t)f)(x, y) = f(x, y + t)$ and

    $$(V(t)f)(x, y) = \begin{cases} \overline{z}f(x + t, y) & \text{for } y > 0, x < 0, x + t > 0 \\ z\, f(x + t, y) & \text{for } y > 0, \ x > 0, \ x + t < 0 \\ f(x + t, y) & \text{otherwise} \end{cases}.$$

    That is, V is the left translation parallel to the x-axis with the modification that the function is multiplied by z when the positive y-axis is crossed.
    Let $\mathcal{R}$ be the C^∞-manifold with boundary obtained by cutting $\mathbb{R}^2$ along the nonnegative y-axis, and let $\mathcal{D}$ be the set of $f \in \mathcal{H}$ such that $f \in C^\infty(\mathcal{R})$ and

    $$\frac{\partial^n f}{\partial x^n}(+0, y) = z\frac{\partial^n f}{\partial x^n}(-0, y) \quad \text{for all } y > 0, \ n \in \mathbb{N}_0.$$

    Let iA and iB be the generators of U and V, respectively.
    a. Show that $\mathcal{D} \subseteq \mathcal{D}(A) \cap \mathcal{D}(B)$, $Af = -i\frac{\partial}{\partial y} f$, and $Bf = -i\frac{\partial}{\partial x} f$ for $f \in \mathcal{D}$.
    b. Show that $A\mathcal{D} \subseteq \mathcal{D}$, $B\mathcal{D} \subseteq \mathcal{D}$, and $ABf = BAf$ for $f \in \mathcal{D}$.
    c. Show that $\mathcal{D}$ is a core for the self-adjoint operators A^n and B^n for $n \in \mathbb{N}$.
    Hint: Use Exercise 3.

d. Show that $(I - V(-t)U(-s)V(t)U(s))f = (1 - z)\chi_{ts} \cdot f$ for $f \in \mathcal{H}$ and $t, s \in [0, +\infty)$, where χ_{ts} is the characteristic function of $[0, t] \times [0, s]$.

e. Show that A and B commute strongly if and only if $z = 1$.
   Hint: Use Proposition 6.15.

Chapter 7
Miscellanea

This chapter gathers together some pertinent topics and tools for the study of closed and self-adjoint operators. Section 7.1 is devoted to the polar decomposition of a densely defined closed operator. In Sect. 7.2 the polar decomposition is used to study the operator relation $A^*A = AA^* + I$. Section 7.3 deals with the bounded transform and its inverse. The usefulness of this transform has been already seen in the proofs of various versions of the spectral theorem in Chap. 5. In Sect. 7.4 we introduce special classes of vectors (analytic vectors, quasi-analytic vectors, Stieltjes vectors) to derive criteria for the self-adjointness and for the strong commutativity of self-adjoint operators. The tensor product of unbounded operators on Hilbert spaces is studied in great detail in Sect. 7.5.

7.1 The Polar Decomposition of a Closed Operator

For any complex number t, there is a real φ such that $t = |t|e^{i\varphi}$, where $|t| = (\bar{t}t)^{1/2}$. In this section we deal with a generalization of this fact to Hilbert space operators. It is easy to guess that $|t|$ should be generalized to $(T^*T)^{1/2}$. The factor $e^{i\varphi}$ cannot be replaced by a unitary in general, because the operator may have a kernel or its range may not be dense. The proper generalization of $e^{i\varphi}$ is that of a partial isometry.

Let $\mathcal{G}_1$ and $\mathcal{G}_2$ be closed subspaces of Hilbert spaces $\mathcal{H}_1$ and $\mathcal{H}_2$, respectively. A linear operator of $\mathcal{H}_1$ into $\mathcal{H}_2$ that maps $\mathcal{G}_1$ isometrically onto $\mathcal{G}_2$ and annihilates $\mathcal{G}_1^\perp$ is called a *partial isometry*. Then $\mathcal{G}_1$ is called the *initial space*, and $\mathcal{G}_2$ is called the *final space* of the partial isometry. Some characterizations of partial isometries are collected in Exercise 1.

Let T be a *densely defined closed linear operator* of a Hilbert space $\mathcal{H}_1$ into a Hilbert space $\mathcal{H}_2$. From Proposition 3.18(ii) it follows that T^*T is a positive self-adjoint operator on $\mathcal{H}_1$, so by Proposition 5.13 it has a unique positive square root

$$|T| := (T^*T)^{1/2}. \tag{7.1}$$

Lemma 7.1 $\mathcal{D}(T) = \mathcal{D}(|T|)$, and $\|Tx\| = \||T|x\|$ for $x \in \mathcal{D}(T)$.

K. Schmüdgen, *Unbounded Self-adjoint Operators on Hilbert Space*,
Graduate Texts in Mathematics 265,
DOI 10.1007/978-94-007-4753-1_7, © Springer Science+Business Media Dordrecht 2012

Proof If $x \in \mathcal{D}(T^*T) = \mathcal{D}(|T|^2)$, then

$$\|Tx\|^2 = \langle Tx, Tx \rangle = \langle T^*Tx, x \rangle = \langle |T|^2x, x \rangle = \||T|x\|^2.$$

By Proposition 3.18(ii), $\mathcal{D}(T^*T)$ is a core for the operator T. By Corollary 4.14, $\mathcal{D}(|T|^2)$ is a core for $|T|$. Hence, the assertion follows by taking the closure of $\mathcal{D}(T^*T) = \mathcal{D}(|T|^2)$ in the graph norm $\|T \cdot \| + \| \cdot \| = \||T| \cdot \| + \| \cdot \|$. $\square$

Theorem 7.2 *Suppose that T is a densely defined closed operator of $\mathcal{H}_1$ into $\mathcal{H}_2$. Then there is a partial isometry U_T with initial space $\mathcal{N}(T)^\perp = \overline{\mathcal{R}(T^*)} = \overline{\mathcal{R}(|T|)}$ and final space $\mathcal{N}(T^*)^\perp = \overline{\mathcal{R}(T)}$ such that*

$$T = U_T|T|. \tag{7.2}$$

If A is a positive self-adjoint operator on $\mathcal{H}_1$ and U is a partial isometry of $\mathcal{H}_1$ into $\mathcal{H}_2$ with initial space $\overline{\mathcal{R}(A)}$ such that $T = UA$, then $A = |T|$ and $U = U_T$.

We call the operator $|T|$ the *modulus* of T, the operator U_T the *phase* of T, and the decomposition (7.2) the *polar decomposition* of T. The reader should notice that a number of properties of the modulus of complex numbers fail (see Exercise 4).

Proof Set $U_T(|T|x) = Tx$ for $x \in \mathcal{D}(|T|)$. By Lemma 7.1, U_T is a well-defined isometric linear map of $\mathcal{R}(|T|)$ onto $\mathcal{R}(T)$. It extends by continuity to an isometry of $\overline{\mathcal{R}(|T|)}$ onto $\overline{\mathcal{R}(T)}$. Setting U_T zero on $\overline{\mathcal{R}(|T|)}^\perp$, U_T becomes a linear operator of $\mathcal{H}_1$ into $\mathcal{H}_2$ which is defined on the whole Hilbert space $\mathcal{H}_1$. Since $\mathcal{N}(T) = \mathcal{N}(|T|)$ by Lemma 7.1, $\overline{\mathcal{R}(|T|)} = \mathcal{N}(|T|)^\perp = \mathcal{N}(T)^\perp = \overline{\mathcal{R}(T^*)}$. Thus, U_T is a partial isometry with initial space $\overline{\mathcal{R}(T^*)}$ and final space $\overline{\mathcal{R}(T)} = \mathcal{N}(T^*)^\perp$. By the definition of U_T, (7.2) holds.

We prove the uniqueness assertion of Theorem 7.2. Because U is bounded, the equality $T = UA$ implies that $T^* = AU^*$ by Proposition 1.7(ii), so we obtain $T^*T = AU^*UA$. Since the initial space of the partial isometry U contains $\mathcal{R}(A)$, we have $U^*UA = A$ and so $T^*T = A^2$. Hence, $|T| = A$ by the uniqueness of the positive square root of T^*T. Since $T = U|T| = U_T|T|$ and $\overline{\mathcal{R}(A)} = \overline{\mathcal{R}(|T|)}$ is the initial space of U and of U_T, we conclude that $U = U_T$. $\square$

Note that the operator U_T is an isometry if and only if $\mathcal{N}(T) = \{0\}$. Further, U_T is a coisometry if and only if $\mathcal{R}(T)$ is dense in $\mathcal{H}_2$, or equivalently, $\mathcal{N}(T^*) = \{0\}$. Therefore, U_T is unitary if and only if $\mathcal{N}(T) = \mathcal{N}(T^*) = \{0\}$.

We state without proof a number of related facts and useful formulas:

$$|T| = (U_T)^*T = T^*U_T, \qquad\qquad \mathcal{R}(|T|) = \mathcal{R}(T^*) = (U_T)^*\mathcal{R}(T),$$

$$|T^*| = U_TT^* = T(U_T)^*, \qquad\qquad \mathcal{R}(|T^*|) = \mathcal{R}(T) = U_T\mathcal{R}(T^*),$$

$$T = |T^*|U_T, \qquad\qquad\qquad\qquad U_T\mathcal{D}(T) = \mathcal{D}(T^*) \cap \mathcal{R}(U_T),$$

$$T^* = (U_T)^*|T^*| = |T|(U_T)^*, \qquad (U_T)^*\mathcal{D}(T^*) = \mathcal{D}(T) \cap \mathcal{R}((U_T)^*).$$

Moreover, $(U_T)^* = U_{T^*}$ and $T^* = (U_T)^*|T^*|$ is the polar decomposition of T^*.

The polar decomposition (7.2) is an important technical tool in operator theory. Often it is used to reduce problems for arbitrary densely defined closed operators to those for positive self-adjoint operators.

Suppose that $\mathcal{H}_1 = \mathcal{H}_2$. Clearly, the normality $T^*T = TT^*$ of the operator T is equivalent to the equality $|T| = |T^*|$.

Assume that T is a *normal operator* on $\mathcal{H}$. Since $T = |T^*|U_T$ as stated above, we have $T = U_T|T| = |T|U_T$, that is, the operators U_T and $|T|$ *commute*. Moreover, $\mathcal{R}(T^*) = \mathcal{R}(|T^*|) = \mathcal{R}(|T|) = \mathcal{R}(T)$. Hence, the restriction of U_T to $\overline{\mathcal{R}(T^*)}$ is a *unitary operator on the Hilbert subspace* $\overline{\mathcal{R}(T^*)}$.

Example 7.1 (*Polar decomposition of a self-adjoint operator*) Let T be a self-adjoint operator on a Hilbert space $\mathcal{H}$, and let $T = U_T|T|$ be its polar decomposition. Since $U_T = U_{T^*} = U_T^*$, U_T is self-adjoint. Let $\mathcal{H}_0 = \mathcal{N}(T)$. Since $\mathcal{N}(T)^\perp = \overline{\mathcal{R}(T^*)} = \overline{\mathcal{R}(T)}$, U_T is an isometric map of $\mathcal{H}_0^\perp$ onto $\mathcal{H}_0^\perp$. That is, $U_T^2 x = U_T^* U_T x = x$ for $x \in \mathcal{H}_0^\perp$. Setting $\mathcal{H}_\pm := \mathcal{N}(U_T \mp I)$, each vector $x \in \mathcal{H}_0^\perp$ can be written as $x = x_+ + x_-$ with $x_\pm \in \mathcal{H}_\pm$, where $x_\pm := \frac{1}{2}(I \pm U_T)x$. Thus, we have the orthogonal decomposition

$$\mathcal{H} = \mathcal{H}_+ \oplus \mathcal{H}_- \oplus \mathcal{H}_0. \tag{7.3}$$

Since $U_T T = T U_T$ and $U_T|T| = T = |T|U_T$ by the above formulas, the decomposition (7.3) reduces the self-adjoint operators T and $|T|$. That is, there are self-adjoint operators T_+, T_-, T_0 on $\mathcal{H}_+, \mathcal{H}_-, \mathcal{H}_0$, respectively, such that

$$T = T_+ \oplus T_- \oplus T_0. \tag{7.4}$$

For $x \in \mathcal{H}_\pm$, $T_\pm x = Tx = |T|U_T x = \pm|T|x$, and so $\langle T_\pm x, x \rangle = \pm\langle|T|x, x \rangle$. Moreover, $\mathcal{N}(T_\pm) = \{0\}$, since $\mathcal{N}(T) = \mathcal{N}(T_0) = \mathcal{H}_0$ by construction. Thus, T_+ is a strictly positive operator, T_- is a strictly negative operator, and $T_0 = 0$. That is, (7.4) is the decomposition of T into its strictly positive part, its strictly negative part, and its kernel. Moreover, the projections P_+, P_-, P_0 of $\mathcal{H}$ onto $\mathcal{H}_+, \mathcal{H}_-, \mathcal{H}_0$, respectively, are given by $P_\pm = \frac{1}{2}(U_T^2 \pm U_T)$, and $P_0 = I - U_T^2$.   ○

7.2 Application to the Operator Relation $AA^* = A^*A + I$

In this section we use the polar decomposition as a tool to study the operator relation

$$AA^* = A^*A + I. \tag{7.5}$$

We shall say that an operator A on a Hilbert space $\mathcal{H}$ satisfies the relation (7.5) if A is densely defined and closed, $\mathcal{D}(AA^*) = \mathcal{D}(A^*A)$, and $AA^*x = A^*Ax + x$ for all $x \in \mathcal{D}(AA^*)$.

First, we develop a model of operators for which (7.5) holds. Let $\mathcal{G}$ be a fixed Hilbert space. We denote by $l^2(\mathbb{N}_0, \mathcal{G})$ the Hilbert space $\bigoplus_{n \in \mathbb{N}_0} \mathcal{G}_n$, where $\mathcal{G}_n = \mathcal{G}$ for all $n \in \mathbb{N}_0$. Elements of $l^2(\mathbb{N}_0, \mathcal{G})$ will be written as sequences $(x_0, x_1, x_2, \ldots)$ or briefly as (x_n). We define two operators $A_\mathcal{G}$ and $A_\mathcal{G}^+$ on $l^2(\mathbb{N}_0, \mathcal{G})$ by

$$A_\mathcal{G}(x_0, x_1, x_2, \ldots) = (x_1, \sqrt{2}\,x_2, \sqrt{3}\,x_3, \ldots), \tag{7.6}$$

$$A_\mathcal{G}^+(x_0, x_1, x_2, \ldots) = (0, x_0, \sqrt{2}\,x_1, \sqrt{3}\,x_2, \ldots) \tag{7.7}$$

for $(x_0, x_1, x_2, \ldots)$ in the domain

$$\mathcal{D}(A_\mathcal{G}) = \mathcal{D}\big(A_\mathcal{G}^+\big) := \big\{(x_n) \in l^2(\mathbb{N}_0, \mathcal{G}) : (\sqrt{n}\,x_n) \in l^2(\mathbb{N}_0, \mathcal{G})\big\}.$$

Formulas (7.6) and (7.7) are usually written as $A_\mathcal{G}(x_n) = (\sqrt{n+1}\,x_{n+1})$ and $A_\mathcal{G}^+(x_n) = (\sqrt{n}\,x_{n-1})$, where $x_{-1} := 0$.

It is not difficult to check that $(A_\mathcal{G})^* = A_\mathcal{G}^+$ and $(A_\mathcal{G}^+)^* = A_\mathcal{G}$. Hence, $A_\mathcal{G}$ is a closed operator, and $A_\mathcal{G}^+$ is its adjoint. The operators $A_\mathcal{G}^+ A_\mathcal{G}$ and $A_\mathcal{G} A_\mathcal{G}^+$ act on the same domain

$$\mathcal{D}\big(A_\mathcal{G}^+ A_\mathcal{G}\big) = \mathcal{D}\big(A_\mathcal{G} A_\mathcal{G}^+\big) = \big\{(x_n) \in l^2(\mathbb{N}_0, \mathcal{G}) : (n x_n) \in l^2(\mathbb{N}_0, \mathcal{G})\big\}$$

as diagonal operators $A_\mathcal{G}^+ A_\mathcal{G}(x_n) = (n x_n)$ and $A_\mathcal{G} A_\mathcal{G}^+(x_n) = ((n+1)x_n)$. From the latter we obtain $A_\mathcal{G} A_\mathcal{G}^+(x_n) = A_\mathcal{G}^+ A_\mathcal{G}(x_n) + (x_n)$ for $(x_n) \in \mathcal{D}(A_\mathcal{G}^+ A_\mathcal{G})$. That is, $A_\mathcal{G}$ is a *densely defined closed operator satisfying the operator relation* (7.5).

The next theorem states that up to unitary equivalence all closed operators satisfying (7.5) are of this form.

Theorem 7.3 *Let A be a densely defined closed operator on a Hilbert space $\mathcal{H}$ which satisfies relation* (7.5). *Then there exist a Hilbert space $\mathcal{G}$ and a unitary operator V of $l^2(\mathbb{N}_0, \mathcal{G})$ onto $\mathcal{H}$ such that $V^* A V = A_\mathcal{G}$ and $V^* A^* V = A_\mathcal{G}^+$.*

Proof The proof is based on the spectral analysis for the positive self-adjoint operator $C := A^* A = |A|^2$.

First, we note that $\mathcal{N}(A^*) = \{0\}$. Indeed, if $x \in \mathcal{N}(A^*)$, then $A^* A x + x = 0$ by (7.5), so $\|Ax\|^2 + \|x\|^2 = \langle A^* A x + x, x \rangle = 0$, and hence $x = 0$.

Let U denote the phase U_A and P_0 the projection of $\mathcal{H}$ onto the closed subspace $\mathcal{N}(A)$. By Theorem 7.2 we have the polar decomposition $A = U|A|$ of A, where U is a partial isometry with initial space $\mathcal{N}(A)^\perp$ and final space $\mathcal{N}(A^*)^\perp$. Therefore,

$$U^* U = I - P_0 \quad \text{and} \quad U U^* = I. \tag{7.8}$$

Recall that $A^* = |A| U^*$ by the formulas in Sect. 7.1. Hence, $A A^* = U |A|^2 U^* = U C U^*$ and $A^* A = |A|^2 = C$, so relation (7.5) is equivalent to

$$U C U^* = C + I. \tag{7.9}$$

Let $x \in \mathcal{D}(C)$. Since $P_0 x \in \mathcal{N}(A) \subseteq \mathcal{N}(C)$, we have $U^* U x = (I - P_0)x \in \mathcal{D}(C)$ by (7.8) and $U C x = U C(I - P_0)x = U C U^* U x = (C + I)U x$ by (7.9). That is, $U C \subseteq (C + I)U$. Hence, $U E_C = E_{C+I} U$ by Proposition 5.15, where E_T denotes the spectral measure of T. Let Δ be a Borel set of $\mathbb{R}$. Since obviously $E_{C+I}(\Delta) = E_C(\Delta - 1)$, we obtain

$$U E_C(\Delta) = E_C(\Delta - 1)U. \tag{7.10}$$

Now we essentially use Proposition 5.10(i). Since $UE_C(\Delta)U^* = E_C(\Delta - 1)$ by (7.8) and (7.10), it follows from Proposition 5.10(i) that $\lambda - 1 \in \sigma(C)$ implies $\lambda \in \sigma(C)$. If $0 \notin \Delta$, then (7.8) and (7.10) yield $E_C(\Delta) = U^*E_C(\Delta - 1)U$, because $P_0\mathcal{H} \subset \mathcal{N}(C) = E_C(\{0\})\mathcal{H}$. Therefore, if $\lambda \neq 0$ is in $\sigma(C)$, then $\lambda - 1 \in \sigma(C)$. Since $C = A^*A \geq 0$ and hence $\sigma(C) \subseteq [0, \infty)$, the two preceding conditions on $\sigma(C)$ are only possible if $\sigma(C) = \mathbb{N}_0$. Therefore, by Corollary 5.11, $\sigma(C)$ consists of eigenvalues $n \in \mathbb{N}_0$ only. The corresponding eigenspaces are $\mathcal{H}_n := E_C(\{n\})\mathcal{H}$, and we have $\mathcal{H} = \bigoplus_{n \in \mathbb{N}_0} \mathcal{H}_n$, since $E_C(\mathbb{N}_0) = I$.

From (7.10), applied with $\Delta = \{n\}$, it follows that $U\mathcal{H}_n \subseteq \mathcal{H}_{n-1}$, where $\mathcal{H}_{-1} := \{0\}$. Applying the adjoint to Eq. (7.10) yields $E_C(\Delta)U^* = U^*E_C(\Delta - 1)$. Setting $\Delta = \{n+1\}$, the latter gives $U^*\mathcal{H}_n \subseteq \mathcal{H}_{n+1}$. From $U\mathcal{H}_{n+1} \subseteq \mathcal{H}_n$ and (7.8) we conclude that $\mathcal{H}_{n+1} = U^*U\mathcal{H}_{n+1} \subseteq U^*\mathcal{H}_n$. Therefore, $U^*\mathcal{H}_n = \mathcal{H}_{n+1}$. Hence, $\mathcal{H}_n = (U^*)^n\mathcal{H}_0$ for $n \in \mathbb{N}_0$.

Set $\mathcal{G} := \mathcal{H}_0$ and define $V(x_n) = ((U^*)^n x_n)$ for $(x_n) \in l^2(\mathbb{N}_0, \mathcal{G})$. Recall that U^* is an isometry by (7.8) and that $\mathcal{H} = \bigoplus_n \mathcal{H}_n$, where $\mathcal{H}_n = (U^*)^n\mathcal{H}_0$ for $n \in \mathbb{N}_0$. Therefore, V is a unitary mapping of $l^2(\mathbb{N}_0, \mathcal{G})$ onto $\mathcal{H}$. For $x \in \mathcal{H}_0$, we have $C(U^*)^n x = n(U^*)^n x$, and hence $|A|(U^*)^n x = \sqrt{n}(U^*)^n x$, because $|A| = C^{1/2}$. Let $(x_n) \in l^2(\mathbb{N}_0, \mathcal{G})$. If $V(x_n) \in \mathcal{D}(A)$, we compute

$$V^*AV(x_n) = V^*U|A|V(x_n) = V^*U\big(\sqrt{n}(U^*)^n x_n\big)$$
$$= V^*\big(\sqrt{n+1}\,(U^*)^n x_{n+1}\big) = (\sqrt{n+1}\,x_{n+1}) = A_{\mathcal{G}}(x_n).$$

This proves that $V^*AV \subseteq A_{\mathcal{G}}$. If $(x_n) \in \mathcal{D}(A_{\mathcal{G}})$, then $(\sqrt{n}\,x_n) \in l^2(\mathbb{N}_0, \mathcal{G})$, and hence $V(x_n) = ((U^*)^n x_n) \in \mathcal{D}(|A|) = \mathcal{D}(A)$, so that $(x_n) \in \mathcal{D}(V^*AV)$. Thus, $V^*AV = A_{\mathcal{G}}$. Applying the adjoint to both sides, we get $V^*A^*V = A_{\mathcal{G}}^+$. $\qquad\square$

Let us return to our model operator $A_{\mathcal{G}}$. If $\mathcal{G}$ is the orthogonal sum of two Hilbert subspaces $\mathcal{G}_1$ and $\mathcal{G}_2$, it follows at once from the definition that $A_{\mathcal{G}}$ is the orthogonal sum $A_{\mathcal{G}_1} \oplus A_{\mathcal{G}_2}$ on $l^2(\mathbb{N}_0, \mathcal{G}) = l^2(\mathbb{N}_0, \mathcal{G}_1) \oplus l^2(\mathbb{N}_0, \mathcal{G}_2)$. Therefore, the operator $A_{\mathcal{G}}$ is not irreducible (see Definition 1.8) if $\dim\mathcal{G} > 1$.

Now we consider the case where $\dim\mathcal{G} = 1$, that is, $\mathcal{G} = \mathbb{C}$. Then the operators $A_{\mathbb{C}}$ and $A_{\mathbb{C}}^+$ act on the standard orthonormal basis $\{e_n\}$ of $l^2(\mathbb{N}_0) = l^2(\mathbb{N}_0, \mathbb{C})$ by

$$A_{\mathbb{C}}e_n = \sqrt{n}\,e_{n-1} \quad \text{and} \quad A_{\mathbb{C}}^+e_n = \sqrt{n+1}\,e_{n+1}, \quad n \in \mathbb{N}_0,\ e_{-1} := 0. \quad (7.11)$$

In quantum mechanics, these operators $A_{\mathbb{C}}$ and $A_{\mathbb{C}}^+$ are called *annihilation operator* and *creation operator*, respectively, and (7.11) is called the *Segal–Bargmann representation* of the *canonical commutation relation* (7.5).

We now prove that the operator $A_{\mathbb{C}}$ is irreducible. Let $A_{\mathbb{C}}$ be a direct sum $A_1 \oplus A_2$ of operators on $l^2(\mathbb{N}_0) = \mathcal{H}_1 \oplus \mathcal{H}_2$. Then $\mathbb{C}e_0 = \mathcal{N}(A_{\mathbb{C}}) = \mathcal{N}(A_1) \oplus \mathcal{N}(A_2)$. Hence, e_0 is in $\mathcal{H}_1$ or in $\mathcal{H}_2$, say $e_0 \in \mathcal{H}_1$. But $A_{\mathbb{C}}^* = A_1^* \oplus A_2^*$. Hence, all vectors e_n are in $\mathcal{H}_1$ by (7.11). Thus, $\mathcal{H}_1 = l^2(\mathbb{N}_0)$ which proves that $A_{\mathbb{C}}$ is irreducible.

The preceding results combined with Theorem 7.3 give the following corollary.

Corollary 7.4 *The operator $A_{\mathcal{G}}$ is irreducible if and only if $\mathcal{G}$ is one-dimensional. Each densely defined closed irreducible operator A satisfying relation (7.5) is unitarily equivalent to $A_{\mathbb{C}}$.*

7.3 The Bounded Transform of a Closed Operator

It is easily seen that $t \to z_t := t(1 + \bar{t}t)^{-1/2}$ is a bijection of the complex plane onto the open unit disc with inverse given by $z \to t_z = z(1 - \bar{z}z)^{-1/2}$. In Sect. 5.2 we defined operator analogs of these mappings; we now continue their study.

For a Hilbert space $\mathcal{H}$, let $\mathcal{C}(\mathcal{H})$ denote the set of all densely defined closed linear operators on $\mathcal{H}$, and $\mathcal{Z}(\mathcal{H})$ the set of all bounded operators Z on $\mathcal{H}$ such that $\|Z\| \leq 1$ and $\mathcal{N}(I - Z^*Z) = \{0\}$. For $T \in \mathcal{C}(\mathcal{H})$ and $Z \in \mathcal{Z}(\mathcal{H})$, we define

$$Z_T = T(I + T^*T)^{-1/2}, \tag{7.12}$$

$$T_Z = Z(I - Z^*Z)^{-1/2}. \tag{7.13}$$

The operators $I + T^*T$ (by Proposition 3.18) and $I - Z^*Z$ are positive and self-adjoint and they have trivial kernels. Hence, $(I + T^*T)^{-1/2}$ and $(I - Z^*Z)^{-1/2}$ are well-defined positive self-adjoint operators for $T \in \mathcal{C}(\mathcal{H})$ and $Z \in \mathcal{Z}(\mathcal{H})$. Since $((I + T^*T)^{-1})^{1/2} = (I + T^*T)^{-1/2}$ (by Exercise 5.9a.), the operator Z_T from (7.12) coincides with the operator Z_T defined by (5.6).

As in Sect. 5.2, Z_T is called the *bounded transform* of the operator T.

Theorem 7.5 *The mapping $T \to Z_T$ is a bijection of $\mathcal{C}(\mathcal{H})$ onto $\mathcal{Z}(\mathcal{H})$ with inverse given by $Z \to T_Z$. Both mappings preserve adjoints, that is, $(Z_T)^* = Z_{T^*}$ and $(T_Z)^* = T_{Z^*}$, and we have $(I + T^*T)^{-1/2} = (I - Z_T^*Z_T)^{1/2}$ for $T \in \mathcal{C}(\mathcal{H})$.*

Proof From Lemma 5.8 we already know that the map $T \to Z_T$ takes $\mathcal{C}(\mathcal{H})$ into $\mathcal{Z}(\mathcal{H})$ and that $Z_T^* = Z_{T^*}$ and $(I + T^*T)^{-1} = I - Z_T^*Z_T$. Therefore, it suffices to prove that $Z \to T_Z$ maps $\mathcal{Z}(\mathcal{H})$ into $\mathcal{C}(\mathcal{H})$ and that it is the inverse of $T \to Z_T$.

Let $T \in \mathcal{C}(\mathcal{H})$. First, we verify that

$$T = Z_T(I + T^*T)^{1/2}. \tag{7.14}$$

As shown in the proof of Lemma 5.8(i), $Z_T = TC_T^{1/2}$, where $C_T = (I + T^*T)^{-1}$, is everywhere defined. Therefore, we have $C_T^{1/2}\mathcal{H} \subseteq \mathcal{D}(T)$, and so $Z_T(C_T^{1/2})^{-1} = Z_T(I + T^*T)^{1/2} \subseteq T$. To prove equality, it suffices to show that $\mathcal{D}(T) \subseteq C_T^{1/2}\mathcal{H}$. Let $x \in \mathcal{D}(T)$. For arbitrary $y \in \mathcal{H}$, we have

$$\langle (C_T^{1/2}Z_T^*T + C_T)x, y \rangle = \langle Tx, Z_T C_T^{1/2}y \rangle + \langle x, C_T y \rangle$$

$$= \langle Tx, T(I + T^*T)^{-1}y \rangle + \langle x, (I + T^*T)^{-1}y \rangle$$

$$= \langle (T^*T + I)x, (I + T^*T)^{-1}y \rangle = \langle x, y \rangle,$$

so $x = C_T^{1/2}(Z_T^*T + C_T^{1/2})x \in C_T^{1/2}\mathcal{H}$. This completes the proof of (7.14).

Since $(I + T^*T)^{-1} = I - Z_T^*Z_T$ by formula (5.7) and hence $(I + T^*T)^{1/2} = (I - Z_T^*Z_T)^{-1/2}$, it follows from (7.14) and (7.13) that

$$T = Z_T(I - Z_T^*Z_T)^{-1/2} = T_{(Z_T)}. \tag{7.15}$$

Now let $Z \in \mathcal{Z}(\mathcal{H})$. Using the definition (7.13), for $x \in \mathcal{D}(T_Z)$, we compute

$$\|T_Z x\|^2 + \|x\|^2 = \langle Z^* Z (I - Z^* Z)^{-1/2} x, (I - Z^* Z)^{-1/2} x \rangle + \langle x, x \rangle$$
$$= -\langle (I - Z^* Z)(I - Z^* Z)^{-1/2} x, (I - Z^* Z)^{-1/2} x \rangle + \langle x, x \rangle$$
$$+ \langle (I - Z^* Z)^{-1/2} x, (I - Z^* Z)^{-1/2} x \rangle$$
$$= \|(I - Z^* Z)^{-1/2} x\|^2. \tag{7.16}$$

We prove that T_Z is closed. Suppose that $x_n \to x$ and $T_Z x_n \to y$ for a sequence of vectors $x_n \in \mathcal{D}(T_Z)$. Then, by (7.16), $((I - Z^* Z)^{-1/2} x_n)$ is a Cauchy sequence, so it converges in $\mathcal{H}$. Since $(I - Z^* Z)^{-1/2}$ is self-adjoint and hence closed, we get $x \in \mathcal{D}((I - Z^* Z)^{-1/2}) = \mathcal{D}(T_Z)$ and $(I - Z^* Z)^{-1/2} x_n \to (I - Z^* Z)^{-1/2} x$. Hence,

$$T_Z x_n = Z(I - Z^* Z)^{-1/2} x_n \to Z(I - Z^* Z)^{-1/2} x = T_Z x = y.$$

This proves that the operator T_Z is closed. Clearly, T_Z is densely defined, because the self-adjoint operator $(I - Z^* Z)^{-1/2}$ is. Therefore, $T_Z \in \mathcal{C}(\mathcal{H})$.

Our next aim is to prove that

$$(I + T_Z^* T_Z)^{-1} = I - Z^* Z. \tag{7.17}$$

First, we note that $Z^* \in \mathcal{Z}(\mathcal{H})$. Obviously, $\|Z^*\| \le 1$. If $x \in \mathcal{N}(I - ZZ^*)$, then $Z^* x \in \mathcal{N}(I - Z^* Z) = \{0\}$, and hence $x = ZZ^* x = 0$. Thus, $Z^* \in \mathcal{Z}(\mathcal{H})$.

Since $Z(Z^* Z)^n = (ZZ^*)^n Z$ for $n \in \mathbb{N}$, we have $Z p(I - Z^* Z) = p(I - ZZ^*)Z$ for any polynomial p. If $(p_n(t))_{n \in \mathbb{N}}$ is a sequence of polynomials converging uniformly on $[0, 1]$ to the function $t^{1/2}$, then the sequences $(p_n(I - Z^* Z))$ and $(p_n(I - ZZ^*))$ converge in the operator norm to $(I - Z^* Z)^{1/2}$ and $(I - ZZ^*)^{1/2}$, respectively, by Corollary 5.5. Therefore,

$$Z(I - Z^* Z)^{1/2} = (I - ZZ^*)^{1/2} Z. \tag{7.18}$$

Using the relation $(T_Z)^* = (Z(I - Z^* Z)^{-1/2})^* \supseteq (I - Z^* Z)^{-1/2} Z^*$ and applying formula (7.18) twice, first to Z and then to $Z^* \in \mathcal{Z}(\mathcal{H})$ in place of Z, we compute

$$T_Z^* T_Z (I - Z^* Z) \supseteq (I - Z^* Z)^{-1/2} Z^* Z (I - Z^* Z)^{-1/2}(I - Z^* Z)$$
$$= (I - Z^* Z)^{-1/2} Z^* (I - ZZ^*)^{1/2} Z$$
$$= (I - Z^* Z)^{-1/2}(I - Z^* Z)^{1/2} Z^* Z = Z^* Z.$$

Since $Z^* Z$ is everywhere defined on $\mathcal{H}$, we have $T_Z^* T_Z (I - Z^* Z) = Z^* Z$, and so $(I + T_Z^* T_Z)(I - Z^* Z) = I$. Multiplying this equation by $(I + T_Z^* T_Z)^{-1}$ from the left, we obtain (7.17).

(7.17) yields $(I + T_Z^* T_Z)^{-1/2} = (I - Z^* Z)^{1/2}$. Hence, by (7.12) and (7.13),

$$Z_{(T_Z)} = T_Z (I + T_Z^* T_Z)^{-1/2} = Z(I - Z^* Z)^{-1/2}(I - Z^* Z)^{1/2} = Z. \tag{7.19}$$

Formulas (7.15) and (7.19) show that the maps $T \to Z_T$ and $Z \to T_Z$ are inverse to each other. $\qquad \square$

7.4 Analytic Vectors, Quasi-analytic Vectors, Stieltjes Vectors, and Self-adjointness of Symmetric Operators

In this section we develop criteria for the self-adjointness of symmetric operators and the strong commutativity of self-adjoint operators by means of analytic vectors, quasi-analytic vectors, and Stieltjes vectors. These are classes of vectors x which are defined by growth conditions on the sequence $(\|T^n x\|)_{n \in \mathbb{N}_0}$.

Let us begin with some basics on *quasi-analytic functions*. Let $(m_n)_{n \in \mathbb{N}_0}$ be a positive sequence, and $\mathcal{J} \subseteq \mathbb{R}$ be an open interval. We denote by $C\{m_n\}$ the set of functions $f \in C^\infty(\mathcal{J})$ for which there exists a constant $K_f > 0$ such that

$$\left| f^{(n)}(t) \right| \le K_f^n m_n \quad \text{for all } t \in \mathcal{J}, \ n \in \mathbb{N}_0. \tag{7.20}$$

A linear subspace C of $C^\infty(\mathcal{J})$ is called a *quasi-analytic class*, provided that the following holds: if $f \in C$ and there is a point $t_0 \in \mathcal{J}$ such that $f^{(n)}(t_0) = 0$ for all $n \in \mathbb{N}_0$, then $f(t) \equiv 0$ on $\mathcal{J}$.

The quasi-analytic classes among the linear subspaces $C\{m_n\}$ are characterized by the following famous *Denjoy–Carleman theorem*.

Proposition 7.6 $C\{m_n\}$ *is a quasi-analytic class if and only if*

$$\sum_{n=1}^{\infty} \left(\inf_{k \ge n} m_k^{1/k} \right)^{-1} = \infty. \tag{7.21}$$

A proof of Proposition 7.6 can be found in [Hr, Theorem 1.3.8]. In the case of log convex sequences a proof is given in [Ru1, Theorem 19.11].

For the results obtained in this section, it suffices to have the following:

Corollary 7.7 *Let* $(m_n)_{n \in \mathbb{N}_0}$ *be a positive sequence such that*

$$\sum_{n=1}^{\infty} m_n^{-1/n} = \infty. \tag{7.22}$$

Suppose that $f \in C^\infty(\mathcal{J})$ *and there is a constant* $K_f > 0$ *such that* (7.20) *is satisfied. If there exists a* $t_0 \in \mathcal{J}$ *such that* $f^{(n)}(t_0) = 0$ *for all* $n \in \mathbb{N}_0$, *then* $f(t) \equiv 0$ *on* $\mathcal{J}$.

Proof Since obviously $m_n^{1/n} \ge \inf_{k \ge n} m_k^{1/k}$, (7.22) implies (7.21), so $C\{m_n\}$ is a quasi-analytic class by Proposition 7.6. This gives the assertion. $\qquad\square$

Example 7.2 $(m_n = n!)_{n \in \mathbb{N}}$ Since $n! \le n^n$, the sequence $(n!)_{n \in \mathbb{N}_0}$ satisfies (7.21) and (7.22), so $C\{n!\}$ is a quasi-analytic class. But any function $f \in C\{n!\}$ has a *holomorphic extension* to some strip $\{z : \operatorname{Re} z \in \mathcal{J}, \ |\operatorname{Im} z| < \delta\}$, $\delta > 0$ (see, e.g., [Ru1, Theorem 19.9]). Using the latter fact, the assertion of Corollary 7.7 is well known in this special case. $\qquad\qquad\circ$

Example 7.3 (*log convex sequences*) Let $(m_n)_{n \in \mathbb{N}_0}$ be a positive sequence such that $m_0 = 1$ and

$$m_n^2 \leq m_{n-1} m_{n+1} \quad \text{for } n \in \mathbb{N}. \tag{7.23}$$

Condition (7.23) means that $(\log m_n)_{n \in \mathbb{N}_0}$ is a convex sequence. It is well known and easily checked that then $m_n^{1/n} \leq m_k^{1/k}$ for $n \leq k$, so $m_n^{1/n} = \inf_{k \geq n} m_k^{1/k}$. Therefore, by Proposition 7.6, in this case $C\{m_n\}$ is a quasi-analytic class if and only if $\sum_n m_n^{-1/n} = \infty$, that is, if (7.22) is satisfied. ○

Next, we define and investigate various classes of vectors for a linear operator T on a Hilbert space $\mathcal{H}$. A vector $x \in \mathcal{H}$ is called a C^∞-*vector* of T if x belongs to

$$\mathcal{D}^\infty(T) := \bigcap_{n=1}^\infty \mathcal{D}(T^n).$$

The reason for this terminology stems from the following fact which follows easily from Proposition 6.1: If T is self-adjoint, then a vector $x \in \mathcal{H}$ belongs to $\mathcal{D}^\infty(T)$ if and only if $t \to e^{itT}x$ is a C^∞-map of $\mathbb{R}$ into the Hilbert space $\mathcal{H}$.

Definition 7.1 Let $x \in \mathcal{D}^\infty(T)$. We write
$x \in \mathcal{D}^b(T)$ and x is called *bounded* for T if there is a constant $B_x > 0$ such that

$$\|T^n x\| \leq B_x^n \quad \text{for } n \in \mathbb{N}, \tag{7.24}$$

$x \in \mathcal{D}^a(T)$ and x is called *analytic* for T if there exists a constant $C_x > 0$ such that

$$\|T^n x\| \leq C_x^n n! \quad \text{for } n \in \mathbb{N}_0, \tag{7.25}$$

$x \in \mathcal{D}^{qa}(T)$ and x is called *quasi-analytic* for T if

$$\sum_{n=1}^\infty \|T^n x\|^{-1/n} = \infty, \tag{7.26}$$

$x \in \mathcal{D}^s(T)$ and x is called a *Stieltjes vector* for T if

$$\sum_{n=1}^\infty \|T^n x\|^{-1/(2n)} = \infty. \tag{7.27}$$

Obviously, eigenvectors of T are bounded vectors. The sets $\mathcal{D}^a(T)$ and $\mathcal{D}^b(T)$ are linear subspaces of $\mathcal{D}^\infty(T)$, and we have

$$\mathcal{D}^b(T) \subseteq \mathcal{D}^a(T) \subseteq \mathcal{D}^{qa}(T) \subseteq \mathcal{D}^s(T). \tag{7.28}$$

The second inclusion in (7.28) follows from the inequality $\|T^n x\|^{-1/n} \geq (C_x n)^{-1}$, since $n! \leq n^n$. A vector x is in $\mathcal{D}^{qa}(T)$ (resp. $\mathcal{D}^s(T)$) if and only if λx is for $\lambda \neq 0$, but the sum of two vectors from $\mathcal{D}^{qa}(T)$ (resp. $\mathcal{D}^s(T)$) is in general not in $\mathcal{D}^{qa}(T)$ (resp. $\mathcal{D}^s(T)$).

Example 7.4 (*Example* 7.3 *continued*) Suppose that the operator T is symmetric and x is a unit vector of $\mathcal{D}^\infty(T)$. Set $m_n := \|T^n x\|$. Then the assumptions of Example 7.3 are fulfilled, since $m_0 = 1$ and

$$m_n^2 = \|T^n x\|^2 = \langle T^n x, T^n x \rangle = \langle T^{n-1} x, T^{n+1} x \rangle \le m_{n-1} m_{n+1} \quad \text{for } n \in \mathbb{N}.$$

Hence, as stated in Example 7.3, $C\{\|T^n x\|\}$ is a quasi-analytic class if and only if $\sum_n \|T^n x\|^{-1/n} = \infty$, that is, $x \in \mathcal{D}^{qa}(T)$.

In Example 7.3 it was also noted that $\|T^{2n+1} x\|^{1/(2n+1)} \ge \|T^{2n} x\|^{1/(2n)}$ for $n \in \mathbb{N}$. This in turn yields that $\sum_k \|T^k x\|^{-1/k} \le 2 \sum_n \|T^{2n} x\|^{-1/(2n)}$. Therefore, $x \in \mathcal{D}^{qa}(T)$ if and only if $\sum_n \|T^{2n} x\|^{-1/(2n)} = \infty$. $\circ$

Example 7.5 (*Examples of analytic vectors*) Let A be a self-adjoint operator on $\mathcal{H}$, and let $\alpha > 0$. Clearly, $e^{-\alpha |A|}$ and $e^{-\alpha A^2}$ are bounded operators on $\mathcal{H}$. Suppose that $u, v \in \mathcal{H}$.

Statement $x := e^{-\alpha |A|} u$ and $y := e^{-\alpha A^2} v$ are analytic vectors for A.

Proof By Stirling's formula [RA, p. 45] there is a null sequence (ε_n) such that

$$n! = \sqrt{2\pi}\, n^{n+1/2} e^{-n} (1 + \varepsilon_n) \quad \text{for } n \in \mathbb{N}. \tag{7.29}$$

Hence, there is a constant $c > 0$ such that $n^{2n} \le c e^{2n} (n!)^2$ for $n \in \mathbb{N}$. By some elementary arguments one shows that $\lambda^{2n} e^{-2\alpha |\lambda|} \le \alpha^{-2n} n^{2n}$ for $\lambda \in \mathbb{R}$. Thus,

$$\int_{\mathbb{R}} \lambda^{2n} e^{-2\alpha|\lambda|} d\langle E_A(\lambda) u, u \rangle \le \int_{\mathbb{R}} \alpha^{-2n} n^{2n} d\langle E_A(\lambda) u, u \rangle \le c \|u\|^2 \alpha^{-2n} e^{2n} (n!)^2.$$

Hence, $u \in \mathcal{D}(A^n e^{-\alpha|A|})$ by (4.26), that is, $x \in \mathcal{D}(A^n)$. Since the first integral is equal to $\|A^n e^{-\alpha|A|} u\|^2 = \|A^n x\|^2$ by (4.33), the preceding shows that $x \in \mathcal{D}^a(A)$.

Since $\lambda^{2n} e^{-2\alpha\lambda^2} \le \lambda^{2n} e^{-2\alpha|\lambda|}$ for $\lambda \in \mathbb{R}$, $|\lambda| \ge 1$, a slight modification of this reasoning yields $y \in \mathcal{D}^a(A)$. $\square$ $\circ$

Proposition 7.8 *Suppose that the operator T is self-adjoint and $x \in \mathcal{D}^a(T)$. For any $z \in \mathbb{C}$, $|z| < C_x^{-1}$, where C_x is given by (7.25), we have $x \in \mathcal{D}(e^{zT})$ and*

$$e^{zT} x = \lim_{n \to \infty} \sum_{k=0}^{n} \frac{z^k}{k!} T^k x. \tag{7.30}$$

Proof Fix $z \in \mathbb{C}$, $|z| < C_x^{-1}$. Let E be the spectral measure of T, and $E_x(\cdot)$ the measure $\langle E(\cdot) x, x \rangle$. Let $n \in \mathbb{N}$. Then $T_n := T E([-n, n])$ is a bounded self-adjoint operator on $\mathcal{H}$ with spectrum contained in $[-n, n]$. Since the series $\sum_{k=0}^{\infty} \frac{z^k}{k!} \lambda^k$ converges to the function $e^{z\lambda}$ uniformly on $[-n, n]$, we have $e^{zT_n} x = \sum_{k=0}^{\infty} \frac{z^k}{k!} T_n^k x$ (see Proposition 4.12(v)). Using properties of the functional calculus, we derive

$$\left(\int_{-n}^{n}\left|e^{z\lambda}\right|^2 dE_x(\lambda)\right)^{1/2} = \left\|e^{zT}E([-n,n])x\right\| = \left\|e^{zT_n}x\right\| = \left\|\sum_{k=0}^{\infty}\frac{z^k}{k!}T_n^k x\right\|$$

$$\leq \sum_{k=0}^{\infty}\frac{|z|^k}{k!}\left\|T_n^k x\right\| \leq \sum_{k=0}^{\infty}\frac{|z|^k}{k!}\left\|T^k x\right\|$$

$$\leq \sum_{k=0}^{\infty}\left|zC_x\right|^k = \left(1 - |zC_x|\right)^{-1}.$$

Letting $n \to \infty$, we get $\int_{\mathbb{R}}|e^{z\lambda}|^2 dE_x(\lambda) < \infty$. Therefore, $x \in \mathcal{D}(e^{zT})$ by (4.26).

Replacing z by $|z|$ and by $-|z|$ in the preceding, we conclude that the functions $e^{2|z|\lambda}$ and $e^{-2|z|\lambda}$ are $L^1(\mathbb{R}, E_x)$. Hence, $e^{2|z\lambda|}$ is in $L^1(\mathbb{R}, E_x)$. The sequence

$$\varphi_n(\lambda) := \left|e^{z\lambda} - \sum_{k=0}^{n}\frac{z^k}{k!}\lambda^k\right|^2, \quad n \in \mathbb{N},$$

converges to zero on $\mathbb{R}$ and has the integrable majorant $e^{2|z\lambda|}$, because

$$\varphi_n(\lambda) = \left|\sum_{k\geq n+1}\frac{z^k}{k!}\lambda^k\right|^2 \leq e^{2|z\lambda|}.$$

Therefore, it follows from Lebesgue's dominated convergence theorem (Theorem B.1) that

$$\left\|e^{zT}x - \sum_{k=0}^{n}\frac{z^k}{k!}T^k x\right\|^2 = \int_{\mathbb{R}}\left|e^{z\lambda} - \sum_{k=0}^{n}\frac{z^k}{k!}\lambda^k\right|^2 dE_x(\lambda) = \int_{\mathbb{R}}\varphi_n(\lambda)\, dE_x(\lambda) \to 0$$

as $n \to \infty$. This proves formula (7.30). $\qquad\square$

Corollary 7.9 *Suppose that T is a self-adjoint operator on $\mathcal{H}$. For any $x \in \mathcal{H}$, the following statements are equivalent:*

(i) $x \in \mathcal{D}^a(T)$.
(ii) *There is a $c > 0$ such that $x \in \mathcal{D}(e^{zT})$ for all $z \in \mathbb{C}$, $|\operatorname{Re} z| < c$.*
(iii) *There is a $c > 0$ such that $x \in \mathcal{D}(e^{c|T|})$.*

Proof Clearly, by formula (4.26), $x \in \mathcal{D}^{\infty}(T)$ if (ii) or (iii) holds. The implication (i)$\to$(ii) was proved by Proposition 7.8. The equivalence of (ii) and (iii) follows at once from (4.26). Suppose that (iii) is satisfied and set $u = e^{c|T|}x$. Then $x = e^{-c|T|}u$, so $x \in \mathcal{D}^a(T)$ by Example 7.5. $\qquad\square$

The following corollary contains some explanation for the terminology "analytic vector"; another reason is given by Example 7.8 below.

Corollary 7.10 *Let T be a self-adjoint operator on $\mathcal{H}$, $y \in \mathcal{H}$, and $x \in \mathcal{D}^a(T)$. Then $f(z) := \langle e^{izT}x, y\rangle$ is a holomorphic function on the strip $\{z : |\operatorname{Im} z| < C_x^{-1}\}$, where C_x is given by (7.25).*

Proof Let $s \in \mathbb{R}$. By formula (7.30), applied with z replaced by $i(z - s)$, the series

$$f(z) = \langle e^{izT} x, y \rangle = \langle e^{(iz-is)T} x, e^{-isT} y \rangle = \sum_{n=0}^{\infty} \frac{(i(z - s))^n}{n!} \langle T^n x, e^{-isT} y \rangle$$

converges if $|z - s| < C_x^{-1}$. Hence, f is holomorphic on the set $\{|\operatorname{Im} z| < C_x^{-1}\}$. $\square$

Before we turn to the main results of this section, we prove three simple lemmas.

Lemma 7.11 *Let $(r_n)_{n \in \mathbb{N}}$ be a positive sequence and set $m_n = \alpha r_n + \beta$ for $n \in \mathbb{N}$, where $\alpha > 0$ and $\beta > 0$. If $\sum_n r_n^{-1/n} = \infty$, then $\sum_n m_n^{-1/n} = \infty$.*

Proof Let M be the set $\{n : \alpha r_n \leq \beta\}$. For $n \in M$, we have $m_n \leq 2\beta$ and

$$m_n^{-1/n} \geq (2\beta)^{-1/n} \geq (1+2\beta)^{-1},$$

so $\sum_n m_n^{-1/n} = \infty$ when the set M is infinite. If M is finite, then there is a $k \in \mathbb{N}$ such that $m_n \leq 2\alpha r_n$ for all $n \geq k$, and so

$$\sum_{n \geq k} m_n^{-1/n} \geq \sum_{n \geq k} (2\alpha)^{-1/n} r_n^{-1/n} \geq \sum_{n \geq k} (1+2\alpha)^{-1} r_n^{-1/n} = \infty. \qquad \square$$

Lemma 7.12 *If T if a symmetric operator, then $\mathcal{D}^s(T) \subseteq \mathcal{D}^s(T+zI)$ for $z \in \mathbb{C}$.*

Proof Let $x \in \mathcal{D}^s(T)$. Since $\lambda x \in \mathcal{D}^s(T)$ for $\lambda \in \mathbb{C}$, we can assume without loss of generality that $Tx \neq 0$ and $\|x\| = 1$. As noted in Examples 7.3 and 7.4, we then have $\|T^l x\|^{1/l} \leq \|T^n x\|^{1/n}$ for $l \leq n$. Using this inequality, we derive for $n \in \mathbb{N}$,

$$\|(T + zI)^n x\| = \left\| \sum_{l=0}^{n} \binom{n}{l} z^{n-l} T^l x \right\| \leq \sum_{l=0}^{n} \binom{n}{l} |z|^{n-l} \|T^l x\|$$

$$\leq \sum_{l=0}^{n} \binom{n}{l} |z|^{n-l} \|T^n x\|^{1/n} = \left(\|T^n x\|^{1/n} + |z| \right)^n$$

$$= \|T^n x\| \left(1 + |z| \|T^n x\|^{-1/n} \right)^n \leq \|T^n x\| \left(1 + |z| \|Tx\|^{-1} \right)^n.$$

Since $x \in \mathcal{D}^s(T)$, we have $\sum_n \|T^n x\|^{-1/(2n)} = \infty$. By the preceding inequality this implies that $\sum_n \|(T+zI)^n x\|^{-1/(2n)} = \infty$, that is, $x \in \mathcal{D}^s(T+zI)$. $\square$

Lemma 7.13 *If T is a self-adjoint operator on $\mathcal{H}$, then $\mathcal{D}^b(T)$ is dense in $\mathcal{H}$.*

Proof Let E be the spectral measure of T. If $a > 0$ and $x \in E([-a, a])\mathcal{H}$, then

$$\|T^n x\|^2 = \int_{-a}^{a} \lambda^{2n} d\langle E(\lambda) x, x \rangle \leq a^{2n} \|x\|^2 \leq \left(a(1 + \|x\|) \right)^{2n}$$

for $n \in \mathbb{N}$, so $x \in \mathcal{D}^b(T)$. For any $y \in \mathcal{H}$, $y = \lim_{a \to \infty} E([-a, a])y$. Therefore, the set $\bigcup_{a>0} E([-a, a])\mathcal{H}$ is contained in $\mathcal{D}^b(T)$ and dense in $\mathcal{H}$. $\square$

The following three theorems are the main results in this section. They all contain sufficient conditions for the self-adjointness of symmetric operators in terms of the density of special classes of vectors. In the proof of Theorem 7.15 we shall use Theorem 10.17 below. Theorems 7.14 and 7.15 are due to A.E. Nussbaum.

Theorem 7.14 *If T is a symmetric operator on a Hilbert space $\mathcal{H}$ such that the linear span of $\mathcal{D}^{qa}(T)$ is dense in $\mathcal{H}$, then T is essentially self-adjoint.*

Proof First, we note that T is densely defined, because the linear span of $\mathcal{D}^{qa}(T)$ is dense and obviously contained in $\mathcal{D}(T)$.

We prove that $\mathcal{N}(T^* - iI) = \{0\}$. Assume to the contrary that there exists a nonzero vector $y \in \mathcal{N}(T^* - iI)$. Let $x \in \mathcal{D}^{qa}(T)$, $x \neq 0$. By Proposition 3.17 the symmetric operator T has a self-adjoint extension A in a possibly larger Hilbert space $\mathcal{G} \supseteq \mathcal{H}$. Define a function f on $\mathbb{R}$ by $f(t) = \langle e^{itA}x, y \rangle - e^t \langle x, y \rangle$.

Since $T \subseteq A$, we have $\mathcal{D}^\infty(T) \subseteq \mathcal{D}^\infty(A)$, so $x \in \mathcal{D}^\infty(A)$. Therefore, by a repeated application of Proposition 6.1(iii), it follows that $f \in C^\infty(\mathbb{R})$ and

$$f^{(n)}(t) = \langle e^{itA}(iA)^n x, y \rangle - e^t \langle x, y \rangle \quad \text{for } t \in \mathbb{R},\ n \in \mathbb{N}. \tag{7.31}$$

Since $T^*y = iy$, we have $\langle T^n x, y \rangle = \langle T^{n-1}x, iy \rangle = \cdots = \langle x, i^n y \rangle$, and hence

$$f^{(n)}(0) = \langle (iA)^n x, y \rangle - \langle x, y \rangle = i^n \langle T^n x, y \rangle - \langle x, y \rangle = 0 \quad \text{for } n \in \mathbb{N}_0. \tag{7.32}$$

Let $a > 0$ and $\mathcal{J} = (-a, a)$. Put $\alpha = \|y\|$, $\beta = e^a \|x\| \|y\|$ and $m_n = \alpha \|T^n x\| + \beta$ for $n \in \mathbb{N}_0$. Since $x \in \mathcal{D}^{qa}(T)$, we have $\sum_n \|T^n x\|^{-1/n} = \infty$, and hence $\sum_n m_n^{-1/n} = \infty$ by Lemma 7.11. From (7.31) we obtain

$$\left| f^{(n)}(t) \right| \leq \|A^n x\| \|y\| + e^a \|x\| \|y\| = \alpha \|T^n x\| + \beta = m_n \quad \text{for } t \in \mathcal{J},\ n \in \mathbb{N}_0,$$

so f satisfies Eq. (7.20) with $K_f = 1$. Further, $f^{(n)}(0) = 0$ for all $n \in \mathbb{N}_0$ by (7.32). By the preceding we have shown that all assumptions of Corollary 7.7 are fulfilled. From Corollary 7.7 we conclude that $f(t) \equiv 0$ on $(-a, a)$ and hence on $\mathbb{R}$, because $a > 0$ was arbitrary. That is, we have $\langle e^{itA}x, y \rangle = e^t \langle x, y \rangle$ for all $t \in \mathbb{R}$.

But the function $\langle e^{itA}x, y \rangle$ is bounded on $\mathbb{R}$, while e^t is unbounded on $\mathbb{R}$. So the latter equality is only possible when $\langle x, y \rangle = 0$. Thus, we have proved that $\mathcal{D}^{qa}(T) \perp y$. Since the linear span of $\mathcal{D}^{qa}(T)$ is dense in $\mathcal{H}$, we get $y = 0$, which is the desired contradiction. Therefore, $\mathcal{N}(T^* - iI) = 0$.

A similar reasoning shows that $\mathcal{N}(T^* + iI) = 0$. Hence, T is essentially self-adjoint by Proposition 3.8.   □

Theorem 7.15 *Suppose that T is a semibounded symmetric operator on $\mathcal{H}$. If the linear span of $\mathcal{D}^s(T)$ is dense in $\mathcal{H}$, then T is essentially self-adjoint.*

Proof The proof follows a similar pattern as the proof of Theorem 7.14. By Lemma 7.12 we can assume without loss of generality that $T \geq I$. Then $0 \in \pi(T)$ by Lemma 3.3. Therefore, by Proposition 3.9, it suffices to prove that $\mathcal{N}(T^*) = \{0\}$. Assume to the contrary that $\mathcal{N}(T^*) \neq \{0\}$ and take a unit vector $y \in \mathcal{N}(T^*)$. Let

$x \in \mathcal{D}^s(T)$, $x \neq 0$. By Theorem 10.17 (which we take for granted for this proof!) there exists a self-adjoint operator $A \geq I$ on $\mathcal{H}$ which extends T.

By Proposition 6.7, $u(t) := (\cos A^{1/2}t)x$ solves the Cauchy problem $u(0) = x$, $u'(0) = 0$ for the wave equation $u''(t) = -Au(t)$ and $u'(t) = -(A^{1/2}\sin A^{1/2}t)x$ for $t \in \mathbb{R}$. Define a function f on $\mathbb{R}$ by $f(t) := \langle u(t), y \rangle - \langle x, y \rangle$.

Since $x \in \mathcal{D}^\infty(T) \subseteq \mathcal{D}^\infty(A)$, from the functional calculus of the self-adjoint operator A we easily derive that

$$u^{(2k)}(t) = (-A)^k u(t) \quad \text{and} \quad u^{(2k-1)}(t) = (-A)^{(k-1)}u'(t) \quad \text{for } k \in \mathbb{N}.$$

Therefore, $u \in C^\infty(\mathbb{R}, \mathcal{H})$, and hence $f \in C^\infty(\mathbb{R})$. Since $\|y\| = 1$ and $A \geq I$, it follows from the preceding and the formulas for u and u' that

$$\left| f^{(2k-1)}(t) \right| = \left| \langle (-A)^{k-1}u'(t), y \rangle \right| \leq \left\| A^{k-1}A^{1/2}x \right\| \leq \left\| A^k x \right\| = \left\| T^k x \right\|,$$
$$\left| f^{(2k)}(t) \right| = \left| \langle (-A)^k u(t), y \rangle \right| \leq \left\| A^k x \right\| = \left\| T^k x \right\|$$

for $k \in \mathbb{N}$ and $|f(t)| \leq 2\|x\|$ on $\mathbb{R}$. That is, setting $m_0 = 2\|x\|$, $m_{2k} = \|T^k x\|$, and $m_{2k-1} = \|T^k x\|$, the preceding shows that $|f^{(n)}(t)| \leq m_n$ for all $n \in \mathbb{N}$. Clearly,

$$\sum_n m_n^{-1/n} \geq \sum_n \left\| T^n x \right\|^{-1/(2n)} = \infty$$

by (7.27). Since $u'(0) = 0$, $u(0) = x$ and $T^*y = 0$, we obtain

$$f^{(2k-1)}(0) = \langle (-A)^k u'(0), y \rangle = 0, \qquad f(0) = \langle u(0), y \rangle - \langle x, y \rangle = 0,$$
$$f^{(2k)}(0) = \langle (-A)^k u(0), y \rangle = \langle (-T)^k x, y \rangle = \langle -(-T)^{k-1}x, T^*y \rangle = 0$$

for $k \in \mathbb{N}$, that is, we have shown that $f^{(n)}(0) = 0$ for all $n \in \mathbb{N}_0$. Therefore, all assumptions of Corollary 7.7 are satisfied, so we conclude that $f(t) \equiv 0$ on $\mathbb{R}$.

Since $f(t) = 0$ and $\cos A^{1/2}t$ is a bounded self-adjoint operator for $t \in \mathbb{R}$, we have

$$\langle x, y \rangle = \langle u(t), y \rangle = \langle (\cos A^{1/2}t)x, y \rangle = \langle x, (\cos A^{1/2}t)y \rangle.$$

Therefore, $\langle x, y - (\cos A^{1/2}t)y \rangle = 0$ for all vectors $x \in \mathcal{D}^s(T)$. By assumption the linear span of $\mathcal{D}^s(T)$ is dense in $\mathcal{H}$. Hence, it follows that $y - (\cos A^{1/2}t)y = 0$. If E is the spectral measure of A, from the functional calculus we derive

$$0 = \left\| y - (\cos A^{1/2}t)y \right\|^2 = \int_1^\infty \left| 1 - \cos(\lambda^{1/2}t) \right|^2 d\langle E(\lambda)y, y \rangle.$$

This implies that $\lambda^{1/2}t \in 2\pi \cdot \mathbb{Z}$ for any λ in the support of the measure $\langle E(\cdot)y, y \rangle$. Because $t \in \mathbb{R}$ was arbitrary, this only possible when this measure is zero. Since $\|y\| = 1$, we arrive at a contradiction. This completes the proof. $\qquad \square$

The following result is the famous *Nelson theorem*.

Theorem 7.16 *Let T be a symmetric operator on $\mathcal{H}$. If the space $\mathcal{D}^a(T)$ of analytic vectors for T is dense in $\mathcal{H}$, then T is essentially self-adjoint. If T is closed, then T is self-adjoint if and only if $\mathcal{D}^a(T)$ is dense in $\mathcal{H}$.*

Proof Since $\mathcal{D}^a(T)$ is a linear space and $\mathcal{D}^b(T) \subseteq \mathcal{D}^a(T) \subseteq \mathcal{D}^{qa}(T)$ by (7.28), the assertions follow immediately from Theorem 7.14 and Lemma 7.13. $\qquad\square$

Nelson's and Nussbaum's theorems are powerful tools for proving self-adjointness by means of growth estimates of $\|T^n x\|$. We illustrate this by an example.

Example 7.6 (*Jacobi operators*) Let $\{e_n : n \in \mathbb{N}_0\}$ be an orthonormal basis of a Hilbert space $\mathcal{H}$. Given a complex sequence $(a_n)_{n \in \mathbb{N}}$ and a real sequence $(b_n)_{n \in \mathbb{N}}$, we define the linear operator T with domain $\mathcal{D}(T) = \text{Lin}\{e_n : n \in \mathbb{N}_0\}$ by

$$T e_n = a_{n-1} e_{n-1} + b_n e_n + \overline{a_n} e_{n+1}, \quad n \in \mathbb{N}_0, \tag{7.33}$$

where $e_{-1} := 0$. It is easily seen that T is symmetric.

Operators of the form (7.33) are called *Jacobi operators*. They play a crucial role in the study of the moment problem in Sect. 16.1, see, e.g., (16.7).

Statement *Suppose that there exist two positive real constants α and β such that $|a_k| \leq \alpha n + \beta$ and $|b_k| \leq \alpha n + \beta$ for all $k \leq n$, $k, n \in \mathbb{N}$. Then we have $\mathcal{D}^a(T) = \mathcal{D}(T)$, and the operator T is essentially self-adjoint.*

Proof The main part of the proof is to show that each vector e_k is analytic for T. From formula (7.33) we conclude $T^n e_k$ is a sum of at most 3^n summands of the form $\gamma_m e_m$, where $m \leq k + n$ and γ_m is a product of n factors a_j, $\overline{a}_j$, or b_j with $j < k + n$. Therefore, by the above assumption, $\|T^n e_k\| \leq 3^n (\alpha(n+k) + \beta)^n$.

Set $c := k + \beta \alpha^{-1}$. Clearly, $c_0 := \sup_n (\frac{n+c}{n})^n < \infty$. As noted in Example 7.5, it follows from Stirling's formula that there is a constant $c_1 > 0$ such that $n^n \leq c_1 e^n n!$ for $n \in \mathbb{N}$. From the preceding we obtain

$$\|T^n e_k\| \leq 3^n (\alpha(n+k) + \beta)^n = (3\alpha)^n (n+c)^n \leq (3\alpha)^n c_0 n^n \leq c_0 c_1 (3\alpha e)^n n!$$

for $n \in \mathbb{N}_0$. This implies that $e_k \in \mathcal{D}^a(T)$. Hence, $\mathcal{D}(T) = \mathcal{D}^a(T)$, because $\mathcal{D}^a(T)$ is a linear space. By Nelson's Theorem 7.16, T is essentially self-adjoint. $\qquad\square$ $\circ$

Example 7.7 (*Annihilation and creation operators*) Recall from (7.11) that the *annihilation operator* A and the *creation operator* A^+ act on the standard orthonormal basis $\{e_n : n \in \mathbb{N}_0\}$ of $\mathcal{H} = l^2(\mathbb{N}_0)$ by

$$A e_n = \sqrt{n}\, e_{n-1} \quad \text{and} \quad A^+ e_n = \sqrt{n+1}\, e_{n+1}, \quad n \in \mathbb{N}_0,$$

where $e_{-1} := 0$. Since $A^* = A^+$, setting

$$P_0 = \frac{1}{\sqrt{2}i}(A - A^+) \quad \text{and} \quad Q_0 = \frac{1}{\sqrt{2}}(A + A^+),$$

P_0 and Q_0 are symmetric operators on $l^2(\mathbb{N}_0)$, and we have

$$P_0 e_n = \frac{1}{\sqrt{2}i}(\sqrt{n}\, e_{n-1} - \sqrt{n+1}\, e_{n+1}),$$

$$Q_0 e_n = \frac{1}{\sqrt{2}}(\sqrt{n}\, e_{n-1} + \sqrt{n+1}\, e_{n+1}).$$

Since the assumptions of the statement in Example 7.6 are satisfied, the restrictions of P_0 and Q_0 to $\mathcal{D}_0 := \mathrm{Lin}\{e_n : n \in \mathbb{N}_0\}$ are essentially self-adjoint. It is easily checked that the operators P_0 and Q_0 satisfy the canonical commutation relation

$$P_0 Q_0 - Q_0 P_0 = -\mathrm{i}I \upharpoonright \mathcal{D}_0. \tag{7.34}$$

The operator $P = -\mathrm{i}\frac{d}{dx}$ and the multiplication operator $Q = M_x$ by the variable x on $L^2(\mathbb{R})$ satisfy this relation as well. It is well known that the Hermite functions

$$h_n(x) := (-1)^n \left(2^n n! \sqrt{\pi}\right)^{-1/2} e^{x^2/2} \frac{d^n}{dx^n} e^{-x^2}, \quad n \in \mathbb{N}_0,$$

form an orthonormal basis of $L^2(\mathbb{R})$. Hence, there is a unitary operator U of $L^2(\mathbb{R})$ onto $l^2(\mathbb{N}_0)$ defined by $Uh_n = e_n$. Then it can be shown that $UPU^*x = P_0 x$ and $UQU^*x = Q_0 x$ for all $x \in \mathcal{D}_0$.   ○

Example 7.8 (*Analytic vectors for the self-adjoint operator* $T = -\mathrm{i}\frac{d}{dx}$ *on* $L^2(\mathbb{R})$)
Let $\mathcal{F}$ denote the Fourier transform, and M_x the multiplication operator by the independent variable x on $L^2(\mathbb{R})$. For this example, we will take for granted that $T = \mathcal{F}^{-1} M_x \mathcal{F}$. (This follows from formula (8.2) in Sect. 8.1.)

Statement $f \in L^2(\mathbb{R})$ *is an analytic vector for* T *if and only if* f *is the restriction to* $\mathbb{R}$ *of a holomorphic function* F *on a strip* $\{z : |\mathrm{Im}\, z| < c\}$ *for some* $c > 0$ *satisfying*

$$\sup_{|y|<c} \int_{\mathbb{R}} |F(x+\mathrm{i}y)|^2 \, dx < \infty. \tag{7.35}$$

Proof Since T and M_x are unitarily equivalent, $f \in \mathcal{D}^a(T)$ if and only if $\hat{f} = \mathcal{F}(f)$ is in $\mathcal{D}^a(M_x)$. Obviously, $\mathcal{D}^a(M_x) = \mathcal{D}^a(|M_x|) = \mathcal{D}^a(M_{|x|})$. Thus, by Corollary 7.9, $f \in \mathcal{D}^a(T)$ if and only if $\hat{f}$ is in $\mathcal{D}(e^c M_{|x|})$, or equivalently, $e^{c|x|} \hat{f}(x)$ is in $L^2(\mathbb{R})$ for some $c > 0$. By a classical theorem of Paley–Wiener [Kz, Theorem 7.1] the latter is equivalent to the existence of a function F as stated above.   □

From this statement it follows that the null vector is the only analytic vector for T contained in $C_0^\infty(\mathbb{R})$. Nevertheless, the restriction of T to $C_0^\infty(\mathbb{R})$ is essentially self-adjoint, since $C_0^\infty(\mathbb{R})$ is a core for T as shown in Example 6.1. However, by Theorem 7.16 this cannot happen if the symmetric operator T is closed.

Let $(\varepsilon_n)_{n\in\mathbb{N}}$ be a positive null sequence. Then the function

$$g(z) := \sum_{n=1}^{\infty} 2^{-n} \frac{1}{z - n - \varepsilon_n \mathrm{i}}$$

is holomorphic on the set $\mathbb{C} \backslash \{n + \varepsilon_n \mathrm{i} : n \in \mathbb{N}\}$. In particular, g is analytic on $\mathbb{R}$. One easily checks that $g \in \mathcal{D}^\infty(T)$. But g is not an analytic vector for T, because condition (7.35) is not fulfilled.   ○

Our next main theorem states that if two commuting symmetric operators have sufficiently many *joint* (!) quasi-analytic vectors, their closures are *strongly commuting* self-adjoint operators. For this, we need a preliminary lemma.

If T is a linear operator with domain $\mathcal{D}$ and $T\mathcal{D} \subseteq \mathcal{D}$, we denote by $\{T\}^c$ the set of all densely defined linear operators S satisfying $\mathcal{D} \subseteq \mathcal{D}(S) \cap \mathcal{D}(S^*)$, $S\mathcal{D} \subseteq \mathcal{D}$, and $T S x = S T x$ for all $x \in \mathcal{D}$.

Lemma 7.17 *Let T be a symmetric operator with domain $\mathcal{D}$ such that $T\mathcal{D} \subseteq \mathcal{D}$. For any operator $S \in \{T\}^c$, we have $S\mathcal{D}^{qa}(T) \subseteq \mathcal{D}^{qa}(T)$. In particular, $\mathcal{D}^{qa}(T)$ is invariant under the operator T.*

Proof Let $x \in \mathcal{D}$. Using the properties of T and $S \in \{T\}^c$, we obtain

$$\left\| T^n S x \right\|^2 = \langle T^{2n} S x, S x \rangle = \langle S T^{2n} x, S x \rangle = \langle T^{2n} x, S^* S x \rangle \leq \left\| T^{2n} x \right\| \left\| S^* S x \right\|.$$

Let $x \in \mathcal{D}^{qa}(T)$. Without loss of generality we can assume that $\|x\| = 1$. Then we have $\sum_n \|T^{2n}x\|^{-1/(2n)} = \infty$ by Example 7.4. The preceding inequality implies that $\|T^n S x\|^{1/n} \leq \|T^{2n}x\|^{1/(2n)}(1 + \|S^* S x\|)$. Hence,

$$\sum_n \left\| T^n S x \right\|^{-1/n} \geq \sum_n \left\| T^{2n} x \right\|^{-1/(2n)} \left(1 + \left\| S^* S x \right\|\right)^{-1} = \infty,$$

so that $S x \in \mathcal{D}^{qa}(T)$. $\qquad\square$

Theorem 7.18 *Let A and B be symmetric operators acting on the same dense domain $\mathcal{D}$ of $\mathcal{H}$ such that $A\mathcal{D} \subseteq \mathcal{D}$, $B\mathcal{D} \subseteq \mathcal{D}$, and $ABx = BAx$ for $x \in \mathcal{D}$. If the linear span $\mathcal{D}_Q$ of*

$$Q := \left\{ S x : S \in \{A\}^c \cap \{B\}^c, \ x \in \mathcal{D}^{qa}(A) \cap \mathcal{D}^{qa}(B) \right\}$$

is dense in $\mathcal{H}$ (in particular, if $\mathcal{D}^a(A) \cap \mathcal{D}^a(B)$ is dense in $\mathcal{H}$), then the closures $\overline{A}$ and $\overline{B}$ are strongly commuting self-adjoint operators on $\mathcal{H}$.

Proof By the definition of the set $\{T\}^c$ we have $Q \subseteq \mathcal{D}$. From Lemma 7.17 it follows that Q is contained in $\mathcal{D}^{qa}(A)$ and $\mathcal{D}^{qa}(B)$. Therefore, $\overline{A}$ and $\overline{B}$ are self-adjoint by Theorem 7.14.

Let A_0 and B_0 denote the restrictions of A and B to $\mathcal{D}_Q$ and $\mathcal{E}_Q := (A - iI)\mathcal{D}_Q$, respectively. Since $Q \subseteq \mathcal{D}^{qa}(A_0)$ has a dense linear span, Theorem 7.14 applies also to A_0, so A_0 is essentially self-adjoint. Hence, $\mathcal{E}_Q = (A_0 - iI)\mathcal{D}_Q = \text{Lin}(A - iI)Q$ is dense in $\mathcal{H}$. But $(A - iI)Q \subseteq \mathcal{D}^{qa}(B)$ by Lemma 7.17, because $Q \subseteq \mathcal{D}^{qa}(B)$ and $(A - iI) \in \{B\}^c$. Thus, $(A - iI)Q \subseteq \mathcal{D}^{qa}(B_0)$, since $B_0 = B \restriction \mathcal{E}_Q$. Hence, Theorem 7.14 applies to B_0 and the set $(A - iI)Q$ as well and implies that B_0 is essentially self-adjoint. Since $B_0 \subseteq B$ and hence $\overline{B_0} \subseteq \overline{B}$, we obtain $\overline{B_0} = \overline{B}$.

Since A and B commute on $\mathcal{D}$, we have $B_0(A - iI)x = (A - iI)Bx$ for $x \in \mathcal{D}_Q$. Setting $y = (A - iI)x$, the latter yields $(\overline{A} - iI)^{-1} B_0 y = B(\overline{A} - iI)^{-1}y$ for $y \in \mathcal{E}_Q = \mathcal{D}(B_0)$. From this relation we derive $(\overline{A} - iI)^{-1}\overline{B_0}\, y = \overline{B}(\overline{A} - iI)^{-1}y$ for $y \in \mathcal{D}(\overline{B_0})$. Since $\overline{B_0} = \overline{B}$ as noted above, we get $(\overline{A} - iI)^{-1}\overline{B} \subseteq \overline{B}(\overline{A} - iI)^{-1}$. Hence, by Proposition 5.27, the self-adjoint operators $\overline{A}$ and $\overline{B}$ strongly commute. $\square$

Completion of the proof of Proposition 5.20 Suppose that v is a generating vector for A. Put $y = e^{-A^2}v$. Let $\mathcal{D}_0$ denote the linear span of vectors $A^n y$, $n \in \mathbb{N}_0$, and let $\mathcal{H}_0$ be its closure in $\mathcal{H}$. To complete the proof, it suffices to show that $\mathcal{H}_0 = \mathcal{H}$.

First, we prove that $v \in \mathcal{H}_0$. Set $p_n(t) := \sum_{k=0}^{n} \frac{t^{2k}}{k!}$ and $\varphi_n(t) := p_n(t)e^{-t^2}$. The sequence $(\varphi_n(t))_{n \in \mathbb{N}}$ converges monotonically to 1 on $\mathbb{R}$. From the functional calculus we therefore conclude that $\varphi_n(A)v = p_n(A)y \to v$ in $\mathcal{H}$. Hence, $v \in \mathcal{H}_0$, since $p_n(A)y \in \mathcal{D}_0$.

Obviously, $A_0 := A \upharpoonright \mathcal{D}_0$ is a symmetric operator on the Hilbert space $\mathcal{H}_0$ and $A_0 \mathcal{D}_0 \subseteq \mathcal{D}_0$. By Example 7.5, we have $y \in \mathcal{D}^a(A)$. Since $A\mathcal{D}^a(A) \subseteq \mathcal{D}^a(A)$, the latter yields $\mathcal{D}_0 \subseteq \mathcal{D}^a(A)$, so that $\mathcal{D}_0 = \mathcal{D}^a(A_0)$. Since $\overline{A_0}$ has a dense set $\mathcal{D}_0$ of analytic vectors, $\overline{A_0}$ is self-adjoint by Theorem 7.16. Hence, Proposition 1.17 applies, and A decomposes as $A = \overline{A_0} \oplus A_1$ with respect to the orthogonal sum $\mathcal{H} = \mathcal{H}_0 \oplus \mathcal{H}_0^\perp$. Clearly, $E_{\overline{A_0}}(M) = E_A(M) \upharpoonright \mathcal{H}_0$. From $v \in \mathcal{H}_0$ we get $E_A(M)v = E_{\overline{A_0}}(M)v \in \mathcal{H}_0$. Since v is a generating vector for A, the span of vectors $E_A(M)v$ is dense in $\mathcal{H}$. Therefore, $\mathcal{H} = \mathcal{H}_0$.   □

7.5 Tensor Products of Operators

Let us adopt the notational convention to denote algebraic tensor products of Hilbert spaces and operators by $\odot$ and their completions and closures, respectively, by $\otimes$.

7.5.1 Tensor Product of Hilbert Spaces

In this subsection we want to define the Hilbert space tensor product $(\mathcal{H}_1 \otimes \mathcal{H}_2, \langle \cdot, \cdot \rangle)$ of two Hilbert spaces $(\mathcal{H}_1, \langle \cdot, \cdot \rangle_1)$ and $(\mathcal{H}_2, \langle \cdot, \cdot \rangle_2)$.

First, let $\mathcal{S}(\mathcal{H}_1, \mathcal{H}_2)$ denote the vector space of all conjugate-bilinear mapping $\varphi : \mathcal{H}_1 \times \mathcal{H}_2 \to \mathbb{C}$. That φ is conjugate-bilinear means that

$$\varphi(\lambda_1 x_1 + \lambda_2 x_2, y) = \overline{\lambda_1} \varphi(x_1, y) + \overline{\lambda_2} \varphi(x_2, y),$$
$$\varphi(x, \lambda_1 y_1 + \lambda_2 y_2) = \overline{\lambda_1} \varphi(x, y_1) + \overline{\lambda_2} \varphi(x, y_2)$$

for $\lambda_1, \lambda_2 \in \mathbb{C}, x_1, x_2, x \in \mathcal{H}_1$, and $y_1, y_2, y \in \mathcal{H}_2$.

For $x \in \mathcal{H}_1$ and $y \in \mathcal{H}_2$, let $x \otimes y$ be the conjugate-bilinear form on $\mathcal{H}_1 \times \mathcal{H}_2$ given by

$$(x \otimes y)(x_1, y_1) = \langle x, x_1 \rangle \langle y, y_1 \rangle, \quad x_1 \in \mathcal{H}_1, \ y_1 \in \mathcal{H}_2. \tag{7.36}$$

These forms $x \otimes y$ are called elementary tensors. Let $\mathcal{H}_1 \odot \mathcal{H}_2$ be the linear subspace of $\mathcal{S}(\mathcal{H}_1, \mathcal{H}_2)$ spanned by elementary tensors $x \otimes y$, where $x \in \mathcal{H}_1$ and $y \in \mathcal{H}_2$. By the linearity of scalar products in the first variables, the definition (7.36) implies the following bilinearity properties of elementary tensors:

$$(x_1 + x_2) \otimes y = x_1 \otimes y + x_2 \otimes y, \quad x \otimes (y_1 + y_2) = x \otimes y_1 + x \otimes y_2,$$
$$\lambda(x \otimes y) = (\lambda x) \otimes y = x \otimes (\lambda y).$$

Since $\lambda(x \otimes y) = (\lambda x) \otimes y$, each element of $\mathcal{H}_1 \odot \mathcal{H}_2$ is a finite sum $\sum_k x_k \otimes y_k$ of elementary tensors. One of the main difficulties in dealing with tensor products is that this representation is highly nonunique. Let $\sum_l x_l' \otimes y_l'$ be another finite sum of elementary tensors. By definition both sums are equal in $\mathcal{H}_1 \odot \mathcal{H}_2$ if and only if they coincide as conjugate-bilinear forms in $\mathcal{S}(\mathcal{H}_1, \mathcal{H}_2)$, that is, if

$$\sum_k \langle x_k, x \rangle_1 \langle y_k, y \rangle_2 = \sum_l \langle x_l', x \rangle_1 \langle y_l', y \rangle_2 \quad \text{for all } x \in \mathcal{H}_1, \ y \in \mathcal{H}_2.$$

For $u = \sum_{k=1}^r x_k \otimes y_k \in \mathcal{H}_1 \odot \mathcal{H}_2$ and $v = \sum_{l=1}^s x_l' \otimes y_l' \in \mathcal{H}_1 \odot \mathcal{H}_2$, we define

$$\langle u, v \rangle = \sum_{k=1}^r \sum_{l=1}^s \langle x_k, x_l' \rangle_1 \langle y_k, y_l' \rangle_2. \tag{7.37}$$

Lemma 7.19 $\langle \cdot, \cdot \rangle$ *is a well-defined scalar product on* $\mathcal{H}_1 \odot \mathcal{H}_2$.

Proof First, we prove that $\langle \cdot, \cdot \rangle$ is well defined, that is, the definition of $\langle u, v \rangle$ does not depend on the particular representations of u and v as sums of elementary tensors. For this, it suffices to show that $\langle u, v \rangle = 0$ when $u = 0$ in $\mathcal{S}(\mathcal{H}_1, \mathcal{H}_2)$ and that $\langle u, v \rangle = 0$ when $v = 0$. If $u = 0$ as a form, using (7.36) and (7.37), we obtain

$$0 = \sum_l u(x_l', y_l') = \sum_{k,l} (x_k \otimes y_k)(x_l', y_l') = \sum_{k,l} \langle x_k, x_l' \rangle_1 \langle y_k, y_l' \rangle_2 = \langle u, v \rangle.$$

Similarly, it follows that $\langle u, v \rangle = 0$ when $v = 0$ as a form. Thus, $\langle \cdot, \cdot \rangle$ is well defined.

Clearly, $\langle \cdot, \cdot \rangle$ is a Hermitian sesquilinear form. It remains to prove that $\langle \cdot, \cdot \rangle$ is positive definite. We choose orthonormal bases $\{e_1, \dots, e_p\}$ of $\mathrm{Lin}\{x_1, \dots, x_r\}$ and $\{f_1, \dots, f_q\}$ of $\mathrm{Lin}\{y_1, \dots, y_r\}$. Using the bilinearity rules for elementary tensors, we get

$$u = \sum_{k,i,j} \langle x_k, e_i \rangle_1 \langle y_k, f_j \rangle_2 \, e_i \otimes f_j. \tag{7.38}$$

Since $\langle \cdot, \cdot \rangle$ is well defined, we can insert the expression (7.38) into (7.37) and get

$$\langle u, u \rangle = \sum_{k,i,i',j,j'} \langle x_k, e_i \rangle_1 \overline{\langle x_k, e_{i'} \rangle_1} \langle y_k, f_j \rangle_2 \overline{\langle y_k, f_{j'} \rangle_2} \langle e_i, e_{i'} \rangle_1 \langle f_j, f_{j'} \rangle_2$$

$$= \sum_k \sum_{i,j} |\langle x_k, e_i \rangle_1|^2 |\langle y_k, f_j \rangle_2|^2 \geq 0.$$

Further, if $\langle u, u \rangle = 0$, then $\langle x_k, e_i \rangle_1 \cdot \langle y_k, f_j \rangle_2 = 0$ for all k, i, j, so $u = 0$. $\square$

Definition 7.2 The Hilbert space $(\mathcal{H}_1 \otimes \mathcal{H}_2, \langle \cdot, \cdot \rangle)$ obtained as a completion of $(\mathcal{H}_1 \odot \mathcal{H}_2, \langle \cdot, \cdot \rangle)$ is called the *tensor product of Hilbert spaces* $\mathcal{H}_1$ and $\mathcal{H}_2$.

Since $\langle x \otimes y, x' \otimes y' \rangle = \langle x, x' \rangle_1 \langle y, y' \rangle_2$ by (7.37), the norm $\| \cdot \|$ of the Hilbert space $\mathcal{H}_1 \otimes \mathcal{H}_2$ has the cross property

$$\|x \otimes y\| = \|x\|_1 \|y\|_2, \quad x \in \mathcal{H}_1, \ y \in \mathcal{H}_2. \tag{7.39}$$

Further, if $\{e_i : i \in I_1\}$ is an orthonormal basis of $\mathcal{H}_1$ and $\{f_j : j \in I_2\}$ is an orthonormal basis of $\mathcal{H}_2$, it is easily seen that the set $\{e_i \otimes f_j : i \in I_1, j \in I_2\}$ is an orthonormal basis of the Hilbert space $\mathcal{H}_1 \otimes \mathcal{H}_2$. That is, the elements of $\mathcal{H}_1 \otimes \mathcal{H}_2$ are precisely the sums

$$u = \sum_{i \in I_1} \sum_{j \in I_2} \alpha_{ij} e_i \otimes f_j$$

with complex numbers α_{ij} satisfying $\sum_{i,j} |\alpha_{ij}|^2 < \infty$.

The next example shows that the L^2-space of a product measure fits nicely into this construction.

Example 7.9 (*Tensor product of two L^2-spaces*) Let (X_1, μ_1) and (X_2, μ_2) be two σ-finite measure spaces, and let $(X_1 \times X_2, \mu)$ be the measure space with the product measure $\mu = \mu_1 \times \mu_2$. Let $\mathcal{H}_1 = L^2(X_1, \mu_1)$ and $\mathcal{H}_2 = L^2(X_2, \mu_2)$. Using the definitions (7.37) of the scalar product and the product measure $\mu_1 \times \mu_2$, one verifies that

$$J : \sum_k f_k \otimes g_k \rightarrow \sum_k f_k(x_1) g_k(x_2), \quad x_1 \in X_1, \ x_2 \in X_2,$$

is a well-defined linear map of $\mathcal{H}_1 \odot \mathcal{H}_2$ onto a dense subspace of $L^2(X_1 \times X_2, \mu)$ that preserves the scalar product. By continuity J extends to a unitary isomorphism of the Hilbert spaces $\mathcal{H}_1 \otimes \mathcal{H}_2 = L^2(X_1, \mu_1) \otimes L^2(X_2, \mu_2)$ and $L^2(X_1 \times X_2, \mu)$.

In particular, the map $f \otimes g \rightarrow f(x_1) g(x_2)$ extends to an isometric isomorphism of $L^2(\mathbb{R}^n) \otimes L^2(\mathbb{R}^m)$ onto $L^2(\mathbb{R}^{n+m})$. Likewise, the map $f \otimes g \rightarrow (f_n g_k)$ has a unique extension to an isometric isomorphism of $l^2(\mathbb{N}) \otimes l^2(\mathbb{N})$ onto $l^2(\mathbb{N}^2)$.   ○

7.5.2 Tensor Product of Operators

Throughout this subsection T_1 and T_2 denote linear operators on Hilbert spaces $(\mathcal{H}_1, \langle \cdot, \cdot \rangle_1)$ and $(\mathcal{H}_2, \langle \cdot, \cdot \rangle_2)$, respectively. Define

$$\mathcal{D}(T_1 \odot T_2) = \left\{ \sum_{k=1}^r x_k \otimes y_k : x_k \in \mathcal{D}(T_1), \ y_k \in \mathcal{D}(T_2), \ r \in \mathbb{N} \right\},$$

$$(T_1 \odot T_2) \left(\sum_{k=1}^r x_k \otimes y_k \right) = \sum_{k=1}^r T_1 x_k \otimes T_2 x_k, \quad x_k \in \mathcal{D}(T_1), \ y_k \in \mathcal{D}(T_2), \ r \in \mathbb{N}.$$

Proposition 7.20

(i) $T_1 \odot T_2$ is a well-defined linear operator on the Hilbert space $\mathcal{H}_1 \otimes \mathcal{H}_2$ with domain $\mathcal{D}(T_1 \odot T_2)$.

(ii) If T_1 and T_2 are bounded, then $T_1 \odot T_2$ is bounded and $\| T_1 \odot T_2 \| = \| T_1 \| \| T_2 \|$.

(iii) If T_1 and T_2 are symmetric operators, so is $T_1 \odot T_2$. If, in addition, T_1 and T_2 are positive, then $T_1 \odot T_2$ is also positive.

Proof Let $u = \sum_{k=1}^{r} x_k \otimes y_k$, where $x_k \in \mathcal{D}(T_1)$, $y_k \in \mathcal{D}(T_2)$. We choose an orthonormal basis $\{e_1, \ldots, e_n\}$ of $\mathrm{Lin}\{x_1, \ldots, x_r\}$ and set $f_i = \sum_k \langle x_k, e_i \rangle_1 y_k$. Then

$$u = \sum_{k,i} \langle x_k, e_i \rangle_1 e_i \otimes y_k = \sum_i e_i \otimes f_i, \qquad (7.40)$$

$$\|u\|^2 = \sum_{i,j} \langle e_i, e_j \rangle_1 \langle f_i, f_j \rangle_2 = \sum_j \|f_j\|_2^2. \qquad (7.41)$$

(i): To prove that $T_1 \odot T_2$ is well defined, we show that $\sum_k T_1 x_k \otimes T_2 y_k = 0$ whenever $u = 0$. If $u = 0$, then all f_i are zero by (7.41), and hence

$$\sum_k T_1 x_k \otimes T_2 y_k = \sum_{k,i} \langle x_k, e_i \rangle_1 T_1 e_i \otimes T_2 y_k = \sum_i T_1 e_i \otimes T_2 f_i = 0.$$

The linearity of $T_1 \odot T_2$ is obvious.

(ii): We write $u \in \mathcal{D}(T_1 \odot T_2)$ as in (7.40). Set $I_1 := I \restriction \mathcal{D}(T_1)$ and $I_2 := I \restriction \mathcal{D}(T_2)$. Using (7.41) twice, first for the element $(I_1 \otimes T_2)u$ and then for u, we derive

$$\|(I_1 \odot T_2)u\|^2 = \left\| \sum_i e_i \otimes T_2 f_i \right\|^2 = \sum_i \|T_2 f_i\|_2^2 \leq \|T_2\|^2 \sum_i \|f_i\|^2 = \|T_2\|^2 \|u\|^2.$$

This proves that $\|I_1 \odot T_2\| \leq \|T_2\|$. Similarly, $\|T_1 \odot I_2\| \leq \|T_1\|$. Hence,

$$\|T_1 \odot T_2\| = \left\| (T_1 \odot I_2)(I_1 \odot T_2) \right\| \leq \|T_1\| \|I_1 \odot T_2\| \leq \|T_1\| \|T_2\|.$$

We prove the reverse inequality. Let $\varepsilon > 0$. We choose unit vectors $x \in \mathcal{D}(T_1)$ and $y \in \mathcal{D}(T_2)$ such that $\|T_1\| \leq \|T_1 x\|_1 + \varepsilon$ and $\|T_2\| \leq \|T_2 y\|_2 + \varepsilon$. Using the cross property (7.39) of the norm, we conclude that

$$\left(\|T_1\| - \varepsilon \right) \left(\|T_2\| - \varepsilon \right) \leq \|T_1 x\|_1 \|T_2 y\|_2 = \|T_1 x \otimes T_2 y\|$$
$$= \left\| (T_1 \otimes T_2)(x \otimes y) \right\| \leq \|T_1 \odot T_2\| \|x \otimes y\| = \|T_1 \odot T_2\|,$$

so $\|T_1\| \|T_2\| \leq \|T_1 \odot T_2\|$ by letting $\varepsilon \to +0$. Thus, $\|T_1 \odot T_2\| = \|T_1\| \|T_2\|$.

(iii): An easy computation shows that $T_1 \odot T_2$ is symmetric. Suppose now in addition that $T_1 \geq 0$ and $T_2 \geq 0$. We will prove that $T_1 \odot T_2 \geq 0$.

Let u be as above. Clearly, $\langle T_1 \cdot, \cdot \rangle_1$ is a Hermitian sesquilinear form on the vector space $G := \mathrm{Lin}\{x_1, \ldots, x_r\}$. From linear algebra we know that this form can be diagonalized. Hence, there exists a basis $\{g_1, \ldots, g_r\}$ of G such that $\langle T_1 g_i, g_j \rangle_1 = 0$ for all i, j, $i \neq j$. As above, we can write $u = \sum_i g_i \otimes h_i$, where $h_i \in \mathcal{D}(T_2)$. Then

$$\langle (T_1 \odot T_2)u, u \rangle = \sum_{i,j} \langle T_1 g_i, g_j \rangle_1 \langle T_2 h_i, h_j \rangle_2 = \sum_i \langle T_1 g_i, g_i \rangle_1 \langle T_2 h_i, h_i \rangle_2 \geq 0. \qquad \square$$

Lemma 7.21 *Suppose that T_1 and T_2 are densely defined and closable. Then $T_1 \odot T_2$ is also densely defined and closable, and we have $(T_1 \odot T_2)^* \supseteq T_1^* \odot T_2^*$.*

Proof Since T_1 and T_2 are densely defined and closable, T_1^* and T_2^* are densely defined by Theorem 1.8(i). Therefore, $T_1^* \odot T_2^*$ is densely defined. A simple computation shows that $(T_1 \odot T_2)^* \supseteq T_1^* \odot T_2^*$. Thus, $(T_1 \odot T_2)^*$ is densely defined. Hence, $T_1 \odot T_2$ is closable again by Theorem 1.8(i). $\square$

From now on we assume that T_1 and T_2 are *densely defined* and *closable*.

Definition 7.3 The closure of the closable operator $T_1 \odot T_2$ (by Lemma 7.21) is denoted by $T_1 \otimes T_2$ and called the *tensor product of the operators* T_1 and T_2.

Lemma 7.22 $\mathcal{D}(T_1 \odot T_2)$ *is a core for* $\overline{T}_1 \odot \overline{T}_2$, *and we have* $\overline{T}_1 \otimes \overline{T}_2 = T_1 \otimes T_2$.

Proof The first assertion follows easily by approximating all vectors $x_i \in \mathcal{D}(\overline{T}_1)$ and $y_i \in \mathcal{D}(\overline{T}_2)$ of $u = \sum_i x_i \otimes y_i \in \mathcal{D}(\overline{T}_1 \odot \overline{T}_2)$ in the graph norms by $\tilde{x}_i \in \mathcal{D}(T_1)$ and $\tilde{y}_i \in \mathcal{D}(T_2)$. Since $\overline{T}_1 \otimes \overline{T}_2$ is the closure of $\overline{T}_1 \odot \overline{T}_2$ and $\mathcal{D}(T_1 \odot T_2)$ is a core for $\overline{T}_1 \odot \overline{T}_2$, it follows that $\overline{T}_1 \otimes \overline{T}_2 = T_1 \otimes T_2$. $\qquad\square$

Theorem 7.23 *Suppose that T_1 and T_2 are self-adjoint. Then the tensor products $P := T_1 \otimes T_2, T_1 \otimes I$, and $I \otimes T_2$ are self-adjoint and the sum $S := T_1 \otimes I + I \otimes T_2$ is essentially self-adjoint. The dense linear subspace*

$$\mathcal{D}_b := \operatorname{Lin}\{x \otimes y : x \in \mathcal{D}^b(T_1),\ y \in \mathcal{D}^b(T_2)\} \tag{7.42}$$

of $\mathcal{H}_1 \otimes \mathcal{H}_2$ is an invariant core and all vectors of $\mathcal{D}_b$ are bounded vectors for each of these four operators.

If T_1 and T_2 are positive, then $T_1 \otimes T_2, T_1 \otimes I, I \otimes T_2$, and S are positive as well.

Proof From Proposition 7.20(iii) it follows that each of the operators $T = T_1 \otimes T_2$, $T_1 \otimes I, I \otimes T_2$, and S is symmetric.

Let $x \in \mathcal{D}^b(T_1)$ and $y \in \mathcal{D}^b(T_2)$. Then there is a constant $C \geq 1$ such that $\|T_1^k x\|_1 \leq C^k$ and $\|T_2^k y\|_2 \leq C^k$ for $k \in \mathbb{N}$. For $n \in \mathbb{N}$, we have

$$\|P^n(x \otimes y)\| = \|T_1^n x \otimes T_2^n y\| \leq \|T_1^n x\|_1 \|T_2^n y\| \leq (C^2)^n, \tag{7.43}$$

$$\|S^n(x \otimes y)\| = \left\| \sum_{l=0}^n \binom{n}{l} T_1^l x \otimes T_2^{n-l} y \right\|$$

$$\leq \sum_{l=0}^n \binom{n}{l} \|T_1^l x\|_1 \|T_2^{n-l} y\|_2 \leq \sum_{l=0}^n \binom{n}{l} C^l C^{n-l} = (2C)^n.$$

Hence, $x \otimes y$ is a bounded vector for P and S. Setting $T_2 = I$ resp. $T_1 = I$ in (7.43) the same reasoning shows that $x \otimes y$ is a bounded vector for $T_1 \otimes I$ and $I \otimes T_2$.

Obviously, sums of bounded vectors are again bounded vectors. Since the sets $\mathcal{D}^b(T_1)$ and $\mathcal{D}^b(T_2)$ are dense by Lemma 7.13, $\mathcal{D}_b$ is a dense set of bounded, hence analytic, vectors for each of these four symmetric operators. Therefore, by Nelson's Theorem 7.16, the closed symmetric operators $P, T_1 \otimes I, I \otimes T_2$ are self-adjoint, and the symmetric operator S is essentially self-adjoint. Since $\mathcal{D}^b(T_j)$ is invariant under T_j, $\mathcal{D}_b$ is invariant under each of these operators.

If T_1 and T_2 are positive, then $T_1 \odot T_2, T_1 \odot I$, and $I \odot T_2$ are positive by Proposition 7.20(iii). Hence, all four operators are positive. $\qquad\square$

Lemma 7.24 *If T_1 and T_2 are self-adjoint operators, then $T_1 \otimes I$ and $I \otimes T_2$ are strongly commuting self-adjoint operators, and the joint spectrum $\sigma(T)$ of the pair $T = \{T_1 \otimes I, I \otimes T_2\}$ is equal to $\sigma(T_1) \times \sigma(T_2)$.*

Proof Let $\lambda_1 \in \rho(T_1)$. Clearly, the bounded operator $R_{\lambda_1}(T_1) \otimes I$ leaves $\mathcal{D}(T_1 \odot I)$ invariant, and its restriction to this domain is the inverse of $T_1 \odot I - \lambda_1 I$. Since $T_1 \otimes I$ is the closure of $T_1 \odot I$, this implies that $\lambda_1 \in \rho(T_1 \otimes I)$ and $R_{\lambda_1}(T_1) \otimes I = R_{\lambda_1}(T_1 \otimes I)$. In particular, $\sigma(T_1 \otimes I) \subseteq \sigma(T_1)$. Similarly, $\sigma(I \otimes T_2) \subseteq \sigma(T_2)$ and $I \otimes R_{\lambda_2}(T_2) = R_{\lambda_2}(I \otimes T_2)$ for $\lambda_2 \in \rho(T_2) \subseteq \rho(I \otimes T_2)$. (Because of Corollary 3.14, this gives a second proof of the self-adjointness of $T_1 \otimes I$ and $I \otimes T_2$ that avoids the use of Nelson's theorem.) Since the resolvents $R_\lambda(T_1) \otimes I$ and $I \otimes R_\lambda(T_2)$ obviously commute, $T_1 \otimes I$ and $I \otimes T_2$ strongly commute by Proposition 5.27. (The strong commutativity would also follow from Theorem 7.18, because $T_1 \odot I$ and $I \odot T_2$ are commuting symmetric operators having a common dense set of analytic vectors by Theorem 7.23.) Thus, we have $\sigma(T) \subseteq \sigma(T_1 \otimes I) \times \sigma(I \otimes T_2) \subseteq \sigma(T_1) \times \sigma(T_2)$, where the first inclusion stems from Proposition 5.24(ii).

To prove the inclusion $\sigma(T_1) \times \sigma(T_2) \subseteq \sigma(T)$ we apply Proposition 5.24(i) twice, first to T_1, T_2 and then to T. Let $j = 1, 2$ and $\lambda_j \in \sigma(T_j)$. Then there is a sequence of unit vectors $x_{jn} \in \mathcal{D}(T_j)$ such that $\lim_n (T_j - \lambda_j) x_{jn} = 0$. Then $y_n := x_{1n} \otimes x_{2n}$ are unit vectors and $\lim_n (T_1 \otimes I - \lambda_1) y_n = \lim_n (I \otimes T_2 - \lambda_2) y_n = 0$. Therefore, $(\lambda_1, \lambda_2) \in \sigma(T)$. This completes the proof of the equality $\sigma(T) = \sigma(T_1) \times \sigma(T_2)$. $\square$

Let us retain the assumptions and the notation of Theorem 7.23 and Lemma 7.24. To any complex polynomial $p(t_1, t_2) = \sum_{k,l} \alpha_{kl} \, t_1^k t_2^l$ the functional calculus (5.32) for the pair $T = (T_1 \otimes I, I \otimes T_2)$ of strongly commuting self-adjoint operators associates a normal operator $p(T)$ which acts on vectors $\sum_i x_i \otimes y_i \in \mathcal{D}_b$ by

$$p(T)\left(\sum_i x_i \otimes y_i \right) = \sum_i \sum_{k,l} \alpha_{kl} T_1^k x_i \otimes T_2^l y_i. \tag{7.44}$$

Combining Proposition 5.25 and Lemma 7.24, we conclude that the *spectrum of the operator $p(T)$ is the closure of the set* $\{p(t_1, t_2) : t_1 \in \sigma(T_1), \ t_2 \in \sigma(T_2)\}$.

For $p_1(t_1, t_2) = t_1 t_2$ and $p_2(t_1, t_2) = t_1 + t_2$, we have $p_1(T) = P$ and $p_2(T) = \overline{S}$. (To see these equalities, it suffices to note that by (7.44) the corresponding operators coincide on $\mathcal{D}_b$ which is a core for the self-adjoint operators P and $\overline{S}$.) We restate the result about the spectrum in these two cases separately as

Corollary 7.25 *If T_1 and T_2 are self-adjoint operators on $\mathcal{H}_1$ and $\mathcal{H}_2$, then*

$$\sigma(T_1 \otimes T_2) = \overline{\sigma(T_1) \cdot \sigma(T_2)}, \qquad \sigma(\overline{T_1 \otimes I + I \otimes T_2}) = \overline{\sigma(T_1) + \sigma(T_2)}, \tag{7.45}$$

where the bars on the right-hand sides mean the closures of the corresponding sets.

In general none of the three bars in (7.45) can be omitted (see Example 7.10 below and Exercise 17). But $\sigma(T_1) \cdot \sigma(T_2)$ is closed if one of the operators is bounded and 0 is not in the spectrum of the other one (Exercise 18.b). Further, $T_1 \otimes I + I \otimes T_2$

and $\sigma(T_1) + \sigma(T_2)$ are closed if T_1 and T_2 are both lower semibounded (see Exercises 17.a and 18.a). In particular, all three bars in (7.45) can be omitted if T_1 and T_2 are *bounded* self-adjoint operators.

It can be shown that both formulas in (7.45) hold (without all bars) if T_1 and T_2 are arbitrary bounded everywhere defined operators [RS4, Theorem XIII.9].

Example 7.10 For $j = 1, 2$, let T_j be the multiplication operator by the variable t_j on $\mathcal{H}_j := L^2(\mathbb{R})$. Then $T_1 \otimes I$, $I \otimes T_2$, and $T_1 \otimes T_2$ are the multiplication operators by t_1, t_2, and $t_1 t_2$, respectively, and the closure of $S := T_1 \otimes I + I \otimes T_2$ is the multiplication operator by $t_1 + t_2$ on $\mathcal{H}_1 \otimes \mathcal{H}_2 = L^2(\mathbb{R}^2)$. For the function

$$f(t_1, t_2) = \chi_{(1, +\infty)}(t_1) \chi_{(-1,1)}(t_1 + t_2) t_1^{-\alpha} \in L^2(\mathbb{R}^2), \quad \text{where } 1/2 < \alpha \le 3/2,$$

we have $(t_1 + t_2) f \in L^2(\mathbb{R}^2)$ and $t_j f \notin L^2(\mathbb{R}^2)$. The latter means that $f \in \mathcal{D}(\overline{S})$ and $f \notin \mathcal{D}(T_1 \otimes I)$, so $f \notin \mathcal{D}(S)$. That is, the operator S is not closed.

Let g denote the characteristic function of the set $\{0 \le t_1 \le t_2^{-\beta}, 1 \le t_2\}$, where $1 < \beta \le 3$. Then g and $t_1 t_2 g$ are in $L^2(\mathbb{R}^2)$, so $g \in \mathcal{D}(T_1 \otimes T_2)$. Since $t_2 g$ is not in $L^2(\mathbb{R}^2)$, we have $g \notin \mathcal{D}(T_2 \otimes I)$, and hence $g \notin \mathcal{D}((T_1 \otimes I)(I \otimes T_2))$. Therefore, $(T_1 \otimes I)(I \otimes T_2) \ne T_1 \otimes T_2$. In fact, $T_1 \otimes T_2$ is the closure of $(T_1 \otimes I)(I \otimes T_2)$.   ∘

Proposition 7.26 *If T_1 and T_2 are densely defined and closable, then*

$$(T_1 \otimes T_2)^* = T_1^* \otimes T_2^*. \tag{7.46}$$

Proof By Lemma 7.22 we can assume that T_1 and T_2 are closed.

From Lemma 7.21 we have $(T_1 \odot T_2)^* \supseteq T_1^* \odot T_2^*$. Hence, it follows that $(T_1 \otimes T_2)^* = (T_1 \odot T_2)^* \supseteq T_1^* \otimes T_2^*$, so the graph $\mathcal{G}(T_1^* \otimes T_2^*)$ of the closed operator $T_1^* \otimes T_2^*$ is a closed subspace of the graph $\mathcal{G}((T_1 \otimes T_2)^*)$. To prove the equality $(T_1 \otimes T_2)^* = T_1^* \otimes T_2^*$, it therefore suffices to show that the orthogonal complement of $\mathcal{G}(T_1^* \otimes T_2^*)$ in the graph Hilbert space $(\mathcal{G}(T), \langle \cdot, \cdot \rangle_T)$ defined by (1.3), where $T := (T_1 \otimes T_2)^*$, is zero.

Suppose that $(u, Tu) \in \mathcal{G}(T)$ is orthogonal to $\mathcal{G}(T_1^* \otimes T_2^*)$. Let $x \in \mathcal{D}(T_1^* T_1)$ and $y \in \mathcal{D}(T_2^* T_2)$. Then $(T_1 \otimes T_2)(x \otimes y) \in \mathcal{D}(T_1^* \otimes T_2^*)$. Since $T \supseteq T_1^* \otimes T_2^*$ as noted in the preceding paragraph, we obtain

$$T(T_1 \otimes T_2)(x \otimes y) = \left(T_1^* \otimes T_2^*\right)(T_1 x \otimes T_2 y) = T_1^* T_1 x \otimes T_2^* T_2 y.$$

From the orthogonality $(u, Tu) \perp \mathcal{G}(T_1^* \otimes T_2^*)$ it follows that

$$\begin{aligned} 0 &= \langle u, (T_1 \otimes T_2)(x \otimes y) \rangle + \langle Tu, T(T_1 \otimes T_2)(x \otimes y) \rangle \\ &= \langle (T_1 \otimes T_2)^* u, x \otimes y \rangle + \langle Tu, T_1^* T_1 x \otimes T_2^* T_2 y \rangle \\ &= \langle Tu, \left(I + T_1^* T_1 \odot T_2^* T_2\right)(x \otimes y) \rangle. \end{aligned}$$

Hence, $Tu \perp \mathcal{E} := (I + T_1^* T_1 \odot T_2^* T_2) \mathcal{D}(T_1^* T_1 \odot T_2^* T_2)$. Since T_1 and T_2 are closed, $T_1^* T_1$ and $T_2^* T_2$ are positive self-adjoint operators by Proposition 3.18. Hence, $T_1^* T_1 \otimes T_2^* T_2$ is self-adjoint and positive by Theorem 7.23. Because the linear operator $T_1^* T_1 \otimes T_2^* T_2$ is the closure of $T_1^* T_1 \odot T_2^* T_2$, it follows from

Proposition 3.9 that $\mathcal{E}$ is dense in $\mathcal{H}_1 \otimes \mathcal{H}_2$. Thus, $Tu \perp \mathcal{E}$ implies that $Tu = 0$. But $(u, Tu) = (u, 0) \perp \mathcal{G}(T_1^* \otimes T_2^*)$ implies that u is orthogonal to the dense domain $\mathcal{D}(T_1^* \odot T_2^*)$. Therefore, $u = 0$. □

Often separations of variables lead in a natural way to the tensor product of operators. We illustrate this by an example from quantum physics.

Example 7.11 (*Separation of a two-particle Hamiltonian*) In quantum mechanics the Hamiltonian of two particles is given by the operator

$$T = -\frac{1}{2m_1}\Delta_{x_1} - \frac{1}{2m_2}\Delta_{x_2} + V(x_1 - x_2) \tag{7.47}$$

on $L^2(\mathbb{R}^6)$, where Δ_{x_j} is the Laplacian in $x_j \in \mathbb{R}^3$, and V is a potential on $\mathbb{R}^6$ describing the interaction of the particles and depending only on the difference $x_1 - x_2$.

To obtain a separation of variables in (7.47), we introduce new variables

$$y_s = (m_1 x_1 + m_2 x_2)M^{-1} \quad \text{and} \quad y_r = x_1 - x_2, \tag{7.48}$$

where we set $M := m_1 + m_2$. That is, we consider the coordinate transformation $y \equiv (y_r, y_s) := \mathfrak{A}x \equiv \mathfrak{A}(x_1, x_2)$ on $\mathbb{R}^6$, where $\mathfrak{A}$ is the block matrix

$$\mathfrak{A} = \begin{pmatrix} 1 & -1 \\ m_1 M^{-1} & m_2 M^{-1} \end{pmatrix}.$$

Since $\mathfrak{A}$ has determinant one, there is a unitary operator U on $L^2(\mathbb{R}^6)$ defined by $(Uf)(y) := f(x) = f(\mathfrak{A}^{-1}y)$. Our aim is to prove that

$$UTU^* = -\frac{1}{2M}\Delta_{y_s} + \left(-\frac{1}{2m}\Delta_{y_r} + V(y_r)\right). \tag{7.49}$$

That is, UTU^* is of the form $T_1 \otimes I + I \otimes T_2$ on $L^2(\mathbb{R}^6) = L^2(\mathbb{R}^3) \otimes L^2(\mathbb{R}^3)$, where $T_1 = -\frac{1}{2M}\Delta_{y_s}$, $T_2 = -\frac{1}{2m}\Delta_{y_r} + V(y_r)$, $m := m_1 m_2 M^{-1}$.

Now let us prove formula (7.49). Set $g = U^* f$. Then $g(x) = f(\mathfrak{A}x)$. Using the substitution (7.48), we compute for $j = 1, 2, 3$,

$$\frac{\partial^2}{\partial x_{1j}^2} g(x) \equiv \frac{\partial^2}{\partial x_{1j}^2} f(\mathfrak{A}x) = \frac{\partial}{\partial x_{1j}}\left(\frac{\partial}{\partial y_{rj}} f(\mathfrak{A}x) + m_1 M^{-1}\frac{\partial}{\partial y_{sj}} f(\mathfrak{A}x)\right)$$

$$= \frac{\partial^2}{\partial y_{rj}^2} f(\mathfrak{A}x) + m_1^2 M^{-2}\frac{\partial^2}{\partial y_{sj}^2} f(\mathfrak{A}x) + 2m_1 M^{-1}\frac{\partial^2}{\partial y_{rj}\partial y_{sj}} f(\mathfrak{A}x),$$

where $x_1 = (x_{11}, x_{12}, x_{13})$, $y_r = (y_{r1}, y_{r2}, y_{r3})$ and $y_s = (y_{s1}, y_{s2}, y_{s3})$. Hence,

$$(\Delta_{x_1}g)(x) = (\Delta_{y_r}f)(\mathfrak{A}x) + m_1^2 M^{-2}(\Delta_{y_s}f)(\mathfrak{A}x) + 2m_1 M^{-1}\sum_j \frac{\partial^2 f}{\partial y_{rj}\partial y_{sj}}(\mathfrak{A}x).$$

In a similar manner we derive

$$(\Delta_{x_2} g)(x) = (\Delta_{y_r} f)(\mathfrak{A}x) + m_2^2 M^{-2}(\Delta_{y_s} f)(\mathfrak{A}x) - 2m_2 M^{-1} \sum_j \frac{\partial^2 f}{\partial y_{rj} \partial y_{sj}}(\mathfrak{A}x).$$

Therefore, using the relations $\frac{1}{2m_1} + \frac{1}{2m_2} = \frac{1}{2m}$, $\frac{m_1}{2M^2} + \frac{m_2}{2M^2} = \frac{1}{2M}$, we get

$$(Tg)(x) = -\frac{1}{2m}(\Delta_{y_r} f)(\mathfrak{A}x) - \frac{1}{2M}(\Delta_{y_s} f)(\mathfrak{A}x) + V(x_1 - x_2)f(\mathfrak{A}x).$$

Applying the unitary U to the latter equation gives

$$(UTU^* f)(y) = (UTg)(y) = -\frac{1}{2m}(\Delta_{y_r} f)(y) - \frac{1}{2M}(\Delta_{y_s} f)(y) + V(y_r)f(y),$$

which completes the proof of formula (7.49).   o

7.6 Exercises

1. (*Partial isometries*)
   Let $T \in \mathbf{B}(\mathcal{H}_1, \mathcal{H}_2)$. Prove that the following six conditions are equivalent:
   (i) T is a partial isometry, (ii) T^* is a partial isometry, (iii) $TT^*T = T$,
   (iv) $T^*TT^* = T^*$, (v) T^*T is a projection, (vi) TT^* is a projection.
   If one of these conditions is fulfilled, then $I - T^*T$ and $I - TT^*$ are the projections on the initial space and final space of T, respectively.
2. Let $T = U_T |T|$ be the polar decomposition of T. Prove that $T^* = (U_T)^* |T^*|$ is the polar decomposition of T^* and $T^* = |T|(U_T)^*$.
3. Determine the polar decompositions of the weighted shift operators R_α and L_α defined in Exercise 12 of Chap. 1.
4. Show that there are linear operators A, B, C acting on the Hilbert space $\mathbb{C}^2$ such that $|A^*| \neq |A|$ and $|B + C| \not\leq |B| + |C|$.
5. Let A_1 and A_2 be densely defined closed operators on Hilbert spaces $\mathcal{H}_1$ and $\mathcal{H}_2$, respectively, satisfying relation (7.5). Prove that the following three statements are equivalent:
   (i) A_1 and A_2 are unitarily equivalent, (ii) $\dim \mathcal{N}(A_1) = \dim \mathcal{N}(A_2)$,
   (iii) $\dim \mathcal{N}((A_1)^* A_1 - nI) = \dim \mathcal{N}((A_2)^* A_2 - nI)$ for some $n \in \mathbb{N}_0$.
6. Let A be a densely defined closed operator on $\mathcal{H}$ satisfying (7.5) and $\lambda \in \mathbb{C}$.
   a. Prove that $\langle Ax, Ay \rangle - \langle A^*x, A^*y \rangle = \langle x, y \rangle$ for $x, y \in \mathcal{D}(A) = \mathcal{D}(A^*)$.
   b. Show that the operator $A - \lambda I$ also satisfies (7.5).
   *c. Suppose that A is irreducible and let $\{e_n : n \in \mathbb{N}_0\}$ be an orthonormal basis of $\mathcal{H}$ such that $Ae_n = \sqrt{n}\, e_{n-1}$, $n \in \mathbb{N}_0$. Determine an orthonormal basis $\{f_n : n \in \mathbb{N}_0\}$ such that $(A - \lambda I)f_n = \sqrt{n}\, f_{n-1}$, where $e_{-1} = f_{-1} := 0$.
      Hint: Compute first a nonzero vector $f_0 \in \mathcal{N}(A - \lambda I)$.
*7. Let T be a closed densely defined symmetric operator, and let Z_T be its bounded transform.
   a. Show that $(Z_T)^*(I - (Z_T)^* Z_T)^{1/2} = (I - (Z_T)^* Z_T)^{1/2} Z_T$.
   b. Show that $W_\pm(T) := Z_T \pm i(I - (Z_T)^* Z_T)^{1/2}$ is an isometry.

8. Let T be a symmetric operator such that $T\mathcal{D}(T) \subseteq \mathcal{D}(T)$. Prove the relations $S\mathcal{D}^s(T) \subseteq \mathcal{D}^s(T)$, $S\mathcal{D}^a(T) \subseteq \mathcal{D}^a(T)$, $S\mathcal{D}^b(T) \subseteq \mathcal{D}^b(T)$ for $S \in \{T\}^c$.
   Hint: See the proof of Lemma 7.17. For $\mathcal{D}^a(T)$, use Stirling's formula (7.29).

9. Let A and B be symmetric operators on the same dense domain $\mathcal{D}$ of $\mathcal{H}$ such that $A\mathcal{D} \subseteq \mathcal{D}$, $B\mathcal{D} \subseteq \mathcal{D}$, and $ABx = BAx$ for $x \in \mathcal{D}$. Suppose that the linear span of $A^n B^k (\mathcal{D}^a(A) \cap \mathcal{D}^a(B))$, $n, k \in \mathbb{N}_0$, is dense in $\mathcal{H}$. Show that $\overline{A}$ and $\overline{B}$ are strongly commuting self-adjoint operators.

10. Let $\mathcal{H}_1$ and $\mathcal{H}_2$ be Hilbert spaces.
    a. Given $x_1, x_2 \in \mathcal{H}_1$ and $y_1, y_2 \in \mathcal{H}_2$, decide when $x_1 \otimes y_1 = x_2 \otimes y_2$.
    b. Show that $\mathcal{H}_1 \odot \mathcal{H}_2 = \mathcal{H}_1 \otimes \mathcal{H}_2$ if and only if $\mathcal{H}_1$ or $\mathcal{H}_2$ is finite-dimensional.
    c. Give an example $T_1 \in \mathbf{B}(\mathcal{H}_1)$ such that $T_1 \odot I \neq T_1 \otimes I$ on $\mathcal{H}_1 \otimes \mathcal{H}_2$.

11. Let T_j and S_j be linear operators on $\mathcal{H}_j$, $j = 1, 2$.
    a. Show that $T_1 S_1 \odot T_2 S_2 \subseteq (T_1 \odot T_2)(S_1 \odot S_2)$. Give an example where equality fails. Show that equality holds if $\mathcal{D}(T_1) = \mathcal{H}_1$ and $\mathcal{D}(T_2) = \mathcal{H}_2$.
    b. Show that $(T_1 + S_1) \odot T_2 = T_1 \odot T_2 + S_1 \odot T_2$.

12. Let T_j be a densely defined closable operator, and let B_j be a bounded operator on $\mathcal{H}_j$ such that $\mathcal{D}(B_j) = \mathcal{H}_j$, $j = 1, 2$. Show that
    $(B_1 \otimes B_2)(T_1 \otimes T_2) \subseteq B_1 T_1 \otimes B_2 T_2$ and $T_1 B_1 \otimes T_2 B_2 \subseteq (T_1 \otimes T_2)(B_1 \otimes B_2)$.

13. Let T_1 and T_2 be densely defined closed operators on $\mathcal{H}_1$ resp. $\mathcal{H}_2$.
    a. Show that $U_{T_1 \otimes T_2} = U_{T_1} \otimes U_{T_2}$ and $|T_1 \otimes T_2| = |T_1| \otimes |T_2|$,
    b. Show that $T_1 \otimes T_2 = (U_{T_1} \otimes U_{T_2})(|T_1| \otimes |T_2|)$ is the polar decomposition of $T_1 \otimes T_2$.

For the following three exercises, we assume that T_1 and T_2 are self-adjoint operators on Hilbert spaces $\mathcal{H}_1$ and $\mathcal{H}_2$, respectively.

14. Show that $\{e^{it T_1} \otimes e^{it T_2} : t \in \mathbb{R}\}$ is a unitary group on $\mathcal{H}_1 \otimes \mathcal{H}_2$ with generator equal to the closure of $i(T_1 \otimes I + I \otimes T_2)$.

15. Suppose that $x \in \mathcal{D}^a(T_1)$ and $y \in \mathcal{D}^a(T_2)$. Show that $x \otimes y \in \mathcal{D}^a(T_1 \otimes T_2)$ and $x \otimes y \in \mathcal{D}^a(T_1 \otimes I + I \otimes T_2)$.

16. Let $p(t_1, t_2) \in \mathbb{C}[t_1, t_2]$ be a polynomial.
    a. Show that each vector of the space $\mathcal{D}_b$ defined by (7.42) is a bounded vector for the operator $p(T_1, T_2)$.
    b. Show that $p(T_1, T_2) \upharpoonright \mathcal{D}_b$ is essentially self-adjoint if $p \in \mathbb{R}[t_1, t_2]$.
    c. Show that $\mathcal{D}_b$ is a core for the operator $p(T_1, T_2)$.
       Hint for c.: Show that $\mathcal{D}_b$ is a core for $p(T_1, T_2)^* p(T_1, T_2)$.

17. a. Let T_1 and T_2 be lower semibounded self-adjoint operators on $\mathcal{H}_1$ and $\mathcal{H}_2$, respectively. Show that $T_1 \otimes I + I \otimes T_2$ is closed and hence self-adjoint.
    b. Find self-adjoint operators T_1 and T_2 for which $\sigma(T_1) + \sigma(T_2)$ is not closed.
       Hint: Take diagonal operators with eigenvalues $-n$ resp. $n + \frac{1}{2n}$, $n \in \mathbb{N}$.
    c. Find self-adjoint operators T_1 and T_2 for which $\sigma(T_1) \cdot \sigma(T_2)$ is not closed.
       Hint: Take diagonal operators with eigenvalues n resp. $\frac{1}{n} + \frac{1}{n^2}$, $n \in \mathbb{N}$.

18. Let M and N be closed subsets of $\mathbb{R}$. Prove the following assertions:
    a. $M + N$ is closed if M is bounded or if M and N are lower bounded.
    b. $M \cdot N$ is closed if M is bounded and $0 \notin N$ or if $0 \notin M$ and $0 \notin N$.

7.7 Notes to Part III

Chapter 6:

Stone's Theorem 6.2 was proved in [St3]. The Hille–Yosida Theorem 6.11 was published almost simultaneously in [Hl] and [Yo]. The Trotter formula was proved in [Tr]. Standard books on operator semigroups are [D1, EN, HP, Pa].

Chapter 7:

Theorem 7.3 is essentially due to Tillmann [Ti]. The bounded transform can be found in [Ka]. It is a useful tool for the study of unbounded operators affiliated with C^*-algebras, see, e.g., [Wo].

Analytic vectors appeared first in Harish-Chandra's work [Hs] on Lie group representations. Nelson's Theorem 7.16 was proved in his pioneering paper [Ne1], while Nussbaum's Theorems 7.14 and 7.15 have been obtained in [Nu].

Part IV
Perturbations of Self-adjointness and Spectra

Chapter 8
Perturbations of Self-adjoint Operators

The sum of two unbounded self-adjoint operators A and B is not self-adjoint in general, but it is if the operator B is "small" with respect to the operator A, or in rigorous terms, if B is relatively A-bounded with A-bound less than one. This chapter deals with "perturbations" $A + B$ of a self-adjoint operator A by a "small" self-adjoint operator B. The main motivation stems from quantum mechanics, where $A = -\Delta$, $B = V$ is a potential, and $A + B = -\Delta + V$ is a Schrödinger operator.

In Sect. 8.1 we treat differential operators with constant coefficients on $\mathbb{R}^d$ and their spectral properties. In Sect. 8.2 the self-adjointness of the sum under relatively bounded perturbations is considered, and the Kato–Rellich theorem is proved. The essential spectrum of self-adjoint operators is studied in Sect. 8.4. This is followed, in Sect. 8.5, by Weyl's result on the invariance of the essential spectrum under relatively compact perturbations. In Sects. 8.3 and 8.6 these operator-theoretic results are applied to Schrödinger operators.

8.1 Differential Operators with Constant Coefficients on $L^2(\mathbb{R}^d)$

The main technical tool in this section is the Fourier transform considered as a unitary operator on $L^2(\mathbb{R}^d)$. We shall use some basic results from Appendix C.

For $\alpha = (\alpha_1, \dots, \alpha_d) \in \mathbb{N}_0^d$, we recall the notations $|\alpha| = \alpha_1 + \cdots + \alpha_d$,

$$x^\alpha := x_1^{\alpha_1} \cdots x_d^{\alpha_d}, \quad \text{and} \quad D^\alpha := (-\mathrm{i})^{|\alpha|} \frac{\partial^{\alpha_1}}{\partial x_1^{\alpha_1}} \cdots \frac{\partial^{\alpha_d}}{\partial x_d^{\alpha_d}}.$$

Let $p(x) = \sum_{|\alpha| \le n} a_\alpha x^\alpha$ be a polynomial with complex coefficients $a_\alpha \in \mathbb{C}$ and consider the linear partial differential expression

$$p(D) := \sum_{|\alpha| \le n} a_\alpha D^\alpha.$$

In Sect. 1.3.2 we associated with such an expression two closed operators $p(D)_{\min}$ and $p(D)_{\max}$ on the Hilbert space $L^2(\mathbb{R}^d)$. Recall that $p(D)_{\min}$ is the closure of the

K. Schmüdgen, *Unbounded Self-adjoint Operators on Hilbert Space*, Graduate Texts in Mathematics 265, DOI 10.1007/978-94-007-4753-1_8, © Springer Science+Business Media Dordrecht 2012

operator $p(D)_0$ with domain $C_0^\infty(\mathbb{R}^d)$ defined by $p(D)_0 f = p(D)f$. The domain $\mathcal{D}(p(D)_{\max})$ consists of all $f \in L^2(\mathbb{R}^d)$ for which the distribution $p(D)f$ is given by a function $g \in L^2(\mathbb{R}^d)$, and $p(D)_{\max}$ acts then by $p(D)_{\max} f = g$. By formula (1.22) we have $(p(D)_{\min})^* = p^+(D)_{\max}$, where $p^+(x) := \sum_{|\alpha| \le n} \overline{a_\alpha} x^\alpha$.

In addition, we define yet another operator $p(D)_{\mathcal{F}}$ by $p(D)_{\mathcal{F}} := \mathcal{F}^{-1} M_p \mathcal{F}$. Recall that $\hat{f} = \mathcal{F}(f)$ denotes the Fourier transform of f (see Appendix C) and M_p is the multiplication operator by the polynomial $p(x)$ on $L^2(\mathbb{R}^d)$.

Proposition 8.1 $(p(D)_{\mathcal{F}})^* = p^+(D)_{\mathcal{F}}$ and $p(D)_{\mathcal{F}} = p(D)_{\min} = p(D)_{\max}$.

Proof Since $(M_p)^* = M_{\overline{p}} = M_{p^+}$ (by Example 1.3) and $\mathcal{F}$ is unitary (by Theorem C.4), we have $(p(D)_{\mathcal{F}})^* = \mathcal{F}^{-1}(M_p)^* \mathcal{F} = \mathcal{F}^{-1} M_{p^+} \mathcal{F} = p^+(D)_{\mathcal{F}}$.

To prove the second assertion, it is convenient to introduce the operator $p(D)_1$ with domain $\mathcal{S}(\mathbb{R}^d)$ defined by $p(D)_1 f = p(D)f$ for $f \in \mathcal{S}(\mathbb{R}^d)$.

For $f \in \mathcal{S}(\mathbb{R}^d)$, we have $\hat{f} \in \mathcal{S}(\mathbb{R}^d)$ and $p(D)f = \mathcal{F}^{-1}(p \cdot \mathcal{F}f)$ by formula (C.3). That is, $p(D)_1 \subseteq p(D)_{\mathcal{F}}$. In particular, $p(D)_1$ is closable, because $p(D)_{\mathcal{F}}$ is closed. Since obviously $p(D)_0 \subseteq p(D)_1$, we obtain $p(D)_{\min} = \overline{p(D)_0} \subseteq \overline{p(D)_{\mathcal{F}}}$.

We prove that $\overline{p(D)_1} \subseteq p(D)_{\min}$. Let $f \in \mathcal{D}(p(D)_1) = \mathcal{S}(\mathbb{R}^d)$. We choose a function $\omega \in C_0^\infty(\mathbb{R}^d)$ such that $w(x) = 1$ for $\|x\| \le 1$. Put $\omega_n(x) := \omega(n^{-1}x)$ and $f_n := \omega_n \cdot f$ for $n \in \mathbb{N}$. Fix $\alpha \in \mathbb{N}_0^d$. Applying the Leibniz rule, we derive

$$D^\alpha f_n = \omega_n D^\alpha f + n^{-1} \sum_{|\beta| < |\alpha|} \eta_{n,\beta} D^\beta f, \qquad (8.1)$$

where $\eta_{n,\beta} \in C_0^\infty(\mathbb{R}^d)$ are functions depending on ω and its derivatives such that $\sup_{n,x} |\eta_{n,\beta}(x)| < \infty$. Hence, it follows from (8.1) that $D^\alpha f_n \to D^\alpha f$ in $L^2(\mathbb{R}^d)$. Therefore, $f_n \to f$ and $p(D)f_n \to p(D)f$ in $L^2(\mathbb{R}^d)$. Since $f_n \in C_0^\infty(\mathbb{R}^d)$ and $p(D)_{\min}$ is the closure of $p(D)_0$, the latter implies that $f \in \mathcal{D}(p(D)_{\min})$ and $p(D)_{\min} f = p(D)_1 f$. This proves that $p(D)_1 \subseteq p(D)_{\min}$. Since $p(D)_{\min}$ is closed, this implies that $\overline{p(D)_1} \subseteq p(D)_{\min}$.

Next, we show that $p(D)_{\mathcal{F}} \subseteq \overline{p(D)_1}$. Let $g \in \mathcal{D}(p(D)_{\mathcal{F}})$. Then $\hat{g} \in \mathcal{D}(M_p)$ by the definition of $p(D)_{\mathcal{F}}$. Let $\varepsilon > 0$. Since $C_0^\infty(\mathbb{R}^d)$ is dense in the Hilbert space $L^2(\mathbb{R}^d, (1 + |p|)^2 dx)$, there exists $\varphi_\varepsilon \in C_0^\infty(\mathbb{R}^d)$ such $\|(1 + |p|)(\hat{g} - \varphi_\varepsilon)\| < \varepsilon$. Then $\psi_\varepsilon := \mathcal{F}^{-1}\varphi_\varepsilon \in \mathcal{S}(\mathbb{R}^d)$ and $\hat{\psi}_\varepsilon = \varphi_\varepsilon$, so $\|g - \psi_\varepsilon\| = \|\hat{g} - \varphi_\varepsilon\| < \varepsilon$, because $\mathcal{F}$ is unitary. Since $p(D)\psi_\varepsilon = \mathcal{F}^{-1}(p \cdot \hat{\psi}_\varepsilon) = \mathcal{F}^{-1}(p \cdot \varphi_\varepsilon)$ by (C.3), we obtain

$$\left\| p(D)_{\mathcal{F}} g - P(D)\psi_\varepsilon \right\| = \left\| \mathcal{F}^{-1}\left(p \cdot (\hat{g} - \varphi_\varepsilon)\right) \right\| = \left\| p(\hat{g} - \varphi_\varepsilon) \right\| < \varepsilon.$$

Thus, we have shown that $g \in \mathcal{D}(\overline{p(D)_1})$ and $p(D)_{\mathcal{F}} g = \overline{p(D)_1} g$.

Combining the assertions of the preceding two paragraphs, we conclude that $p(D)_{\mathcal{F}} \subseteq p(D)_{\min}$. Since $p(D)_{\min} \subseteq p(D)_{\mathcal{F}}$ as noted above, $p(D)_{\mathcal{F}} = p(D)_{\min}$.

Applying the adjoint to the latter equality, we obtain $p^+(D)_{\mathcal{F}} = p^+(D)_{\max}$. Replacing now p by p^+, we get $p(D)_{\mathcal{F}} = p(D)_{\max}$. $\qquad\square$

In what follows we denote the operator $p(D)_{\mathcal{F}} = p(D)_{\min} = p(D)_{\max}$ simply by $p(D)$. Then

$$p(D) = \mathcal{F}^{-1} M_p \mathcal{F}. \qquad (8.2)$$

The latter equality means that $\mathcal{D}(p(D)) = \{f \in L^2(\mathbb{R}^d) : p(x)\hat{f}(x) \in L^2(\mathbb{R}^d)\}$ and $p(D)f = \mathcal{F}^{-1}(p \cdot \hat{f})$ for $f \in \mathcal{D}(p(D))$. Since the differential operator $p(D)$ is unitarily equivalent to the multiplication operator M_p, properties of $p(D)$ can be easily derived from those of M_p. By Example 2.1, $\sigma(M_p)$ is the closure $\overline{p(\mathbb{R}^d)}$ of the range of the polynomial p, so we have

$$\sigma(p(D)) = \overline{p(\mathbb{R}^d)}. \tag{8.3}$$

Proposition 8.2 *If $p(x) = \sum_\alpha a_\alpha x^\alpha$ is a polynomial with real coefficients a_α, then:*

(i) *$p(D)$ is a self-adjoint operator on $L^2(\mathbb{R}^d)$.*
(ii) *If g is a Borel function on $\sigma(p(D))$, then $g(p(D)) = \mathcal{F}^{-1}M_{g \circ p}\mathcal{F}$.*
(iii) *If g is a Borel function on $\sigma(p(D))$ such that $g \circ p \in L^2(\mathbb{R}^d)$, then $g(p(D))$ is a convolution operator on $L^2(\mathbb{R}^d)$ with kernel $\mathcal{F}^{-1}(g \circ p)$, that is,*

$$\big(g(p(D))f\big)(x) = (2\pi)^{-d/2} \int_{\mathbb{R}^d} \big(\mathcal{F}^{-1}(g \circ p)\big)(x-y)f(y)\,dy. \tag{8.4}$$

Proof (i) and (ii) follow at once from the unitary equivalence (8.2).

(iii): Let $f \in L^2(\mathbb{R}^d)$. By (ii) we have

$$g(p(D))f = \mathcal{F}^{-1}\big((g \circ p) \cdot \mathcal{F}(f)\big) = \mathcal{F}^{-1}\big(\mathcal{F}(\mathcal{F}^{-1}(g \circ p)) \cdot \mathcal{F}(f)\big). \tag{8.5}$$

Since $\mathcal{F}^{-1}(g \circ p) \in L^2(\mathbb{R}^d)$, the convolution Theorem C.6 applies to the right-hand side of (8.5) and yields $(2\pi)^{-d/2}\mathcal{F}^{-1}(g \circ p) * f$, which gives formula (8.4). $\square$

The most important special case is of course the Laplace operator on $\mathbb{R}^d$. We collect some properties of this operator in the following example.

Example 8.1 (*Laplace operator on $L^2(\mathbb{R}^d)$*) Let $p(x) := \|x\|^2 = x_1^2 + \cdots + x_d^2$ and $p_k(x) := x_k$. Then $p(D) = -\Delta$ and $p_k(D) = D_k \equiv -i\frac{\partial}{\partial x_k}$. Since $M_p = \sum_k M_{p_k}^2$, we have $-\Delta = \sum_{k=1}^{d} D_k^2$. By (8.2) and Proposition 8.2, $-\Delta$ *is the self-adjoint operator on $L^2(\mathbb{R}^d)$ defined by*

$$\mathcal{D}(-\Delta) = \big\{f \in L^2(\mathbb{R}^d) : \|x\|^2\hat{f}(x) \in L^2(\mathbb{R}^d)\big\}, \qquad (-\Delta)f = \mathcal{F}^{-1}(\|x\|^2\hat{f}), \tag{8.6}$$

with spectrum $\sigma(-\Delta) = \overline{p(\mathbb{R}^d)} = [0, \infty)$, and $C_0^\infty(\mathbb{R}^d)$ is a core for $-\Delta$.

(That $C_0^\infty(\mathbb{R}^d)$ is a core follows from the facts that $p(D)_{\min} = -\Delta$ as shown above and that $p(D)_{\min}$ is defined as the closure of $p(D)_0$.)

By Theorem D.3 the domain $\mathcal{D}(-\Delta)$ is just the Sobolev space $H^2(\mathbb{R}^d) = H_0^2(\mathbb{R}^d)$, and we have $\mathcal{D}((-\Delta)^m) = H^{2m}(\mathbb{R}^d) \subseteq C^k(\mathbb{R}^d)$ for $2m > k + d/2$, $k, m \in \mathbb{N}_0$. Note that, in particular, $\mathcal{D}(-\Delta) \subseteq C^1(\mathbb{R})$ if $d = 1$ and $\mathcal{D}(-\Delta) \subseteq C(\mathbb{R}^d)$ if $d = 2, 3$. In general, functions of $\mathcal{D}(-\Delta)$ are not even continuous if $d > 4$.

Next we consider the convolution formula (8.4) in some interesting cases. We only state the formulas without carrying out the corresponding computations.

Let $d \in \mathbb{N}$ and $\lambda \in \mathbb{C}$, $\mathrm{Re}\,\lambda > 0$. Put $g(y) = e^{-\lambda y}$. Then $(g \circ p)(x) = e^{-\lambda \|x\|^2}$ is in $L^2(\mathbb{R}^d)$, and one can compute that $(\mathcal{F}^{-1}(g \circ p))(x) = (2\lambda)^{-d/2} e^{-\|x\|^2/4\lambda}$. Therefore, formula (8.4) yields

$$\left(e^{-\lambda(-\Delta)} f\right)(x) = (4\pi\lambda)^{-d/2} \int_{\mathbb{R}^d} e^{-\|x-y\|^2/4\lambda} f(y)\,dy, \quad f \in L^2(\mathbb{R}^d).$$

For real t and $g(y) = e^{-ity}$, the function $g \circ p$ is obviously not in $L^2(\mathbb{R}^d)$. But nevertheless it can be shown that

$$\left(e^{-it(-\Delta)} f\right)(x) = \lim_{R \to \infty} (4\pi it)^{-d/2} \int_{\|y\| \leq R} e^{i\|x-y\|^2/4t} f(y)\,dy, \quad f \in L^2(\mathbb{R}^d),$$

where lim means the limit in $L^2(\mathbb{R}^d)$. In quantum physics the kernel function

$$P(x, y; t) = (4\pi it)^{-d/2} e^{i\|x-y\|^2/4t}$$

is called the *free propagator*.

These formulas describe the *contraction semigroup* $(0, +\infty) \ni \lambda \to e^{-\lambda(-\Delta)}$ and the *unitary group* $\mathbb{R} \ni t \to e^{it(-\Delta)}$ of the positive self-adjoint operator $-\Delta$.

Now suppose that $d \leq 3$ and $\lambda \in \rho(-\Delta) = \mathbb{C}\setminus[0, \infty)$. Set $g(y) = (y - \lambda)^{-1}$. Then $(g \circ p)(x) = (\|x\|^2 - \lambda)^{-1}$ is in $L^2(\mathbb{R}^d)$, so formula (8.4) applies. The function $G_d(x, y; \lambda) := (\mathcal{F}^{-1}(g \circ p))(x - y)$ is called the *free Green function*.

In dimensions $d = 1$ and $d = 3$ one can compute (see also Exercise 3)

$$G_1(x, y; \lambda) = \frac{i}{2\sqrt{\lambda}} e^{i\sqrt{\lambda}|x-y|}, \qquad G_3(x, y; \lambda) = \frac{e^{i\sqrt{\lambda}\|x-y\|}}{4\pi\|x-y\|}, \tag{8.7}$$

where $\sqrt{\lambda}$ denotes the square root of λ with positive imaginary part. For the resolvent of $-\Delta$ on $L^2(\mathbb{R}^3)$, we therefore obtain from formula (8.4),

$$\left((-\Delta - \lambda I)^{-1} f\right)(x) = \frac{1}{4\pi} \int_{\mathbb{R}^3} \frac{e^{i\sqrt{\lambda}\,\|x-y\|}}{\|x-y\|} f(y)\,dy, \quad f \in L^2(\mathbb{R}^3). \qquad \circ$$

8.2 Relatively Bounded Perturbations of Self-adjoint Operators

Let A and B be linear operators on a Hilbert space $\mathcal{H}$.

Definition 8.1 We say that B is *relatively A-bounded* if $\mathcal{D}(B) \supseteq \mathcal{D}(A)$ and there exist nonnegative real numbers a and b such that

$$\|Bx\| \leq a\|Ax\| + b\|x\| \quad \text{for all } x \in \mathcal{D}(A). \tag{8.8}$$

The infimum of all $a \geq 0$ for which there exists a number $b \geq 0$ such that (8.8) holds is called the *A-bound* of B and denoted by $\alpha_A(B)$.

It is easily seen that B is relatively A-bounded if and only if $\mathcal{D}(B) \supseteq \mathcal{D}(A)$ and B maps the normed space $\mathcal{D}_A := (\mathcal{D}(A), \|\cdot\|_A)$ continuously in $\mathcal{H}$.

If the operator B is bounded, then B is obviously relatively A-bounded, and (8.8) holds with $a = \alpha_A(B) = 0$ and $b = \|B\|$. However, the following simple lemma, applied to an unbounded operator B, shows that in general there is no constant b such that (8.8) is satisfied for $a = \alpha_A(B)$.

Lemma 8.3 *Each symmetric operator B is relatively B^2-bounded with B^2-bound equal to zero.*

Proof Obviously, $\mathcal{D}(B) \supseteq \mathcal{D}(B^2)$. For $\varepsilon > 0$ and $x \in \mathcal{D}(B^2)$, we have

$$\left\langle (\varepsilon^2 B^2 - I)x, (\varepsilon^2 B^2 - I)x \right\rangle = \varepsilon^4 \|B^2 x\|^2 - 2\varepsilon^2 \|Bx\|^2 + \|x\|^2 \geq 0,$$

so $\varepsilon^2 \|Bx\|^2 \leq (\varepsilon^2 \|B^2 x\| + \|x\|)^2$, and hence $\|Bx\| \leq \varepsilon \|B^2 x\| + \varepsilon^{-1} \|x\|$. $\qquad\square$

Lemma 8.4 *If A is a closed operator and B is a densely defined closable operator on a Hilbert space $\mathcal{H}$ such that $\mathcal{D}(B) \supseteq \mathcal{D}(A)$, then B is relatively A-bounded.*

Proof Since A is closed, $\mathcal{D}_A = (\mathcal{D}(A), \|\cdot\|_A)$ is a Hilbert space by Proposition 1.4.

We show that B, considered as a mapping of $\mathcal{D}_A$ into $\mathcal{H}$, is closed. Suppose that $x_n \to x$ in $\mathcal{D}_A$ and $Bx_n \to y$ in $\mathcal{H}$. Then $x \in \mathcal{D}(A) \subseteq \mathcal{D}(B)$ and for $u \in \mathcal{D}(B^*)$,

$$\langle Bx_n, u \rangle = \langle x_n, B^* u \rangle \to \langle y, u \rangle = \langle x, B^* u \rangle = \langle Bx, u \rangle.$$

Since B is closable, $\mathcal{D}(B^*)$ is dense in $\mathcal{H}$. Hence, the equality $\langle y, u \rangle = \langle Bx, u \rangle$ for $u \in \mathcal{D}(B^*)$ implies that $y = Bx$. This proves that $B : \mathcal{D}_A \to \mathcal{H}$ is closed.

From the closed graph theorem we conclude that $B : \mathcal{D}_A \to \mathcal{H}$ is continuous. Hence, B is relatively A-bounded. $\qquad\square$

The next result is the celebrated *Kato–Rellich theorem*. It requires an A-bound strictly *less than one*.

Theorem 8.5 *Let A be a self-adjoint operator on $\mathcal{H}$. Suppose that B is a relatively A-bounded symmetric operator on $\mathcal{H}$ with A-bound $a_A(B) < 1$. Then:*

(i) *The operator $A + B$ on $\mathcal{D}(A + B) = \mathcal{D}(A)$ is self-adjoint.*
(ii) *If A is essentially self-adjoint on $\mathcal{D} \subseteq \mathcal{D}(A)$, so is $A + B$ on $\mathcal{D}$.*

Proof (i): By assumption, there are positive constants a, b such that $a < 1$ and (8.8) holds. Suppose that $\beta \in \mathbb{R}$, $\beta \neq 0$. Since A is symmetric, by formula (3.2),

$$\left\| (A - i\beta I)x \right\|^2 = \|Ax\|^2 + |\beta|^2 \|x\|^2$$

for $x \in \mathcal{D}(A)$, so that $\|Ax\| \leq \|(A - i\beta I)x\|$ and $\|x\| \leq |\beta|^{-1} \|(A - i\beta I)x\|$. Combined with (8.8), we therefore obtain

$$\|Bx\| \leq \left(a + b|\beta|^{-1}\right) \|(A - i\beta I)x\|, \quad x \in \mathcal{D}(A). \tag{8.9}$$

Since A is self-adjoint, $i\beta \in \rho(A)$ by Corollary 3.14, and $x := (A - i\beta I)^{-1}y$ is in $\mathcal{D}(A)$ for any $y \in \mathcal{H}$. Inserting this into (8.9), we get

$$\left\| B(A - i\beta I)^{-1} y \right\| \leq \left(a + b|\beta|^{-1}\right) \|y\|, \quad y \in \mathcal{H}.$$

Since $a < 1$, $a + b|\beta|^{-1} < 1$ for large $|\beta|$. Hence, $C := B(A - i\beta I)^{-1}$ is a bounded operator with domain $\mathcal{D}(C) = \mathcal{H}$ and norm less than 1. Therefore, $I + C$ has a bounded inverse defined on $\mathcal{H}$ (given by the series $(I + C)^{-1} = \sum_{n=0}^{\infty}(-C)^n$), and hence $\mathcal{R}(I + C) = \mathcal{H}$. Since $i\beta \in \rho(A)$, $\mathcal{R}(A - i\beta I) = \mathcal{H}$. By the definition of C we have $(I + C)(A - i\beta I)x = (A + B - i\beta I)x$ for $x \in \mathcal{D}(A)$. Hence, we obtain $\mathcal{R}(A + B - i\beta I) = \mathcal{H}$ for large $|\beta|$, so $A + B$ is self-adjoint by Proposition 3.13.

(ii): We have to show that $\mathcal{D}$ is a core for $A + B$, provided that $\mathcal{D}$ is a core for A. Let $x \in \mathcal{D}(A)$. Since $\mathcal{D}$ is a core for A, there exists a sequence $(x_n)_{n \in \mathbb{N}}$ from $\mathcal{D}$ such that $x_n \to x$ and $Ax_n \to Ax$. From (8.8) we obtain

$$\left\|(A + B)(x_n - x)\right\| \leq (1 + a)\left\|A(x_n - x)\right\| + \|x_n - x\|,$$

so that $(A + B)x_n \to (A + B)x$. This shows that $\mathcal{D}$ is also a core for $A + B$. $\qquad\square$

If the A-bound is one, then assertion (i) of Theorem 8.5 is no longer true. For instance, if A is an unbounded self-adjoint operator and $B = -A$, then $A + B = 0 \upharpoonright \mathcal{D}(A)$ is not closed and so not self-adjoint, but it is e.s.a. The next result, due to R. Wüst, says that the operator $A + B$ is always e.s.a. if (8.8) holds with $a = 1$.

Proposition 8.6 *Suppose that A is an essentially self-adjoint operator and B is a symmetric operator on $\mathcal{H}$ such that $\mathcal{D}(B) \supseteq \mathcal{D}(A)$. Suppose there is a constant $b > 0$ such that*

$$\|Bx\| \leq \|Ax\| + b\|x\| \quad \textit{for } x \in \mathcal{D}(A). \tag{8.10}$$

Then $A + B$ is essentially self-adjoint on $\mathcal{D}(A)$ and on any core $\mathcal{D} \subseteq \mathcal{D}(A)$ for A.

Proof Let $(c_n)_{n \in \mathbb{N}}$ be a sequence from the interval $(0, 1)$ such that $\lim_{n \to \infty} c_n = 1$. Set $T := (A + B)\upharpoonright \mathcal{D}$. Suppose that $x \in \mathcal{R}(T + iI)^{\perp}$. We will show that $x = 0$.

Since $c_n B$ is relatively A-bounded with A-bound $a_A(c_n B) \leq c_n < 1$ by (8.10) and $\mathcal{D}$ is a core for A, the symmetric operator $T_n := (A + c_n B)\upharpoonright \mathcal{D}$ is e.s.a. by Theorem 8.5(ii). Therefore, $\mathcal{R}(T_n + iI)$ is dense in $\mathcal{H}$ by Proposition 3.8. Hence, there exists a sequence $(x_n)_{n \in \mathbb{N}}$ from $\mathcal{D}(T_n) = \mathcal{D}$ such that

$$x = \lim_{n \to \infty} (T_n + iI)x_n. \tag{8.11}$$

Since T_n is symmetric, $\|T_n x_n\| \leq \|(T_n + iI)x_n\|$ and $\|x_n\| \leq \|(T_n + iI)x_n\|$ by formula (3.2). Using assumption (8.10) and these inequalities, we estimate

$$\begin{aligned}
\left\|(T_n - T)x_n\right\| &= (1 - c_n)\|Bx_n\| \leq \|Ax_n\| - c_n\|Bx_n\| + b\|x_n\| \\
&\leq \left\|(A + c_n B)x_n\right\| + b\|x_n\| = \|T_n x_n\| + b\|x_n\| \\
&\leq (b + 1)\left\|(T_n + iI)x_n\right\|.
\end{aligned}$$

Combined with (8.11) the preceding implies that

$$\limsup_{n \to \infty} \|x_n\| \leq \|x\| \quad \text{and} \quad \limsup_{n \to \infty}\left\|(T_n - T)x_n\right\| \leq (b + 1)\|x\|. \tag{8.12}$$

Since $\mathcal{D}$ is dense in $\mathcal{H}$, given $\varepsilon > 0$, there exists $y_\varepsilon \in \mathcal{D} = \mathcal{D}(T_n) = \mathcal{D}(T)$ such that $\|x - y_\varepsilon\| < \varepsilon$. Using (8.11), the facts that $x \perp (T + \mathrm{i}I)x_n$ and $\lim_{n\to\infty} c_n = 1$, and both inequalities from (8.12), we get

$$
\begin{aligned}
\|x\|^2 &= \lim_{n\to\infty} \langle x, (T_n + \mathrm{i}I)x_n\rangle = \lim_{n\to\infty} \big(\langle x, (T_n-T)x_n\rangle + \langle x, (T + \mathrm{i}I)x_n\rangle\big) \\
&= \lim_{n\to\infty} \big(\langle x-y_\varepsilon, (T_n-T)x_n\rangle + \langle y_\varepsilon, (T_n-T)x_n\rangle\big) \\
&\le \|x-y_\varepsilon\|\limsup_{n\to\infty}\|(T_n-T)x_n\| + \limsup_{n\to\infty}\big(\|(T_n-T)y_\varepsilon\|\|x_n\|\big) \\
&\le \varepsilon(b+1)\|x\| + \lim_{n\to\infty}(1-c_n)\|By_\varepsilon\|\|x\| = \varepsilon(b+1)\|x\|,
\end{aligned}
$$

so $\|x\| \le \varepsilon(b+1)$. Since $\varepsilon > 0$ is arbitrary, $x = 0$. This proves that $\mathcal{R}(T + \mathrm{i}I)^\perp = \{0\}$. Similarly, $\mathcal{R}(T-\mathrm{i}I)^\perp = \{0\}$. Hence, $T = (A + B)\!\restriction\!\mathcal{D}$ is e.s.a. by Proposition 3.8. $\qquad\square$

8.3   Applications to Schrödinger Operators: Self-adjointness

One of the most important classes of operators in nonrelativistic quantum mechanics are the so-called *Schrödinger operators*. These are operators of the form

$$
H = -\frac{\hbar^2}{2m}\Delta + V(x),
$$

acting on the Hilbert space $L^2(\mathbb{R}^d)$, where Δ denotes the Laplace operator and $V(x)$ is a real-valued measurable function on $\mathbb{R}^d$, called the *potential*. Schrödinger operators describe the energy of quantum systems, such as atoms, molecules and nuclei. The symbol m refers to the mass, and $\hbar$ is Planck's constant. Upon scaling coordinates we can assume that the constant in front of the Laplacian is -1. The potential acts as a multiplication operator, that is, $(Vf)(x) = V(x)f(x)$ for f in the domain $\mathcal{D}(V) = \{f \in L^2(\mathbb{R}^d) : V\cdot f \in L^2(\mathbb{R}^d)\}$.

The next proposition is about the domain of the operator $-\Delta$. It is the main technical ingredient for our first result (Theorem 8.8 below) on the self-adjointness of Schrödinger operators. In its proof we essentially use the Fourier transform.

For $p \in [1, \infty]$, let $\|\cdot\|_p$ denote the norm of $L^p(\mathbb{R}^d)$. Recall that $\|\cdot\|_2 = \|\cdot\|$.

Proposition 8.7 *We set $q = \infty$ for $d \le 3$ and assume that $2 \le q < +\infty$ for $d = 4$ and $2 \le q < 2d/(d-4)$ for $d \ge 5$. Then, for each $a > 0$, there exists a constant $b > 0$, depending on a, d, and q, such that $f \in L^q(\mathbb{R}^d)$ and*

$$
\|f\|_q \le a\|-\Delta f\| + b\|f\| \quad \text{for all } f \in \mathcal{D}(-\Delta). \tag{8.13}
$$

Proof In the first part of the proof we derive inequality (8.13) with some constants, while in the second part the constant $a > 0$ will be made arbitrary small.

Let $f \in \mathcal{D}(-\Delta)$ and set $g = \mathcal{F}(f)$. By (8.6), we have $\|x\|^2 g(x) \in L^2(\mathbb{R}^d)$ and $-\Delta f = \mathcal{F}^{-1}(\|x\|^2 g(x))$. Since the Fourier transform $\mathcal{F}$ is unitary, it follows that

$$
\|-\Delta f\| = \big\|\|x\|^2 g(x)\big\| \quad \text{and} \quad \|f\| = \|g\|. \tag{8.14}
$$

First, assume that $d \leq 3$. Then $c := \|(1 + \|x\|^2)^{-1}\| = (\int_{\mathbb{R}^d} (1 + \|x\|^2)^{-2} dx)^{1/2}$ is finite. Applying the Hölder inequality (B.5), we derive

$$\|g\|_1 = \|(1 + \|x\|^2)^{-1}(1 + \|x\|^2)g(x)\|_1$$
$$\leq c\|(1 + \|x\|^2)g(x)\|_2 \leq c\|\|x\|^2 g(x)\| + c\|g\|. \tag{8.15}$$

Thus, $g \in L^1(\mathbb{R}^d)$. Therefore, we have $f = \mathcal{F}^{-1}g \in L^\infty(\mathbb{R}^d)$ and $\|f\|_\infty \leq \|g\|_1$. Combining the latter and (8.14) with (8.15), we get

$$\|f\|_\infty \leq c\|-\Delta f\| + c\|f\|. \tag{8.16}$$

Next, suppose that $d \geq 4$. First, let $q > 2$. We define p and r by $p^{-1} + q^{-1} = 2^{-1}$, so $p = 2q/(q-2)$, and $p^{-1} + 2^{-1} = r^{-1}$. In the case $d \geq 5$ we have $q < 2d/(d-4)$ by assumption. This yields $p > d/2$. Since $q > 2$, this also holds for $d = 4$. Since $p > d/2$, we have $c := \|(1 + \|x\|^2)^{-1}\|_p < \infty$. By Hölder's inequality (B.5),

$$\|g\|_r = \|(1 + \|x\|^2)^{-1}(1 + \|x\|^2)g\|_r \leq c\|(1 + \|x\|^2)g\|_2 \leq c\|\|x\|^2 g(x)\| + c\|g\|.$$

Note that $q^{-1} + r^{-1} = 1$ and $q > 2$. Therefore, by the Hausdorff–Young Theorem C.5, $\mathcal{F}$ maps $L^r(\mathbb{R}^d)$ continuously into $L^q(\mathbb{R}^d)$, so there exists a constant $c_1 > 0$ such that $\|\mathcal{F}g\|_q \leq c_1\|g\|_r$. Since $f(x) = (\mathcal{F}^{-1}g)(x) = (\mathcal{F}g)(-x)$, this yields $\|f\|_q \leq c_1\|g\|_r$. Inserting this into the above inequality and using (8.14), we obtain

$$\|f\|_q \leq c_1 c\|-\Delta f\| + c_1 c\|f\|. \tag{8.17}$$

If $c_1 c = 1$, inequality (8.17) is trivially true for $q = 2$, which was excluded. Summarizing, (8.16) and (8.17) show that (8.13) is valid for *some* constants.

To complete the proof, we have to make the constants in front of $\|-\Delta f\|$ *small*. For this reason, we scale the function f by setting $f_\gamma(x) = f(\gamma x)$ for $d \leq 3$ and $f_\gamma(x) = \gamma^{d/q} f(\gamma x)$ for $d \geq 4$, where $\gamma > 0$. Clearly, $\Delta f_\gamma = \gamma^2 (\Delta f)_\gamma$. Then

$$\|f_\gamma\|_\infty = \|f\|_\infty, \qquad \|-\Delta f_\gamma\| = \gamma^{2-d/2}\|-\Delta f\|,$$
$$\|f_\gamma\| = \gamma^{-d/2}\|f\| \quad \text{if } d \leq 3,$$
$$\|f_\gamma\|_q = \|f\|_q, \qquad \|-\Delta f_\gamma\| = \gamma^{2+d/q-d/2}\|-\Delta f\|,$$
$$\|f_\gamma\| = \gamma^{d/q-d/2}\|f\| \quad \text{if } d \geq 4.$$

Replacing f by f_γ in inequalities (8.16) and (8.17) therefore yields

$$\|f\|_\infty \leq c\gamma^{2-d/2}\|-\Delta f\| + c\gamma^{-d/2}\|f\| \quad \text{for } d \leq 3,$$
$$\|f\|_q \leq c_1 c\gamma^{2+d/q-d/2}\|-\Delta f\| + c_1 c\gamma^{d/q-d/2}\|f\| \quad \text{for } d \geq 4.$$

If $d \leq 3$, then $2 - d/2 > 0$. If $d \geq 4$, then $q < 2d/(d-4)$, and so $2 + d/q - d/2 > 0$. Hence, for sufficiently small $\gamma > 0$, the coefficients in front of $\|-\Delta f\|_2$ are smaller than any given positive a. This completes the proof of the proposition. $\qquad\square$

If $d \leq 3$, then $\mathcal{D}(-\Delta)$ is contained in the space $C_0(\mathbb{R}^d)$ of continuous functions on $\mathbb{R}^d$ vanishing at infinity. This follows from the Riemann–Lebesgue Lemma C.1, since $\mathcal{F}f = g \in L^1(\mathbb{R}^d)$ for $f \in \mathcal{D}(-\Delta)$ by (8.15). It can be also easily derived from (8.13). Indeed, since $\mathcal{S}(\mathbb{R}^d)$ is a core for $-\Delta$, each $f \in \mathcal{D}(-\Delta)$ is the limit of a sequence (f_n) from $\mathcal{S}(\mathbb{R}^d)$ in the graph norm of $-\Delta$. From (8.13), applied to $f_n - f_k$, it follows that (f_n) is a Cauchy sequence in the uniform convergence on $\mathbb{R}^d$. Because $f_n \in C_0(\mathbb{R}^d)$ and $(C_0(\mathbb{R}^d), \| \cdot \|_\infty)$ is complete, f is in $C_0(\mathbb{R}^d)$.

Definition 8.2 A *Kato–Rellich potential* is a Borel function V on $\mathbb{R}^d$ such that

$$V \in L^2(\mathbb{R}^d) + L^\infty(\mathbb{R}^d) \quad \text{if } d = 1, 2, 3,$$
$$V \in L^p(\mathbb{R}^d) + L^\infty(\mathbb{R}^d) \quad \text{for some } p > d/2 \text{ if } d \geq 4.$$

Clearly, $V \in L^{p_1}(\mathbb{R}^d) + L^{p_2}(\mathbb{R}^d)$ means that V can be written as $V = V_1 + V_2$ with $V_1 \in L^{p_1}(\mathbb{R}^d)$ and $V_2 \in L^{p_2}(\mathbb{R}^d)$.

The reason for this terminology is that this is a natural class of potentials for which the Kato–Rellich theorem, Theorem 8.5, applies with $A = -\Delta$ and $B = V$. The proof of the next theorem even shows that each Kato–Rellich potential is relatively $(-\Delta)$-bounded with $(-\Delta)$-bound zero.

Theorem 8.8 *Let V be a real-valued Kato–Rellich potential. Then $-\Delta + V$ is self-adjoint on $\mathcal{D}(-\Delta)$ and essentially self-adjoint on each core for $-\Delta$.*

Proof We write V as $V = V_1 + V_2$ with $V_2 \in L^\infty(\mathbb{R}^d)$ and $V_1 \in L^p(\mathbb{R}^d)$, where $p = 2$ for $d \leq 3$ and $p > d/2$ for $d \geq 4$.

First, suppose that $d \leq 3$. Applying (8.13) with $q = \infty$, we get for $f \in \mathcal{D}(-\Delta)$,

$$\|Vf\| \leq \|V_1\|_2 \|f\|_\infty + \|V_2\|_\infty \|f\| \leq a\|V_1\|_2 \|-\Delta f\| + \left(b\|V_1\|_2 + \|V_2\|_\infty\right)\|f\|.$$

Now let $d \geq 4$. Define q by $q^{-1} + p^{-1} = 2^{-1}$, that is, $q = 2p/(p - 2)$. If $d = 4$, then $p > 2$, and so $q \geq 2$. For $d \geq 5$, we have $p > d/2$ which implies that $2 \leq q < 2d/(d-4)$. That is, q satisfies the assumptions of Proposition 8.7. Using first the Hölder inequality (B.5) and then (8.13), we derive for $f \in \mathcal{D}(-\Delta)$,

$$\|V_1 f\|_2 \leq \|V_1\|_p \|f\|_q \leq a\|V_1\|_p \|-\Delta f\| + b\|V_1\|_p \|f\|. \tag{8.18}$$

Since $a > 0$ was arbitrary, these inequalities show that V has $(-\Delta)$-bound zero. Hence, the assertions follow from the Kato–Rellich theorem. □

Example 8.2 Let $\alpha \in [0, \infty)$. Suppose that $\alpha < d/2$ if $d \leq 3$ and $\alpha < 2$ if $d \geq 4$. Then, for any $c \in \mathbb{R}$, the function

$$V(x) = \frac{c}{\|x\|^\alpha}$$

on $\mathbb{R}^d$ is a Kato–Rellich potential, and hence $-\Delta + V$ is self-adjoint on $\mathcal{D}(-\Delta)$.

Indeed, let χ_n be the characteristic function of $\{x : \|x\| < n\}$ for $n \in \mathbb{N}$ and set $V_{1,n} := V \cdot \chi_n$. Then $V_{1,n} \in L^2(\mathbb{R}^d)$ if $d < 3$ and $V_{1,n} \in L^p(\mathbb{R}^d)$ if $d \geq 4$, where p is chosen such that $d/2 < p < d/\alpha$. Obviously, $V_{2,n} := V - V_{1,n} \in L^\infty(\mathbb{R}^d)$.   ∘

Example 8.3 The Schrödinger operator of a particle in an electric field with potential V and a magnetic field with vector potential $\mathfrak{a} = (a_1, a_2, a_3)$ is

$$T = \sum_{k=1}^{3} (D_k - a_k)^2 + V(x).$$

(Here some physical constants have been set equal to 1.) We easily compute

$$T = -\Delta - 2 \sum_{k=1}^{3} a_k D_k + V_0(x), \quad \text{where } V_0(x) := \mathrm{i}\,\mathrm{div}\, a + a_1^2 + a_2^2 + a_3^2 + V(x).$$

Statement *Let $V \in L^2(\mathbb{R}^3) + L^\infty(\mathbb{R}^3)$, and $a_1, a_2, a_3 \in C^1(\mathbb{R}^3)$ be real-valued functions such that $a_1, a_2, a_3, \mathrm{div}\, a \in L^\infty(\mathbb{R}^3)$. Then T is self-adjoint on $\mathcal{D}(-\Delta)$.*

Proof The domain of $\mathcal{D}(T)$ is $\mathcal{D}(-\Delta)$. Fix $\varepsilon > 0$. For $f \in \mathcal{D}(T)$, we derive

$$\|D_k f\|^2 = \langle D_k f, D_k f \rangle = \langle D_k^2 f, f \rangle \leq \langle -\Delta f, f \rangle$$
$$\leq \|-\Delta f\| \, \|f\| \leq \left(\varepsilon \|-\Delta f\| + \varepsilon^{-1} \|f\|\right)^2. \tag{8.19}$$

Since $V_0 \in L^2(\mathbb{R}^3) + L^\infty(\mathbb{R}^3)$ by the assumptions, V_0 has the $(-\Delta)$-bound zero as shown in the proof of Theorem 8.8. Since $a_1, a_2, a_3 \in L^\infty(\mathbb{R}^3)$, it follows from (8.19) that $a_k D_k$ has the $(-\Delta)$-bound zero. These facts imply that the symmetric operator $B := -2 \sum_{k=1}^{3} a_k D_k + V_0$ is relatively $(-\Delta)$-bounded with $(-\Delta)$-bound zero. Therefore, by Theorem 8.5, the operator $T = -\Delta + B$ is self-adjoint on $\mathcal{D}(-\Delta)$ and e.s.a. on any core for $-\Delta$.   ☐ ∘

Lemma 8.9 *If V is a Borel function on $\mathbb{R}^d$ such that $\mathcal{D}(V) \supseteq \mathcal{D}(-\Delta)$, then we have $V \in L^2_{\mathrm{loc}}(\mathbb{R}^d)$ and*

$$\delta_V := \sup_{c \in \mathbb{R}^d} \int_{\mathcal{B}(c)} |V(x)|^2 \, dx < \infty, \tag{8.20}$$

where $\mathcal{B}(c) := \{x \in \mathbb{R}^d : \|x - c\| \leq 1\}$ is the closed unit ball centered at c.

Proof Since $\mathcal{D}(V) \supseteq \mathcal{D}(-\Delta)$, it follows from Lemma 8.4 that V is $(-\Delta)$-bounded, that is, there are positive constants a and b such that

$$\|Vf\| \leq a \|-\Delta f\| + b \|f\|, \quad f \in \mathcal{D}(-\Delta).$$

We choose $\eta \in C_0^\infty(\mathbb{R}^d)$ such that $\eta = 1$ on the unit ball $\mathcal{B}(0)$. For $c \in \mathbb{R}^d$, we put $\eta_c(x) = \eta(x - c)$. Then $\eta_c = 1$ on $\mathcal{B}(c)$, and

$$\int_{\mathcal{B}(c)} |V(x)|^2 \, dx \leq \|V\eta_c\|^2 \leq \left(a \|-\Delta \eta_c\| + b \|\eta_c\|\right)^2 = \left(a \|-\Delta \eta\| + b \|\eta\|\right)^2.$$

Since the right-hand side does not depend on c, the supremum of the left-hand side over $c \in \mathbb{R}^d$ is finite, that is, $\delta_V < \infty$.   □

Lemma 8.9 says that the condition $\delta_V < \infty$ is necessary for a function V on $\mathbb{R}^d$ to be relatively $(-\Delta)$-bounded. The next proposition shows (among others) that for $d = 1$, this condition is also sufficient. With a slightly more involved proof, this assertion holds in dimensions $d \leq 3$ (see, e.g., [RS4], Theorem XIII.96).

Proposition 8.10 *For a Borel function V on $\mathbb{R}$, the following are equivalent:*

(i) *V is relatively $(-\frac{d^2}{dx^2})$-bounded with $(-\frac{d^2}{dx^2})$-bound zero.*

(ii) *V is relatively $(-\frac{d^2}{dx^2})$-bounded.*

(iii) *$\mathcal{D}(V) \supseteq \mathcal{D}(-\frac{d^2}{dx^2})$.*

(iv) *$\delta_V = \sup_{c \in \mathbb{R}} \int_{c-1}^{c+1} |V(x)|^2 \, dx < \infty$.*

Proof The implications (i) $\rightarrow$ (ii) $\rightarrow$ (iii) are trivial, and (iii) $\rightarrow$ (iv) follows from Lemma 8.9. It remains to prove that (iv) implies (i).

Recall that $\mathcal{D}(-\frac{d^2}{dx^2}) = H^2(\mathbb{R})$. Fix $c \in \mathbb{R}$. First, let $f \in H^2(\mathbb{R})$ be real-valued. Note that $H^2(\mathbb{R}) \subseteq C^1(\mathbb{R})$. For arbitrary $x, y \in \mathcal{B}(c) = [c-1, c+1]$, we have

$$f(x)^2 - f(y)^2 = \int_y^x 2f(t) f'(t) \, dt \leq \int_{c-1}^{c+1} f'(t)^2 \, dt + \int_{c-1}^{c+1} f(t)^2 \, dt.$$

Since f is real-valued, by the mean value theorem for integrals we can choose a number $y \in [c-1, c+1]$ such that $2f(y)^2 = \int_{c-1}^{c+1} f(t)^2 \, dt$. Then

$$\left| f(x) \right|^2 \leq \int_{c-1}^{c+1} \left| f'(t) \right|^2 \, dt + 2 \int_{c-1}^{c+1} \left| f(t) \right|^2 \, dt \quad \text{for } x \in [c-1, c+1].$$

$$(8.21)$$

By decomposing f into real and imaginary parts (8.21) remains valid for complex-valued $f \in H^2(\mathbb{R})$. Multiplying by $|V(x)|^2$ and integrating over $[c-1, c+1]$ yields

$$\int_{c-1}^{c+1} \left| V(x) f(x) \right|^2 \, dx \leq \delta_V \int_{c-1}^{c+1} \left| f'(t) \right|^2 \, dt + 2\delta_V \int_{c-1}^{c+1} \left| f(t) \right|^2 \, dt.$$

By summing over $c \in 2\mathbb{Z}$ we derive

$$\|Vf\|^2 \leq \delta_V \|f'\|^2 + 2\delta_V \|f\|^2 \leq \delta_V \left(\|f'\| + 2\|f\| \right)^2. \qquad (8.22)$$

Let $\varepsilon > 0$. By the proof of Lemma 8.3, applied to the operator $B = -i\frac{d}{dx}$, we have

$$\|f'\| = \|Bf\| \leq \varepsilon \|B^2 f\| + \varepsilon^{-1} \|f\| = \varepsilon \|-f''\| + \varepsilon^{-1} \|f\|. \qquad (8.23)$$

Combining (8.22) and (8.23), we obtain

$$\|Vf\| \leq \varepsilon \delta_V^{1/2} \|-f''\| + (2 + \varepsilon^{-1}) \delta_V^{1/2} \|f\|. \qquad (8.24)$$

This proves that V is relatively $(-\frac{d^2}{dx^2})$-bounded with $(-\frac{d^2}{dx^2})$-bound zero.   □

8.4 Essential Spectrum of a Self-adjoint Operator

In this section, A denotes a *self-adjoint* operator on $\mathcal{H}$ with spectral measure E_A.

Definition 8.3 The *discrete spectrum* $\sigma_d(A)$ of A is the set of all eigenvalues of A of finite multiplicities which are isolated points of the spectrum $\sigma(A)$. The complement set $\sigma_{\mathrm{ess}}(A) := \sigma(A)\backslash\sigma_d(A)$ is called the *essential spectrum* of A.

By Corollary 5.11, an isolated point of the spectrum $\sigma(A)$ is always an eigenvalue of A. Therefore, a number belongs to the essential spectrum $\sigma_{\mathrm{ess}}(A)$ of A if and only if it is an accumulation point of $\sigma(A)$ or an eigenvalue of infinite multiplicity.

Remark Definition 8.3 is the standard definition of the essential spectrum of a self-adjoint operator used in most books, see, e.g., [RS1, BS, We]. Let T be a *closed* operator on a Hilbert space $\mathcal{H}$. There are various definitions of the essential spectrum of T in the literature (see [EE, I.3 and IX.1] for a detailed discussion), but all of them coincide with Definition 8.3 when T is self-adjoint [EE, Theorem IX.1.6].

Let $\rho_F(T)$ (resp. $\rho_{SF}(T)$) denote the set of all complex numbers λ such that $T - \lambda I$ is a Fredholm operator (resp. semi-Fredholm operator), that is, $\mathcal{R}(T-\lambda I)$ is closed and $\dim \mathcal{N}(T-\lambda I) < \infty$ and (resp. or) $\dim \mathcal{H}/\mathcal{R}(T-\lambda I) < \infty$. Then the set $\sigma_{\mathrm{ess}}(T) := \mathbb{C}\backslash\rho_F(T)$ is called the *essential spectrum* of the closed operator T. Note that in [K2] the essential spectrum is defined by the smaller set $\mathbb{C}\backslash\rho_{SF}(T)$.

Let us illustrate Definition 8.3 by a very simple example.

Example 8.4 Let $(\alpha_n)_{n\in\mathbb{N}}$ be a real sequence and define a self-adjoint operator A by $A(\varphi_n) = (\alpha_n\varphi_n)$ with domain $\mathcal{D}(A) = \{(\varphi_n)\in l^2(\mathbb{N}) : (\alpha_n\varphi_n) \in l^2(\mathbb{N})\}$ on $l^2(\mathbb{N})$. As in Example 2.2, $\sigma(A)$ is the closure of the set $\{\alpha_n : n \in \mathbb{N}\}$, and each α_n is an eigenvalue of A. The essential spectrum $\sigma_{\mathrm{ess}}(A)$ consists of all $\lambda \in \mathbb{R}$ such that the set $\{n \in \mathbb{N} : \alpha_n \in (\lambda-\varepsilon, \lambda + \varepsilon)\}$ is infinite for each $\varepsilon > 0$. For instance, if the sequence (α_n) converges and $\alpha := \lim \alpha_n$, then $\sigma(A) = \{\alpha, \alpha_n : n \in \mathbb{N}\}$ and $\sigma_{\mathrm{ess}}(A) = \{\alpha\}$. ∘

The next proposition contains *Weyl's criterion* for the essential spectrum. It is based on the notion of a *singular sequence*.

Definition 8.4 Let $\lambda \in \mathbb{R}$. A *singular sequence for A at λ* is a sequence $(x_n)_{n\in\mathbb{N}}$ of vectors $x_n \in \mathcal{D}(A)$ such that

$$\liminf_{n\to\infty} \|x_n\| > 0, \qquad \text{w-}\lim_{n\to\infty} x_n = 0, \qquad \lim_{n\to\infty} (A - \lambda I)x_n = 0.$$

Recall that w-$\lim_{n\to\infty} x_n = 0$ means that $\lim_{n\to\infty}\langle x_n, y\rangle = 0$ for all $y \in \mathcal{H}$.

Proposition 8.11 *For any $\lambda \in \mathbb{R}$, the following statements are equivalent:*

(i) $\lambda \in \sigma_{\mathrm{ess}}(A)$.

(ii) *There exists an orthonormal singular sequence for A at λ.*
(iii) *There exists a singular sequence for A at λ.*
(iv) $\dim(E_A(\lambda + \varepsilon) - E_A(\lambda - \varepsilon))\mathcal{H} = \infty$ *for each $\varepsilon > 0$.*

Proof (i) $\rightarrow$ (ii): If λ is an eigenvalue of infinite multiplicity, then any orthonormal sequence from $\mathcal{N}(A - \lambda I)$ is obviously a singular sequence for A at λ.

Now let λ be an accumulation point of $\sigma(A)$. Then there is a sequence $(t_n)_{n \in \mathbb{N}}$ of points $t_n \in \sigma(A)$ such that $t_n \neq t_k$ if $n \neq k$ and $\lim_n t_n = \lambda$. We choose a positive null sequence (ε_n) such that the intervals $\mathcal{J}_n = (t_n - \varepsilon_n, t_n + \varepsilon_n)$ are pairwise disjoint. Since $t_n \in \sigma(A)$, $E_A(\mathcal{J}_n) \neq 0$ by Proposition 5.10(i). Hence, we can find unit vectors $x_n \in E_A(\mathcal{J}_n)\mathcal{H}$. Since $\mathcal{J}_n \cap \mathcal{J}_k = \emptyset$ if $n \neq k$, the sequence (x_n) is orthonormal, and hence w-$\lim_n x_n = 0$ by Bessel's inequality. Then

$$\left\| (A - \lambda I)x_n \right\|^2 = \left\| (A - \lambda I)E_A(\mathcal{J}_n)x_n \right\|^2 = \int_{\mathcal{J}_n} (t - \lambda)^2 d\langle E_A(t)x_n, x_n \rangle$$

$$\leq \int_{\mathcal{J}_n} \left(|t_n - \lambda| + \varepsilon_n \right)^2 d\langle E_A(t)x_n, x_n \rangle \leq \left(|t_n - \lambda| + \varepsilon_n \right)^2 \to 0$$

as $n \to \infty$. Thus, $(x_n)_{n \in \mathbb{N}}$ is an orthonormal singular sequence for A at λ.

(ii) $\rightarrow$ (iii) is trivial.

(iii) $\rightarrow$ (iv): Let $(x_n)_{n \in \mathbb{N}}$ be a singular sequence for A at λ and assume to the contrary that $\dim(E_A(\lambda + \varepsilon) - E_A(\lambda - \varepsilon))\mathcal{H} < \infty$ for some $\varepsilon > 0$. We compute

$$\varepsilon^2 \|x_n\|^2 = \int_{(\lambda - \varepsilon, \lambda + \varepsilon]} \varepsilon^2 d\langle E_A(t)x_n, x_n \rangle + \int_{\mathbb{R} \setminus (\lambda - \varepsilon, \lambda + \varepsilon]} \varepsilon^2 d\langle E_A(t)x_n, x_n \rangle$$

$$\leq \int_{(\lambda - \varepsilon, \lambda + \varepsilon]} \varepsilon^2 d\langle E_A(t)x_n, x_n \rangle + \int_{\mathbb{R}} (t - \lambda)^2 d\langle E_A(t)x_n, x_n \rangle$$

$$= \varepsilon^2 \left\| (E_A(\lambda + \varepsilon) - E_A(\lambda - \varepsilon))x_n \right\|^2 + \left\| (A - \lambda I)x_n \right\|^2. \qquad (8.25)$$

Since w-$\lim_n x_n = 0$ and the finite rank operator $E_A(\lambda + \varepsilon) - E_A(\lambda - \varepsilon)$ is compact, $\lim_n (E_A(\lambda + \varepsilon) - E_A(\lambda - \varepsilon))x_n = 0$. Hence, the right-hand side of (8.25) tends to zero, which contradicts the condition $\liminf_n \|x_n\| > 0$ in Definition 8.4.

(iv) $\rightarrow$ (i): Assume that $\lambda \notin \sigma_{\mathrm{ess}}(A)$. Then, either $\lambda \in \rho(A)$, or λ is an eigenvalue of finite multiplicity which is an isolated point of $\sigma(A)$. By Proposition 5.10 this implies that $\dim E_A(\{\lambda\})\mathcal{H} < \infty$ and $E_A(t)$ is constant on the intervals $(\lambda - 2\varepsilon, \lambda)$ and $(\lambda, \lambda + 2\varepsilon)$ for some $\varepsilon > 0$. Therefore, $\dim(E_A(\lambda - \varepsilon) - E_A(\lambda + \varepsilon))\mathcal{H} < \infty$, that is, condition (iv) does not hold. $\qquad \square$

8.5 Relatively Compact Perturbations of Self-adjoint Operators

In this section we prove some results about the invariance of the essential spectrum under compact perturbations. All of them are derived from the following theorem.

Theorem 8.12 *Let A_1 and A_2 be self-adjoint operators on a Hilbert space $\mathcal{H}$. Suppose that there exists a number $\mu \in \rho(A_1) \cap \rho(A_2)$ such that*

$$C_\mu := (A_2 - \mu I)^{-1} - (A_1 - \mu I)^{-1} \qquad (8.26)$$

is a compact operator on $\mathcal{H}$. Then we have $\sigma_{\mathrm{ess}}(A_1) = \sigma_{\mathrm{ess}}(A_2)$.

Proof Let $\lambda \in \sigma_{\mathrm{ess}}(A_1)$. We apply Weyl's criterion (Proposition 8.11). Then there exists a singular sequence $(x_n)_{n \in \mathbb{N}}$ for A_1 at λ. In order to prove that $\lambda \in \sigma_{\mathrm{ess}}(A_2)$, it suffices to show that

$$y_n := (A_2 - \mu I)^{-1}(A_1 - \mu I)x_n, \quad n \in \mathbb{N},$$

defines a singular sequence for A_2 at λ. Clearly, $y_n \in \mathcal{D}(A_2)$. We have

$$
\begin{aligned}
y_n - x_n &= (A_2 - \mu I)^{-1}(A_1 - \mu I)x_n - x_n \\
&= \left(C_\mu + (A_1 - \mu I)^{-1}\right)(A_1 - \mu I)x_n - x_n \\
&= C_\mu(A_1 - \lambda I)x_n + (\lambda - \mu)C_\mu x_n. \qquad (8.27)
\end{aligned}
$$

Since w-$\lim_n x_n = 0$ and C_μ is compact, $C_\mu x_n \to 0$ in $\mathcal{H}$. From $(A_1 - \lambda I)x_n \to 0$ it follows that $C_\mu(A_1 - \lambda I)x_n \to 0$. Hence, by (8.27) we obtain $y_n - x_n \to 0$. Therefore, since $\liminf_n \|x_n\| > 0$ and w-$\lim_n x_n = 0$, we conclude that $\liminf_n \|y_n\| > 0$ and w-$\lim_n y_n = 0$. Further,

$$(A_2 - \lambda I)y_n = (A_2 - \mu I)y_n + (\mu - \lambda)y_n = (A_1 - \lambda I)x_n + (\mu - \lambda)(y_n - x_n).$$

Since $(A_1 - \lambda I)x_n \to 0$ and $y_n - x_n \to 0$ as just shown, $(A_2 - \lambda I)x_n \to 0$. This proves that $(y_n)_{n \in \mathbb{N}}$ is a singular sequence for A_2 at λ.

Therefore, $\lambda \in \sigma_{\mathrm{ess}}(A_2)$ and $\sigma_{\mathrm{ess}}(A_1) \subseteq \sigma_{\mathrm{ess}}(A_2)$. Interchanging A_1 and A_2, we also have $\sigma_{\mathrm{ess}}(A_2) \subseteq \sigma_{\mathrm{ess}}(A_1)$. Thus, $\sigma_{\mathrm{ess}}(A_2) = \sigma_{\mathrm{ess}}(A_1)$. $\qquad \square$

Corollary 8.13 *Let T be a symmetric operator on $\mathcal{H}$ such that $d_+(T) < \infty$. If A_1 and A_2 are self-adjoint extensions of T on $\mathcal{H}$, then $\sigma_{\mathrm{ess}}(A_1) = \sigma_{\mathrm{ess}}(A_2)$.*

Proof Since $T \subseteq A_2$ and $T \subseteq A_1$, the resolvents $(A_2 - iI)^{-1}$ and $(A_1 - iI)^{-1}$ coincide on $\mathcal{R}(T - iI)$. Hence, the range of $C_i := (A_2 - iI)^{-1} - (A_1 - iI)^{-1}$ is contained in $\mathcal{R}(T - iI)^\perp$. Since $d_+(T) = \dim \mathcal{R}(T - iI)^\perp < \infty$, C_i is a finite rank operator and hence compact. Thus, $\sigma_{\mathrm{ess}}(A_1) = \sigma_{\mathrm{ess}}(A_2)$ by Theorem 8.12. $\qquad \square$

Definition 8.5 Let A be a closed operator on a Hilbert space $\mathcal{H}$. A linear operator C on $\mathcal{H}$ is called *relatively A-compact* if $\mathcal{D}(C) \supseteq \mathcal{D}(A)$ and C is a compact mapping of the Hilbert space $\mathcal{D}_A = (\mathcal{D}(A), \|\cdot\|_A)$ into the Hilbert space $\mathcal{H}$.

Recall that the compactness of $C : \mathcal{D}_A \to \mathcal{H}$ means that for each sequence $(x_n)_{n \in \mathbb{N}}$ of $\mathcal{D}(A)$ satisfying $\sup_n(\|x_n\| + \|Ax_n\|) < \infty$, the sequence $(C(x_n))_{n \in \mathbb{N}}$ has a subsequence $(C(x_{n_k}))_{k \in \mathbb{N}}$ which converges in $\mathcal{H}$.

For instance, if C is a compact operator on $\mathcal{H}$ with $\mathcal{D}(C) = \mathcal{H}$, then C is obviously relatively A-compact for any closed operator A.

Proposition 8.14 *Let A be a closed operator, and C a linear operator on $\mathcal{H}$.*

(i) *Suppose that $\rho(A)$ is not empty. Then C is relatively A-compact if and only if $\mathcal{D}(C) \supseteq \mathcal{D}(A)$ and $C(A - \mu I)^{-1}$ is a compact operator on $\mathcal{H}$ for some (and then for all) $\mu \in \rho(A)$.*

(ii) *If A is densely defined and C is relatively A-compact, then C is relatively A-bounded with A-bound zero.*

Proof (i): If C is relatively A-compact, then $C(A - \mu I)^{-1}$ is the composition of the continuous operator $(A - \mu I)^{-1} : \mathcal{H} \to \mathcal{D}_A$ and the compact operator $C : \mathcal{D}_A \to \mathcal{H}$ and hence compact. Conversely, if $C(A - \mu I)^{-1}$ is a compact operator on $\mathcal{H}$, then $C : \mathcal{D}_A \to \mathcal{H}$ is compact as a product of the continuous operator $A - \mu I : \mathcal{D}_A \to \mathcal{H}$ and the compact operator $C(A - \mu I)^{-1} : \mathcal{H} \to \mathcal{H}$.

(ii): Suppose that C is relatively A-compact. Then $C : \mathcal{D}_A \to \mathcal{H}$ is compact and hence continuous. This means that C is relatively A-bounded. We prove that $\alpha_A(C) = 0$. Assume to the contrary that $\alpha_A(C) > 0$. Then there is an $a > 0$ for which there is no $b > 0$ such that (8.8) is valid (with B replaced by C). Hence, there exists a sequence $(u_n)_{n \in \mathbb{N}}$ of vectors $u_n \in \mathcal{D}(A)$ such that $\|Cu_n\| > a\|Au_n\| + n\|u_n\|$. Upon replacing u_n by $\|Cu_n\|^{-1} u_n$ we can assume that $\|Cu_n\| = 1$. Then $\|u_n\| < 1/n$ and $\|Au_n\| < a^{-1}$ for all $n \in \mathbb{N}$. Hence, $u_n \to 0$ in $\mathcal{H}$. The bounded sequence (Au_n) has a weakly converging subsequence in $\mathcal{H}$, say w-$\lim_k A(u_{n_k}) = y$. Then

$$\langle A(u_{n_k}), x \rangle = \langle u_{n_k}, A^* x \rangle \to \langle y, x \rangle = \langle 0, A^* x \rangle = 0$$

for all $x \in \mathcal{D}(A^*)$. Since A is closed, $\mathcal{D}(A^*)$ is dense in $\mathcal{H}$, and hence $y = 0$. Therefore, (u_{n_k}) is a weak null sequence in the Hilbert space $\mathcal{D}_A$. The compact operator $C : \mathcal{D}_A \to \mathcal{H}$ maps this sequence into a null sequence $(C(u_{n_k}))$ of $\mathcal{H}$. Since $\|Cu_n\| = 1$ by construction, this is the desired contradiction. □

Theorem 8.15 *Let A be a self-adjoint operator, and C a symmetric relatively A-compact operator on $\mathcal{H}$. Then $A + C$ is self-adjoint, and $\sigma_{\text{ess}}(A + C) = \sigma_{\text{ess}}(A)$.*

Proof Because C has the A-bound zero by Proposition 8.14(ii), the operator $A + C$ is self-adjoint by the Kato–Rellich theorem (Theorem 8.5). Since $\mathcal{D}(A + C) = \mathcal{D}(A)$, the first resolvent identity (2.4) applies with $T = A + C$, $S = A$ and yields

$$(A + C - \lambda i I)^{-1} - (A - \lambda i I)^{-1} = -(A + C - \lambda i I)^{-1} C(A - \lambda i I)^{-1}$$

$$(8.28)$$

for nonzero $\lambda \in \mathbb{R}$. By Proposition 8.14(i), $C(A - \lambda i I)^{-1}$ is compact. Therefore, the difference of resolvents in (8.28) is compact. Hence, the hypothesis of Theorem 8.12 is fulfilled, and we obtain $\sigma_{\text{ess}}(A) = \sigma_{\text{ess}}(A + C)$. □

The special case of Theorem 8.15 when C is an "ordinary" compact operator on $\mathcal{H}$ is Weyl's original classical theorem. We state this separately as the following:

Corollary 8.16 *If A is a self-adjoint operator and C is a compact self-adjoint operator on a Hilbert space $\mathcal{H}$, then $\sigma_{\mathrm{ess}}(A + C) = \sigma_{\mathrm{ess}}(A)$.*

Let A be a self-adjoint operator, C a compact self-adjoint operator, and U a unitary operator on $\mathcal{H}$. Set $B = U(A + C)U^{-1}$. Since the essential spectrum is obviously preserved under unitary transformations, by Corollary 8.16 we have

$$\sigma_{\mathrm{ess}}(A) = \sigma_{\mathrm{ess}}(A + C) = \sigma_{\mathrm{ess}}\big(U(A + C)U^{-1}\big) = \sigma_{\mathrm{ess}}(B).$$

A result of von Neumann (see [AG, No. 94]) states a converse of this assertion:

If A and B are bounded self-adjoint operators on a separable Hilbert space $\mathcal{H}$ with equal essential spectra $\sigma_{\mathrm{ess}}(A) = \sigma_{\mathrm{ess}}(B)$, then there exist a compact operator C and a unitary operator U on $\mathcal{H}$ such that $B = U(A + C)U^{-1}$.

8.6 Applications to Schrödinger Operators: Essential Spectrum

First, we develop some preliminaries on integral operators with convolution kernels. For any Borel function ψ on $\mathbb{R}$ we define in accordance with (8.2),

$$\psi(D) = \mathcal{F}^{-1}M_\psi \mathcal{F}.$$

Lemma 8.17

(i) *If $\psi \in L^2(\mathbb{R}^d) \cap L^\infty(\mathbb{R}^d)$, then the operator $\psi(D)$ acts by the convolution multiplication $\psi(D)f = (2\pi)^{-d/2}(\mathcal{F}^{-1}\psi) * f$ for $f \in L^2(\mathbb{R}^d)$.*

(ii) *If $V, \psi \in L^2(\mathbb{R}^d) \cap L^\infty(\mathbb{R}^d)$, then $T := V(t)\psi(D)$ is an integral operator on $\mathcal{D}(T) = L^2(\mathbb{R}^d)$ with kernel $K(x, y) := (2\pi)^{-d/2}V(x)(\mathcal{F}^{-1}\psi)(x - y)$ belonging to $L^2(\mathbb{R}^{2d})$. In particular, T is a compact operator.*

(iii) *Let $V \in L^2(\mathbb{R}^d)$ and $d \leq 3$. Then the operator $T := V(x)(-\Delta + I)^{-1}$ is compact. It is an integral operator with domain $\mathcal{D}(T) = L^2(\mathbb{R}^d)$ and kernel $K(x, y) := V(x)h(x - y) \in L^2(\mathbb{R}^{2d})$, where $h := (2\pi)^{-d/2}\mathcal{F}^{-1}((\|x\|^2 + 1)^{-1})$.*

Proof (i): From the definition of $\psi(D)$ and the convolution theorem (Theorem C.6), we obtain $\psi(D)f = \mathcal{F}^{-1}(\psi \cdot (\mathcal{F}f)) = (2\pi)^{-d/2}(\mathcal{F}^{-1}\psi) * f$ for $f \in L^2(\mathbb{R}^d)$.

(ii): Since $V, \psi \in L^\infty(\mathbb{R}^d)$, the operators V and $\psi(D)$ are bounded, so $\mathcal{D}(T) = L^2(\mathbb{R}^d)$. Further, $h := (2\pi)^{-d/2}\mathcal{F}^{-1}\psi \in L^2(\mathbb{R}^d)$, since $\psi \in L^2(\mathbb{R}^d)$. By (i),

$$(Tf)(x) = V(x)(h * f)(x) = \int_{\mathbb{R}^d} V(x)h(x - y)f(y)\, dy = \int_{\mathbb{R}^d} K(x, y)f(y)\, dy,$$

$$\int_{\mathbb{R}^d} \int_{\mathbb{R}^d} |K(x, y)|^2\, dy\, dx = \int_{\mathbb{R}^d} \int_{\mathbb{R}^d} |V(x)h(x - y)|^2\, dy\, dx$$

$$= \int_{\mathbb{R}^d} \int_{\mathbb{R}^d} |V(x)h(y')|^2\, dy'\, dx = \|h\|_2 \|V\|_2 < \infty.$$

This shows that T is an integral operator with kernel $K \in L^2(\mathbb{R}^{2d})$. Therefore, by Theorem A.6, T is a Hilbert–Schmidt operator. In particular, T is compact.

(iii): Let $g \in L^2(\mathbb{R}^d)$. Then $f := (-\Delta + I)^{-1} g \in \mathcal{D}(-\Delta)$. Hence, $f \in L^\infty(\mathbb{R}^d)$ by Proposition 8.7, since $d \leq 3$. Therefore, $V \cdot f \in L^2(\mathbb{R}^d)$, that is, $g \in \mathcal{D}(T)$. This shows that $\mathcal{D}(T) = L^2(\mathbb{R}^d)$. Using again that $d \leq 3$, it follows that the function $\psi(x) := (\|x\|^2 + 1)^{-1}$ is in $L^2(\mathbb{R}^d)$. The remaining assertions follow as in (ii). □

Lemma 8.18 *Let* $p \in [2, \infty]$ *and* $V, \psi \in L^p(\mathbb{R}^d)$.

(i) *The operator* $V(x)\psi(D)$ *is bounded, and there is a constant* $c > 0$ *such that*

$$\left\| V(x)\psi(D)f \right\| \leq c\|V\|_p \|\psi\|_p \|f\| \quad \text{for } f \in \mathcal{D}\big(V(x)\psi(D)\big). \quad (8.29)$$

(ii) *If* $d \geq 4$ *and* $p > d/2$, *then* $T := V(x)(-\Delta + I)^{-1}$ *is a compact operator on* $\mathcal{D}(T) = L^2(\mathbb{R}^d)$.

Proof (i): Define $q \in [2, \infty]$ and $r \in [1, 2]$ by $p^{-1} + q^{-1} = 2^{-1}$ and $q^{-1} + r^{-1} = 1$. By the Hausdorff–Young Theorem C.5 there exists a constant $c > 0$ such that

$$\left\| \mathcal{F}^{-1}g \right\|_q \leq c\|g\|_r \quad \text{for } g \in L^r(\mathbb{R}^d). \quad (8.30)$$

Let $f \in \mathcal{D}(V(x)\psi(D))$. Note that $p^{-1} + 2^{-1} = r^{-1}$. Applying the Hölder inequality (B.5), Eq. (8.2), inequality (8.30), and again the Hölder inequality, we derive

$$\left\| V(x)\psi(D)f \right\|_2 \leq \|V\|_p \left\| \psi(D)f \right\|_q = \|V\|_p \left\| \mathcal{F}^{-1}(\psi \cdot \mathcal{F}f) \right\|_q$$
$$\leq c\|V\|_p \|\psi \cdot \mathcal{F}f\|_r \leq c\|V\|_p \|\psi\|_p \|f\|_2.$$

(ii): Let $g \in L^2(\mathbb{R}^d)$. Then $f := (-\Delta + I)^{-1}g \in \mathcal{D}(-\Delta)$. Since $V \in L^p(\mathbb{R}^d)$ and $p > d/2$, inequality (8.18) derived in the proof of Theorem 8.8 implies that $V \cdot f \in L^2(\mathbb{R}^d)$, that is, $g \in \mathcal{D}(T)$. This proves that $\mathcal{D}(T) = L^2(\mathbb{R}^d)$.

We choose a sequence (V_n) of functions $V_n \in L^2(\mathbb{R}^d) \cap L^\infty(\mathbb{R}^d)(\subseteq L^p(\mathbb{R}^d))$ such that $V = \lim_{n \to \infty} V_n$ in $L^p(\mathbb{R}^d)$. Let χ_n be the characteristic function of the set $\{x \in \mathbb{R}^d : \|x\|^2 \leq n\}$ and put $\psi_n(x) = (\|x\|^2 + 1)^{-1}\chi_n(x)$, $n \in \mathbb{N}$. Then $\psi_n \in L^2(\mathbb{R}^d) \cap L^\infty(\mathbb{R}^d)$, and it follows from Lemma 8.17(ii) that the operator

$$T_n := V_n(x)\psi_n(D) = V_n(x)(-\Delta + I)^{-1}\chi_n(-\Delta)$$

is compact and $\mathcal{D}(T_n) = L^2(\mathbb{R}^d)$. Since $p > d/2$, the function $\psi(x) := (\|x\|^2 + 1)^{-1}$ is in $L^p(\mathbb{R}^d)$. Obviously, V_n and ψ_n are in $L^p(\mathbb{R}^d)$. Using the definitions of the corresponding operators and finally inequality (8.29), we obtain

$$\|T - T_n\| = \left\| (V - V_n)(-\Delta + I)^{-1} + V_n(-\Delta + I)^{-1}\big(I - \chi_n(-\Delta)\big) \right\|$$
$$\leq \left\| (V - V_n)(x)\psi(D) \right\| + \left\| V_n(x)(\psi - \psi_n)(D) \right\|$$
$$\leq c\|V - V_n\|_p \|\psi\|_p + c\|V_n\|_p \|\psi - \psi_n\|_p.$$

Since $V = \lim_n V_n$ and $\psi = \lim_n \psi_n$ in $L^p(\mathbb{R}^d)$, it follows from the preceding that $\lim_n \|T - T_n\| = 0$. Since the operators T_n are compact, so is T. □

Let $V(x)$ be a Borel function on $\mathbb{R}^d$. We shall write $V \in L^p(\mathbb{R}^d) + L^\infty(\mathbb{R}^d)_\varepsilon$ if for each $\varepsilon > 0$, there exist functions $V_{1,\varepsilon} \in L^p(\mathbb{R}^d)$ and $V_{2,\varepsilon} \in L^\infty(\mathbb{R}^d)$ such that $V = V_{1,\varepsilon} + V_{2,\varepsilon}$ and $\|V_{2,\varepsilon}\|_\infty < \varepsilon$.

Clearly, $V \in L^p(\mathbb{R}^d) + L^\infty(\mathbb{R}^d)_\varepsilon$ if and only if there is a sequence $(V_n)_{n \in \mathbb{N}}$ of functions $V_n \in L^p(\mathbb{R}^d)$ such that $V - V_n \in L^\infty(\mathbb{R}^d)$ and $\lim_{n \to \infty} \|V - V_n\|_\infty = 0$. The following theorem is the main result of this section.

Theorem 8.19 *Let* $V \in L^p(\mathbb{R}^d) + L^\infty(\mathbb{R}^d)_\varepsilon$ *be a real-valued function, where* $p = 2$ *if* $d \leq 3$ *and* $p > d/2$ *if* $d \geq 4$.

Then V *is relatively* $(-\Delta)$-*compact. The operator* $-\Delta + V$ *is self-adjoint on* $\mathcal{D}(-\Delta + V) = \mathcal{D}(-\Delta)$, *and* $\sigma_{\mathrm{ess}}(-\Delta + V) = [0, +\infty)$.

Proof Since $V \in L^p(\mathbb{R}^d) + L^\infty(\mathbb{R}^d)_\varepsilon$, we can choose a sequence of functions $V_n \in L^p(\mathbb{R}^d)$ such that $V - V_n \in L^\infty(\mathbb{R}^d)$ and $\lim_{n \to \infty} \|V - V_n\|_\infty = 0$. We set

$$T := V(x)(-\Delta + I)^{-1} \quad \text{and} \quad T_n := V_n(x)(-\Delta + I)^{-1}.$$

By Lemma 8.17(iii) for $d \leq 3$ and Lemma 8.18(ii) for $d \geq 4$, T_n is compact, and $\mathcal{D}(T_n) = L^2(\mathbb{R}^d)$. Since $V - V_n \in L^\infty(\mathbb{R}^d)$, we have $\mathcal{D}(T) = L^2(\mathbb{R}^d)$. From

$$\|T - T_n\| = \left\| (V - V_n)(-\Delta + I)^{-1} \right\| \leq \|V - V_n\|_\infty \left\| (-\Delta + I)^{-1} \right\| \to 0$$

it follows that T is compact. Hence, V is relatively $(-\Delta)$-compact by Proposition 8.14(i). Since $\sigma_{\mathrm{ess}}(-\Delta) = [0, +\infty)$, all assertions follow from Theorem 8.15. $\square$

Example 8.5 (*Example* 8.2 *continued*) Let V be as in Example 8.2. For the decomposition of $V = V_{1,n} + V_{2,n}$ given in Example 8.2, we have $\lim_{n \to \infty} \|V_{2,n}\|_\infty = 0$. Hence, $V \in L^2(\mathbb{R}^d) + L^\infty(\mathbb{R}^d)_\varepsilon$, so that $\sigma_{\mathrm{ess}}(-\Delta + V) = [0, +\infty)$ by Theorem 8.19. $\circ$

The following result is the counterpart of Proposition 8.10. It characterizes those potentials V on $\mathbb{R}$ which are relatively $(-\frac{d^2}{dx^2})$-compact.

Proposition 8.20 *Let* $V \in L^2_{\mathrm{loc}}(\mathbb{R})$. *The multiplication operator by* V *is relatively* $(-\frac{d^2}{dx^2})$-*compact on* $L^2(\mathbb{R})$ *if and only if*

$$\lim_{|c| \to +\infty} \int_{c-1}^{c+1} |V(x)|^2 \, dx = 0. \tag{8.31}$$

Proof Let T denote the operator $-i\frac{d}{dx}$ with domain $\mathcal{D}(T) = H^1(\mathbb{R})$. Then we have $T^2 = -\frac{d^2}{dx^2}$ and $\mathcal{D}(T^2) = H^2(\mathbb{R})$.

First, we suppose that (8.31) holds. Let $f \in H^2(\mathbb{R})$. From (8.21) it follows that

$$|f(x)|^2 \leq \|Tf\|^2 + 2\|f\|^2 \quad \text{for } x \in \mathbb{R}. \tag{8.32}$$

Since $H^2(\mathbb{R})$ is dense in $H^1(\mathbb{R})$, (8.32) remains valid for $f \in H^1(\mathbb{R})$. Applying (8.32) to f and f' and using that $\|Tf\|^2 = \langle T^2 f, f \rangle \leq \|T^2 f\|^2 + \|f\|^2$, we derive

$$|f(x)|^2 + |f'(x)|^2 \leq 2\|T^2 f\|^2 + 5\|f\|^2 \quad \text{for } f \in H^2(\mathbb{R}). \tag{8.33}$$

Let $(f_n)_{n\in\mathbb{N}}$ be a sequence from $\mathcal{D}(T^2)$ such that

$$M := \sup_n \left(\|T^2 f_n\| + \|f_n\| \right) < \infty. \tag{8.34}$$

From (8.33) it follows that the sequence (f_n) is uniformly bounded and equicontinuous on $\mathbb{R}$. Therefore, by the Arzelà–Ascoli theorem, for any compact interval, it has a uniformly convergent subsequence. By passing further to subsequences we can find a subsequence (f_{n_k}) which converges uniformly on *each* compact interval.

Let $\varepsilon > 0$ be given. By (8.31) there exists $b > 0$ such that

$$\int_{c-1}^{c+1} |V(x)|^2 \, dx \le \varepsilon^2 \quad \text{if } |c| \ge b. \tag{8.35}$$

Set $V_b(x) = V(x)$ if $|x| < b$ and $V_b(x) = 0$ if $|x| \ge b$. Clearly, (8.35) implies that $\delta_{V-V_b} < \varepsilon^2$, where δ_{V-V_b} was defined in Proposition 8.10. By (8.24), applied to $V - V_b$ with $\varepsilon = 1$, and (8.34) we obtain

$$\left\| (V - V_b)(f_{n_k} - f_{n_l}) \right\| \le \varepsilon \left\| T^2(f_{n_k} - f_{n_l}) \right\| + 3\varepsilon \|f_{n_k} - f_{n_l}\| \le \varepsilon 6M$$

for $k, l \in \mathbb{N}$. Since the sequence (f_{n_k}) converges uniformly on $[-b, b]$ and

$$\left\| V_b(f_{n_k} - f_{n_l}) \right\| \le \|V_b\|_{L^2(-b,b)} \|f_{n_k} - f_{n_l}\|_{L^\infty(-b,b)},$$

the two preceding inequalities imply that $(V f_{n_k} = (V - V_b) f_{n_k} + V_b f_{n_k})_{k\in\mathbb{N}}$ is a Cauchy sequence in $L^2(\mathbb{R})$. This proves that V is relatively T^2-compact.

To prove the converse direction, we suppose that (8.31) does not hold. Then there exist a $\gamma > 0$ and a real sequence $(c_n)_{n\in\mathbb{N}}$ such that $|c_{n+1}| \ge |c_n| + 4$ and $\int_{c_n-1}^{c_n+1} |V(x)|^2 \, dx \ge \gamma$. Take $\eta \in C_0^\infty(\mathbb{R})$ such that $\eta(x) = 1$ if $|x| \le 1$ and $\eta(x) = 0$ if $|x| \ge 2$, and set $\eta_n(x) = \eta(x - c_n)$. Since $\|T^2 \eta_n\| + \|\eta_n\| = \|T^2 \eta\| + \|\eta\|$, the sequence $(\eta_n)_{n\in\mathbb{N}}$ is bounded in $\mathcal{D}_{T^2}$. But for $n \ne k$, we have

$$\left\| V(\eta_n - \eta_k) \right\|^2 = \int_{c_n-2}^{c_n+2} |V(x)\eta_n(x)|^2 \, dx + \int_{c_k-2}^{c_k+2} |V(x)\eta_k(x)|^2 \, dx \ge 2\gamma,$$

so $(V \eta_n)_{n\in\mathbb{N}}$ cannot have a convergent subsequence in $L^2(\mathbb{R})$. Therefore, V is not relatively T^2-compact. $\qquad\square$

8.7 Exercises

1. Let $p(x) = \sum_\alpha a_\alpha x^\alpha$ be a polynomial with real coefficients, and let M be a Borel subset of $\mathbb{R}$. Prove that the spectral projection $E_{p(D)}(M)$ of the self-adjoint operator $p(D)$ on $L^2(\mathbb{R}^d)$ is $\mathcal{F}^{-1}\chi_{p^{-1}(M)}\mathcal{F}$, where $\chi_{p^{-1}(M)}$ is the characteristic function of the set $p^{-1}(M) = \{x \in \mathbb{R}^d : p(x) \in M\}$.

2. Let $T = -i\frac{d}{dx}$ and $a, b, c \in \mathbb{R}$, $a < b$, $c \ge 0$. Show that the spectral projections $E_T((a, b))$ and $E_{T^2}((0, c))$ of the self-adjoint operators T and T^2 on $L^2(\mathbb{R})$ are

$$\left(E_T((a, b)) f \right)(x) = (2\pi)^{-1} i \int_{\mathbb{R}} \frac{e^{ia(x-y)} - e^{ib(x-y)}}{x - y} f(y) \, dy, \quad f \in L^2(\mathbb{R}),$$

$$\left(E_{T^2}((0, c)) f \right)(x) = \pi^{-1} \int_{\mathbb{R}} \frac{\sin c(x - y)}{x - y} f(y) \, dy, \quad f \in L^2(\mathbb{R}).$$

3. Prove formula (8.7) for the free Green function G_3.

Hints: Introduce first spherical coordinates to compute the Fourier transform of $(\|x\|^2 - \lambda)^{-1}$. Use the residue method to evaluate the integral along the real line by integrating over a rectangle with vertices $-r, r, r + i\sqrt{r}, -r + i\sqrt{r}$, where $r > 0$.

4. Prove that the self-adjoint operator $-\Delta$ on $L^2(\mathbb{R}^d)$ has no eigenvalues.

5. Let A and B be linear operators on $\mathcal{H}$ such that $\mathcal{D}(B) \supseteq \mathcal{D}(A)$. Prove that B is relatively A-bounded if and only if there are positive numbers a, b such that $\|Bx\|^2 \leq a\|Ax\|^2 + b\|x\|^2$ for all $x \in \mathcal{D}(A)$.

6. Let A be a self-adjoint operator, and B a linear operator such that $\mathcal{D}(B) \supseteq \mathcal{D}(A)$.

   a. Prove that B is relatively A-bounded if and only if $B(A - icI)^{-1}$ is bounded for one, hence all, $c \in \mathbb{R}$, $c \neq 0$.

   b. Suppose that B is relatively A-bounded. Prove that the A-bound of B is

   $$a_A(B) = \lim_{|c| \to +\infty} \left\| B(A - icI)^{-1} \right\|.$$

   Hint: Use some arguments from the proof of Theorem 8.5.

7. Let A and B be linear operators on $\mathcal{H}$ such that B is relatively A-bounded with A-bound $\alpha_A(B) < 1$.

   a. Prove that $A + B$ is closed if and only if A is closed.

   b. Suppose that A is closable. Prove that $A + B$ is closable. Show that there is an extension $\widetilde{B}$ of B such that $\alpha_{\overline{A}}(\widetilde{B}) = \alpha_A(B)$ and $\overline{A + B} = \overline{A} + \widetilde{B}$.

8. Let A be a lower semibounded self-adjoint operator, and B a symmetric operator on $\mathcal{H}$ such that $\mathcal{D}(B) \supseteq \mathcal{D}(A)$. Suppose that there are positive numbers $a < 1$ and b such that $\|Bx\| \leq a\|Ax\| + b\|x\|$ for all $x \in \mathcal{D}(A)$.

Prove that the self-adjoint operator $A + B$ is lower semibounded with lower bound

$$m := m_A - \max\{b(1-a)^{-1}, a|m_A| + b\}.$$

Hints: Let $\lambda < m$. Set $C := B(A - \lambda I)^{-1}$. Mimic the proof of Theorem 8.5 to show that $\|C\| < 1$, $(A + B - \lambda I)^{-1} = (A - \lambda I)^{-1}(I + C)^{-1}$ and $\lambda \in \rho(A + B)$.

9. Let V be a Borel function on $\mathbb{R}^d$ such that $c := \limsup_{|x| \to \infty} |V(x)| < \infty$ and $\mathcal{D}(V) \supseteq \mathcal{D}(-\Delta)$. Prove that $V \in L^2(\mathbb{R}^d) + L^\infty(\mathbb{R}^d)$.

10. Suppose that $0 < \alpha < 1/2$. Define the function V on $\mathbb{R}$ by $V(x) = |x - 2n|^{-\alpha}$ for $n - 1 < x \leq n + 1$ and $n \in \mathbb{Z}$.

   a. Prove that $\delta_V < \infty$ and $-\frac{d^2}{dx^2} + V(x)$ is self-adjoint on $\mathcal{D}(-\frac{d^2}{dx^2})$.

   b. Prove that V is not a Kato–Rellich potential on $\mathbb{R}$.

11. (*Self-adjointness of Jacobi operators*)

Let $(b_n)_{n \in \mathbb{N}_0}$ be a real sequence, and $(a_n)_{n \in \mathbb{N}_0}$ a complex sequence. Define operators A and B on $l^2(\mathbb{N}_0)$ by $(B(\varphi_n))_k = a_{k-1}\varphi_{k-1} + \overline{a}_k\varphi_{k+1}$ and $A(\varphi_n) = (b_n\varphi_n)$ with domains $\mathcal{D}(A) = \{(\varphi_n) \in l^2(\mathbb{N}_0) : (b_n\varphi_n) \in l^2(\mathbb{N}_0)\}$ and $\mathcal{D}(B) = \{(\varphi_n) \in l^2(\mathbb{N}_0) : (a_{n-1}\varphi_{n-1} + \overline{a}_n\varphi_{n+1}) \in l^2(\mathbb{N}_0)\}$, where $a_{-1} := 0$.

   a. Suppose that B is relatively A-bounded. Show that for any $\alpha > \alpha_A(B)$, there is a number $\beta_\alpha > 0$ such that $|a_{n-1}^2| + |a_{n+1}|^2 \leq \alpha^2|b_n|^2 + \beta_\alpha$ for $n \in \mathbb{N}$.

b. Suppose that there exist constants $\alpha > 0$ and $\beta > 0$ such that

$$2|a_{n-1}|^2 + 2|a_{n+1}|^2 \leq \alpha^2 |b_n|^2 + \beta \quad \text{for } n \in \mathbb{N}. \tag{8.36}$$

Show that B is relatively A-bounded with A-bound $\alpha_A(B) \leq \alpha$.

c. Suppose that (8.36) holds and $\alpha < 1$. Show that the operator $A + B$ is self-adjoint on $\mathcal{D}(A + B) = \mathcal{D}(A)$.

12. Find examples of self-adjoint operators A for which $\sigma_{\mathrm{ess}}(A) = \emptyset$, $\sigma_{\mathrm{ess}}(A) = \mathbb{Z}$, and $\sigma_{\mathrm{ess}}(A) = \mathbb{R}$.

13. Prove that $\sigma_{\mathrm{ess}}(A) \neq \emptyset$ for each bounded self-adjoint operator A acting on an infinite-dimensional Hilbert space.

14. Let A be a self-adjoint operator, and B a symmetric operator on $\mathcal{H}$ which is relatively A-bounded with A-bound $\alpha_A(B) < 1$. Suppose that C is a relatively A-compact operator on $\mathcal{H}$. Prove that C is relatively $(A + B)$-compact.

15. Derive Weyl's classical theorem (Corollary 8.16) directly from Proposition 8.11 without using Theorem 8.12.

Chapter 9
Trace Class Perturbations of Spectra of Self-adjoint Operators

Let A and B be self-adjoint operators on a Hilbert space $\mathcal{H}$ such that the closure of their difference is of trace class. *Krein's spectral shift* associated with such a pair $\{B, A\}$ is a real function ξ on $\mathbb{R}$ which enters into an integral representation formula (9.58) for the trace of the difference of functions of the operators A and B. Spectral shift and this trace formula are fundamental tools in perturbation theory of spectra. An introduction into this topic is given in Sect. 9.6 thereby following M.G. Krein's original approach based on perturbation determinants. Some basics on infinite determinants and perturbation determinants are developed in Sects. 9.4 and 9.5, respectively.

Sections 9.2 and 9.3 deal with the case where $B = A + \alpha \langle \cdot, u \rangle u$ is a rank one perturbation of A, where $\alpha \in \mathbb{R}$ and $u \in \mathcal{H}$. In these sections boundary values on $\mathbb{R}$ of the holomorphic function $F(z) = \langle R_z(A)u, u \rangle$ play a crucial role. In Sect. 9.3 the spectral shift is defined in the rank one case. Section 9.2 contains the beautiful Aronszajn–Donoghue spectral theory of the operators B. The decomposition of a self-adjoint operator and its spectrum into a pure point part, a singularly continuous part, and an absolutely continuous part is treated in Sect. 9.1.

9.1 Parts of the Spectrum of a Self-adjoint Operator

To explain the main idea, let μ be a positive regular Borel measure on $\mathbb{R}$, and A the multiplication operator $(Af)(x) = xf(x)$ on $L^2(\mathbb{R}, \mu)$. Then the spectrum of A is the support of μ, see Example 5.4. However, if μ is the Lebesgue measure or if μ is the sum of positive multiples of delta measures at the rationals, in both cases we obtain the same set $\operatorname{supp} \mu = \mathbb{R}$. In the second case, A has a complete set of eigenvectors, while in the first case, A has no eigenvector. Thus, the spectrum is a rather coarse invariant of the operator A. Finer invariants are obtained by the decomposition $\mu = \mu_p + \mu_{sc} + \mu_{ac}$ of μ (see Proposition B.8) into a pure point part μ_p, a singularly continuous part μ_{sc}, and an absolutely continuous part μ_{ac}. These parts are mutually singular, so we have $L^2(\mathbb{R}, \mu) = L^2(\mathbb{R}, \mu_p) \oplus L^2(\mathbb{R}, \mu_{sc}) \oplus L^2(\mathbb{R}, \mu_{ac})$,

K. Schmüdgen, *Unbounded Self-adjoint Operators on Hilbert Space*,
Graduate Texts in Mathematics 265,
DOI 10.1007/978-94-007-4753-1_9, © Springer Science+Business Media Dordrecht 2012

and each of these spaces reduces the operator A. In this section we derive a similar decomposition (achieved by formula (9.1)) for an arbitrary self-adjoint operator.

Assume that A is a self-adjoint operator on a Hilbert space $\mathcal{H}$ with spectral measure E_A. Our main aim are the decompositions (9.1) and (9.2) of the operator A.

We denote by $\mathcal{H}_p = \mathcal{H}_p(A)$ the closed linear span of all eigenspaces of A. If A has no eigenvalue, we set $\mathcal{H}_p = \{0\}$. Further, let $\mathcal{H}_c = \mathcal{H}_c(A)$ be the set of all vectors $x \in \mathcal{H}$ for which the function $\lambda \to \langle E_A(\lambda)x, x \rangle$ is continuous on $\mathbb{R}$.

Proposition 9.1

(i) *A vector $x \in \mathcal{H}$ belongs to $\mathcal{H}_p(A)$ if and only if there is an at most countable subset N of $\mathbb{R}$ such that $E_A(N)x = x$.*

(ii) *A vector $x \in \mathcal{H}$ is in $\mathcal{H}_c(A)$ if and only if $E_A(N)x = 0$ for each one point subset and hence for each countable subset N of $\mathbb{R}$.*

(iii) *$\mathcal{H}_p(A)$ and $\mathcal{H}_c(A)$ are closed subspaces of $\mathcal{H}$ and $\mathcal{H} = \mathcal{H}_p(A) \oplus \mathcal{H}_c(A)$.*

Proof (i): Let $x \in \mathcal{H}_p$. Then x is of the form $x = \sum_{n=1}^{\infty} x_n$, where $Ax_n = \lambda_n x_n$ with $\lambda_n \in \mathbb{R}$. Put $N = \{\lambda_n : n \in \mathbb{N}\}$. Since $E_A(\{\lambda_n\})x_n = x_n$ by Proposition 5.10(ii), we have $E_A(N)x = \sum_n E_A(N)x_n = \sum_n E_A(\{\lambda_n\})x_n = \sum_n x_n = x$.

Conversely, let N be an at most countable set of pairwise different elements λ_n such that $E_A(N)x = x$. Set $x_n := E_A(\{\lambda_n\})x$. Then, $Ax_n = \lambda_n x_n$ by Proposition 5.10(ii), and $x = E_A(N)x = \sum_n E_A(\{\lambda_n\})x = \sum_n x_n$, so $x \in \mathcal{H}_p$.

(ii): The monotone increasing function $\langle E_A(\lambda)x, x \rangle$ is continuous if and only if it has no jumps, that is, if $\langle E_A(\{\lambda\})x, x \rangle = 0$, or equivalently, if $E_A(\{\lambda\})x = 0$ for each $\lambda \in \mathbb{R}$. By the σ-additivity of the spectral measure E_A the latter holds if and only if $E_A(N)x = 0$ for each countable subset N of $\mathbb{R}$.

(iii): Let $x \in (\mathcal{H}_p)^{\perp}$. Since $x_\lambda := E_A(\{\lambda\})x \in \mathcal{N}(A - \lambda I) \subseteq \mathcal{H}_p$ for λ in $\mathbb{R}$ again by Proposition 5.10(ii), we have $0 = \langle x_\lambda, x \rangle = \langle E_A(\{\lambda\})x, x \rangle$. Hence, $E_A(\{\lambda\})x = 0$, so that $x \in \mathcal{H}_c$ by (ii).

Conversely, suppose that $x \in \mathcal{H}_c$. Let $y \in \mathcal{H}_p$. By (i), there is an at most countable set N satisfying $E_A(N)y = y$. Then, $E_A(N)x = 0$ by (ii), and hence $\langle x, y \rangle = \langle x, E_A(N)y \rangle = \langle E_A(N)x, y \rangle = 0$. That is, $x \in (\mathcal{H}_p)^{\perp}$.

Thus, we have shown that $(\mathcal{H}_p)^{\perp} = \mathcal{H}_c$. In particular, $\mathcal{H}_c$ is closed. Note that $\mathcal{H}_p$ is closed by definition. $\qquad \square$

Now we turn to another decomposition of $\mathcal{H}$. Let $\mathcal{H}_{ac} = \mathcal{H}_{ac}(A)$ be the set of all vectors $x \in \mathcal{H}$ for which the measure $\mu_x(\cdot) = \langle E_A(\cdot)x, x \rangle$ on $\mathcal{B}(\mathbb{R})$ is *absolutely continuous* with respect to the Lebesgue measure on $\mathbb{R}$, that is, $\mu_x(N) = 0$ (or equivalently, $E_A(N) = 0$) for each Lebesgue null set N.

Let $\mathcal{H}_{sing} = \mathcal{H}_{sing}(A)$ (resp. $\mathcal{H}_{sc} = \mathcal{H}_{sc}(A)$) denote the set of $x \in \mathcal{H}$ (resp. $x \in \mathcal{H}_c$) for which the measure μ_x is *singular* with respect to the Lebesgue measure on $\mathbb{R}$, that is, there exists a Lebesgue null subset N of $\mathbb{R}$ such that $\mu_x(\mathbb{R}\backslash N) = 0$ (or equivalently, $E_A(\mathbb{R}\backslash N)x = 0$, that is, $x = E_A(N)x$). Clearly, $\mathcal{H}_{sc} = \mathcal{H}_c \cap \mathcal{H}_{sing}$.

Proposition 9.2 *$\mathcal{H}_{ac}(A)$, $\mathcal{H}_{sing}(A)$, and $\mathcal{H}_{sc}(A)$ are closed linear subspaces of $\mathcal{H}$ such that $\mathcal{H} = \mathcal{H}_{ac}(A) \oplus \mathcal{H}_{sing}(A)$ and $\mathcal{H}_{sing}(A) = \mathcal{H}_p(A) \oplus \mathcal{H}_{sc}(A)$.*

Proof First, we prove that $\mathcal{H}_{\text{sing}}$ is a closed linear subspace of $\mathcal{H}$. Let $(x_n)_{n \in \mathbb{N}}$ be a sequence of $\mathcal{H}_{\text{sing}}$ converging to $x \in \mathcal{H}$. By the definition of $\mathcal{H}_{\text{sing}}$ there exist Lebesgue null sets N_n such that $E_A(N_n)x_n = x_n$. Then $N := \bigcup_n N_n$ is also a Lebesgue null set, and $E_A(N)x_n = E_A(N_n)x_n = x_n$ for all n. Letting $n \to \infty$, we obtain $E_A(N)x = x$. That is, $x \in \mathcal{H}_{\text{sing}}$, which proves that $\mathcal{H}_{\text{sing}}$ is closed.

We verify that $\mathcal{H}_{\text{sing}}$ is a linear subspace. Obviously, $\lambda x \in \mathcal{H}_{\text{sing}}$ for $x \in \mathcal{H}_{\text{sing}}$ and $\lambda \in \mathbb{C}$. Let $x_1, x_2 \in \mathcal{H}_{\text{sing}}$. Then there are Lebesgue null sets N_1 and N_2 such that $E_A(N_1)x_1 = x_1$ and $E_A(N_2)x_2 = x_2$. Clearly, $N = N_1 \cup N_2$ is a Lebesgue null set, and $E_A(N)x_j = E_A(N_j)x_j = x_j$ for $j = 1, 2$. Hence, we have $E_A(N)(x_1 + x_2) = x_1 + x_2$, that is, $x_1 + x_2 \in \mathcal{H}_{\text{sing}}$.

We prove that $\mathcal{H}_{\text{ac}} = (\mathcal{H}_{\text{sing}})^{\perp}$. First, suppose that $x \in \mathcal{H}_{\text{ac}}$. Let $y \in \mathcal{H}_{\text{sing}}$. By the definitions of $\mathcal{H}_{\text{sing}}$ and $\mathcal{H}_{\text{ac}}$ there is a Lebesgue null set N such that $E_A(N)y = y$ and $E_A(N)x = 0$. Therefore, $\langle x, y \rangle = \langle x, E_A(N)y \rangle = \langle E_A(N)x, y \rangle = 0$. That is, $x \in (\mathcal{H}_{\text{sing}})^{\perp}$. Conversely, let $x \in (\mathcal{H}_{\text{sing}})^{\perp}$. Let N be a Lebesgue null set. For $y \in \mathcal{H}$, we have $u_y := E_A(N)y \in \mathcal{H}_{\text{sing}}$, and so $\langle E_A(N)x, y \rangle = \langle x, E_A(N)y \rangle = \langle x, u_y \rangle = 0$ for all $y \in \mathcal{H}$. Hence, $E_A(N)x = 0$, that is, $x \in \mathcal{H}_{\text{ac}}$. Together with the preceding paragraph, we have shown that $\mathcal{H} = \mathcal{H}_{\text{ac}} \oplus \mathcal{H}_{\text{sing}}$.

Since at most countable sets have Lebesgue measure zero, it follows from Proposition 9.1(i) that $\mathcal{H}_{\text{p}} \subseteq \mathcal{H}_{\text{sing}}$. Since $\mathcal{H}_{\text{sc}} = \mathcal{H}_{\text{c}} \cap \mathcal{H}_{\text{sing}}$, the decomposition $\mathcal{H} = \mathcal{H}_{\text{p}} \oplus \mathcal{H}_{\text{c}}$ implies that $\mathcal{H}_{\text{sing}} = \mathcal{H}_{\text{p}} \oplus \mathcal{H}_{\text{sc}}$. $\qquad \square$

Proposition 9.3 *Each of the closed linear subspaces* $\mathcal{H}_{\text{p}}(A)$, $\mathcal{H}_{\text{c}}(A)$, $\mathcal{H}_{\text{ac}}(A)$, $\mathcal{H}_{\text{sing}}(A)$, *and* $\mathcal{H}_{\text{sc}}(A)$ *reduces the self-adjoint operator* A.

Proof Let P_s denote the projection of $\mathcal{H}$ on $\mathcal{H}_{\text{sing}}$. Suppose that $x \in \mathcal{H}_{\text{sing}}$. Then there is a Lebesgue null set N such that $E_A(N)x = x$. For $\lambda \in \mathbb{R}$, we have $E_A(N)E_A(\lambda)x = E_A(\lambda)E_A(N)x = E_A(\lambda)x$, and so $E_A(\lambda)x \in \mathcal{H}_{\text{sing}}$. That is, $E_A(\lambda)\mathcal{H}_{\text{sing}} \subseteq \mathcal{H}_{\text{sing}}$, which implies that $E_A(\lambda)P_s = P_s E_A(\lambda)$. Therefore, $P_s A \subseteq A P_s$ by Proposition 5.15. Hence, $\mathcal{H}_{\text{sing}}$ reduces A by Proposition 1.15.

Replacing the null set N by an at most countable set, the same reasoning shows that $\mathcal{H}_{\text{p}}$ reduces A. The other subspaces are reducing, because they are orthogonal complements of reducing subspaces. $\qquad \square$

The restrictions A_{p}, A_{c}, A_{ac}, A_{sing}, and A_{sc} of A to the reducing subspaces $\mathcal{H}_{\text{p}}(A)$, $\mathcal{H}_{\text{c}}(A)$, $\mathcal{H}_{\text{ac}}(A)$, $\mathcal{H}_{\text{sing}}(A)$, and $\mathcal{H}_{\text{sc}}(A)$, respectively, are called the (*spectrally*) *discontinuous, continuous, absolutely continuous, singular,* and *singularly continuous parts* of A, respectively. The *continuous spectrum* $\sigma_{\text{c}}(A)$, the *absolutely continuous spectrum* $\sigma_{\text{ac}}(A)$, the *singular spectrum* $\sigma_{\text{sing}}(A)$, and the *singularly continuous spectrum* $\sigma_{\text{sc}}(A)$ of A are defined as the spectra of the self-adjoint operators A_{c}, A_{as}, A_{sing}, and A_{sc}, respectively. By Propositions 9.1, 9.2, 9.3 we have the decompositions

$$A = A_{\text{p}} \oplus A_{\text{sc}} \oplus A_{\text{ac}}, \qquad \sigma(A) = \sigma(A_{\text{p}}) \cup \sigma_{\text{sc}}(A) \cup \sigma_{\text{ac}}(A), \qquad (9.1)$$

$$A = A_{\text{sing}} \oplus A_{\text{ac}}, \qquad \sigma(A) = \sigma_{\text{sing}}(A) \cup \sigma_{\text{ac}}(A). \qquad (9.2)$$

Note that $\sigma(A_{\text{p}})$ is the closure of the set $\sigma_{\text{p}}(A)$ of eigenvalues of A.

The operator A is said to have a *purely point spectrum, purely continuous spectrum, purely absolutely continuous spectrum, purely singular spectrum,* and *purely singularly continuous spectrum* if $\mathcal{H}_p$, $\mathcal{H}_c$, $\mathcal{H}_{ac}$, $\mathcal{H}_{sing}$, and $\mathcal{H}_{sc}$ is equal to $\mathcal{H}$, respectively. We shall say that A has one of these properties on an interval J when it holds for the restriction of A to the reducing subspace $E_A(J)\mathcal{H}$.

Example 9.1 (*Multiplication operators by the independent variable on* $\mathbb{R}$) Let μ be a positive regular Borel measure on $\mathbb{R}$. Let A be the multiplication operator $(Af)(x) = xf(x)$ on $\mathcal{H} = L^2(\mathbb{R}, \mu)$. As noted above, by Proposition B.8 there is a unique decomposition $\mu = \mu_p + \mu_{sc} + \mu_{ac}$ of μ as a sum of measures, where μ_p is pure point, μ_{sc} is singularly continuous, and μ_{ac} is absolutely continuous. Then

$$\mathcal{H}_p(A) = L^2(\mathbb{R}, \mu_p), \qquad \mathcal{H}_{sc}(A) = L^2(\mathbb{R}, \mu_{sc}), \qquad \mathcal{H}_{ac}(A) = L^2(\mathbb{R}, \mu_{ac}),$$

and the operators A_p, A_{sc}, and A_{ac} act as multiplication operators by the variable x on the corresponding L^2-spaces. Clearly,

- $\mathcal{H}_p(A) = \mathcal{H}$ iff μ is supported by a countable set,
- $\mathcal{H}_{sc}(A) = \mathcal{H}$ iff μ is supported by a Lebesgue null set and $\mu(\{t\}) = 0$ for all $t \in \mathbb{R}$,
- $\mathcal{H}_{ac}(A) = \mathcal{H}$ iff μ is absolutely continuous with respect to the Lebesgue measure, that is, $\mu(N) = 0$ for each Lebesgue null set N.   ∘

Example 9.2 (*Multiplication operators by functions on* $\mathbb{R}^d$) Let φ be a real-valued Borel function on an open subset Ω of $\mathbb{R}^d$. Let A_φ denote the multiplication operator by the function φ on $\mathcal{H} = L^2(\Omega)$, that is, $(A_\varphi f)(x) = \varphi(x)f(x)$ for $f \in \mathcal{D}(A_\varphi) = \{f \in \mathcal{H} : \varphi \cdot f \in \mathcal{H}\}$.

By Example 5.3 the spectral measure of A_φ acts by $E(M)f = \chi_{\varphi^{-1}(M)} \cdot f$. Hence,

$$\left\| E(M)f \right\|^2 = \int_\Omega \chi_{\varphi^{-1}(M)}(x)\left| f(x) \right|^2 dx = \int_{\varphi^{-1}(M)} \left| f(x) \right|^2 dx \qquad (9.3)$$

for $M \in \mathcal{B}(\mathbb{R})$ and $f \in \mathcal{H}$. From (9.3) it follows that A_φ has purely absolutely continuous spectrum if and only if $\varphi^{-1}(N)$ has Lebesgue measure zero for each Lebesgue null set N. The problem of when this condition is fulfilled is difficult even when $d = 1$ or $\varphi \in C^\infty(\Omega)$. However, there is the following important result:

Statement *If* $\varphi \in C^1(\Omega)$ *and* $(\text{grad}\,\varphi)(x) \neq 0$ *a.e. on* Ω, *then* A_φ *has a purely absolutely continuous spectrum. In particular,* A_φ *has a purely absolutely continuous spectrum when* φ *is a nonconstant real polynomial.*

Proof A proof of this assertion is based on the *co-area formula* from geometric measure theory [Fe]. Indeed, combining Theorems 3.2.3 and 3.2.12 in [Fe], one has

$$\int_{\varphi^{-1}(M)} f(x)\left| (\text{grad}\,\varphi)(x) \right| dx = \int_M \left(\int_{\varphi^{-1}(s)} f(t)\, dH^{d-1}(t) \right) ds \qquad (9.4)$$

for $M \in \mathfrak{B}(\mathbb{R})$ and each nonnegative Borel function f on Ω, where H^{d-1} denotes a $(d-1)$-dimensional Hausdorff measure.

Suppose that M is a Lebesgue null set. Then the right-hand side of (9.4) is zero. Let f be the characteristic function of the set $K_n := \{x \in \Omega : |(\mathrm{grad}\,\varphi)(x)| \geq 1/n\}$. It follows then from (9.4) that the Lebesgue measure of $K_n \cap \varphi^{-1}(M)$ is zero. Since $\mathrm{grad}\,\varphi \neq 0$ a.e., we conclude that $\varphi^{-1}(M)$ is a Lebesgue null set.   □

In the special case $\varphi(x) = \|x\|^2$ on $\Omega = \mathbb{R}^d$, the assertion follows by splitting the integral (9.3) into a radial and a surface part over the unit sphere S^{d-1}, that is,

$$\|E(M)f\|^2 = \int_{\{x:\|x\|^2 \in M\}} |f(x)|^2\, dx$$

$$= c_d \int_{\{r:\sqrt{r} \in M\}} \left(\int_{S^{d-1}} r^{d-1} |f(tr)|^2\, d\sigma(t) \right) dr,$$

where σ is the surface measure on S^{d-1}, and c_d is a constant. Therefore, if M is a Lebesgue null set, so is $\{r : \sqrt{r} \in M\}$, and hence $E(M)f = 0$.   ○

Corollary 9.4 *Let $p(x)$ be a nonconstant polynomial in d variables with real coefficients. Then the self-adjoint operator $p(D)$ on $L^2(\mathbb{R}^d)$ defined by (8.2) has a purely absolutely continuous spectrum.*

Proof Since the multiplication operator $A_{p(x)}$ has a purely absolutely continuous spectrum by Example 9.2, so has the operator $p(D) = \mathcal{F}^{-1} A_{p(x)} \mathcal{F}$.   □

The next proposition contains an interesting characterization of the space $\mathcal{H}_c(A)$.

Proposition 9.5 *A vector $x \in \mathcal{H}$ belongs to the subspace $\mathcal{H}_c(A)$ if and only if*

$$\lim_{T \to \infty} (2T)^{-1} \int_T^T |\langle e^{-itA}x, x\rangle|^2\, dt = 0. \tag{9.5}$$

Proof For a finite positive Borel measure μ on $\mathbb{R}$, we set $F_\mu(t) = \int_\mathbb{R} e^{-its}\, d\mu(s)$. A classical theorem of N. Wiener ([Wr], see e.g. [RS3, Theorem XI.114]) states that

$$\lim_{T \to \infty} (2T)^{-1} \int_T^T |F_\mu(t)|^2\, dt = \sum_{s \in \mathbb{R}} \mu(\{s\}). \tag{9.6}$$

(In particular, this means that the limit in (9.6) exists and is finite. Note that $\mu(\{s\})$ can be nonzero only for at most countably many numbers $s \in \mathbb{R}$.)

Setting $\mu := \langle E_A(\cdot)x, x\rangle$, we have $F_\mu(t) = \int_\mathbb{R} e^{-it\lambda}\, d\langle E_A(\lambda)x, x\rangle = \langle e^{-itA}x, x\rangle$. By comparison of (9.5) and (9.6) it follows that condition (9.5) is satisfied if and only if $\langle E_A(\{s\})x, x\rangle = 0$, or equivalently, if $E_A(\{s\})x = 0$ for all $s \in \mathbb{R}$, that is, $x \in \mathcal{H}_c$ by Proposition 9.1(ii).   □

9.2 Aronszajn–Donoghue Theory of Rank One Perturbations

Throughout this section, A denotes a (possibly unbounded) self-adjoint operator on a Hilbert space $\mathcal{H}$, and u is a fixed unit vector of $\mathcal{H}$. Our aim is to study spectral properties of the one-parameter family of self-adjoint operators

$$A_\alpha := A + \alpha \langle \cdot, u \rangle u, \quad \text{where } \alpha \in \mathbb{R}. \tag{9.7}$$

One main result that will be proved below is the following theorem.

Theorem 9.6 *For any $\alpha \in \mathbb{R}$, the absolutely continuous parts of A_α and A are unitarily equivalent. In particular, $\sigma_{\mathrm{ac}}(A_\alpha) = \sigma_{\mathrm{ac}}(A)$.*

First, we develop some preliminaries. Let $\alpha \in \mathbb{R}$. As usual, E_{A_α} denotes the spectral measure of A_α. We define the measure μ_α on $\mathbb{R}$ and the function F_α by

$$\mu_\alpha(\cdot) := \langle E_{A_\alpha}(\cdot)u, u \rangle \quad \text{and} \quad F_\alpha(z) := \langle R_z(A_\alpha)u, u \rangle = \int_{\mathbb{R}} \frac{d\mu_\alpha(\lambda)}{\lambda - z}, \quad z \in \rho(A_\alpha).$$

Obviously, $A = A_0$. In the case $\alpha = 0$, we write $\mu = \mu_0$ and $F = F_0$.

Lemma 9.7 *For $\alpha, \beta \in \mathbb{R}$ and $z \in \mathbb{C} \backslash \mathbb{R}$, we have $1 + (\alpha - \beta)F_\beta(z) \neq 0$,*

$$F_\alpha(z) = \frac{F_\beta(z)}{1 + (\alpha - \beta)F_\beta(z)}, \quad \mathrm{Im}\, F_\alpha(z) = \frac{\mathrm{Im}\, F_\beta(z)}{|1 + (\alpha - \beta)F_\beta(z)|^2}. \tag{9.8}$$

Proof From the resolvent identity (2.4) and Eq. (9.7) we conclude that

$$R_z(A_\alpha) - R_z(A_\beta) = R_z(A_\alpha)(A_\beta - A_\alpha)R_z(A_\beta)$$
$$= (\beta - \alpha)\langle R_z(A_\beta)\cdot, u \rangle R_z(A_\alpha)u. \tag{9.9}$$

Applying both sides of (9.9) to the vector u, it follows that

$$R_z(A_\alpha)u - R_z(A_\beta)u = (\beta - \alpha)F_\beta(z)R_z(A_\alpha)u. \tag{9.10}$$

Taking the scalar product with u yields $F_\alpha(z) - F_\beta(z) = (\beta - \alpha)F_\beta(z)F_\alpha(z)$, so $F_\alpha(z)(1 + (\alpha - \beta)F_\beta(z)) = F_\beta(z)$. Since $F_\beta(z) \neq 0$, we have $1 + (\alpha - \beta)F_\beta(z) \neq 0$, and we obtain the first equality of (9.8). The second equality of (9.8) follows at once from the first one. $\qquad \square$

Let $\mathcal{H}_\alpha$ denote the closure of $\mathcal{D}_\alpha := \mathrm{Lin}\{R_z(A_\alpha)u : z \in \mathbb{C}\backslash\mathbb{R}\}$. By Lemma 5.17, $\mathcal{H}_\alpha$ is the smallest reducing subspace for the self-adjoint operator A_α containing the vector u. Formula (9.10) implies that $\mathcal{D}_\beta \subseteq \mathcal{D}_\alpha$. Interchanging the role of α and β, it follows that $\mathcal{D}_\beta = \mathcal{D}_\alpha$ and $\mathcal{H}_\beta = \mathcal{H}_\alpha$. In particular, $\mathcal{H}_0 = \mathcal{H}_\alpha$. This proves the following:

Lemma 9.8 *For any $\alpha \in \mathbb{R}$, $\mathcal{H}_0$ is the smallest reducing subspace for A_α containing u. The vector u is generating for A_α if and only if it is generating for A.*

The subspace $\mathcal{H}_0$ contains u and is reducing for A_α and A. By (9.7) the parts of A_α and A on the subspace $(\mathcal{H}_0)^\perp$ coincide. To study the spectrum of the operators A_α, we can therefore assume that u *is a generating vector for* A.

In what follows, the main technical tools are the functions F and G defined by

$$F(z) = \int_{\mathbb{R}} \frac{d\mu(\lambda)}{\lambda - z} \quad \text{and} \quad G(t) = \int_{\mathbb{R}} \frac{d\mu(\lambda)}{(\lambda - t)^2}, \quad t \in \mathbb{R}. \tag{9.11}$$

Clearly, $F(z)$ is defined on $\mathbb{C}\backslash\mathbb{R}$, but it might also exist for certain real numbers. From Theorem F.3 it follows that the limit

$$F(t + i0) := \lim_{\varepsilon \to +0} F(t + i\varepsilon)$$

exists and is finite a.e. on $\mathbb{R}$.

Lemma 9.9 *Let* $t \in \mathbb{R}$. *Suppose that* $G(t) < \infty$. *Then* $F(t) \in \mathbb{R}$ *exists, and*

$$F(t) = F(t + i0), \qquad iG(t) = \lim_{\varepsilon \to +0} \varepsilon^{-1}\big(F(t + i\varepsilon) - F(t)\big). \tag{9.12}$$

Proof Put $f(\lambda) := (\lambda - t)^{-1}$. Since $f \in L^2(\mathbb{R}, \mu)$ (by $G(t) < \infty$) and the measure μ is finite, $f \in L^1(\mathbb{R}, \mu)$ by Hölder's inequality, that is, $F(t) \in \mathbb{R}$ exists. Define

$$f_\varepsilon(\lambda) := \big(\lambda - (t + i\varepsilon)\big)^{-1} \quad \text{and} \quad g_\varepsilon(\lambda) := \varepsilon^{-1}\big((\lambda - (t + i\varepsilon))^{-1} - (\lambda - t)^{-1}\big).$$

Clearly, $f_\varepsilon(\lambda) \to f(\lambda)$ and $g_\varepsilon(\lambda) \to if(\lambda)^2$ a.e. on $\mathbb{R}$ as $\varepsilon \to +0$. It is easily checked that $|f_\varepsilon(\lambda)| \le |f(\lambda)|$ and $|g_\varepsilon(\lambda)| \le |f(\lambda)|^2$ a.e. on $\mathbb{R}$. Therefore, since f and f^2 are in $L^1(\mathbb{R}, \mu)$, the dominated convergence theorem (Theorem B.1) applies and yields

$$F(t + i\varepsilon) = \int_{\mathbb{R}} f_\varepsilon(\lambda)\, d\mu(\lambda) \to \int_{\mathbb{R}} f(\lambda)\, d\mu(\lambda) = F(t),$$

$$\varepsilon^{-1}\big(F(t + i\varepsilon) - F(t)\big) = \int_{\mathbb{R}} g_\varepsilon(\lambda)\, d\mu(\lambda) \to \int_{\mathbb{R}} if(\lambda)^2\, d\mu(\lambda) = iG(t). \qquad \square$$

Remark It may happen that the limit $F(t + i0)$ exists, but $F(t)$ does not. For instance, if $F(z) = \int_{-1}^{1}(\lambda - z)^{-1}d\lambda$, then $F(0 + i0) = \lim_{\varepsilon \to +0} F_\alpha(i\varepsilon) = i\pi$, but $F(0)$ does not exist.

The next theorem collects the main results of the *Aronszajn–Donoghue theory*. It gives a detailed spectral analysis of the self-adjoint operators A_α in terms of the functions F and G defined by (9.11) which depend only on the measure $\mu(\cdot) = \langle E_A(\cdot)u, u \rangle$. For this analysis, Theorem F.6 from Appendix F plays a crucial role.

Theorem 9.10 *Suppose that* u *is a generating vector for the self-adjoint operator* A. *Let* $\alpha, \beta, \beta_1, \beta_2 \in \mathbb{R}$ *and suppose that* $\alpha \ne 0$.

(i) *A real number t is an eigenvalue of A_α if and only if t belongs to the set*

$$P_\alpha := \left\{ t \in \mathbb{R} : F(t) = \int_\mathbb{R} (\lambda - t)^{-1} d\mu(\lambda) \in \mathbb{R} \text{ exists, } F(t) = -\alpha^{-1}, \ G(t) < \infty \right\}$$
$$= \left\{ t \in \mathbb{R} : F(t+i0) = -\alpha^{-1}, \ G(t) < \infty \right\}.$$

The pure point part $(\mu_\alpha)_\mathrm{p}$ of the measure μ_α is supported by the set P_α.

(ii) *For $t \in P_\alpha$, we have $\mu_\alpha(\{t\}) \equiv \langle E_{A_\alpha}(\{t\})u, u \rangle = \alpha^{-2} G(t)^{-1}$.*

(iii) *The singularly continuous part $(\mu_\alpha)_\mathrm{sc}$ of the measure μ_α is supported by*

$$S_\alpha = \left\{ t \in \mathbb{R} : F(t+i0) = -\alpha^{-1}, \ G(t) = \infty \right\}.$$

(iv) *The measures $(\mu_{\beta_1})_\mathrm{sing}$ and $(\mu_{\beta_2})_\mathrm{sing}$ are mutually singular if $\beta_1 \neq \beta_2$.*

(v) *The absolutely continuous part $(\mu_\alpha)_\mathrm{ac}$ of the measure μ_α is supported by*

$$L := \left\{ t \in \mathbb{R} : (\operatorname{Im} F)(t+i0) \neq 0 \right\} = \left\{ t \in \mathbb{R} : (\operatorname{Im} F)(t+i0) > 0 \right\}.$$

(vi) *The absolutely continuous parts $(A_\alpha)_\mathrm{ac}$ and A_ac of A_α and A, respectively, are unitarily equivalent.*

The three sets P_α, S_α, and L defined above are mutually disjoint.

Proof (i): Both descriptions of the set P_α are equal, since $F(t) = F(t+i0)$ by (9.12) when $G(t) < \infty$.

Since u is a generating vector, A has a simple spectrum, and by Proposition 5.18 we can assume that, up to unitary equivalence, A is the multiplication operator $(Af)(\lambda) = \lambda f(\lambda)$ and u is the function $u(\lambda) \equiv 1$ in $\mathcal{H} = L^2(\mathbb{R}, \mu)$.

Let t be an eigenvalue of A_α with eigenvector f in $L^2(\mathbb{R}, \mu)$. Then we have

$$(A_\alpha f)(\lambda) = \lambda f(\lambda) + \alpha \langle f, u \rangle = t f(\lambda) \quad \mu\text{-a.e. on } \mathbb{R}. \tag{9.13}$$

We prove that $\langle f, u \rangle \neq 0$ and $\mu(\{t\}) = 0$. Assume to the contrary that $\langle f, u \rangle = 0$. Then, by (9.13), f is a multiple of $\chi_{\{t\}}$, say $f = c\chi_{\{t\}}$, so $\mu(\{t\}) \neq 0$. Then

$$0 = \bar{c} \langle f, u \rangle = \bar{c} \int_\mathbb{R} c \chi_{\{t\}}(\lambda) 1 \, d\mu(\lambda) = \bar{c} c \mu(\{t\}) = \|f\|^2 \neq 0,$$

which is a contradiction. Thus, $\langle f, u \rangle \neq 0$. From (9.13) it follows that $\alpha \langle f, u \rangle = 0$ on the singleton $\{t\}$. Because $\alpha \langle f, u \rangle \neq 0$, this is only possible when $\mu(\{t\}) = 0$.

Since $\mu(\{t\}) = 0$, we have $f(\lambda) = -\alpha \langle f, u \rangle (\lambda - t)^{-1}$ μ-a.e. by (9.13), and so

$$\langle f, u \rangle = -\alpha \, \langle f, u \rangle \int_\mathbb{R} (\lambda - t)^{-1} d\mu(\lambda),$$

$$\|f\|^2 = \alpha^2 |\langle f, u \rangle|^2 \int_\mathbb{R} (\lambda - t)^{-2} d\mu(\lambda) < \infty.$$

Since $\langle f, u \rangle \neq 0$, these relations imply that $F(t) = \int (\lambda - t)^{-1} d\mu(\lambda) \in \mathbb{R}$ exists, $F(t) = -\alpha^{-1}$, and $G(t) < \infty$. That is, $t \in P_\alpha$.

Conversely, suppose that $t \in P_\alpha$. Set $f_t(\lambda) := -\alpha(\lambda - t)^{-1}$. Then f_t belongs to $L^2(\mathbb{R}, \mu)$ by $G(t) < \infty$. Further,

$$\langle f_t, u \rangle = -\alpha \int_{\mathbb{R}} (\lambda - t)^{-1} d\mu(\lambda) = -\alpha F(t) = 1.$$

Since $\mu(\{t\}) = 0$ because of $G(t) < \infty$, we have

$$\lambda f_t(\lambda) + \alpha \langle f_t, u \rangle = \lambda f_t(\lambda) + \alpha = t f_t(\lambda) \quad \mu\text{-a.e. on } \mathbb{R},$$

which means that $A_\alpha f_t = t f_t$ in $L^2(\mathbb{R}, \mu)$. Hence, t is an eigenvalue of A_α.

By definition $(\mu_\alpha)_p$ is supported by the set of atoms of μ. Obviously, this set coincides with the set P_α of eigenvalues of A_α.

(ii): We prove the formula for $\mu_\alpha(\{t\})$. Clearly, $\|f_t\|^2 = \alpha^2 G(t)$. Since u is a generating vector for A_α by Lemma 9.8, all eigenvalues of A_α are simple by Corollary 5.19. Therefore, $E_{A_\alpha}(\{t\})$ is the rank one projection $\|f_t\|^{-2} \langle \cdot, f_t \rangle f_t$. Since $\langle f_t, u \rangle = 1$, we get

$$\mu_\alpha(\{t\}) = \langle E_{A_\alpha}(\{t\}) u, u \rangle = \|f_t\|^{-2} |\langle f_t, u \rangle|^2 = \alpha^{-2} G(t)^{-1}.$$

We shall give another proof of (ii) which is based on (9.8). Indeed, using (9.8) and the equality $F(t) = -\alpha^{-1}$ (by $t \in P_\alpha$), we obtain the identity

$$\varepsilon \operatorname{Im} F_\alpha(t + i\varepsilon) = \frac{\varepsilon \operatorname{Im} F(t + i\varepsilon)}{|1 + \alpha F(t + i\varepsilon)|^2} = \frac{\operatorname{Im} \varepsilon^{-1}(F(t + i\varepsilon) - F(t))}{\alpha^2 |\varepsilon^{-1}(F(t + i\varepsilon) - F(t))|^2}. \quad (9.14)$$

Letting $\varepsilon \to +0$, the right-hand side of (9.14) tends to $\frac{G(t)}{(\alpha G(t))^2} = \alpha^{-2} G(t)^{-1}$ by (9.12), while the left-hand side of (9.14) tends to $\mu_\alpha(\{t\})$ by formula (F.5).

(iii): For arbitrary $\beta \in \mathbb{R}$, by Theorem F.6(i), $(\mu_\beta)_{\text{sing}}$ is supported by

$$S'_\beta := \left\{ t \in \mathbb{R} : (\operatorname{Im} F_\beta)(t + i0) \equiv \lim_{\varepsilon \to +0} (\operatorname{Im} F_\beta)(t + i\varepsilon) = +\infty \right\}, \quad (9.15)$$

that is, $(\mu_\beta)_{\text{sing}}(\mathbb{R} \backslash S'_\beta) = 0$. Here we apply this result for $\beta = \alpha \neq 0$. By (i) the points of P_α are eigenvalues of A_α and hence atoms of μ_α, so that $(\mu_\alpha)_{\text{sc}}(P_\alpha) = 0$. Thus, $(\mu_\alpha)_{\text{sc}}$ is supported by $S'_\alpha \backslash P_\alpha$. To prove that $(\mu_\alpha)_{\text{sc}}$ is supported by S_α, it therefore suffices to show that $S'_\alpha \backslash P_\alpha \subseteq S_\alpha$.

Indeed, from identity (9.8) we obtain

$$F(t + i\varepsilon) + \alpha^{-1} = \frac{F_\alpha(t + i\varepsilon)}{1 - \alpha F_\alpha(t + i\varepsilon)} + \alpha^{-1} = \frac{\alpha^{-1}}{1 - \alpha F_\alpha(t + i\varepsilon)}. \quad (9.16)$$

Let $t \in S'_\alpha \backslash P_\alpha$. Then $|F_\alpha(t + i\varepsilon)| \to +\infty$, and hence $F(t + i\varepsilon) + \alpha^{-1} \to 0$ by (9.16) as $\varepsilon \to +0$, so that $F(t + i0) = -\alpha^{-1}$. Since $t \notin P_\alpha$, we have $G(t) = +\infty$, and therefore $t \in S_\alpha$. This proves that $S'_\alpha \backslash P_\alpha \subseteq S_\alpha$.

(iv): For any positive regular Borel measure ν, we have $\nu_{\text{sing}} = \nu_p + \nu_{\text{sc}}$. Therefore, by (i) and (iii), $(\mu_\alpha)_{\text{sing}}$ is supported by the set

$$P_\alpha \cup S_\alpha = \left\{ t \in \mathbb{R} : F(t + i0) = -\alpha^{-1} \right\}.$$

Recall that $\beta_1 \neq \beta_2$. Hence, if $\beta_1 \neq 0$ and $\beta_2 \neq 0$, then $(\mu_{\beta_1})_{\text{sing}}$ and $(\mu_{\beta_2})_{\text{sing}}$ are supported by the disjoint sets $P_{\beta_1} \cup S_{\beta_1}$ and $P_{\beta_2} \cup S_{\beta_2}$. Now assume that one number, say β_1, is zero. By formula (9.15), applied with $\beta = 0$, it follows that $(\mu_0)_{\text{sing}}$ is supported by $S_0' := \{t \in \mathbb{R} : (\text{Im } F)(t + i0) = +\infty\}$. Since $S_0' \cap (P_{\beta_2} \cup S_{\beta_2})$ is obviously empty, $(\mu_0)_{\text{sing}}$ and $(\mu_{\beta_2})_{\text{sing}}$ are also supported by disjoint sets.

(v): By Theorem F.6(ii) the absolutely continuous part $(\mu_\beta)_{\text{ac}}$ of μ_β is given by

$$d(\mu_\beta)_{\text{ac}}(\lambda) = h_\beta(\lambda)\, d\lambda, \quad \text{where } h_\beta(\lambda) := \pi^{-1}(\text{Im } F_\beta)(\lambda + i0).$$

Set $L_\beta := \{\lambda \in \mathbb{R} : h_\beta(\lambda) \neq 0\}$. The second formula (9.8) implies that the sets L_β and $L_0 = L$ coincide up to a Lebesgue null set. Hence, $(\mu_\beta)_{\text{ac}}$ is supported by L.

(vi): By Lemma 9.8, u is a generating vector for A_β. Hence, A_β is (unitarily equivalent to) the multiplication operator by the variable λ on the Hilbert space $L^2(\mathbb{R}, \mu_\beta)$ by Proposition 5.18. Then, as noted in Example 9.1, $(A_\beta)_{\text{ac}}$ is the multiplication operator by λ in $L^2(\mathbb{R}, (\mu_\beta)_{\text{ac}})$. Clearly, $h_\beta(\lambda) = \pi^{-1}(\text{Im } F_\beta)(\lambda) \geq 0$ a.e. on $\mathbb{R}$. We shall use the preceding facts twice, for the given β and also for $\beta = 0$. Recall that $\mu = \mu_0$ and $A = A_0$. It is easy to check that the map U defined by $(U(f))(\lambda) = (h_0^{-1} h_\beta)^{1/2}(\lambda) f(\lambda)$ is a unitary isomorphism of $L^2(\mathbb{R}, \mu_{\text{ac}})$ onto $L^2(\mathbb{R}, (\mu_\beta)_{\text{ac}})$ such that $U A_{\text{ac}} U^{-1} = (A_\beta)_{\text{ac}}$. $\square$

Having Theorem 9.10, it is easy to prove Theorem 9.6. Indeed, by Lemma 9.8, $\mathcal{H}_0$ is a reducing subspace for A and A_α. Because u is a generating vector for the part of A on $\mathcal{H}_0$, the absolutely continuous parts of the parts of A_α and A on $\mathcal{H}_0$ are unitarily equivalent by Theorem 9.10(vi). Since the parts of A_α and A on the orthogonal complement of $\mathcal{H}_0$ coincide, this completes the proof of Theorem 9.6.

We close this section by developing four interesting examples in detail. Examples 9.3 and 9.4 deal with the behavior of eigenvalues, while Examples 9.5 and 9.6 show that the pure point spectrum and the singularly continuous spectrum may dramatically change when passing from A to the perturbed operator A_α.

Example 9.3 (*Diagonal operator with purely discrete spectrum*) Suppose that $(\lambda_n)_{n \in \mathbb{N}}$ is a real sequence such that $\lim_{n \to \infty} |\lambda_n| = +\infty$. Let A be the diagonal operator on $\mathcal{H} = l^2(\mathbb{N})$ with domain $\mathcal{D}(A) = \{(x_n) \in l^2(\mathbb{N}) : (\lambda_n x_n) \in l^2(\mathbb{N})\}$ defined by $A(x_n) = (\lambda_n x_n)$. Then the spectral measure E_A of A acts by $E_A(M)x = \sum_{n \in M} x_n$ for $x = (x_n) \in \mathcal{H}$.

We fix a unit vector $u = (u_n)$ of $\mathcal{H}$ such that $u_n \neq 0$ for all $n \in \mathbb{N}$. The latter assumption implies that $\text{Lin}\{E_A(M)u\}$ is dense, so u is a generating vector for A. Since $\mu(M) = \langle E_A(M)u, u \rangle = \sum_{n \in M} u_n^2$, the functions F and G are given by

$$F(z) = \sum_{n=1}^\infty \frac{u_n^2}{\lambda_n - z}, \quad z \in \mathbb{C} \setminus \mathbb{R}, \quad \text{and} \quad G(t) = \sum_{n=1}^\infty \frac{u_n^2}{(\lambda_n - t)^2}, \quad t \in \mathbb{R}.$$

Suppose that $\alpha \in \mathbb{R}$, $\alpha \neq 0$. Let $t \in \mathbb{R}$. Since $\lim_n |\lambda_n| = +\infty$, $G(t) = +\infty$ if and only if $t = \lambda_k$ for some k. If $t \neq \lambda_k$ for all k, then $F(t + i0) = F(t) \in \mathbb{R}$ by Lemma 9.9. If $t = \lambda_k$ for some k, then $\text{Im } F(t + i0) = +\infty$. These facts imply that $L = \{\lambda_n : n \in \mathbb{N}\}$, so L is a Lebesgue null set, and hence $\sigma_{\text{ac}}(A_\alpha) = \emptyset$, and that $S_\alpha = \emptyset$, so $\sigma_{\text{sc}}(A_\alpha) = \emptyset$.

By Theorem 9.10(i), $t \in \mathbb{R}$ is an eigenvalue of A_α if and only if $t \in P_\alpha$, that is, $t \neq \lambda_k$ for all k and $F(t) = -\alpha^{-1}$. In order to analyze the latter conditions more in detail, we now assume in addition that $\lambda_n < \lambda_{n+1}$ for all $n \in \mathbb{N}$.

Suppose that $\alpha > 0$. If $t < \lambda_1$, then $F(t) > 0$, so the equation $F(t) = -\alpha^{-1}$ has no solution on $(-\infty, \lambda_1)$. The function $F(t)$ is strictly increasing on $(\lambda_k, \lambda_{k+1})$, $F(t) \to -\infty$ as $t \to \lambda_k + 0$, and $F(t) \to +\infty$ as $t \to \lambda_{k+1} - 0$. Therefore, the equation $F(t) = -\alpha^{-1}$ has a unique solution, denoted by $v(\alpha)_k$, in the interval $(\lambda_k, \lambda_{k+1})$. In the case $\alpha < 0$ the equation $F(t) = -\alpha^{-1}$ has unique solutions $v(\alpha)_1$ in $(-\infty, \lambda_1)$ and $v(\alpha)_{k+1}$ in $(\lambda_k, \lambda_{k+1})$.

Summarizing, we have shown that the pure point spectrum $\sigma_p(A_\alpha)$ is the set $P_\alpha = \{v(\alpha)_k : k \in \mathbb{N}\}$ of eigenvalues of A_α for all $\alpha \neq 0$. Moreover, we have

$$\lambda_k < v(\alpha)_k < \lambda_{k+1} \quad \text{if } \alpha > 0, \qquad v(\alpha)_k < \lambda_k < v(\alpha)_{k+1} \quad \text{if } \alpha < 0. \quad (9.17)$$

○

Example 9.4 (*Embedded eigenvalue*) Let μ be the sum of the Lebesgue measure on $[a, b]$ and the delta measure δ_c, where $a < c < b$. As above, A is the multiplication operator $(Af)(\lambda) = \lambda f(\lambda)$ on $L^2(\mathbb{R}, \mu)$. We then compute

$$\operatorname{Im} F(t + i0) = \begin{cases} \pi & \text{if } a < t < b, t \neq c \\ 0 & \text{if } t \notin [a, b] \\ \pi/2 & \text{if } t = a, b \\ +\infty & \text{if } t = c \end{cases}. \quad (9.18)$$

Thus, $L = [a, b]$, so that $\sigma_{ac}(A_\alpha) = [a, b]$ for all $\alpha \in \mathbb{R}$ in accordance with Theorem 9.6. Obviously, $G(t) = +\infty$ if and only if $t \in [a, b]$. Since $\operatorname{Im} F(t + i0) \neq 0$ on $[a, b]$, the sets S_α and $\sigma_{sc}(A_\alpha)$ are empty for all α.

Clearly, A has a single eigenvalue $t = c$. Since $F(t)$ does not exist for $t \in [a, b]$, A_α has no eigenvalue in $[a, b]$ for $\alpha \neq 0$. For $t \in \mathbb{R} \setminus [a, b]$, we have $G(t) < \infty$ and

$$F(t) = F(t + i0) = \log \frac{|b - t|}{|a - t|} + (c - t)^{-1}.$$

For $\alpha \neq 0$, the equation $F(t) = -\alpha^{-1}$ has a unique solution $t_\alpha \in \mathbb{R} \setminus [a, b]$. Hence, A_α has the single eigenvalue t_α. Note that $t_\alpha < a$ if $\alpha < 0$ and $t_\alpha > b$ if $\alpha > 0$.   ○

Example 9.5 (*Singularly continuous spectrum is not invariant*) We replace in Example 9.4 the measure δ_c by a finite singular measure v on $[a, b]$. Then $\sigma_{sc}(A) \neq \emptyset$. In fact, the singularly continuous part A_{sc} of A is the multiplication operator $(A_{sc}f)(\lambda) = \lambda f(\lambda)$ on $L^2(\mathbb{R}, v)$.

We show that $\sigma_{sc}(A_\alpha)$ is empty for all $\alpha \neq 0$. We have $F = H_1 + H_2$, where $H_1(z) := \int_a^b (\lambda - z)^{-1} d\lambda$ and $H_2(z) := \int_a^b (\lambda - z)^{-1} dv(\lambda)$. Then $\operatorname{Im} H_1(t + i0) = \pi$ if $a < t < b$, $\operatorname{Im} H_1(a + i0) = \operatorname{Im} H_1(b + i0) = \pi/2$ and $\operatorname{Im} H_2(t + i\varepsilon) \geq 0$ if $\varepsilon > 0$, so it follows that $\operatorname{Im} F(t + i0) \geq \pi/2$ and hence $1 + \alpha F(t + i0) \neq 0$ for all $t \in [a, b]$. Obviously, $G(t) < +\infty$ for $t \in \mathbb{R} \setminus [a, b]$. These facts imply that $S_\alpha = \emptyset$. Hence, the singularly continuous spectrum $\sigma_{sc}(A_\alpha)$ is empty for $\alpha \neq 0$.   ○

Example 9.6 (*Purely point spectrum versus singularly continuous spectrum*) Let $(a_n)_{n \in \mathbb{N}}$ and $(x_n)_{n \in \mathbb{N}}$ be real sequences such that $(a_n) \in l^1(\mathbb{N})$ and $a_n > 0$ for all n. Then $\mu := \sum_{n=1}^{\infty} a_n \delta_{x_n}$ is a finite positive Borel measure on $\mathbb{R}$. The multiplication operator A by the variable λ on $L^2(\mathbb{R}, \mu)$ has a purely point spectrum.

Let us assume for a moment that

$$G(t) \equiv \int_{\mathbb{R}} (\lambda - t)^{-2} d\mu(\lambda) = +\infty \quad \text{for all } t \in \mathbb{R}. \tag{9.19}$$

Then $P_\alpha = \emptyset$, and so $\sigma_p(A_\alpha) = \emptyset$ for $\alpha \neq 0$. Also, $\sigma_{ac}(A_\alpha) = \sigma(A) = \emptyset$ by Theorem 9.6. Hence, A_α has a purely singularly continuous spectrum for $\alpha \neq 0$.

Now we show the existence of a measure μ of the above form satisfying (9.19). First, we note that it suffices to find sequences (b_k) and (y_k) such that $(b_k) \in l^1(\mathbb{N})$, $b_k > 0$ and $y_k \in [0, 1]$ for all k, and $\int (\lambda - t)^{-2} d\nu(\lambda) = +\infty$ for all $t \in [0, 1]$, where $\nu = \sum_k b_k \delta_{y_k}$. Then a measure μ with the desired properties is obtained by

$$\mu := \sum_{n \in \mathbb{Z}} \sum_{k \in \mathbb{N}} 2^{-|n|} b_k \delta_{y_k - n}.$$

To construct such sequences (b_k) and (y_k), we use a trick which goes back to M. Riesz. Take a positive sequence $(b_k) \in l^1(\mathbb{N})$ such that $(b_k^{1/2}) \notin l^1(\mathbb{N})$ and define $y_k \in [0, 1]$ by $y_k \equiv s_k \mod 1$, where $s_k := \sum_{l=1}^{k} b_l^{1/2}$. Set $\nu := \sum_k b_k \delta_{y_k}$.

Fix $t \in [0, 1]$. Since $\lim_k s_k = +\infty$ and $\lim_k b_k^{1/2} = 0$, it is easily verified that given $n \in \mathbb{N}$, there exists $k > n$ such that $|y_k - t| \leq b_k^{1/2}$. Thus, there exists a subsequence (y_{k_n}) such that $|y_{k_n} - t| \leq b_{k_n}^{1/2}$ for $n \in \mathbb{N}$. Hence,

$$\int_{\mathbb{R}} (\lambda - t)^{-2} d\nu(\lambda) \geq \sum_n (y_{k_n} - t)^{-2} b_{k_n} \geq \sum_n b_{k_n}^{-1} b_{k_n} = +\infty.$$

This completes the construction of a measure μ satisfying (9.19).   ∘

9.3 Krein's Spectral Shift for Rank One Perturbations

First, we consider the spectral shift and the trace formula in the simplest case.

Example 9.7 (*Spectral shift and trace formula on finite-dimensional Hilbert spaces*) Let A be a self-adjoint operator on a finite-dimensional Hilbert space $\mathcal{H}$ of dimension d. Let $\lambda_1 < \cdots < \lambda_k$ denote the eigenvalues of A, and $P_1, \ldots, P_k$ the projections on the corresponding eigenspaces. The spectral projections $E_A(\lambda)$ of A are

$$E_A(\lambda) = P_1 + \cdots + P_j \quad \text{if } \lambda_j \leq \lambda < \lambda_{j+1},$$

$$E_A(\lambda) = 0 \quad \text{if } \lambda < \lambda_1, \qquad E_A(\lambda) = I \quad \text{if } \lambda_k \leq \lambda.$$

Therefore, setting $n_j := \dim P_j \mathcal{H}$ and $\lambda_{k+1} := +\infty$, we have

$$\operatorname{Tr} E_A(\lambda) = \sum_{j=1}^{k} (n_1 + \cdots + n_j) \chi_{[\lambda_j, \lambda_{j+1})}(\lambda).$$

For any function $f \in C^1(\mathbb{R})$ such that $\lim_{\lambda \to +\infty} f(\lambda) = 0$, we compute

$$\int_{\mathbb{R}} f'(\lambda) \operatorname{Tr} E_A(\lambda) \, d\lambda = \sum_{j=1}^{k} (n_1 + \cdots + n_j)\big(f(\lambda_{j+1}) - f(\lambda_j)\big)$$

$$= -\sum_{j=1}^{k} n_j f(\lambda_j) = -\operatorname{Tr} f(A). \qquad (9.20)$$

Let B be another self-adjoint operator on $\mathcal{H}$. If we define the "spectral shift" by

$$\xi(\lambda) = \operatorname{Tr}\big[E_A(\lambda) - E_B(\lambda)\big], \quad \lambda \in \mathbb{R}, \qquad (9.21)$$

then we obtain from (9.20) the "trace formula"

$$\int_{\mathbb{R}} f'(\lambda)\xi(\lambda) \, d\lambda = \operatorname{Tr}\big[f(B) - f(A)\big]. \qquad (9.22)$$

∘

If B is a rank one or even a trace class perturbation of A, then the operator $E_A(\lambda) - E_B(\lambda)$ may fail to be of trace class. In Sect. 9.6 the spectral shift will be therefore defined by means of perturbation determinants. However, Corollary 9.21 below shows that formula (9.21) remains valid under certain assumptions.

In this section we treat only the case of a rank one perturbation and assume that B is the operator A_α defined by (9.7), that is, $B \equiv A_\alpha = A + \alpha \langle \cdot, u \rangle u$. We remain the notation introduced at the beginning of Sect. 9.2 and define

$$\Delta_{B/A}(z) := 1 + \alpha F(z) = 1 + \alpha \int_{\mathbb{R}} \frac{d\mu(\lambda)}{\lambda - z}. \qquad (9.23)$$

Lemma 9.11 *For $z \in \mathbb{C} \backslash \mathbb{R}$, we have $1 + \alpha F(z) \neq 0$ and*

$$\operatorname{Tr}\big[R_z(A) - R_z(B)\big] = \frac{\alpha}{1 + \alpha F(z)} \langle R_z(A)^2 u, u \rangle = \frac{d}{dz} \log \Delta_{B/A}(z). \quad (9.24)$$

Proof The proof is based on identities (9.10) and (9.9) applied with $\beta = 0$. Then $A = A_\beta$ and $B = A_\alpha$. From (9.10) we obtain $R_z(B)u = (1 + \alpha F(z))^{-1} R_z(A)u$. Inserting this equation into (9.9), we derive

$$R_z(A) - R_z(B) = \alpha\big(1 + \alpha F(z)\big)^{-1}\langle \cdot, R_{\bar{z}}(A)u \rangle R_z(A)u.$$

Taking the trace on both sides and recalling that a rank one operator $\langle \cdot, \varphi \rangle \psi$ has the trace $\langle \psi, \varphi \rangle$, this gives the first equality of formula (9.24).

Using that $\frac{d}{dz} F(z) \equiv \frac{d}{dz}\langle R_z(A)u, u \rangle = \langle R_z(A)^2 u, u \rangle$ as easily checked (see Exercise 2.10), the second equality follows by differentiating $\log \Delta_{B/A}(z)$. $\square$

Remark While $\log \Delta_{B/A}(z)$ depends of course on the chosen single-valued branch for this holomorphic function, the derivative $\frac{d}{dz} \log \Delta_{B/A}(z)$ does not on $\mathbb{C} \backslash \mathbb{R}$, since any two branches differ only by a constant on connected open sets. That is, whenever an expression $\frac{d}{dz} \log \Delta_{B/A}(z)$ occurs in what follows, there is no need to specify the branch of $\log \Delta_{B/A}(z)$.

Theorem 9.12 *There exists a function* $\xi \in L^1(\mathbb{R})$ *such that*

$$\xi(\lambda) = \pi^{-1} \arg \Delta_{B/A}(\lambda + i0) := \pi^{-1} \lim_{\varepsilon \to +0} \arg \Delta_{B/A}(\lambda + i\varepsilon) \quad \textit{a.e. on } \mathbb{R},$$

$$(9.25)$$

$$\log \Delta_{B/A}(z) = \int_{\mathbb{R}} \frac{\xi(\lambda)}{\lambda - z} d\lambda, \quad z \in \mathbb{C}\backslash\mathbb{R}, \tag{9.26}$$

$$0 \le (\mathrm{sign}\,\alpha)\xi(\lambda) \le 1 \; \textit{a.e. on } \mathbb{R},$$

$$\int_{\mathbb{R}} \xi(\lambda)\, d\lambda = \alpha, \qquad \int_{\mathbb{R}} |\xi(\lambda)|\, d\lambda = |\alpha|, \tag{9.27}$$

$$\mathrm{Tr}\big[R_z(B) - R_z(A)\big] = -\int_{\mathbb{R}} \frac{\xi(\lambda)}{(\lambda - z)^2} d\lambda, \quad z \in \mathbb{C}\backslash\mathbb{R}. \tag{9.28}$$

In (9.26) *the symbol "log" means the principal branch of the log function, that is,* $\log w = \log |w| + i \arg w$, *where* $-\pi < \arg w < \pi$.

Proof Assume first that $\alpha > 0$. Obviously, $\mathrm{Im}\, \Delta_{B/A}(iy) = \alpha\,\mathrm{Im}\, F(iy) > 0$ if $y > 0$ by (9.23). Hence, the function $C(z) := \log \Delta_{B/A}(z)$ is holomorphic with positive imaginary part $\mathrm{Im}\, C(z) = \arg \Delta_{B/A}(z)$ on the upper half-plane.

Employing the dominated convergence theorem (Theorem B.1), we derive

$$-iy\big(\Delta_{B/A}(iy) - 1\big) = \alpha \int_{\mathbb{R}} \frac{-i\lambda y + y^2}{\lambda^2 + y^2} d\mu(\lambda) \to \alpha \int_{\mathbb{R}} d\mu(\lambda) = \alpha \|u\|^2 = \alpha$$

for $y \in \mathbb{R}$ and $y \to \infty$. From this we easily conclude that

$$\lim_{y\to\infty} -iyC(iy) = \lim_{y\to\infty} -iy \log \Delta_{B/A}(iy) = \alpha. \tag{9.29}$$

Hence, $C(z)$ is a Nevanlinna function that fulfills the assumptions of Theorem F.7. Therefore, since $\mathrm{Im}\, C(\lambda + i\varepsilon) = \arg \Delta_{B/A}(\lambda + i\varepsilon)$, it follows that the limit in (9.25) exists a.e. on $\mathbb{R}$ and that it defines a function $\xi \in L^1(\mathbb{R})$ satisfying (9.26).

Since $\arg \Delta_{B/A}(\lambda + i\varepsilon) \in (0, \pi]$, we have $\xi(\lambda) \in [0, 1]$ a.e. Using (9.29), (9.27), and the dominated convergence theorem (because of $\xi \in L^1(\mathbb{R})$), we obtain

$$\alpha = \lim_{y\to\infty} -iyC(iy) = \lim_{y\to\infty} \int_{\mathbb{R}} \frac{-iy\xi(\lambda)}{\lambda - iy} d\lambda = \int_{\mathbb{R}} \xi(\lambda)\, d\lambda.$$

Since $\xi(\lambda) \in [0, 1]$ a.e., the latter gives $\int |\xi|\, d\lambda = |\alpha|$. This proves (9.27).

In the case $\alpha < 0$, we replace z by $\bar{z}$ in the preceding proof. We then have $\arg \Delta_{B/A}(z) \in [-\pi, 0)$ if $\mathrm{Im}\, z > 0$. Hence, we get $\xi(\lambda) \in [-1, 0]$ by (9.26).

Formula (9.28) follows by differentiating both sides of (9.26) and using (9.24). Differentiation and integration on the right-hand side of (9.26) can be interchanged by applying again the dominated convergence theorem.   □

The function ξ defined by (9.25) is called the *spectral shift* of the pair $\{B, A\}$. It will be generalized to trace class perturbations and studied more in detail in Sect. 9.6. Formula (9.28) is the trace formula (9.22) for the function $f_z(\lambda) = (\lambda - z)^{-1}$. In Sect. 9.7 this formula will be derived for a much larger class of functions.

9.4 Infinite Determinants

Throughout this section, $\mathcal{H}$ is an infinite-dimensional separable Hilbert space, and T and S are *trace class operators* on $\mathcal{H}$ (see Appendix A and Definition A.3).

Let $(\lambda_n(T))_{n \in \mathbb{N}}$ denote the sequence defined in Theorem A.10. That is, $\lambda_n(T)$ are the eigenvalues of T counted with multiplicities; if T has no eigenvalue or only finitely many eigenvalues, the remaining numbers $\lambda_n(T)$ are set equal to zero.

Definition 9.1 The *determinant* $\det(I + T)$ is defined by

$$\det(I + T) = \prod_{n=1}^{\infty} \left(1 + \lambda_n(T)\right). \tag{9.30}$$

Note that the infinite product in (9.30) converges absolutely, since $T \in \mathbf{B}_1(\mathcal{H})$ and hence the sequence $(\lambda_n(T))_{n \in \mathbb{N}}$ is in $l^1(\mathbb{N})$ by (A.5).

Definition 9.1 is used in [GK]. It is justified by the result from linear algebra that the determinant of a linear transformation on a finite-dimensional complex vector space is equal to the product of its eigenvalues. We briefly mention another elegant approach to determinants that is developed in [RS4] and [Sm2]. It is based on formula (9.31) below.

For a Hilbert space $\mathcal{H}$ and $n \in \mathbb{N}$, let $\otimes^n \mathcal{H}$ be the n-fold tensor product Hilbert space $\mathcal{H} \otimes \cdots \otimes \mathcal{H}$. We denote by $\Lambda^n(\mathcal{H})$ the closed linear subspace of $\otimes^n \mathcal{H}$ which is generated by all elements of the form

$$x_1 \wedge \cdots \wedge x_n := \frac{1}{\sqrt{n!}} \sum_{\tau} \epsilon(\tau) x_{\tau(1)} \otimes \cdots \otimes x_{\tau(n)},$$

where the summation is over all permutations of $\{1, \ldots, n\}$, $\epsilon(\tau)$ is the sign of the permutation τ, and $x_1, \ldots, x_n \in \mathcal{H}$. If $\{x_k : k \in \mathbb{N}\}$ is an orthonormal basis of $\mathcal{H}$, then $\{x_{k_1} \wedge \cdots \wedge x_{k_n} : k_1 < k_2 < \cdots < k_n\}$ is an orthonormal basis of $\Lambda^n(\mathcal{H})$.

Let $A \in \mathbf{B}(\mathcal{H})$. Obviously, the operator $A \otimes \cdots \otimes A$ on $\otimes^n \mathcal{H}$ leaves $\Lambda^n(\mathcal{H})$ invariant. Its restriction to $\Lambda^n(\mathcal{H})$ is denoted by $\Lambda^n(A)$. From linear algebra we know that $\dim \Lambda^n(\mathbb{C}^n) = 1$ and $\Lambda^n(A)$ acts on $\Lambda^n(\mathbb{C}^n)$ as multiplication by $\det A$.

Then, for any $T \in \mathbf{B}_1(\mathcal{H})$, we have $\Lambda^n(T) \in \mathbf{B}_1(\Lambda^n(\mathcal{H}))$ for all $n \in \mathbb{N}$ and

$$\det(I + T) = 1 + \sum_{n=1}^{\infty} \operatorname{Tr} \Lambda^n(T). \tag{9.31}$$

The equality of both definitions (9.30) and (9.31) is proved in [RS4, Th. XIII.106].

The determinant is continuous with respect to the trace norm. This means that for any sequence $(T_n)_{n \in \mathbb{N}}$ of operators $T_n \in \mathbf{B}_1(\mathcal{H})$, we have

$$\det(I + T_n) \to \det(I + T) \quad \text{if } \|T_n - T\|_1 \to 0 \text{ as } n \to \infty. \tag{9.32}$$

This important fact is proved in [GK, p. 160]; we sketch a proof in Exercise 8. The result (9.32) follows also from the following inequality [RS4, p. 324], [GGK, p. 6]:

$$\left|\det(I + T) - \det(I + S)\right| \le \|T - S\|_1 \exp\left(1 + \|T\|_1 + \|S\|_1\right). \tag{9.33}$$

We now develop a number of further properties of determinants.

Lemma 9.13 *Let* $(A_n)_{n\in\mathbb{N}}$ *and* $(B_n)_{n\in\mathbb{N}}$ *be sequences of operators* $A_n, B_n \in \mathbf{B}(\mathcal{H})$ *such that* s-$\lim_{n\to\infty} A_n = I$ *and* s-$\lim_{n\to\infty} B_n = I$. *Then for* $T \in \mathbf{B}_1(\mathcal{H})$, *we have* $\lim_{n\to\infty} \|A_n T B_n^* - T\|_1 = 0$.

Proof The Banach–Steinhaus theorem yields $c := \sup_{n\in\mathbb{N}}(\|A_n\| + \|B_n^*\|) < \infty$. Let $\varepsilon > 0$ be given. The finite rank operators are dense in $(\mathbf{B}_1(\mathcal{H}), \|\cdot\|_1)$, so there is a finite rank operator $T_\varepsilon = \sum_{k=1}^m \langle\cdot, u_k\rangle v_k$ such that $\|T - T_\varepsilon\|_1 (c+1)^2 \le \varepsilon$. Since

$$\|A_n T_\varepsilon B_n^* - T_\varepsilon\|_1 = \left\| \sum_{k=1}^m \left(\langle\cdot, (B_n - I)u_k\rangle A_n v_k + \langle\cdot, u_k\rangle(A_n - I)v_k \right) \right\|_1$$

$$\le \sum_{k=1}^m \left(\|(B_n - I)u_k\| c \|v_k\| + \|u_k\| \|(A_n - I)v_k\| \right),$$

the assumptions imply that $\|A_n T_\varepsilon B_n^* - T_\varepsilon\|_1 \le \varepsilon$ for $n \ge n(\varepsilon)$. For those n,

$$\|A_n T B_n^* - T\|_1 = \left\| A_n (T - T_\varepsilon)(B_n^* - I) + (A_n - I)(T - T_\varepsilon) + A_n T_\varepsilon B_n^* - T_\varepsilon \right\|_1$$

$$\le \|T - T_\varepsilon\|_1 \left(\|A_n\| \|B_n^* - I\| + \|A_n - I\| \right) + \|A_n T_\varepsilon B_n^* - T_\varepsilon\|_1$$

$$\le \|T - T_\varepsilon\|_1 \left(c(c+1) + (c+1) \right) + \varepsilon \le 2\varepsilon. \qquad\square$$

Corollary 9.14 *Let* $\{x_n : n \in \mathbb{N}\}$ *be an orthonormal basis of* $\mathcal{H}$. *Then*

$$\det(I + T) = \lim_{n\to\infty} \det\left(\delta_{kj} + \langle Tx_k, x_j\rangle \right)_{j,k=1}^n \quad \text{for } T \in \mathbf{B}_1(\mathcal{H}).$$

Proof Let P_n be the projection on the linear span of $x_1, \ldots, x_n$. Then we have $P_n T P_n = \sum_{k=1}^n \langle\cdot, x_k\rangle T x_k$. Clearly, $\det(\delta_{kj} + \langle Tx_k, x_j\rangle)_{j,k=1}^n$ is the determinant of $I + P_n T P_n$. Since $\lim_n P_n x = x$ for $x \in \mathcal{H}$, we obtain $\lim_n \|P_n T P_n - T\|_1 = 0$ by Lemma 9.13, and hence $\lim_n \det(I + P_n T P_n) = \det(I + T)$ by (9.32). $\qquad\square$

The following formulas are easily derived from the finite-dimensional results by applying Corollary 9.14 and (9.32); details can be found in [GK, Chap. IV, § 1]:

$$\left| \det(I + T) \right| \le e^{\|T\|_1}, \tag{9.34}$$

$$\overline{\det(I + T)} = \det(I + T^*), \tag{9.35}$$

$$\det(I + T)(I + S) = \det(I + T)\det(I + S), \tag{9.36}$$

$$\det(I + AT) = \det(I + TA) \quad \text{for } T, S \in \mathbf{B}_1(\mathcal{H}), \ A \in \mathbf{B}(\mathcal{H}). \tag{9.37}$$

Lemma 9.15 *Suppose that* $T \in \mathbf{B}_1(\mathcal{H})$.

(i) $I + T$ *has an inverse in* $\mathbf{B}(\mathcal{H})$ *if and only if* $\det(I + T) \ne 0$.

(ii) *If* $\|T\|_1 < 1$, *then* $\det(I + T) = \exp(\mathrm{Tr}\log(I + T))$, *where* $\log(I + T)$ *denotes the trace class operator defined by the convergent series* $\log(I + T) = \sum_{k=1}^\infty \frac{(-1)^{n+1}}{n} T^n$ *in the Banach space* $(\mathbf{B}_1(\mathcal{H}), \|\cdot\|_1)$.

Proof (i): If $I + T$ is invertible in $\mathbf{B}(\mathcal{H})$, then $S := -T(I + T)^{-1} \in \mathbf{B}_1(\mathcal{H})$ and $(I + S)(I + T) = I$, so $\det(I + T)\det(I + S) = \det I = 1$ by (9.36), and hence $\det(I + T) \neq 0$.

Suppose that $I + T$ is not invertible. Then $0 \in \sigma(I + T)$, so $-1 \in \sigma(T)$. Since T is compact, -1 is one of the eigenvalues $\lambda_n(T)$. Hence, $\det(I + T) = 0$ by (9.30).

(ii): Since $\|T^n\|_1 \leq \|T\|_1 \|T^{n-1}\| \leq \|T\|_1 \|T\|^{n-1} \leq \|T\|_1^n$ by (A.2) and $\|T\|_1 < 1$, the series converges in the norm $\|\cdot\|_1$ and defines an operator $\log(I + T) \in \mathbf{B}_1(\mathcal{H})$. From the spectral mapping theorem (or from the series expansion) it follows that the operator $\log(I + T)$ has the eigenvalues $\log(1 + \lambda_n(T))$. Therefore, $\operatorname{Tr}\log(I + T) = \sum_n \log(1 + \lambda_n(T))$ by Lidskii's Theorem A.5 and

$$\det(I + T) = e^{\log(\prod_n(1+\lambda_n(T)))} = e^{(\sum_n \log(1+\lambda_n(T)))} = e^{\operatorname{Tr}\log(I+T)}. \qquad \square$$

The following technical lemma will be used in the proof of the formula (9.44) below.

Lemma 9.16 *Suppose that $z \to T(z)$ is a holomorphic mapping of an open subset Ω of $\mathbb{C}$ into the Banach space $(\mathbf{B}_1(\mathcal{H}), \|\cdot\|_1)$ such that $(I + T(z))^{-1} \in \mathbf{B}(\mathcal{H})$ for all $z \in \Omega$. Then $\det(I + T(z))$ is a holomorphic function on Ω, and*

$$\frac{d}{dz}\log\det(I + T(z)) = \operatorname{Tr}\left[(I + T(z))^{-1}\frac{dT(z)}{dz}\right], \quad z \in \Omega. \qquad (9.38)$$

Proof We fix an orthonormal basis $\{x_n : n \in \mathbb{N}\}$ of $\mathcal{H}$. Let P_n be the projection on the span of $x_1, \dots, x_n$. We abbreviate

$$T_n(z) := P_n T(z) P_n, \qquad \Delta_n(z) := \det(I + T_n(z)), \qquad \Delta(z) := \det(I + T(z)).$$

Let $\mathcal{I}_n$ denote the unit matrix of type n. Clearly,

$$\Delta_n(z) = \det(\mathcal{I}_n + \mathcal{T}_n(z)), \quad \text{where } \mathcal{T}_n(z) := (t_{jk}(z) := \langle T(z)x_k, x_j \rangle)_{j,k=1}^n.$$

Since the matrix entries of $\mathcal{I}_n + \mathcal{T}_n(z)$ are holomorphic functions on Ω, so is the determinant $\Delta_n(z)$. Let $\gamma_{nrs}(z)$ denote the cofactor of the element $\delta_{rs} + t_{rs}(z)$ in the matrix $\mathcal{I}_n + \mathcal{T}_n$. Assume for a moment that the matrix $\mathcal{I}_n + \mathcal{T}_n(z)$ is invertible. Using the formulas for the expansion of $\Delta_n(z)$ and for the matrix inverse, we derive

$$\frac{d}{dz}\log\det(I + T_n(z)) = \Delta_n(z)^{-1}\Delta_n'(z) = \sum_{k,j=1}^n \Delta_n(z)^{-1}\gamma_{nkj}(z)t_{kj}'(z)$$

$$= \sum_{k,j=1}^n \left((\mathcal{I}_n + \mathcal{T}_n(z))^{-1}\right)_{jk} t_{kj}'(z)$$

$$= \operatorname{Tr}\left[(I + T_n(z))^{-1}\frac{dT_n(z)}{dz}\right]. \qquad (9.39)$$

Here the last equality holds, because the operator in square brackets leaves $P_n\mathcal{H}$ invariant and is zero on $(I - P_n)\mathcal{H}$. In fact, (9.39) is the assertion (9.38) in the finite-dimensional case. In the rest of this proof we show that one can pass to the limit in (9.39), thus proving the assertion in the general case.

From Lemma 9.13, $\lim_n \|T_n(z) - T(z)\|_1 = 0$, and so $\lim_n \Delta_n(z) = \Delta(z)$ by (9.32). Since $\Delta(z) \neq 0$ by Lemma 9.15(i), $\Delta_n(z) \neq 0$, and hence $\mathcal{I}_n + \mathcal{T}_n(z)$ is invertible for large n. Since $|\Delta_n(z)| \leq \exp(\|T(z)\|_1)$ by (9.34), the sequence $(\Delta_n(z))_{n \in \mathbb{N}}$ of holomorphic functions is uniformly bounded on compact subsets of Ω. Therefore, by Montel's theorem (see, e.g., [BG], 2.2.8) it has a subsequence which converges uniformly (to $\Delta(z)$) on each compact subset of Ω. For notational simplicity, we assume that the sequence (Δ_n) has already this property. Hence, $\Delta(z)$ is holomorphic on Ω. Writing $\Delta_n'(z)$ by the Cauchy formula as an integral of $(2\pi i)^{-1}\Delta_n(\zeta)(\zeta - z)^{-2}$ along some circle around z and passing to the limit, it follows that $\Delta_n'(z) \to \Delta'(z)$. Hence, the left-hand side of (9.39) tends to $\Delta^{-1}(z)\Delta'(z) = \frac{d}{dz}\log(I + T(z))$, which is the left-hand side of (9.38).

We have $(I + T(z))^{-1} \in \mathbf{B}(\mathcal{H})$ and $\lim_n \|T_n(z) - T(z)\| = 0$, since $\|\cdot\| \leq \|\cdot\|_1$. Hence, because the inversion operation in $\mathbf{B}(\mathcal{H})$ is continuous in the operator norm, $(I + T_n(z))^{-1} \in \mathbf{B}(\mathcal{H})$ for large n, and $\lim_n \|(I + T_n(z))^{-1} - (I + T(z))^{-1}\| = 0$. For notational simplicity, we write $T'(z)$ for $\frac{dT(z)}{dz}$, etc. Then $T'(z) \in \mathbf{B}_1(\mathcal{H})$ (because $T(z)$ is a holomorphic map into $(\mathbf{B}_1(\mathcal{H}), \|\cdot\|_1)$), and obviously we have $P_n T'(z) P_n = T_n'(z)$. Hence, $\lim_n \|T_n'(z) - T'(z)\|_1 = 0$ by Lemma 9.13. Thus,

$$\left\|(I + T_n(z))^{-1}T_n'(z) - (I + T(z))^{-1}T'(z)\right\|_1$$
$$= \left\|\left((I + T_n(z))^{-1} - (I + T(z))^{-1}\right)T_n'(z) + (I + T(z))^{-1}\left(T_n'(z) - T'(z)\right)\right\|_1$$
$$\leq \left\|(I + T_n(z))^{-1} - (I + T(z))^{-1}\right\|\left\|T_n'(z)\right\|_1$$
$$+ \left\|(I + T(z))^{-1}\right\|\left\|T_n'(z) - T'(z)\right\|_1.$$

The latter expression tends to zero, since $\|T_n'(z)\|_1 = \|P_n T'(z) P_n\|_1 \leq \|T'(z)\|_1$. Therefore, by the continuity of the trace in the trace norm, the right-hand side of (9.39) tends to the right-hand side of (9.38). This completes the proof of (9.38). $\square$

9.5 Perturbation Determinants

Throughout this section we assume that A is a *closed operator with nonempty resolvent set* and D is a *trace class operator* on $\mathcal{H}$. We define the operator $B := A + D$ on $\mathcal{D}(B) = \mathcal{D}(A)$ and consider B as the "perturbation" of A by D.

Definition 9.2 The *perturbation determinant* $\Delta_{B/A}$ of the ordered pair $\{B, A\}$ of operators is defined by

$$\Delta_{B/A}(z) = \det(I + D(A - zI)^{-1}) = \det((B - zI)(A - zI)^{-1}), \quad z \in \rho(A).$$
$$(9.40)$$

Example 9.8 Let D be the rank one operator $\alpha \langle \cdot, u \rangle u$, where $u \in \mathcal{H}$ and $\alpha \in \mathbb{C}$. Then $DR_z(A) = \alpha \langle R_z(A) \cdot, u \rangle u$ has the eigenvalues $\lambda_1 = \alpha \langle R_z(A)u, u \rangle$ and $\lambda_n = 0$ if $n > 1$. Therefore, by (9.40) and (9.30), we obtain

$$\Delta_{B/A}(z) = 1 + \alpha \langle R_z(A)u, u \rangle = 1 + \alpha F(z),$$

that is, the expression in (9.23) is the perturbation determinant $\Delta_{B/A}(z)$. ○

Proposition 9.17 $\Delta_{B/A}(z)$ is a holomorphic function on the resolvent set $\rho(A)$.

Proof We apply Corollary 9.14 with $T := DR_z(A)$. Since the determinant of the operator $I + P_n D R_z(A) P_n$ is $\det(\delta_{kj} + \langle DR_z(A)x_k, x_j \rangle)^n_{j,k=1}$, it is holomorphic on $\rho(A)$ (by Proposition 2.9) and converges to $\det(I + DR_z(A))$. By (9.34),

$$\det\!\left(I + P_n D R_z(A) P_n\right) \le e^{\|P_n D R_z(A) P_n\|_1} \le e^{\|D\|_1 \|R_z(A)\|}. \tag{9.41}$$

The resolvent formula (5.12) implies that $\|R_z(A)\|$ is bounded on compact subsets of $\rho(A)$. Therefore, by (9.41), the sequence $(\det(I + P_n D R_z(A) P_n))_{n \in \mathbb{N}}$ of holomorphic functions is uniformly bounded on compact subsets of $\rho(A)$. Hence, its limit $\Delta_{B/A}(z) = \det(I + DR_z A)$ is also holomorphic on $\rho(A)$, see, e.g., [BG, 2.2.8]. □

Proposition 9.18 Let D_0 be another trace class operator and set $C := B + D_0$. For any $z \in \rho(A) \cap \rho(B)$, we have

$$\Delta_{C/A}(z) = \Delta_{C/B}(z) \Delta_{B/A}(z), \tag{9.42}$$

$$\Delta_{B/A}(z) \Delta_{A/B}(z) = 1, \tag{9.43}$$

$$\frac{d}{dz} \log \Delta_{B/A}(z) \equiv \Delta_{B/A}(z)^{-1} \frac{d}{dz} \Delta_{B/A}(z) = \text{Tr}\big[R_z(A) - R_z(B)\big]. \tag{9.44}$$

Proof Using formulas (9.40) and (9.36), we compute

$$\begin{aligned}
\Delta_{C/A}(z) &= \det\!\big((C - zI)(B - zI)^{-1}(B - zI)(A - zI)^{-1}\big) \\
&= \det\!\big((I + D_0)(B - zI)^{-1}(I + D)(A - zI)^{-1}\big) \\
&= \det(I + D_0)(B - zI)^{-1} \det(I + D)(A - zI)^{-1} = \Delta_{C/B}(z) \Delta_{B/A}(z).
\end{aligned}$$

By (9.42) we have $\Delta_{B/A}(z) \Delta_{A/B}(z) = \Delta_{A/A}(z) = 1$, which gives (9.43).

Now we prove (9.44). First, we verify that the assumptions of Lemma 9.16 are satisfied with $T(z) := DR_z(A)$ and $\Omega := \rho(A)$. Suppose that $z, z_0 \in \rho(A)$ and $|z - z_0| < \|R_{z_0}(A)\|^{-1}$. Applying the bounded operator D to the series (2.6) for the resolvent $R_z(A)$ (which converges in the operator norm), we obtain

$$DR_z(A) = \sum_{n=0}^{\infty} (z - z_0)^n D R_{z_0}(A)^{n+1}. \tag{9.45}$$

Since $\|DR_{z_0}(A)^{n+1}\|_1 \leq \|D\|_1 \|R_{z_0}(A)\|^{n+1}$, the series (9.45) converges also in the stronger norm $\|\cdot\|_1$. Thus, $z \to DR_z(A)$ is a holomorphic map of $\rho(A)$ into the Banach space $(\mathbf{B}_1(\mathcal{H}), \|\cdot\|_1)$. Since $\Delta_{B/A}(z) \neq 0$ by (9.43), $I + DR_z(A)$ is boundedly invertible by Lemma 9.15(i). It is easily checked that

$$\left(I + DR_z(A)\right)^{-1} = (A - zI)(B - zI)^{-1} = I - D(B - zI)^{-1} \in \mathbf{B}(\mathcal{H}). \quad (9.46)$$

Applying (9.38), (9.46), the trace property $\operatorname{Tr} T_1 T_2 = \operatorname{Tr} T_2 T_1$, and finally the resolvent identity (2.4), we derive

$$\begin{aligned}
\frac{d}{dz} \log \Delta_{B/A}(z) &= \frac{d}{dz} \log \det\left(I + DR_z(A)\right) \\
&= \operatorname{Tr}\left[\left(I + DR_z(A)\right)^{-1} \frac{d}{dz} DR_z(A)\right] \\
&= \operatorname{Tr}\left[(A - zI)(B - zI)^{-1} DR_z(A)^2\right] \\
&= \operatorname{Tr}\left[R_z(A)(A - zI)(B - zI)^{-1} DR_z(A)\right] \\
&= \operatorname{Tr} R_z(B) DR_z(A) = \operatorname{Tr}\left[R_z(A) - R_z(B)\right]. \quad \square
\end{aligned}$$

Proposition 9.19 *Suppose that A and D, hence B, are self-adjoint operators. Then*

$$\overline{\Delta_{B/A}(z)} = \Delta_{B/A}(\overline{z}) \quad \text{for } z \in \rho(A), \quad (9.47)$$

$$e^{-\|D\|_1 |\operatorname{Im} z|^{-1}} \leq \left|\Delta_{B/A}(z)\right| \leq e^{\|D\|_1 |\operatorname{Im} z|^{-1}} \quad \text{for } z \in \mathbb{C} \setminus \mathbb{R}. \quad (9.48)$$

Moreover, $\lim_{|\operatorname{Im} z| \to \infty} \Delta_{B/A}(z) = 1$ *uniformly on* $\mathbb{C}$.

Proof Using equalities (9.40), (9.35), and (9.37), we obtain for $z \in \rho(A)$,

$$\begin{aligned}
\overline{\Delta_{B/A}(z)} &= \overline{\det\left(I + DR_z(A)\right)} = \det\left(I + R_{\overline{z}}(A)D\right) \\
&= \det\left(I + DR_{\overline{z}}(A)\right) = \Delta_{B/A}(\overline{z}).
\end{aligned}$$

Let $z \in \mathbb{C} \setminus \mathbb{R}$. Then $\|R_z(A)\| \leq |\operatorname{Im} z|^{-1}$ by Proposition 3.10, and hence

$$\left\|DR_z(A)\right\|_1 \leq \|D\|_1 \left\|R_z(A)\right\| \leq \|D\|_1 |\operatorname{Im} z|^{-1} \quad \text{for } z \in \mathbb{C} \setminus \mathbb{R}. \quad (9.49)$$

Combined with (9.34) the latter yields

$$\left|\Delta_{B/A}(z)\right| = \left|\det\left(I + DR_z(A)\right)\right| \leq e^{\|D\|_1 |\operatorname{Im} z|^{-1}},$$

which is the second inequality in (9.48). Since $B = A + (-D)$, A and B can be interchanged, and the same inequality holds for $\Delta_{A/B}(z)$. But $\Delta_{B/A}(z) = (\Delta_{A/B}(z))^{-1}$ by (9.43). This implies the first inequality in (9.48).

From (9.48) we conclude that $\lim_{|\operatorname{Im} z| \to \infty} \Delta_{B/A}(z) = 1$ uniformly on $\mathbb{C}$. (Note that $\lim_{|\operatorname{Im} z| \to \infty} \Delta_{B/A}(z) = 1$ follows also from (9.49) combined with (9.32).) $\square$

Proposition 9.20 *Let A and D be self-adjoint operators. Let $\lambda_0 \in \mathbb{R}$ be in the resolvent set or an isolated point of the spectrum of A and B. Suppose that*

$$n_A := \dim \mathcal{N}(A - \lambda_0 I) < \infty \quad \text{and} \quad n_B := \dim \mathcal{N}(B - \lambda_0 I) < \infty.$$

Then $\Delta_{B/A}(z)$ has a zero of order $n_B - n_A$ if $n_B - n_A > 0$, resp. a pole of order $n_A - n_B$ if $n_B - n_A < 0$ at λ_0, and $\Delta_{B/A}(\lambda_0 + 0) \neq 0$ if $n_A = n_B$.

Proof Throughout this proof we freely use the assertions of Proposition 5.10.

First, let us suppose that $\lambda_0 \in \mathbb{R}$ is in the resolvent set or an isolated point of the spectrum of a self-adjoint operator C such that $n_C := \dim \mathcal{N}(C - \lambda_0 I) < \infty$. Then $R_z(C) E_C(\{\lambda_0\}) = (\lambda_0 - z)^{-1} E_C(\{\lambda_0\})$, and hence

$$\operatorname{Tr} R_z(C) E_C(\{\lambda_0\}) = -n_C (z - \lambda_0)^{-1}. \tag{9.50}$$

By the assumption there exists an $\varepsilon > 0$ such that $M \setminus \{\lambda_0\} \subseteq \rho(C)$, where $M := (\lambda_0 - \varepsilon, \lambda_0 + \varepsilon)$. Then the spectral projection of the set M for the self-adjoint operator $C_M := C(I - E_C(M))$ is zero. Therefore, $M \subseteq \rho(C_M)$, and for $z \in \mathbb{C} \setminus \mathbb{R}$,

$$R_z(C)\big(I - E_C(\{\lambda_0\})\big) = R_z(C)\big(I - E_C(M)\big) = R_z(C_M). \tag{9.51}$$

By Proposition 2.9 this operator-valued function has a holomorphic extension to the subset M of $\rho(C_M)$. In particular, it is holomorphic in λ_0.

We now apply this result and Eq. (9.50) to the operators A and B. Recall that by the assumptions, λ_0 is an isolated singularity of $R_z(A)$ and $R_z(B)$, so of $\frac{d}{dz} \log \Delta_{B/A}(z)$ by (9.44) and hence of $\Delta_{B/A}(z)$. Using (9.44) and (9.50), we get

$$\begin{aligned}
g(z) &:= \frac{d}{dz} \log\big[\Delta_{B/A}(z)(z - \lambda_0)^{n_A - n_B}\big] \\
&= \frac{d}{dz} \log \Delta_{B/A}(z) + (n_A - n_B)(z - \lambda_0)^{-1} \\
&= \operatorname{Tr}\big[R_z(A) - R_z(B)\big] + (n_A - n_B)(z - \lambda_0)^{-1} \\
&= \operatorname{Tr}\big[R_z(A)\big(I - E_A(\{\lambda_0\})\big) - R_z(B)\big(I - E_B(\{\lambda_0\})\big)\big].
\end{aligned}$$

Since the operator-valued function (9.51) is holomorphic in λ_0, so is $g(z)$. Therefore, $h(z) := \Delta_{B/A}(z)(z - \lambda_0)^{n_A - n_B}$ is holomorphic in λ_0 and $h(\lambda_0) \neq 0$, since otherwise g is not holomorphic in λ_0. Then $\Delta_{B/A}(z) = (z - \lambda_0)^{n_B - n_A} h(z)$ with $h(\lambda_0) \neq 0$, which gives the assertion. $\qquad\square$

9.6 Krein's Spectral Shift for Trace Class Perturbations

Throughout this section we assume that A and D are *self-adjoint operators* on $\mathcal{H}$ such that D is *trace class* and $B := A + D$.

The following theorem extends the spectral shift and the basic formulas (9.25), (9.26), (9.27), (9.28) proved in Theorem 9.12 for the rank one case to trace class perturbations.

Theorem 9.21 *Then there exists a function $\xi \in L^1(\mathbb{R})$ such that*

$$\Delta_{B/A}(z) = \exp \int_{\mathbb{R}} \frac{\xi(\lambda)}{\lambda - z} \, d\lambda, \quad z \in \mathbb{C} \backslash \mathbb{R}, \tag{9.52}$$

$$\int_{\mathbb{R}} \xi(\lambda) \, d\lambda = \operatorname{Tr} D, \qquad \int_{\mathbb{R}} |\xi(\lambda)| \, d\lambda \le \|D\|_1, \tag{9.53}$$

and Eq. (9.28) holds. The function $\xi \in L^1(\mathbb{R})$ is uniquely determined by (9.52).

Before we turn to the proof of this theorem, we note a simple lemma which follows at once by applying Lebesgue's dominated convergence theorem.

Lemma 9.22 *For any function ξ from $L^1(\mathbb{R})$, we have*

$$\lim_{|\operatorname{Im} z| \to \infty} \int_{\mathbb{R}} \frac{\xi(\lambda)}{\lambda - z} \, d\lambda = 0. \tag{9.54}$$

Proof of Theorem 9.21 Since $D = D^* \in \mathbf{B}_1(\mathcal{H})$ by assumption, by Corollary A.9 there exist an orthonormal basis $\{u_k : k \in \mathbb{N}\}$ of $\mathcal{H}$ and a real sequence $(\alpha_k)_{k \in \mathbb{N}}$ such that

$$D = \sum_{k=1}^{\infty} \alpha_k \langle \cdot, u_k \rangle u_k, \qquad \operatorname{Tr} D = \sum_{k=1}^{\infty} \alpha_k, \qquad \|D\|_1 = \sum_{k=1}^{\infty} |\alpha_k| < \infty. \tag{9.55}$$

Set $A_0 := A$ and $A_n := A_{n-1} + \alpha_n \langle \cdot, u_n \rangle u_n$ for $n \in \mathbb{N}$.

Let ξ_n be the function from Proposition 9.12 applied to $B = A_n$, $A = A_{n-1}$. Since $\int |\xi_n| \, d\lambda = |\alpha_n|$ by (9.27), it follows from the third equality of (9.55) that the series $\sum_n \xi_n$ converges in $L^1(\mathbb{R})$ and therefore defines a function $\xi \in L^1(\mathbb{R})$. The formulas (9.53) follow then immediately by combining (9.27) and (9.55).

Now we prove (9.52). Let $z \in \mathbb{C} \backslash \mathbb{R}$ and $n \in \mathbb{N}$. Applying the multiplicativity rule (9.42) and formula (9.26) for $B = A_n$, $A = A_{n-1}$, we obtain

$$\Delta_{A_n, A}(z) = \prod_{k=1}^{n} \Delta_{A_n, A_{n-1}}(z) = \prod_{k=1}^{n} \exp \int_{\mathbb{R}} \frac{\xi_k(\lambda)}{\lambda - z} \, d\lambda = \exp \int_{\mathbb{R}} \frac{\sum_{k=1}^{n} \xi_k(\lambda)}{\lambda - z} \, d\lambda. \tag{9.56}$$

Set $D_n := A_n - A = \sum_{k=1}^{n} \alpha_k \langle \cdot, u_k \rangle u_k$. Then we have $\lim_n \|D - D_n\|_1 = 0$, and hence $\lim_n \|D R_z(A) - D_n R_z(A)\|_1 = 0$, so that $\lim_n \Delta_{A_n, A}(z) = \Delta_{B, A}(z)$ by (9.32). Since $\xi \in L^1(\mathbb{R})$, the dominated convergence theorem (Theorem B.1) yields

$$\int_{\mathbb{R}} \frac{\sum_{k=1}^{n} \xi_k(\lambda)}{\lambda - z} \, d\lambda \to \int_{\mathbb{R}} \frac{\sum_{k=1}^{\infty} \xi_k(\lambda)}{\lambda - z} \, d\lambda = \int_{\mathbb{R}} \frac{\xi(\lambda)}{\lambda - z} \, d\lambda \quad \text{as } n \to \infty.$$

Letting $n \to \infty$ in (9.56) and using the preceding facts, we obtain (9.52).

To prove (9.28), we differentiate (9.52). By Lebesgue's Theorem B.1, differentiation and integration can be interchanged, since $\xi \in L^1(\mathbb{R})$. Thus, we derive

$$\Delta_{B/A}(z)^{-1} \frac{d}{dz} \Delta_{B/A}(z) = \int_{\mathbb{R}} \frac{\xi(\lambda)}{(\lambda - z)^2} \, d\lambda, \quad z \in \mathbb{C} \backslash \mathbb{R}.$$

Comparing the left-hand side of this formula with (9.44), we obtain (9.28).

Finally, we prove the uniqueness assertion. Let $\zeta_1, \zeta_2 \in L^1(\mathbb{R})$ be two functions satisfying (9.52) on $\mathbb{C} \setminus \mathbb{R}$. Put $f_k(z) = \int_{\mathbb{R}} \frac{\zeta_k(\lambda)}{\lambda - z} d\lambda$ for $k = 1, 2$. Since $e^{f_1(z)} = e^{f_2(z)}$ by (9.52) for $z \in \mathbb{C} \setminus \mathbb{R}$, there exists an integer $k(z)$ such that $f_1(z) = f_2(z) + 2\pi i k(z)$. Since f_1 and f_2 are continuous on $\mathbb{C} \setminus \mathbb{R}$, so is $k(z)$. Hence, $k(z)$ is constant on both half-planes. Letting $|\operatorname{Im} z| \to +\infty$ and using Lemma 9.22, we conclude that these constants are zero, that is, $f_1(z) = f_2(z)$ on $\mathbb{C} \setminus \mathbb{R}$. This means that the Stieltjes transforms of the complex regular Borel measures ν_1 and ν_2 coincide, where $d\nu_k(\lambda) := \zeta_k(\lambda) d\lambda$. By Theorem F.2 the Stieltjes transform uniquely determines the corresponding measure. Therefore, $\zeta_1 = \zeta_2$ a.e. on $\mathbb{R}$. $\qquad \square$

Corollary 9.23 *If the operator D has precisely n_+ positive (resp. n_- negative) eigenvalues (counted with multiplicities), then $\xi(\lambda) \leq n_+$ (resp. $\xi(\lambda) \geq -n_-$) a.e. on $\mathbb{R}$. In particular, if $D \geq 0$, then we have $\xi(\lambda) \geq 0$ a.e. on $\mathbb{R}$.*

Proof Suppose that D has precisely n_+ positive eigenvalues. Then n_+ numbers α_k are positive, and all others are negative or zero. Then, by (9.27), n_+ summands ξ_k have values in $[0, 1]$, while all other summands have values in $[-1, 0]$. Hence, $\xi = \sum_k \xi_k \leq n_+$. A similar reasoning shows that $\xi \geq -n_-$. $\qquad \square$

Definition 9.3 The function $\xi(\lambda; B, A) := \xi(\lambda)$ of $L^1(\mathbb{R})$ from Theorem 9.21 is called the *spectral shift* of the ordered pair $\{B, A\}$.

Let us reformulate Eq. (9.52). Suppose that f is a holomorphic function on an open subset Ω of $\mathbb{C}$. Recall from complex analysis (see, e.g., [BG, 1.6.28]) that a holomorphic function g on Ω is called a *branch* of the function $\log f$ on Ω if $e^{g(z)} = f(z)$ for all $z \in \Omega$. Clearly, on connected components of Ω, two branches differ by $2\pi i k$ for some constant $k \in \mathbb{Z}$. Thus, Eq. (9.52) means that the function $\int \frac{\xi(\lambda)}{\lambda - z} d\lambda$ is a branch of $\log \Delta_{B/A}(z)$ on $\mathbb{C} \setminus \mathbb{R}$. Further, Eq. (9.54) holds, since $\xi \in L^1(\mathbb{R})$, and $\int \frac{\xi(\lambda)}{\lambda - z} d\lambda$ is the unique branch of $\log \Delta_{B/A}(z)$ which tends to zero as $|\operatorname{Im} z| \to \infty$. In what follows, we fix this particular branch of $\log \Delta_{B/A}(z)$ and define $\arg \Delta_{B/A}(z) := \operatorname{Im}(\log \Delta_{B/A}(z))$ on $\mathbb{C} \setminus \mathbb{R}$. Then both formulas (9.26) and (9.25) remain valid in this case.

Some further basic properties of the spectral shift are collected in Exercise 9.

Next, we investigate the function ξ on the discrete spectrum of A and B. Formula (9.57) shows that ξ has a possible jump at isolated eigenvalues with finite multiplicities of A and B and that ξ behaves like an index. This formula and the two formulas in Corollary 9.25 below give some justification for the name "spectral shift".

Proposition 9.24

(i) *Let $\mathcal{J}$ be an open interval of $\mathbb{R}$ contained in $\rho(A) \cap \rho(B)$. Then there is an integer $n \in \mathbb{Z}$ such that $\xi(\lambda) = n$ for all $\lambda \in \mathcal{J}$.*

(ii) *Let $\lambda_0 \in \mathbb{R}$ be in the resolvent set or an isolated point of the spectrum of A and B such that $n_A := \dim \mathcal{N}(A - \lambda_0 I) < \infty$ and $n_B := \dim \mathcal{N}(B - \lambda_0 I) < \infty$. Then*

$$\xi(\lambda_0 + 0) - \xi(\lambda_0 - 0) = n_A - n_B. \tag{9.57}$$

Proof (i): By Eq. (9.47), $\Delta_{B/A}(z)$ is real on $\mathcal{J}$, so ξ is integer-valued by (9.25). Since $\Delta_{B/A}(z)$ is holomorphic on $\mathcal{J} \subseteq \rho(A) \cap \rho(B)$, ξ is constant on $\mathcal{J}$.

(ii): From Proposition 9.20 it follows that there is a holomorphic function h at λ_0 such that $\Delta_{B/A}(z) = (z - \lambda_0)^{n_B - n_A} h(z)$ and $h(\lambda_0) \neq 0$. We choose an open disk M centered at λ_0 such that h is holomorphic and nonzero on M. Let $C_r \subseteq M$ be a circle with radius $r > 0$ centered at λ_0 oriented counterclockwise. By the argument principle of complex analysis [BG, p. 55] the change of arg $\Delta_{B/A}(z)$ along C_r is $2\pi(n_A - n_B)$. Since $\overline{\Delta_{B/A}(z)} = \Delta_{B/A}(\overline{z})$ by (9.47), the change of arg $\Delta_{B/A}(z)$ along the part of C_r in the upper half-plane is $\pi(n_A - n_B)$. By (9.26) this means that

$$\xi(\lambda_0 + r) - \xi(\lambda_0 - r) = \pi^{-1}\left[\arg \Delta_{B/A}(\lambda_0 + r) - \arg \Delta_{B/A}(\lambda_0 - r)\right] = n_A - n_B.$$

Letting $r \to +0$, we obtain (9.57). $\qquad\square$

We say that the spectrum of a self-adjoint operator A is *discrete* in an open interval $\mathcal{J}$ if $\sigma_{\mathrm{ess}}(A) \cap \mathcal{J} = \emptyset$, or equivalently, $\sigma(A) \cap \mathcal{J}$ is empty or consists of eigenvalues of finite multiplicities which have no accumulation point belonging to $\mathcal{J}$.

Corollary 9.25

(i) *Suppose that A has a discrete spectrum in the interval (a, b), where $-\infty \leq a < b \leq +\infty$. If $a < c < d < b$, then*

$$\xi(d - 0) - \xi(c + 0) = \mathrm{Tr}\left[E_A\big((c, d)\big) - E_B\big((c, d)\big)\right]$$
$$= \dim E_A\big((c, d)\big)\mathcal{H} - \dim E_B\big((c, d)\big)\mathcal{H}.$$

(ii) *If A is lower semibounded and has a discrete spectrum in $(-\infty, b)$, then for all $\lambda < b$, we have*

$$\xi(\lambda - 0) = \mathrm{Tr}\left[E_A(\lambda - 0) - E_B(\lambda - 0)\right]$$
$$= \dim E_A\big((-\infty, \lambda)\big)\mathcal{H} - \dim E_B\big((-\infty, \lambda)\big)\mathcal{H}.$$

(All spectral projections occurring in these formulas have finite rank.)

Proof We only carry out the proof of (i); the proof of (ii) is similar.

Since $B = A + D$ and D is trace class and hence compact, B has also a discrete spectrum in (a, b) by Weyl's theorem (Corollary 8.16). Let C denote A or B. Then $\sigma(C) \cap (c, d)$ is empty or it consists of finitely many eigenvalues $\lambda_1 < \cdots < \lambda_k$ of finite multiplicities $n_C(\lambda_1), \ldots, n_C(\lambda_k)$. Hence, it follows from Proposition 5.10 that

$$\mathrm{Tr}\, E_C\big((c, d)\big) = \dim E_C\big((c, d)\big)\mathcal{H} = \sum_{j=1}^{k} n_C(\lambda_j).$$

Since ξ is constant on open intervals contained in $\rho(A) \cap \rho(B)$ by Proposition 9.24(i), the assertion is obtained at once from (9.57). $\qquad\square$

If C denotes one of the operators A or B in Corollary 9.25, then $\operatorname{Tr} E_C((c,d))$ (resp. $\operatorname{Tr} E_C(\lambda - 0)$) is just the sum of multiplicities of eigenvalues of C contained in the interval (c,d) (resp. $(-\infty, \lambda)$).

Example 9.9 (*Example* 9.3 *continued*) Let B be the self-adjoint operator A_α from Example 9.3. Assume that $\alpha > 0$ and $\lambda_n < \lambda_{n+1}$ for $n \in \mathbb{N}$. Combining Corollary 9.25 and formula (9.17), in this case the spectral shift $\xi(\cdot) \equiv \xi(\cdot; B, A)$ is given by

$$\xi(\lambda) = 1 \quad \text{if } \lambda_k < \lambda < \nu(\alpha)_k, \qquad \xi(\lambda) = 0 \quad \text{if } \lambda < \lambda_1 \text{ or } \nu(\alpha)_k < \lambda < \lambda_{k+1}. \quad \circ$$

9.7 Krein's Trace Formula

We retain the assumptions and the notation from the preceding section. Recall that $\xi(\cdot) = \xi(\cdot; B, A)$ denotes the spectral shift function of the ordered pair $\{B, A\}$.

Let $f \in C^1(\mathbb{R})$. We shall say that the *Krein trace formula* holds for the function f if the operator $f(B) - f(A)$ is of trace class and

$$\operatorname{Tr}\big[f(B) - f(A) \big] = \int_{\mathbb{R}} \xi(\lambda) f'(\lambda) \, d\lambda. \tag{9.58}$$

In particular, formula (9.28) in Theorem 9.21 says that the Krein trace formula is valid for each function $f(\lambda) = (\lambda - z)^{-1}$, where $z \in \mathbb{C} \backslash \mathbb{R}$.

Theorem 9.27 below shows that the trace formula holds for a rather large class of functions. The crucial technical step is contained in the following lemma.

Lemma 9.26 *The Krein trace formula holds for* $f_t(\lambda) = e^{-it\lambda}$, *where* $t \in \mathbb{R}$, *and*

$$\big\| f_t(B) - f_t(A) \big\|_1 \equiv \big\| e^{-itB} - e^{-itA} \big\|_1 \le |t| \, \|D\|_1 \quad \text{for } t \in \mathbb{R}. \tag{9.59}$$

Proof Let $x \in \mathcal{D}(A) = \mathcal{D}(B)$. Then $e^{-isA} x \in \mathcal{D}(A) = \mathcal{D}(B)$ for all $s \in \mathbb{R}$. Therefore, using the Leibniz rule, we compute

$$\frac{d}{ds}\big(I - e^{isB} e^{-isA}\big)x = -e^{isB} iB e^{-isA} x - e^{isB}(-iA) e^{-isA} x = -i e^{isB} D e^{-isA} x,$$

so that

$$\big(I - e^{itB} e^{-itA}\big)x = -i \int_0^t e^{isB} D e^{-isA} x \, ds. \tag{9.60}$$

Since $D \in \mathbf{B}_1(\mathcal{H})$, $e^{isB} D e^{-isA} \in \mathbf{B}_1(\mathcal{H})$ for $s \in \mathbb{R}$. First, we show that the map $\mathbb{R} \ni s \to e^{isB} D e^{-isA} \in (\mathbf{B}_1(\mathcal{H}), \| \cdot \|_1)$ is continuous. For this, we write D as in formula (9.55). Set $T_k = \langle \cdot, u_k \rangle u_k$ and $D_n = \sum_{k=1}^n \alpha_k T_k$. For $s, s' \in \mathbb{R}$,

$$\big\| e^{isB} T_k e^{-isA} - e^{is'B} T_k e^{-is'A} \big\|_1$$
$$= \big\| \langle \cdot, \big(e^{isA} - e^{is'A}\big) u_k \rangle e^{isB} u_k + \langle \cdot, e^{is'A} u_k \rangle \big(e^{isB} - e^{is'B}\big) u_k \big\|_1$$
$$\le \big\| \big(e^{isA} - e^{-is'A}\big) u_k \big\| + \big\| \big(e^{isB} - e^{-is'B}\big) u_k \big\|.$$

Hence, the map $s \to e^{isB}T_k e^{-isA}$ is $\|\cdot\|_1$-continuous, and so is $s \to e^{isB}D_n e^{-isA}$. The $\|\cdot\|_1$-continuity of the map $s \to e^{isB}D e^{-isA}$ follows now from

$$\left\| e^{isB} D e^{-isA} - e^{isB} D_n e^{-isA} \right\|_1 = \left\| e^{isB}(D - D_n)e^{-isA} \right\|_1 \leq \|D - D_n\|_1 \to 0.$$

Next, we consider the integral $\int_0^t e^{isB} D e^{-isA}\, ds$ of the continuous function $e^{isB} D e^{-isA}$ with values in the Banach space $(\mathbf{B}_1(\mathcal{H}), \|\cdot\|_1)$. Since (9.60) holds for all x from the dense subset $\mathcal{D}(A)$, this integral is $i(I - e^{itB}e^{-itA})$. Therefore,

$$e^{-itB} - e^{-itA} = e^{-itB}\left(I - e^{itB}e^{-itA}\right) = -ie^{-itB}\int_0^t e^{isB} D e^{-isA}\, ds, \quad (9.61)$$

where the integral is meant in the norm $\|\cdot\|_1$. Taking the trace norm in (9.61), we get that the operator $e^{-itB} - e^{-itA}$ is of trace class and (9.59) holds.

Next, we prove (9.58) for f_t. Fix $\varepsilon > 0$. Let $\lambda \in \mathbb{R}$. If C is a self-adjoint operator, formula (6.18), applied to the unitary group $t \to e^{-itC}$ with generator $-iC$, yields

$$R_{\lambda+i\varepsilon}(C) = \left(C - (\lambda + i\varepsilon)I\right)^{-1} = -i\left(-iC - (\varepsilon - i\lambda)I\right)^{-1}$$
$$= i\int_0^\infty e^{-(\varepsilon - i\lambda)t} e^{-itC}\, dt.$$

Therefore, if χ_+ denotes the characteristic function of $[0, \infty)$, we have

$$R_{\lambda+i\varepsilon}(B) - R_{\lambda+i\varepsilon}(A) = i\int_{\mathbb{R}} e^{(i\lambda - \varepsilon)t}\chi_+(t)\left(e^{-itB} - e^{-itA}\right) dt.$$

As above, the integral can be considered with respect to the trace norm. Taking the traces of both sides and interchanging integral and trace, we obtain

$$\mathrm{Tr}\left[R_{\lambda+i\varepsilon}(B) - R_{\lambda+i\varepsilon}(A)\right] = i\int_{\mathbb{R}} e^{it\lambda} e^{-\varepsilon t}\chi_+(t)\mathrm{Tr}\left[e^{-itB} - e^{-itA}\right] dt.$$

This shows that $\lambda \to (2\pi)^{-1/2}\,\mathrm{Tr}[R_{\lambda+i\varepsilon}(B) - R_{\lambda+i\varepsilon}(A)]$ is the inverse Fourier transform of the function $t \to ie^{-\varepsilon t}\chi_+(t)\mathrm{Tr}[e^{-itB} - e^{-itA}]$ from $L^1(\mathbb{R}) \cap L^2(\mathbb{R})$. By the formula (C.2) for the inverse Fourier transform we have

$$ie^{-\varepsilon t}\mathrm{Tr}\left[e^{-itB} - e^{-itA}\right] = (2\pi)^{-1}\int_{\mathbb{R}} e^{-it\lambda}\mathrm{Tr}\left[R_{\lambda+i\varepsilon}(B) - R_{\lambda+i\varepsilon}(A)\right] d\lambda$$

for $t \geq 0$. Inserting formula (9.28) for $z = \lambda + i\varepsilon$ on the right-hand side, we get

$$ie^{-\varepsilon t}\mathrm{Tr}\left[e^{-itB} - e^{-itA}\right] = -(2\pi)^{-1}\int_{\mathbb{R}} e^{-it\lambda}\left(\int_{\mathbb{R}} \xi(s)\left(s - (\lambda + i\varepsilon)\right)^{-2} ds\right) d\lambda.$$

Since $\xi \in L^1(\mathbb{R})$, by Fubini's Theorem B.5 we can interchange the order of integrations. Computing then the interior integral by the residue method, we derive

$$\int_{\mathbb{R}} e^{-it\lambda}\left(s - (\lambda + i\varepsilon)\right)^{-2} d\lambda = -2\pi t e^{-\varepsilon t} e^{-it\lambda}.$$

Inserting the latter into the preceding formula and dividing by $ie^{-\varepsilon t}$, we obtain the trace formula for the function $f_t(\lambda) = e^{-it\lambda}$ when $t \geq 0$. Passing to the complex conjugates on both sides, we obtain the corresponding formula for $t \leq 0$. $\qquad\square$

Let $W_1(\mathbb{R})$ denote the set of all functions f on $\mathbb{R}$ of the form

$$f(\lambda) := c + \int_{\mathbb{R}} (e^{-it\lambda} - 1) it^{-1} d\mu(t), \quad \lambda \in \mathbb{R}, \tag{9.62}$$

where $c \in \mathbb{C}$, and μ is a (finite) complex regular Borel measure on $\mathbb{R}$. That is, $W_1(\mathbb{R})$ is the set of functions $f \in C^1(\mathbb{R})$ for which there exists a (finite) complex regular Borel measure μ on $\mathbb{R}$ such that

$$f'(\lambda) = \int_{\mathbb{R}} e^{-it\lambda} d\mu(t), \quad \lambda \in \mathbb{R}. \tag{9.63}$$

The equivalence of both descriptions is easily shown by writing μ as a combination of finite positive measures and interchanging differentiation and integration in (9.62) by using Lebesgue's dominated convergence theorem (Theorem B.1).

Theorem 9.27 *Let $f \in W_1(\mathbb{R})$. Then the operator $f(B) - f(A)$ is of trace class and Krein's trace formula (9.58) is satisfied for f. Moreover, if (9.62) holds, then*

$$\|f(B) - f(A)\|_1 \leq |\mu|(\mathbb{R}) \|D\|_1, \tag{9.64}$$

where $|\mu|$ is the total variation of μ (see, e.g., (B.2) in Appendix B).

Proof We can assume without loss of generality that μ is a finite positive measure. Then functional calculus and integration in (9.62) can be interchanged, so that

$$f(B) - f(A) = \int_{\mathbb{R}} (e^{-itB} - e^{-itA}) it^{-1} d\mu(t). \tag{9.65}$$

From formula (9.59) it follows that $t \to e^{-itB} - e^{-itA}$ is a continuous map of $\mathbb{R}$ into the Banach space $(\mathbf{B}_1(\mathcal{H}), \|\cdot\|_1)$. Hence, the integral (9.65) converges also in the trace norm, and the operator $f(B) - f(A)$ is of trace class. By (9.59) we obtain

$$\|f(B) - f(A)\|_1 \leq \int_{\mathbb{R}} \|e^{-itB} - e^{-itA}\|_1 |t|^{-1} d\mu(t)$$

$$\leq \int_{\mathbb{R}} \|D\|_1 d\mu(t) = \mu(\mathbb{R}) \|D\|_1.$$

Further, using first (9.65), then the trace formula (9.58) for $f_t(\lambda) = e^{-it\lambda}$ (by Lemma 9.26), and finally (9.63), we derive

$$\mathrm{Tr}[f(B) - f(A)] = \int_{\mathbb{R}} \mathrm{Tr}[e^{-itB} - e^{-itA}] it^{-1} d\mu(t)$$

$$= \int_{\mathbb{R}} \left(\int_{\mathbb{R}} \xi(\lambda)(-ite^{-it\lambda}) d\lambda \right) it^{-1} d\mu(t)$$

$$= \int_{\mathbb{R}} \xi(\lambda) \left(\int_{\mathbb{R}} e^{-it\lambda} d\mu(t) \right) d\lambda$$

$$= \int_{\mathbb{R}} \xi(\lambda) f'(\lambda) d\lambda.$$

Since $\xi \in L^1(\mathbb{R})$ and $\mu(\mathbb{R}) < \infty$, by Fubini's Theorem B.5 the change of integration order in the line before last is justified. This proves that the trace formula (9.58) is valid for the function f. $\qquad\square$

The class $W_1(\mathbb{R})$ of functions for which the trace formula (9.58) holds is rather large. Clearly, $W_1(\mathbb{R})$ contains the Schwartz space $\mathcal{S}(\mathbb{R})$. Indeed, if $f \in \mathcal{S}(\mathbb{R})$, the inverse Fourier transform $\mathcal{F}^{-1}(f')$ of f' is in $L^1(\mathbb{R})$, and (9.63) holds for the complex regular Borel measure μ given by $d\mu(\lambda) = (2\pi)^{1/2}\mathcal{F}^{-1}(f')(\lambda)\,d\lambda$.

Taking the measures $\mu = \delta_0$, $\mu = -is\delta_s$ for $s \in \mathbb{R}$, and $d\mu(t) = \frac{1}{2}e^{-|t|}\,dt$ in (9.63), we obtain the functions $f(\lambda) = \lambda$, $f_s(\lambda) = e^{-is\lambda}$, and $f(\lambda) = \arctan\lambda$, respectively, belonging to the class $W_1(\mathbb{R})$.

The above assumption that the operator $D = \overline{B - A}$ is of trace class does not cover all interesting cases. For instance, if T is a symmetric operator with finite deficiency indices and A and B are self-adjoint extensions of T, this assumption is not fulfilled in general, but the difference of the resolvents of A and B is a finite rank operator and hence of trace class. This suggests the following definition.

Definition 9.4 Two self-adjoint operators A and B on $\mathcal{H}$ are called *resolvent comparable* if the difference $R_z(B) - R_z(A)$ of resolvents is of trace class for one (and hence for all) $z \in \rho(A) \cap \rho(B)$.

The following fundamental result is due Krein [Kr5]; proofs can be found in [Kr6, BY], and [Yf, Chap. 8, § 7].

Theorem 9.28 *Let A and B be resolvent comparable self-adjoint operators on a Hilbert space $\mathcal{H}$. Then there exists a real-valued function ξ on $\mathbb{R}$ such that $(1 + \lambda^2)^{-1}\xi(\lambda) \in L^1(\mathbb{R})$, and formula (9.28) holds for all $z \in \rho(A) \cap \rho(B)$. This function ξ is uniquely determined up to an additive constant by these properties. It can be derived from the generalized perturbation determinant $\widetilde{\Delta}_{A/B}$, that is, there is a real constant c such that*

$$\xi(\lambda) = \pi^{-1}\arg\widetilde{\Delta}_{B/A}(\lambda + i0) + c := \pi^{-1}\lim_{\varepsilon \to +0}\widetilde{\Delta}_{B/A}(\lambda + i\varepsilon) + c \quad \text{a.e. on } \mathbb{R}.$$

Here $\widetilde{\Delta}_{A/B}$ is defined by $\widetilde{\Delta}_{A/B}(z) := \Delta_{R_{z_0}(B)/R_{z_0}(A)}(\zeta)$, where $\zeta = (z - z_0)^{-1}$ for some fixed $z_0 \in \rho(A) \cap \rho(B)$. If another such number z_0 is chosen, $\widetilde{\Delta}_{A/B}(z)$ is multiplied by a constant, and hence a constant is added to $\arg\widetilde{\Delta}_{B/A}(\lambda + i0)$.

In Theorem 9.21 the spectral shift from Theorem 9.12 was defined for trace class perturbations. Likewise Theorem 9.6 extends to trace class perturbations as well.

The assertions of the following theorem under assumptions (i) and (ii) are called the *Kato–Rosenblum theorem* and *Kuroda–Birman theorem*, respectively.

Theorem 9.29 *Let A and B be self-adjoint operators on $\mathcal{H}$. Suppose that one of the following assumptions is fulfilled:*

(i) There is a trace class operator D on $\mathcal{H}$ such that $B = A + D$.

(ii) $(B - zI)^{-1} - (A - zI)^{-1}$ is of trace class for one (hence all) $z \in \rho(A) \cap \rho(B)$.

Then the absolutely continuous parts A_{ac} and B_{ac} of A and B, respectively, are unitarily equivalent. In particular, $\sigma_{ac}(A) = \sigma_{ac}(B)$.

Theorem 9.29 is usually proved in scattering theory, where the unitary equivalence of A_{ac} and B_{ac} is provided by means of the *wave operators* $\Omega_{\pm}(A, B)$. We do not carry out the proof and refer to [RS3, Theorems XI.8 and XI.9].

Let $P_{ac}(A)$ and $P_{ac}(B)$ be the orthogonal projections of $\mathcal{H}$ onto the absolutely continuous subspaces $\mathcal{H}_{ac}(A)$ and $\mathcal{H}_{ac}(B)$, respectively, and define

$$\mathcal{D}\big(\Omega_{\pm}(A, B)\big) = \Big\{ x \in \mathcal{H} : \lim_{t \to \pm\infty} e^{itA} e^{-itB} P_{ac}(B)x \text{ exists} \Big\},$$

$$\Omega_{\pm}(A, B)x = \lim_{t \to \pm\infty} e^{itA} e^{-itB} P_{ac}(B)x, \quad x \in \mathcal{D}\big(\Omega_{\pm}(A, B)\big).$$

Then, under the assumptions of Theorem 9.29, $U := \Omega_{\pm}(A, B) \restriction P_{ac}(B)$ is an isometric linear operator of $P_{ac}(B)\mathcal{H}$ onto $P_{ac}(A)\mathcal{H}$, and we have $A_{ac} = U B_{ac} U^*$.

9.8 Exercises

1. Let A be a self-adjoint operator. Show that the operator A_c has no eigenvalue.

2. Find a self-adjoint operator A such that $\sigma(A_c)$ contains an eigenvalue of A.
   Hint: Consider a multiplication operator on some measure space.

3. Define $(Af)(t) = t^n f(t)$, $n \in \mathbb{N}$, on $L^2(0, 1)$ and $(Bf)(t) = t^2 f(t)$ on $L^2(-1, 1)$. Prove (without using the result of Example 9.2) that the self-adjoint operators A and B are purely absolutely continuous.

4. Use formula (9.3) to describe the subspaces $\mathcal{H}_p(A_{\varphi})$, $\mathcal{H}_{sing}(A_{\varphi})$, $\mathcal{H}_{sc}(A_{\varphi})$, and $\mathcal{H}_{ac}(A_{\varphi})$ for the self-adjoint operator A_{φ} in Example 9.2.

5. Replace in Example 9.4 the point measure δ_c by a finite sum $\sum_j \delta_{c_j}$ of point measures δ_{c_j}, where $c_j \in (a, b)$, and determine the three parts of the spectrum.

6. Let A be a self-adjoint operator, and D a self-adjoint finite rank operator on $\mathcal{H}$. Prove that $\sigma_{ac}(A + D) = \sigma_{ac}(A)$.
   Hint: Decompose D as a sum of rank one operators and apply Theorem 9.6.

*7. Suppose that $T \in \mathbf{B}_1(\mathcal{H})$ and define $D_T(z) := \det(I + zT)$ for $z \in \mathbb{C}$.
   a. Prove that $D_T(z)$ is an entire function satisfying $|D_T(z)| \le e^{\|T\|_1 |z|}$, $z \in \mathbb{C}$.
   b. Let $z \in \mathbb{C}$, $z \neq 0$, and let $\mathcal{C}_z$ be a smooth curve in $\mathbb{C}$ from 0 to z such that $-\lambda_n(T)^{-1} \notin \mathcal{C}_z$ for all nonzero $\lambda_n(T)$. Prove that

$$D_T(z) = \exp\Big(\int_{\mathcal{C}_z} \text{Tr}\, T(I + \zeta T)^{-1} \, d\zeta \Big). \tag{9.66}$$

   Hint: Show that $D_T(z)^{-1} \frac{d}{dz} D_T(z) = \text{Tr}\, T(I + zT)^{-1}$ when $-z^{-1} \in \rho(T)$.

*8. Prove the continuity of $\det(I + T)$ in the $\| \cdot \|_1$-norm, that is, for any $\varepsilon > 0$, there exists $\delta > 0$ such that $|\det(I + T) - \det(I + S)| < \varepsilon$ when $\|T - S\|_1 < \delta$.

Hints: First note that $\det(I + T) = D_T(1)$. Verify the inequality

$$\left\| T(I + \zeta T)^{-1} - S(I + \zeta S)^{-1} \right\|_1 \le \left\| (I + \zeta S)^{-1} \right\| \left\| T - S \right\|_1 \left\| (I + \zeta T)^{-1} \right\|$$

and use it to show that

$$\mathcal{J}(T, S) := \int_{\mathcal{C}_1} \mathrm{Tr}\left[T(I + \zeta T)^{-1} - S(I + \zeta S)^{-1} \right] d\zeta$$

is small when $\| T - S \|_1$ is small. Then apply (9.66) with $z = 1$ and write

$$\det(I + T) - \det(I + S) = \big(\det(I + T) \big) \big(1 - e^{\mathcal{J}(T,S)} \big).$$

9. Let A be a self-adjoint operator on $\mathcal{H}$. Let D and D_1 be self-adjoint trace class operators on $\mathcal{H}$. Set $B = A + D$ and $B_1 = A + D_1$.
   a. Prove that $\xi(\cdot; B, A) = -\xi(\cdot; A, B)$ a.e. on $\mathbb{R}$.
   b. Prove that $\xi(\cdot; B_1, B) + \xi(\cdot; B, A) = \xi(\cdot; B_1, A)$ a.e. on $\mathbb{R}$.
   c. Prove that $\| \xi(\cdot; B_1, A) - \xi(\cdot; B, A) \|_{L^1(\mathbb{R})} \le \| D_1 - D \|_1$.
   d. Prove that $D_1 \ge D$ implies that $\xi(\cdot; B_1, A) \ge \xi(\cdot; B, A)$ a.e. on $\mathbb{R}$.
   Hints: Use (9.44) for a., (9.42) for b., (9.53) and b. for c., and Corollary 9.23 for d.
10. Compute the spectral shift in Example 9.9 in the case $\alpha < 0$.

9.9 Notes to Part IV

Chapter 8:

The Kato–Rellich theorem 8.5 was proved by Rellich [Re]. Its applications to Schrödinger operators by Kato [K1] were a milestone in mathematical physics. Proposition 8.6 was proved by Wüst [Wt], and Proposition 8.10 is a special case of a result due to Strichartz [Str]. The invariance of the essential spectrum under compact perturbations was discovered by Weyl [W1].

Chapter 9:

The Aronszajn–Donoghue theory was developed in [Ar] and [Dn2]; see, e.g., [AD, Sm3, SW] for further developments.

Standard references for infinite determinants are [GK, Sm1, Sm2], [RS4, Theorem XIII.17], and [GGK].

The notion of a spectral shift was invented at a formal level by the theoretical physicist Lifshits [Lf]. The mathematical theory of the spectral shift is due to Krein [Kr4, Kr5]. A very readable exposition is given in Krein's lectures [Kr6]. An excellent presentation of the theory can be found in Chap. 8 of the monograph [Yf]. A nice survey and essay around the spectral shift is [BY]. The papers [Kr5, Yf], and [BY] treat the case of resolvent comparable self-adjoint operators in detail. There is now an extensive literature about the spectral shift and its applications. We mention only a few important papers such as [BK, BP, GMN, Kp, Sm2].

Part V
Forms and Operators

Chapter 10
Semibounded Forms and Self-adjoint Operators

This chapter and the next are concerned with operators associated with forms and with the interplay between forms and operators. Forms are basic tools for the definition and the study of semibounded self-adjoint operators and of m-sectorial operators. Often forms are easier to define and to deal with than operators, and form domains are less sensitive and larger than operator domains.

In Sect. 10.2 basic concepts on forms are introduced. In Sect. 10.2 a one-to-one correspondence between lower semibounded self-adjoint operators and lower semibounded densely defined closed forms is established and studied in detail. Abstract boundary problems and variational problems are investigated in this context. Section 10.3 is devoted to the order relation of lower semibounded operators defined by forms, while Sect. 10.4 deals with the Friedrichs extension of a lower semibounded densely defined symmetric operator. Examples of differential operators defined by forms are developed in Sects. 10.5 and 10.6. Section 10.7 is concerned with form perturbations and form sums of lower semibounded self-adjoint operators. Form sums allow one to treat Schrödinger operators having more singular potentials.

10.1 Closed and Closable Lower Semibounded Forms

Let us begin with a number of general notions and operations with forms.

Definition 10.1 A *sesquilinear form* (or briefly, a *form*) on a linear subspace $\mathcal{D}(\mathfrak{t})$ of a Hilbert space $\mathcal{H}$ is a mapping $\mathfrak{t}[\cdot,\cdot]: \mathcal{D}(\mathfrak{t}) \times \mathcal{D}(\mathfrak{t}) \to \mathbb{C}$ such that

$$\mathfrak{t}[\alpha x + \beta y, z] = \alpha \mathfrak{t}[x, z] + \beta \mathfrak{t}[y, z], \qquad \mathfrak{t}[z, \alpha x + \beta y] = \overline{\alpha} \mathfrak{t}[z_1 x] + \overline{\beta} \mathfrak{t}[z, y]$$

for $\alpha, \beta \in \mathbb{C}$ and $x, y, z \in \mathcal{D}(\mathfrak{t})$. The subspace $\mathcal{D}(\mathfrak{t})$ is called the *domain* of $\mathfrak{t}$.

The *quadratic form* $\mathfrak{t}[\cdot]: \mathcal{D}(\mathfrak{t}) \to \mathbb{C}$ associated with a form $\mathfrak{t}$ is defined by

$$\mathfrak{t}[x] := \mathfrak{t}[x, x]. \tag{10.1}$$

K. Schmüdgen, *Unbounded Self-adjoint Operators on Hilbert Space*,
Graduate Texts in Mathematics 265,
DOI 10.1007/978-94-007-4753-1_10, © Springer Science+Business Media Dordrecht 2012

The sesquilinear form can be always uniquely recovered from its quadratic form by the *polarization identity*

$$4\mathfrak{t}[x, y]$$
$$= \mathfrak{t}[x + y] - \mathfrak{t}[x - y] + i\mathfrak{t}[x + iy] - i\mathfrak{t}[x - iy], \quad x, y \in \mathcal{D}(\mathfrak{t}). \quad (10.2)$$

If $\mathfrak{s}$ and $\mathfrak{t}$ are forms on $\mathcal{H}$ and $\alpha \in \mathbb{C}$, the *sum* $\mathfrak{s} + \mathfrak{t}$ and the *scalar multiple* $\alpha\mathfrak{s}$ are

$$(\mathfrak{s} + \mathfrak{t})[x, y] = \mathfrak{s}[x, y] + \mathfrak{t}[x, y], \quad x, y \in \mathcal{D}(\mathfrak{s} + \mathfrak{t}) := \mathcal{D}(\mathfrak{s}) \cap \mathcal{D}(\mathfrak{t}),$$
$$(\alpha\mathfrak{s})[x, y] = \alpha\mathfrak{s}[x, y], \quad x, y \in \mathcal{D}(\alpha\mathfrak{s}) := \mathcal{D}(\mathfrak{s}).$$

We denote by α the form $\alpha \langle \cdot, \cdot \rangle$. In particular, $\mathfrak{t} + \alpha$ is the form given by

$$(\mathfrak{t} + \alpha)[x, y] := \mathfrak{t}[x, y] + \alpha \langle x, y \rangle, \quad x, y \in \mathcal{D}(\mathfrak{t} + \alpha) = \mathcal{D}(\mathfrak{t}).$$

A form $\mathfrak{t}$ is called *symmetric*, or *Hermitian*, if

$$\mathfrak{t}[x, y] = \overline{\mathfrak{t}[y, x]} \quad \text{for } x, y \in \mathcal{D}(\mathfrak{t}).$$

A symmetric form $\mathfrak{t}$ is called *lower semibounded* if there is an $m \in \mathbb{R}$ such that

$$\mathfrak{t}[x] \geq m \|x\|^2 \quad \text{for } x \in \mathcal{D}(\mathfrak{t}).$$

In this case we write $\mathfrak{t} \geq m$ and call m a *lower bound* for $\mathfrak{t}$. The *largest lower bound* of $\mathfrak{t}$ is denoted by $m_\mathfrak{t}$ and given by the infimum of $\mathfrak{t}[x]\|x\|^{-2}$ over all nonzero $x \in \mathcal{D}(\mathfrak{t})$. The form $\mathfrak{t}$ is said to be *positive* if $\mathfrak{t} \geq 0$, that is, if $\mathfrak{t}[x] \geq 0$ for all $x \in \mathcal{D}(\mathfrak{t})$.

In the rest of this section we assume that $\mathfrak{t}$ is a *lower semibounded symmetric form*. Let m be a lower bound of $\mathfrak{t}$. Since $\mathfrak{t} \geq m$, the equation

$$\langle x, y \rangle_\mathfrak{t} := \mathfrak{t}[x, y] + (1 - m)\langle x, y \rangle, \quad x, y \in \mathcal{D}(\mathfrak{t}), \quad (10.3)$$

defines a scalar product $\langle \cdot, \cdot \rangle_\mathfrak{t}$ on $\mathcal{D}(\mathfrak{t})$, and the corresponding norm

$$\|x\|_\mathfrak{t} := \langle x, x \rangle_\mathfrak{t}^{1/2} = \left(\mathfrak{t}[x] + (1 - m)\|x\|^2\right)^{1/2}, \quad x \in \mathcal{D}(\mathfrak{t}), \quad (10.4)$$

satisfies $\| \cdot \|_\mathfrak{t} \geq \| \cdot \|$ on $\mathcal{D}(\mathfrak{t})$. From the Cauchy–Schwarz inequality, applied to the positive form $\mathfrak{t} - m$, and the relation $\| \cdot \| \leq \| \cdot \|_\mathfrak{t}$ we easily derive that

$$\left|\mathfrak{t}[x, y]\right| \leq \left(1 + |m|\right) \|x\|_\mathfrak{t} \|y\|_\mathfrak{t}, \quad x, y \in \mathcal{D}(\mathfrak{t}). \quad (10.5)$$

That is, $\mathfrak{t}$ is a bounded form on the pre-Hilbert space $(\mathcal{D}(\mathfrak{t}), \langle \cdot, \cdot \rangle_\mathfrak{t})$.

The following definition contains three basic concepts.

Definition 10.2 Let $\mathfrak{t}$ be a lower semibounded form. We say that $\mathfrak{t}$ is *closed* if $(\mathcal{D}(\mathfrak{t}), \| \cdot \|_\mathfrak{t})$ is complete, that is, if $(\mathcal{D}(\mathfrak{t}), \langle \cdot, \cdot \rangle_\mathfrak{t})$ is a Hilbert space, and that $\mathfrak{t}$ is *closable* if there exists a closed lower semibounded form which is an extension of $\mathfrak{t}$. A linear subspace $\mathcal{D}$ of $\mathcal{D}(\mathfrak{t})$ is called a *core* of $\mathfrak{t}$ if $\mathcal{D}$ is dense in $(\mathcal{D}(\mathfrak{t}), \| \cdot \|_\mathfrak{t})$.

Obviously, these notions do not depend on the particular lower bound m of $\mathfrak{t}$ appearing in (10.3), because another lower bound yields an equivalent norm $\| \cdot \|_\mathfrak{t}$.

In applications to differential operators the space $(\mathcal{D}(\mathfrak{t}), \| \cdot \|_\mathfrak{t})$ is often a Sobolev space with some equivalent norm. In this case the form is closed, because Sobolev spaces are complete.

The following two propositions collect various characterizations of closedness and closability of a lower semibounded form. They can be considered as counter-parts of Propositions 1.4 and 1.5 for forms.

Recall that a function $f : X \to \mathbb{R} \cup \{+\infty\}$ on a metric space X is called *lower semicontinuous* if for each convergent sequence $(x_n)_{n \in \mathbb{N}}$ in X, we have

$$f\left(\lim_{n \to \infty} x_n \right) \leq \lim_{n \to \infty} \inf f(x_n). \tag{10.6}$$

An equivalent requirement is that the inverse image $f^{-1}((a, +\infty])$ of each interval $(a, +\infty]$, $a \in \mathbb{R}$, is open in X. From (10.6) it follows at once that the pointwise supremum of any set of lower semicontinuous functions is again lower semicontinuous. Hence, monotone limits of continuous functions are lower semicontinuous.

We now extend the quadratic form $\mathfrak{t}$ to a mapping $\mathfrak{t}' : \mathcal{H} \to \mathbb{R} \cup \{+\infty\}$ by setting $\mathfrak{t}'[x] = \mathfrak{t}[x]$ if x is in $\mathcal{D}(\mathfrak{t})$ and $\mathfrak{t}'[x] = +\infty$ if x is not in $\mathcal{D}(\mathfrak{t})$.

Proposition 10.1 *The following four conditions are equivalent:*

(i) $\mathfrak{t}$ *is closed.*

(ii) *If* $(x_n)_{n \in \mathbb{N}}$ *is a sequence from* $\mathcal{D}(\mathfrak{t})$ *such that* $\lim_{n \to \infty} x_n = x$ *in* $\mathcal{H}$ *for some* $x \in \mathcal{H}$ *and* $\lim_{n,k \to \infty} \mathfrak{t}[x_n - x_k] = 0$, *then* $x \in \mathcal{D}(\mathfrak{t})$ *and* $\lim_{n \to \infty} \mathfrak{t}[x_n - x] = 0$.

(iii) $\mathfrak{t}'$ *is a lower semicontinuous function on* $\mathcal{H}$.

(iv) *If* $(x_n)_{n \in \mathbb{N}}$ *is a sequence from* $\mathcal{D}(\mathfrak{t})$ *such that* $\lim_{n \to \infty} x_n = x$ *in* $\mathcal{H}$ *for some* $x \in \mathcal{H}$ *and the set* $\{\mathfrak{t}[x_n] : n \in \mathbb{N}\}$ *is bounded, then we have* $x \in \mathcal{D}(\mathfrak{t})$ *and* $\mathfrak{t}[x] \leq \liminf_{n \to \infty} \mathfrak{t}[x_n]$.

Proof (i) $\leftrightarrow$ (ii): Since Cauchy sequences in $(\mathcal{D}(\mathfrak{t}), \|\cdot\|_{\mathfrak{t}})$ converge in $\mathcal{H}$, (ii) is only a formulation of the fact that each Cauchy sequence of $(\mathcal{D}(\mathfrak{t}), \|\cdot\|_{\mathfrak{t}})$ converges.

(iii) $\to$ (iv): Let (x_n) be a sequence as in (iv). Then $\mathfrak{t}'[x] \leq \liminf \mathfrak{t}[x_n]$ by the lower semicontinuity of $\mathfrak{t}'$. Since the set $\{\mathfrak{t}[x_n] : n \in \mathbb{N}\}$ is bounded, $\mathfrak{t}'[x] < \infty$, and hence $x \in \mathcal{D}(\mathfrak{t})$.

(iv) $\to$ (ii): Let (x_n) be a sequence as in (ii). Then (x_n) is a Cauchy sequence in $(\mathcal{D}(\mathfrak{t}), \|\cdot\|_{\mathfrak{t}})$. Hence, the set $\{\|x_n\|_{\mathfrak{t}}\}$ is bounded. By (ii), given $\varepsilon > 0$, there exists $n(\varepsilon)$ such that $\mathfrak{t}[x_k - x_n] < \varepsilon$ for $n, k \geq n(\varepsilon)$. Fix $k \geq n(\varepsilon)$ and consider the sequence $(x_n' := x_k - x_n)_{n \in \mathbb{N}}$. Since $\mathfrak{t}[x_n'] \leq \|x_n'\|_{\mathfrak{t}} \leq \|x_k\|_{\mathfrak{t}} + \mathfrak{t}[x_n] + (1 - m)\|x_n\|^2$ and $\lim_n x_n' = x_k - x$, this sequence also satisfies the conditions in (iv). Therefore, by (iv) we obtain $x_k - x \in \mathcal{D}(\mathfrak{t})$ and $\mathfrak{t}[x_k - x] \leq \liminf_{n \to \infty} \mathfrak{t}[x_k - x_n] \leq \varepsilon$. This proves that $x \in \mathcal{D}(\mathfrak{t})$ and $\lim_k \mathfrak{t}[x_k - x] = 0$.

The implication (i) $\to$ (iii) will be proved after Theorem 10.7 below. $\square$

Corollary 10.2 *The sum of finitely many lower semibounded closed forms of* $\mathcal{H}$ *is also closed.*

Proof The sum of finitely many lower semicontinuous functions is obviously lower semicontinuous, so the assertion follows from Proposition 10.1, (i) $\leftrightarrow$ (iii). $\square$

Let $\mathcal{H}_t$ denote the Hilbert space obtained by completing the pre-Hilbert space $(\mathcal{D}(t), \langle\cdot,\cdot\rangle_t)$. Since the embedding map $(\mathcal{D}(t), \|\cdot\|_t) \to (\mathcal{H}, \|\cdot\|)$ is continuous, it has a unique extension $\mathcal{I}_t$ to a continuous linear mapping from $\mathcal{H}_t$ into $\mathcal{H}$.

Proposition 10.3 *The following statements are equivalent:*

(i) t *is closable.*

(ii) *For any sequence* $(x_n)_{n\in\mathbb{N}}$ *of vectors* $x_n \in \mathcal{D}(t)$ *such that* $\lim_{n\to\infty} x_n = 0$ *in* $\mathcal{H}$ *and* $\lim_{n,k\to\infty} t[x_n - x_k] = 0$, *we have* $\lim_{n\to\infty} t[x_n] = 0$.

(iii) $\mathcal{I}_t : \mathcal{H}_t \to \mathcal{H}$ *is injective.*

Proof (i) $\to$ (ii): Since t is closable, it has a closed extension $\mathfrak{s}$. Let m be a lower bound for $\mathfrak{s}$. Then $\|\cdot\|_t = \|\cdot\|_{\mathfrak{s}}$ on $\mathcal{D}(t)$. Let (x_n) be a sequence as in (ii). By (10.4), (x_n) is a Cauchy sequence in $(\mathcal{D}(t), \|\cdot\|_t)$ and so in $(\mathcal{D}(\mathfrak{s}), \|\cdot\|_{\mathfrak{s}})$. Since $\mathfrak{s}$ is closed, there exists $x \in \mathcal{D}(\mathfrak{s})$ such that $\lim_n \|x_n - x\|_{\mathfrak{s}} = 0$. Then $\lim_n \|x_n - x\| = 0$, and hence $x = 0$, because $\lim_n x_n = 0$ in $\mathcal{H}$. From $\lim_n \|x_n\|_t = \lim_n \|x_n\|_{\mathfrak{s}} = 0$ we obtain $\lim_n t[x_n] = 0$.

(ii) $\to$ (iii): Let $x \in \mathcal{N}(\mathcal{I}_t)$. Then there is a sequence (x_n) from $\mathcal{D}(t)$ such that $\lim_n x_n = x$ in $\mathcal{H}_t$ and $\lim_n x_n = \lim_n \mathcal{I}_t(x_n) = \mathcal{I}_t(x) = 0$ in $\mathcal{H}$. Because (x_n) is a Cauchy sequence in $\mathcal{H}_t$, $\lim_{n,k} t[x_n - x_k] = 0$. Hence, $\lim_n t[x_n] = 0$ by (ii). This and the fact that $\lim_n x_n = 0$ in $\mathcal{H}$ imply that $\lim_n x_n = 0$ in $\mathcal{H}_t$. Therefore, $x = 0$.

(iii) $\to$ (i): Since $\mathcal{I}_t$ is injective, we can identify $x \in \mathcal{H}_t$ and $\mathcal{I}_t(x) \in \mathcal{H}$. Then $\mathcal{H}_t$ becomes a linear subspace of $\mathcal{H}$ which contains $\mathcal{D}(t)$. By (10.5), t is bounded on $\mathcal{D}(t)$, so it extends by continuity to a lower semibounded form, denoted by $\bar{t}$, on $\mathcal{H}_t$. The scalar product of $\mathcal{H}_t$ and $\langle\cdot,\cdot\rangle_{\bar{t}}$ are both continuous extensions of the scalar product of $\mathcal{D}(t)$, so they coincide. Hence, $\bar{t}$ is closed, because $\mathcal{H}_t$ is complete. $\square$

The form $\bar{t}$ in the preceding proof is called the *closure* of the closable form t.

By construction, $\bar{t}$ is the smallest closed extension of t. The domain $\mathcal{D}(\bar{t})$ consists of all vectors $x \in \mathcal{H}$ which are limits in $\mathcal{H}$ (!) of a Cauchy sequence $(x_n)_{n\in\mathbb{N}}$ in $(\mathcal{D}(t), \|\cdot\|_t)$, that is, $x \in \mathcal{H}$ is in $\mathcal{D}(\bar{t})$ if and only if there is a sequence $(x_n)_{n\in\mathbb{N}}$ from $\mathcal{D}(t)$ such that $\lim_{n\to\infty} x_n = x$ in $\mathcal{H}$ and $\lim_{n,k\to\infty} t[x_n - x_k] = 0$. If (x_n) and (y_n) are such sequences for x and y in $\mathcal{D}(\bar{t})$, respectively, then

$$\bar{t}[x, y] = \lim_{n\to\infty} t[x_n, y_n].$$

The latter implies in particular that t and $\bar{t}$ have the same greatest lower bounds.

Example 10.1 (*Nonclosable forms*) Let $a, b \in \mathbb{R}$, $a < b$, $c \in [a, b]$, and $\mathcal{H} = L^2(a, b)$. Define positive forms by

$$t_1[f, g] = f(c)\overline{g(c)},$$
$$t_2[f, g] = \langle f, g \rangle + f(c)\overline{g(c)}, \quad f, g \in \mathcal{D}(t_1) = \mathcal{D}(t_2) := C([a, b]).$$

For any $\alpha \in \mathbb{C}$, there is a sequence $(f_n)_{n\in\mathbb{N}}$ from $\mathcal{D}(t_1) = \mathcal{D}(t_2)$ such $f_n(c) = \alpha$ for all $n \in \mathbb{N}$ and $\lim_n f_n = 0$ in $\mathcal{H}$. Then $t_1[f_n] = |\alpha|^2$ and $\lim_n t_2[f_n] = |\alpha|^2$. Hence, *both forms* t_1 *and* t_2 *are not closable* by Proposition 10.3 (i) $\leftrightarrow$ (iii). It is not difficult to see that $\mathcal{H}_{t_j} = \mathcal{H} \oplus \mathbb{C}$ and $\mathcal{I}_{t_j}(f, \alpha) = f$ for $f \in \mathcal{H}$, $\alpha \in \mathbb{C}$, $j = 1, 2$.

However, the positive form t_3 defined by

$$t_3[f, g] = \langle f', g' \rangle + f(c)\overline{g(c)}, \quad f, g \in \mathcal{D}(t_3) := H^1(a, b),$$

is *closed*. Indeed, by Lemma 1.12, the functional $f \to f(c)$ is continuous on the Sobolev space $H^1(a, b)$. Hence, the form norm $\| \cdot \|_{t_3}$ is equivalent to the norm of $H^1(a, b)$. Since $H^1(a, b)$ is complete, t_3 is closed.   ○

10.2 Semibounded Closed Forms and Self-adjoint Operators

In this section we assume that A is a *self-adjoint operator* on $\mathcal{H}$. Let E_A denote its spectral measure. First, we associate a form t_A with A. Define

$$\mathcal{D}(t_A) \equiv \mathcal{D}[A] := \mathcal{D}(|A|^{1/2}) = \left\{ x \in \mathcal{H} : \int_{\mathbb{R}} |\lambda| \, d\langle E_A(\lambda)x, x \rangle < \infty \right\}, \quad (10.7)$$

$$t_A[x, y] \equiv A[x, y] = \int_{\mathbb{R}} \lambda \, d\langle E_A(\lambda)x, y \rangle \quad \text{for } x, y \in \mathcal{D}(t_A). \quad (10.8)$$

(By the definition of $\mathcal{D}[A]$ the existence of the integral (10.8) follows at once from formula (4.19) in Lemma 4.8 applied with $f(\lambda) = |\lambda|^{1/2}$ and $g(\lambda) = f(\lambda)\,\mathrm{sign}\,\lambda$.)

It is easily seen that t_A is a densely defined symmetric form on $\mathcal{H}$.

Definition 10.3 We call t_A the *form associated with the self-adjoint operator* A and $\mathcal{D}[A]$ the *form domain* of A.

It should be emphasized that "squared brackets" as in $\mathcal{D}[A]$ and $A[\cdot, \cdot]$ always refer to the form domain of A and the form associated with A, respectively.

Note that $\mathcal{D}(A) = \mathcal{D}(|A|) \subseteq \mathcal{D}(|A|^{1/2}) = \mathcal{D}[A]$. If the operator A is unbounded, the form domain $\mathcal{D}[A]$ is strictly larger than the domain $\mathcal{D}(A)$ of the operator A.

The next proposition gives a description of t_A avoiding the use of the spectral representation and shows how the operator A can be recovered from its form t_A.

Proposition 10.4

(i) $A[x, y] = \langle U_A |A|^{1/2}x, |A|^{1/2}y \rangle$ *for* $x, y \in \mathcal{D}[A]$, *where* U_A *is the partial isometry from the polar decomposition* $A = U_A|A|$ *of* A.

(ii) $\mathcal{D}(A)$ *is the set of all* $x \in \mathcal{D}[A]$ *for which there exists a vector* $u \in \mathcal{H}$ *such that* $A[x, y] = \langle u, y \rangle$ *for all* $y \in \mathcal{D}[A]$. *If this holds, then* $u = Ax$, *and hence*

$$A[x, y] = \langle Ax, y \rangle \quad \text{for all } x \in \mathcal{D}(A) \text{ and } y \in \mathcal{D}[A]. \quad (10.9)$$

(iii) A *is lower semibounded if and only if* t_A *is lower semibounded. In this case*

$$m_A = m_{t_A} = \inf\{\lambda : \lambda \in \sigma(A)\}. \quad (10.10)$$

Proof (i): Set $f(\lambda) := \text{sign}\,\lambda$ and $g(\lambda) := |\lambda|^{1/2}$. Then we have $f(A) = U_A$ and $g(A) = |A|^{1/2}$ (see, e.g., Example 7.1). If $x, y \in \mathcal{D}[A]$, then $x \in \mathcal{D}((fg)(A))$ and $y \in \mathcal{D}(g(A))$, so formula (4.32) in Proposition 4.15 applies, and we compute

$$A[x, y] = \int \lambda\, d\langle E_A(\lambda)x, y\rangle = \int (fg)(\lambda)g(\lambda)\, d\langle E_A(\lambda)x, y\rangle$$
$$= \langle (fg)(A)x, g(A)y\rangle = \langle f(A)g(A)x, g(A)y\rangle = \langle U_A|A|^{1/2}x, |A|^{1/2}y\rangle.$$

(ii): Let $x \in \mathcal{D}(A)$ and $y \in \mathcal{D}[A]$. Being functions of A, the operators U_A and $|A|^{1/2}$ commute and $U_A|A|^{1/2}x \in \mathcal{D}(|A|^{1/2})$ by the functional calculus. Using the self-adjointness of $|A|^{1/2}$, the polar decomposition $A = U_A|A|$, and (i), we obtain

$$A[x, y] = \langle U_A|A|^{1/2}x, |A|^{1/2}y\rangle = \langle |A|^{1/2}U_A|A|^{1/2}x, y\rangle$$
$$= \langle U_A|A|^{1/2}|A|^{1/2}x, y\rangle = \langle Ax, y\rangle.$$

Conversely, assume that for $x \in \mathcal{D}[A]$, there exists a vector $u \in \mathcal{H}$ such that $\langle U_A|A|^{1/2}x, |A|^{1/2}y\rangle \equiv A[x, y] = \langle u, y\rangle$ for all $y \in \mathcal{D}[A] = \mathcal{D}(|A|^{1/2})$. Since $|A|^{1/2}$ is self-adjoint, we get $u = |A|^{1/2}U_A|A|^{1/2}x = U_A|A|^{1/2}|A|^{1/2}x = Ax$.

(iii): Since $\sigma(A) = \text{supp}\, E_A$ by Proposition 5.10, $m := \inf \sigma(A) = \inf \text{supp}\, E_A$. Hence, $A \geq mI$ and $\mathfrak{t}_A \geq m$ by (10.8), so that $m_A \geq m$ and $m_{\mathfrak{t}_A} \geq m$.

Let $\alpha > m$. Since $m = \inf \text{supp}\, E_A$, there is a unit vector $x \in E_A([m, \alpha])\mathcal{H}$. Then $\langle Ax, x\rangle \leq \alpha$ and $\mathfrak{t}_A[x] \leq \alpha$ by (10.8), so $m_A \leq m$ and $m_{\mathfrak{t}_A} \leq m$. $\square$

The next two propositions relate properties of forms to properties of operators.

Proposition 10.5 *Let $A \geq mI$ be a lower semibounded self-adjoint operator.*

(i) *For $x, y \in \mathcal{D}[A] = \mathcal{D}((A - mI)^{1/2})$, we have*
$$A[x, y] = \langle (A - mI)^{1/2}x, (A - mI)^{1/2}y\rangle + m\langle x, y\rangle. \tag{10.11}$$

(ii) *$\mathfrak{t}_A$ is a (lower semibounded) closed form.*

(iii) *A linear subspace of $\mathcal{D}[A]$ is a core for the form $\mathfrak{t}_A$ if and only if it is a core for the operator $(A - mI)^{1/2}$.*

(iv) *$\mathcal{D}(A)$ is a core for the form $\mathfrak{t}_A$.*

(v) *Let $\mathcal{D}$ be a core of $\mathfrak{t}_A$. If B is a linear operator such that $\mathcal{D}(B) \subseteq \mathcal{D}[A]$ and $A[x, y] = \langle Bx, y\rangle$ for all $x \in \mathcal{D}(B)$ and $y \in \mathcal{D}$, then $B \subseteq A$.*

Proof (i): Clearly, $\mathcal{D}[A] = \mathcal{D}(|A|^{1/2}) = \mathcal{D}(|A - mI|^{1/2})$ by (4.26). Since $A \geq m \cdot I$, we have $A = \int_m^\infty \lambda\, dE_A(\lambda)$. Therefore, using (10.7) and (4.32), we derive

$$A[x, y] - m\langle x, y\rangle = \int_m^\infty (\lambda - m)\, d\langle E_A(\lambda)x, y\rangle = \langle (A - mI)^{1/2}x, (A - mI)^{1/2}y\rangle.$$

(ii) and (iii): From (i) we conclude that the graph norm $\|\cdot\|_{(A-mI)^{1/2}}$ (defined by (1.4)) and the form norm $\|\cdot\|_{\mathfrak{t}_A}$ (defined by (10.4)) are equivalent on the domain $\mathcal{D}((A - mI)^{1/2}) = \mathcal{D}[A]$. Hence, since the operator $(A - mI)^{1/2}$ is closed, the form $\mathfrak{t}_A$ is closed, and the operator $(A - mI)^{1/2}$ and the form $\mathfrak{t}_A$ have the same cores.

(iv) follows from (iii), since $\mathcal{D}(A)$ is a core for $(A - mI)^{1/2}$ by Corollary 4.14.

(v): Since the form t_A and the scalar product of $\mathcal{H}$ are $\|\cdot\|_{t_A}$-continuous and $\mathcal{D}$ is $\|\cdot\|_{t_A}$-dense in $\mathcal{D}[A]$, the equality $A[x, y] = \langle Bx, y\rangle$, $x \in \mathcal{D}(B)$, extends by continuity to all $y \in \mathcal{D}[A]$. Now Proposition 10.4(ii) yields $Bx = Ax$ for $x \in \mathcal{D}(B)$. That is, $B \subseteq A$. □

If the self-adjoint operator A is *positive*, then by Proposition 10.5(i),

$$t_A[x, y] \equiv A[x, y] = \langle A^{1/2}x, A^{1/2}y\rangle \quad \text{for } x, y \in \mathcal{D}[A] = \mathcal{D}(A^{1/2}). \tag{10.12}$$

For positive operators A, we may take (10.12) as the definition of the form t_A. In this case most proofs are essentially shorter; in particular, the use of the polar decomposition and of formula (4.32) can be avoided in the preceding proofs.

The following proposition is similar to Proposition 5.12.

Proposition 10.6 *Suppose that A is a lower semibounded self-adjoint operator and $m < m_A$. Then the following assertions are equivalent:*

(i) *The embedding map $\mathcal{I}_{t_A} : (\mathcal{D}[A], \|\cdot\|_{t_A}) \to (\mathcal{H}, \|\cdot\|)$ is compact.*
(ii) *$(A - mI)^{-1/2}$ is compact.*
(iii) *The resolvent $R_\lambda(A)$ is compact for one, hence for all, $\lambda \in \rho(A)$.*
(iv) *A has a purely discrete spectrum.*

Proof Note that $\|(A - mI)^{1/2}x\| \geq (m_A - m)^{1/2}\|x\|$ by the functional calculus. Hence, it follows from (10.11) that $\|\cdot\|'_{t_A} := \|(A - mI)^{1/2}\cdot\|$ defines a norm on $\mathcal{D}[A] = \mathcal{D}((A - mI)^{1/2})$ which is equivalent to the form norm $\|\cdot\|_{t_A}$. Then we have $\|(A - mI)^{-1/2}y\|'_{t_A} = \|y\|$ for $y \in \mathcal{H}$. The equivalence of (i) and (ii) will be derived from the two preceding norm equalities.

(i) → (ii): Let M be a bounded subset of $\mathcal{H}$. Then $(A - mI)^{-1/2}M$ is bounded in $\mathcal{D}[A]$ by the second equality and hence relatively compact in $\mathcal{H}$, because $\mathcal{I}_{t_A}$ is compact by (i). This shows that $(A - mI)^{-1/2}$ is compact.

(ii) → (i): Let N be a bounded set in $\mathcal{D}[A]$. Then $M = (A - mI)^{1/2}N$ is bounded in $\mathcal{H}$ by the first equality. Since $(A - mI)^{-1/2}$ is compact, $N = (A - mI)^{-1/2}M$ is compact in $\mathcal{H}$. This proves that $\mathcal{I}_{t_A}$ is compact.

(ii) ↔ (iii): The bounded self-adjoint operator $(A - mI)^{-1/2}$ is compact if and only if its square $R_m(A) = (A - mI)^{-1} = ((A - mI)^{-1/2})^2$ is compact. By the resolvent identity (2.5), the latter implies that $R_\lambda(A)$ is compact for all $\lambda \in \rho(A)$.

The equivalence of (iii) and (iv) was already contained in Proposition 5.12. □

Now we proceed in reverse order by associating an operator with an *arbitrary densely defined* form.

Definition 10.4 Let t be a densely defined form on $\mathcal{H}$. The *operator A_t associated with the form t* is defined by $A_t x := u_x$ for $x \in \mathcal{D}(A_t)$, where

$$\mathcal{D}(A_t) = \{x \in \mathcal{D}(t) : \text{There exists } u_x \in \mathcal{H} \text{ such that } t[x, y] = \langle u_x, y\rangle \text{ for } y \in \mathcal{D}(t)\}.$$

Since $\mathcal{D}(\mathfrak{t})$ is dense in $\mathcal{H}$, the vector u_x is uniquely determined by x, and the operator $A_\mathfrak{t}$ is therefore well defined and linear. By Definition 10.4, $\mathcal{D}(A_\mathfrak{t}) \subseteq \mathcal{D}(\mathfrak{t})$, and

$$\mathfrak{t}[x, y] = \langle A_\mathfrak{t}x, y \rangle \quad \text{for } x \in \mathcal{D}(A_\mathfrak{t}) \text{ and } y \in \mathcal{D}(\mathfrak{t}). \tag{10.13}$$

Clearly, we have $A_{\mathfrak{t}+\lambda} = A_\mathfrak{t} + \lambda I$ for any $\lambda \in \mathbb{C}$.

The above definition is suggested by Proposition 10.4(ii) which says that each self-adjoint operator A is just the operator $A_{(\mathfrak{t}_A)}$ associated with its form $\mathfrak{t}_A$.

The following representation theorem is the main result of this section.

Theorem 10.7 (*Representation theorem for semibounded forms*) *If $\mathfrak{t}$ is a densely defined lower semibounded closed form on $\mathcal{H}$, then the operator $A_\mathfrak{t}$ is self-adjoint, and $\mathfrak{t}$ is equal to the form $\mathfrak{t}_{(A_\mathfrak{t})}$ associated with $A_\mathfrak{t}$.*

Proof Since $A_{\mathfrak{t}+\lambda} = A_\mathfrak{t} + \lambda I$ for $\lambda \in \mathbb{R}$, we can assume that $\mathfrak{t} \geq 1$. Then we can set $m = 1$ in (10.3), so we have $\langle \cdot, \cdot \rangle_\mathfrak{t} = \mathfrak{t}[\cdot, \cdot]$.

First, we prove that $\mathcal{R}(A_\mathfrak{t}) = \mathcal{H}$. Let $u \in \mathcal{H}$. Since $\| \cdot \| \leq \| \cdot \|_\mathfrak{t}$, the mapping $y \to \langle y, u \rangle$ is a continuous linear functional on $(\mathcal{D}(\mathfrak{t}), \langle \cdot, \cdot \rangle_\mathfrak{t})$. Because $\mathfrak{t}$ is closed, $(\mathcal{D}(\mathfrak{t}), \langle \cdot, \cdot \rangle_\mathfrak{t})$ is a Hilbert space, so by Riesz' theorem there exists a vector $x \in \mathcal{D}(\mathfrak{t})$ such that $\langle y, u \rangle = \langle y, x \rangle_\mathfrak{t}$ for $y \in \mathcal{D}(\mathfrak{t})$. That is, $\mathfrak{t}[x, y] = \langle u, y \rangle$ for $y \in \mathcal{D}(\mathfrak{t})$. From Definition 10.4 it follows that $x \in \mathcal{D}(A_\mathfrak{t})$ and $A_\mathfrak{t}(x) = u$. Thus, $\mathcal{R}(A_\mathfrak{t}) = \mathcal{H}$.

Because the form $\mathfrak{t}$ is symmetric, the operator $A_\mathfrak{t}$ is symmetric by (10.13). Since $\mathcal{R}(A_\mathfrak{t}) = \mathcal{H}$, $A_\mathfrak{t}$ is self-adjoint by Corollary 3.12.

Next, we show that $\mathcal{D}(A_\mathfrak{t})$ is dense in the Hilbert space $(\mathcal{D}(\mathfrak{t}), \langle \cdot, \cdot \rangle_\mathfrak{t})$. Suppose that $y \in \mathcal{D}(\mathfrak{t})$ is orthogonal to $\mathcal{D}(A_\mathfrak{t})$ in $(\mathcal{D}(\mathfrak{t}), \langle \cdot, \cdot \rangle_\mathfrak{t})$. Since $\mathcal{R}(A_\mathfrak{t}) = \mathcal{H}$, there is a vector $x \in \mathcal{D}(A_\mathfrak{t})$ such that $A_\mathfrak{t}x = y$. Using formula (10.13), we obtain $\langle y, y \rangle = \langle A_\mathfrak{t}x, y \rangle = \mathfrak{t}[x, y] = \langle x, y \rangle_\mathfrak{t} = 0$. Hence, $y = 0$, so $\mathcal{D}(A_\mathfrak{t})$ is dense in $(\mathcal{D}(\mathfrak{t}), \langle \cdot, \cdot \rangle_\mathfrak{t})$.

By (10.13) we have $A_\mathfrak{t} \geq I$. Hence, $\mathfrak{s} := \mathfrak{t}_{(A_\mathfrak{t})}$ is a lower semibounded closed form by Proposition 10.5(ii). From (10.13) and (10.9) we obtain $\mathfrak{t}[x, y] = \langle A_\mathfrak{t}x, y \rangle = \mathfrak{s}[x, y]$, and so $\langle x, y \rangle_\mathfrak{t} = \langle x, y \rangle_\mathfrak{s}$ for $x, y \in \mathcal{D}(A_\mathfrak{t})$. Recall that $\mathcal{D}(A_\mathfrak{t})$ is dense in $(\mathcal{D}(\mathfrak{t}), \langle \cdot, \cdot \rangle_\mathfrak{t})$ as shown in the preceding paragraph. Since $\mathfrak{s}$ is the form associated with $A_\mathfrak{t}$, $\mathcal{D}(A_\mathfrak{t})$ is dense in $(\mathcal{D}(\mathfrak{s}), \langle \cdot, \cdot \rangle_\mathfrak{s})$ by Proposition 10.5(iv). Hence, it follows that $(\mathcal{D}(\mathfrak{t}), \langle \cdot, \cdot \rangle_\mathfrak{t}) = (\mathcal{D}(\mathfrak{s}), \langle \cdot, \cdot \rangle_\mathfrak{s})$ and $\mathfrak{t} = \mathfrak{s}$. That is, $\mathfrak{t} = \mathfrak{t}_{(A_\mathfrak{t})}$. □

The next corollary follows by combining Proposition 10.4 and Theorem 10.7.

Corollary 10.8 *The mapping $A \to \mathfrak{t}_A$ is a bijection of the set of lower semibounded self-adjoint operators on a Hilbert space $\mathcal{H}$ on the set of densely defined lower semibounded closed forms on $\mathcal{H}$. The inverse of this mapping is given by $\mathfrak{t} \to A_\mathfrak{t}$.*

Having the form representation theorem (Theorem 10.7), it is now easy to provide the missing

Proof of Proposition 10.1 (i) $\to$ (iii). Upon replacing $\mathcal{H}$ by the closure of $\mathcal{D}(\mathfrak{t})$ we can assume that $\mathfrak{t}$ is densely defined. Since $\mathfrak{t}$ is lower semibounded and closed by (i), A_t is a lower semibounded self-adjoint operator by Theorem 10.7. Let $A_t = \int_m^\infty \lambda \, dE(\lambda)$ be its spectral decomposition and set $A(n) = \int_m^{m+n} \lambda \, dE(\lambda)$, $n \in \mathbb{N}$, $n > m$. Then t' is the limit of the monotone sequence of continuous quadratic forms $t_{A(n)}$ on $\mathcal{H}$. Hence, t' is lower semicontinuous. $\qquad\square$

The following theorem deals with a variational problem associated with forms.

Theorem 10.9 *Let t be a densely defined lower semibounded closed form on $\mathcal{H}$, and let $\lambda \in \mathbb{R}$, $\lambda < m_t$. For $u \in \mathcal{H}$, let J_u denote the functional on $\mathcal{D}(t)$ defined by*

$$J_u(x) = t[x] - \lambda \|x\|^2 - 2\operatorname{Re}\langle u, x\rangle, \quad x \in \mathcal{D}(t). \tag{10.14}$$

The (nonlinear) functional J_u attains its minimum over $\mathcal{D}(t)$ at precisely one element $x_u \in \mathcal{D}(t)$, and we have $x_u = (A_t - \lambda I)^{-1}u$. The minimum of J_u is

$$-\langle (A_t - \lambda I)^{-1}u, u\rangle = -t[x_u] + \lambda \|x_u\|^2. \tag{10.15}$$

Proof Upon replacing t by $t - \lambda$ and A_t by $A_t - \lambda I$ we can assume without loss of generality that $\lambda = 0$. Since then $t \geq m_t > 0$ and $A_t \geq m_t \cdot I$ by Proposition 10.4(iii), we have $A_t^{-1} \in \mathbf{B}(\mathcal{H})$. For $x \in \mathcal{D}(t)$, we compute

$$\begin{aligned}
\left\| A_t^{1/2}\big(x - A_t^{-1}u\big) \right\|^2 &= \big\langle A_t^{1/2}\big(x - A_t^{-1}u\big), A_t^{1/2}\big(x - A_t^{-1}u\big)\big\rangle \\
&= \big\langle A_t^{1/2}x, A_t^{1/2}x\big\rangle - \big\langle A_t^{-1/2}u, A_t^{1/2}x\big\rangle \\
&\quad - \big\langle A_t^{1/2}x, A_t^{-1/2}u\big\rangle + \big\langle A_t^{-1/2}u, A_t^{-1/2}u\big\rangle \\
&= t[x] - \langle u, x\rangle - \langle x, u\rangle + \big\langle A_t^{-1}u, u\big\rangle = J_u(x) + \big\langle A_t^{-1}u, u\big\rangle.
\end{aligned}$$

From this equation it follows that J_u attains its minimum at x if and only if $\left\| A_t^{1/2}(x - A_t^{-1}u) \right\| = 0$, that is, if $x = x_u := A_t^{-1}u$. Further, this minimum is equal to $-\langle A_t^{-1}u, u\rangle = -\langle x_u, A_t x_u\rangle = -\|A_t^{1/2}x_u\|^2 = -t[x_u]$. $\qquad\square$

Suppose that t is a densely defined lower semibounded closed form on $\mathcal{H}$ and $\lambda < m_t$. Then $\lambda \in \rho(A_t)$. We consider the following *abstract boundary problem*: Given $u \in \mathcal{H}$, find $x \in \mathcal{D}(t)$ such that

$$t[x, y] - \lambda\langle x, y\rangle = \langle u, y\rangle \quad \text{for all } y \in \mathcal{D}(t). \tag{10.16}$$

By Definition 10.4, Eq. (10.16) holds if and only if $x \in \mathcal{D}(A_t)$ and $A_t x = \lambda x + u$. Since $\lambda \in \rho(A_t)$, problem (10.16) has a unique solution $x_u = (A_t - \lambda I)^{-1}u$.

Theorem 10.9 states that this abstract boundary problem (10.16) is equivalent to the following *variational problem*:

Given $u \in \mathcal{H}$, minimize the (nonlinear) functional

$$J_u(x) = t[x] - \lambda \|x\|^2 - 2\operatorname{Re}\langle u, x\rangle \quad \text{for } x \in \mathcal{D}(t). \tag{10.17}$$

Thus, *both problems* (10.16) *and* (10.17) *are equivalent reformulations of the equation $A_t x - \lambda x = u$ in terms of the form t, and they have the same unique solution*

$$x_u = (A_t - \lambda I)^{-1}u.$$

That both problems (10.16) and (10.17) (and so their special cases considered in Sect. 10.6 below) are uniquely solvable and that their solutions coincide follows already from elementary Hilbert space theory without using the operator A_t.

Indeed, since $\lambda < m_t$ and t is closed, $\mathcal{D}(t)$ is a Hilbert space $\mathcal{D}_t$ with scalar product $(\cdot, \cdot)_t := (t - \lambda)[\cdot, \cdot]$ and norm $|\cdot|_t$. Since the linear functional $F_u(\cdot) := \langle \cdot, u \rangle$ is continuous on $\mathcal{D}_t$, by Riesz' theorem it is of the form $(\cdot, x_u)_t$ for some unique vector $x_u \in \mathcal{D}_t$. This means that x_u is the unique solution of the boundary problem (10.16). Further, we have

$$|x - x_u|_t^2 = |x_u|_t^2 + |x|_t^2 - 2\operatorname{Re}(x, x_u)_t = |x_u|_t^2 + (t - \lambda)[x] - 2\operatorname{Re}\langle x, u \rangle$$
$$= |x_u|_t^2 + J_u(x).$$

Hence, the minimum of J_u is attained at x_u, and it is $-|x_u|_t^2 = -t[x_u] + \lambda \|x_u\|^2$. That is, x_u is the unique solution of the variational problem (10.17).

10.3 Order Relations for Self-adjoint Operators

In this section we use forms to define an order relation $\geq$ for self-adjoint operators.

Definition 10.5 Let A and B be symmetric operators on Hilbert spaces $\mathcal{G}$ and $\mathcal{K}$, respectively, such that $\mathcal{G}$ is a subspace of $\mathcal{K}$. We shall write $A \succeq B$ (and $B \preceq A$) if $\mathcal{D}(A) \subseteq \mathcal{D}(B)$ and $\langle Ax, x \rangle \geq \langle Bx, x \rangle$ for all $x \in \mathcal{D}(A)$.

If A and B are self-adjoint operators, we write $A \geq B$ (and $B \leq A$) if $\mathcal{D}[A] \subseteq \mathcal{D}[B]$ and $A[x] \geq B[x]$ for all $x \in \mathcal{D}[A]$.

Obviously, for bounded self-adjoint operators defined on the same Hilbert space $\mathcal{G} = \mathcal{K}$, both relations $\geq$ and $\succeq$ are the same.

If A is self-adjoint, it follows at once from Proposition 10.4(iii) that the relations $A \geq m \cdot I$ and $A \succeq m \cdot I$ are equivalent.

If A and B are *positive* self-adjoint operators, by (10.12) we have

$$A \geq B \quad \text{if and only if} \quad \mathcal{D}\left(A^{1/2}\right) \subseteq \mathcal{D}\left(B^{1/2}\right) \text{ and}$$
$$\left\|A^{1/2}x\right\| \geq \left\|B^{1/2}x\right\|, \; x \in \mathcal{D}\left(A^{1/2}\right).$$

Lemma 10.10 *If A and B are lower semibounded self-adjoint operators, then:*

(i) $A \geq B$ *if and only if* $\mathcal{D}(A) \subseteq \mathcal{D}[B]$ *and* $A[y] \geq B[y]$ *for all* $y \in \mathcal{D}(A)$.
(ii) *If* $A \succeq B$, *then* $A \geq B$.

Proof Upon adding a multiple of the identity we can assume that $A \geq 0$ and $B \geq 0$.
(i): Since $\mathcal{D}(A) \subseteq \mathcal{D}[A]$, the only if implication is trivial.
We prove the converse direction. From the above condition it follows that

$$\left\|A^{1/2}y\right\| \geq \left\|B^{1/2}y\right\| \quad \text{for } y \in \mathcal{D}(A) \subseteq \mathcal{D}[B]. \tag{10.18}$$

Let $x \in \mathcal{D}(A^{1/2})$. Since $\mathcal{D}(A)$ is a core for $A^{1/2}$, there is a sequence $(x_n)_{n \in \mathbb{N}}$ from $\mathcal{D}(A)$ such that $\lim_n x_n = x$ and $\lim_n A^{1/2}x_n = A^{1/2}x$. Applying (10.18) to $y = x_n - x_k$, it follows that $(B^{1/2}x_n)$ is a Cauchy sequence. Since $B^{1/2}$ is closed, we have $x \in \mathcal{D}(B^{1/2})$ and $\lim_n B^{1/2}x_n = B^{1/2}x$. Setting $y = x_n$ in (10.18) and passing to the limit, we get $\|A^{1/2}x\| \geq \|B^{1/2}x\|$. This proves that $A \geq B$.

(ii): Since $A \succeq B$, we have $\|A^{1/2}y\|^2 = \langle Ay, y \rangle \geq \langle By, y \rangle = \|B^{1/2}y\|^2$ for y in $\mathcal{D}(A) \subseteq \mathcal{D}(B) \subseteq \mathcal{D}[B]$, that is, (10.18) holds. Hence, $A \geq B$ by (i). $\square$

For semibounded self-adjoint operators, the order relation $\geq$ is finer and more important than the relation $\succeq$. We illustrate this by the following simple example.

Example 10.2 Let $\mathcal{T}$ be the set of self-adjoint extensions of a densely defined positive symmetric operator T on $\mathcal{H}$. If $A, B \in \mathcal{T}$ and $A \succeq B$, then $\mathcal{D}(A) \subseteq \mathcal{D}(B)$, and hence $A = B$, because A and B are restrictions of T^*. That is, in the set $\mathcal{T}$ the relation $\succeq$ is trivial, since $A \succeq B$ is equivalent to $A = B$. We shall see in Sects. 13.4 and 14.8 that the relation $\geq$ is not trivial in $\mathcal{T}$! For instance, the Friedrichs extension defined in the next section is the largest element of $\mathcal{T}$ with respect to $\geq$.

The technical ingredient for our next results is the following lemma which relates the form of a positive invertible self-adjoint operator to the form of its inverse.

Lemma 10.11 *If T is a positive self-adjoint operator such that $\mathcal{N}(T) = \{0\}$, then*

$$\mathfrak{t}'_{T^{-1}}[x] = \sup\{|\langle x, y \rangle|^2 T[y]^{-1} : y \in \mathcal{D}[T], \ y \neq 0\} \quad \text{for } x \in \mathcal{H}, \quad (10.19)$$

where $\mathfrak{t}'_{T^{-1}}[x] := T^{-1}[x]$ if $x \in \mathcal{D}[T^{-1}]$ and $\mathfrak{t}'_{T^{-1}}[x] := +\infty$ if $x \notin \mathcal{D}[T^{-1}]$.

Proof First, let us note that T^{-1} is self-adjoint by Corollary 1.9, and $\mathcal{D}[T^{-1}] = \mathcal{D}(T^{-1/2})$, because $(T^{-1})^{1/2} = T^{-1/2}$ (see Exercise 5.9). Fix a nonzero $x \in \mathcal{H}$.

Let $y \in \mathcal{D}[T] = \mathcal{D}(T^{1/2})$, $y \neq 0$. If $x \in \mathcal{D}[T^{-1}]$, we have

$$|\langle x, y \rangle|^2 = |\langle T^{-1/2}x, T^{1/2}y \rangle|^2 \leq \|T^{-1/2}x\|^2 \|T^{1/2}y\|^2 = \mathfrak{t}'_{T^{-1}}[x]T[y].$$
(10.20)

Inequality (10.20) holds trivially if $x \notin \mathcal{D}[T^{-1}]$, because then $\mathfrak{t}'_{T^{-1}}[x] = +\infty$ and $T[y] = \|T^{1/2}y\|^2 \neq 0$ by the assumption $\mathcal{N}(T) = \{0\}$.

Put $E_n := E_T((n^{-1}, n))$ for $n \in \mathbb{N}$, where E_T is the spectral measure of T. Then $E_n x \in \mathcal{D}(T^{-1})$. Setting $y_n := T^{-1}E_n x$, we have $T[y_n] = \|T^{-1/2}E_n x\|^2$ and $|\langle x, y_n \rangle|^2 = \|T^{-1/2}E_n x\|^4$. Since $x \neq 0$, $T[y_n] \neq 0$ for large n, and hence

$$|\langle x, y_n \rangle|^2 T[y_n]^{-1} = \|T^{-1/2}E_n x\|^2 \nearrow \mathfrak{t}'_{T^{-1}}[x] \quad \text{as } n \to \infty. \quad (10.21)$$

Equality (10.19) follows by combining (10.20) and (10.21). $\square$

Corollary 10.12 *Let A and B be self-adjoint operators on $\mathcal{H}$ such that $A \geq B \geq 0$ and $\mathcal{N}(B) = \{0\}$. Then we have $\mathcal{N}(A) = \{0\}$ and $B^{-1} \geq A^{-1}$.*

Proof Let $x \in \mathcal{N}(A)$. Then, $0 = \langle Ax, x \rangle = \|A^{1/2}x\|^2 \geq \|B^{1/2}x\|^2$, so that we have $x \in \mathcal{N}(B^{1/2}) \subseteq \mathcal{N}(B)$, and therefore $x = 0$. Thus, $\mathcal{N}(A) = \{0\}$.

By formula (10.19), the relation $A \geq B$ implies that $\mathcal{D}[B^{-1}] \subseteq \mathcal{D}[A^{-1}]$ and $B^{-1}[x] \geq A^{-1}[x]$ for $x \in \mathcal{D}[B^{-1}]$, that is, $B^{-1} \geq A^{-1}$. $\square$

The following corollary characterizes the order relation $A \geq B$ in terms of the usual order relation for bounded self-adjoint operators of the resolvents.

Corollary 10.13 *Let A and B be lower semibounded self-adjoint operators on a Hilbert space $\mathcal{H}$, and let $\lambda \in \mathbb{R}$, $\lambda < m_A$, and $\lambda < m_B$. Then $\lambda \in \rho(A) \cap \rho(B)$, and the relation $A \geq B$ holds if and only if $(B - \lambda I)^{-1} \geq (A - \lambda I)^{-1}$.*

Proof By Proposition 10.4(iii), we have $\lambda \in \rho(A) \cap \rho(B)$. Since $t_{A-\lambda I} = t_A - \lambda$ and $t_{B-\lambda I} = t_B - \lambda$, the relation $A \geq B$ is equivalent to $A - \lambda I \geq B - \lambda I$. Hence, we can assume without loss of generality that $\lambda = 0$. Then we have $A \geq B \geq 0$ and $0 \in \rho(A) \cap \rho(B)$, so by Corollary 10.12, $A \geq B$ is equivalent to $B^{-1} \geq A^{-1}$. $\square$

Theorem 10.9 provides an elegant proof for the only if part in Corollary 10.13. Indeed, let J_u^A and J_u^B denote the functional from Theorem 10.9 for t_A resp. t_B. If $A \geq B$, we have $J_u^A \geq J_u^B$ by Definition 10.5, and hence by Theorem 10.9,

$$-\langle (A - \lambda I)^{-1}u, u \rangle = \min J_u^A \geq \min J_u^B = -\langle (B - \lambda I)^{-1}u, u \rangle$$

for arbitrary $u \in \mathcal{H}$, so that $(B - \lambda I)^{-1} \geq (A - \lambda I)^{-1}$.

The following assertion is usually called the *Heinz inequality*.

Proposition 10.14 *Let A and B be self-adjoint operators on $\mathcal{H}$. If $A \geq B \geq 0$, then $A^\alpha \geq B^\alpha$ for any $\alpha \in (0, 1)$.*

Proof Let T be a positive self-adjoint operator. By Proposition 5.16 we have

$$\left\| T^{\alpha/2}x \right\|^2 = \pi^{-1} \sin \pi\alpha \int_0^\infty t^{\alpha-1} \langle T(T + tI)^{-1}x, x \rangle \, dt \qquad (10.22)$$

for $x \in \mathcal{D}(T^{\alpha/2})$ and a vector $x \in \mathcal{H}$ is in $\mathcal{D}(T^{\alpha/2})$ if and only and if the integral in (10.22) is finite.

Since $A \geq B \geq 0$, for any $t > 0$, we have by Corollary 10.13,

$$A(A + tI)^{-1} = I - t(A + tI)^{-1} \geq I - t(B + tI)^{-1} = B(B + tI)^{-1}.$$

Therefore, by the preceding result, applied to $T = A$ and $T = B$, we conclude that $\mathcal{D}(A^{\alpha/2}) \subseteq \mathcal{D}(B^{\alpha/2})$ and $\|A^{\alpha/2}x\| \geq \|B^{\alpha/2}x\|$ for $x \in \mathcal{D}(A^{\alpha/2})$. By Definition 10.5 the latter means that $A^\alpha \geq B^\alpha$. $\square$

Example 10.3 ($A \geq B \geq 0$ *does* not *imply* $A^2 \geq B^2$) For the operators A and B on the Hilbert space $\mathbb{C}^2$ given by the matrices

$$A = \begin{pmatrix} 2 & 1 \\ 1 & 1 \end{pmatrix} \quad \text{and} \quad B = \begin{pmatrix} 1 & 0 \\ 0 & 0 \end{pmatrix},$$

it is obvious that $A \geq B \geq 0$, but $A^2 \not\geq B^2$, since

$$A^2 - B^2 = \begin{pmatrix} 4 & 3 \\ 3 & 2 \end{pmatrix}.$$

It can be even shown that $A^\alpha \not\geq B^\alpha$ for each $\alpha > 1$.   o

Proposition 10.14 is a result from the theory of *operator monotone functions*. Detailed treatments of this theory can be found in [Dn1] and [Ao].

Recall that a *Nevanlinna function* is a holomorphic function f on the upper half-plane $\mathbb{C}_+$ with nonnegative imaginary part (see Appendix F). By Theorem F.1 a function f on $\mathbb{C}_+$ is a Nevanlinna function if and only if there are a finite positive regular Borel measure μ on $\mathbb{R}$ and numbers $a \in \mathbb{R}$ and $b \geq 0$ such that

$$f(z) = a + bz + \int_{\mathbb{R}} \frac{1 + tz}{t - z} d\mu(t) \tag{10.23}$$

for $z \in \mathbb{C}_+$. The measure μ and the numbers a, b are uniquely determined by f.

Let $\mathcal{J}$ be an open interval on $\mathbb{R}$. A measurable function f on $\mathcal{J}$ is called *operator monotone on $\mathcal{J}$* if $f(A) \leq f(B)$ for all self-adjoint operators A and B on a Hilbert space $\mathcal{H}$ such that $\sigma(A) \subseteq \mathcal{J}$, $\sigma(B) \subseteq \mathcal{J}$, and $A \leq B$. (In fact it suffices to assume this for Hermitian matrices A and B of arbitrary size with eigenvalues in $\mathcal{J}$.)

A complete characterization of operator monotone functions is given by the following result which is *Löwner's theorem* [Lö]; see [Dn1, pp. 26 and 85] for a proof.

Theorem 10.15 *A function f on an open interval $\mathcal{J}$ is operator monotone on $\mathcal{J}$ if and only if there are numbers $a \in \mathbb{R}$, $b \geq 0$, and a finite positive regular Borel measure μ on $\mathbb{R}$ satisfying $\mu(\mathcal{J}) = 0$ such that $f(z)$ is given by (10.23) for $z \in \mathcal{J}$.*

Since $\mu(\mathcal{J}) = 0$, the holomorphic function f on $\mathbb{C}_+$ given by (10.23) admits then an analytic continuation across the interval $\mathcal{J}$ into the lower half-plane.

Theorem 10.15 implies in particular that for each operator monotone function on $\mathbb{R}$, we have $\mu(\mathbb{R}) = 0$, so it is of the form $f(z) = a + bz$ with $a \in \mathbb{R}$ and $b \geq 0$.

10.4   The Friedrichs Extension of a Semibounded Symmetric Operator

In this section we use the form representation theorem (Theorem 10.7) to construct a distinguished self-adjoint extension, called the *Friedrichs extension*, of a densely defined lower semibounded symmetric operator T. This extension is the largest with respect to the order relation $\geq$ among all lower semibounded self-adjoint extensions of T. The crucial step of this construction is contained in the following lemma.

Lemma 10.16 *If T is a densely defined lower semibounded symmetric operator, then the form $\mathfrak{s}_T$ defined by $\mathfrak{s}_T[x, y] = \langle Tx, y \rangle$, $x, y \in \mathcal{D}(\mathfrak{s}_T):=\mathcal{D}(T)$, is closable.*

Proof We write $\mathfrak{s}$ for $\mathfrak{s}_T$. Let $\mathcal{H}_\mathfrak{s}$ denote the completion of $(\mathcal{D}(\mathfrak{s}), \langle \cdot, \cdot \rangle_\mathfrak{s})$. By Proposition 10.3, it suffices to show that the embedding $\mathcal{I}_\mathfrak{s} : \mathcal{H}_\mathfrak{s} \to \mathcal{H}$ is injective.

Let $x \in \mathcal{N}(\mathcal{I}_\mathfrak{s})$. Since $\mathcal{D}(\mathfrak{s})$ is dense in $\mathcal{H}_\mathfrak{s}$, there is a sequence $(x_n)_{n\in\mathbb{N}}$ from $\mathcal{D}(\mathfrak{s})$ such that $\lim_n x_n = x$ in $\mathcal{H}_\mathfrak{s}$ and $\lim_n \mathcal{I}_\mathfrak{s}(x_n) = \lim_n x_n = \mathcal{I}_\mathfrak{s}(x) = 0$ in $\mathcal{H}$. Using that T is symmetric for arbitrary $y \in \mathcal{D}(T)$, we compute

$$\langle x, y \rangle_\mathfrak{s} = \lim_{n\to\infty} \langle x_n, y \rangle_\mathfrak{s} = \lim_{n\to\infty} \left(\mathfrak{s}[x_n, y] + (1 - m)\langle x_n, y \rangle \right)$$
$$= \lim_{n\to\infty} \left(\langle Tx_n, y \rangle + (1 - m)\langle x_n, y \rangle \right) = \lim_{n\to\infty} \langle x_n, (T + (1 - m)I)y \rangle = 0.$$

Since $\mathcal{D}(T) = \mathcal{D}(\mathfrak{s})$ is dense in $\mathcal{H}_\mathfrak{s}$, $x = 0$. Thus, $\mathcal{I}_\mathfrak{s}$ is injective, and $\mathfrak{s}$ is closed. $\square$

The closure $\mathfrak{t} := \overline{\mathfrak{s}_T}$ of $\mathfrak{s}_T$ is a densely defined lower semibounded *closed* form. By Theorem 10.7, the operator $A_\mathfrak{t}$ associated with this form is self-adjoint.

Definition 10.6 The operator $A_\mathfrak{t}$ is called the *Friedrichs extension* of T and denoted by T_F.

The first assertion of the following theorem says that T_F is indeed an extension of the symmetric operator T.

Theorem 10.17 *Let T be a densely defined lower semibounded symmetric operator on a Hilbert space $\mathcal{H}$.*

(i) *T_F is a lower semibounded self-adjoint extension of T which has the same greatest lower bound as T.*

(ii) *If S is another lower semibounded self-adjoint extension of T, then $T_F \geq S$ and $\mathfrak{t}_{T_F} = \overline{\mathfrak{s}_T} \subseteq \mathfrak{t}_S = \overline{\mathfrak{s}_S}$.*

(iii) *$\mathcal{D}(T_F) = \mathcal{D}(T^*) \cap \mathcal{D}(\overline{\mathfrak{s}_T})$ and $T_F = T^* \upharpoonright \mathcal{D}(\overline{\mathfrak{s}_T})$. Moreover, T_F is the only self-adjoint extension of T with domain contained in $\mathcal{D}(\overline{\mathfrak{s}_T})$.*

(iv) *$(T + \lambda I)_F = T_F + \lambda I$ for $\lambda \in \mathbb{R}$.*

Proof (i): If $x \in \mathcal{D}(T)$, then $\mathfrak{t}_{T_F}[x, y] = \overline{\mathfrak{s}_T}[x, y] = \langle Tx, y \rangle$ for all $y \in \mathcal{D}(T)$. From Proposition 10.5(v), applied to the core $\mathcal{D}(T) = \mathcal{D}(\mathfrak{s}_T)$ of $\mathfrak{t}_{T_F} = \overline{\mathfrak{s}_T}$, we obtain $T \subseteq T_F$. Clearly, the greatest lower bound of T coincides with the greatest lower bound of $\mathfrak{s}_F$, so of $\overline{\mathfrak{s}_F}$, and hence of T_F by Proposition 10.4(iii).

(ii): Let S be a lower semibounded self-adjoint extension of T. By (i), S_F is a self-adjoint extension of S, and so $S_F = S$. Hence, $\overline{\mathfrak{s}_S} = \mathfrak{t}_{S_F} = \mathfrak{t}_S$ by the definition of S_F. From $S \supseteq T$ we obtain $\mathfrak{s}_S \supseteq \mathfrak{s}_T$, and hence $\mathfrak{t}_S = \overline{\mathfrak{s}_S} \supseteq \overline{\mathfrak{s}_T} = \mathfrak{t}_{T_F}$. The relation $\mathfrak{t}_{T_F} \subseteq \mathfrak{t}_S$ implies that $T_F \geq S$ according to Definition 10.5.

(iii): Since $T \subseteq T_F$ by (i), we have $(T_F)^* = T_F \subseteq T^*$, so $T_F = T^* \upharpoonright \mathcal{D}(T_F)$ and $\mathcal{D}(T_F) \subseteq \mathcal{D}(T^*) \cap \mathcal{D}(\overline{\mathfrak{s}_T}) =: \mathcal{D}(R)$. Let $R := T^* \upharpoonright \mathcal{D}(R)$ and $x \in \mathcal{D}(R)$. Using the symmetry of $\mathfrak{t}_{T_F}$, formula (10.9), and the relations $\overline{\mathfrak{s}_T} = \mathfrak{t}_{T_F}$ and $T \subseteq T_F$, we obtain

$$\mathfrak{t}_{T_F}[x, y] = \overline{\mathfrak{t}_{T_F}[y, x]} = \overline{\langle T_F y, x \rangle} = \langle x, Ty \rangle = \langle T^*x, y \rangle = \langle Rx, y \rangle$$

for $y \in \mathcal{D}(T)$. Applying again Proposition 10.5(v) with $\mathcal{D} = \mathcal{D}(T)$, we get $R \subseteq T_F$, and so $\mathcal{D}(R) \subseteq \mathcal{D}(T_F)$. Thus, we have shown that $\mathcal{D}(T_F) = \mathcal{D}(T^*) \cap \mathcal{D}(\overline{\mathfrak{s}_T})$.

Let S be an arbitrary self-adjoint extension of T such that $\mathcal{D}(S) \subseteq \mathcal{D}(\overline{\mathfrak{s}_T})$. Clearly, $T \subseteq S$ implies that $S = S^* \subseteq T^*$, so $\mathcal{D}(S) \subseteq \mathcal{D}(T^*) \cap \mathcal{D}(\overline{\mathfrak{s}_F}) = \mathcal{D}(T_F)$. Since $S = T^* \upharpoonright \mathcal{D}(S)$ and $T_F = T^* \upharpoonright \mathcal{D}(T_F)$, we get $S \subseteq T_F$. Hence, $S = T_F$.

(iv): Since $\mathfrak{s}_{T+\lambda I} = \mathfrak{s}_T + \lambda$, $\mathfrak{t}_{(T+\lambda I)_F} = \overline{\mathfrak{s}_{T+\lambda I}} = \overline{\mathfrak{s}_T} + \lambda = \mathfrak{t}_{T_F} + \lambda = \mathfrak{t}_{T_F + \lambda I}$, and hence $(T + \lambda I)_F = T_F + \lambda I$. $\qquad\square$

10.5 Examples of Semibounded Forms and Operators

Let us begin with a simple but standard example for the Friedrichs extension.

Example 10.4 (*Friedrichs extension of $A = -\frac{d^2}{dx^2}$ on $\mathcal{D}(A) = H_0^2(a,b)$, $a,b \in \mathbb{R}$*)
From Example 1.4 we know that A is the square T^2 of the symmetric operator $T = -i\frac{d}{dx}$ with domain $\mathcal{D}(T) = H_0^1(a,b)$ in $L^2(a,b)$ and $A^* = -\frac{d^2}{dx^2}$ on $\mathcal{D}(A^*) = H^2(a,b)$. Thus, A is a densely defined positive symmetric operator, and

$$\mathfrak{s}_A[f] = \langle Af, f\rangle = \|Tf\|^2 = \|f'\|^2 \quad \text{for } f \in \mathcal{D}(\mathfrak{s}_A) = \mathcal{D}(A) = H_0^2(a,b).$$

Hence, the form norm of $\mathfrak{s}_A$ coincides with the norm of the Sobolev space $H^1(a,b)$. Since $C_0^\infty(a,b) \subseteq H_0^2(a,b)$, the completion $\mathcal{D}(\overline{\mathfrak{s}_A})$ of $(\mathcal{D}(\mathfrak{s}_A), \|\cdot\|_{\mathfrak{s}_A})$ is $H_0^1(a,b)$. Therefore, by Theorem 10.17(iii), the Friedrichs extension A_F of A is the restriction of A^* to $\mathcal{D}(\overline{\mathfrak{s}_A})$, that is,

$$A_F f = -f'' \quad \text{for } f \in \mathcal{D}(A_F) = \{f \in H^2(a,b) : f(a) = f(b) = 0\}.$$

By Theorem D.1, the embedding of $H_0^1(a,b) = \mathcal{D}(\mathfrak{t}_{A_F})$ into $L^2(a,b)$ is compact. It follows therefore from Proposition 10.6 that A_F has a purely discrete spectrum. One easily computes that A_F has the orthonormal basis $\{f_n : n \in \mathbb{N}\}$ of eigenfunctions:

$$A_F f_n = \pi^2 (b-a)^{-2} n^2 f_n, \quad \text{where } f_n(x) = \sqrt{2}(b-a)^{-1/2} \sin \pi n \frac{x-a}{b-a}, \ n \in \mathbb{N}.$$

The smallest eigenvalue of A_F is $\pi^2(b-a)^{-2}$, so we have $A_F \geq \pi^2(b-a)^{-2} \cdot I$. Hence, $\mathfrak{t}_{A_F} \geq \pi^2(b-a)^{-2}$, which yields the Poincaré inequality (see, e.g., (D.1))

$$\|f'\| \geq \pi(b-a)^{-1}\|f\| \quad \text{for } f \in \mathcal{D}[A_F] = H_0^1(a,b). \tag{10.24}$$

$\circ$

Before we continue with other ordinary differential operators, we treat a simple operator-theoretic example.

Example 10.5 Let T be a linear operator of a Hilbert space $(\mathcal{H}_1, \langle\cdot,\cdot\rangle_1)$ into a Hilbert space $(\mathcal{H}_2, \langle\cdot,\cdot\rangle_2)$. Define a positive form $\mathfrak{t}$ by

$$\mathfrak{t}[x,y] = \langle Tx, Ty\rangle_2, \quad x, y \in \mathcal{D}(\mathfrak{t}) := \mathcal{D}(T). \tag{10.25}$$

Taking $m = 0$ as a lower bound of $\mathfrak{t}$, by (10.4) and (1.4), we have

$$\|x\|_{\mathfrak{t}}^2 = \|Tx\|_2^2 + \|x\|_1^2 = \|x\|_T^2, \quad x \in \mathcal{D}(\mathfrak{t}),$$

that is, form norm $\| \cdot \|_t$ and graph norm $\| \cdot \|_T$ coincide on $\mathcal{D}(t) = \mathcal{D}(T)$. Therefore, the form t is closed if and only the operator T is closed. Likewise, t is closable if and only if T is. In the latter case, we have $\bar{t}[x, y] = \langle \overline{T}x, \overline{T}y \rangle_2$ for $x, y \in \mathcal{D}(\bar{t}) = \mathcal{D}(\overline{T})$. Further, a linear subspace of $\mathcal{D}(t)$ is a core of t if and only if it is a core of T.

Statement *If $\mathcal{D}(T)$ is dense in $\mathcal{H}_1$ and the operator T is closed, then $A_t = T^*T$.*

Proof For $x \in \mathcal{D}(A_t)$ and $y \in \mathcal{D}(T)$, we have $\langle Tx, Ty \rangle_2 = t[x, y] = \langle A_t x, y \rangle_1$. This implies that $Tx \in \mathcal{D}(T^*)$ and $T^*Tx = A_t x$. Thus, $A_t \subseteq T^*T$. Since A_t is self-adjoint and T^*T is obviously symmetric, we get $A_t = T^*T$. $\qquad \square$

By Proposition 10.4(iv), $\mathcal{D}(A_t) = \mathcal{D}(T^*T)$ is a core for t and so for T. Thus, we have given another proof of Proposition 3.18(ii), which states that *the operator T^*T is self-adjoint and $\mathcal{D}(T^*T)$ is a core for the densely defined closed operator T.* ∘

Example 10.6 We apply the preceding example to the closed operators $T = -i\frac{d}{dx}$ with *different domains* in the Hilbert space $L^2(a, b)$, where $a, b \in \mathbb{R}, a < b$. The form t is always given by the same expression

$$t[f, g] = \langle -if', -ig' \rangle \quad \text{for } f, g \in \mathcal{D}(t) := \mathcal{D}(T).$$

We define the operators $T_0 = -i\frac{d}{dx}$ and $T_{(\alpha, \beta)} = -i\frac{d}{dx}$ for $(\alpha, \beta) \in \mathbb{C}^2$ with domains

$$\mathcal{D}(T_0) := H_0^1(a, b) \quad \text{and} \quad \mathcal{D}(T_{(\alpha, \beta)}) = \{ f \in H^1(a, b) : \alpha f(a) = \beta f(b) \}.$$

Since the functionals $f(a)$ and $f(b)$ on $H^1(a, b)$ are continuous by Lemma 1.12, the operators T_0 and $T_{(\alpha, \beta)}$ are closed. Similarly as in Example 1.4, we derive

$$T_{(0,0)}^* = T_0, \qquad T_0^* = T_{(0,0)}, \quad \text{and} \quad T_{(\alpha, \beta)}^* = T_{(\overline{\beta}, \overline{\alpha})} \quad \text{for } (\alpha, \beta) \neq (0, 0).$$

Therefore, we obtain

$$\mathcal{D}(T_0^* T_0) = \{ f \in H^2(a, b) : f(a) = f(b) = 0 \},$$
$$\mathcal{D}(T_0 T_0^*) = \{ f \in H^2(a, b) : f'(a) = f'(b) = 0 \},$$
$$\mathcal{D}(T_{(\alpha, \beta)}^* T_{(\alpha, \beta)}) = \{ f \in H^2(a, b) : \alpha f(a) = \beta f(b), \overline{\beta} f'(a) = \overline{\alpha} f'(b) \}$$

for $(\alpha, \beta) \neq (0, 0)$. By Example 10.5, the self-adjoint operator A_t associated with the form t on $\mathcal{D}(t) = \mathcal{D}(T)$ is T^*T. That is, $A_t f = -\frac{d^2}{dx^2}$ with $\mathcal{D}(A_t) = \mathcal{D}(T^*T)$, where T is T_0, T_0^* or $T_{(\alpha, \beta)}$. In particular, $T_0^* T_0$ is the differential operator $-\frac{d^2}{dx^2}$ with *Dirichlet boundary conditions* $f(a) = f(b) = 0$, while $T_0 T_0^*$ is the operator $-\frac{d^2}{dx^2}$ with the *Neumann boundary conditions* $f'(a) = f'(b) = 0$.

Clearly, $T_0^* T_0$ is the Friedrichs extension A_F of the operator A from Example 10.4. Since A has finite deficiency indices, by Corollary 8.13 all other self-adjoint extensions of A, in particular all above operators T^*T, have purely discrete spectra as well. Let $\lambda_n(T^*T)$, $n \in \mathbb{N}$, denote the eigenvalues of T^*T enumerated with multiplicities in increasing order. Since $T^*T \leq A_F$ by Theorem 10.17(ii), from Corollary 12.3 below we conclude that $\lambda_n(T^*T) \leq \pi^2(b - a)^{-2}n^2$ for $n \in \mathbb{N}$. ∘

Example 10.7 In this example we keep the domain $\mathcal{D}(\mathfrak{t}) = H^1(\mathbb{R})$ fixed, and we change the form $\mathfrak{t}$ by adding a quadratic form of point evaluations.

Let $a, b \in \mathbb{R}$, $a < b$, $\alpha, \gamma \in \mathbb{R}$, $\beta \in \mathbb{C}$. We define the form $\mathfrak{t}$ on $\mathcal{D}(\mathfrak{t}) = H^1(\mathbb{R})$ by

$$\mathfrak{t}[f, g] = \langle -if', -ig' \rangle + \alpha f(a)\overline{g(a)} + \beta f(a)\overline{g(b)} + \overline{\beta}\, f(b)\overline{g(a)} + \gamma f(b)\overline{g(b)}.$$

From the continuity of the functionals $f(a)$ and $f(b)$ on $H^1(\mathbb{R})$ (by Lemma 1.12) it follows easily that the form $\mathfrak{t}$ is lower semibounded and that the form norm of $\mathfrak{t}$ is equivalent to the Hilbert space norm of $H^1(\mathbb{R})$. Hence, $\mathfrak{t}$ is a closed form.

Statement $A_\mathfrak{t}$ *is the operator* $-\frac{d^2}{dx^2}$ *with domain*

$$\mathcal{D}(A_\mathfrak{t}) = \big\{ f \in H^2(-\infty, a) \oplus H^2(a, b) \oplus H^2(b, +\infty) : f(a+) = f(a-),$$
$$f(b+) = f(b-), f'(b+) - f'(b-) = \beta f(a) + \gamma f(b),$$
$$f'(a+) - f'(a-) = \alpha f(a) + \overline{\beta} f(b) \big\}.$$

Proof Define $T_0 = -\frac{d^2}{dx^2}$ on $\mathcal{D}(T_0) = \{ f \in H^2(\mathbb{R}) : f(a) = f(b) = f'(a) = f'(b) = 0 \}$. As in Example 1.6, it follows that the adjoint operator T_0^* acts by $T_0^* f = -f''$ for $f \in \mathcal{D}(T_0^*) = H^2(-\infty, a) \oplus H^2(a, b) \oplus H^2(b, +\infty)$.

Let $g \in H^1(\mathbb{R})$ and $f \in \mathcal{D}(T_0^*)$. Since $f'(+\infty) = f'(-\infty) = 0$ by Lemma 1.11, we obtain by partial integration

$$\mathfrak{t}[f, g] - \langle -f'', g \rangle = f'(a-)\overline{g(a)} + f'(b-)\overline{g(b)} - f'(a+)\overline{g(a)} - f'(b+)\overline{g(b)}$$
$$+ \alpha f(a)\overline{g(a)} + \beta f(a)\overline{g(b)} + \overline{\beta}\, f(b)\overline{g(a)} + \gamma f(b)\overline{g(b)}.$$

We use this equality two times. First, if $f \in \mathcal{D}(T_0)$, it yields $\mathfrak{t}[f, g] = \langle T_0 f, g \rangle$ for all $g \in \mathcal{D}(\mathfrak{t})$, so that $A_\mathfrak{t} f = T_0 f$. That is, we have $T_0 \subseteq A_\mathfrak{t}$, and hence $A_\mathfrak{t} \subseteq T_0^*$. Since $\mathcal{D}(A_\mathfrak{t}) \subseteq \mathcal{D}(\mathfrak{t}) = H^1(\mathbb{R})$, all functions from $\mathcal{D}(A_\mathfrak{t})$ are continuous on $\mathbb{R}$.

Now let $f \in \mathcal{D}(T_0^*)$. Then $f \in \mathcal{D}(A_\mathfrak{t})$ if and only if the expression after the equality sign is zero. Since the numbers $g(a)$ and $g(b)$ are arbitrary if g runs through $H^1(\mathbb{R})$, this is true if and only if the coefficients of $g(a)$ and $g(b)$ vanish, that is, if f satisfies the boundary conditions of the domain $\mathcal{D}(A_\mathfrak{t})$. $\qquad \square$ o

Example 10.8 (*Forms associated with Sturm–Liouville operators*) Let $\mathcal{J}$ be an open interval of $\mathbb{R}$, and $\mathcal{H} = L^2(\mathcal{J})$. Let p and q be real functions from $L^\infty(\mathcal{J})$. Suppose that there is a constant $c > 0$ such that $p(x) \geq c$ a.e. on $\mathcal{J}$.

We define the symmetric form $\mathfrak{t}$ on $\mathcal{D}(\mathfrak{t}) = H^1(\mathcal{J})$ by

$$\mathfrak{t}[f, g] = \int_{\mathcal{J}} p(x) f'(x) \overline{g'(x)}\, dx + \int_{\mathcal{J}} q(x) f(x) \overline{g(x)}\, dx. \tag{10.26}$$

If M is a constant such that $|p(x)| \leq M$ and $|q(x)| \leq M$ a.e. on $\mathcal{J}$, we estimate

$$M \|f\|^2_{H^1(\mathcal{J})} \geq \mathfrak{t}[f] \geq c \|f'\|^2 - M \|f\|^2 = c \|f\|^2_{H^1(\mathcal{J})} - (M + c) \|f\|^2.$$

From these inequalities we conclude that $\mathfrak{t}$ is lower semibounded and that the form norm (10.4) on $\mathcal{D}(\mathfrak{t})$ is equivalent to the norm of $H^1(\mathcal{J})$. Therefore, $\mathfrak{t}$ is closed. By Theorem 10.7 the operator $A_\mathfrak{t}$ is self-adjoint and lower semibounded.

For all functions p and q as above, the form domain $\mathcal{D}(t) = \mathcal{D}[A_t]$ is $H^1(\mathcal{J})$. But as we shall see from formula (10.27), the domains of the corresponding operators A_t can be quite different! We will meet this phenomenon again in Sect. 14.8.

Now suppose that $q = 0$ and $\mathcal{J} = (a, b)$, $a, b \in \mathbb{R}$. We then describe the operator A_t "explicitly". Indeed, using integration by parts, it can be shown that

$$A_t f = -\big(pf'\big)',$$
$$\mathcal{D}(A_t) = \big\{f \in H^1(\mathcal{J}) : pf' \in H^1(\mathcal{J}), f'(a) = f'(b) = 0\big\}. \tag{10.27}$$

That the operator A_t from (10.27) is self-adjoint can be also derived from Lemma 10.18 below. We sketch this argument. Indeed, we define $T := -\mathrm{i}\frac{d}{dx}$ on $\mathcal{D}(T) = H^1(\mathcal{J})$ and $S := p^{1/2}Tp^{1/2}$. Since T is closed, S is closed, and we have $S^* = p^{1/2}T^*p^{1/2}$ by Lemma 10.18, applied with $B = p^{1/2}$. Hence, S^*S is self-adjoint, and so is $p^{-1/2}S^*Sp^{-1/2}$ by Lemma 10.18, now applied with $B = p^{-1/2}$. Since $T^* = T_0 = -\mathrm{i}\frac{d}{dx}$ on $\mathcal{D}(T_0) = H_0^1(\mathcal{J})$, we have $p^{-1/2}S^*Sp^{-1/2} = T_0 pT$, and this is just the operator A_t given by (10.27). ○

Lemma 10.18 *Suppose that $B \in \mathbf{B}(\mathcal{H})$ has a bounded inverse $B^{-1} \in \mathbf{B}(\mathcal{H})$. If T is a densely defined operator on $\mathcal{H}$, then $(B^*TB)^* = B^*T^*B$. If T is closed or self-adjoint, so is B^*TB.*

Proof Put $S := B^*TB$. Clearly, $S^* \supseteq B^*T^*B$. Let $x \in \mathcal{D}(S^*)$. It suffices to prove that $x \in \mathcal{D}(B^*T^*B)$. If $y \in \mathcal{D}(T)$, then $y' := B^{-1}y \in \mathcal{D}(TB) = \mathcal{D}(S)$, and

$$\langle Ty, Bx\rangle = \langle B^*Ty, x\rangle = \langle Sy', x\rangle = \langle y', S^*x\rangle = \langle B^{-1}y, S^*x\rangle = \langle y, \big(B^{-1}\big)^*S^*x\rangle.$$

From this it follows that $Bx \in \mathcal{D}(T^*)$. Thus, $x \in \mathcal{D}(T^*B) = \mathcal{D}(B^*T^*B)$.

If T is closed, then $T^{**} = T$, and hence $(B^*TB)^{**} = B^*T^{**}B = B^*TB$ by the preceding, that is, B^*TB is also closed. □

10.6 Dirichlet and Neumann Laplacians on Domains of $\mathbb{R}^d$

Let Ω be an open subset of $\mathbb{R}^d$. In this section we apply the theory of forms to develop two operator realizations on the Hilbert space $\mathcal{H} = L^2(\Omega)$ of the Laplacian

$$\mathcal{L} = -\Delta = -\frac{\partial^2}{\partial x_1^2} - \cdots - \frac{\partial^2}{\partial x_d^2}.$$

For this, some results on Sobolev spaces on Ω from Appendix D are essentially used. Recall that the minimal operator $L_{\min}$ and the maximal operator $L_{\max}$ associated with $\mathcal{L}$ have been defined in Sect. 1.3.2.

10.6.1 Laplacian with Dirichlet Boundary Condition

To begin with, we define a positive form t with domain $\mathcal{D}(t) = H_0^1(\Omega)$ by

$$t[u, v] = \int_\Omega \nabla u \cdot \overline{\nabla v} \, dx \equiv \int_\Omega (\partial_1 u \overline{\partial_1 v} + \cdots + \partial_d u \overline{\partial_d v}) \, dx. \tag{10.28}$$

Then $\|u\|_t^2 = t[u] + \|u\|^2 = \|u\|_{H_0^1(\Omega)}^2$, that is, the form norm $\|\cdot\|_t$ is just the Hilbert space norm of $H_0^1(\Omega)$. Therefore, $(\mathcal{D}(t), \|\cdot\|_t)$ is complete, and hence t is closed.

To derive the next formula (10.29), we argue as in the proof of Lemma 1.13. Let $u, v \in \mathcal{D}(L_{\min}) = C_0^\infty(\Omega)$. We choose a bounded open set $\widetilde{\Omega} \subseteq \Omega$ with C^∞-boundary such that $\operatorname{supp} u \subseteq \widetilde{\Omega}$ and $\operatorname{supp} v \subseteq \widetilde{\Omega}$. Applying the Green formula (D.5) twice to $\widetilde{\Omega}$ and using that all boundary terms on $\partial\widetilde{\Omega}$ vanish, we compute

$$\langle L_{\min} u, v \rangle = \langle -\Delta u, v \rangle = t[u, v] = \langle u, -\Delta v \rangle = \langle u, L_{\min} v \rangle. \tag{10.29}$$

This shows that $L_{\min}$ is a positive symmetric operator on $\mathcal{H}$ and that its form $\mathfrak{s}_{L_{\min}}$ (defined in Lemma 10.16) is the restriction of the form t to $C_0^\infty(\Omega)$. Since $C_0^\infty(\Omega)$ is dense in $H_0^1(\Omega)$, t is the closure of $\mathfrak{s}_{L_{\min}}$. Therefore, by Definition 10.6, the associated operator A_t is the Friedrichs extension of $L_{\min}$, and $\mathcal{D}[A_t] = \mathcal{D}(t) = H_0^1(\Omega)$. We call the positive self-adjoint operator $A_t = (L_{\min})_F$ the *Dirichlet Laplacian* on Ω and denote it by $-\Delta_D$.

For general open sets Ω, the Dirichlet Laplacian $-\Delta_D$ refers to vanishing boundary values at $\partial\Omega$ in the sense that $C_0^\infty(\Omega)$ is a form core for $-\Delta_D$. If the open set Ω is C^1, the trace map γ_0 is defined on $H^1(\Omega)$ and $H_0^1(\Omega) = \{u \in H^1(\Omega) : \gamma_0(u) = 0\}$ by Theorem D.6. In this case any function f from the domain $\mathcal{D}(-\Delta_D) \subseteq \mathcal{D}[A_t] = H_0^1(\Omega)$ has vanishing boundary values $\gamma_0(f) \equiv f \upharpoonright \partial\Omega = 0$ at $\partial\Omega$.

Recall from formula (1.22) that $(L_{\min})^* = L_{\max}$, where $L_{\max} f = -\Delta f$ for f in $\mathcal{D}(L_{\max}) = \{f \in L^2(\Omega) : \Delta f \in L^2(\Omega)\}$. Therefore, by Theorem 10.17(iii),

$$\mathcal{D}(-\Delta_D) = \mathcal{D}(L_{\max}) \cap H_0^1(\Omega),$$
$$-\Delta_D f = L_{\max} f \quad \text{for } f \in \mathcal{D}(-\Delta_D). \tag{10.30}$$

If $f \in H^2(\Omega)$, then obviously $\Delta f \in L^2(\Omega)$, so that $f \in \mathcal{D}(L_{\max})$. That is,

$$H^2(\Omega) \cap H_0^1(\Omega) \subseteq \mathcal{D}(-\Delta_D). \tag{10.31}$$

In general, the domain $\mathcal{D}(L_{\max})$ is much larger than $H^2(\Omega)$, see, e.g., Exercise 11.a.

Now we turn to the two fundamental problems (10.16) and (10.17) of Sect. 10.2. Suppose that $\lambda < m_t$. Then $\lambda \in \rho(-\Delta_D)$. Inserting the definition (10.28) of the form t and replacing h by $\overline{h}$ in (10.16), these problems are the following:

Given $g \in L^2(\Omega)$, to find $f \in H_0^1(\Omega)$ such that

$$\int_\Omega \nabla f \cdot \nabla h \, dx - \lambda \int_\Omega f h \, dx = \int_\Omega g h \, dx \quad \text{for all } h \in H_0^1(\Omega), \tag{10.32}$$

or to find $f \in H_0^1(\Omega)$ which minimizes the functional

$$\int_\Omega \left(|\nabla f|^2 - \lambda |f|^2 \right) dx - \int_\Omega (g\overline{f} + f\overline{g}) \, dx. \tag{10.33}$$

These problems are referred to as the *weak Dirichlet problem* and the *variational Dirichlet problem* for the Laplacian on Ω, respectively. It should be emphasized that in (10.32) and (10.33) only first-order derivatives occur. From Sect. 10.2 we know that both problems have the same unique solution given by

$$f = (A_t - \lambda I)^{-1} g = (-\Delta_D - \lambda I)^{-1} g. \tag{10.34}$$

This function f is called a *weak solution* of the equation $(-\Delta_D - \lambda)f = g$. The word "weak" stems from the fact that $-\Delta_D$ acts on $f \in \mathcal{D}(-\Delta_D) = H_0^1(\Omega) \cap \mathcal{D}(L_{\max})$ in distributional sense. The question of when a weak solution is "smooth enough", and thus a classical solution is the so-called *regularity problem* which is treated in many books (see, e.g., [Ag, Br, LM]). A typical result reads as follows:

If Ω is bounded and of class C^m for sufficiently large m, $g \in H^n(\Omega)$, $n, k \in \mathbb{N}_0$, and $n + 2 - k > d/2$, then $f \in H^{n+2}(\Omega) \subseteq C^k(\Omega)$.

From now on we assume that Ω is *bounded*. Then the embedding of $H_0^1(\Omega)$ into $L^2(\Omega)$ is compact (by Theorem D.1) and $(H_0^1(\Omega), \|\cdot\|_{H_0^1(\Omega)}) = (\mathcal{D}(t), \|\cdot\|_t)$. Therefore, it follows from Proposition 10.6 that the operator $-\Delta_D$ has a purely discrete spectrum.

Further, by the Poincaré inequality (D.1), we have $t \geq c$, and hence $-\Delta_D \geq c \cdot I$ for some constant $c = c_\Omega > 0$. Therefore, $\sigma(-\Delta_D) \subseteq [c, +\infty)$. In particular, we have $0 \in \rho(-\Delta_D)$, so we can set $\lambda = 0$ in problems (10.32) and (10.33).

From now on we suppose that Ω is *bounded and of class C^2* (see Appendix D). Then, by Theorem D.10(i), for each $g \in L^2(\Omega)$, the solution $f = (-\Delta_D + I)^{-1} g$ of problem (10.32) with $\lambda = -1$ is in $H^2(\Omega)$. Thus, $\mathcal{D}(-\Delta_D) \subseteq H^2(\Omega)$. Hence, combining (10.30) and (10.31), it follows that

$$\mathcal{D}(-\Delta_D) = H^2(\Omega) \cap H_0^1(\Omega). \tag{10.35}$$

Next, we prove that on $\mathcal{D}(-\Delta_D)$ the graph norm of $-\Delta_D$ and the norm $\|\cdot\|_{H^2(\Omega)}$ are equivalent. Since the trace map γ_0 is continuous on $H^1(\Omega)$ and hence on $H^2(\Omega)$ and $H_0^1(\Omega) = \mathcal{N}(\gamma_0)$ by Theorem D.6, $H^2(\Omega) \cap H_0^1(\Omega)$ is a closed subspace of $H^2(\Omega)$. Hence, $\mathcal{D}(-\Delta_D) = H^2(\Omega) \cap H_0^1(\Omega)$ is complete in the graph norm of $-\Delta_D$ and also in the norm $\|\cdot\|_{H^2(\Omega)}$. Clearly, $\|-\Delta_D \cdot\| \leq \|\cdot\|_{H^2(\Omega)}$. Therefore, by the closed graph theorem, both norms are equivalent.

By definition, $C_0^\infty(\Omega)$ is dense in $\mathcal{D}(L_{\min})$ with respect to the graph norm of $-\Delta_D$ and in $H_0^2(\Omega)$ with respect to the norm $\|\cdot\|_{H^2(\Omega)}$. The equivalence of both norms implies that $\mathcal{D}(L_{\min}) = H_0^2(\Omega)$.

We summarize some of the preceding results in the following theorem.

Theorem 10.19 *Suppose that $\Omega \subseteq \mathbb{R}^d$ is an open bounded set of class C^2. Then the Dirichlet Laplacian $-\Delta_D$ is the positive self-adjoint operator on $L^2(\Omega)$ defined by*

$$(-\Delta_D)f = -\Delta f \quad \text{for } f \in \mathcal{D}(-\Delta_D) = H^2(\Omega) \cap H_0^1(\Omega).$$

It is the Friedrichs extension of the operator $L_{\min}$ with domain $\mathcal{D}(L_{\min}) = H_0^2(\Omega)$. The graph norm of $-\Delta_D$ is equivalent to the norm $\|\cdot\|_{H^2(\Omega)}$ on $\mathcal{D}(-\Delta_D)$.

The form of $-\Delta_D$ *is given by* (10.28), *and the form Hilbert space* $\mathcal{D}[-\Delta_D]$ *is the Sobolev space* $H_0^1(\Omega)$. *The operator* $-\Delta_D$ *has a purely discrete spectrum contained in* $[c_\Omega, +\infty)$, *where* $c_\Omega > 0$ *is the constant occurring in the Poincaré inequality* (D.1).

10.6.2 Laplacian with Neumann Boundary Condition

In this subsection we assume that $\Omega \subseteq \mathbb{R}^d$ is a *bounded open set of class* C^2 (see Appendix D). Let ν denote the outward normal unit vector to $\partial\Omega$. By Theorem D.7 any function $u \in H^2(\Omega)$ has well-defined boundary values $\gamma_1(u) = \frac{\partial u}{\partial \nu} \restriction \partial\Omega$.

As in the preceding subsection, we define a positive form t by (10.28) but now on the domain $\mathcal{D}(\mathfrak{t}) = H^1(\Omega)$. Then the form norm of $\mathfrak{t}$ is the norm of $H^1(\Omega)$. Hence, $\mathfrak{t}$ is closed. The positive self-adjoint operator $A_\mathfrak{t}$ associated with this form $\mathfrak{t}$ is called the *Neumann Laplacian* on Ω and denoted by $-\Delta_N$. This name and this notation will be justified by formula (10.39) for the domain $\mathcal{D}(-\Delta_N)$ given below.

Let us begin with the problems (10.16) and (10.17). We now have $m_\mathfrak{t} = 0$ and hence $m_{-\Delta_N} = 0$, since $\mathfrak{t}[1,1] = 0$. Hence, in contrast to the Dirichlet Laplacian, we cannot take $\lambda = 0$. Assume that $\lambda < 0$. The problems (10.16) and (10.17) are as follows:

Given $g \in L^2(\Omega)$, to find $f \in H^1(\Omega)$ such that

$$\int_\Omega \nabla f \cdot \nabla h \, dx - \lambda \int_\Omega f h \, dx = \int_\Omega g h \, dx \quad \text{for all } h \in H^1(\Omega), \quad (10.36)$$

or to find $f \in H^1(\Omega)$ such that f minimizes the functional

$$\int_\Omega \left(|\nabla f|^2 - \lambda |f|^2 \right) dx - \int_\Omega (g\overline{f} + f\overline{g}) \, dx. \quad (10.37)$$

These problems are called the *weak Neumann problem* and the *variational Neumann problem* for the Laplacian on Ω. Both problems have the same unique solution

$$f = (A_\mathfrak{t} - \lambda I)^{-1} g = (-\Delta_N - \lambda I)^{-1} g.$$

Now we turn to the description of the domain of the self-adjoint operator $-\Delta_N$.

Arguing as in the preceding subsection by using the Green formula (D.5), we derive $\mathfrak{t}[u,v] = \langle L_{\min} u, v \rangle$ for $u \in C_0^\infty(\Omega)$ and $v \in H^1(\Omega)$. Therefore, by the definition of $A_\mathfrak{t}$ we obtain $L_{\min} \subseteq A_\mathfrak{t} = -\Delta_N$. Hence, $-\Delta_N \subseteq (L_{\min})^* = L_{\max}$.

Next, we prove that a function $f \in H^2(\Omega)$ is in the domain $\mathcal{D}(-\Delta_N)$ if and only if it satisfies the *Neumann boundary condition* $\gamma_1(f) \equiv \frac{\partial f}{\partial \nu} = 0$ on $\partial\Omega$. Indeed, since $f \in H^2(\Omega)$, we have $f \in \mathcal{D}(L_{\max})$, and the Green formula (D.5) implies

$$\mathfrak{t}[f,v] - \langle L_{\max} f, v \rangle = \int_{\partial\Omega} \frac{\partial f}{\partial \nu} \overline{v} \, d\sigma \quad \text{for } v \in \mathcal{D}(\mathfrak{t}) = H^1(\Omega), \quad (10.38)$$

where $d\sigma$ is the surface measure on $\partial\Omega$. Therefore, by Definition 10.4, f belongs to $\mathcal{D}(A_\mathfrak{t}) = \mathcal{D}(-\Delta_N)$ if and only if the integral in (10.38) vanishes for all

$v \in H^1(\Omega)$. From the definition of the class C^2 (see Appendix D) it follows easily that for each point $x \in \partial\Omega$, there is a neighborhood U_x such that $\gamma_0(H^1(\Omega))$ is dense in $L^2(\partial\Omega \cap U_x, d\sigma)$. Hence, $\gamma_0(H^1(\Omega))$ is dense in $L^2(\partial\Omega, d\sigma)$. Therefore, by (10.38), $f \in \mathcal{D}(-\Delta_N)$ if and only if $\gamma_1(f) \equiv \frac{\partial f}{\partial \nu} = 0$ σ-a.e. on $\partial\Omega$.

In order to derive a complete description of the domain $\mathcal{D}(-\Delta_N)$, we apply the regularity Theorem D.10(ii) for the Neumann problem to conclude that any solution $f = (-\Delta_N + I)^{-1}g$ of problem (10.36), where $g \in L^2(\Omega)$, belongs to $H^2(\Omega)$. Hence, $\mathcal{D}(-\Delta_N) \subseteq H^2(\Omega)$. Therefore, by the result of the preceding paragraph,

$$\mathcal{D}(-\Delta_N) = \left\{ f \in H^2(\Omega) : \gamma_1(f) \equiv \frac{\partial f}{\partial \nu} \upharpoonright \partial\Omega = 0 \right\}. \tag{10.39}$$

Since the embedding of $H^1(\Omega) = (\mathcal{D}(\mathfrak{t}), \|\cdot\|_{\mathfrak{t}})$ into $L^2(\Omega)$ is compact according to Theorem D.4, the operator $A_{\mathfrak{t}} = -\Delta_N$ has a purely discrete spectrum by Proposition 10.6.

Summarizing the preceding, we have proved the following theorem.

Theorem 10.20 *Let $\Omega \subseteq \mathbb{R}^d$ be an open bounded set of class C^2. Then the Neumann Laplacian $-\Delta_N$ is the positive self-adjoint operator on $L^2(\Omega)$ which acts by $(-\Delta_N)f = -\Delta f$ for f in the domain (10.39). The form Hilbert space $\mathcal{D}[-\Delta_N]$ is the Sobolev space $H^1(\Omega)$, and the form of $-\Delta_N$ is given by (10.28). The operator $-\Delta_N$ has a purely discrete spectrum contained in $[0, +\infty)$.*

The Dirichlet Laplacian was defined as the Friedrichs extension of the minimal operator $L_{\min}$. It can be shown that the Neumann Laplacian can be obtained in a similar manner as the Friedrichs extension of the operator $L^N = -\Delta$ with domain

$$\mathcal{D}(L^N) = \left\{ f \in C^\infty(\overline{\Omega}) : \frac{\partial f}{\partial \nu} \upharpoonright \partial\Omega = 0 \right\},$$

where $C^\infty(\overline{\Omega})$ denotes the set of $f \in C^\infty(\Omega)$ all of whose partial derivatives can be extended continuously to the closure $\overline{\Omega}$ of Ω.

Indeed, for $f, g \in \mathcal{D}(L^N)$, the boundary integral in the Green formula (D.5) vanishes, and we obtain $\langle L^N f, g \rangle = \mathfrak{t}[f, g]$. Hence, L^N is a densely defined positive symmetric operator whose form $\mathfrak{s}_{L^N}$ (defined by Lemma 10.16) is the restriction to $\mathcal{D}(L^N)$ of $\mathfrak{t}$. It can be shown that $\mathcal{D}(L^N)$ is dense in the Sobolev space $H^1(\Omega)$. Taking this for granted, it follows that $\mathcal{D}(L^N)$ is a core for the form $\mathfrak{t}$. Therefore, the Friedrichs extension $(L^N)_F$ coincides with the operator $A_{\mathfrak{t}} = -\Delta_N$.

The one-dimensional Examples 10.6 and 10.7 suggest that other types of boundary conditions can be treated in this manner. For instance, there are *mixed boundary conditions* (that is, a Dirichlet condition on a part Γ_0 of the boundary and a Neumann condition on the rest) and *Robin conditions* (that is, $af + \frac{\partial f}{\partial \nu} = 0$ on $\partial\Omega$ for some function $a \in L^\infty(\partial\Omega, d\sigma)$). In the first case one takes the form $\mathfrak{t}$ defined by (10.28) but on the domain $\{u \in H^1(\Omega) : \gamma_0(u) = 0$ on $\Gamma_0\}$, while in the second case the integral $\int_{\partial\Omega} au\overline{v}\, d\sigma$ has to be added in the definition (10.28) of $\mathfrak{t}$. For details, we refer to the literature on elliptic differential operators.

10.7 Perturbations of Forms and Form Sums

In this section we develop the form counterpart of the Kato–Rellich perturbation Theorem 8.5, and we use it to define form sums of self-adjoint operators.

We begin with some definitions on relatively bounded forms.

Definition 10.7 Let $\mathfrak{t}$ be a lower semibounded symmetric form on a Hilbert space $\mathcal{H}$. A form $\mathfrak{s}$ on $\mathcal{H}$ is called *relatively $\mathfrak{t}$-bounded* if $\mathcal{D}(\mathfrak{t}) \subseteq \mathcal{D}(\mathfrak{s})$ and there exist nonnegative numbers a, b such that

$$\left|\mathfrak{s}[x]\right| \leq a\left|\mathfrak{t}[x]\right| + b\|x\|^2 \quad \text{for all } x \in \mathcal{D}(\mathfrak{t}). \tag{10.40}$$

The $\mathfrak{t}$-*bound* $\beta_{\mathfrak{t}}(\mathfrak{s})$ is then the infimum over all $a \geq 0$ for which there exists a number $b \geq 0$ such that (10.40) is satisfied.

Let A and B be self-adjoint operators on $\mathcal{H}$ such that A is lower semibounded. We say that B is *relatively A-form bounded* if the form $\mathfrak{t}_B$ is relatively $\mathfrak{t}_A$-bounded. Then the number $\beta_{\mathfrak{t}_A}(\mathfrak{t}_B)$ is called the A-*form bound* of B and denoted by $\beta_A(B)$.

It is easily checked that $\mathfrak{s}$ is relatively $\mathfrak{t}$-bounded if and only if $\mathfrak{s}$ is relatively $(\mathfrak{t} + c)$-bounded for any real number c and that $\beta_{\mathfrak{t}+c}(\mathfrak{s}) = \beta_{\mathfrak{t}}(\mathfrak{s})$. Hence, we can assume without loss of generality that $\mathfrak{t}$ and A are positive.

The next theorem is usually called the *KLMN theorem* named after T. Kato, P. Lax, J. Lions, A. Milgram, and E. Nelson.

Theorem 10.21 (*KLMN theorem*) *Let A be a positive self-adjoint operator on a Hilbert space $\mathcal{H}$. Suppose that $\mathfrak{s}$ is a relatively $\mathfrak{t}_A$-bounded symmetric form on $\mathcal{H}$ with $\mathfrak{t}_A$-bound $\beta_{\mathfrak{t}_A}(\mathfrak{s}) < 1$, that is, $\mathcal{D}[A] \subseteq \mathcal{D}(\mathfrak{s})$, and there are nonnegative real numbers $a < 1$ and b such that*

$$\left|\mathfrak{s}[x]\right| \leq aA[x] + b\|x\|^2 \quad \text{for } x \in \mathcal{D}[A]. \tag{10.41}$$

Then there exists a unique self-adjoint operator C, called the form sum *of A and $\mathfrak{s}$ and denoted by $A \dot{+} \mathfrak{s}$, such that*

$$\mathcal{D}[A] = \mathcal{D}[C] \quad \text{and} \quad C[x, y] = A[x, y] + \mathfrak{s}[x, y] \quad \text{for } x, y \in \mathcal{D}[C].$$

The operator C has a lower bound $m_C \geq (1 - a)m_A - b$. Any core for the form $\mathfrak{t}_A$, in particular, any core for the operator A, is a core for the form $\mathfrak{t}_C$.

Proof We define the form $\tilde{\mathfrak{t}}$ by $\tilde{\mathfrak{t}}[x, y] = A[x, y] + \mathfrak{s}[x, y]$ for $x, y \in \mathcal{D}(\tilde{\mathfrak{t}}) := \mathcal{D}[A]$. Let $x \in \mathcal{D}(\tilde{\mathfrak{t}})$. Since $A - m_A I \geq 0$, $(A - m_A I)[x] \geq 0$. Using (10.41), we derive

$$\tilde{\mathfrak{t}}[x] = A[x] + \mathfrak{s}[x] \geq A[x] - a\,A[x] - b\|x\|^2$$

$$= (1 - a)\,(A - m_A I)[x] + \big((1 - a)m_A - b\big)\|x\|^2 \geq \big((1 - a)m_A - b\big)\|x\|^2.$$

Hence, $\tilde{\mathfrak{t}}$ is a lower semibounded symmetric form.

Using assumption (10.41) twice, we obtain for $x \in \mathcal{D}(\tilde{t})$,

$$(1 - a)A[x] = A[x] - \left(aA[x] + b\|x\|^2\right) + b\|x\|^2$$
$$\leq A[x] + \mathfrak{s}[x] + b\|x\|^2 = \tilde{t}[x] + b\|x\|^2 \leq (1 + a)A[x] + 2b\|x\|^2.$$

These inequalities show that the form norms $\|\cdot\|_{\tilde{t}}$ and $\|\cdot\|_{t_A}$ defined by (10.4) are equivalent. Hence, $\tilde{t}$ is closed, because t_A is closed. By the form representation Theorem 10.7, there is a unique self-adjoint operator C such that $\tilde{t} = t_C$. Since the form norms of $\tilde{t}$ and t_A are equivalent, both forms have the same cores.

The number $(1 - a)m_A - b$ is a lower bound of $\tilde{t} = t_C$ by the first inequality proved above and hence of C by Proposition 10.4(iii).   □

Proposition 10.22 *Let A and B be lower semibounded self-adjoint operators such that $\mathcal{D}[A] \cap \mathcal{D}[B]$ is dense in the underlying Hilbert space $\mathcal{H}$.*

(i) *There is a unique self-adjoint operator C, called the* form sum *of A and B and denoted by $A \dotplus B$, such that $\mathcal{D}[C] = \mathcal{D}[A] \cap \mathcal{D}[B]$ and $t_C = t_A + t_B$. The operator C is also lower semibounded, and $m_C \geq \min{(m_A, m_B)}$.*

(ii) *If $x \in \mathcal{D}(A) \cap \mathcal{D}(B)$, then $x \in \mathcal{D}(C) = \mathcal{D}(A \dotplus B)$ and $(A + B)x = (A \dotplus B)x = Cx$. That is, $A + B \subseteq A \dotplus B$.*

Proof (i): Since the lower semibounded form $t_A + t_B$ is densely defined by assumption and closed by Corollary 10.2, the assertion follows at once from Theorem 10.7.

(ii): For $x \in \mathcal{D}(A) \cap \mathcal{D}(B)$ and $y \in \mathcal{D}[C]$, we have

$$C[x, y] = A[x, y] + B[x, y] = \langle Ax, y \rangle + \langle Bx, y \rangle = \langle(A + B)x, y\rangle.$$

Therefore, by Proposition 10.5(i), $x \in \mathcal{D}(C)$ and $Cx = (A + B)x$.   □

The next proposition contains an application of form sums to Schrödinger operators.

Proposition 10.23 *Let $\mathcal{F}$ be a finite subset of $\mathbb{R}^d$, and V a nonnegative Borel function on $\mathbb{R}^d$ such that $V \in L^1(K)$ for each compact set $K \subseteq \mathbb{R}^d$ satisfying $K \cap \mathcal{F} = \emptyset$. Then the form sum $(-\Delta) \dotplus V(x)$ is a positive self-adjoint operator on $L^2(\mathbb{R}^d)$.*

Proof Obviously, $\mathcal{D} := C_0^\infty(\mathbb{R}^d \backslash \mathcal{F})$ is dense in $L^2(\mathbb{R}^d)$. Let $f \in \mathcal{D}$. Then supp f is compact, and (supp f) $\cap \mathcal{F} = \emptyset$, so by the assumption we have

$$\int_{\text{supp} f} V(x)\big|f(x)\big|^2 dx = \int_{\mathbb{R}^d} \big|V^{1/2}(x)f(x)\big|^2 dx < \infty,$$

that is, $f \in \mathcal{D}(V^{1/2}) = \mathcal{D}[V]$. Hence, $\mathcal{D}[V] \cap \mathcal{D}[-\Delta]$ contains the dense subset $\mathcal{D}$ of $L^2(\mathbb{R}^d)$, so Proposition 10.22 applies.   □

We illustrate the preceding proposition by two examples.

Example 10.9 Let $\alpha_1, \ldots, \alpha_k \in \mathbb{R}$, $s_1, \ldots, s_k \in \mathbb{R}^d$, and let $f_1, \ldots, f_k$ be nonnegative functions from $L^\infty(\mathbb{R}^d)$. Then the potential

$$V(x) = \frac{f_1(x)}{\|x - s_1\|^{\alpha_1}} + \cdots + \frac{f_k(x)}{\|x - s_k\|^{\alpha_k}}$$

satisfies the assumptions of Proposition 10.23. Hence, the positive self-adjoint operator $(-\Delta) \dotplus V(x)$ on $L^2(\mathbb{R}^d)$ exists. ○

Example 10.10 (*Form sum versus operator sum*) Let $\omega \in C_0^\infty(\mathbb{R})$ be a nonnegative function such that $\operatorname{supp}\omega \subseteq [-2, 2]$ and $\omega(x) = 1$ for $x \in [-1, 1]$. Let $\{r_n : n \in \mathbb{N}\}$ be an enumeration of the rational numbers, and $1/2 \le \alpha < 1$. We define $V(x) = 0$ if x is rational and

$$V(x) = \sum_{n=1}^\infty 2^{-n} \frac{\omega(x - r_n)}{|x - r_n|^\alpha}$$

if x is irrational. Since the integral of $|\omega(x - r_n)||x - r_n|^{-\alpha}$ over $\mathbb{R}$ is finite and does not depend on n, Fatou's Lemma B.3 implies that $V \in L^1(\mathbb{R})$. Thus, by Proposition 10.23, the *form sum* $(-\frac{d^2}{dx^2}) \dotplus V(x)$ is a positive self-adjoint operator on $L^2(\mathbb{R})$.

But the domain of the *operator sum* $(-\frac{d^2}{dx^2}) + V(x)$ consists only of the zero vector! Indeed, assume to the contrary that there is an $f \ne 0$ in $\mathcal{D}(-\frac{d^2}{dx^2}) \cap \mathcal{D}(V)$. Since $\mathcal{D}(-\frac{d^2}{dx^2}) \subseteq C(\mathbb{R})$, there exist numbers $n \in \mathbb{N}$, $1 > \varepsilon > 0$, and $\delta > 0$ such that $|f(x)| \ge \delta$ on $\mathcal{I} := (r_n - \varepsilon, r_n + \varepsilon)$. Since $2\alpha \ge 1$, we have

$$\int_{\mathbb{R}} |V(x)f(x)|^2 \, dx \ge 2^{-2n} \int_{\mathcal{I}} \delta^2 |x - r_n|^{-2\alpha} \, dx = +\infty.$$

Hence, f is not in the domain $\mathcal{D}(V)$, which is a contradiction. ○

A large class of potentials on $\mathbb{R}^3$ for which the form sum $-\Delta \dotplus V$ exists by the KLMN theorem is the *Rollnik class* $\mathfrak{R}$, that is, the set of real Borel functions V on $\mathbb{R}^3$ satisfying

$$\|V\|_{\mathfrak{R}} := \int_{\mathbb{R}^6} \frac{|V(x)V(y)|}{\|x - y\|^2} \, dx \, dy < \infty. \tag{10.42}$$

It can be shown (see [RS2]) that if $V \in \mathfrak{R} + L^\infty(\mathbb{R}^3)$, then V is relatively $(-\Delta)$-form bounded with $(-\Delta)$-form bound zero. Hence, the form sum $-\Delta \dotplus V$ exists by the KLMN Theorem 10.21. Further, we have $L^{3/2}(\mathbb{R}^3) \subseteq \mathfrak{R}$.

For instance, if $\gamma \in \mathbb{R}$ and $\alpha < 2$, then $V(x) := \gamma \|x\|^{-\alpha} \in L^{3/2}(\mathbb{R}^3) \subseteq \mathfrak{R}$, and hence the form sum $-\Delta \dotplus V(x)$ exists. Recall that the operator sum $-\Delta + V(x)$ is a self-adjoint operator on $L^2(\mathbb{R}^3)$ by the Kato–Rellich Theorem 8.5 when $\alpha < \frac{3}{2}$. That is, the KLMN theorem allows one to treat "stronger singularities" than the Kato–Rellich theorem.

All operators A_t in Example 10.7 are in fact form sums of the operator $-\frac{d^2}{dx^2}$ and some forms of point evaluations. Here we reconsider only the simplest case.

Example 10.11 (*Form sum* $-\frac{d^2}{dx^2} \dotplus \alpha\delta$) Let $T = -i\frac{d}{dx}$ and $A := T^2 = -\frac{d^2}{dx^2}$ be the self-adjoint operators on $L^2(\mathbb{R})$ from Example 1.7. Then $\mathfrak{t}[f, g] = \langle Tf, Tg \rangle$ for $f, g \in \mathcal{D}(\mathfrak{t}) := \mathcal{D}(T)$ defines a positive closed form, and we have $A_\mathfrak{t} = A$ by Example 10.5. For $\alpha \in \mathbb{R}$, we define the form

$$\mathfrak{s}_\alpha[f, g] = \alpha f(0)\overline{g(0)} \quad \text{for } f, g \in \mathcal{D}(\mathfrak{s}_\alpha) := \mathcal{D}(T) = H^1(\mathbb{R}).$$

From Lemma 1.12 it follows that the form $\mathfrak{s}_\alpha$ is relatively $\mathfrak{t}$-bounded with $\mathfrak{t}$-bound zero. Therefore, by Theorem 10.21 the self-adjoint operator $C_\alpha := A \dotplus \mathfrak{s}_\alpha$ exists. We write symbolically $A \dotplus \alpha\delta$ for this operator C_α. By the definition of C_α we have

$$C_\alpha[f, g] = A[f, g] + \mathfrak{s}_\alpha[f, g] = \langle -if', -ig' \rangle + \alpha f(0)\overline{g(0)}.$$

The latter form is the special case of Example 10.7 with $a = 0$ and $\beta = \gamma = 0$, so the statement therein describes the operator $A \dotplus \alpha\delta$ explicitly in terms of boundary conditions. That is, we have $(A \dotplus \alpha\delta)f = -f''$ for f in the domain

$$\mathcal{D}(A \dotplus \alpha\delta) = \left\{ f \in H^2(-\infty, 0) \oplus H^2(0, +\infty) : \right.$$
$$\left. f(+0) = f(-0), f'(+0) - f'(-0) = \alpha f(0) \right\}. \qquad \circ$$

We close this section by developing an operator-theoretic example in detail.

Example 10.12 (*Rank one form perturbations of self-adjoint operators*) Throughout this example we assume that A is a self-adjoint operator on $\mathcal{H}$ such that $A \geq I$. Then $B := A^{-1}$ is a bounded positive self-adjoint operator.

Let F be a linear functional on $\mathcal{D}[A] = \mathcal{D}(A^{1/2})$. We define the form $\mathfrak{s}_F$ by

$$\mathfrak{s}_F[x, y] := F(x)\overline{F(y)} \quad \text{for } x, y \in \mathcal{D}(\mathfrak{s}_F) := \mathcal{D}[A].$$

Statement 1 $\mathfrak{s}_F$ *is relatively* $\mathfrak{t}_A$-*bounded if and only if there is a vector* $u \in \mathcal{H}$ *such that* $F(x) = \langle A^{1/2}x, u \rangle$, $x \in \mathcal{D}[A]$. *If this is true, the* $\mathfrak{t}_A$-*bound of* $\mathfrak{s}_F$ *is zero.*

Proof The if assertion is easily verified. We carry out the proof of the only if part.

Assume that $\mathfrak{s}_F$ is relatively $\mathfrak{t}_A$-bounded. Then there are positive constants a, b such that (10.40) is satisfied. Since $A \geq I$, (10.40) implies that

$$|\mathfrak{s}_F[x]| = |F(x)|^2 \leq (a + b)\|A^{1/2}x\|^2 \quad \text{for } x \in \mathcal{D}[A],$$

so F is a continuous linear functional on the Hilbert space $(\mathcal{D}[A], \langle A^{1/2}\cdot, A^{1/2}\cdot \rangle)$. By the Riesz' theorem there is a $w \in \mathcal{D}[A]$ such that $F(x) = \langle A^{1/2}x, A^{1/2}w \rangle$ for all $x \in \mathcal{D}[A]$. Setting $u = A^{1/2}w$, we get $F(\cdot) = \langle A^{1/2}\cdot, u \rangle$.

Let $\varepsilon > 0$ be given. Since the domain $\mathcal{D}(A^{1/2})$ is dense in $\mathcal{H}$, we can find a vector $v \in \mathcal{D}(A^{1/2})$ such that $\|u - v\| < \sqrt{\varepsilon/2}$. Then we obtain

$$|\mathfrak{s}_F[x]| = |F(x)|^2 = |\langle A^{1/2}x, u \rangle|^2 = |\langle A^{1/2}x, u - v \rangle + \langle A^{1/2}x, A^{-1/2}A^{1/2}v \rangle|^2$$
$$\leq \left(\|A^{1/2}x\|\sqrt{\varepsilon/2} + \|x\|\|A^{1/2}v\| \right)^2 \leq \varepsilon\|A^{1/2}x\|^2 + 2\|A^{1/2}v\|^2\|x\|^2.$$

This proves that $\mathfrak{s}_F$ has the $\mathfrak{t}_A$-bound zero. $\qquad\square$

The vector u in Statement 1 is uniquely determined by F, since $A^{1/2}\mathcal{D}[A] = \mathcal{H}$.

Now suppose that $\mathfrak{s}_F$ is relatively $\mathfrak{t}_A$-bounded and $\alpha \in \mathbb{R}$. Then the $\mathfrak{t}_A$-bound of $\alpha \mathfrak{s}_F$ is zero, hence, the form sum $A \dotplus \alpha \mathfrak{s}_F$ exists by the KLMN Theorem 10.21. Let $F(\cdot) = \langle A^{1/2}\cdot, u \rangle$ be as in Statement 1 and assume in addition that $\|u\| = 1$ and

$$u \notin \mathcal{D}[A] = \mathcal{D}(A^{1/2}). \tag{10.43}$$

Let T be the restriction of A to $\mathcal{D}(T) := \mathcal{D}(A) \cap \mathcal{N}(F) = \{x \in \mathcal{D}(A) : F(x) = 0\}$, and let P denote the projection of $\mathcal{H}$ onto $\mathbb{C} \cdot B^{1/2}u$. Since $F(\cdot) = \langle A^{1/2}\cdot, u \rangle$ is obviously continuous in the graph norm of A, T is a closed symmetric operator.

Statement 2 $\mathcal{D}(T) = B(I - P)\mathcal{H}$ is dense in $\mathcal{H}$, $\mathcal{D}(T^*) = B\mathcal{H} + \mathbb{C} \cdot B^{1/2}u$, and

$$T^*\left(By + \lambda B^{1/2}u\right) = y \quad \text{for } y \in \mathcal{H}, \ \lambda \in \mathbb{C}. \tag{10.44}$$

Proof Clearly, a vector $x \in \mathcal{H}$ is in $\mathcal{D}(T)$ if and only if $x = By$ for some $y \in \mathcal{H}$ and $F(x) = \langle A^{1/2}x, u \rangle = \langle y, B^{1/2}u \rangle = 0$, or equivalently, $x = B(I - P)y$. This proves that $\mathcal{D}(T) = B(I - P)\mathcal{H}$.

If $v \perp \mathcal{D}(T)$, then $\langle v, B(I - P)y \rangle = \langle Bv, (I - P)y \rangle = 0$ for all $y \in \mathcal{H}$, and hence $Bv \in \mathbb{C} \cdot B^{1/2}u$. Since $u \notin \mathcal{D}[A] = B^{1/2}\mathcal{H}$ by the assumption (10.43), the latter is only possible when $v = 0$. Therefore, $\mathcal{D}(T)$ is dense in $\mathcal{H}$.

A vector $x \in \mathcal{H}$ is in $\mathcal{D}(T^*)$ if and only if there is a $y(= T^*x) \in \mathcal{H}$ such that

$$\langle TB(I - P)v, x \rangle \equiv \langle (I - P)v, x \rangle = \langle B(I - P)v, y \rangle$$

for all $v \in \mathcal{H}$, or equivalently, $x - By \in P\mathcal{H}$, that is, $x \in B\mathcal{H} + \mathbb{C} \cdot B^{1/2}u$. Then $y = T^*x = T^*(By + \lambda B^{1/2}u)$. $\qquad\square$

Put $v_\alpha := (1 + \alpha)Bu - \alpha \langle B^{1/2}u, u \rangle B^{1/2}u$. Let C_α denote the restriction of T^* to the domain $\mathcal{D}(C_\alpha) := \mathcal{D}(T) + \mathbb{C} \cdot v_\alpha = B(I - P)\mathcal{H} + \mathbb{C} \cdot v_\alpha$. By (10.44),

$$C_\alpha\left(B(I - P)y + \lambda v_\alpha\right) = (I - P)y + \lambda(1 + \alpha)u \quad \text{for } y \in \mathcal{H}, \lambda \in \mathbb{C}.$$

Statement 3 C_α is a self-adjoint operator and equal to the form sum $A \dotplus \alpha \mathfrak{s}_F$.

Proof In this proof we abbreviate $\beta := \langle B^{1/2}u, u \rangle$, $\gamma := 1 + \alpha$, and $\delta := -\alpha\beta$. Let $x = B(I - P)y + \lambda v_\alpha$ and $x' = B(I - P)y' + \lambda' v_\alpha$ be vectors of $\mathcal{D}(C_\alpha)$.

Using the preceding definitions and the fact that $(I - P)\mathcal{H} \perp B^{1/2}u$, we compute

$$\langle C_\alpha x, x' \rangle = \langle (I - P)y + \lambda\gamma u, B(I - P)y' + \lambda'(\gamma Bu + \delta B^{1/2}u) \rangle$$
$$= \langle (I - P)y + \lambda\gamma u, B((I - P)y' + \lambda'\gamma u) \rangle + \lambda\overline{\lambda'}\gamma\delta\beta = \langle x, C_\alpha x' \rangle.$$

This shows that C_α is symmetric. Clearly, the graph $\mathcal{G}(C_\alpha)$ of C_α is equal to the sum $\mathcal{G}(T) + \mathbb{C} \cdot (v_\alpha, C_\alpha v_\alpha)$. Therefore, since T is closed, C_α is closed.

Next, we note that $v_\alpha \notin \mathcal{D}(T) = B(I - P)\mathcal{H}$. Indeed, if $\alpha \neq 0$, this follows at once from (10.43). If $\alpha = 0$, then $v_\alpha = Bu \in B(I - P)\mathcal{H}$ would imply that $u \perp B^{1/2}u$, which is impossible, since $B \geq I$. Hence, $\dim \mathcal{D}(C_\alpha)/\mathcal{D}(T) = 1$. Since $\mathcal{D}(T^*) = \mathcal{D}(T) + \mathrm{Lin}\{BP\mathcal{H}, B^{1/2}u\}$ by Statement 2, $\dim \mathcal{D}(T^*)/\mathcal{D}(T) \leq 2$. From

$T \neq C_\alpha$ we get $T^* \neq C_\alpha^*$, because C_α is closed. Since $T \subseteq C_\alpha \subseteq C_\alpha^* \subseteq T^*$, the preceding facts imply that $C_\alpha = C_\alpha^*$. That is, C_α is self-adjoint.

We prove that $\mathcal{D}(C_\alpha)$ is a core for $A^{1/2}$. Since $A^{1/2} \geq I$, it suffices to show that the orthogonal complement of $\mathcal{D}(C_\alpha)$ in the Hilbert space $(\mathcal{D}[A], \langle A^{1/2}\cdot, A^{1/2}\cdot \rangle)$ is trivial. Let $x \in D[A]$ be orthogonal to $\mathcal{D}(C_\alpha)$. Then $\langle A^{1/2}x, A^{1/2}B(I - P)y \rangle = 0$ for all $y \in \mathcal{H}$. This implies that $x = \lambda B^{1/2}u$ for some $\lambda \in \mathbb{C}$. Further, we have $0 = \langle A^{1/2}x, A^{1/2}v_\alpha \rangle = \lambda \langle u, \gamma B^{1/2}u + \delta u \rangle = \lambda(\gamma\beta + \delta) = \lambda\beta$, so $\lambda = 0$ and $x = 0$.

Now we compute

$$
\begin{aligned}
A[x, x'] &= \langle A^{1/2}x, A^{1/2}x' \rangle \\
&= \langle B^{1/2}(I - P)y + \lambda(\gamma B^{1/2}u + \delta u), B^{1/2}(I - P)y' + \lambda'(\gamma B^{1/2}u + \delta u) \rangle \\
&= \langle (I - P)y + \lambda\gamma u, B((I - P)y' + \lambda'\gamma u) \rangle + \lambda\overline{\lambda'}(2\gamma\delta\beta + \delta^2), \\
F(x) &= \langle A^{1/2}x, u \rangle = \langle A^{1/2}(B(I - P)y + \lambda v_\alpha), u \rangle \\
&= \langle (I - P)y, B^{1/2}u \rangle + \lambda \langle \gamma B^{1/2}u + \delta u, u \rangle = 0 + \lambda(\gamma\beta + \delta) = \lambda\beta,
\end{aligned}
$$

and similarly $F(x') = \lambda'\beta$, so that

$$
\alpha \mathfrak{s}_F[x, x'] = \alpha F(x) \overline{F(x')} = \lambda\overline{\lambda'}\alpha\beta^2.
$$

Since $\alpha\beta^2 + 2\gamma\delta\beta + \delta^2 = \gamma\delta\beta + \alpha\beta^2 - (1 + \alpha)\alpha\beta^2 + (\alpha\beta)^2 = \gamma\delta\beta$, from the preceding formulas for $\langle C_\alpha x, x' \rangle$, $A[x, x']$, and $\alpha\mathfrak{s}_F[x, x']$ we conclude that

$$
(\mathfrak{t}_A + \alpha\mathfrak{s}_F)[x, x'] = A[x, x'] + \alpha\mathfrak{s}_F[x, x'] = \langle C_\alpha x, x' \rangle, \quad x, x' \in \mathcal{D}(C_\alpha). \tag{10.45}
$$

As shown above, $\mathcal{D}(C_\alpha)$ is a core for $A^{1/2}$, hence it is a core for the form $\mathfrak{t}_A$. Since $\alpha\mathfrak{s}_F$ is relatively $\mathfrak{t}_A$-bounded, $\mathcal{D}(C_\alpha)$ is also a core for the form $\mathfrak{t}_A + \alpha\mathfrak{s}_F$. Therefore, it follows from (10.45) and Proposition 10.5(v) that $C_\alpha \subseteq A \dotplus \alpha\mathfrak{s}_F$. But both operators C_α and $A \dotplus \alpha\mathfrak{s}_F$ are self-adjoint. Hence, $C_\alpha = A \dotplus \alpha\mathfrak{s}_F$.   □ ∘

10.8 Exercises

1. Let A be the multiplication operator by a Borel function φ on an interval $\mathcal{J}$ on $\mathcal{H} = L^2(\mathcal{J})$, that is, $Af = f \cdot \varphi$ for $f \in \mathcal{D}(A) := \{f \in L^2(\mathcal{J}) : f \cdot \varphi \in L^2(\mathcal{J})\}$.
   a. Describe $\mathcal{D}[A]$ and show that $\mathcal{D}[A] = \mathcal{D}(A)$ if and only if $\varphi \in L^\infty(\mathcal{J})$.
   b. In the cases $\varphi(x) = x$, $\mathcal{J} = \mathbb{R}$ and $\varphi(x) = \log x$, $\mathcal{J} = (0, 2)$, find elements of $\mathcal{D}[A]$ which are not in $\mathcal{D}(A)$.

2. Let A be a self-adjoint operator on $\mathcal{H}$. Let A_+ and A_- be its strictly positive and strictly negative parts (see Example 7.1), respectively. Prove that

$$
A[x, y] = \langle A_+^{1/2}x, A_+^{1/2}y \rangle - \langle (-A_-)^{1/2}x, (-A_-)^{1/2}y \rangle \quad \text{for } x, y \in \mathcal{D}(\mathfrak{t}_A).
$$

3. Describe the domains of the operators $A_{\mathfrak{t}_1}$ and $A_{\mathfrak{t}_2}$ associated with the forms $\mathfrak{t}_1$ and $\mathfrak{t}_2$ in Example 10.1.

4. Let $\mathfrak{t}[f, g] = \langle -if', -ig' \rangle + \alpha f(1)\overline{g(1)}$. Describe the operator $A_\mathfrak{t}$ in terms of boundary conditions in the following six cases:

    a. $\mathcal{D}(t) = H^1(0, \infty)$,
    b. $\mathcal{D}(t) = H_0^1(0, \infty)$,
    c. $\mathcal{D}(t) = H^1(0, 2)$,
    d. $\mathcal{D}(t) = H_0^1(0, 2)$,
    e. $\mathcal{D}(t) = H^1(1, 2)$,
    f. $\mathcal{D}(t) = H_0^1(1, 2)$.

5. Prove relation (10.27) in Example 10.8.

6. a. Let A and B be symmetric operators. Show that $A \succeq B$ and $B \succeq A$ imply that $A = B$.

   b. Let A and B be lower semibounded self-adjoint operators. Show that the relations $A \geq B$ and $B \geq A$ imply that $A = B$.

7. Let A and B be positive self-adjoint operators on $\mathcal{H}$.

  *a. Show $\mathcal{D}(A) \subseteq \mathcal{D}(B)$ implies that $\mathcal{D}[A] = \mathcal{D}(A^{1/2}) \subseteq \mathcal{D}(B^{1/2}) = \mathcal{D}[B]$.
     Sketch of proof. Use Lemma 8.4 to prove that there exists a $c > 0$ such that $\|Bx\| \leq c\|(I + A)x\|$ for $x \in \mathcal{D}(A)$. Then $B^2 \leq (c(I + A))^2$ by Lemma 10.10(i), and $B \leq c(I + A)$ by Proposition 10.14, which implies the assertion.

   b. Show that $\mathcal{D}(A) = \mathcal{D}(B)$ implies that $\mathcal{D}[A] = \mathcal{D}[B]$.

   c. Suppose that $\mathcal{N}(B) = \{0\}$ and $A \geq B$. Prove that $\log A \geq \log B$. (That is, the function $f(x) = \log x$ is operator monotone on the interval $(0, +\infty)$.) Hint: Use Proposition 10.14 and Exercise 5.9.b.

8. Let A and B be positive self-adjoint operators on $\mathcal{H}$. Prove that $A \leq B$ if and only if there is an operator $C \in \mathbf{B}(\mathcal{H})$ such that $\|C\| \leq 1$ and $CB^{1/2} \subseteq A^{1/2}$. Show that the operator C can be chosen such that $\mathcal{N}(B) \subseteq \mathcal{N}(C)$ and that C is then uniquely determined by these properties.

9. Find positive self-adjoint operators A and B on some Hilbert space $\mathcal{H}$ such that $\mathcal{D}[A] \cap \mathcal{D}[B]$ is dense in $\mathcal{H}$, but $\mathcal{D}(A) \cap \mathcal{D}(B) = \{0\}$. Hint: See Example 10.10.

10. Define operators $T = -\frac{d^2}{dx^2}$ and $S = -\frac{d^2}{dx^2}$ on $L^2(a, b)$, $a, b \in \mathbb{R}$, by

$$\mathcal{D}(T) = \left\{ f \in H^2(a, b) : f'(a) = f'(b) = 0, \ f(a) = f(b) \right\},$$
$$\mathcal{D}(S) = \left\{ f \in H^2(a, b) : f(a) = f'(b) = 0, \ f'(a) = f(b) \right\}.$$

   a. Show that T and S are densely defined closed positive symmetric operators.
   b. Show that $\mathcal{D}(T^*) = \{f \in H^2(a, b) : f'(a) = f'(b)\}$.
   c. Show that $\mathcal{D}(S^*) = \{f \in H^2(a, b) : f(a) = f'(b)\}$.
   d. Show that $\mathcal{D}(T_F) = \{f \in H^2(a, b) : f(a) = f(b), f'(a) = f'(b)\}$.
   e. Show that $\mathcal{D}(S_F) = \{f \in H^2(a, b) : f(a) = f'(b) = 0\}$.

11. Let $\mathcal{L} = -\Delta$, $\Omega_0 := \{(x, y) \in \mathbb{R}^2 : x^2 + y^2 < 1/4\}$ and $\Omega := \Omega_0 \setminus \{(0, 0)\}$.

   a. Prove that $f(x, y) := \log \sqrt{x^2 + y^2}$ is in $\mathcal{D}(L_{\max})$ on Ω, but $f \notin H^1(\Omega)$. Hint: f is harmonic on Ω. Verify that $f \in L^2(\Omega)$ and $f_x \notin L^2(\Omega)$.

   b. Prove that f is not in $\mathcal{D}(L_{\max})$ on Ω_0. Hint: $(-\Delta)f$ is a multiple of the delta distribution δ_0 on Ω_0.

   c. Prove that $g(x, y) := \log|\log\sqrt{x^2 + y^2}|$ is in $H^1(\Omega_0)$ and in $H^1(\Omega)$.
     (Note that g is neither continuous on Ω_0 nor bounded on Ω.)

*12. Let $d \in \mathbb{N}$. Prove that the linear subspace $\mathcal{D} := \{f \in \mathcal{S}(\mathbb{R}^d) : f(0) = 0\}$ is a
    core for the self-adjoint operator $-\Delta$ on $L^2(\mathbb{R}^d)$ if and only if $d \geq 4$.
    (The assertion remains valid if $\mathcal{S}(\mathbb{R}^d)$ is replaced by $C_0^\infty(\mathbb{R}^d)$, see, e.g.,
    [RS2, p. 161].)
    Sketch of proof. Since $\mathcal{S}(\mathbb{R}^d)$ is invariant under the Fourier transform,
    by Example 8.1, $\mathcal{D}$ is a core for $-\Delta$ if and only if $\mathcal{E} := \{g \in \mathcal{S}(\mathbb{R}^d) :$
    $\int g(x)\,dx = 0\}$ is a core for the multiplication operator by $\|x\|^2$ on $L^2(\mathbb{R}^d)$.
    Note that $\mathcal{E}$ is the null space of the linear functional F defined by $F(g) =$
    $\int g\,dx$. By Exercise 1.6.a, $\mathcal{E}$ is *not* a core for $\|x\|^2$ if and only if F
    is continuous in the graph norm, or equivalently, by Riesz' theorem, F
    is of the form $F(g) = \int gh(1 + \|x\|^4)\,dx$ for some function h satisfy-
    ing $\int |h|^2(1 + \|x\|^4)\,dx < \infty$. Since $F(g) = \int g\,dx$, we obtain $h(x) =$
    $(1 + \|x\|^4)^{-1}$. But $\int |h|^2(1 + \|x\|^4)\,dx < \infty$ if and only if $d \leq 3$. Therefore,
    $\mathcal{E}$ is not a core fore for $\|x\|^2$ if and only if $d \leq 3$.

Chapter 11
Sectorial Forms and m-Sectorial Operators

In this short chapter we are dealing with nonsymmetric forms. Our main aim are two representation theorems, one for bounded coercive forms on densely and continuously embedding Hilbert spaces in Sect. 11.1 and another one for densely defined closed sectorial forms in Sect. 11.2. An application to second-order elliptic differential operators is given in Sect. 11.3.

11.1 Bounded Coercive Forms on Embedded Hilbert Spaces

First, we need some more terminology. Given a form t, the *adjoint form* t^*, the *real part* $\operatorname{Re} t$, and the *imaginary part* $\operatorname{Im} t$ of t are defined by

$$t^*[x, y] = \overline{t[y, x]} \quad \text{for } x, y \in \mathcal{D}(t^*) := \mathcal{D}(t),$$
$$\operatorname{Re} t = \frac{1}{2}(t + t^*), \quad \operatorname{Im} t = \frac{1}{2i}(t - t^*). \tag{11.1}$$

Clearly, both forms $\operatorname{Re} t$ and $\operatorname{Im} t$ are symmetric, and we have $t = \operatorname{Re} t + i \operatorname{Im} t$. Since the quadratic form of a symmetric form takes only real values, $(\operatorname{Re} t)[x]$ and $(\operatorname{Im} t)[x]$ are indeed real and imaginary parts of the complex number $t[x]$.

Let $(V, \|\cdot\|_V)$ be a normed space which contains the domain $\mathcal{D}(t)$ as a linear subspace. We say that the form t is *bounded* on V if there is a constant $C > 0$ such that

$$\left|t[u, v]\right| \leq C \|u\|_V \|v\|_V \quad \text{for } u, v \in \mathcal{D}(t) \tag{11.2}$$

and that t is *coercive* on V if there exists a constant $c > 0$ such that

$$\left|t[u]\right| \geq c \|u\|_V^2 \quad \text{for } u \in \mathcal{D}(t). \tag{11.3}$$

The following simple lemma shows that it suffices to check the boundedness condition on quadratic forms only.

Lemma 11.1 *A form t is bounded on V if and only if there is constant $M > 0$ such that $|t[u]| \leq M \|u\|_V^2$ for $u \in \mathcal{D}(t)$.*

K. Schmüdgen, *Unbounded Self-adjoint Operators on Hilbert Space*,
Graduate Texts in Mathematics 265,
DOI 10.1007/978-94-007-4753-1_11, © Springer Science+Business Media Dordrecht 2012

Proof The only if direction is trivial by setting $u = v$ in (11.2). Hence, it suffices to verify the if part. If $\|u\|_V \leq 1$ and $\|v\|_V \leq 1$, then we have $\mathfrak{t}[u + \tau v] \leq 4M$ for $\tau = 1, -1, i, -i$, and hence $4|\mathfrak{t}[u, v]| \leq 16M$ by the polarization identity (10.2). This implies that $|\mathfrak{t}[u, v]| \leq 4M\|u\|_V \|v\|_V$ for arbitrary $u, v \in \mathcal{D}(\mathfrak{t})$.   □

The considerations in this section are essentially based on the following result.

Lemma 11.2 (*Lax–Milgram lemma*) *Let* $\mathfrak{t}$ *be a bounded coercive form on a Hilbert space* $(V, \langle \cdot, \cdot \rangle_V)$ *such that* $\mathcal{D}(\mathfrak{t}) = V$. *Then there exists an operator* $B \in \mathbf{B}(V)$ *with inverse* $B^{-1} \in \mathbf{B}(V)$ *satisfying*

$$\mathfrak{t}[u, v] = \langle Bu, v \rangle_V \quad \text{for } u, v \in V.$$

Moreover, if c *and* C *are positive constants such that* (11.2) *and* (11.3) *hold, then*

$$c \leq \|B\|_V \leq C \quad \text{and} \quad C^{-1} \leq \|B^{-1}\|_V \leq c^{-1}.$$

Proof Let $u \in V$. By (11.2), $F_u(\cdot) := \overline{\mathfrak{t}[u, \cdot]}$ is a bounded linear functional on V with $\|F_u\| \leq C\|u\|_V$. From the Riesz representation theorem it follows that there is a unique element $u' \in V$ such that $F_u(v) = \langle v, u' \rangle_V$, $v \in V$, and we have $\|u'\|_V \leq C\|u\|_V$. Defining $Bu := u'$, we obtain a linear operator $B \in \mathbf{B}(V)$ such that $\langle Bu, v \rangle_V = \langle u', v \rangle_V = \overline{F_u(v)} = \mathfrak{t}[u, v]$ for $u, v \in V$ and $\|B\|_V \leq C$.

Next we show that B has a bounded inverse $B^{-1} \in \mathbf{B}(V)$. Using (11.3), we have

$$\|Bu\|_V \|u\|_V \geq |\langle Bu, u \rangle_V| = |\mathfrak{t}[u]| \geq c\|u\|_V^2,$$

so that $\|Bu\|_V \geq c\|u\|_V$ for all $u \in V$. The latter implies that B is injective and $\mathcal{R}(B)$ is closed in V. If $u_0 \in V$ is orthogonal to $\mathcal{R}(B)$, then

$$0 = \langle Bu_0, u_0 \rangle_V = \mathfrak{t}[u_0, u_0] \geq c\|u_0\|_V^2,$$

and hence $u_0 = 0$. Therefore, $\mathcal{R}(B) = V$. Since $\|Bu\|_V \geq c\|u\|_V$ as noted above, $B^{-1} \in \mathbf{B}(V)$ and $\|B^{-1}\|_V \leq c^{-1}$. The two remaining norm inequalities are easily checked.   □

In the rest of this section we assume that $(V, \langle \cdot, \cdot \rangle_V)$ is a Hilbert space which is *densely* and *continuously* embedded into the Hilbert space $(\mathcal{H}, \langle \cdot, \cdot \rangle)$. This means the following: V is a dense linear subspace of $\mathcal{H}$, and the embedding $J : V \to \mathcal{H}$, $Jv = v$ for $v \in V$, is continuous, that is, there exists a number $a > 0$ such that

$$\|v\| \leq a\|v\|_V \quad \text{for all } v \in V. \tag{11.4}$$

(A typical example is a Sobolev space $V = H^n(\Omega)$ embedded into $\mathcal{H} = L^2(\Omega)$.)

If no confusion can arise, we identify each element v of V with its image Jv in $\mathcal{H}$. Then any form $\mathfrak{t}$ on V becomes a form on $\mathcal{H}$, which is also denoted by $\mathfrak{t}$. (Strictly speaking, the form on $\mathcal{H}$ is $\tilde{\mathfrak{t}}[x, y] := \mathfrak{t}[J^{-1}x, J^{-1}y]$, $x, y \in \mathcal{D}(\tilde{\mathfrak{t}}) := J(\mathcal{D}(\mathfrak{t}))$, and we should write $\mathfrak{t}[J^{-1}x, J^{-1}y] = \langle A_{\tilde{\mathfrak{t}}}x, y \rangle$ for $x, y \in \mathcal{D}(\tilde{\mathfrak{t}})$.)

The main result of this section is the following form representation theorem.

Theorem 11.3 *Let $\mathfrak{t}$ be a bounded coercive form on the Hilbert space $(V, \langle \cdot, \cdot \rangle_V)$. Then the operator $A_\mathfrak{t}$ associated with $\mathfrak{t}$, considered as a form on $\mathcal{H}$ with domain $\mathcal{D}(\mathfrak{t}) = V$, according to Definition 10.4 is a densely defined closed operator on $\mathcal{H}$ with bounded inverse $A_\mathfrak{t}^{-1} \in \mathbf{B}(\mathcal{H})$. Moreover, we have:*

(i) *$\mathcal{D}(A_\mathfrak{t})$ is a dense linear subspace of the Hilbert space $(V, \| \cdot \|_V)$.*
(ii) *$(A_\mathfrak{t})^* = A_{\mathfrak{t}^*}$.*
(iii) *If the embedding $J : V \to \mathcal{H}$ is compact, then $R_\lambda(A_\mathfrak{t})$ is compact for all $\lambda \in \rho(A_\mathfrak{t})$, and the operator $A_\mathfrak{t}$ has a purely discrete spectrum.*

Proof The scalar product $\mathfrak{s} := \langle \cdot, \cdot \rangle_V$ of V is a positive form on $\mathcal{H}$. Since the form norm of $\mathfrak{s}$ and the norm of V are equivalent by (11.4) and V is a Hilbert space, $\mathfrak{s}$ is closed. Let $A_\mathfrak{s}$ be the operator associated with $\mathfrak{s}$ according to Definition 10.4. By Theorem 10.7, $A_\mathfrak{s}$ is a positive self-adjoint operator on $\mathcal{H}$ and $\mathfrak{s} = \mathfrak{t}_{A_\mathfrak{s}}$. Hence,

$$\|y\|_V = \left\| A_\mathfrak{s}^{1/2} y \right\| \quad \text{and}$$

$$\langle x, y \rangle_V = \langle A_\mathfrak{s} x, y \rangle, \quad x \in \mathcal{D}(A_\mathfrak{s}), \ y \in V = \mathcal{D}(\mathfrak{s}) = \mathcal{D}\left(A_\mathfrak{s}^{1/2}\right).$$

From the inequality $a^{-1} \|y\| \leq \|y\|_V = \left\| A_\mathfrak{s}^{1/2} y \right\|$ for $y \in \mathcal{D}(A_\mathfrak{s}^{1/2})$ by (11.4) it follows that $0 \in \rho(A_\mathfrak{s}^{1/2})$ by Proposition 3.10. Hence, $A_\mathfrak{s}^{-1/2} \in \mathbf{B}(\mathcal{H})$.

Let B denote the operator for the form $\mathfrak{t}$ on V given by the Lax–Milgram Lemma 11.2. We will prove that $A_\mathfrak{t} = A_\mathfrak{s} B$.

By Lemma 11.2 we have $\|B^{-1}\|_V \leq c^{-1}$ and $\mathfrak{t}[u, v] = \langle Bu, v \rangle_V$ for $u, v \in V$. Setting $S := B^{-1} A_\mathfrak{s}^{-1/2}$, we derive for $x \in \mathcal{H}$,

$$\|Sx\| = \left\| B^{-1} A_\mathfrak{s}^{-1/2} x \right\| \leq a \left\| B^{-1} A_\mathfrak{s}^{-1/2} x \right\|_V \leq ac^{-1} \left\| A_\mathfrak{s}^{-1/2} x \right\|_V = ac^{-1} \|x\|,$$

so $S \in \mathbf{B}(\mathcal{H})$, and hence $T := S A_\mathfrak{s}^{-1/2} = B^{-1} A_\mathfrak{s}^{-1} \in \mathbf{B}(\mathcal{H})$. Clearly, $T(\mathcal{H}) \subseteq V$, since $\mathcal{R}(A_\mathfrak{s}^{-1}) = \mathcal{D}(A_\mathfrak{s}) \subseteq \mathcal{D}(A_\mathfrak{s}^{1/2}) = V$. Putting the preceding together, we get

$$\mathfrak{t}[Tx, y] = \mathfrak{t}\left[B^{-1} A_\mathfrak{s}^{-1} x, y \right] = \left\langle A_\mathfrak{s}^{-1} x, y \right\rangle_V = \langle x, y \rangle, \quad x \in \mathcal{H}, \ y \in V = \mathcal{D}(\mathfrak{t}). \tag{11.5}$$

If $Tx = 0$, then $\langle x, y \rangle = 0$ for all $y \in V$ by (11.5), and hence $x = 0$, because V is dense in $\mathcal{H}$ by assumption. Thus, T is invertible.

Next, we prove that $T^{-1} = A_\mathfrak{t}$. Let $u \in \mathcal{D}(T^{-1})$. Setting $x = T^{-1} u$ in (11.5), we obtain $\mathfrak{t}[u, y] = \langle T^{-1} u, y \rangle$ for $y \in \mathcal{D}(\mathfrak{t})$. By the definition of $A_\mathfrak{t}$ this implies that $u \in \mathcal{D}(A_\mathfrak{t})$ and $T^{-1} u = A_\mathfrak{t} u$. Thus, $T^{-1} \subseteq A_\mathfrak{t}$. Since $\mathcal{D}(T) = \mathcal{H}$, T^{-1} is surjective. If $A_\mathfrak{t} x = 0$, then $\mathfrak{t}[x] = \langle A_\mathfrak{t} x, x \rangle = 0$, and so $x = 0$ by (11.3). That is, $A_\mathfrak{t}$ is injective. Therefore, Lemma 1.3 applies and yields $T^{-1} = A_\mathfrak{t}$. Hence, $A_\mathfrak{t}^{-1} = T \in \mathbf{B}(\mathcal{H})$, and $A_\mathfrak{t}$ is closed. In particular, $A_\mathfrak{t} = T^{-1} = (B^{-1} A_\mathfrak{s}^{-1})^{-1} = A_\mathfrak{s} B$.

(i): Let $v \in V$ be orthogonal to $\mathcal{D}(A_\mathfrak{t}) = \mathcal{R}(T)$ in $(V, \langle \cdot, \cdot \rangle_V)$. By (11.5) we have

$$\left\langle x, \left(B^{-1} \right)^* v \right\rangle = \left\langle A_\mathfrak{s}^{-1} x, \left(B^{-1} \right)^* v \right\rangle_V = \left\langle B^{-1} A_\mathfrak{s}^{-1} x, v \right\rangle_V = \langle Tx, v \rangle_V = 0$$

for all $x \in \mathcal{H}$, where $(B^{-1})^*$ is the adjoint of $B^{-1} \in \mathbf{B}(V)$. Thus, we obtain $(B^*)^{-1} v = (B^{-1})^* v = 0$, and hence $v = 0$. This shows that $\mathcal{D}(A_\mathfrak{t})$ is dense in V. In particular, the latter implies that the domain $\mathcal{D}(A_\mathfrak{t})$ is dense in $\mathcal{H}$.

(ii): Obviously, t^* is also a bounded coercive form on V. Let $x \in \mathcal{D}(A_t)$ and $y \in \mathcal{D}(A_{t^*})$. Then $x, y \in \mathcal{D}(t) = \mathcal{D}(t^*)$, and from (11.1) we obtain

$$\langle A_t x, y \rangle = t[x, y] = \overline{t^*[y, x]} = \overline{\langle A_{t^*} y, x \rangle} = \langle x, A_{t^*} y \rangle.$$

From this equality we conclude that $A_{t^*} \subseteq (A_t)^*$. Since the inverse of the operator associated with a bounded coercive form is in $\mathbf{B}(\mathcal{H})$ as shown above, A_t and A_{t^*} are surjective. Hence, $(A_t)^*$ is injective. Therefore, $A_{t^*} = (A_t)^*$ by Lemma 1.3.

(iii): Recall that $\|A_s^{1/2} y\| = \|y\|_V$ for $y \in V = \mathcal{D}(A_s^{1/2})$. From this relation it follows easily that $J : V \to \mathcal{H}$ is compact if and only if $A_s^{-1/2}$ is compact (see, e.g., Proposition 10.6, (i) $\leftrightarrow$ (ii)), or equivalently, if $(A_s^{-1/2})^2 = A_s^{-1} = R_0(A_t)$ is compact. The other assertions of (iii) follow from Proposition 2.11. $\qquad\square$

Corollary 11.4 *Let* t_1 *and* t_2 *be bounded coercive forms on Hilbert spaces* $(V_1, \langle \cdot, \cdot \rangle_{V_1})$ *and* $(V_2, \langle \cdot, \cdot \rangle_{V_2})$ *which are densely and continuously embedded into the Hilbert space* $\mathcal{H}$. *If* $A_{t_1} = A_{t_2}$, *then* $V_1 = V_2$ (*as vector spaces*) *and* $t_1 = t_2$.

Proof Since $t_1[x, y] = \langle A_{t_1} x, y \rangle = \langle A_{t_2} x, y \rangle = t_2[x, y]$ for all $x, y \in \mathcal{D} := \mathcal{D}(A_{t_1}) = \mathcal{D}(A_{t_2})$, both forms t_1 and t_2 coincide on $\mathcal{D}$. Therefore, from

$$c_k \|u\|_{V_k}^2 \leq |t_k[u]| \leq C_k \|u\|_{V_k}^2, \quad u \in \mathcal{D}, \; k = 1, 2,$$

by (11.3) and (11.2) we infer that the norms $\| \cdot \|_{V_1}$ and $\| \cdot \|_{V_2}$ are equivalent on $\mathcal{D}$. By Theorem 11.3(i), $\mathcal{D}$ is dense in V_1 and in V_2. Hence, the Hilbert spaces V_1 and V_2 coincide as vector subspaces of $\mathcal{H}$, and they have equivalent norms. Because each form t_k is continuous on V_k by assumption, we conclude that $t_1 = t_2$. $\qquad\square$

Corollary 11.5 *Let* t *be a bounded coercive form on the Hilbert space* $(V, \langle \cdot, \cdot \rangle_V)$. *Then* A_t *is self-adjoint if and only if* t *is symmetric.*

Proof By Theorem 11.3(ii), A_t is self-adjoint if and only if $A_t = A_{t^*}$. By Corollary 11.4, $A_t = A_{t^*}$ is equivalent to $t = t^*$. $\qquad\square$

It might be instructive to look at the preceding from a slightly different perspective which is often used in the literature (see, e.g., [Tn]).

Let V^* denote the antidual of V, that is, V^* is the space of continuous conjugate linear functionals on the Hilbert space V. Since the embedding $J : \mathcal{H} \to V$ is continuous, the adjoint $J^* : V^* \to \mathcal{H}^*$ is also continuous. Since J has a dense range in $\mathcal{H}$, J^* is injective. Because J is injective, J^* has a dense range in V^*. We identify $\mathcal{H}$ with its antidual $\mathcal{H}^*$ by identifying $x \in \mathcal{H}$ with the conjugate linear functional $\langle x, \cdot \rangle$ by Riesz' theorem. Summarizing, we have a *triplet of Hilbert spaces*

$$V \to \mathcal{H} \sim \mathcal{H}^* \to V^*, \tag{11.6}$$

where J and J^* are *continuous embeddings with dense ranges*. Note that $J^*(x)$ is just the functional $\langle x, J \cdot \rangle$ on V.

Let $u \in V$. Since the form t is bounded on V, $t[u, \cdot]$ is a continuous conjugate linear functional on V and so an element $u' \in V^*$. Defining $\mathcal{A}_t(u) = u'$, we obtain a linear mapping $\mathcal{A}_t : V \to V^*$ such that $\mathcal{A}_t(u)(v) = t[u, v]$ for $u, v \in V$. Since t is bounded and coercive, $\mathcal{A}_t$ is a Hilbert space isomorphism of V and V^*. It follows at once from Definition 10.4 that the operator A_t on $\mathcal{H}$ is the restriction of $\mathcal{A}_t$ to the space of all elements $u \in V$ for which $\mathcal{A}_t(u)$ is in $\mathcal{H}$ rather than in V^*, that is,

$$\mathcal{D}(A_t) = \{u \in V : \mathcal{A}_t(u) \in \mathcal{H}\}, \qquad A_t(u) = \mathcal{A}_t(u) \quad \text{for } u \in \mathcal{D}(A_t).$$

11.2 Sectorial Forms

In this section we investigate forms t for which the *numerical range*

$$\Theta(t) := \{t[x] : x \in \mathcal{D}(t), \|x\| = 1\}$$

of t is contained in a sector with opening angle less than π.

Definition 11.1 A form t on $\mathcal{H}$ is called *sectorial* if there are numbers $c \in \mathbb{R}$ and $\theta \in [0, \pi/2)$ such that its numerical range $\Theta(t)$ lies in the sector

$$S_{c,\theta} = \{\lambda \in \mathbb{C} : |\operatorname{Im}\lambda| \leq \tan\theta(\operatorname{Re}\lambda - c)\} = \{\lambda \in \mathbb{C} : \arg(\lambda - c) \leq \theta\}. \ (11.7)$$

A sectorial form t is said to be *closed* if $\operatorname{Re} t$ is closed. It is called *closable* if there is a closed sectorial form which extends t. A *core* of a sectorial form t is a dense linear subspace of the pre-Hilbert space $(\mathcal{D}(t), \| \cdot \|_{\operatorname{Re} t})$.

Obviously, if t is sectorial, so are t^* and $t + \lambda$ for $\lambda \in \mathbb{C}$.

The next proposition provides some characterizations of sectorial forms. Recall that the closedness was defined only for lower semibounded forms, but the real part of a sectorial form t is indeed lower semibounded as shown by Proposition 11.6(ii).

Proposition 11.6 *For any form t on $\mathcal{H}$, the following assertions are equivalent:*

(i) t *is sectorial.*
(ii) $\operatorname{Re} t$ *is lower semibounded, and* $\operatorname{Im} t$ *is bounded on* $(\mathcal{D}(t), \| \cdot \|_{\operatorname{Re} t})$.
(iii) $\operatorname{Re} t$ *is lower semibounded, and t is bounded on* $(\mathcal{D}(t), \| \cdot \|_{\operatorname{Re} t})$.
(iv) *There is a real number c such that $\operatorname{Re} t \geq c$ and $t + 1 - c$ is a bounded coercive form on* $(\mathcal{D}(t), \| \cdot \|_{\operatorname{Re} t})$.

Further, if $\Theta(t) \subseteq S_{c,\theta}$, then $\operatorname{Re} t \geq c$, and the form $t + 1 - c$ is bounded and coercive on $(\mathcal{D}(t), \| \cdot \|_{\operatorname{Re} t})$.

Proof To prove the boundedness of a form, by Lemma 11.1 it suffices to check the continuity condition for the corresponding quadratic form. Further, we recall that $\operatorname{Im} t[x] = (\operatorname{Im} t)[x]$ and $\operatorname{Re} t[x] = (\operatorname{Re} t)[x]$ for $x \in \mathcal{D}(t)$.

(i) $\to$ (ii): Assume that $\Theta(t)$ is contained in a sector $S_{c,\theta}$. By (11.7) this yields $|\operatorname{Im} t[x]| \leq \tan\theta(\operatorname{Re} t[x] - c\|x\|^2)$ for all $x \in \mathcal{D}(t)$. Hence, $\operatorname{Re} t \geq c$.

Since $(\operatorname{Re} t)[x] - c\|x\|^2 \leq \|x\|_{\operatorname{Re} t}^2$ by (10.4), we get $|(\operatorname{Im} t)[x]| \leq \tan\theta \|x\|_{\operatorname{Re} t}^2$ for $x \in \mathcal{D}(t)$, that is, $\operatorname{Im} t$ is bounded on $(\mathcal{D}(t), \| \cdot \|_{\operatorname{Re} t})$.

(ii) $\rightarrow$ (iii): Applying (10.5) to the lower semibounded (!) form $\operatorname{Re} \mathfrak{t}$, we conclude that $\operatorname{Re} \mathfrak{t}$ is always bounded on $(\mathcal{D}(\mathfrak{t}), \|\cdot\|_{\operatorname{Re} \mathfrak{t}})$. Hence, $\mathfrak{t}$ is bounded by (ii).

(iii) $\rightarrow$ (iv): By (iii) there exists $c \in \mathbb{R}$ such that $\operatorname{Re} \mathfrak{t} \geq c$. Hence, by (10.4),

$$\|x\|_{\operatorname{Re} \mathfrak{t}}^2 = (\operatorname{Re} \mathfrak{t} + 1 - c)[x] = \operatorname{Re}\big((\mathfrak{t} + 1 - c)[x]\big) \leq \big|(\mathfrak{t} + 1 - c)[x]\big|, \quad x \in \mathcal{D}(\mathfrak{t}),$$

whence $\mathfrak{t} + 1 - c$ is coercive on $(\mathcal{D}(\mathfrak{t}), \|\cdot\|_{\operatorname{Re} \mathfrak{t}})$. By (iii), $\mathfrak{t}$ is bounded, and so is $\mathfrak{t} + 1 - c$.

(iv) $\rightarrow$ (i): By (iv) we choose a $c \in \mathbb{R}$ such that $\operatorname{Re} \mathfrak{t} \geq c$ and $\mathfrak{t} + 1 - c$ is bounded. Hence, $\operatorname{Im} \mathfrak{t}$ is also bounded, that is, there exists a constant $C > 0$ such that

$$\big|\operatorname{Im} \mathfrak{t}[x]\big| \leq C\|x\|_{\operatorname{Re} \mathfrak{t}}^2 = C(\operatorname{Re} \mathfrak{t} + 1 - c)[x] = C\big(\operatorname{Re} \mathfrak{t}[x] + (1 - c)\|x\|^2\big)$$

for all $x \in \mathcal{D}(\mathfrak{t})$. Choosing an angle $\theta \in [0, \pi/2)$ such that $\tan \theta = C$, this means that $\Theta(\mathfrak{t}) \subseteq S_{c-1,\theta}$.

The last assertion was shown in the proofs of (i) $\rightarrow$ (ii) and (iii) $\rightarrow$ (iv). $\qquad\square$

The next proposition is the counterpart to Lemma 10.16.

Proposition 11.7 *Let T be a (not necessary densely defined) sectorial operator. Then the form $\mathfrak{s}_T$ defined by $\mathfrak{s}_T[x, y] = \langle Tx, y\rangle, x, y \in \mathcal{D}(\mathfrak{s}_T) := \mathcal{D}(T)$, is closable. There is a unique closed sectorial extension $\overline{\mathfrak{s}_T}$ of $\mathfrak{s}_T$, and $\Theta(T)$ is dense in $\Theta(\overline{\mathfrak{s}_T})$.*

Proof Let us abbreviate $\mathfrak{s} = \mathfrak{s}_T$. Since obviously $\Theta(T) = \Theta(\mathfrak{s})$, $\mathfrak{s}$ is sectorial, so $\operatorname{Re} \mathfrak{s}$ is lower semibounded. By adding some constant we can assume without loss of generality that $\operatorname{Re} \mathfrak{s} \geq 1$. Then $\|\cdot\|_{\operatorname{Re} \mathfrak{s}}^2 = \operatorname{Re} \mathfrak{s}[\cdot]$. By Proposition 11.6(iii), there exists a constant $C > 0$ such that $|\mathfrak{s}[x, y]| \leq C\|x\|_{\operatorname{Re} \mathfrak{s}}\|y\|_{\operatorname{Re} \mathfrak{s}}$ for all $x, y \in \mathcal{D}(\mathfrak{s})$.

We prove that $\operatorname{Re} \mathfrak{s}$ is closable and apply Proposition 10.3(iii). Let $x \in \mathcal{N}(\mathcal{I}_{\operatorname{Re} \mathfrak{s}})$. Then there is a sequence $(x_n)_{n \in \mathbb{N}}$ from $\mathcal{D}(\mathfrak{s})$ such that $\lim_n x_n = x$ in $\mathcal{H}_{\operatorname{Re} \mathfrak{s}}$ and $\lim_n x_n = 0$ in $\mathcal{H}$. Since the converging sequence (x_n) in $\mathcal{H}_{\operatorname{Re} \mathfrak{s}}$ is bounded, we have $K := \sup_n \|x_n\|_{\operatorname{Re} \mathfrak{s}} < \infty$. For $n, k \in \mathbb{N}$, from the preceding we obtain

$$\begin{aligned}
\|x_n\|_{\operatorname{Re} \mathfrak{s}}^2 = \operatorname{Re} \mathfrak{s}[x_n] &\leq \big|\mathfrak{s}[x_n, x_n]\big| \leq \big|\mathfrak{s}[x_n, x_n - x_k]\big| + \big|\mathfrak{s}[x_n, x_k]\big| \\
&\leq C\|x_n\|_{\operatorname{Re} \mathfrak{s}}\|x_n - x_k\|_{\operatorname{Re} \mathfrak{s}} + \big|\langle Tx_n, x_k\rangle\big| \\
&\leq CK\|x_n - x_k\|_{\operatorname{Re} \mathfrak{s}} + \big|\langle Tx_n, x_k\rangle\big|.
\end{aligned} \tag{11.8}$$

Given $\varepsilon > 0$, there exists an N_ε such that $CK\|x_n - x_k\|_{\operatorname{Re} \mathfrak{s}} \leq \varepsilon$ for $n, k \geq N_\varepsilon$. Fix $n \geq N_\varepsilon$. Letting $k \rightarrow \infty$ in (11.8) and using that $\lim_k x_k = 0$ in $\mathcal{H}$, we get $\|x_n\|_{\operatorname{Re} \mathfrak{s}}^2 \leq \varepsilon$ for $n \geq N_\varepsilon$. This proves that $\lim_n \|x_n\|_{\operatorname{Re} \mathfrak{s}} = 0$. Hence, $x = 0$, since $\lim_n x_n = x$ in $\mathcal{H}_{\operatorname{Re} \mathfrak{s}}$. Thus, $\mathcal{I}_{\operatorname{Re} \mathfrak{s}}$ is injective, and $\operatorname{Re} \mathfrak{s}$ is closable by Proposition 10.3.

Let $\mathfrak{h}$ be the closure of $\operatorname{Re} \mathfrak{s}$. Since $\operatorname{Im} \mathfrak{s}$ is continuous on $(\mathcal{D}(\mathfrak{s}), \|\cdot\|_{\operatorname{Re} \mathfrak{s}})$, it extends by continuity to a symmetric form $\mathfrak{k}$ on $\mathcal{D}(\mathfrak{h})$. Clearly, $\overline{\mathfrak{s}} := \mathfrak{h} + i\mathfrak{k}$ is the unique closed sectorial extension of $\mathfrak{s}$, and $\Theta(T) = \Theta(\mathfrak{s})$ is dense in $\Theta(\overline{\mathfrak{s}})$. $\qquad\square$

Theorem 11.8 (*Representation theorem for sectorial forms*) *Suppose that $\mathfrak{t}$ is a densely defined closed sectorial form on $\mathcal{H}$. Then the operator $A_\mathfrak{t}$ is m-sectorial, and we have:*

(i) $\mathcal{D}(A_t)$ *is a dense linear subspace of the Hilbert space* $(\mathcal{D}(t), \|\cdot\|_{\mathrm{Re}\,t})$.

(ii) $(A_t)^* = A_{t^*}$.

(iii) *If the embedding of* $(\mathcal{D}(t), \|\cdot\|_{\mathrm{Re}\,t})$ *into* $\mathcal{H}$ *is compact, then the resolvent* $R_\lambda(A_t)$ *is compact for all* $\lambda \in \rho(A_t)$, *and* A_t *has a purely discrete spectrum.*

(iv) *If* $\Theta(t)$ *is contained in a sector* $S_{c,\theta}$ *for* $c \in \mathbb{R}$, $\theta \in [0, \pi/2)$, *then* $\sigma(A_t) \subseteq S_{c,\theta}$.

(v) *Let* B *be a linear operator on* $\mathcal{H}$ *such that* $\mathcal{D}(B) \subseteq \mathcal{D}(t)$, *and let* $\mathcal{D}$ *be a core for* t. *If* $t[x, y] = \langle Bx, y\rangle$ *for all* $x \in \mathcal{D}(B)$ *and* $y \in \mathcal{D}$, *then* $B \subseteq A_t$.

Proof Because t is closed, $V := (\mathcal{D}(t), \|\cdot\|_{\mathrm{Re}\,t})$ is a Hilbert space. Since $\mathcal{D}(t)$ is dense in $\mathcal{H}$ and $\|\cdot\| \leq \|\cdot\|_{\mathrm{Re}\,t}$, the Hilbert space V is densely and continuously embedded into $\mathcal{H}$. By Proposition 11.6 there is a $c \in \mathbb{R}$ such that $\Theta(T) \subseteq S_{c,\theta}$ and $t - (c - 1)$ is a bounded coercive form on $(\mathcal{D}(t), \|\cdot\|_{\mathrm{Re}\,t})$. Therefore, Theorem 11.3 applies and shows that the closed operator $A_{t-(c-1)} = A_t - (c - 1)I$ has a bounded inverse in $\mathbf{B}(\mathcal{H})$. Hence, $c - 1 \in \rho(A_t)$, so A_t is m-sectorial by Proposition 3.19(ii).

(i)–(iii) restate the corresponding assertions of Theorem 11.3. Since obviously $\Theta(A_t) \subseteq \Theta(t)$, by Proposition 3.19 the spectrum $\sigma(A_t)$ is contained in $S_{c,\theta}$ when $\Theta(t)$ is. Using Definition 11.1 of a core and the boundedness of t (by Proposition 11.6(iii)), the proof of (v) is the same as the proof of Proposition 10.5(v). $\qquad\square$

Corollary 11.9 *The map* $t \to A_t$ *gives a one-to-one correspondence between densely defined closed sectorial forms and m-sectorial operators on a Hilbert space.*

Proof The existence and injectivity of this mapping follow from Theorem 11.8 and Corollary 11.4. It remains to prove the surjectivity.

Let T be an m-sectorial operator, and let t be the closure of the form $\mathfrak{s}_T$ from Proposition 11.7. By Proposition 3.19(iii), T and hence t are densely defined. Since $t[x, y] = \mathfrak{s}_T[x, y] = \langle Tx, y\rangle$ for $x, y \in \mathcal{D}(T)$ and $\mathcal{D}(T)$ is a core for $t = \overline{\mathfrak{s}_T}$, we conclude that $T \subseteq A_t$ by Theorem 11.8(v). Since T and A_t are m-sectorial, Proposition 3.24 yields $T = A_t$. $\qquad\square$

In the next section sectorial forms are applied to elliptic linear partial differential operators. The following definition fits to these applications. As above, $(V, \langle\cdot,\cdot\rangle_V)$ is a Hilbert space which is densely and continuously embedded into $\mathcal{H}$.

Definition 11.2 A form t with domain $\mathcal{D}(t) = V$ is called *elliptic* if there exist constants $C > 0$, $\gamma > 0$, and $c \in \mathbb{R}$ such that

$$|t[u, v]| \leq C\|u\|_V\|v\|_V \quad \text{for } u, v \in V, \tag{11.9}$$

$$(\mathrm{Re}\,t)[u] - c\|u\|^2 \geq \gamma\|u\|_V^2 \quad \text{for } u \in V. \tag{11.10}$$

Condition (11.9) means that the form t is bounded on the Hilbert space $(V, \langle\cdot,\cdot\rangle_V)$, while (11.10) is called the *abstract Gårding inequality*. Obviously, (11.10) implies that the form $\mathrm{Re}\,t$ is lower semibounded with lower bound c.

Let $\mathfrak{t}$ be an elliptic form on V, and let $v \in V$. Using formulas (11.10), (10.4), (10.5), and (11.4), we derive

$$\gamma \|u\|_V^2 \leq (\operatorname{Re}\mathfrak{t})[u] + (1-c)\|u\|^2 = \|u\|_{\operatorname{Re}\mathfrak{t}}^2$$
$$\leq |\mathfrak{t}[u]| + (1-c)\|u\|^2 \leq \left(C + \left(1 + |c|\right)a^2\right)\|u\|_V^2.$$

This shows that the form norm $\|\cdot\|_{\operatorname{Re}\mathfrak{t}}$ of $\operatorname{Re}\mathfrak{t}$ and the Hilbert space norm $\|\cdot\|_V$ of V are equivalent. Hence, $\operatorname{Re}\mathfrak{t}$ is closed, since V is a Hilbert space.

Because the norms $\|\cdot\|_{\operatorname{Re}\mathfrak{t}}$ and $\|\cdot\|_V$ are equivalent, it follows from Proposition 11.6(iii) that each elliptic form is sectorial. Since elliptic forms are densely defined (because V is dense in $\mathcal{H}$) and closed, Theorem 11.8 applies to each elliptic form.

Conversely, any densely defined closed sectorial form $\mathfrak{t}$ is elliptic on the Hilbert space $(\mathcal{D}(\mathfrak{t}), \|\cdot\|_{\operatorname{Re}\mathfrak{t}})$ which is densely and continuously embedded into $\mathcal{H}$. Indeed, by Proposition 11.6, (iii) and (iv), the form $\mathfrak{t}$ is bounded, and $\mathfrak{t} + 1 - c$ is coercive on $(\mathcal{D}(\mathfrak{t}), \|\cdot\|_{\operatorname{Re}\mathfrak{t}})$. The latter implies that the Gårding inequality (11.10) holds.

Summarizing the preceding, we have shown that *elliptic forms are in fact the same as densely defined closed sectorial forms.*

The formulation of the *abstract boundary problem* (10.16) carries over verbatim to the case of an elliptic form $\mathfrak{t}$ on $\mathcal{H}$: Given $u \in \mathcal{H}$, to find $x \in \mathcal{D}(\mathfrak{t})$ such that

$$\mathfrak{t}[x, y] - \lambda\langle x, y\rangle = \langle u, y\rangle \quad \text{for all } y \in \mathcal{D}(\mathfrak{t}). \tag{11.11}$$

By the definition of $A_{\mathfrak{t}}$, (11.11) holds if and only if $x \in \mathcal{D}(A_{\mathfrak{t}})$ and $A_{\mathfrak{t}}x = \lambda x + u$. Therefore, if $\lambda \in \rho(A_{\mathfrak{t}})$, this problem has a unique solution $x_u = (A_{\mathfrak{t}} - \lambda I)^{-1}u$.

Remark A powerful tool for the study of sectorial forms are holomorphic semigroups of operators. If $\mathfrak{t}$ is a densely defined closed sectorial form on $\mathcal{H}$ such that $\Theta(\mathfrak{t}) \subseteq S_{0,\theta}$, it can be shown that the operator $-A_{\mathfrak{t}}$ is the infinitesimal generator of a strongly continuous contraction semigroup $e^{-\mathfrak{t}A_{\mathfrak{t}}}$ on $\mathcal{H}$ which is holomorphic in the sector $\{z \in \mathbb{C} : |\arg z| < \frac{\pi}{2} - \theta\}$. In this book we do not cover this topic and refer to [Ka] or [Tn] for detailed treatments of this matter.

11.3 Application to Second-Order Elliptic Differential Operators

Let Ω be an open subset of $\mathbb{R}^d$. Let $a_{kl}(x), b_k(x), c_k(x), q(x) \in L^\infty(\Omega)$, $k, l = 1, \ldots, d$, be given. Throughout this section we assume that there is a constant $\alpha > 0$ such that

$$\operatorname{Re} \sum_{k,l=1}^{d} a_{kl}(x)\xi_l\overline{\xi_k} \geq \alpha \sum_{k=1}^{d} |\xi_k|^2 \quad \text{for all } \xi = (\xi_1, \ldots, \xi_d) \in \mathbb{C}^d, \; x \in \Omega.$$

$$\tag{11.12}$$

This condition means that the differential expression $\mathcal{L}$ defined by Eq. (11.16) below is *uniformly elliptic* on Ω.

Let V be a closed subspace of the Hilbert space $H^1(\Omega)$ such that $H_0^1(\Omega) \subseteq V$. Then the Hilbert space $(V, \| \cdot \|_{H^1(\Omega)})$ is densely and continuously embedded into $L^2(\Omega)$. We define a (not necessarily symmetric) form t with domain $\mathcal{D}(t) = V$ by

$$t[u, v] = \int_\Omega \sum_{k,l} a_{kl} \partial_l u \overline{\partial_k v} \, dx + \int_\Omega \sum_k (b_k \partial_k u \, \overline{v} + c_k u \overline{\partial_k v}) \, dx + \int_\Omega q u \overline{v} \, dx.$$

Recall that ∂_k denotes the partial derivative $\frac{\partial}{\partial x_k}$.

Proposition 11.10 *The form t on V is an elliptic form according to Definition 11.2. More precisely, for any $\gamma < \alpha$, there exists a real constant c_γ such that*

$$\mathrm{Re}\, t[u] - c_\gamma \|u\|^2 \geq \gamma \|u\|^2_{H^1(\Omega)}, \quad u \in V. \tag{11.13}$$

Proof Since $a_{kl}, b_k, c_k, q \in L^\infty(\Omega)$, it follows at once from the Cauchy–Schwarz inequality that t is bounded on V, that is, there is a constant $C > 0$ such that

$$|t[u, v]| \leq C \|u\|^2_{H^1(\Omega)} \|v\|^2_{H^1(\Omega)}, \quad u, v \in V.$$

Thus, condition (11.9) is satisfied.

Now we prove the Gårding inequality (11.13). Recall that $\| \cdot \|$ is the norm of $L^2(\Omega)$. Let $u \in V$. Setting $\xi_k = \partial_k u(x)$ in (11.12) and integrating over Ω, we get

$$\mathrm{Re} \int_\Omega \sum_{k,l} a_{kl} \partial_l u \overline{\partial_k u} \, dx \geq \alpha \int_\Omega \sum_k |\partial_k u|^2 \, dx = \alpha \big(\|u\|^2_{H^1(\Omega)} - \|u\|^2 \big).$$

$$\tag{11.14}$$

Fix $\varepsilon > 0$. Because the functions b_k, c_k, q are in $L^\infty(\Omega)$, there is a constant $M > 0$ such that $|b_k|, |c_k|, |q| \leq M$ a.e. on Ω. Using again the Cauchy–Schwarz inequality and the elementary inequality $2ab \leq \varepsilon a^2 + \varepsilon^{-1} b^2$, $a, b \in \mathbb{R}$, it follows that

$$\left| \int_\Omega \sum_k (b_k \partial_k u \overline{u} + c_k u \overline{\partial_k u}) \, dx + \int_\Omega q u \overline{u} \, dx \right|$$

$$\leq M \|u\|^2 + 2M \sum_k \|\partial_k u\| \|u\| \leq M d\varepsilon \|u\|^2_{H^1(\Omega)} + M \big(1 + d\varepsilon^{-1} \big) \|u\|^2.$$

Combining the latter inequality and (11.14) with the definition of Re t, we derive

$$(\mathrm{Re}\, t)[u] = \mathrm{Re}\, t[u] \geq (\alpha - M d\varepsilon) \|u\|^2_{H^1(\Omega)} - \big(\alpha + M \big(1 + d\varepsilon^{-1} \big) \big) \|u\|^2.$$

Choosing $\varepsilon > 0$ such that $M d\varepsilon \leq \alpha - \gamma$ and setting $c_\gamma = -(\alpha + M(1 + d\varepsilon^{-1}))$, we obtain the Gårding inequality (11.13). This proves that t is an elliptic form.   □

Since each elliptic form is densely defined, closed, and sectorial, Theorem 11.8 applies to the form t defined above. Hence, the corresponding operator A_t is m-sectorial, and its spectrum is contained in a sector (11.7). If, in addition, the embedding map of V into $L^2(\Omega)$ is compact, then A_t has a purely discrete spectrum.

From its definition it is clear that the form $\mathfrak{t}$ is symmetric, provided that

$$a_{kl}(x) = \overline{a_{lk}(x)}, \qquad b_k(x) = \overline{c_k(x)}, \qquad q(x) = \overline{q(x)} \quad \text{a.e. on } \Omega, \ k, l = 1, \dots, d. \tag{11.15}$$

In this case the symmetric form $\mathfrak{t} = \operatorname{Re} \mathfrak{t}$ is lower semibounded (by the Gårding inequality (11.13)) and closed (since $\mathfrak{t}$ is elliptic); hence, the operator $A_{\mathfrak{t}}$ is self-adjoint and lower semibounded by Theorem 10.7.

Though the form $\mathfrak{t}$ is given by a fairly explicit and simple formula, for general L^∞-coefficients, it seems to be very difficult to describe the domain $\mathcal{D}(A_{\mathfrak{t}})$ and the action of operator $A_{\mathfrak{t}}$ explicitly, say in terms of differentiations.

In the remaining part of this section we assume that all functions a_{kl}, b_k, c_k, q are in $C^\infty(\Omega) \cap L^\infty(\Omega)$. In this case we can say more about the operator $A_{\mathfrak{t}}$.

Let us consider the partial differential expression

$$\mathcal{L} = -\sum_{k,l=1}^{d} \partial_k a_{kl} \partial_l + \sum_{k=1}^{d} (b_k \partial_k - \partial_k c_k) + q. \tag{11.16}$$

Since all coefficients are in $C^\infty(\Omega)$, we know from Sect. 1.3.2 that the formal adjoint $\mathcal{L}^+$ and the operators $L_{\min}$ and $L_{\max}$ are well defined. We obtain

$$\mathcal{L}^+ = -\sum_{k,l=1}^{d} \partial_l \overline{a_{kl}} \partial_k + \sum_{k=1}^{d} (\overline{c_k} \partial_k - \partial_k \overline{b_k}) + \overline{q}.$$

Recall that $L_{\min}$ is the closure of L_0, where $L_0 u = \mathcal{L}u$ for $u \in \mathcal{D}(L_0) = C_0^\infty(\Omega)$, and $L_{\max} v = \mathcal{L}v$ for $v \in \mathcal{D}(L_{\max}) = \{f \in L^2(\Omega) : \mathcal{L}f \in L^2(\Omega)\}$.

Lemma 11.11 $L_{\min} \subseteq A_{\mathfrak{t}} \subseteq L_{\max}$.

Proof Let $u \in C_0^\infty(\Omega)$. We argue in a similar manner as in the proof of Eq. (10.29). We choose a bounded open set $\widetilde{\Omega} \subseteq \Omega$ of class C^2 such that $\operatorname{supp} u \subseteq \widetilde{\Omega}$ and apply the Green formula (D.5) to $\widetilde{\Omega}$. Since $u \restriction \partial \widetilde{\Omega} = 0$, we then obtain

$$\langle L_0 u, v \rangle = \mathfrak{t}[u, v] \quad \text{for } v \in V. \tag{11.17}$$

Hence, $L_0 u = A_{\mathfrak{t}} u$ by the definition of the operator $A_{\mathfrak{t}}$. That is, $L_0 \subseteq A_{\mathfrak{t}}$. Since $L_{\min}$ is the closure of L_0 and $A_{\mathfrak{t}}$ is closed, the latter implies that $L_{\min} \subseteq A_{\mathfrak{t}}$.

Replacing $\mathcal{L}$ by $\mathcal{L}^+$ and $\mathfrak{t}$ by $\mathfrak{t}^*$, we obtain $(L^+)_0 \subseteq A_{\mathfrak{t}^*}$. Applying the adjoint to this inclusion and using that $A_{\mathfrak{t}^*} = (A_{\mathfrak{t}})^*$ (by Theorem 11.8(ii)) and $((L^+)_0)^* = L_{\max}$ (by formula (1.22)), we derive $A_{\mathfrak{t}} = (A_{\mathfrak{t}^*})^* \subseteq ((L^+)_0)^* = L_{\max}$. $\qquad \square$

The following propositions give descriptions of the operator $A_{\mathfrak{t}}$ in the two extreme cases $V = H_0^1(\Omega)$ and $V = H^1(\Omega)$. In the first case $V = H_0^1(\Omega)$ we say that $A_{\mathfrak{t}}$ is the *Dirichlet operator* of $\mathcal{L}$ and denote it by L_D, while in the second case $V = H^1(\Omega)$ the operator $A_{\mathfrak{t}}$ is called the *Neumann operator* of $\mathcal{L}$ and denoted by L_N. The corresponding boundary problems (11.11) are called *weak Dirichlet problem* resp. *weak Neumann problem* for $\mathcal{L}$. This terminology will be justified by the boundary conditions occurring in Propositions 11.12 and 11.14. For the Laplacian $\mathcal{L} = -\Delta$, these (and more) results have been already developed in Sect. 10.6.

Proposition 11.12 *Let $V = H_0^1(\Omega)$ and $a_{kl}, b_k, c_k, q \in C^\infty(\Omega) \cap L^\infty(\Omega)$ for all k, l. Assume that the ellipticity assumption (11.12) holds. Then $A_{\mathfrak{t}}$ is the restriction of the maximal operator $L_{\max}$ to the domain $\mathcal{D}(A_{\mathfrak{t}}) = H_0^1(\Omega) \cap \mathcal{D}(L_{\max})$.*

Proof Since $A_{\mathfrak{t}} \subseteq L_{\max}$ by Lemma 11.11 and $\mathcal{D}(A_{\mathfrak{t}}) \subseteq \mathcal{D}[A_{\mathfrak{t}}] = H_0^1(\Omega)$, we have $\mathcal{D}(A_{\mathfrak{t}}) \subseteq H_0^1(\Omega) \cap \mathcal{D}(L_{\max})$.

Let $u \in \mathcal{D}(L_{\max}) \cap H_0^1(\Omega)$. Using once more identity (11.17), but with L and $\mathfrak{t}$ replaced by L^+ and $\mathfrak{t}^*$, and the equality $L_{\max} = (L^+)_0^*$ (by (1.22)), we obtain

$$\mathfrak{t}[u, v] = \overline{\mathfrak{t}^*[v, u]} = \overline{\langle (L^+)_0 v, u \rangle} = \langle u, (L^+)_0 v \rangle = \langle L_{\max} u, v \rangle \qquad (11.18)$$

for $v \in C_0^\infty(\Omega)$. Since $C_0^\infty(\Omega)$ is dense in $H_0^1(\Omega)$ and hence a core for the form $\mathfrak{t}$, we conclude from (11.18) and Theorem 11.8(v) that $u \in \mathcal{D}(A_{\mathfrak{t}})$.

The preceding proves the equality $\mathcal{D}(A_{\mathfrak{t}}) = H_0^1(\Omega) \cap \mathcal{D}(L_{\max})$. $\square$

Let $V = H_0^1(\Omega)$, and let conditions (11.15) be satisfied. Then the form $\mathfrak{t}$ and hence the operator $L_0 = (L^+)_0$ are symmetric and lower semibounded by (11.13). Since $C_0^\infty(\Omega)$ is dense in $\mathcal{D}(\mathfrak{t}) = H_0^1(\Omega)$, $\mathfrak{t}$ is the closure of the form $\mathfrak{s}_{L_0}$ defined in Lemma 10.16. Hence, the self-adjoint operator $A_{\mathfrak{t}}$ is just the Friedrichs extension of L_0. The next corollary restates Proposition 11.12 in this case.

Corollary 11.13 *Retain the assumptions of Proposition 11.12 and suppose that the conditions (11.15) hold. Then the Friedrichs extension $(L_0)_F$ of the lower semibounded symmetric operator L_0 and of its closure $L_{\min}$ are given by*

$$\mathcal{D}((L_0)_F) = H_0^1(\Omega) \cap \mathcal{D}(L_{\max}) \quad and \quad (L_0)_F f = L_{\max} f \quad for \ f \in \mathcal{D}((L_0)_F).$$

Now we turn to the second extreme case $V = H^1(\Omega)$.

Proposition 11.14 *Suppose that Ω is of class C^2. For the operator L_N, we have $L_{\min} \subseteq L_N \subseteq L_{\max}$. A function $f \in H^2(\Omega)$ belongs to $\mathcal{D}(L_N)$ if and only if it satisfies the boundary condition*

$$\sum_{k=1}^d v_k \left(\sum_{l=1}^d a_{kl} \partial_l f + c_k f \right) = 0 \quad \sigma\text{-a.e. on } \partial\Omega, \qquad (11.19)$$

where $v(x) = (v_1, \ldots, v_d)$ denotes the outward unit normal vector at $x \in \partial\Omega$, and σ is the surface measure of $\partial\Omega$.

Proof The relations $L_{\min} \subseteq L_N \subseteq L_{\max}$ are already contained in Lemma 11.11.

Let $f \in H^2(\Omega)$. Clearly, $f \in \mathcal{D}(L_{\max})$. For $v \in V = H^1(\Omega)$, we compute

$$\mathfrak{t}[f, v] - \langle L_{\max} f, v \rangle = \int_\Omega \sum_{k,l} \left((a_{kl} \partial_l f) \overline{\partial_k v} + (\partial_k a_{kl} \partial_l f) \overline{v} \right) dx$$

$$+ \int_\Omega \sum_k \left((c_k f) \overline{\partial_k v} + (\partial_k c_k f) \overline{v} \right) dx$$

$$= \int_{\partial\Omega} \sum_k v_k \left(\sum_l a_{kl} \partial_l f + c_k f \right) \overline{v} \, d\sigma. \qquad (11.20)$$

For the first equality, only the definitions of $\mathfrak{t}$ and $\mathcal{L}$ have been inserted, while the second equality follows by applying the Gauss formula (D.4) to both integrals.

Since $L_N \subseteq L_{\max}$, we have $f \in \mathcal{D}(L_N)$ if and only if $\mathfrak{t}[f, v] = \langle L_{\max} f, v \rangle$, or equivalently, the integral (11.20) vanishes for all $v \in V = H^1(\Omega)$. Because the range $\gamma_0(H^1(\Omega))$ of the trace map γ_0 (see Appendix D) is dense in $L^2(\Omega, d\sigma)$, the latter is equivalent to the boundary condition (11.19). □

In Sect. 10.6 much stronger results on the domains $\mathcal{D}(L_D)$ and $\mathcal{D}(L_N)$ were proved for the Laplacian $\mathcal{L} = -\Delta$, see (10.35) and (10.39). These descriptions have been essentially derived from the regularity theorem (Theorem D.10) for weak solutions. Under certain assumptions concerning the coefficients, similar results are valid for general elliptic differential operators, see, e.g., [Ag, LM].

11.4 Exercises

1. Let (X, μ) be a measure space, and M_φ the multiplication operator by a measurable function φ on $\mathcal{H} = L^2(X, \mu)$ (see Example 3.8). Define a form by $\mathfrak{t}[f, g] = \int_X \varphi f \overline{g} \, d\mu$ for $f, g \in \mathcal{D}(\mathfrak{t}) = \{f \in L^2(X, \mu) : f |\varphi|^{1/2} \in L^2(X, \mu)\}$.
   a. Determine $\Theta(\mathfrak{t})$.
   b. Suppose that $\sigma(M_\varphi) \subseteq S_{c,\theta}$ for some $c \in \mathbb{R}$, $\theta \in [0, \pi/2)$. Show that $\mathfrak{t}$ is a closed sectorial form and $A_\mathfrak{t} = M_\varphi$.
2. Let $\mathfrak{s}$ and $\mathfrak{t}$ be symmetric forms such that $\mathcal{D}(\mathfrak{s}) = \mathcal{D}(\mathfrak{t})$ and $|\mathfrak{t}[u]| \leq \mathfrak{s}[u]$ for all $u \in \mathcal{D}(\mathfrak{s})$. Prove that $|\mathfrak{t}[u, v]|^2 \leq \mathfrak{s}[u] \mathfrak{s}[v]$ for $u, v \in \mathcal{D}(\mathfrak{s})$.
   (Compare this result with Lemma 11.1.)
   Hints: Replacing u by zu for some $|z| = 1$, one can assume that $\mathfrak{t}[u, v]$ is real. Then $4\mathfrak{t}[u, v] = \mathfrak{t}[u + v] - \mathfrak{t}[u + v]$. Consider $\mathfrak{t}[u, v] = \mathfrak{t}[cu, c^{-1}v]$ for some $c > 0$.
3. Let $\alpha \in \mathbb{C}$ and define $\mathfrak{t}[f, g] = \langle f', g' \rangle + \alpha f(0) \overline{g(0)}$ for $f, g \in \mathcal{D}(\mathfrak{t}) := H^1(\mathbb{R})$. Show that $\mathfrak{t}$ is a sectorial form.
4. Let $(a_n)_{n \in \mathbb{N}}$ be a positive sequence, and $(b_n)_{n \in \mathbb{N}}$ a complex sequence. Define the form $\mathfrak{t}$ by $\mathfrak{t}[(\varphi_n), (\psi_n)] = \sum_{n=1}^\infty (a_n \varphi_n \overline{\psi}_n + b_n \varphi_n \overline{\psi}_{n+1})$ on $\mathcal{H} = l^2(\mathbb{N})$.
   a. Find sufficient conditions on (a_n) and (b_n) such that $\mathfrak{t}$ becomes a (densely defined closed) sectorial form on $\mathcal{H}$ and describe the domain $\mathcal{D}(\mathfrak{t})$.
   b. Describe the corresponding m-sectorial operator $A_\mathfrak{t}$.
   c. Is $\mathcal{D}_0 = \{(\varphi_1, \ldots, \varphi_n, 0, \ldots) : \varphi_k \in \mathbb{C}, n \in \mathbb{N}\}$ a core for $\mathfrak{t}$ or $A_\mathfrak{t}$?

The following four exercises develop a number of simple examples of sectorial forms in detail. Throughout $\mathcal{J}$ is an open interval, and $\mathcal{H} = L^2(\mathcal{J})$.

5. Let φ, ψ, η be bounded continuous functions on $\mathcal{J}$ such that φ, ψ are real-valued, $m_\varphi := \inf\{\varphi(x) : x \in \mathcal{J}\} > 0$, and $m_\psi := \inf\{\psi(x) : x \in \mathcal{J}\} > 0$. Further, put $M_\eta := \sup\{|\eta(x)| : x \in \mathcal{J}\}$. Suppose that $4m_\varphi m_\psi - M_\eta^2 > 0$. Define

$$\mathfrak{t}[f,g] = \langle \varphi f', g' \rangle + \langle \eta f', g \rangle + \langle \psi f, g \rangle, \quad f, g \in \mathcal{D}(\mathfrak{t}) := H^1(\mathcal{J}).$$

   a. Show that $\mathfrak{t}$ is a bounded coercive form on the Hilbert space $H^1(\mathcal{J})$.
   b. Consider $\mathfrak{t}$ as a form on $L^2(\mathcal{J})$ and determine $\mathfrak{t}^*$, $\mathrm{Re}\,\mathfrak{t}$, and $\mathrm{Im}\,\mathfrak{t}$.
   c. Show that $\mathfrak{t}$ is a sectorial form on the Hilbert space $L^2(\mathcal{J})$ and determine a sector $S_{c,\theta}$ such that $\Theta(\mathfrak{t}) \subseteq S_{c,\theta}$.

Let $\mathcal{D}(\mathfrak{t})$ be a linear subspace of $H^1(\mathcal{J})$ which contains $H_0^1(\mathcal{J})$, and let $\alpha, \beta, \gamma \in \mathbb{C}$. Define the form $\mathfrak{t}$ by

$$\mathfrak{t}[f,g] = \langle f', g' \rangle + \alpha \langle f', g \rangle + \beta \langle f, g' \rangle + \gamma \langle f, g \rangle, \quad f, g \in \mathcal{D}(\mathfrak{t}).$$

6. a. Determine $\mathfrak{t}^*$, $\mathrm{Re}\,\mathfrak{t}$, and $\mathrm{Im}\,\mathfrak{t}$.
   b. Show that the form norm of $\mathrm{Re}\,\mathfrak{t}$ is equivalent to the norm of $H^1(\mathcal{J})$.
   c. Show that $\mathrm{Re}\,\mathfrak{t}$ is lower semibounded and that $\mathfrak{t}$ is a closed sectorial form.
      Hint: Use $|\langle f', f \rangle| = |\langle f, f' \rangle| \leq \|f'\| \, \|f\| \leq \varepsilon \|f'\|^2 + \varepsilon^{-1} \|f\|^2$ for $\varepsilon > 0$.
7. Let $\mathcal{J} = (a, b)$, where $a, b \in \mathbb{R}$, $a < b$.
   a. Let $\mathcal{D}(\mathfrak{t}) = \{f \in H^1(\mathcal{J}) : zf(a) = wf(b)\}$, where $z, w \in \mathbb{C}$, $(z, w) \neq (0, 0)$. Show that $\mathcal{D}(A_{\mathfrak{t}}) = \{f \in H^2(\mathcal{J}) : \bar{z}(f' + \beta f)(b) = \bar{w}(f' + \beta f)(a)\}$.
   b. Let $\mathcal{D}(\mathfrak{t}) = H^1(\mathcal{J})$. Show that $\mathcal{D}(A_{\mathfrak{t}}) = \{f \in H^2(\mathcal{J}) : (f' + \beta f)(b) = (f' + \beta f)(a) = 0\}$.
   c. Let $\mathcal{D}(\mathfrak{t}) = H_0^1(\mathcal{J})$. Show that $\mathcal{D}(A_{\mathfrak{t}}) = \mathcal{D}(A_{\mathfrak{t}^*}) = H^2(\mathcal{J})$.
   d. Suppose that $\mathcal{D}(\mathfrak{t}) \neq H_0^1(\mathcal{J})$ and $\gamma \in \mathbb{R}$. Show that $\mathcal{D}(A_{\mathfrak{t}}) = \mathcal{D}(A_{\mathfrak{t}^*})$ if and only if $\alpha = \bar{\beta}$, or equivalently, if $\mathfrak{t} = \mathfrak{t}^*$.
8. Let $\mathcal{J} = (a, \infty)$, where $a \in \mathbb{R}$.
   a. Show that $\mathcal{D}(A_{\mathfrak{t}}) = \{f \in H^2(\mathcal{J}) : (f' + \beta f)(a) = 0\}$ if $\mathcal{D}(\mathfrak{t}) = H^1(\mathcal{J})$ and $\mathcal{D}(A_{\mathfrak{t}}) = H^2(\mathcal{J})$ if $\mathcal{D}(\mathfrak{t}) = H_0^1(\mathcal{J})$.
   b. Deduce that if $\mathcal{D}(\mathfrak{t}) = H^1(\mathcal{J})$, then $\mathcal{D}(A_{\mathfrak{t}}) = \mathcal{D}(A_{\mathfrak{t}^*})$ if and only if $\alpha = \bar{\beta}$.

Chapter 12
Discrete Spectra of Self-adjoint Operators

This chapter is concerned with eigenvalues of self-adjoint operators. The Fischer–Courant min–max principle is an important and useful tool for studying and comparing discrete spectra of lower semibounded self-adjoint operators. In Sect. 12.1 we prove this result and derive a number of applications. Section 12.2 contains some results about the existence of negative or positive eigenvalues of Schrödinger operators. In Sect. 12.3 we develop Weyl's classical asymptotic formula for the eigenvalues of the Dirichlet Laplacian on a bounded open subset of $\mathbb{R}^d$.

12.1 The Min–Max Principle

Let A be a *lower semibounded* self-adjoint operator on an *infinite-dimensional* Hilbert space $\mathcal{H}$. The *min–max principle* gives a variational characterization of the eigenvalues below the bottom of the essential spectrum of A. It is crucial that only the values $\langle Ax, x \rangle$ or $A[x]$ enter into this description but not the eigenvectors of A.

To formulate the min–max principle, we define three sequences $(\mu_n(A))_{n\in\mathbb{N}}$, $(\tilde{\mu}_n(A))_{n\in\mathbb{N}}$, and $(\lambda_n(A))_{n\in\mathbb{N}}$ associated with A. Let $\mathfrak{F}_n = \mathfrak{F}_n(\mathcal{H})$ denote the set of linear subspaces of $\mathcal{H}$ of dimension at most n. For $n \in \mathbb{N}$, we set

$$\mu_n(A) = \sup_{\mathcal{D}\in\mathfrak{F}_{n-1}} \inf_{x\in\mathcal{D}(A),\, \|x\|=1,\, x\perp\mathcal{D}} \langle Ax, x \rangle, \tag{12.1}$$

$$\tilde{\mu}_n(A) = \sup_{\mathcal{D}\in\mathfrak{F}_{n-1}} \inf_{x\in\mathcal{D}[A],\, \|x\|=1,\, x\perp\mathcal{D}} A[x]. \tag{12.2}$$

By these definitions, $\mu_1(A) = m_A$ and $\tilde{\mu}_1(A) = m_{\mathfrak{t}_A}$. Hence, by formula (10.10),

$$\mu_1(A) = \tilde{\mu}_1(A) = \inf\{\lambda : \lambda \in \sigma(A)\}. \tag{12.3}$$

For the definition of the third sequence $(\lambda_n(A))_{n\in\mathbb{N}}$, we consider two cases.

Case 1 ($\sigma_{\mathrm{ess}}(A) = \emptyset$) Then $\sigma(A)$ has no accumulation points and consists of eigenvalues of finite multiplicities. Let $\lambda_n(A)$, $n \in \mathbb{N}$, be the eigenvalues of A

K. Schmüdgen, *Unbounded Self-adjoint Operators on Hilbert Space*,
Graduate Texts in Mathematics 265,
DOI 10.1007/978-94-007-4753-1_12, © Springer Science+Business Media Dordrecht 2012

counted according to their multiplicities and arranged in increasing order, that is, $\lambda_n(A) \leq \lambda_{n+1}(A)$ for $n \in \mathbb{N}$.

Case 2 ($\sigma_{\mathrm{ess}}(A) \neq \emptyset$) Let $\alpha = \inf\{\lambda : \lambda \in \sigma_{\mathrm{ess}}(A)\}$ be the bottom of the essential spectrum of A. The set of all eigenvalues $\lambda_n(A)$ of A less than α will be enumerated as in Case 1. If there are precisely $k \in \mathbb{N}$ eigenvalues less than α, we set $\lambda_n(A) = \alpha$ for $n > k$. If there are no eigenvalues of A less than α, we put $\lambda_n(A) = \alpha$ for all $n \in \mathbb{N}$.

From these definitions it follows at once that A has a purely discrete spectrum if and only if $\lim_{n \to \infty} \lambda_n(A) = +\infty$.

Theorem 12.1 (*Min–max principle*) *Let A be a lower semibounded self-adjoint operator on an infinite-dimensional Hilbert space $\mathcal{H}$, and let E_A denote its spectral measure. Then we have*

$$\lambda_n(A) = \mu_n(A) = \widetilde{\mu}_n(A) = \inf_{\dim E_A((-\infty,\lambda))\mathcal{H} \geq n} \lambda \quad \text{for } n \in \mathbb{N}. \tag{12.4}$$

In the proof we use the following simple lemma.

Lemma 12.2 *Set $\mathcal{E}_\lambda := E_A((-\infty, \lambda))\mathcal{H}$ and $d(\lambda) := \dim \mathcal{E}_\lambda$ for $\lambda \in \mathbb{R}$. Then*

$$d(\lambda) < n \quad \text{if } \lambda < \mu_n(A), \tag{12.5}$$

$$d(\lambda) \geq n \quad \text{if } \lambda > \mu_n(A). \tag{12.6}$$

Proof Assume to the contrary that (12.5) is false, that is, $d(\lambda) \geq n$ and $\lambda < \mu_n(A)$. Let $\mathcal{D} \in \mathfrak{F}_{n-1}$ be given. Since $d(\lambda) = \dim \mathcal{E}_\lambda \geq n$, there exists a unit vector $x \in \mathcal{E}_\lambda$ such that $x \perp \mathcal{D}$. (Indeed, by choosing bases of $\mathfrak{F}_{n-1}$ and $\mathcal{E}_\lambda$, the requirement $x \perp \mathcal{D}$ leads to at most $n - 1$ linear equations for at least n variables, so there is always a nontrivial solution.) Since A is bounded from below and $x = E_A((-\infty, \lambda))x$, it follows from the functional calculus that $x \in \mathcal{D}(A)$ and $\langle Ax, x \rangle \leq \lambda \|x\|^2 = \lambda$. This implies that $\mu_n(A) \leq \lambda$, which is a contradiction.

Suppose now that (12.6) is not true, that is, $d(\lambda) < n$ and $\lambda > \mu_n(A)$. Then $\mathcal{E}_\lambda \in \mathfrak{F}_{n-1}$. If $x \in \mathcal{D}(A)$, $\|x\| = 1$, and $x \perp \mathcal{E}_\lambda$, then $E_A((-\infty, \lambda))x = 0$, and hence $\langle Ax, x \rangle \geq \lambda$ by the spectral theorem. Thus, $\mu_n(A) \geq \lambda$, which is a contradiction. $\square$

Proof of Theorem 12.1 Let us abbreviate $\mu_n = \mu_n(A)$ and $\lambda_n = \lambda_n(A)$. From (12.5) and (12.6) it follows at once that $\mu_n = \inf\{\lambda : d(\lambda) \geq n\}$ for all $n \in \mathbb{N}$.

To prove that $\mu_n = \lambda_n$, we proceed by induction. By (12.3) we have $\mu_1 = \lambda_1$. Now suppose that $\mu_1 = \lambda_1, \ldots, \mu_{n-1} = \lambda_{n-1}$. We shall prove that $\mu_n = \lambda_n$.

First, we show that $\mu_n \in \sigma(A)$. Assume to the contrary that $\mu_n \notin \sigma(A)$. Then $E_A((\mu_n - \varepsilon, \mu_n + \varepsilon)) = 0$ for some $\varepsilon > 0$ by Proposition 5.10(i), and hence $\mathcal{E}_{\mu_n + \varepsilon} = \mathcal{E}_{\mu_n - \varepsilon/2}$. Therefore, $n \leq d(\mu_n + \varepsilon) = d(\mu_n - \varepsilon/2) < n$ by (12.6) and (12.5), which is a contradiction. Thus, $\mu_n \in \sigma(A)$.

To complete the proof, we consider the following two distinct cases:

Case I: $d(\mu_n + \varepsilon_0) < \infty$ for some $\varepsilon_0 > 0$.

Since $\dim E_A((\mu_n - \varepsilon, \mu_n + \varepsilon))\mathcal{H} \le d(\mu_n + \varepsilon_0) < \infty$ for $0 < \varepsilon < \varepsilon_0$ and $\mu_n \in \sigma(A)$, Proposition 5.10(i) implies that μ_n is an isolated eigenvalue of finite multiplicity. Hence, there exists $\delta > 0$ such that $\sigma(A) \cap (\mu_n - \delta, \mu_n + \delta)$ contains only μ_n. Therefore, by Proposition 5.10(i), $E_A((\mu_n, \mu_n + \delta)) = 0$, so that $\dim E_A((-\infty, \mu_n])\mathcal{H} = d(\mu_n + \delta) \ge n$ by (12.6). Hence, A has at least n eigenvalues (counted with multiplicities) in $(-\infty, \mu_n]$. These are $\lambda_1 \le \ldots \le \lambda_{n-1} \le \lambda_n$. If λ_n would be less than μ_n, then $\dim E_A((-\infty, \lambda))\mathcal{H} = d(\lambda) \ge n$ for $\lambda_n < \lambda < \mu_n$. This contradicts (12.5). Thus, $\mu_n = \lambda_n$.

Case II: $d(\mu_n + \varepsilon) = \infty$ for all $\varepsilon > 0$.

Since $d(\mu_n - \varepsilon) < n$ by (12.5), it follows that $\dim E_A((\mu_n - \varepsilon, \mu_n + \varepsilon))\mathcal{H} = \infty$ for all $\varepsilon > 0$. Therefore, $\mu_n \in \sigma_{\mathrm{ess}}(A)$ by Proposition 8.11. Suppose that $\lambda < \mu_n$ and choose $\delta > 0$ such that $\lambda + \delta < \mu_n$. Then $d(\lambda + \delta) < n$ by (12.5), and hence $\dim E_A((\lambda - \delta, \lambda + \delta))\mathcal{H} < n$, so $\lambda \notin \sigma_{\mathrm{ess}}(A)$ again by Proposition 8.11. This proves that $\mu_n = \inf\{\mu : \mu \in \sigma_{\mathrm{ess}}(A)\}$. By the definition of λ_n (Case 2) we therefore have $\lambda_n \le \mu_n$.

Assume that $\lambda_n < \mu_n$. Then we have $\dim E_A((-\infty, \lambda))\mathcal{H} = d(\lambda) \ge n$ for $\lambda_n < \lambda < \mu_n$, which contradicts (12.5). The preceding proves that $\lambda_n = \mu_n$.

The equality $\tilde{\mu}_n(A) = \lambda_n(A)$ is proved by a similar reasoning. $\square$

Theorem 12.1 has a number of important applications, both for theoretical investigations and for practical computations. As above, $\mathcal{H}$ is an infinite-dimensional Hilbert space. Our first corollary shows that inequalities for (not necessarily commuting!) operators imply inequalities for the corresponding eigenvalues. The relations "$\ge$" and "$\succeq$" have been defined in Definition 10.5. For $\lambda \in \mathbb{R}$, we denote by $N(A; \lambda)$ the number of $n \in \mathbb{N}$ for which $\lambda_n(A) < \lambda$.

Corollary 12.3 *Let A and B be lower semibounded self-adjoint operators on (infinite-dimensional) Hilbert spaces $\mathcal{G}$ and $\mathcal{H}$, respectively, such that $\mathcal{G}$ is a subspace of $\mathcal{H}$. Suppose that $A \ge B$ or $A \succeq B$. Then we have $\lambda_n(A) \ge \lambda_n(B)$ for all $n \in \mathbb{N}$ and $N(A; \lambda) \le N(B; \lambda)$ for $\lambda \in \mathbb{R}$. If B has a purely discrete spectrum, so has A.*

Proof From the relation $A \ge B$ resp. $A \succeq B$ it follows immediately that $\tilde{\mu}_n(A) \ge \tilde{\mu}_n(B)$ resp. $\mu_n(A) \ge \mu_n(B)$ for $n \in \mathbb{N}$, so $\lambda_n(A) \ge \lambda_n(B)$ by equality (12.4). Obviously, the latter implies that $N(A; \lambda) \le N(B; \lambda)$.

As noted above, a lower semibounded self-adjoint operator C has a purely discrete spectrum if and only if $\lim_n \lambda_n(C) = +\infty$. Hence, A has a purely discrete spectrum if B has. $\square$

Corollary 12.4 *Let A be a positive self-adjoint operator, and B a self-adjoint operator on $\mathcal{H}$. Suppose that B is relatively A-bounded with A-bound zero and $\sigma_{\mathrm{ess}}(A + \beta B) = [0, +\infty)$ for $\beta \ge 0$. Then the function $\beta \to \lambda_n(A + \beta B)$ on $[0, +\infty)$ is nonincreasing for each $n \in \mathbb{N}$.*

Proof First, we note that by the Kato–Rellich Theorem 8.5 the operator $A + \beta B$ is self-adjoint on $\mathcal{D}(A + \beta B) = \mathcal{D}(A)$. From the assumption $\sigma_{\mathrm{ess}}(A + \beta B) = [0, +\infty)$ and the above definition of $\lambda_n(\cdot)$ it follows that $\lambda_n(A + \beta B) \leq 0$ for all $n \in \mathbb{N}$. Recall that $\lambda_n(A + \beta B) = \mu_n(A + \beta B)$ by (12.4). Therefore,

$$\mu_n(A + \beta B) = \sup_{\mathcal{D} \in \mathfrak{F}_{n-1}} \inf_{x \in \mathcal{D}(A),\, \|x\|=1,\, x \perp \mathcal{D}} \min\{0, \langle Ax, x \rangle + \beta \langle Bx, x \rangle\}.$$

Since $\langle Ax, x \rangle \geq 0$, the function $\beta \to \min\{0, \langle Ax, x \rangle + \beta \langle Bx, x \rangle\}$ is nonincreasing on $[0, \infty)$ as easily seen, so is $\beta \to \mu_n(A + \beta B) = \lambda_n(A + \beta B)$. $\qquad\square$

In perturbation theory of quantum mechanics the variable β in the sum $A + \beta B$ is called the *coupling constant*. Then Corollary 12.4 says that the negative eigenvalues $\lambda_n(A + \beta B)$ are nonincreasing functions of the coupling constant $\beta \in [0, +\infty)$. Hence, the number of negative eigenvalues of $A + \beta B$ is increasing. It can also be shown that the function $\beta \to \lambda_n(A + \beta B)$ is continuous on $[0, +\infty)$ and strictly monotone at all point β where $\lambda_n(A + \beta B) < 0$.

Remarks 1. Exercises 1–4 contain additional facts concerning Theorem 12.1. Exercise 1 shows that in Eqs. (12.1) and (12.2) supremum and infimum can be replaced by maximum and minimum, respectively, while Exercises 2 and 3 state "dual" versions to Theorem 12.1 ("max–min principles").

2. Suppose that the Hilbert space $\mathcal{H}$ has dimension $d \in \mathbb{N}$. Then the numbers $\mu_n(A)$ and $\widetilde{\mu}_n(A)$ are not defined for $n > d$, since there is no unit vector x such that $x \perp \mathcal{H}$. But, as the corresponding proofs show, the assertions of Theorem 12.1 and of Corollaries 12.3 and 12.4 remain valid for all $n = 1, \ldots, d$.

Proposition 12.5 *Let A be a lower semibounded self-adjoint operator on $\mathcal{H}$. Let V be a d-dimensional subspace of $\mathcal{D}(A)$, and $A_V := PA\lceil V$ the compression of A to V, where P is the orthogonal projection onto V. If $\lambda_1(A_V), \ldots, \lambda_d(A_V)$ denote the eigenvalues of the self-adjoint operator A_V on V enumerated in increasing order, then*

$$\lambda_1(A) \leq \lambda_1(A_V), \quad \lambda_2(A) \leq \lambda_2(A_V), \quad \ldots, \quad \lambda_d(A) \leq \lambda_d(A_V). \tag{12.7}$$

Proof Let $k \in \{1, \ldots, d\}$. Since $\lambda_k(A) = \mu_k(A)$ and $\lambda_k(A_V) = \mu_k(A_V)$ by (12.4) (see, e.g., Remark 2.), it suffices to prove that $\mu_k(A_V) \geq \mu_k(A)$. By (12.1),

$$\mu_k(A_V) = \sup_{\mathcal{D} \in \mathfrak{F}_{k-1}(V)} \inf_{v \in V,\, \|v\|=1,\, v \perp \mathcal{D}} \langle A_V v, v \rangle$$

$$= \sup_{\mathcal{D} \in \mathfrak{F}_{k-1}(\mathcal{H})} \inf_{v \in V,\, \|v\|=1,\, v \perp P\mathcal{D}} \langle Av, v \rangle$$

$$= \sup_{\mathcal{D} \in \mathfrak{F}_{k-1}(\mathcal{H})} \inf_{v \in V,\, \|v\|=1,\, v \perp \mathcal{D}} \langle Av, v \rangle$$

$$\geq \sup_{\mathcal{D} \in \mathfrak{F}_{k-1}(\mathcal{H})} \inf_{x \in \mathcal{D}(A),\, \|x\|=1,\, x \perp \mathcal{D}} \langle Ax, x \rangle = \mu_k(A).$$

For the third equality, we used the fact that for $v \in V$, we have $v \perp P\mathcal{D}$ if and only if $v \perp \mathcal{D}$, since $\langle v, Py \rangle = \langle Pv, y \rangle = \langle v, y \rangle$ for any $y \in \mathcal{H}$. $\qquad \square$

Inequalities (12.7) are estimates of the eigenvalues of A from *above* by finite-dimensional approximations. They are the basic tools for the *Rayleigh–Ritz method* which is useful for practical calculations of eigenvalues, see, e.g., [CH].

Derivation of lower bounds for the eigenvalues is more subtle. A simple result of this kind is the next proposition, which contains a *lower* bound for the first eigenvalue.

Proposition 12.6 (*Temple's inequality*) *Let A be a lower semibounded self-adjoint operator such that $\lambda_1(A) < \lambda_2(A)$, and let $\alpha \in \mathbb{R}$, $x \in \mathcal{D}(A^2)$, $\|x\| = 1$. Suppose that $\langle Ax, x \rangle < \alpha < \lambda_2(A)$. Then*

$$\lambda_1(A) \geq \langle Ax, x \rangle - \frac{\langle A^2 x, x \rangle - \langle Ax, x \rangle^2}{\alpha - \langle Ax, x \rangle} = \langle Ax, x \rangle - \frac{\langle (A - \langle Ax, x \rangle)^2 x, x \rangle}{\alpha - \langle Ax, x \rangle}.$$

Proof Set $\lambda_1 := \lambda_1(A)$. Since $\lambda_1(A) < \alpha < \lambda_2(A)$, we have $\sigma(A) \cap (\lambda_1, \alpha) = \emptyset$. Hence,

$$\langle (A - \lambda_1 I)(A - \alpha I)x, x \rangle = \int_{\sigma(A)} (t - \lambda_1)(t - \alpha) \, d\langle E_A(t)x, x \rangle \geq 0,$$

which in turn implies that

$$\langle (A - \alpha I)Ax, x \rangle \geq \lambda_1 \langle (A - \alpha I)x, x \rangle. \tag{12.8}$$

By assumption, $\langle (A - \alpha I)x, x \rangle = \langle Ax, x \rangle - \alpha < 0$. Therefore, (12.8) yields

$$\lambda_1 \geq \frac{\langle (A - \alpha I)Ax, x \rangle}{\langle (A - \alpha I)x, x \rangle} = \langle Ax, x \rangle - \frac{\langle A^2 x, x \rangle - \langle Ax, x \rangle^2}{\alpha - \langle Ax, x \rangle}. \qquad \square$$

Corollary 12.3 can be used to derive results by comparing eigenvalues of Schrödinger operators with different potentials. A sample is the following proposition.

Proposition 12.7 *Let V be a nonnegative function of $L^\infty_{\text{loc}}(\mathbb{R}^d)$ such that $\lim_{\|x\| \to \infty} V(x) = +\infty$. Then the form sum $A = -\Delta \dotplus V$ is a positive self-adjoint operator with purely discrete spectrum.*

Proof First, we note that the form sum $A = -\Delta \dotplus V$ is a well-defined positive self-adjoint operator by Proposition 10.23, since $V \in L^\infty_{\text{loc}}(\mathbb{R}^d)$.

Let $M > 0$ be given. Since $\lim_{\|x\| \to +\infty} V(x) = +\infty$, there is an $r > 0$ such that $V(x) \geq M$ if $\|x\| \geq r$. Set $U_r(x) = -M$ if $\|x\| \leq r$, $U_r(x) = 0$ if $\|x\| > r$, and define the form sum $B_r = -\Delta \dotplus (U_r + M)$. By construction, $V(x) \geq U_r(x) + M$, and hence $A \geq B_r$. Therefore, $\lambda_n(A) \geq \lambda_n(B_r)$ for $n \in \mathbb{N}$ by Corollary 12.3.

Since U_r is bounded and of compact support, $-\Delta + U_r$ is a self-adjoint operator such that $\sigma_{\text{ess}}(-\Delta + U_r) = [0, +\infty)$ by Theorem 8.19. Obviously, $-\Delta + U_r$ is lower semibounded. Hence, there exists n_0 such $\lambda_n(-\Delta + U_r) \geq -1$ for $n \geq n_0$. Since the operator sum $-\Delta + (U_r + M)$ is a self-adjoint operator by Theorem 8.8, it coincides with the form sum B_r by Proposition 10.22(ii). Thus, we have

$$\lambda_n(B_r) = \lambda_n\big(-\Delta + (U_r + M)\big) = M + \lambda_n(-\Delta + U_r).$$

By the preceding, for $n \geq n_0$, we have

$$\lambda_n(A) \geq \lambda_n(B_r) = M + \lambda_n(-\Delta + U_r) \geq M - 1.$$

This proves that $\lim_{n \to \infty} \lambda_n(A) = +\infty$, so A has a purely discrete spectrum. $\square$

12.2 Negative or Positive Eigenvalues of Schrödinger Operators

Our first proposition collects some simple criteria concerning the existence of negative eigenvalues. As usual, eigenvalues are counted with their multiplicities.

Proposition 12.8 *Let A be a self-adjoint operator on a Hilbert space $\mathcal{H}$ such that $\sigma_{\mathrm{ess}}(A) \subseteq [0, +\infty)$.*

(i) *A has a negative eigenvalue if and only if $\langle Ax, x \rangle < 0$ for some $x \in \mathcal{D}(A)$.*

(ii) *A has only a finite number of negative eigenvalues if and only if there exists a finite-dimensional linear subspace $\mathcal{D}$ of $\mathcal{H}$ such that $\langle Ax, x \rangle \geq 0$ for all x in $\mathcal{D}(A)$, $x \perp \mathcal{D}$. The number of negative eigenvalues is then less than or equal to $\dim \mathcal{D}$.*

(iii) *A has infinitely many negative eigenvalues if and only if there is an infinite-dimensional subspace $\mathcal{E}$ of $\mathcal{D}(A)$ such that $\langle Ax, x \rangle < 0$ for all $x \in \mathcal{E}$, $x \neq 0$.*

Proof Let E_A denote the spectral measure of A. We freely use the description of the spectrum given in Proposition 5.10. Since $\sigma_{\mathrm{ess}}(A) \subseteq [0, +\infty)$, all points of $\sigma(A) \cap (-\infty, 0)$ are eigenvalues of finite multiplicities which have no accumulation point in $(-\infty, 0)$, and $\dim E_A((-\infty, 0))\mathcal{H}$ is the number of negative eigenvalues.

(i): A has no negative eigenvalues if and only if $E_A((-\infty, 0)) = 0$, or equivalently, if $A \geq 0$.

(ii): First, suppose that A has n negative eigenvalues. Then $\mathcal{D} := E_A((-\infty, 0))\mathcal{H}$ has dimension n. If $x \in \mathcal{D}(A)$ and $x \perp \mathcal{D}$, then $x \in E_A([0, +\infty))\mathcal{H}$, and hence $\langle Ax, x \rangle \geq 0$ by the functional calculus.

Conversely, suppose that there is a finite-dimensional subspace $\mathcal{D}$ such that $\langle Ax, x \rangle \geq 0$ for $x \in \mathcal{D}(A) \cap \mathcal{D}^\perp$. If $\dim E_A((-\infty, 0))\mathcal{H}$ would be larger than $\dim \mathcal{D}$, then there exists a nonzero vector $y \in E_A((-\infty, 0))\mathcal{H} \cap \mathcal{D}(A)$ such that $y \perp \mathcal{D}$. But then $\langle Ay, y \rangle < 0$, contradicting $\langle Ay, y \rangle \geq 0$ by the assumption.

(iii): If A has infinitely many negative eigenvalues, $\mathcal{E} := E_A((-\infty, 0))\mathcal{D}(A)$ is infinite-dimensional and has the desired property.

Conversely, assume that $\mathcal{E}$ is such a subspace. Let $\mathcal{D}$ be an arbitrary finite-dimensional subspace of $\mathcal{H}$. Since $\mathcal{E}$ has infinite dimension, we can find a nonzero $y \in \mathcal{E}$ such that $y \perp \mathcal{D}$. Hence, the condition in (ii) cannot hold, since otherwise we would have $\langle Ay, y \rangle \geq 0$. This contradicts $\langle Ay, y \rangle < 0$ by the condition in (iii). Therefore, by (ii), A has infinitely many negative eigenvalues. $\quad\square$

Now we turn to the existence of *negative* eigenvalues of Schrödinger operators. In quantum mechanics the corresponding eigenfunctions are called *bound states*. If a particle is in a bound state, it cannot leave the system without additional energy.

Proposition 12.9 *Let V be a real-valued function of $L^2(\mathbb{R}^3) + L^\infty(\mathbb{R}^3)_\varepsilon$. Suppose that there exist positive constants c, R_0, and δ such that $\delta < 2$ and*

$$V(x) \leq -c\|x\|^{-\delta} \quad \text{for } \|x\| \geq R_0. \tag{12.9}$$

Then the self-adjoint operator $-\Delta + V$ has infinitely many negative eigenvalues.

Proof By Theorems 8.8 and 8.19, $A := -\Delta + V$ is a self-adjoint operator on $L^2(\mathbb{R}^3)$ and $\sigma_{\text{ess}}(A) = [0, +\infty)$. Hence, the assumptions of Proposition 12.8 are satisfied.

Let us fix a function $\varphi \in C_0^\infty(\mathbb{R}^3)$ such that $\operatorname{supp}\varphi \subseteq \{x : 1 < \|x\| < 2\}$ and $\|\varphi\| = 1$. Put $\varphi_\alpha(x) = \alpha^{-3/2}\varphi(\alpha^{-1}x)$ for $\alpha > 0$. Then we have

$$\langle \Delta\varphi_\alpha, \varphi_\alpha \rangle = \alpha^{-2}\langle \Delta\varphi, \varphi \rangle \quad \text{and} \quad \langle \|x\|^{-\delta}\varphi_\alpha, \varphi_\alpha \rangle = \alpha^{-\delta}\langle \|x\|^{-\delta}\varphi, \varphi \rangle. \tag{12.10}$$

Since $\operatorname{supp}\varphi_\alpha \subseteq \{x : \alpha < \|x\| < 2\alpha\}$, for $\alpha > R_0$, it follows from (12.9) and (12.10) that

$$\langle A\varphi_\alpha, \varphi_\alpha \rangle = \langle -\Delta\varphi_\alpha, \varphi_\alpha \rangle + \langle V\varphi_\alpha, \varphi_\alpha \rangle \leq \alpha^{-2}\langle -\Delta\varphi, \varphi \rangle - c\alpha^{-\delta}\langle \|x\|^{-\delta}\varphi, \varphi \rangle.$$

Therefore, since $\delta < 2$, there exists $\beta > 0$ such that $\langle A\varphi_\alpha, \varphi_\alpha \rangle < 0$ for all $\alpha > \beta$.

Set $\psi_n := \varphi_{2^n\beta}$. Clearly, $(\psi_n)_{n \in \mathbb{N}}$ is an orthonormal sequence of functions with disjoints supports, so that $\langle A\psi_n, \psi_k \rangle = 0$ for $n \neq k$. Hence, the linear span $\mathcal{E}$ of vectors ψ_n is an infinite-dimensional linear subspace of $\mathcal{D}(A)$ such that $\langle A\psi, \psi \rangle < 0$ for all $\psi \in \mathcal{E}$, $\psi \neq 0$. Therefore, by Proposition 12.8(iii), A has infinitely many negative eigenvalues. $\quad\square$

The condition $\delta < 2$ is crucial in Proposition 12.9. It turns out that the potential $V_0(x) = \|x\|^{-2}$ is the borderline case for having *infinitely many* negative eigenvalues. In fact, the following result (see [RS4], Theorem XIII.6) is true:

Let $V \in L^2(\mathbb{R}^3) + L^\infty(\mathbb{R}^3)_\varepsilon$ be real-valued. If there are positive constants R_0 and b such that $b < 1/4$ and

$$V(x) \geq -b\|x\|^{-2} \quad \text{for } \|x\| \geq R_0,$$

then $-\Delta + V$ has only a finite number of negative eigenvalues.

Another application of the min–max principle are estimates for the numbers of negative eigenvalues of Schrödinger operators. This matter is treated in great detail in [RS4, Chap. XIII, Sect. 3]. Here we state only the theorem about the *Birman–Schwinger bound* (12.11) without proof (see [RS4, Theorem XIII.10]):

Let V be a real-valued Borel function on $\mathbb{R}^3$ of the Rollnik class $\mathfrak{R}$ (see Sect. 10.7). If $N(V)$ denotes the number of negative eigenvalues of the self-adjoint operator $-\Delta \dotplus V$, then

$$N(V) \le (4\pi)^{-2} \|V\|_{\mathfrak{R}}, \tag{12.11}$$

where $\|V\|_{\mathfrak{R}}$ is given by (10.42). In particular, $N(V)$ is finite.

The next result contains a criterion for the nonexistence of *positive* eigenvalues.

Proposition 12.10 *Let V be a real-valued Borel function on $\mathbb{R}^d$ which is relatively $(-\Delta)$-bounded with $(-\Delta)$-bound less than one. Suppose that there exists a number $\beta \in (0, 2)$ such that*

$$V(ax) = a^{-\beta} V(x) \quad \text{for } a \in (0,1) \text{ and } x \in \mathbb{R}^d, \ x \neq 0. \tag{12.12}$$

Then the self-adjoint operator $-\Delta + V$ has no eigenvalue contained in $[0, +\infty)$.

Proof By the Kato–Rellich Theorem 8.5, the operator $-\Delta + V$ is self-adjoint on $\mathcal{D}(-\Delta)$. Let λ be an eigenvalue of $-\Delta + V$ with eigenvector f, $\|f\| = 1$. Put $f_a(x) = f(ax)$ for $a \in (0, 1)$. Using assumption (12.12) and the equation $-(\Delta f)(x) = -V(x)f(x) + \lambda f(x)$, we derive

$$-(\Delta f_a)(x) = -a^2 (\Delta f)(ax) = -a^2 V(ax)f(ax) + a^2 \lambda f(ax)$$
$$= -a^{2-\beta} V(x) f_a(x) + a^2 \lambda f_a(x).$$

Using this relation and the fact that the operator $-\Delta$ is symmetric, we obtain

$$a^2 \lambda \langle f_a, f \rangle - a^{2-\beta} \langle V f_a, f \rangle = \langle -\Delta f_a, f \rangle = \langle f_a, -\Delta f \rangle = \langle f_a, -Vf + \lambda f \rangle$$
$$= -\langle V f_a, f \rangle + \lambda \langle f_a, f \rangle,$$

which leads to the equation

$$\lambda \langle f_a, f \rangle = \frac{a^{2-\beta} - 1}{a^2 - 1} \langle V f_a, f \rangle.$$

Letting $a \to 1 - 0$, we conclude that $\lambda = \lambda \langle f, f \rangle = \frac{2-\beta}{2} \langle Vf, f \rangle$. Hence,

$$\langle -\Delta f, f \rangle = \langle -Vf + \lambda f, f \rangle = -2(2 - \beta)^{-1}\lambda + \lambda = \lambda \beta (\beta - 2)^{-1}. \tag{12.13}$$

Since $\langle -\Delta f, f \rangle > 0$ and $\beta(\beta - 2) < 0$, it follows from (12.13) that $\lambda < 0$. $\square$

We illustrate the preceding results by the standard example of potentials.

Example 12.1 (*Coulomb potential $V(x) = -\gamma \|x\|^{-1}$ on $L^2(\mathbb{R}^3)$*) First, suppose that $\gamma \in \mathbb{R}$. Recall from Example 8.5 that $V \in L^2(\mathbb{R}^3) + L^\infty(\mathbb{R}^3)_\varepsilon$, so V has the $(-\Delta)$-bound zero, and $\sigma_{\text{ess}}(-\Delta + V) = [0, +\infty)$. Thus, by Corollary 12.4, the function $\beta \to \lambda_n(-\Delta + \beta V)$ is nonincreasing on $[0, +\infty)$.

The potential V satisfies the homogeneity condition (12.12) with $\beta = 1$. Hence, by Proposition 12.10, the self-adjoint operator $-\Delta + V$ has no eigenvalue in $[0, +\infty)$.

Further, let λ be an eigenvalue of $-\Delta + V$ with normalized eigenfunction f. Since $\frac{2-\beta}{2} = \frac{1}{2}$ and hence $\lambda = \frac{1}{2}\langle Vf, f \rangle$ by the above proof, (12.13) yields

$$-\langle -\Delta f, f \rangle = \frac{1}{2}\langle Vf, f \rangle = \lambda. \tag{12.14}$$

This assertion is called *virial theorem* (see, e.g., [Th], 4.1.4). Since $\langle -\Delta f, f \rangle > 0$, Eq. (12.14) implies that $\lambda < 0$.

Now suppose that $\gamma > 0$. Then (12.9) is fulfilled with $\delta = 1$, $\gamma = c$, $R_0 = 1$. Therefore, by Proposition 12.9, the operator $-\Delta + V$ has *infinitely many negative* eigenvalues. It can be shown that these are precisely the eigenvalues $\frac{-\gamma^2}{4k^2}$, where $k \in \mathbb{N}$, with multiplicities k^2 and the corresponding eigenfunctions are built from spherical harmonics and Laguerre functions. This is carried out in detail in standard books on quantum mechanics. $\circ$

12.3 Asymptotic Distribution of Eigenvalues of the Dirichlet Laplacian

Let Ω be an open subset of $\mathbb{R}^d$. The Dirichlet Laplacian $-\Delta_{D,\Omega}$ and the Neumann Laplacian $-\Delta_{N,\Omega}$ on Ω have been studied extensively in Sect. 10.6. Here we only recall that $-\Delta_{D,\Omega}$ and $-\Delta_{N,\Omega}$ are positive self-adjoint operators on $L^2(\Omega)$ which are defined by their quadratic forms $\mathfrak{t}_{D,\Omega}$ and $\mathfrak{t}_{N,\Omega}$, where

$$\begin{aligned}
\mathcal{D}(\mathfrak{t}_{D,\Omega}) = \mathcal{D}[-\Delta_{D,\Omega}] := H_0^1(\Omega), \\
\mathcal{D}(\mathfrak{t}_{N,\Omega}) = \mathcal{D}[\Delta_{N,\Omega}] := H^1(\Omega),
\end{aligned} \tag{12.15}$$

and both forms are given by the same expression

$$\mathfrak{t}[f, g] = \int_{\Omega} (\nabla f)(x) \cdot \overline{(\nabla g)(x)} \, dx. \tag{12.16}$$

By (12.16) the form norm of $-\Delta_{D,\Omega}$ coincides with the norm of $H_0^1(\Omega)$.

Suppose that Ω is *bounded*. Then the embedding of $H_0^1(\Omega)$ into $L^2(\Omega)$ is compact by Theorem D.1. Hence $-\Delta_{D,\Omega}$ has a purely discrete spectrum by Proposition 10.6. If, in addition, Ω is of class C^1 (see Appendix D), the embedding of $H^1(\Omega)$ into $L^2(\Omega)$ is compact by Theorem D.4, so $-\Delta_{N,\Omega}$ has also a purely discrete spectrum. The spectrum of $-\Delta_{D,\Omega}$ is formed by an increasing positive sequence $\lambda_n(-\Delta_{D,\Omega})$, $n \in \mathbb{N}$, of eigenvalues counted with multiplicities and converging to $+\infty$. The main result of this section (Theorem 12.14) describes the asymptotic behavior of this sequence for some "nice" open bounded sets Ω. To achieve this goal, a number of preliminaries are needed.

Lemma 12.11 *Let Ω and $\widetilde{\Omega}$ be open subsets of $\mathbb{R}^d$.*

(i) *If $\Omega \subseteq \widetilde{\Omega}$, then $-\Delta_{D,\widetilde{\Omega}} \leq -\Delta_{D,\Omega}$.*
(ii) $-\Delta_{N,\Omega} \leq -\Delta_{D,\Omega}$.

Proof Both inequalities follow at once from the definitions of the order relation "$\leq$" (Definition 10.5) and the corresponding forms (12.15) and (12.16). □

Lemma 12.12 *Let $\Omega_1, \ldots, \Omega_q$ be pairwise disjoint open subsets of $\mathbb{R}^d$.*

(i) *If $\Omega = \bigcup_{j=1}^q \Omega_j$, then $-\Delta_{D,\Omega} = \bigoplus_{j=1}^q -\Delta_{D,\Omega_j}$ on $L^2(\Omega) = \bigoplus_{j=1}^q L^2(\Omega_j)$.*

(ii) *Let Ω be an open subset of $\mathbb{R}^d$ such that $\bigcup_{j=1}^q \Omega_j \subseteq \Omega$ and $\Omega \setminus \bigcup_{j=1}^q \Omega_j$ is a Lebesgue null set. Then $\bigoplus_{j=1}^q -\Delta_{N,\Omega_j} \leq -\Delta_{N,\Omega}$ on $L^2(\Omega) = \bigoplus_{j=1}^q L^2(\Omega_j)$.*

Proof For a function f on Ω we denote its restriction to Ω_j by f_j.

(i): Since Ω is the union of the disjoint sets Ω_j, for $f, g \in C_0^\infty(\Omega)$, we have

$$t_{D,\Omega}[f, g] = \int_\Omega \nabla f \cdot \overline{\nabla g}\, dx = \sum_{j=1}^q \int_{\Omega_j} \nabla f_j \cdot \overline{\nabla g_j}\, dx$$

$$= \sum_{j=1}^q t_{D,\Omega_j}[f_j, g_j] \tag{12.17}$$

and $C_0^\infty(\Omega) = \bigoplus_{j=1}^q C_0^\infty(\Omega_j)$. Hence, by continuity Eq. (12.17) extends to the closures of the corresponding forms and yields $t_{D,\Omega}[f, g] = \sum_{j=1}^q t_{D,\Omega_j}[f_j, g_j]$ for $f, g \in \mathcal{D}(t_{D,\Omega}) = \bigoplus_{j=1}^q \mathcal{D}(t_{D,\Omega_j})$. Taking the operators associated with the forms $t_{D,\Omega}$ and t_{D,Ω_j}, it follows that $-\Delta_{D,\Omega} = \bigoplus_{j=1}^q -\Delta_{D,\Omega_j}$.

(ii): Let $f, g \in \mathcal{D}(t_{N,\Omega}) = H^1(\Omega)$. Then $f_j, g_j \in \mathcal{D}(t_{N,\Omega_j}) = H^1(\Omega_j)$. Since $\Omega \setminus \bigcup_{j=1}^q \Omega_j$ has Lebesgue measure zero, we obtain

$$t_{N,\Omega}[f, g] = \int_\Omega \nabla f \cdot \overline{\nabla g}\, dx = \sum_{j=1}^q \int_{\Omega_j} \nabla f_j \cdot \overline{\nabla g_j}\, dx$$

$$= \sum_{j=1}^q t_{N,\Omega_j}[f_j, g_j]. \tag{12.18}$$

Clearly, the operator associated with the form on the right-hand side of (12.18) is $\bigoplus_{j=1}^q -\Delta_{N,\Omega_j}$ on $L^2(\Omega) = \bigoplus_{j=1}^q L^2(\Omega_j)$. Since $\mathcal{D}(t_{N,\Omega}) \subseteq \bigoplus_{j=1}^q \mathcal{D}(t_{N,\Omega_j})$, (12.18) implies that $\bigoplus_{j=1}^q -\Delta_{N,\Omega_j} \leq -\Delta_{N,\Omega}$. □

For a self-adjoint operator A on $\mathcal{H}$ and $\lambda \in \mathbb{R}$ we define

$$N'(A; \lambda) := \dim E_A\big((-\infty, \lambda)\big)\mathcal{H}.$$

If $A = \bigoplus_{k=1}^n A_j$ is an orthogonal sum of self-adjoint operators A_j, we obviously have $E_A(\cdot) = \bigoplus_{j=1}^n E_{A_j}(\cdot)$, and hence

$$N'(A; \lambda) = \sum_{k=1}^n N'(A_k; \lambda), \quad \lambda \in \mathbb{R}. \tag{12.19}$$

If A has a purely discrete spectrum, $N'(A, \lambda)$ is just the number $N(A, \lambda)$ of eigenvalues of A less than λ which appeared in Corollary 12.3. Further, we abbreviate

$$N_{D,\Omega}(\lambda) := N(-\Delta_{D,\Omega}; \lambda) \quad \text{and} \quad N_{N,\Omega}(\lambda) := N(-\Delta_{N,\Omega}; \lambda).$$

That is, if $-\Delta_{D,\Omega}$ resp. $-\Delta_{N,\Omega}$ has a purely discrete spectrum, then $N_{D,\Omega}(\lambda)$ resp. $N_{N,\Omega}(\lambda)$ is the number of eigenvalues of $-\Delta_{D,\Omega}$ resp. $-\Delta_{N,\Omega}$ less than λ.

Example 12.2 ($\Omega = (a, a+l), d = 1, l > 0$) In this case, $-\Delta_{D,\Omega}$ is the differential operator $-\frac{d^2}{dx^2}$ with boundary conditions $f(a) = f(a + l) = 0$. Easy computations show that $-\Delta_{D,\Omega}$ has the spectrum

$$\sigma(-\Delta_{D,\Omega}) = \left\{n^2\pi^2 l^{-2} : n \in \mathbb{N}\right\}$$

consisting of simple eigenvalues $n^2\pi^2 l^{-2}$ and the corresponding eigenfunctions φ_n, $n \in \mathbb{N}$, are

$$\varphi_{2k}(x) = \sin 2k\pi l^{-1}(x - a - l/2),$$
$$\varphi_{2k+1}(x) = \cos(2k + 1)\pi l^{-1}(x - a - l/2).$$

Clearly, $-\Delta_{N,\Omega}$ is the operator $-\frac{d^2}{dx^2}$ with boundary conditions $f'(a) = f'(a + l) = 0$. Hence, the spectrum of $-\Delta_{N,\Omega}$ is formed by simple eigenvalues $n^2\pi^2 l^{-2}$ with eigenfunctions ψ_n, $n \in \mathbb{N}_0$. They are given by

$$\sigma(-\Delta_{N,\Omega}) = \left\{n^2\pi^2 l^{-2} : n \in \mathbb{N}_0\right\},$$
$$\psi_{2k}(x) = \cos 2k\pi l^{-1}(x - a - l/2),$$
$$\psi_{2k+1}(x) = \sin(2k + 1)\pi l^{-1}(x - a - l/2).$$

$\circ$

Example 12.3 ($\Omega = (a_1, a_1 + l) \times \cdots \times (a_d, a_d + l) \subseteq \mathbb{R}^d, l > 0$) By separation of variables we easily determine the spectra and the eigenfunctions of the operators $-\Delta_{D,\Omega}$ and $-\Delta_{N,\Omega}$ and obtain

$$\sigma(-\Delta_{D,\Omega}) = \left\{(n_1^2 + \ldots + n_d^2)\pi^2 l^{-2} : n_1, \ldots, n_d \in \mathbb{N}\right\}, \tag{12.20}$$
$$\varphi_n(x) = \varphi_{n_1}(x_1) \cdots \varphi_{n_d}(x_d), \quad n = (n_1, \ldots, n_d) \in \mathbb{N}^d,$$
$$\sigma(-\Delta_{N,\Omega}) = \left\{(n_1^2 + \ldots + n_d^2)\pi^2 l^{-2} : n_1, \ldots, n_d \in \mathbb{N}_0\right\}, \tag{12.21}$$
$$\psi_n(x) = \psi_{n_1}(x_1) \cdots \psi_{n_d}(x_d), \quad n = (n_1, \ldots, n_d) \in \mathbb{N}_0^d.$$

$\circ$

Let us denote by ω_d the volume of the unit ball in $\mathbb{R}^d$, that is,

$$\omega_d = \pi^{d/2}\Gamma(d/2 + 1), \quad d \in \mathbb{N}. \tag{12.22}$$

Lemma 12.13 *Let* $\Omega = (a_1, a_1 + l) \times \cdots \times (a_d, a_d + l) \subseteq \mathbb{R}^d, l > 0$. *For* $\lambda > 0$,

$$\left| N_{D,\Omega}(\lambda) - \lambda^{d/2}\omega_d l^d (2\pi)^{-d} \right| \leq \sum_{j=0}^{d-1} \lambda^{j/2}\omega_j l^j (2\pi)^{-j}, \qquad (12.23)$$

$$\left| N_{N,\Omega}(\lambda) - \lambda^{d/2}\omega_d l^d (2\pi)^{-d} \right| \leq \sum_{j=0}^{d-1} \lambda^{j/2}\omega_j l^j (2\pi)^{-j}. \qquad (12.24)$$

Proof From Example 12.3 and formula (12.20) it follows that $N_{D,\Omega}(\lambda)$ is the number of points $n = (n_1, \ldots, n_d) \in \mathbb{N}^d$ such that $(n_1^2 + \ldots + n_d^2)\pi^2 l^{-2} < \lambda$, or equivalently, the number of $n \in \mathbb{N}^d$ lying inside the ball B_r centered at the origin with radius $r := \sqrt{\lambda}\, l\pi^{-1}$. Similarly, formula (12.21) implies that $N_{D,\Omega}(\lambda)$ is the number of points $n \in \mathbb{N}_0^d$ inside the ball B_r.

For $n \in \mathbb{N}^d$ and $k \in \mathbb{N}_0^d$, we define the unit cubes $\mathcal{Q}_n$ and $\mathcal{P}_k$ by

$$\mathcal{Q}_n = \{x : n_j - 1 \leq x_j < n_j, j = 1, \ldots, d\},$$
$$\mathcal{P}_k = \{x : k_j \leq x_j < k_j + 1, j = 1, \ldots, d\}.$$

Let $B_{r,+} = \{x \in B_r : x_1 \geq 0, \ldots, x_d \geq 0\}$ be the intersection of the positive octant with B_r. The union of all cubes $\mathcal{Q}_n$, where $n \in \mathbb{N}^d \cap B_r$, is disjoint and contained in $B_{r,+}$, while the union of cubes $\mathcal{P}_k$, where $k \in \mathbb{N}_0^d \cap B_r$, is a disjoint cover of $B_{r,+}$. Therefore, by the preceding, we have

$$N_{D,\Omega}(\lambda) \leq |B_{r,+}| = 2^{-d}\omega_d r^d = \lambda^{\frac{d}{2}}\omega_d l^d (2\pi)^{-d} \leq N_{N,\Omega}(\lambda). \qquad (12.25)$$

Obviously, the difference $N_{N,\Omega}(\lambda) - N_{D,\Omega}(\lambda)$ is just the number of points of $\mathbb{N}_0^d \setminus \mathbb{N}^d$ within B_r. This is the union of sets $\mathcal{R}_j$, $j = 0, \ldots, d-1$, where $\mathcal{R}_j$ denotes the set of all $n \in \mathbb{N}^d \cap B_r$ for which precisely j of the numbers $n_1, \ldots, n_d$ are nonzero. Repeating the above reasoning, we conclude that the number of points of $\mathcal{R}_j$ does not exceed the volume of the positive octant $B_{r,+,j}$ of the ball centered at the origin with radius r in j-dimensional space. Since $|B_{r,+,j}| = 2^{-j}\omega_j r^j$, we get

$$0 \leq N_{N,\Omega}(\lambda) - N_{D,\Omega}(\lambda) \leq \sum_{j=0}^{d-1} 2^{-j}\omega_j r^j = \sum_{j=0}^{d-1} \lambda^{j/2}\omega_j l^j (2\pi)^{-j}. \qquad (12.26)$$

Now the assertions (12.23) and (12.24) follow by combining inequalities (12.25) and (12.26).   □

In the preceding proof the points of $\mathbb{Z}^d$ inside a ball played a crucial role. Estimating the number of such lattice points is an important problem in number theory.

In order to formulate our main theorem, we recall the notion of the Jordan content. Let $\mathcal{M}$ be a *bounded* subset of $\mathbb{R}^d$. Then there are numbers $a, b \in \mathbb{R}$ such that $\mathcal{M}$ is contained in the closed cube $Q = [a, b] \times \ldots \times [a, b]$ in $\mathbb{R}^d$. Let $n \in \mathbb{N}$ and $l := (b - a)/n$. To any n-tuple $\mathfrak{k} = (k_1, \ldots, k_n)$ of numbers $k_j \in \{1, \ldots, n\}$ we associate a cube

$$R_{\mathfrak{k}} = \left[a + (k_1 - 1)l, a + k_1 l\right] \times \cdots \times \left[a + (k_d - 1)l, a + k_d l\right].$$

We call the set of all such cubes a partition of the cube Q and l its length.

We denote by $\underline{J}_n(\mathcal{M})$ the volume of those cubes $R_{\mathfrak{k}}$ contained in $\mathcal{M}$ and by $\overline{J}_n(\mathcal{M})$ the volume of those $R_{\mathfrak{k}}$ that intersect $\mathcal{M}$. The bounded set $\mathcal{M}$ is said to be *Jordan measurable* if $\sup_n \underline{J}_n(\mathcal{M}) = \inf_n \overline{J}_n(\mathcal{M})$; this common value is then denoted by $J(\mathcal{M})$ and called the (d-dimensional) *Jordan content* of $\mathcal{M}$.

Bounded open subsets with piecewise smooth boundary are Jordan measurable. If $\mathcal{M}$ is Jordan measurable, it is Lebesgue measurable and $J(\mathcal{M})$ coincides with the Lebesgue measure of $\mathcal{M}$; see, e.g., [Ap] for more about the Jordan content.

Theorem 12.14 (*Weyl's asymptotic formula*) *Let Ω be a bounded open Jordan measurable subset of $\mathbb{R}^d$. Then we have*

$$\lim_{\lambda \to +\infty} N_{D,\Omega}(\lambda)\lambda^{-\frac{d}{2}} = (2\pi)^{-d}\omega_d J(\Omega). \tag{12.27}$$

Proof Let us fix a cube Q which contains Ω and a partition of Q of length l.

We denote by $\mathcal{Q}_1, \ldots, \mathcal{Q}_q$ the interiors of those cubes $R_{\mathfrak{k}}$ contained in Ω. Since $\mathcal{Q} := \bigcup_{j=1}^q \mathcal{Q}_j \subseteq \Omega$, it follows from Lemmas 12.11(i) and 12.12(i) that

$$-\Delta_{D,\Omega} \leq -\Delta_{D,\mathcal{Q}} = \bigoplus_{j=1}^q -\Delta_{D,\mathcal{Q}_j}. \tag{12.28}$$

Obviously, the volume of $\mathcal{Q}$ is $|\mathcal{Q}| = ql^d$. Applying first Corollary 12.3 and then formula (12.19) to (12.28), and finally (12.23), we derive

$$N_{D,\Omega}(\lambda) \geq N_{D,\mathcal{Q}}(\lambda) = \sum_j N_{D,\mathcal{Q}_j}(\lambda) \geq q\left(\lambda^{\frac{d}{2}}\omega_d l^d (2\pi)^{-d} - \sum_{j=0}^{d-1}\lambda^{\frac{j}{2}}\omega_j l^j (2\pi)^{-j}\right)$$

$$= \lambda^{\frac{d}{2}}|\mathcal{Q}|\omega_d (2\pi)^{-d} - q\sum_{j=0}^{d-1}\lambda^{\frac{j}{2}}\omega_j l^j (2\pi)^{-j}. \tag{12.29}$$

Let $\mathcal{P}_1, \ldots, \mathcal{P}_p$ be the interiors of all cubes $R_{\mathfrak{k}}$ which intersect Ω and let $\mathcal{P}$ be the interior of the union $\bigcup_{j=1}^p \overline{\mathcal{P}_j}$. Then $\Omega \subseteq \mathcal{P}$ and $\mathcal{P}\backslash\bigcup_{j=1}^p \mathcal{P}_j$ is a Lebesgue null set. Therefore, applying Lemmas 12.11, (i) and (ii), and 12.12(ii), we obtain

$$-\Delta_{D,\Omega} \geq -\Delta_{D,\mathcal{P}} \geq -\Delta_{N,\mathcal{P}} \geq \bigoplus_{k=1}^p -\Delta_{N,\mathcal{P}_k}. \tag{12.30}$$

Clearly, $|\mathcal{P}| = pl^p$. Employing again Corollary 12.3 and formula (12.19) (now applied to (12.30)) and finally (12.24), we get

$$N_{D,\Omega}(\lambda) \leq \sum_k N_{N,\mathcal{P}_k}(\lambda) \leq \lambda^{\frac{d}{2}}|\mathcal{P}|\omega_d (2\pi)^{-d} + p\sum_{j=0}^{d-1}\lambda^{\frac{j}{2}}\omega_j l^j (2\pi)^{-j}.$$

$$\tag{12.31}$$

The highest order terms for λ on the right-hand sides of (12.29) and (12.31) are $\lambda^{d/2}$. From (12.29) and (12.31) we therefore conclude that

$$(2\pi)^{-d}\omega_d|\mathcal{Q}| \leq \liminf_{\lambda \to +\infty} N_{D,\Omega}(\lambda)\lambda^{-\frac{d}{2}} \leq \limsup_{\lambda \to +\infty} N_{D,\Omega}(\lambda)\lambda^{-\frac{d}{2}} \leq (2\pi)^{-d}\omega_d|\mathcal{P}|.$$

By definition, $|\mathcal{Q}| = \underline{J}_n(\Omega)$ and $|\mathcal{P}| = \overline{J}_n(\Omega)$. Since Ω is Jordan measurable, the differences $|\mathcal{P}| - J(\Omega)$ and $J(\Omega) - |\mathcal{Q}|$ can be made arbitrary small. Hence, it follows from the preceding inequalities that the limit $\lim_{\lambda \to +\infty} N_{D,\Omega}(\lambda)\lambda^{-d/2}$ exists and is equal to $(2\pi)^{-d}\omega_d J(\Omega)$. This proves Eq. (12.27). $\square$

The function $N_{D,\Omega}(\lambda)$ describes the eigenvalue distribution of the Dirichlet Laplacian $-\Delta_{D,\Omega}$. Formula (12.27) says that its asymptotic behavior is given by

$$N_{D,\Omega}(\lambda) \sim \lambda^{\frac{d}{2}}(2\pi)^{-d}\omega_d J(\Omega) \quad \text{as } \lambda \to +\infty.$$

To put Theorem 12.14 into another perspective, let us think of the region Ω as a (d-dimensional) membrane that is held fixed along the boundary of $\overline{\Omega}$. The transverse vibration of the membrane is described by a function $u(x,t)$ on $\Omega \times [0,+\infty)$ satisfying the wave equation $\Delta u = c^2 u_{tt}$, where c is the quotient of the density of the membrane by the tension. Solutions of the form $u(x,t) = v(x)e^{i\mu t}$ are called normal modes. They represent the pure tones of the membrane and the numbers μ are the "pitches". Clearly, such a function $u(x,t)$ is a normal mode if and only if

$$-\Delta v = c^2\mu^2 v, \qquad v\lceil\partial\overline{\Omega} = 0.$$

In operator-theoretic terms this means that $c^2\mu^2$ is an eigenvalue of the Dirichlet Laplacian $-\Delta_{D,\Omega}$.

If we could listen to the membrane vibrations with a perfect ear and "hear" all eigenvalues, or equivalently the spectrum of $-\Delta_{D,\Omega}$, what geometrical properties of Ω could we get out of this? Or considering Ω as a drum and repeating a famous phrase of Kac [Kc], "Can we hear the shape of a drum?" By Weyl's formula (12.27) we can "hear" the dimension d and the volume $J(\Omega)$ of Ω. Further, the spectrum determines the perimeter of Ω as also shown by Weyl (1912).

This question has an analog for the Laplace–Beltrami operator on a compact Riemannian manifold: What geometrical or topological properties of the manifold are encoded in the spectrum of this operator? This is the subject of a mathematical discipline which is called *spectral geometry*, see, e.g., [Bd].

12.4 Exercises

In Exercises 1, 2, 3, 4, $\mathcal{H}$ denotes an infinite-dimensional separable Hilbert space.

*1. Let A be a lower semibounded self-adjoint operator on $\mathcal{H}$ with purely discrete spectrum. Let $(\lambda_n(A))_{n\in\mathbb{N}}$ be the *nondecreasing* sequence of eigenvalues of A counted with multiplicities, and let $\{x_n : n \in \mathbb{N}\}$ be an orthonormal basis of corresponding eigenvectors of A, that is, $Ax_n = \lambda_n(A)x_n$. Put $\mathcal{D}_0 = \{0\}$ and $\mathcal{D}_n = \mathrm{Lin}\{x_1, \ldots, x_n\}$.

Show that the supremum in Eqs. (12.1) and (12.2) is attained for $\mathcal{D} = \mathcal{D}_{n-1}$ and that in this case the infimum becomes a minimum.

(This means that the assertion of Theorem 12.1 remains valid if "sup" and "inf" in (12.1) and (12.2) are replaced by "max" and "min," respectively. This justifies the name "min–max principle".)

*2. (*Max–min principle*)

Let A and $(\lambda_n(A))_{n\in\mathbb{N}}$ be as in Exercise 12.1. Show that for any $n \in \mathbb{N}$,

$$\lambda_n(A) = \inf_{\mathcal{D}\in\mathfrak{F}_n\cap\mathcal{D}(A)} \sup_{x\in\mathcal{D},\,\|x\|=1} \langle Ax, x\rangle = \inf_{\mathcal{D}\in\mathfrak{F}_n\cap\mathcal{D}[A]} \sup_{x\in\mathcal{D},\,\|x\|=1} A[x].$$

Sketch of proof: Denote the number after the first equality sign by ν_n. Verify that $\sup\{\langle Ax, x\rangle : x \in \mathcal{D}_n, \|x\| = 1\} = \lambda_n(A)$. This implies that $\nu_n \le \lambda_n(A)$. Conversely, if $\mathcal{D} \in \mathfrak{F}_n$, choose a unit vector $x \in \mathcal{D}$ such that $x \perp \mathcal{D}_{n-1}$. Then $\langle Ax, x\rangle \ge \lambda_n(A)$, and hence $\nu_n \ge \lambda_n(A)$.

3. (*Another variant of the max–min principle*)

Let A be a positive self-adjoint compact operator on $\mathcal{H}$, and let $(\widetilde{\lambda}_n(A))_{n\in\mathbb{N}}$ be the *nonincreasing* sequence of eigenvalues of A counted with multiplicities. Show that

$$\widetilde{\lambda}_n(A) = \inf_{\mathcal{D}\in\mathfrak{F}_{n-1}} \sup_{x\perp\mathcal{D},\,\|x\|=1} \langle Ax, x\rangle \quad \text{for } n \in \mathbb{N}.$$

4. Let A be a lower semibounded self-adjoint operator on $\mathcal{H}$, and $C \in \mathbf{B}(\mathcal{H})$. Prove that $|\lambda_n(A + C) - \lambda_n(A)| \le \|C\|$ for $n \in \mathbb{N}$.

Hint: Note that $|\langle (A + C)x, x\rangle - \langle Ax, x\rangle| \le \|C\|$ for $x \in \mathcal{D}(A)$, $\|x\| = 1$.

5. Let T be an operator on a Hilbert space $\mathcal{H}$, and $x \in \mathcal{H}$, $x \ne 0$. Prove that the *Rayleigh quotient*

$$\frac{\langle Tx, x\rangle}{\|x\|^2}$$

is a unique complex number where $\inf\{\|Tx - \lambda x\| : \lambda \in \mathbb{C}\}$ is attained.

6. Let A be a self-adjoint operator with purely discrete spectrum. Let $p \in \mathbb{C}[t]$ and $x \in \mathcal{D}(p(A))$ be such that $\langle p(A)x, x\rangle > 0$. Show that the set $\{t \in \mathbb{R} : p(t) > 0\}$ contains at least one eigenvalue of A.

7. (*Existence of negative eigenvalues*)

Let V be a real-valued Borel function on $\mathbb{R}$ such that

$$\lim_{|c|\to\infty} \int_c^{c+1} |V(x)|^2 \, dx = 0.$$

Define $T = -\frac{d^2}{dx^2} + V(x)$ on $\mathcal{D}(T) = H^2(\mathbb{R})$. Recall that T is a self-adjoint operator and $\sigma_{\mathrm{ess}}(H) = [0, +\infty)$ by Proposition 8.20 and Theorem 8.15.

a. Suppose that there exists an $\varepsilon > 0$ such that $\int_{\mathbb{R}} V(x)e^{-2\varepsilon x^2} dx < -\sqrt{\varepsilon\pi/2}$. Prove that T has at least one negative eigenvalue.

Hint: Set $\varphi_\varepsilon(x) = e^{-\varepsilon x^2}$ and compute $\langle -\varphi_\varepsilon'', \varphi_\varepsilon\rangle = \|\varphi_\varepsilon'\|^2 = \sqrt{\varepsilon\pi/2}$.

   b. Suppose that there is a Borel set M such that $\int_M V(x)\,dx$ exists, is negative, and $V(x) \le 0$ on $\mathbb{R} \setminus M$. Prove that T has at least one negative eigenvalue.
   Hint: Use the fact that $\int_M V(x)e^{-2\varepsilon x^2}\,dx \to \int_M V(x)\,dx$ as $\varepsilon \to +0$.

8. Let $V(x)$ be a real-valued Borel function on $\mathbb{R}$ of compact support such that $\int_{\mathbb{R}} V(x)\,dx < 0$. Show that the self-adjoint operator $T = -\frac{d^2}{dx^2} + V(x)$ on $L^2(\mathbb{R})$ has at least one negative eigenvalue.

9. Find the counterpart of Exercise 7.a on $L^2(\mathbb{R}^d)$ for arbitrary $d \in \mathbb{N}$.
   Hint: Compute $\langle -\Delta\varphi_\varepsilon, \varphi_\varepsilon \rangle = d\varepsilon \sqrt{\pi^d/(2\varepsilon)^d}$.

10. Let $a, b \in \mathbb{R}$, $a < b$, and let $T = -\frac{d^2}{dx^2}$ denote the operator with domain $\mathcal{D}(T) = \{f \in H^2(a, b) : f'(a) = f(b) = 0\}$ in $L^2(a, b)$.
    a. Show that T is a self-adjoint operator with purely discrete spectrum.
    b. Determine the eigenvalues of T and the corresponding eigenfunctions.

12.5 Notes to Part V

Chapters 10 and 11:

That densely defined lower semibounded symmetric operators have self-adjoint extensions with the same lower bounds was conjectured by von Neumann [vN1], p. 103, and proved by Stone [St2], p. 388, and Friedrichs [F1]. Friedrichs' work [F1] was the starting point for the theory of closed symmetric forms. General forms have been studied by various authors [LM, K3, Ls]. The Heinz inequality was obtained in [Hz]. The KLMN Theorem 10.21 bears his name after Kato [K3], Lax and Milgram [LaM], Lions [Ls], and Nelson [Ne2].

The standard reference for the theory of forms is Kato's book [K2]. Our Theorems 11.8 and 10.7 correspond to the first and second form representation theorems therein. General treatments of forms are given in [EE, D2, RS2, BS].

Applications of forms and Hilbert space operators to differential equations are studied in many books such as [Ag, D2, EE, Gr, LM, LU, Tl].

Example 10.11 is the simplest example of "point interaction potentials" studied in quantum mechanics, see, e.g., [AGH] and the references therein.

Chapter 12:

The min–max principle was first stated by Fischer [Fi] in the finite-dimensional case. Its power for applications was discovered by Courant [Cr, CH] and Weyl [W3]. Weyl's asymptotic formula was proved in [W3], see, e.g., [W4] for some historical reminiscences.

Part VI
Self-adjoint Extension Theory of Symmetric Operators

Chapter 13
Self-adjoint Extensions: Cayley Transform and Krein Transform

The subject of this chapter and the next is the self-adjoint extension theory of densely defined symmetric operators. The classical approach, due to J. von Neumann, is based on the Cayley transform. It reduces the problem of self-adjoint extensions of a symmetric operator to the problem of finding unitary extensions of its Cayley transform. This theory is investigated in Sects. 13.1 and 13.2. A similar problem is describing all positive self-adjoint extensions of a densely defined positive symmetric operator. One approach, due to M.G. Krein, uses the Krein transform and is treated in Sect. 13.4. A theorem of Ando and Nishio characterizes when a (not necessarily densely defined!) positive symmetric operator has a positive self-adjoint extension and shows that in this case a smallest positive self-adjoint extension exists. This result and some of its applications are developed in Sect. 13.3. The final Sect. 13.5 deals with two special situations for the construction of self-adjoint extensions. These are symmetric operators commuting with a conjugation or anticommuting with a symmetry.

Throughout this chapter, λ denotes a fixed complex number such that $\operatorname{Im} \lambda > 0$.

13.1 The Cayley Transform of a Symmetric Operator

Let $\mathbb{T}$ denote the unit circle $\{z \in \mathbb{C} : |z| = 1\}$. It is well known that the Möbius transformation

$$t \to (t - \lambda)(t - \bar{\lambda})^{-1}$$

maps the real axis onto $\mathbb{T}\backslash\{1\}$ and the upper (resp. lower) half-plane onto the set inside (resp. outside) of $\mathbb{T}$. The Cayley transform is an operator analog of this transformation. It maps densely defined symmetric operators onto isometric operators V for which $\mathcal{R}(I - V)$ is dense, and it relates both classes of operators.

First, we develop some simple facts on isometric operators. An *isometric operator* is a linear operator V on a Hilbert space $\mathcal{H}$ such that $\|Vx\| = \|x\|$ for all $x \in \mathcal{D}(V)$. By the polarization formula (1.2) we then have $\langle Vx, Vy \rangle = \langle x, y \rangle$

K. Schmüdgen, *Unbounded Self-adjoint Operators on Hilbert Space*,
Graduate Texts in Mathematics 265,
DOI 10.1007/978-94-007-4753-1_13, © Springer Science+Business Media Dordrecht 2012

for $x, y \in \mathcal{D}(V)$. The domain $\mathcal{D}(V)$ can be an arbitrary linear subspace of $\mathcal{H}$. Obviously, if V is isometric, it is invertible, and its inverse V^{-1} is also isometric. Clearly, V is closed if and only if $\mathcal{D}(V)$ is a closed linear subspace of $\mathcal{H}$.

Let V be an isometric operator on $\mathcal{H}$. From the inequality

$$\left\| (V - \mu I)x \right\| \geq \left| \|Vx\| - |\mu|\|x\| \right| = \left| 1 - |\mu| \right| \|x\| \quad \text{for } x \in \mathcal{D}(V) \text{ and } \mu \in \mathbb{C}$$

it follows that $\mathbb{C}\backslash\mathbb{T} \subseteq \pi(T)$. Hence, by Proposition 2.4, the defect numbers of V are constant on the interior of $\mathbb{T}$ and on the exterior of $\mathbb{T}$. The cardinal numbers

$$d^i(V) := d_\mu(V) = \mathcal{R}(V - \mu I)^\perp \quad \text{for } |\mu| < 1, \tag{13.1}$$

$$d^e(V) := d_\mu(V) = \mathcal{R}(V - \mu I)^\perp \quad \text{for } |\mu| > 1 \tag{13.2}$$

are called the *deficiency indices* of the isometric operator V. A nice description of these numbers is given by the next lemma.

Lemma 13.1 $d^i(V) = \dim \mathcal{R}(V)^\perp$ *and* $d^e(V) = \dim \mathcal{D}(V)^\perp$.

Proof By (13.1), we have $d^i(V) = d_0(V) = \dim \mathcal{R}(V)^\perp$.

Fix $\mu \in \mathbb{C}$, $0 < |\mu| < 1$. Since $(V^{-1} - \mu)Vx = (I - \mu V)x = -\mu(V - \mu^{-1})x$ for $x \in \mathcal{D}(V)$, we obtain $\mathcal{R}(V^{-1} - \mu I) = \mathcal{R}(V - \mu^{-1}I)$, and therefore by (13.2),

$$d^e(V) = \dim \mathcal{R}\big(V - \mu^{-1}I\big)^\perp = \dim \mathcal{R}\big(V^{-1} - \mu I\big)^\perp$$

$$= d^i\big(V^{-1}\big) = \dim \mathcal{R}\big(V^{-1}\big)^\perp = \dim \mathcal{D}(V)^\perp. \qquad \square$$

Lemma 13.2 *If V is an isometric operator on $\mathcal{H}$ and $\mathcal{R}(I - V)$ is dense in $\mathcal{H}$, then $\mathcal{N}(I - V) = \{0\}$.*

Proof Let $x \in \mathcal{N}(I - V)$. For $v \in \mathcal{D}(V)$, we have

$$\langle (I - V)v, x \rangle = \langle v, x \rangle - \langle Vv, x \rangle = \langle v, x \rangle - \langle Vv, Vx \rangle = \langle v, x \rangle - \langle v, x \rangle = 0.$$

Hence, $x = 0$, since $\mathcal{R}(I - V)$ is dense. $\qquad \square$

Let T be a *densely defined symmetric linear operator* on $\mathcal{H}$. Since $\text{Im}\,\lambda > 0$, we have $\bar{\lambda} \in \pi(T)$ by Proposition 3.2(i). Hence, $T - \bar{\lambda}I$ is invertible. The operator

$$V_T = (T - \lambda I)(T - \bar{\lambda}I)^{-1} \quad \text{with domain } \mathcal{D}(V_T) = \mathcal{R}(T - \bar{\lambda}I) \tag{13.3}$$

is called the *Cayley transform* of T. That is, V_T is defined by

$$V_T(T - \bar{\lambda}I)x = (T - \lambda I)x \quad \text{for } x \in \mathcal{D}(T). \tag{13.4}$$

Some useful properties of the Cayley transform are collected in the next proposition.

Proposition 13.3

(i) *The Cayley transform V_T is an isometric operator on $\mathcal{H}$ with domain $\mathcal{D}(V_T) = \mathcal{R}(T - \bar{\lambda}I)$ and range $\mathcal{R}(V_T) = \mathcal{R}(T - \lambda I)$.*

(ii) $\mathcal{R}(I - V_T) = \mathcal{D}(T)$ *and* $T = (\lambda I - \bar{\lambda}V_T)(I - V_T)^{-1}$.

(iii) T is closed if and only if V_T is closed.
(iv) If S is another symmetric operator on $\mathcal{H}$, then $T \subseteq S$ if and only if $V_T \subseteq V_S$.
(v) $d^i(V_T) = d_-(T)$ and $d^e(V_T) = d_+(T)$.

Proof (i): We write $\lambda = \alpha + i\beta$, where $\alpha, \beta \in \mathbb{R}$. Let $x \in \mathcal{D}(T)$. Using that the operator T is symmetric, we compute

$$\langle (T - \alpha I)x \pm i\beta x, (T - \alpha I)x \pm i\beta x \rangle$$
$$= \|(T - \alpha I)x\|^2 + |\beta|^2 \|x\|^2 \pm i\beta \langle x, (T - \alpha I)x \rangle \mp i\beta \langle (T - \alpha I)x, x \rangle$$
$$= \|(T - \alpha I)x\|^2 + |\beta|^2 \|x\|^2.$$

(The same computation appeared already in formula (3.2).) Setting $y := (T - \bar{\lambda}I)x$, we have $V_T y = (T - \lambda I)x$, and hence

$$\|V_T y\| = \|(T - \alpha I)x - i\beta x\| = \|(T - \alpha I)x + i\beta x\| = \|y\|,$$

so V_T is isometric. The equality $\mathcal{R}(V_T) = \mathcal{R}(T - \lambda I)$ follows from (13.4).

(ii): Recall that $\lambda - \bar{\lambda} \neq 0$. Since $(I - V_T)y = (\lambda - \bar{\lambda})x$, $\mathcal{R}(I - V_T) = \mathcal{D}(T)$. Hence, $\mathcal{N}(I - V_T) = \{0\}$ by Lemma 13.2. (Of course, this follows also directly, since $(I - V_T)y = (\lambda - \bar{\lambda})x = 0$ implies that $x = 0$ and therefore $y = 0$.) Further, from the equations $(I - V_T)y = (\lambda - \bar{\lambda})x$ and $(\lambda I - \bar{\lambda} V_T)y = (\lambda - \bar{\lambda})Tx$ it follows that $Tx = (\lambda I - \bar{\lambda} V_T)(I - V_T)^{-1}x$ for $x \in \mathcal{D}(T)$. If $x \in \mathcal{D}((\lambda I - \bar{\lambda} V_T)(I - V_T)^{-1})$, then in particular $x \in \mathcal{D}((I - V_T)^{-1}) = \mathcal{R}(I - V_T) = \mathcal{D}(T)$. By the preceding we have proved that $T = (\lambda I - \bar{\lambda} V_T)(I - V_T)^{-1}$.

(iii): By Proposition 2.1, T is closed if and only if $\mathcal{D}(V_T) = \mathcal{R}(T - \bar{\lambda}I)$ is closed in $\mathcal{H}$, or equivalently, the bounded operator V_T is closed.

(iv) follows at once from formula (13.4).

(v): From Lemma 13.1 and from the descriptions of $\mathcal{D}(V_T)$ and $\mathcal{R}(V_T)$ we obtain

$$d^i(V_T) = \dim \mathcal{R}(V_T)^\perp = \dim \mathcal{R}(T - \lambda I)^\perp = d_-(T),$$
$$d^e(V_T) = \dim \mathcal{D}(V_T)^\perp = \dim \mathcal{R}(T - \bar{\lambda}I)^\perp = d_+(T). \qquad \square$$

Now we proceed in reverse direction and suppose that V is an isometric operator on $\mathcal{H}$ such that $\mathcal{R}(I - V)$ is dense. Since then $\mathcal{N}(I - V) = \{0\}$ by Lemma 13.2, $I - V$ is invertible. The operator

$$T_V = (\lambda I - \bar{\lambda}V)(I - V)^{-1} \quad \text{with domain } \mathcal{D}(T_V) = \mathcal{R}(I - V) \qquad (13.5)$$

is called the *inverse Cayley transform* of V. By this definition we have

$$T_V(I - V)y = (\lambda I - \bar{\lambda}V)y \quad \text{for } y \in \mathcal{D}(V). \qquad (13.6)$$

Proposition 13.4 T_V *is a densely defined symmetric operator which has the Cayley transform* V.

Proof Let $x \in \mathcal{D}(T_V)$. Then $x = (I - V)y$ for some $y \in \mathcal{D}(V)$. Using the assumption that V is isometric, we compute

$$\langle T_V x, x \rangle = \langle T_V(I - V)y, (I - V)y \rangle = \langle (\lambda I - \overline{\lambda}V)y, (I - V)y \rangle$$
$$= \lambda \|y\|^2 + \overline{\lambda}\|Vy\|^2 - \lambda \langle y, Vy \rangle - \overline{\lambda} \langle Vy, y \rangle$$
$$= 2 \operatorname{Re} \lambda \|y\|^2 - 2 \operatorname{Re} \lambda \langle y, Vy \rangle.$$

Hence, $\langle T_V x, x \rangle$ is real. Therefore, T_V is symmetric. Since $\mathcal{D}(T_V) = \mathcal{R}(I - V)$ is dense by assumption, T_V is densely defined. By (13.6) we have

$$(T_V - \overline{\lambda}I)(I - V)y = (\lambda - \overline{\lambda})y, \qquad (T_V - \lambda I)(I - V)y = (\lambda - \overline{\lambda})Vy.$$

From these two equations we derive $Vy = (T_V - \lambda I)(T_V - \overline{\lambda}I)^{-1}y$ for $y \in \mathcal{D}(V)$. Thus, we have shown that $V \subseteq (T_V - \lambda I)(T_V - \overline{\lambda}I)^{-1}$. Since

$$\mathcal{D}\big((T_V - \lambda I)(T_V - \overline{\lambda}I)^{-1}\big) \subseteq \mathcal{D}\big((T_V - \overline{\lambda}I)^{-1}\big) = \mathcal{R}(T_V - \overline{\lambda}I) = \mathcal{D}(V),$$

we get $V = (T_V - \lambda I)(T_V - \overline{\lambda}I)^{-1}$, that is, V is the Cayley transform of T_V.   □

Combining Propositions 13.3(i) and (ii), and 13.4 we obtain the following theorem.

Theorem 13.5 *The Cayley transform $T \to V_T = (T - \lambda I)(T - \overline{\lambda}I)^{-1}$ is a bijective mapping of the set of densely defined symmetric operators on $\mathcal{H}$ onto the set of all isometric operators V on $\mathcal{H}$ for which $\mathcal{R}(I - V)$ is dense in $\mathcal{H}$. Its inverse is the inverse Cayley transform $V \to T_V = (\lambda I - \overline{\lambda}V)(I - V)^{-1}$.*

We derive a number of corollaries to Theorem 13.5.

Corollary 13.6 *A densely defined symmetric operator T is self-adjoint if and only if its Cayley transform V_T is unitary.*

Proof By Proposition 3.12, the symmetric operator T is self-adjoint if and only if $\mathcal{R}(T - \overline{\lambda}I) = \mathcal{H}$ and $\mathcal{R}(I - \lambda I) = \mathcal{H}$. Since $\mathcal{D}(V_T) = \mathcal{R}(T - \overline{\lambda}I)$ and $\mathcal{R}(V_T) = \mathcal{R}(T - \lambda I)$, this holds if and only if the isometric operator V_T is unitary.   □

Corollary 13.7 *A unitary operator V is the Cayley transform of a self-adjoint operator if and only if $\mathcal{N}(I - V) = \{0\}$.*

Proof By Theorem 13.5 and Corollary 13.6, V is the Cayley transform of a self-adjoint operator if and only if $\mathcal{R}(I - V)$ is dense. Since V is unitary, it is easily checked that the latter is equivalent to the relation $\mathcal{N}(I - V) = \{0\}$.   □

Corollary 13.8 *If one of the deficiency indices $d_+(T)$ or $d_-(T)$ of a densely defined symmetric operator T on $\mathcal{H}$ is finite, then each symmetric operator S on $\mathcal{H}$ satisfying $S \supseteq \overline{T}$ is closed.*

Proof We assume without loss of generality that $d_+(T) < \infty$ (otherwise, we replace T by $-T$ and use the relation $d_\pm(T) = d_\mp(-T)$). By Proposition 13.3(iv), the Cayley transform V_S of a symmetric extension S of T is an extension of $V_{\overline{T}}$. Let us abbreviate $\mathcal{N}_\lambda := \mathcal{N}(T^* - \lambda I)$. Since $\mathcal{D}(V_{\overline{T}}) = \mathcal{R}(\overline{T} - \overline{\lambda}I)$ is closed by Proposition 13.3(iii), we have $\mathcal{D}(V_{\overline{T}}) \oplus \mathcal{N}_\lambda = \mathcal{H}$ by (1.7), so there is a linear subspace $\mathcal{E}$ of $\mathcal{N}_\lambda$ such that $\mathcal{D}(V_S) = \mathcal{D}(V_{\overline{T}}) \oplus \mathcal{E}$. Since $d_+(T) = \dim \mathcal{N}_\lambda < \infty$, $\mathcal{E}$ is finite-dimensional. Hence, $\mathcal{D}(V_S)$ is also closed, and so is S by Proposition 13.3(iii). $\square$

The Cayley transform V_T defined above depends on the complex number λ from the upper half plane. In applications it is often convenient to take $\lambda = i$. We restate the corresponding formulas in the special case $\lambda = i$:

$$V_T = (T - iI)(T + iI)^{-1}, \qquad \mathcal{D}(V_T) = \mathcal{R}(T + iI), \qquad \mathcal{R}(V_T) = \mathcal{R}(T - iI),$$

$$T = i(I + V_T)(I - V_T)^{-1}, \qquad \mathcal{D}(T) = \mathcal{R}(I - V_T), \qquad \mathcal{R}(T) = \mathcal{R}(I + V_T),$$

$$y = (T + iI)x, \qquad V_T y = (T - iI)x, \qquad 2ix = (I - V_T)y,$$

$$2Tx = (I + V_T)y \quad \text{for } x \in \mathcal{D}(T) \text{ and } y \in \mathcal{D}(V_T).$$

Recall that an operator $V \in \mathbf{B}(\mathcal{H})$ is called a *partial isometry* if there is a closed linear subspace $\mathcal{H}_1$ of $\mathcal{H}$ such that $\|Vx\| = \|x\|$ for $x \in \mathcal{H}_1$ and $Vz = 0$ for $z \in \mathcal{H}_1^\perp$. The subspace $\mathcal{H}_1$ is called the *initial space* of V, and the space $V(\mathcal{H}_1)$ is called the *final space* of V. Some facts on partial isometries are collected in Exercise 7.1.

Suppose now that the symmetric operator T is *closed*. Then, by Proposition 13.3, V_T is an isometric operator defined on the closed linear subspace $\mathcal{D}(V_T)$ of $\mathcal{H}$. It is natural to extend V_T to an everywhere defined bounded operator on $\mathcal{H}$, denoted also by V_T with a slight abuse of notation, by setting $V_T v = 0$ for $v \in \mathcal{D}(V_T)^\perp$. That is, *the operator V_T becomes a partial isometry with initial space* $\mathcal{D}(V_T) = \mathcal{R}(T - \overline{\lambda}I)$ *and final space* $\mathcal{R}(V_T) = \mathcal{R}(T - \lambda I)$. Then Theorem 13.4 can be restated as follows: *The Cayley transform $T \to V_T$ is a bijective mapping of the densely defined closed symmetric linear operators on $\mathcal{H}$ onto the set of partial isometries V on $\mathcal{H}$ for which $(I - V)\mathcal{H}_V$ is dense in $\mathcal{H}$, where $\mathcal{H}_V$ denotes the initial space of V.*

Let $P_\pm$ be the orthogonal projection of $\mathcal{H}$ onto $\mathcal{N}(T^* \mp iI)$. If we consider the Cayley transform V_T as a partial isometry, then we have in the case $\lambda = i$:

$$\mathcal{D}(T) = (I - V_T)(I - P_+)\mathcal{H}, \qquad V_T^* V_T = I - P_+, \qquad V_T V_T^* = I - P_-, \quad (13.7)$$

$$T\big((I - V_T)(I - P_+)y\big) = i(I + V_T)(I - P_+)y, \tag{13.8}$$

$$(T + iI)x = 2i(I - P_+)y, \qquad (T - iI)x = 2iV_T(I - P_+)y \tag{13.9}$$

for $y \in \mathcal{H}$ and $x = (I - V_T)(I - P_+)y$.

13.2   Von Neumann's Extension Theory of Symmetric Operators

By a *symmetric* (resp. *self-adjoint*) *extension* of T we mean a symmetric (resp. self-adjoint) operator S acting on the same Hilbert space $\mathcal{H}$ such that $T \subseteq S$. It is clear that a self-adjoint operator S on $\mathcal{H}$ is an extension of a densely defined symmetric operator T if and only if S is a restriction of the adjoint operator T^*. (Indeed, $T \subseteq S$

implies that $S = S^* \subseteq T^*$. Conversely, if $S \subseteq T^*$, then $S = S^* \supseteq T^{**} = \overline{T} \supseteq T$.) Therefore, any self-adjoint extension S of T is completely described by its domain $\mathcal{D}(S)$, since $S = T^* {\upharpoonright} \mathcal{D}(S)$.

The Cayley transform allows us to reduce the (difficult) problem of describing all symmetric extensions of T to the (easier) problem of finding all isometric extensions of the Cayley transform V_T. If S is a symmetric extension of T, then V_S is an isometric extension of V_T. Conversely, let V be an isometric operator on $\mathcal{H}$ such that $V_T \subseteq V$. Since $\mathcal{R}(I - V_T) \subseteq \mathcal{R}(I - V)$ and $\mathcal{R}(I - V_T)$ is dense, so is $\mathcal{R}(I - V)$. Hence, V is the Cayley transform of a symmetric operator S and $T \subseteq S$.

This procedure leads to a parameterization of all closed symmetric extensions. Recall that the symbol $\dotplus$ denotes the direct sum of vector spaces.

Theorem 13.9 *Let T be a densely defined symmetric operator on $\mathcal{H}$. Suppose that $\mathcal{G}_+ \subseteq \mathcal{N}(T^* - \lambda I)$ and $\mathcal{G}_- \subseteq \mathcal{N}(T^* - \overline{\lambda} I)$ are closed linear subspaces of $\mathcal{H}$ such that $\dim \mathcal{G}_+ = \dim \mathcal{G}_-$ and U is an isometric linear mapping of $\mathcal{G}_+$ onto $\mathcal{G}_-$. Define*

$$\mathcal{D}(T_U) = \mathcal{D}(\overline{T}) \dotplus (I - U)\mathcal{G}_+ \quad and \quad T_U = T^* {\upharpoonright} \mathcal{D}(T_U),$$

that is, $T_U(x + (I - U)y) = \overline{T}x + \lambda y - \overline{\lambda} U y$ for $x \in \mathcal{D}(\overline{T})$ and $y \in \mathcal{G}_+$.

Then T_U is a closed symmetric operator such that $T \subseteq T_U$. Any closed symmetric extension of T on $\mathcal{H}$ is of this form. Moreover, $d_\pm(T) = d_\pm(T_U) + \dim \mathcal{G}_\pm$.

Proof Since any closed extension of T is an extension of $\overline{T}$ and $(\overline{T})^* = T^*$, we can assume that T is closed. Then, by Proposition 13.3, $\mathcal{D}(V_T)$ is closed and closed symmetric extensions of T are in one-to-one correspondence with isometric extensions V of V_T for which $\mathcal{D}(V)$ is closed. Clearly, the latter are given by $\mathcal{D}(V) = \mathcal{D}(V_T) \oplus \mathcal{G}_+$, $Vy = V_T y$ for $y \in \mathcal{D}(V_T)$ and $Vy = Uy$ for $y \in \mathcal{G}_+$, where $\mathcal{G}_\pm$ and U are as in the theorem. Put $T_U := (\lambda I - \overline{\lambda} V)(I - V)^{-1}$. Then

$$\mathcal{D}(T_U) = (I - V)\big(\mathcal{D}(V_T) \oplus \mathcal{G}_+\big) = (I - V_T)\mathcal{D}(V_T) \dotplus (I - U)\mathcal{G}_+$$
$$= \mathcal{D}(T) \dotplus (I - U)\mathcal{G}_+.$$

The two last sums in the preceding equation are direct, because $I - V$ is injective by Lemma 13.2. Since $T \subseteq T_U$ and T_U is symmetric, $T_U \subseteq (T_U)^* \subseteq T^*$, and hence $T_U(x + (I - U)y) = T^*(x + (I - U)y) = \overline{T}x + \lambda y - \overline{\lambda} U y$ for $x \in \mathcal{D}(\overline{T})$, $y \in \mathcal{G}_+$.

By Lemma 13.1 and Proposition 13.3(v), $\dim \mathcal{D}(V)^\perp = d^e(V) = d_+(T_U)$ and $\dim \mathcal{D}(V_T)^\perp = d^e(V_T) = d_+(T)$. Since $\mathcal{D}(V) = \mathcal{D}(V_T) \oplus \mathcal{G}_+$, we have $\mathcal{D}(V_T)^\perp = \mathcal{D}(V)^\perp \oplus \mathcal{G}_+$, and hence $d_+(T) = d_+(T_U) + \dim \mathcal{G}_+$. Similarly, $d_-(T) = d_-(T_U) + \dim \mathcal{G}_-$. $\qquad\square$

The next result is *J. von Neumann's theorem* on self-adjoint extensions.

Theorem 13.10 *A densely defined symmetric operator T on $\mathcal{H}$ possesses a self-adjoint extension on $\mathcal{H}$ if and only if $d_+(T) = d_-(T)$.*

If $d_+(T) = d_-(T)$, then all self-adjoint extensions of T are given by the operators T_U, where U is an isometric linear mapping of $\mathcal{N}(T^* - \lambda I)$ onto $\mathcal{N}(T^* - \overline{\lambda} I)$, and

$$\mathcal{D}(T_U) = \mathcal{D}(\overline{T}) \dotplus (I - U)\mathcal{N}(T^* - \lambda I),$$

$$T_U(x + (I - U)z) = \overline{T}x + \lambda z - \overline{\lambda}Uz$$

for $x \in \mathcal{D}(T)$ and $z \in \mathcal{N}(T^* - \lambda I)$.

Proof By Corollary 13.6, an operator T_U from Theorem 13.9 is self-adjoint if and only if its Cayley transform is unitary, or equivalently, if $\mathcal{G}_+ = \mathcal{N}(T^* - \lambda I)$ and $\mathcal{G}_- = \mathcal{N}(T^* - \overline{\lambda}I)$ in Theorem 13.9. Hence, the description of all self-adjoint extensions of T stated above follows at once from Theorem 13.9.

Clearly, an isometric operator of a closed subspace $\mathcal{H}_1$ onto another closed subspace $\mathcal{H}_2$ exists if and only if $\dim \mathcal{H}_1 = \dim \mathcal{H}_2$. Hence, T has a self-adjoint extension on $\mathcal{H}$, that is, there exists an isometric map of $\mathcal{N}(T^* - \lambda I)$ onto $\mathcal{N}(T^* - \overline{\lambda}I)$, if and only if $d_+(T) \equiv \dim \mathcal{N}(T^* - \lambda I)$ is equal to $d_-(T) \equiv \dim \mathcal{N}(T^* - \overline{\lambda}I)$. $\square$

As an easy application of Theorem 13.10, we give another proof of Proposition 3.17 on self-adjoint extensions in larger spaces.

Corollary 13.11 *Each densely defined symmetric operator T on $\mathcal{H}$ has a self-adjoint extension acting on a possibly larger Hilbert space.*

Proof Define a symmetric operator $S := T \oplus (-T)$ on the Hilbert space $\mathcal{K} := \mathcal{H} \oplus \mathcal{H}$. Since $S^* = T^* \oplus (-T^*)$, we have $\mathcal{N}(S^* \pm iI) = \mathcal{N}(T^* \pm iI) \oplus \mathcal{N}(T^* \mp iI)$. Therefore, S has equal deficiency indices, so S has a self-adjoint extension A on $\mathcal{K}$ by Theorem 13.10. If we identify $\mathcal{H}$ with the subspace $\mathcal{H} \oplus \{0\}$ of $\mathcal{K}$, we obviously have $T \subseteq S \subseteq A$. $\square$

Summarizing, for a densely defined symmetric operator T, there are precisely three possible cases:

Case 1 $[d_+(T) = d_-(T) = 0]$ Then $\overline{T}$ is the unique self-adjoint extension of T on $\mathcal{H}$, so T is essentially self-adjoint.

Case 2 $[d_+(T) = d_-(T) \neq 0]$ Then there are infinitely many isometric mappings of $\mathcal{N}(T^* - \lambda I)$ onto $\mathcal{N}(T^* - \overline{\lambda}I)$ and hence infinitely many self-adjoint extensions of T on $\mathcal{H}$.

Case 3 $[d_+(T) \neq d_-(T)]$ Then there is no self-adjoint extension of T on $\mathcal{H}$.

We close this section by reconsidering the closed symmetric operator $T = -i\frac{d}{dx}$ on $\mathcal{D}(T) = H_0^1(\mathcal{J})$ on the Hilbert space $L^2(\mathcal{J})$, where $\mathcal{J}$ is a either bounded interval (a, b) or $(0, +\infty)$ or $\mathbb{R}$. The deficiency indices of T have been computed in Example 3.2. If $\mathcal{J} = \mathbb{R}$, then $d_+(T) = d_-(T) = 0$, so we are in Case 1, and T is self-adjoint. If $\mathcal{J} = (0, +\infty)$, we have $d_+(T) = 1$ and $d_-(T) = 0$; hence we are in Case 3, and T has no self-adjoint extension on $\mathcal{H}$. Finally, if $\mathcal{J} = (a, b)$, we are in

Case 2, since $d_+(T) = d_-(T) = 1$. The following example illustrates how all self-adjoint extensions of T on $L^2(a, b)$ can be derived by von Neumann's method. The same result is obtained in Example 14.3 below by a much shorter argument using a boundary triplet.

Example 13.1 (*Examples* 1.4 *and* 1.5 *continued*) Let us abbreviate $\mathcal{N}_{\pm i} := \mathcal{N}(T^* \mp iI)$ and recall from Example 3.2 that $\mathcal{N}_i = \mathbb{C} \cdot e^{-x}$ and $\mathcal{N}_{-i} = \mathbb{C} \cdot e^x$. We apply Theorem 13.10 with $\lambda = i$.

Since $e^{a+b-x} \in \mathcal{N}_i$ and $e^x \in \mathcal{N}_{-i}$ have equal norms in $L^2(a, b)$, the isometric mappings of $\mathcal{N}_i$ onto $\mathcal{N}_{-i}$ are parameterized by $w \in \mathbb{T}$ and determined by $U_w(e^{a+b-x}) = we^x$. Let us write T_w for T_{U_w}. By Theorem 13.10, each $f \in \mathcal{D}(T_w)$ is of the form $f(x) = f_0(x) + \alpha(I - U_w)e^{a+b-x}$, where $f_0 \in \mathcal{D}(T)$ and $\alpha \in \mathbb{C}$. In particular, $f \in H^1(a, b)$. Since $f(b) = \alpha(e^a - we^b)$ and $f(a) = \alpha(e^b - we^a)$, the function $f(x)$ fulfills the boundary condition

$$f(b) = z(w)f(a), \quad \text{where } z(w) := \left(e^a - we^b\right)\left(e^b - we^a\right)^{-1}. \quad (13.10)$$

Conversely, if $f \in H^1(a, b)$ satisfies (13.10), setting $\alpha := f(b)(e^a - we^b)^{-1}$ and $f_0(x) := f(x) - \alpha(I - U_w)e^{a+b-x}$, one verifies that $f_0(a) = f_0(b) = 0$. Hence, $f \in \mathcal{D}(T_w)$. By the preceding we have proved that

$$\mathcal{D}(T_w) = \left\{ f \in H^1(a, b) : f(b) = z(w)f(a) \right\}. \quad (13.11)$$

It is easily checked that the map $w \to z(w)$ is a bijection of $\mathbb{T}$ and that its inverse $z \to w(z)$ is given by the same formula, that is, $w(z) := (e^a - ze^b)(e^b - ze^a)^{-1}$.

By (13.11), T_w is just the operator $S_{z(w)}$ from Example 1.5. Since $w(z(w)) = w$, we also have $T_{w(z)} = S_z$. In particular, this shows that the operators S_z, $z \in \mathbb{T}$, from Example 1.5 exhaust all self-adjoint extensions of T on $L^2(a, b)$. ○

13.3 Existence of Positive Self-adjoint Extensions and Krein–von Neumann Extensions

Throughout this section, T is a *positive symmetric operator* on a Hilbert space $\mathcal{H}$. We do not assume that T is closed or that T is densely defined.

If T is densely defined, we know from Theorem 10.17 that T admits a positive self-adjoint extension on $\mathcal{H}$ and that there is a *largest* such extension, the Friedrichs extension T_F. However, if T is not densely defined, it may happen that T has no positive self-adjoint extension on $\mathcal{H}$; see Example 13.2 and Exercise 9.b below.

The next theorem gives a necessary and sufficient condition for the *existence* of a positive self-adjoint extension on $\mathcal{H}$, and it states that in the affirmative case there is always a *smallest* positive self-adjoint extension.

To formulate this theorem, we introduce some notation. For $y \in \mathcal{H}$, we define

$$v_T(y) := \sup_{x \in \mathcal{D}(T)} \frac{|\langle Tx, y \rangle|^2}{\langle Tx, x \rangle}, \quad (13.12)$$

where we set $\frac{0}{0} := 0$. We denote by $\mathcal{E}(T)$ the set of vectors $y \in \mathcal{H}$ for which $\nu_T(y) < \infty$, or equivalently, there exists a number $c_y \geq 0$ such that

$$|\langle Tx, y \rangle|^2 \leq c_y \langle Tx, x \rangle \quad \text{for all } x \in \mathcal{D}(T). \tag{13.13}$$

Clearly, $\nu_T(y)$ is the smallest number c_y satisfying (13.13) for $y \in \mathcal{E}(T)$. Some properties of ν_T and $\mathcal{E}(T)$ can be found in Exercise 8.

Theorem 13.12 (*Ando–Nishio theorem*) *A positive symmetric operator T on $\mathcal{H}$ admits a positive self-adjoint extension on $\mathcal{H}$ if and only if $\mathcal{E}(T)$ is dense in $\mathcal{H}$. If this is true, there exists a unique smallest (according to Definition 10.5) among all positive self-adjoint extensions of T on $\mathcal{H}$. It is called the* Krein–von Neumann extension *of T and denoted by T_N. Then*

$$\begin{aligned}
\mathcal{D}[T_N] &= \mathcal{D}\big(T_N^{1/2}\big) = \mathcal{E}(T), \\
T_N[y] &= \big\| T_N^{1/2} y \big\|^2 = \nu_T(y), \quad y \in \mathcal{E}(T).
\end{aligned} \tag{13.14}$$

In the proof of this theorem we use the following simple lemma.

Lemma 13.13 $\mathcal{D}(A) \subseteq \mathcal{E}(T)$ *for any positive symmetric extension A of T.*

Proof By the Cauchy–Schwarz inequality, for $x \in \mathcal{D}(T)$ and $y \in \mathcal{D}(A)$,

$$|\langle Tx, y \rangle|^2 = |\langle Ax, y \rangle|^2 \leq \langle Ax, x \rangle \langle Ay, y \rangle = \langle Ay, y \rangle \langle Tx, x \rangle,$$

so that $\nu_T(y) \leq \langle Ay, y \rangle < \infty$, that is, $y \in \mathcal{E}(T)$. $\square$

Proof of Theorem 13.12 If T has a positive self-adjoint extension A on $\mathcal{H}$, then $\mathcal{E}(T)$ is dense in $\mathcal{H}$, since it contains the dense set $\mathcal{D}(A)$ by Lemma 13.13.

Conversely, suppose that $\mathcal{E}(T)$ is dense. First, we define an auxiliary Hilbert space $\mathcal{K}_T$. If $Tx = Tx'$ and $Ty = Ty'$ for $x, x', y, y' \in \mathcal{D}(T)$, using that T is symmetric, we obtain $\langle Tx, y \rangle - \langle Tx', y' \rangle = \langle T(x-x'), y \rangle + \langle x', T(y-y') \rangle = 0$. Hence, there is a well-defined positive semidefinite sesquilinear form $\langle \cdot, \cdot \rangle'$ on $\mathcal{R}(T)$ given by

$$\langle Tx, Ty \rangle' := \langle Tx, y \rangle, \quad x, y \in \mathcal{D}(T). \tag{13.15}$$

If $\langle Tx, Tx \rangle' \equiv \langle Tx, x \rangle = 0$, it follows from (13.13) that $Tx \perp y$ for $y \in \mathcal{E}(T)$. Thus, $Tx = 0$, because $\mathcal{E}(T)$ is dense in $\mathcal{H}$. Hence, $\langle \cdot, \cdot \rangle'$ is a scalar product. We denote by $\mathcal{K}_T$ the Hilbert space completion of $(\mathcal{R}(T), \langle \cdot, \cdot \rangle')$ and by $\| \cdot \|'$ the norm of $\mathcal{K}_T$.

Next, we define the linear operator $J : \mathcal{K}_T \to \mathcal{H}$ with domain $\mathcal{D}(J) = \mathcal{R}(T)$ by $J(Tx) = Tx$ for $x \in \mathcal{D}(T)$. A vector $y \in \mathcal{H}$ is in the domain $\mathcal{D}(J^*)$ if and only if the linear functional $Tx \to \langle J(Tx), y \rangle \equiv \langle Tx, y \rangle$ on $\mathcal{R}(T)$ is $\| \cdot \|'$-continuous. Since $\langle Tx, x \rangle^{1/2} = \| Tx \|'$, this holds if and only if (13.13) is satisfied for some $c_y \geq 0$, that is, $y \in \mathcal{E}(T)$. Thus, $\mathcal{D}(J^*) = \mathcal{E}(T)$. Note that $J^* : \mathcal{H} \to \mathcal{K}_T$. By assumption, $\mathcal{E}(T) = \mathcal{D}(J^*)$ is dense in $\mathcal{H}$, so the operator J is closable. Hence, $T_N := \overline{J} J^* \equiv J^{**} J^*$ is a positive self-adjoint operator by Proposition 3.18(ii).

We show that $T \subseteq T_N$. Let $y \in \mathcal{D}(T)$. Since $\langle J(Tx), y \rangle = \langle Tx, y \rangle = \langle Tx, Ty \rangle'$ for $x \in \mathcal{D}(T)$, we conclude that $y \in \mathcal{D}(J^*)$ and $J^*y = Ty$. Hence, we have $Ty = J(Ty) = \overline{J}(Ty) = \overline{J}J^*y = T_N y$. This proves that $T \subseteq T_N$.

Now we verify formula (13.14). Let $y \in \mathcal{E}(T)$. From Lemma 7.1 and from the relations $T_N^{1/2} = (J^{**}J^*)^{1/2} = |J^*|$ it follows that

$$\mathcal{D}(T_N^{1/2}) = \mathcal{D}(|J^*|) = \mathcal{D}(J^*) = \mathcal{E}(T), \qquad \|T_N^{1/2}y\| = \| |J^*|y\| = \|J^*y\|'. \tag{13.16}$$

We have $|\langle Tx, y \rangle| = |\langle JTx, y \rangle| = |\langle Tx, J^*y \rangle'|$ for $x \in \mathcal{D}(T)$. Therefore, since $\mathcal{S} := \{Tx : \|Tx\|' \equiv \langle Tx, x \rangle^{1/2} = 1, \ x \in \mathcal{D}(T)\}$ is the unit sphere of the dense normed subspace $(\mathcal{R}(T), \|\cdot\|')$ of the Hilbert space $\mathcal{K}_T$, we have

$$\nu_T(y) = \sup\{|\langle Tx, y \rangle| : Tx \in \mathcal{S}\} = \sup\{|\langle Tx, J^*y \rangle'| : Tx \in \mathcal{S}\} = \|J^*y\|'.$$

Combined with (13.16), the latter implies (13.14).

Finally, let A be an arbitrary positive self-adjoint extension of T on $\mathcal{H}$. Fix a vector $y \in \mathcal{D}[A] = \mathcal{D}(A^{1/2})$. For $x \in \mathcal{D}(T)$, we deduce that

$$|\langle Tx, y \rangle|^2 = |\langle A^{1/2}x, A^{1/2}y \rangle|^2 \le \|A^{1/2}y\|^2 \langle A^{1/2}x, A^{1/2}x \rangle = \|A^{1/2}y\|^2 \langle Tx, x \rangle.$$

Therefore, $y \in \mathcal{E}(T) = \mathcal{D}[T_N] = \mathcal{D}(T_N^{1/2})$ and $\|T_N^{1/2}y\| = \nu_T(y) \le \|A^{1/2}y\|$ by (13.14). Thus, $T_N \le A$ according to Definition 10.5. This proves that T_N is the smallest positive self-adjoint extension of T on $\mathcal{H}$. $\square$

Remark Krein [Kr3] called the Friedrichs extension T_F the *hard extension* and the Krein–von Neumann extension T_N the *soft extension* of T.

The proof of Theorem 13.12 gives an interesting factorization of the Krein–von Neumann extension T_N. That is, by the above proof and formula (13.16) therein,

$$T_N = \overline{J}J^* \equiv J^{**}J^* \quad \text{and} \quad T_N[x, y] = \langle J^*x, J^*y \rangle' \quad \text{for } x, y \in \mathcal{D}[T_N] = \mathcal{D}(J^*),$$

where J is the embedding operator, that is, $J(Tx) = Tx$ for $x \in \mathcal{D}(J) = \mathcal{R}(T)$, from the auxiliary Hilbert space $\mathcal{K}_T$ defined in the preceding proof into $\mathcal{H}$. This description of the Krein–von Neumann extension T_N is of interest in itself.

We now develop a similar factorization for the Friedrichs extension thereby giving a *second approach to the Friedrichs extension*.

Let T be a *densely defined* positive symmetric operator on $\mathcal{H}$. Then $\mathcal{E}(T)$ is dense in $\mathcal{H}$, since $\mathcal{E}(T) \supseteq \mathcal{D}(T)$ by Lemma 13.13. Therefore, by the above proof of Theorem 13.12, the Hilbert space $\mathcal{K}_T$ is well defined, the operator $J : \mathcal{K}_T \to \mathcal{H}$ defined by $J(Tx) = Tx$ for $x \in \mathcal{D}(T) = \mathcal{R}(T)$ is closable, $\mathcal{D}(T) \subseteq \mathcal{D}(J^*)$, and $J^*x = Tx$ for $x \in \mathcal{D}(T)$.

Now we define $Q := J^* \restriction \mathcal{D}(T)$. By the properties stated in the preceding paragraph we have $Q : \mathcal{H} \to \mathcal{K}_T$ and $Qx = Tx$ for $x \in \mathcal{D}(Q) = \mathcal{D}(T)$. Since $Q^* \supseteq J^{**} \supseteq J$ and $\mathcal{D}(J) = \mathcal{R}(T)$ is dense in $\mathcal{K}_T$, the operator Q is closable.

Proposition 13.14 *The Friedrichs extension T_F of the densely defined positive symmetric operator T and its form are given by*

$$T_F = Q^*\overline{Q} \equiv Q^*Q^{**} \quad and \quad T_F[x, y] = \langle \overline{Q}x, \overline{Q}y \rangle' \quad for \ x, y \in \mathcal{D}[T_F] = \mathcal{D}(\overline{Q}).$$

Proof Since $\mathcal{D}(T) = \mathcal{D}(Q)$ is dense, $\overline{Q}$ is a densely defined closed operator of $\mathcal{H}$ into $\mathcal{K}_T$. Let $\mathfrak{t}$ be the corresponding closed form defined in Example 10.5, that is,

$$\mathfrak{t}[x, y] = \langle \overline{Q}x, \overline{Q}y \rangle' \quad for \ x, y \in \mathcal{D}(\mathfrak{t}) := \mathcal{D}(\overline{Q}).$$

Let $x \in \mathcal{D}(T)$. Then $x \in \mathcal{D}(T_F)$. Using the above formulas, we obtain

$$T_F x = Tx = JTx = Q^*Tx = Q^*Qx = Q^*\overline{Q}x,$$

so that

$$\mathfrak{t}_{T_F}[x] = \left\| T_F^{1/2} x \right\|^2 = \langle T_F x, x \rangle = \langle Q^*\overline{Q}x, x \rangle = \left(\| \overline{Q}x \|' \right)^2 = \mathfrak{t}[x]. \quad (13.17)$$

By the definition of the Friedrichs extension, $\mathcal{D}(T)$ is a core for the form $\mathfrak{t}_{T_F}$. Clearly, $\mathcal{D}(T) = \mathcal{D}(Q)$ is a core for $\overline{Q}$ and hence for the form $\mathfrak{t}$. Therefore, (13.17) implies that the closed forms $\mathfrak{t}_{T_F}$ and $\mathfrak{t}$ coincide, that is, $T_F[x, y] = \langle \overline{Q}x, \overline{Q}y \rangle'$ for $x, y \in \mathcal{D}[T_F] = \mathcal{D}(\overline{Q})$. Hence, the operators associated with $\mathfrak{t}_{T_F}$ and $\mathfrak{t}$ are equal. These are T_F (see Corollary 10.8) and $Q^*\overline{Q}$ (as shown in Example 10.5), respectively. Thus, $T_F = Q^*\overline{Q}$. $\square$

Now we derive a few corollaries from Theorem 13.12.

Corollary 13.15 *Let T be a densely defined positive symmetric operator on $\mathcal{H}$. Then the Friedrichs extension T_F is the* largest, *and the Krein–von Neumann extension T_N is the* smallest *among all positive self-adjoint extensions of T on $\mathcal{H}$. That is, $T_N \leq A \leq T_F$ for any positive self-adjoint extension A of T on $\mathcal{H}$.*

Proof Since $\mathcal{E}(T) \supseteq \mathcal{D}(T)$ by Lemma 13.13, $\mathcal{E}(T)$ is dense in $\mathcal{H}$, so by Theorem 13.12 the Krein–von Neumann extension T_N exists and is the smallest positive self-adjoint extension. By Theorem 10.17(ii), the Friedrichs extension T_F is the largest such extension. $\square$

Corollary 13.16 *Let S be a given positive self-adjoint operator on $\mathcal{H}$. Then T has a positive self-adjoint extension A on $\mathcal{H}$ satisfying $A \leq S$ if and only if*

$$\left| \langle Tx, y \rangle \right|^2 \leq \langle Tx, x \rangle \langle Sy, y \rangle \quad for \ all \ x \in \mathcal{D}(T), \ y \in \mathcal{D}(S). \quad (13.18)$$

Proof First, we assume that (13.18) holds. Then $\mathcal{D}(S) \subseteq \mathcal{E}(T)$, so $\mathcal{E}(T)$ is dense and the Krein–von Neumann extension T_N exists by Theorem 13.12. Using (13.14), (13.12), and (13.18), we conclude that $T_N[y] = v_T(y) \leq \langle Sy, y \rangle = S[y]$ for y in $\mathcal{D}(S) \subseteq \mathcal{E}(T) = \mathcal{D}[T_N]$. Then $T_N \leq S$ by Lemma 10.10(i), so $A := T_N$ is a self-adjoint extension of T satisfying $A \leq S$.

Conversely, suppose that there exists a positive self-adjoint extension A of T such that $A \leq S$. Then $T_N \leq A \leq S$. Therefore, if $y \in \mathcal{D}(S)$, then $y \in \mathcal{D}[S] \subseteq \mathcal{D}[T_N]$

and $v_T(y) = T_N[y] \le S[y] = \langle Sy, y \rangle$ by (13.14). Inserting the definition (13.12) of $v_T(y)$, the latter implies (13.18). $\square$

The special case of Corollary 13.16 when $S = \gamma I$ and γ is a positive number gives the following:

Corollary 13.17 *The operator T has a bounded positive self-adjoint extension A on $\mathcal{H}$ such that $\|A\| \le \gamma$ if and only if $\|Tx\|^2 \le \gamma \langle Tx, x \rangle$ for all $x \in \mathcal{D}(T)$.*

The next results relate the inverse of the Friedrichs (resp. Krein–von Neumann) extension of T to the Krein–von Neumann (resp. Friedrichs) extension of T^{-1}.

Proposition 13.18 *Suppose that $\mathcal{D}(T)$ is dense in $\mathcal{H}$ and $\mathcal{N}(T) = \{0\}$. Then T^{-1} has a positive self-adjoint extension on $\mathcal{H}$ if and only if $\mathcal{N}(T_F) = \{0\}$. If this is fulfilled, then $(T_F)^{-1} = (T^{-1})_N$.*

Proof Obviously, T^{-1} is also positive and symmetric. Suppose that S is a positive self-adjoint extension of T^{-1}. Since $\mathcal{R}(S)$ contains the dense set $\mathcal{D}(T)$, we have $\mathcal{N}(S) = \mathcal{R}(S)^{\perp} = \{0\}$. By Corollary 1.9, S^{-1} is self-adjoint. Clearly, S^{-1} is a positive self-adjoint extension of T. Since T_F is the largest positive self-adjoint extension of T, we have $S^{-1} \le T_F$. Hence, $\mathcal{N}(T_F) = \{0\}$ and $(T_F)^{-1} \le S$ by Corollary 10.12. This shows that $(T_F)^{-1}$ is the *smallest* positive self-adjoint extension of T^{-1}, that is, $(T_F)^{-1} = (T^{-1})_N$.

Conversely, if $\mathcal{N}(T_F) = \{0\}$, then $\mathcal{N}(T) = \{0\}$ and $(T_F)^{-1}$ is self-adjoint (by Corollary 1.9). Hence, $(T_F)^{-1}$ is a positive self-adjoint extension of T^{-1}. $\square$

Proposition 13.19 *Suppose that $\mathcal{E}(T)$ is dense in $\mathcal{H}$. Then we have $\mathcal{N}(T_N) = \{0\}$ if and only if $\mathcal{R}(T)$ is dense in $\mathcal{H}$. If this is true, then T^{-1} is a densely defined positive symmetric operator and $(T^{-1})_F = (T_N)^{-1}$.*

Proof Since $T_N = J^{**}J^*$, we have $\mathcal{N}(T_N) = \mathcal{N}(J^*) = \mathcal{R}(J)^{\perp} = \mathcal{R}(T)^{\perp}$. Therefore, $\mathcal{N}(T_N)$ is trivial if and only if $\mathcal{R}(T)$ is dense in $\mathcal{H}$.

Suppose that $\mathcal{R}(T)$ is dense in $\mathcal{H}$. Then, since T is symmetric, $\mathcal{N}(T) = \{0\}$ and T^{-1} is a densely defined positive symmetric operator on $\mathcal{H}$.

Let S be an arbitrary positive self-adjoint extension of T on $\mathcal{H}$. Since $\mathcal{R}(S) \supseteq \mathcal{R}(T)$ is dense, we have $\mathcal{N}(S) = \{0\}$. Then S^{-1} is a positive self-adjoint extension of T^{-1}, and hence $S^{-1} \le (T^{-1})_F$ by Theorem 10.17(ii). From Corollary 10.12 it follows that $S \ge ((T^{-1})_F)^{-1}$. It is obvious that $((T^{-1})_F)^{-1}$ is an extension of T. Thus, we have shown that $((T^{-1})_F)^{-1}$ is the *smallest* positive self-adjoint extension of T on $\mathcal{H}$. Therefore, $((T^{-1})_F)^{-1} = T_N$, and hence $(T^{-1})_F = (T_N)^{-1}$. $\square$

For the next example, we need the following simple technical lemma.

Lemma 13.20 *Let A be a positive self-adjoint operator on a Hilbert space $\mathcal{H}$ such that $\mathcal{N}(A) = \{0\}$. A vector $y \in \mathcal{H}$ belongs to $\mathcal{D}(A^{-1/2})$ if and only if*

$$\varphi_A(y) := \sup_{x \in \mathcal{D}(A^{1/2}), x \neq 0} \frac{|\langle x, y \rangle|}{\|A^{1/2}x\|} < \infty.$$

If $\varphi_A(y) < \infty$, then $\varphi_A(y) = \|A^{-1/2}y\|$.

Proof Throughout this proof all suprema are over $x \in \mathcal{D}(A^{1/2})$, $x \neq 0$.

First, suppose that $y \in \mathcal{D}(A^{-1/2})$. Since $\mathcal{N}(A) = \{0\}$, we have $\mathcal{N}(A^{1/2}) = \{0\}$. Therefore, $\mathcal{R}(A^{1/2})$ is dense in $\mathcal{H}$, and hence

$$\varphi_A(y) = \sup \frac{|\langle x, y \rangle|}{\|A^{1/2}x\|} = \sup \frac{|\langle A^{1/2}x, A^{-1/2}y \rangle|}{\|A^{1/2}x\|} = \|A^{-1/2}y\| < \infty.$$

Conversely, assume that $\varphi_A(y) < \infty$. Let $A = \int_0^{+\infty} \lambda \, dE(\lambda)$ be the spectral decomposition of A. Set $M_n := [\frac{1}{n}, n]$ and $y_n = E(M_n)y$ for $n \in \mathbb{N}$. Then we have $y_n \in \mathcal{D}(A^{-1/2})$ by (4.26) and $\|A^{1/2}E(M_n)x\| \leq \|A^{1/2}x\|$ for $x \in \mathcal{D}(A^{1/2})$. Since $\varphi_A(y_n) = \|A^{-1/2}y_n\|$ as shown in the preceding paragraph,

$$\|A^{-1/2}y_n\| = \varphi_A(y_n) = \sup \frac{|\langle x, y_n \rangle|}{\|A^{1/2}x\|} \leq \sup \frac{|\langle E(M_n)x, y \rangle|}{\|A^{1/2}E(M_n)x\|} \leq \varphi_A(y).$$

Since $E(\{0\}) = 0$ (by $\mathcal{N}(A) = \{0\}$) and $\lambda^{-1}\chi_{M_n}(\lambda) \to \lambda^{-1}$ for $\lambda \in (0, +\infty)$, using Fatou's Lemma B.3 and the functional calculus, we derive

$$\int_0^{+\infty} \lambda^{-1} d\langle E(\lambda)y, y \rangle \leq \lim_{n \to \infty} \int_0^{+\infty} \lambda^{-1}\chi_{M_n}(\lambda) \, d\langle E(\lambda)y, y \rangle$$
$$= \lim_{n \to \infty} \|A^{-1/2}y_n\|^2 \leq \varphi_A(y)^2 < \infty.$$

Therefore, $y \in \mathcal{D}(A^{-1/2})$ by (4.26). $\square$

Example 13.2 Let $\mathcal{H}_1$ be a closed linear subspace of $\mathcal{H}$, and $\mathcal{H}_2 := \mathcal{H}_1^{\perp}$. Let A be a bounded positive self-adjoint operator on $\mathcal{H}_1$ such that $\mathcal{N}(A) = \{0\}$, and let B be a bounded operator from $\mathcal{H}_1$ into $\mathcal{H}_2$. Define an operator T by $Tx_1 = Ax_1 + Bx_1$ for $x_1 \in \mathcal{D}(T) := \mathcal{H}_1$. It is easily checked that T is bounded, symmetric, and positive.

Let C be a self-adjoint operator on $\mathcal{H}_2$. Since the operators A and B are bounded, it is straightforward to show that the operator matrix

$$S = \begin{pmatrix} A & B^* \\ B & C \end{pmatrix}$$

defines a self-adjoint extension S of T on $\mathcal{H} = \mathcal{H}_1 \oplus \mathcal{H}_2$ with domain $\mathcal{D}(S) = \mathcal{H}_1 \oplus \mathcal{D}(C)$ and that any self-adjoint extension of T on $\mathcal{H}$ is of this form.

Statement *The self-adjoint operator S is positive if and only if*

$$B^*x_2 \in \mathcal{D}(A^{-1/2}) \quad \text{and} \quad \|A^{-1/2}B^*x_2\|^2 \leq \langle Cx_2, x_2 \rangle \quad \text{for } x_2 \in \mathcal{D}(C).$$

$$(13.19)$$

Proof Clearly, S is positive if and only if

$$\langle S(x_1 + \lambda x_2), x_1 + \lambda x_2 \rangle = \langle Ax_1, x_1 \rangle + 2 \operatorname{Re} \lambda \langle B^* x_2, x_1 \rangle + |\lambda|^2 \langle Cx_2, x_2 \rangle \geq 0$$

for $x_1 \in \mathcal{H}_1, x_2 \in \mathcal{D}(C)$, and all (!) complex numbers λ, or equivalently,

$$|\langle x_1, B^* x_2 \rangle|^2 \leq \langle Ax_1, x_1 \rangle \langle Cx_2, x_2 \rangle \quad \text{for } x_1 \in \mathcal{H}_1, x_2 \in \mathcal{D}(C),$$

or equivalently, $\varphi_A(B^* x_2)^2 \leq \langle Cx_2, x_2 \rangle$ for all $x_2 \in \mathcal{D}(C)$. Now Lemma 13.20 yields the assertion.   □

Having the preceding statement, we can easily construct interesting examples:
- Suppose that the operator $A^{-1/2} B^*$ from $\mathcal{H}_2$ into $\mathcal{H}_1$ is not densely defined. Since $\mathcal{D}(C)$ is dense in $\mathcal{H}_2$, it follows then from the above statement that T has *no positive* self-adjoint extension on $\mathcal{H}$. This situation can be achieved if A^{-1} is unbounded and B^* is a rank one operator with range $\mathbb{C} \cdot u$, where $u \notin \mathcal{D}(A^{-1/2})$.
- Let $\mathcal{H}_1 \cong \mathcal{H}_2$ and define $Bx_1 = x_1$ for $x_1 \in \mathcal{H}_1$. From the above statement and Lemma 10.10(i) we then conclude that S is positive if and only if $A^{-1/2} \leq C^{1/2}$. Therefore, if A^{-1} is unbounded, C has to be unbounded; then T has no *bounded* positive self-adjoint extension on $\mathcal{H}$, but it has *unbounded* ones (for instance, by taking $C = A^{-1}$).   ○

13.4 Positive Self-adjoint Extensions and Krein Transform

Positive self-adjoint extensions of positive symmetric operators can be also studied by means of the Krein transform which is the real counterpart of the Cayley transform. The Krein transform reduces the construction of unbounded *positive* self-adjoint extensions to *norm preserving* self-adjoint extensions of (not necessarily densely defined !) bounded symmetric operators.

13.4.1 Self-adjoint Extensions of Bounded Symmetric Operators

Let B be a *bounded* symmetric operator on a Hilbert space $\mathcal{H}$. We do not assume that the domain $\mathcal{D}(B)$ is closed or that it is dense. We want to extend B to a bounded self-adjoint operator on $\mathcal{H}$ having the same norm $\|B\|$ as B.

Clearly, $B_+ := \|B\| \cdot I + B$ and $B_- := \|B\| \cdot I - B$ are positive symmetric operators. For $x \in \mathcal{D}(B) = \mathcal{D}(B_\pm)$, we have

$$\|B_\pm x\|^2 = \|B\|^2 \|x\|^2 + \|Bx\|^2 \pm 2\|B\| \langle Bx, x \rangle$$
$$\leq 2\|B\| \langle (\|B\| \cdot I \pm B)x, x \rangle = 2\|B\| \langle B_\pm x, x \rangle.$$

Therefore, by Corollary 13.17, the operator $B_\pm$ admits a bounded positive self-adjoint extension $D_\pm$ on $\mathcal{H}$ such that $\|D_\pm\| \leq 2\|B\|$. Then, by Theorem 13.12, there exists a *smallest* positive self-adjoint extension $C_\pm$ of $B_\pm$ on $\mathcal{H}$. Since we have $0 \le C_{\pm} \le D_{\pm}$ and $\|D_{\pm}\| \le 2\|B\|$, the operator $C_{\pm}$ is bounded and satisfies $C_{\pm} \le 2\|B\| \cdot I$.

Proposition 13.21 $C_m := C_+ - \|B\| \cdot I$ *is the smallest, and* $C_M := \|B\| \cdot I - C_-$ *is the largest among all bounded self-adjoint extensions of* B *on* $\mathcal{H}$ *which have the same norm as* B.

Proof By construction, $C_m \supseteq B_+ - \|B\| \cdot I = B$ and $C_M \supseteq \|B\| \cdot I - B_- = B$. Clearly, the relations $0 \le C_{\pm} \le 2\|B\| \cdot I$ imply that $-\|B\| \cdot I \le C_m \le \|B\| \cdot I$ and $-\|B\| \cdot I \le C_M \le \|B\| \cdot I$. Since trivially $\|B\| \le \|C_m\|$ and $\|B\| \le \|C_M\|$, we therefore have $\|C_m\| = \|C_M\| = \|B\|$.

Now let C be an arbitrary bounded self-adjoint extension of B with norm $\|B\|$. Then $\|B\| \cdot I \pm C$ is a positive self-adjoint extension of $B_{\pm} = \|B\| \cdot I \pm B$. Since $C_{\pm}$ was the smallest positive self-adjoint extension of $B_{\pm}$, we have $C_{\pm} \le \|B\| I \pm C$ and hence $C_m = C_+ - \|B\| \cdot I \le C \le \|B\| \cdot I - C_- = C_M$. $\qquad\square$

13.4.2 The Krein Transform of a Positive Symmetric Operator

The Krein transform and its inverse are operator analogs of the bijective mappings

$$[0, \infty) \ni \lambda \to t = (\lambda - 1)(\lambda + 1)^{-1} \in [-1, 1),$$

$$[-1, 1) \ni t \to \lambda = (1 + t)(1 - t)^{-1} \in [0, \infty).$$

It plays a similar role for positive symmetric operators as the Cayley transform does for arbitrary symmetric operators.

Let $\mathcal{P}(\mathcal{H})$ denote the set of all positive symmetric operators on a Hilbert space $\mathcal{H}$, and let $\mathcal{S}_1(\mathcal{H})$ be the set of all bounded symmetric operators B on $\mathcal{H}$ such that $\|B\| \le 1$ and $\mathcal{N}(I - B) = \{0\}$. Note that operators in $\mathcal{P}(\mathcal{H})$ and $\mathcal{S}_1(\mathcal{H})$ are in general neither densely defined nor closed.

If $S \in \mathcal{P}(\mathcal{H})$, then obviously $\mathcal{N}(S + I) = \{0\}$. The operator

$$B_S := (S - I)(S + I)^{-1} \quad \text{with domain } \mathcal{D}(B_S) = \mathcal{R}(S + I)$$

is called the *Krein transform* of S.

Let $B \in \mathcal{S}_1(\mathcal{H})$. Since $\mathcal{N}(I - B) = \{0\}$, $I - B$ is invertible. The operator

$$S_B := (I + B)(I - B)^{-1} \quad \text{with domain } \mathcal{D}(S_B) = \mathcal{R}(I - B)$$

is called the *inverse Krein transform* of B. This is justified by the following:

Proposition 13.22 *The Krein transform* $S \to B_S$ *is a bijective mapping of* $\mathcal{P}(\mathcal{H})$ *onto* $\mathcal{S}_1(\mathcal{H})$. *Its inverse is the inverse Krein transform* $B \to S_B$.

Proof Let $S \in \mathcal{P}(\mathcal{H})$. We first show that $B_S \in \mathcal{S}_1(\mathcal{H})$. Let $x, x' \in \mathcal{D}(S)$ and put $y = (S + I)x$, $y' = (S + I)x'$. Then $B_S y = (S - I)x$ and $B_S y' = (S - I)x'$. If

$y \in \mathcal{N}(I - B_S)$, then $(I - B_S)y = 2x = 0$, and so $y = 0$. Thus, $\mathcal{N}(I - B_S) = \{0\}$. Using that S is symmetric and positive, we compute

$$\langle B_S y, y' \rangle = \langle (S - I)x, (S + I)x' \rangle = \langle Sx, Sx' \rangle - \langle x, x' \rangle$$
$$= \langle (S + I)x, (S - I)x' \rangle = \langle y, B_S y' \rangle,$$

$$\|y\|^2 = \|Sx\|^2 + \|x\|^2 + 2\langle Sx, x \rangle \geq \|Sx\|^2 + \|x\|^2 - 2\langle Sx, x \rangle = \|B_S y\|^2.$$

The preceding shows that $B_S \in \mathcal{S}_1(\mathcal{H})$.

We have $2Sx = (I + B_S)y$ and $2x = (I - B_S)y$ for $x \in \mathcal{D}(S)$. Hence, $\mathcal{D}(S) = \mathcal{R}(I - B_S) = \mathcal{D}(S_{(B_S)})$ and $Sx = (I + B_S)(I - B_S)^{-1}x = S_{(B_S)}x$. This proves that S is the inverse Krein transform $S_{(B_S)}$ of B_S.

Suppose now that $B \in \mathcal{S}_1(\mathcal{H})$. Let $y \in \mathcal{D}(B)$. Setting $x := (I - B)y$, we have $S_B x = (I + B)y$. Using the symmetry of B and the fact that $\|B\| \leq 1$, we derive

$$\langle S_B x, x \rangle = \langle (I + B)y, (I - B)y \rangle = \|y\|^2 - \|By\|^2 \geq 0.$$

In particular, $\langle S_B x, x \rangle$ is real for $x \in \mathcal{D}(S_B)$. Hence, S_B is symmetric. Thus, S_B is in $\mathcal{P}(\mathcal{H})$. The relations $(S_B + I)x = 2y$ and $(S_B - I)x = 2By$ imply that $\mathcal{D}(B) = \mathcal{R}(S_B + I) = \mathcal{D}(B_{(S_B)})$ and $By = (S_B - I)(S_B + I)^{-1}y$. This shows that B is the Krein transform $B_{(S_B)}$ of S_B. $\qquad \square$

Proposition 13.23 *For $S, S_1, S_2 \in \mathcal{P}(\mathcal{H})$ we have:*

(i) *S is self-adjoint if and only if $\mathcal{D}(B_S) = \mathcal{H}$, that is, if B_S is self-adjoint.*
(ii) *$S_1 \subseteq S_2$ if and only if $B_{S_1} \subseteq B_{S_2}$.*
(iii) *Suppose that S_1 and S_2 are self-adjoint. Then $S_1 \geq S_2$ if and only if $B_{S_1} \geq B_{S_2}$.*
(iv) *If S is unbounded, then $\|B_S\| = 1$.*

Proof (i): Since $S \geq 0$, Proposition 3.2(i) implies that -1 is in $\pi(S)$. Hence, S is self-adjoint if and only if $\mathcal{R}(S + I) \equiv \mathcal{D}(B_S)$ is $\mathcal{H}$ by Proposition 3.11.

(ii) is obvious.

(iii): By Proposition 10.13, $S_1 \geq S_2$ if and only if $(S_2 + I)^{-1} \geq (S_1 + I)^{-1}$. Since $B_S = I - 2(S + I)^{-1}$, the latter is equivalent to $B_{S_1} \geq B_{S_2}$.

(iv): Because S is unbounded, there is a sequence $(x_n)_{n \in \mathbb{N}}$ of unit vectors x_n in $\mathcal{D}(S)$ such that $\lim_{n \to \infty} \|Sx_n\| = +\infty$. Set $y_n := (S + I)x_n$. Then we have $B_S y_n = Sx_n - x_n$, and hence

$$1 \geq \|B_S\| \geq \|B_S(y_n)\| \|y_n\|^{-1} \geq (\|Sx_n\| - 1)(\|Sx_n\| + 1)^{-1} \to 1$$

as $n \to \infty$. This proves that $\|B_S\| = 1$. $\qquad \square$

The next theorem summarizes some of the results obtained in this section.

Theorem 13.24 *Let S be a densely defined positive symmetric operator on $\mathcal{H}$. If C is a bounded self-adjoint extension of the Krein transform B_S such that $\|C\| \leq 1$, then $C \in \mathcal{S}_1(\mathcal{H})$, and the inverse Krein transform S_C is a positive self-adjoint extension of S. Any positive self-adjoint extension of S is of this form.*

Proof Since C is symmetric, $\mathcal{N}(I - C) \perp \mathcal{R}(I - C)$. Therefore, since $\mathcal{D}(S)$ is dense and $\mathcal{D}(S) = \mathcal{R}(I - B_S) \subseteq \mathcal{R}(I - C)$, it follows that $\mathcal{N}(I - C) = \{0\}$. Thus, $C \in \mathcal{S}_1(\mathcal{H})$. Because $C = B_{(S_C)}$ is a self-adjoint extension of B_S, S_C is a positive self-adjoint extension of S by Propositions 13.22 and 13.23, (i) and (ii).

Conversely, suppose that T is a positive self-adjoint extension of S. Then B_T is a bounded self-adjoint extension of B_S, $\|B_T\| \leq 1$, and $T = S_{(B_T)}$ by Propositions 13.22 and 13.23. $\qquad\qquad\qquad\qquad\qquad\qquad\qquad\qquad\qquad\qquad\qquad\qquad\qquad\qquad\square$

Suppose that S is a *densely defined positive symmetric operator* on $\mathcal{H}$.

If S is bounded, the closure $\overline{S}$ is obviously the unique (positive) self-adjoint extension of S on $\mathcal{H}$.

Suppose that S is *unbounded*. Then $\|B_S\| = 1$ by Proposition 13.23(iv). Proposition 13.21 implies that the set of all bounded self-adjoint extensions C of B_S on $\mathcal{H}$ having the norm $\|B_S\| = 1$ contains a smallest operator C_m and a largest operator C_M. From Theorem 13.24 it follows that $C_m, C_M \in \mathcal{S}_1(\mathcal{H})$ and

$$S_m = (I + C_m)(I - C_m)^{-1} \quad \text{and} \quad S_M = (I + C_M)(I - C_M)^{-1}$$

are positive self-adjoint extensions of S. Let A be an arbitrary positive self-adjoint extension of S on $\mathcal{H}$. Then B_A is a self-adjoint extension of B_S, and $\|B_A\| = \|B_S\| = 1$, so $C_m \leq B_A \leq C_M$ by the definitions of C_m and C_M, and hence $S_m \leq A \leq S_M$ by Proposition 13.23(iii). Thus, we have shown that S_m is the smallest and S_M is the largest among all positive self-adjoint extensions of S on $\mathcal{H}$. Therefore, by Corollary 13.15, S_m is the *Krein–von Neumann extension* S_N, and S_M is the *Friedrichs extension* S_F of the operator S.

13.5 Self-adjoint Extensions Commuting with a Conjugation or Anticommuting with a Symmetry

In this section we investigate two particular cases of self-adjoint extensions.

Definition 13.1 A *conjugation* on a Hilbert space $\mathcal{H}$ is a mapping J of $\mathcal{H}$ into itself that is an antiunitary involution, that is, for $\alpha, \beta \in \mathbb{C}$ and $x, y \in \mathcal{H}$,

$$J(\alpha x + \beta y) = \overline{\alpha} J(x) + \overline{\beta} J(y), \langle Jx, Jy \rangle = \langle y, x \rangle, \qquad J^2 x = x. \quad (13.20)$$

An operator T on $\mathcal{H}$ is called *J-real* if $J\mathcal{D}(T) \subseteq \mathcal{D}(T)$ and $JTx = TJx$ for $x \in \mathcal{D}(T)$.

Example 13.3 The standard example of a conjugation J is the complex conjugation of functions on a Hilbert space $L^2(X, \mu)$, that is, $(Jf)(t) = \overline{f(t)}$. Obviously, a multiplication operator M_φ is J-real if and only if $\varphi(t)$ is real μ-a.e. on X. $\qquad$ o

Proposition 13.25 *Let J be a conjugation, and let T be a densely defined symmetric operator on a Hilbert space $\mathcal{H}$. Suppose that T is J-real. Then we have:*

(i) T^* is J-real.

(ii) If T_U is a self-adjoint extension of T obtained by Theorem 13.10, then T_U is J-real if and only if $JUJUx = x$ for all $x \in \mathcal{N}(T^* - \lambda I)$.

(iii) T has a J-real self-adjoint extension on $\mathcal{H}$.

(iv) If $d_+(T) = 1$, then each self-adjoint extension of T on $\mathcal{H}$ is J-real.

Proof In this proof we abbreviate $\mathcal{N}_\lambda := \mathcal{N}(T^* - \lambda I)$ and $\mathcal{N}_{\bar\lambda} := \mathcal{N}(T^* - \bar\lambda I)$.

(i): For $u, v \in \mathcal{H}$, it follows from (13.20) that

$$\langle u, Jv \rangle = \langle J^2 u, Jv \rangle = \langle v, Ju \rangle. \tag{13.21}$$

Suppose that $y \in \mathcal{D}(T^*)$. Let $x \in \mathcal{D}(T)$. Then $Jx \in \mathcal{D}(T)$ and $TJx = JTx$. Applying (13.21) twice, with $u = Tx, v = y$ and with $u = T^*y, v = x$, we derive

$$\langle Tx, Jy \rangle = \langle y, JTx \rangle = \langle y, TJx \rangle = \langle T^*y, Jx \rangle = \langle x, JT^*y \rangle.$$

This implies that $Jy \in \mathcal{D}(T^*)$ and $T^*Jy = JT^*y$, which proves that T^* is J-real.

(ii): First, we verify that $J\mathcal{N}_\lambda = \mathcal{N}_{\bar\lambda}$. Let $u \in \mathcal{N}_\lambda$. Using that T^* is J-real by (i) and the first equation of (13.20), we obtain $T^*Ju = JT^*u = J(\lambda u) = \bar\lambda Ju$, so $Ju \in \mathcal{N}_{\bar\lambda}$ and $J\mathcal{N}_\lambda \subseteq \mathcal{N}_{\bar\lambda}$. Applying the latter to $\bar\lambda$ instead of λ, we obtain $\mathcal{N}_{\bar\lambda} = J^2\mathcal{N}_{\bar\lambda} \subseteq J\mathcal{N}_\lambda$. Hence, $J\mathcal{N}_\lambda = \mathcal{N}_{\bar\lambda}$.

Now suppose that T_U is a J-real self-adjoint extension of T. Let $x \in \mathcal{N}_\lambda$. Then $(I - U)x \in \mathcal{N}_\lambda + \mathcal{N}_{\bar\lambda}$. Since $J\mathcal{N}_\lambda = \mathcal{N}_{\bar\lambda}$ as just shown and hence $\mathcal{N}_\lambda = J^2\mathcal{N}_\lambda = J\mathcal{N}_{\bar\lambda}$, it follows that $J(I - U)x \in \mathcal{N}_\lambda + \mathcal{N}_{\bar\lambda}$. On the other hand, $(I - U)x \in \mathcal{D}(T_U)$, and hence $J(I - U)x \in \mathcal{D}(T_U)$, because T_U is J-real. The formula for $\mathcal{D}(T_U)$ in Theorem 13.10 implies that $\mathcal{D}(T_U) \cap (\mathcal{N}_\lambda + \mathcal{N}_{\bar\lambda}) \subseteq (I - U)\mathcal{N}_\lambda$. Hence, there is a $y \in \mathcal{N}_\lambda$ such that $J(I-U)x = (I-U)y$. Using once again the equalities $J\mathcal{N}_\lambda = \mathcal{N}_{\bar\lambda}$ and $J\mathcal{N}_{\bar\lambda} = \mathcal{N}_\lambda$, we conclude that $Jx = -Uy$ and $-JUx = y$. Thus, $JUJUx = JU(-y) = JJx = x$.

Conversely, assume that $JUJUx = x$ for $x \in \mathcal{N}_\lambda$. Then we have $UJUx = Jx$, and hence $J(I - U)x = (I - U)(-JUx)$. Since $-JUx \in J\mathcal{N}_{\bar\lambda} = \mathcal{N}_\lambda$, this shows that $J(I - U)\mathcal{N}_\lambda \subseteq (I - U)\mathcal{N}_\lambda$. Because T is J-real and hence $J\mathcal{D}(T) \subseteq \mathcal{D}(T)$, we conclude that $J\mathcal{D}(T_U) \subseteq \mathcal{D}(T_U)$. Since T^* is J-real, we therefore obtain $JT_Ux = JT^*x = T^*Jx = T_UJx$ for $x \in \mathcal{D}(T_U)$, that is, T_U is J-real.

(iii): It suffices to prove that there exists a U such that the self-adjoint operator T_U is J-real. Let $\{e_i\}$ be an orthonormal basis of $\mathcal{N}_\lambda$. Then $\{Je_i\}$ is an orthonormal basis of $\mathcal{N}_{\bar\lambda}$. We define an isometric linear map U of $\mathcal{N}_\lambda$ onto $\mathcal{N}_{\bar\lambda}$ by $U(e_i) = Je_i$, that is, $U(\sum_i \alpha_i e_i) = \sum_i \alpha_i Je_i$. Then we have $JUJUe_i = JUJJe_i = JUe_i = JJe_i = e_i$, and hence $JUJUx = x$ for all $x \in \mathcal{N}_\lambda$. Therefore, T_U is J-real by (ii).

(iv): Let T_U be an arbitrary self-adjoint extension of T on $\mathcal{H}$. By the assumption $d_+(T) = 1$ we have $\mathcal{N}_\lambda = \mathbb{C} \cdot x$ for some $x \in \mathcal{N}_\lambda, x \neq 0$. Since $JUx \in J\mathcal{N}_{\bar\lambda} = \mathcal{N}_\lambda$, there is $z \in \mathbb{C}$ such that $JUx = zx$. Since U is isometric and J is antiunitary, $z \in \mathbb{T}$. Then $JUJUx = JU(zx) = \bar z JUx = \bar z z x = x$. Hence, T_U is J-real by (ii). $\qquad\square$

Example 13.4 (*Examples* 1.4 *and* 1.5 *continued*) Let J be the conjugation on $\mathcal{H} = L^2(-a, a), a > 0$, given by $(Jf)(x) = \overline{f(-x)}$. Then the symmetric operator

$T = -\mathrm{i}\frac{d}{dx}$ with domain $\mathcal{D}(T) = H_0^1(-a, a)$ is J-real. Indeed, if $f \in \mathcal{D}(T)$, then $Jf \in \mathcal{D}(T)$, and

$$(JTf)(x) = \overline{(Tf)(-x)} = \mathrm{i}\,\overline{\frac{d}{dx} f(-x)} = -\mathrm{i}\,\overline{f'(-x)} = (TJf)(x).$$

As shown in Example 13.1, all self-adjoint extensions of T on $\mathcal{H}$ are the operators $S_z = -\mathrm{i}\frac{d}{dx}$ with domains $\mathcal{D}(S_z) = \{f \in H^1(-a, a) : f(a) = zf(-a)\}$, where $z \in \mathbb{T}$, see, e.g., Example 1.5. Since $|z| = 1$, we have $J\mathcal{D}(S_z) \subseteq \mathcal{D}(S_z)$. Therefore, since T^* is J-real, each self-adjoint operator S_z is J-real. This is in accordance with Proposition 13.25(iv), because $d_{\pm}(T) = 1$.   ∘

Definition 13.2 A *symmetry* of a Hilbert space $\mathcal{H}$ is a self-adjoint operator Q on $\mathcal{H}$ such that $Q^2 = I$.

Let Q be a symmetry. Clearly, $\sigma(Q) \subseteq \{-1, 1\}$. If $P_{\pm}$ denotes the projection on $\mathcal{N}(Q \mp I)$, then $P_+ + P_- = I$ and $Q = P_+ - P_- = 2P_+ - I = I - 2P_-$.

Conversely, if P_+ is an orthogonal projection on $\mathcal{H}$, then $Q := 2P_+ - I$ is a symmetry such that $\mathcal{N}(Q - I) = P_+\mathcal{H}$ and $\mathcal{N}(Q + I) = (I - P_+)\mathcal{H}$.

Proposition 13.26 *Let Q be a symmetry, and T be a densely defined symmetric linear operator on $\mathcal{H}$ such that $QT \subseteq -TQ$. Let $P_{\pm} := \frac{1}{2}(I \pm Q)$.*

Then there exist unique self-adjoint extensions T_+ and T_- of T on $\mathcal{H}$ such that $P_{\mp}x \in \mathcal{D}(\overline{T})$ for $x \in \mathcal{D}(T_{\pm})$. Moreover, $QT_{\pm} \subseteq -T_{\pm}Q$, $P_{\pm}\mathcal{D}(T_{\pm}) = P_{\pm}\mathcal{D}(T^*)$, $P_{\mp}\mathcal{D}(T_{\pm}) = P_{\mp}\mathcal{D}(\overline{T})$, and

$$\mathcal{D}(T_{\pm}) = \big\{x \in \mathcal{D}(T^*) : P_{\mp}x \in \mathcal{D}(\overline{T})\big\}. \qquad (13.22)$$

Proof Let $\mathcal{H}_{\pm} := P_{\pm}\mathcal{H}$. The assumption $QT \subseteq -TQ$ yields $(I \pm Q)T \subseteq T(I \mp Q)$, and so $P_{\pm}T \subseteq TP_{\mp}$. Hence, $P_{\pm}\mathcal{D}(T) \subseteq \mathcal{D}(T)$, and $S_{\mp} := T{\restriction}P_{\pm}\mathcal{D}(T)$ maps $\mathcal{H}_{\pm}$ into $\mathcal{H}_{\mp}$. Since $P_+ + P_- = I$, it follows from the preceding that the operators Q and T act as operator block matrices on $\mathcal{H} = \mathcal{H}_+ \oplus \mathcal{H}_-$:

$$Q = \begin{pmatrix} I & 0 \\ 0 & -I \end{pmatrix}, \quad T = \begin{pmatrix} 0 & S_+ \\ S_- & 0 \end{pmatrix}.$$

Since T is densely defined and symmetric, $\mathcal{D}(S_{\mp}) = P_{\pm}\mathcal{D}(T)$ is dense in $\mathcal{H}_{\pm}$, and $S_{\pm} \subseteq (S_{\mp})^*$. It is easily checked that the operators $\overline{T}$ and T^* acts by the matrices

$$\overline{T} = \begin{pmatrix} 0 & \overline{S_+} \\ \overline{S_-} & 0 \end{pmatrix} \quad \text{and} \quad T^* = \begin{pmatrix} 0 & S_-^* \\ S_+^* & 0 \end{pmatrix}.$$

Using the relations $(S_{\pm})^{**} = \overline{S_{\pm}}$ and $S_{\pm} \subseteq (S_{\mp})^*$, it follows that the operators

$$T_+ := \begin{pmatrix} 0 & \overline{S_+} \\ S_+^* & 0 \end{pmatrix} \quad \text{and} \quad T_- := \begin{pmatrix} 0 & S_-^* \\ \overline{S_-} & 0 \end{pmatrix}$$

are self-adjoint on $\mathcal{H}$ and extensions of T. Clearly, $P_{\pm}\mathcal{D}(T_{\pm}) = \mathcal{D}(S_{\pm}^*) = P_{\pm}\mathcal{D}(T^*)$ and $P_{\mp}\mathcal{D}(T_{\pm}) = \mathcal{D}(\overline{S_{\pm}}) = P_{\mp}\mathcal{D}(\overline{T})$. By a simple computation we verify that

$QT_{\pm} \subseteq -T_{\pm}Q$. A vector $x \in \mathcal{D}(T^*)$ belongs to $\mathcal{D}(T_{\pm})$ if and only if $P_{\mp}x \in \mathcal{D}(\overline{S_{\pm}})$, or equivalently, $P_{\mp}x \in \mathcal{D}(\overline{T})$. This proves (13.22).

We verify the uniqueness assertion. Fix $\tau \in \{+, -\}$ and let A be a self-adjoint extension of T on $\mathcal{H}$ such that $P_{-\tau}x \in \mathcal{D}(\overline{T})$ for $x \in \mathcal{D}(A)$. Since $A \subseteq T^*$, we then have $\mathcal{D}(A) \subseteq \mathcal{D}(T_{\tau})$ by the definition of T_{τ}. Because T_{τ} and A are self-adjoint restrictions of T^*, the latter yields $A \subseteq T_{\tau}$, and hence $A = T_{\tau}$.   □

Example 13.5 ("*Doubling trick*") Let S be a densely defined closed symmetric operator on $\mathcal{H}_0$, and let $z \in \mathbb{T}$. Define a closed symmetric operator T and a symmetry Q_z on $\mathcal{H} = \mathcal{H}_0 \oplus \mathcal{H}_0$ by

$$T = \begin{pmatrix} S & 0 \\ 0 & -S \end{pmatrix}, \qquad Q_z = \begin{pmatrix} 0 & z \\ \bar{z} & 0 \end{pmatrix}.$$

Clearly, $Q_z T \subseteq -T Q_z$, so the assumptions of Proposition 13.26 are fulfilled. Since

$$T^* = \begin{pmatrix} S^* & 0 \\ 0 & -S^* \end{pmatrix} \quad \text{and} \quad P_- = \frac{1}{2} \begin{pmatrix} 1 & -z \\ -\bar{z} & 1 \end{pmatrix},$$

by (13.22) the self-adjoint operator T_+ acts by $T_+(x, y) = (S^*x, -S^*y)$ on

$$\mathcal{D}(T_+) = \{(x, y) : x, y \in \mathcal{D}(S^*), \ x - zy \in \mathcal{D}(S)\}.$$

This general construction remains valid if the number $z \in \mathbb{T}$ is replaced by a unitary operator U on $\mathcal{H}_0$ satisfying $S = USU^*$.

We illustrate the preceding by a simple example. Let S be the operator $-i\frac{d}{dx}$ with domain $\mathcal{D}(S) = H_0^1(\mathcal{J})$ on $\mathcal{H}_0 = L^2(\mathcal{J})$ for an open interval $\mathcal{J}$. From Sect. 1.3.1 we know that $S^* = -i\frac{d}{dx}$ and $\mathcal{D}(S^*) = H^1(\mathcal{J})$. Hence, the operator T_+ acts by $T_+(f, g) = (-if', ig')$. We still have to describe the domain $\mathcal{D}(T_+)$.

Let $\mathcal{J} = (a, b)$, where $a, b \in \mathbb{R}$, $a < b$. Then the domain of T_+ is given by

$$\mathcal{D}(T_+) = \{(f, g) : f, g \in H^1(a, b), \ f(a) = zg(a), \ f(b) = zg(b)\}.$$

Let $\mathcal{J} = (0, +\infty)$. Then $\mathcal{D}(T_+) = \{(f, g) : f, g \in H^1(0, +\infty), f(0) = zg(0)\}$. In this case if z runs through $\mathbb{T}$, these operators T_+ exhaust all self-adjoint extensions of T on $\mathcal{H}$. We now slightly reformulate this example.

Let R denote the operator $-i\frac{d}{dx}$ on $\mathcal{D}(R) = H_0^1(-\infty, 0)$ on the Hilbert space $\mathcal{H}_1 = L^2(-\infty, 0)$. Clearly, $(Uf)(x) = f(-x)$ is a unitary transformation of $\mathcal{H}_1$ onto $\mathcal{H}_0 = L^2(0, +\infty)$ such that $URU^* = -S$ and $UR^*U^* = -S^*$. Therefore, if we identify $f \in \mathcal{H}_1$ with $Uf \in \mathcal{H}_0$, then (up to unitary equivalence) the operators T and T_+ act as $-i\frac{d}{dx}$ with domains $\mathcal{D}(T) = H_0^1(-\infty, 0) \oplus H_0^1(0, +\infty)$ and

$$\mathcal{D}(T_+) = \{f \in H^1(-\infty, 0) \oplus H^1(0, +\infty) : f(+0) = zf(-0)\}. \qquad ○$$

13.6 Exercises

1. (*Restrictions of self-adjoint operators*)
   Let A be a self-adjoint operator on $\mathcal{H}$, and let $V = (A - iI)(A + iI)^{-1}$ be its Cayley transform. Let $\mathcal{K}$ be a closed linear subspace of $\mathcal{H}$. Let $\mathcal{K}_n$ be the

closure of $\mathcal{K} + V^*\mathcal{K} + \cdots + (V^*)^{(n-1)}\mathcal{K}$. We denote by T the restriction of A to the domain $\mathcal{D}(T) = (I - V)\mathcal{K}^\perp$. Let $n, k \in \mathbb{N}, k < n$.

a. Show that T is a closed symmetric operator on $\mathcal{H}$.

b. Show that $\mathcal{D}(T^n)$ is dense in $\mathcal{H}$ if and only if $\mathcal{D}(A) \cap \mathcal{K}_n = \{0\}$. In particular, $\mathcal{D}(T)$ is dense in $\mathcal{H}$ if and only if $\mathcal{D}(A) \cap \mathcal{K} = \{0\}$.

*c. Show that $\mathcal{D}(T^n)$ is a core for T^k if and only if $\mathcal{D}(A^{n-k}) \cap \mathcal{K}_n \subseteq (A - iI)^{k-n}\mathcal{K}_k$.

d. Show that if $\mathcal{K}$ is finite-dimensional, then $\mathcal{D}(T^n)$ is a core for $T^k, k < n$.

*2. Let A be a densely defined closed operator on a separable Hilbert space $\mathcal{H}$. Suppose that A is *not bounded*. Show that there exists a closed linear subspace $\mathcal{K}$ of $\mathcal{H}$ such that

$$\mathcal{D}(A) \cap \mathcal{K} = \mathcal{D}(A) \cap \mathcal{K}^\perp = \{0\}. \qquad (13.23)$$

Sketch of proof. (See [Sch2]) Upon replacing A by $|A|$ there is no loss of generality to assume that A is self-adjoint. Since A is not bounded, use the spectral theorem to show that there is an orthonormal basis $\{x_{kn} : k \in \mathbb{N}, n \in \mathbb{Z}\}$ such that for all $k \in \mathbb{N}$, there exists $n_k \in \mathbb{N}$ such that

$$\left\| (A - iI)^{-1} x_{kn} \right\| \leq \frac{1}{|n|!} \quad \text{when } |n| \geq n_k. \qquad (13.24)$$

Let $\mathcal{K}$ denote the linear subspace of all vectors $y \in \mathcal{H}$ such that for all $k \in \mathbb{Z}$,

$$\sum_{n \in \mathbb{Z}} \langle y, x_{kn} \rangle e^{int} = 0 \quad \text{a.e. on } [0, \pi].$$

Note that $(\langle y, x_{kn} \rangle)_{n \in \mathbb{Z}} \to \sum_n \langle y, x_{kn} \rangle e^{int}$ is a unitary map of the closed subspace $\mathcal{K}_k$ spanned by the vectors $x_{kn}, n \in \mathbb{Z}$, on the subspace of functions f of $L^2(-\pi, \pi)$ for which $f = 0$ a.e. on $[0, \pi]$. This implies that $\mathcal{K}$ is closed and that $\mathcal{K}^\perp$ is the set of all vectors $y \in \mathcal{H}$ such that $\sum_{n \in \mathbb{Z}} \langle y, x_{kn} \rangle e^{int} = 0$ a.e. on $[-\pi, 0]$ for all $k \in \mathbb{N}$.

Suppose that $y \in \mathcal{D}(A)$. Writing $y = (A + iI)^{-1}x$ and using (13.24), it follows that $f_k(z) = \sum_{n \in \mathbb{Z}} \langle y, x_{kn} \rangle z^n$ is a holomorphic function on $\mathbb{C} \setminus \{0\}$. If y is in $\mathcal{K}$ or in $\mathcal{K}^\perp$, then $f_k = 0$ a.e. on a half of the unit circle $\mathbb{T}$. Hence, $f_k \equiv 0$ on $\mathbb{T}$ for all k, which implies that $y = 0$.

3. (*A classical result of J. von Neumann* [vN2]) Let A and $\mathcal{H}$ be as in Exercise 2. Show that there exists a unitary operator on $\mathcal{H}$ such that $\mathcal{D}(A) \cap \mathcal{D}(UAU^*) = \{0\}$. Hint: Set $U = I - 2P$, where P is the projection on a subspace satisfying (13.23).

4. (*Restrictions of unbounded self-adjoint operators continued*) Let A be an (unbounded) self-adjoint operator on $\mathcal{H}$ with Cayley transform $V = (A - iI)(A + iI)^{-1}$. Suppose that $\mathcal{K}$ is a closed linear subspace of $\mathcal{H}$ satisfying (13.23). Let T_1 and T_2 denote the restrictions of A to the domains $\mathcal{D}(T_1) = (I - V)\mathcal{K}$ and $\mathcal{D}(T_2) = (I - V)\mathcal{K}^\perp$.

Show that T_1 and T_2 are densely defined closed symmetric operators such that

$$\mathcal{D}(T_1) \cap \mathcal{D}(T_2) = \{0\} \quad \text{and} \quad \mathcal{D}(T_1^2) = \mathcal{D}(T_2^2) = \{0\}.$$

Hint: Use Exercise 1.

(Details and more results concerning Exercises 1–4 are given in [Sch2].)

*5. (*Cayley transform and bounded transform of a symmetric operator*)

Let T be a densely defined closed symmetric operator on $\mathcal{H}$. Let $P_\pm$ denote the projection of $\mathcal{H}$ on $\mathcal{N}(T^* \mp iI)$, and let V_T be the Cayley transform of T considered as a partial isometry with initial space $(I - P_+)\mathcal{H}$ and final space $(I - P_-)\mathcal{H}$; see (13.7) and (13.9). Let $Z_T = T(I + T^*T)^{-1/2}$ be the bounded transform of T. Set $W_\pm(T) := Z_T \pm i(I - (Z_T)^*Z_T)^{1/2}$; see, e.g., Exercise 7.7.

a. Show that $4(I + T^*T)^{-1} = (I - V_T - P_-)^*(I - V_T - P_-)$.

b. Show that $4T(I + T^*T)^{-1} = i(V_T - (V_T)^* - P_+ + P_-)$.

c. Show that $V_T = W_-(T)W_+(T)^*$ and $P_\pm = I - W_\pm(T)W_\pm(T)^*$.

   (See, e.g., [WN, p. 365].)

6. Let $T = -\frac{d^2}{dx^2}$ with domain $\mathcal{D}(T) = H_0^2(0, +\infty)$ on $\mathcal{H} = L^2(0, +\infty)$.

a. Show that $d_\pm(T) = 1$ and determine the deficiency spaces $\mathcal{N}(T^* \mp iI)$.

b. Use von Neumann's Theorem 13.10 to describe all self-adjoint extension of T in terms of boundary conditions at 0.

7. (*Another formally normal operator without normal extension* [Sch4])

Let S be the unilateral shift operator (see, e.g., Example 5.5), $B := S + S^*$, and $A := i(S + I)(I - S)^{-1}$. Show that $T := A + iB$ is a formally normal operator which has no normal extension in a possibly larger Hilbert space.

Hint: Use the fact that $\dim \mathcal{D}(T^*)/\mathcal{D}(T) = 1$.

8. Let T be a (not necessarily closed or densely defined) positive symmetric operator on $\mathcal{H}$. Let v_T be defined by (13.12), and $\mathcal{E}(T) = \{y \in \mathcal{H} : v_T(y) < \infty\}$.

a. Show that $\mathcal{E}(T)$ is a linear subspace of $\mathcal{H}$ which contains $\mathcal{D}(T)$.

b. Show that $v_T(x) = \langle Tx, x \rangle$ for $x \in \mathcal{D}(T)$.

9. Let $a, b \in \mathbb{C}$. Define a linear operator T on the Hilbert space $\mathcal{H} = \mathbb{C}^2$ with domain $\mathcal{D}(T) = \{(x, 0) : x \in \mathbb{C}\}$ by $T(x, 0) = (ax, bx)$.

a. Show that T is a positive symmetric operator if and only if $a \geq 0$.

b. Suppose that $b \neq 0$ and $a \geq 0$. Show that T has a positive self-adjoint extension on $\mathcal{H}$ if and only if $a > 0$.

   (If $a = 0$ and $b = 1$, then T is a positive symmetric operator which has no positive self-adjoint extension on $\mathcal{H}$.)

10. Let A be a bounded self-adjoint operator on a Hilbert space $\mathcal{H}_0$. Define a linear operator T on $\mathcal{H} = \mathcal{H}_0 \oplus \mathcal{H}_0$ by $Tx = (Ax, x)$ for $x \in \mathcal{D}(T) := \mathcal{H}_0$.

a. Let $\gamma \geq 0$. Show that T has a bounded self-adjoint extension S on $\mathcal{H}$ such that $\|S\| \leq \gamma$ if and only if $A^2 + I - \gamma A \leq 0$.

b. Suppose that $\sigma(A) \subseteq [c, 1]$ for some $c \in (0, 1)$. Show that the smallest possible norm of a bounded self-adjoint extension of T on $\mathcal{H}$ is $c + c^{-1}$.

   Hint: See Example 13.2 and use Corollary 13.17.

11. Retain the notation of Example 13.2. Prove that $|A^{-1/2}B^*|^2$ is the smallest positive self-adjoint operator C for which the block matrix S is positive.

*12. (*Geometric mean of two positive operators*)

Let A, B, and X be positive self-adjoint operators on $\mathcal{H}$ such that A is invertible and $A^{-1} \in \mathbf{B}(\mathcal{H})$. Consider the operator matrix on $\mathcal{H} \oplus \mathcal{H}$ defined by

$$T_X = \begin{pmatrix} A & X \\ X & B \end{pmatrix}.$$

a. Show that the largest operator $X \geq 0$ such that $T_X \geq 0$ is

$$A \bowtie B := A^{1/2}\left(A^{-1/2}BA^{-1/2}\right)^{1/2}A^{-1/2}.$$

(The operator $A \bowtie B$ is called the *geometric mean* of A and B.)

Hints: Prove that $T_X \geq 0$ if and only if $B \geq XA^{-1}X$. Conclude that the largest X is the (unique) positive solution of the equation $B = XA^{-1}X$.

b. Show that $A \bowtie B = (AB)^{1/2}$ when $AB = BA$.

c. Show that $C(A \bowtie B)C^* = (CAC^*) \bowtie (CBC^*)$ for any operator $C \in \mathbf{B}(\mathcal{H})$ such that $C^{-1} \in \mathbf{B}(\mathcal{H})$.

d. Suppose in addition that $B^{-1} \in \mathbf{B}(\mathcal{H})$. Show that X is equal to $A \bowtie B$ if and only if $B^{-1/2}XA^{-1/2}$ (or equivalently, $A^{-1/2}XB^{-1/2}$) is unitary.

(More on geometric means can be found in [ALM].)

*13. (*Positive operator block matrices*)

Let A, B, $C \in \mathbf{B}(\mathcal{H})$. Show that the operator block matrix

$$T = \begin{pmatrix} A & B^* \\ B & C \end{pmatrix} \tag{13.25}$$

on $\mathcal{H} \oplus \mathcal{H}$ is positive if and only if $A \geq 0$, $C \geq 0$, and there exists an operator $D \in \mathbf{B}(\mathcal{H})$ such that $\|D\| \leq 1$ and $B = C^{1/2}DA^{1/2}$.

*14. (*Extremal characterization of the Schur complement*)

Let A, B, $C \in \mathbf{B}(\mathcal{H})$. Suppose that the operator (13.25) is positive on $\mathcal{H} \oplus \mathcal{H}$ and define $S(T) := A^{1/2}(I - D^*D)A^{1/2}$, where D is as in Exercise 13.

a. Show that $S(T)$ is the largest positive operator $X \in \mathbf{B}(\mathcal{H})$ such that

$$\begin{pmatrix} A - X & B^* \\ B & C \end{pmatrix} \geq 0.$$

(The operator $S(T)$ is called the *Schur complement* of C in the matrix T.)

b. Suppose that the operator C is invertible with $C^{-1} \in \mathbf{B}(\mathcal{H})$. Show that $S(T) = A - B^*C^{-1}B$.

(For detailed proofs and more on Exercises 13 and 14, we refer to [Dr].)

15. Let T be a positive symmetric operator on $\mathcal{H}$ which has a positive self-adjoint extension on $\mathcal{H}$. Suppose that $\dim \mathcal{D}(T) < \infty$. Show that $T_N \in \mathbf{B}(\mathcal{H})$, $\mathcal{R}(T_N) = \mathcal{R}(T)$, and $\dim \mathcal{R}(T_N) = \dim \mathcal{D}(T) - \dim \mathcal{D}(T) \cap \mathcal{R}(T)^\perp$.

16. Let A and B be densely defined operators on $\mathcal{H}$ such that $A \subseteq B^*$ and $B \subseteq A^*$, or equivalently, $\langle Ax, y \rangle = \langle x, By \rangle$ for $x \in \mathcal{D}(A)$ and $y \in \mathcal{D}(B)$. In this case we say that the operators A and B form an *adjoint pair*. Define the operator T with domain $\mathcal{D}(T) = \{(x, y) : x \in \mathcal{D}(A), y \in \mathcal{D}(B)\}$ on $\mathcal{H} \oplus \mathcal{H}$ by $T(x, y) = (By, Ax)$.

a. Show that T is a densely defined symmetric operator on $\mathcal{H} \oplus \mathcal{H}$.

b. Find a self-adjoint extension of T on $\mathcal{H} \oplus \mathcal{H}$.

c. Suppose that J_0 is a conjugation on $\mathcal{H}$ such that A and B are J_0-real. Show that T has a J-real self-adjoint extension on $\mathcal{H} \oplus \mathcal{H}$, where $J := J_0 \oplus J_0$.

17. Let Ω be an open subset of $\mathbb{R}^d$, and $V \in L^2_{\text{loc}}(\Omega)$ be real-valued. Let J be the "standard" conjugation on $L^2(\Omega)$ defined by $(Jf)(x) = \overline{f(x)}$.

Show that the symmetric operator $T = -\Delta + V$ on $\mathcal{D}(T) = C_0^\infty(\Omega)$ has a J-real self-adjoint extension on $L^2(\Omega)$.

Chapter 14
Self-adjoint Extensions: Boundary Triplets

This chapter is devoted to a powerful approach to the self-adjoint extension theory which is useful especially for differential operators. It is based on the notion of a boundary triplet for the adjoint of a symmetric operator T. In this context extensions of the operator T are parameterized by linear relations on the boundary space. In Sect. 14.1 we therefore develop some basics on linear relations on Hilbert spaces.

Boundary triplets and the corresponding extensions of T are developed in Sect. 14.2. Gamma fields and Weyl functions are studied in Sect. 14.5, and various versions of the Krein–Naimark resolvent formula are derived in Sect. 14.6. Sect. 14.7 deals with boundary triplets for semibounded operators, while the Krein–Birman–Vishik theory of positive self-adjoint extensions is treated in Sect. 14.8.

Examples of boundary triplets are investigated in Sects. 14.3 and 14.4; further examples will appear in the next chapters.

14.1 Linear Relations

Let $\mathcal{H}_1$ and $\mathcal{H}_2$ be Hilbert spaces with scalar products $\langle \cdot, \cdot \rangle_1$ and $\langle \cdot, \cdot \rangle_2$, respectively.

If T is a linear operator from $\mathcal{H}_1$ into $\mathcal{H}_2$, then an element (x, y) of $\mathcal{H}_1 \oplus \mathcal{H}_2$ belongs to the graph $\mathcal{G}(T)$ of T if and only if $x \in \mathcal{D}(T)$ and $y = Tx$. Thus, the action of the operator T can be completely recovered from the linear subspace $\mathcal{G}(T)$ of $\mathcal{H}_1 \oplus \mathcal{H}_2$. In what follows we will identify the operator T with its graph $\mathcal{G}(T)$.

A linear subspace T of $\mathcal{H}_1 \oplus \mathcal{H}_2$ will be called a *linear relation* from $\mathcal{H}_1$ into $\mathcal{H}_2$. For such a linear relation T, we define the *domain* $\mathcal{D}(T)$, the *range* $\mathcal{R}(T)$, the *kernel* $\mathcal{N}(T)$, and the *multivalued part* $\mathcal{M}(T)$ by

$$\mathcal{D}(T) = \left\{ x \in \mathcal{H}_1 : (x, y) \in T \text{ for some } y \in \mathcal{H}_2 \right\},$$
$$\mathcal{R}(T) = \left\{ y \in \mathcal{H}_2 : (x, y) \in T \text{ for some } x \in \mathcal{H}_1 \right\},$$
$$\mathcal{N}(T) = \left\{ x \in \mathcal{H}_1 : (x, 0) \in T \right\},$$
$$\mathcal{M}(T) = \left\{ y \in \mathcal{H}_2 : (0, y) \in T \right\}.$$

K. Schmüdgen, *Unbounded Self-adjoint Operators on Hilbert Space*, Graduate Texts in Mathematics 265, DOI 10.1007/978-94-007-4753-1_14, © Springer Science+Business Media Dordrecht 2012

The *closure* $\overline{T}$ is just the closure of the linear subspace T, and T is called *closed* if $\overline{T} = T$. Further, we define the *inverse* T^{-1} and the *adjoint* T^* by

$$T^{-1} = \{(y, x) : (x, y) \in T\},$$
$$T^* = \{(y, x) : \langle v, y \rangle_2 = \langle u, x \rangle_1 \text{ for all } (u, v) \in T\}.$$

If T is (the graph of) a linear operator, one easily checks that these notions coincide with the corresponding notions for operators, provided that inverse, adjoint, and closure, respectively, of the operator T exist. It should be emphasized that in contrast to operators for a linear relation inverse, adjoint, and closure always exist! They are again linear relations. For instance, if T is a nonclosable operator, then the closure $\overline{T}$ is only a linear relation, but not (the graph of) an operator (see, e.g., Example 1.1).

If S and T are linear relations from $\mathcal{H}_1$ into $\mathcal{H}_2$, R is a linear relation from $\mathcal{H}_2$ into $\mathcal{H}_3$, and $\alpha \in \mathbb{C} \setminus \{0\}$, we define αT, the *sum* $S + T$, and the *product* RT by

$$\alpha T = \{(x, \alpha y) : (x, y) \in T\},$$
$$S + T = \{(x, u + v) : (x, u) \in S \text{ and } (x, v) \in T\},$$
$$RT = \{(x, y) : (x, v) \in T \text{ and } (v, y) \in R \text{ for some } v \in \mathcal{H}_2\}.$$

A larger number of known facts for operators remains valid for linear relations as well. For arbitrary (!) linear relations S and R, we have

$$(T^*)^* = \overline{T}, \qquad\qquad (T^*)^{-1} = (T^{-1})^*,$$
$$(RT)^{-1} = T^{-1}R^{-1}, \qquad T^*R^* \subseteq (RT)^*, \qquad\qquad (14.1)$$
$$\mathcal{R}(T)^\perp = \mathcal{N}(T^*), \qquad \mathcal{R}(T^*)^\perp = \mathcal{N}(\overline{T}), \qquad\qquad (14.2)$$
$$\mathcal{D}(T)^\perp = \mathcal{M}(T^*), \qquad \mathcal{D}(T^*)^\perp = \mathcal{M}(\overline{T}). \qquad\qquad (14.3)$$

The proofs of the equalities in (14.1) and (14.2) can be given in a similar manner as in the operator case. We do not carry out these proofs and leave them as exercises.

The action set of $x \in \mathcal{D}(T)$ is defined by $T(x) = \{y \in \mathcal{H}_2 : (x, y) \in T\}$. If T is an operator, then $T(x)$ consists of a single element. Clearly, $T(0) = \{0\}$ if and only if T is (the graph of) an operator. In general the multivalued part $\mathcal{M}(T) = T(0)$ of a linear relation T can be an arbitrary linear subspace of $\mathcal{H}_2$. We may therefore interpret linear relations as "multivalued linear mappings."

Linear relations have become now basic objects in modern operator theory. They occur when nondensely defined operators T are studied, since then $\mathcal{M}(T^*) \neq \{0\}$ by (14.3), that is, T^* is *not* (the graph of) an operator. Our main motivation is that they appear in the parameterization of self-adjoint extensions in Sect. 14.2.

Let T be a *closed* linear relation. We define two closed subspaces of $\mathcal{H}_1 \oplus \mathcal{H}_2$ by

$$T_\infty := \{(0, y) \in T\} = \{0\} \oplus \mathcal{M}(T) \quad \text{and} \quad T_s := T \ominus T_\infty.$$

Then $(0, y) \in T_s$ implies that $y = 0$. Therefore, by Lemma 1.1, T_s is (the graph of) a linear operator which is closed, since T_s is closed. The closed operator T_s is called the *operator part* of T. We then have the orthogonal decomposition $T = T_s \oplus T_\infty$.

This equality is a useful tool to extend results which are known for operators to closed linear relations.

Spectrum and resolvent are defined in a similar manner as in the operator case.

Definition 14.1 Let T be a closed relation on a Hilbert space $\mathcal{H}$. The *resolvent set* $\rho(T)$ is the set of all $\lambda \in \mathbb{C}$ such that $(T - \lambda I)^{-1}$ is (the graph of) an operator of $B(\mathcal{H})$. This operator is called the *resolvent* of T at $\lambda \in \rho(T)$. The *spectrum* of T is the set $\sigma(T) = \mathbb{C} \backslash \rho(T)$.

Note that $T - \lambda I$ is not necessarily the graph of an operator when $\lambda \in \rho(T)$. For instance, if $T = \{0\} \oplus \mathcal{H}$, then T^{-1} is the graph of the null operator and $0 \in \rho(T)$.

Proposition 14.1 *Let T be a closed relation on $\mathcal{H}$. A complex number λ is in $\rho(T)$ if and only if $\mathcal{N}(T - \lambda I) = \{0\}$ and $\mathcal{R}(T - \lambda I) = \mathcal{H}$.*

Proof This proof is based on two simple identities which are easily verified:

$$\mathcal{N}(T - \lambda I) = \mathcal{M}\big((T - \lambda I)^{-1}\big), \qquad \mathcal{R}(T - \lambda I) = \mathcal{D}\big((T - \lambda I)^{-1}\big). \quad (14.4)$$

Let $\lambda \in \rho(T)$. Since $(T - \lambda I)^{-1}$ is an operator defined on $\mathcal{H}$, $\mathcal{M}((T - \lambda I)^{-1}) = \{0\}$ and $\mathcal{D}((T - \lambda I)^{-1}) = \mathcal{H}$, so the assertion follows from (14.4).

Conversely, assume that $\mathcal{N}(T - \lambda I) = \{0\}$ and $\mathcal{R}(T - \lambda I) = \mathcal{H}$. Then, by (14.4), $\mathcal{M}((T - \lambda I)^{-1}) = \{0\}$ and $\mathcal{D}((T - \lambda I)^{-1}) = \mathcal{H}$. Hence, $(T - \lambda I)^{-1}$ is the graph $\mathcal{G}(S)$ of an operator S with domain $\mathcal{H}$. Since the relations T and hence $T - \lambda I$ are closed, so is $(T - \lambda I)^{-1}$. Hence, $\mathcal{G}(S)$ is closed, that is, S is a closed operator defined on $\mathcal{H}$. Therefore, by the closed graph theorem, S is bounded. This proves that $\lambda \in \rho(T)$. $\qquad\square$

A linear relation T on a Hilbert space $\mathcal{H}$ is called *symmetric* if $T \subseteq T^*$, that is, $\langle x, v \rangle = \langle y, u \rangle$ for all $(x, y), (u, v) \in T$.

Let T be a symmetric relation. For $\alpha \in \mathbb{R}$, we write $T \geq \alpha I$ if $\langle y, x \rangle \geq \alpha \langle x, x \rangle$ for all $(x, y) \in T$. In this case T is said to be *lower semibounded*. In particular, T is called *positive* if $\langle y, x \rangle \geq 0$ for $(x, y) \in T$.

A linear relation T on $\mathcal{H}$ is called *self-adjoint* if $T = T^*$.

Let $\mathcal{S}(\mathcal{H})$ denote the set of all self-adjoint operators B acting on a closed linear subspace $\mathcal{H}_B$ of $\mathcal{H}$. Note that $\mathcal{R}(B) \subseteq \mathcal{H}_B$ for all $B \in \mathcal{S}(\mathcal{H})$. Throughout we shall denote the projection of $\mathcal{H}$ onto $\mathcal{H}_B$ by P_B.

Proposition 14.2 *There is a one-to-one correspondence between the sets of operators $B \in \mathcal{S}(\mathcal{H})$ and of self-adjoint relations $\mathcal{B}$ on $\mathcal{H}$ given by*

$$\mathcal{B} = \mathcal{G}(B) \oplus \big(\{0\} \oplus (\mathcal{H}_B)^\perp\big) \equiv \big\{(x, Bx + y) : x \in \mathcal{D}(B), \ y \in (\mathcal{H}_B)^\perp\big\}, \quad (14.5)$$

where B is the operator part $\mathcal{B}_s$, and $(\mathcal{H}_B)^\perp$ is the multivalued part $\mathcal{M}(\mathcal{B})$ of $\mathcal{B}$. Moreover, $\mathcal{B} \geq 0$ if and only if $B \geq 0$.

Proof It is easily verified that (14.5) defines a self-adjoint relation $\mathcal{B}$ for $B \in \mathcal{S}(\mathcal{H})$.

Conversely, suppose that $\mathcal{B}$ is a self-adjoint relation on $\mathcal{H}$. Let $B := \mathcal{B}_s$ and $\mathcal{H}_B := \mathcal{M}(\mathcal{B})^\perp$. Since $\mathcal{D}(B)^{\perp\perp} = \mathcal{D}(\mathcal{B})^{\perp\perp} = \mathcal{M}(\mathcal{B})^\perp = \mathcal{H}_B$ by (14.3), $\mathcal{D}(B)$ is a dense linear subspace of $\mathcal{H}_B$. Since $\mathcal{B}_s = \mathcal{G}(B) \perp \{0\} \oplus \mathcal{M}(\mathcal{B})$, we have $\mathcal{R}(B) \subseteq \mathcal{M}(\mathcal{B})^\perp = \mathcal{H}_B$. Thus, B is a densely defined linear operator on $\mathcal{H}_B$. Using that $\mathcal{B}$ is self-adjoint, it is easy to check that B is self-adjoint on $\mathcal{H}_B$.

Let $(x, Bx + y) \in \mathcal{B}$. Since $x \perp y$, we have $\langle Bx + y, x \rangle = \langle Bx, x \rangle$. Hence, $\mathcal{B} \geq 0$ is equivalent to $B \geq 0$.   □

Lemma 14.3 *Let $\mathcal{B}$ be a linear relation on $\mathcal{H}$ such that $\mathcal{B}^{-1}$ is the graph of a positive self-adjoint operator A on $\mathcal{H}$. Then $\mathcal{B}$ is a positive self-adjoint relation.*

Proof Since $\mathcal{N}(A)$ is a reducing subspace for A, we can write $A = C \oplus 0$ on $\mathcal{H} = \mathcal{H}_B \oplus \mathcal{N}(A)$, where C is a positive self-adjoint operator with trivial kernel on $\mathcal{H}_B := \mathcal{N}(A)^\perp$. Then $B := C^{-1} \in \mathcal{S}(\mathcal{H})$ and $\mathcal{B} = A^{-1} = \mathcal{G}(B) \oplus (\{0\} \oplus (\mathcal{H}_B)^\perp)$. Thus, the relation $\mathcal{B}$ is of the form (14.5) and hence self-adjoint by Proposition 14.2. Since $A \geq 0$, we have $C \geq 0$, hence $B \geq 0$, and so $\mathcal{B} \geq 0$.   □

Suppose that $\mathcal{B}$ is a self-adjoint relation. Let B denote its operator part. The *form* $\mathfrak{t}_{\mathcal{B}}[\cdot,\cdot] = \mathcal{B}[\cdot,\cdot]$ associated with $\mathcal{B}$ is defined by

$$\mathcal{B}[x, x'] := B[x, x'] \quad \text{for } x, x' \in \mathcal{D}[\mathcal{B}] := \mathcal{D}[B]. \tag{14.6}$$

As in the operator case, $\mathcal{D}(\mathcal{B}) = \mathcal{D}(B)$ is a core for the form $\mathfrak{t}_{\mathcal{B}}$.

The next result extends the Cayley transform to self-adjoint relations.

Proposition 14.4 *A linear relation $\mathcal{B}$ on $\mathcal{H}$ is self-adjoint if and only if there is a unitary operator V on $\mathcal{H}$ such that*

$$\mathcal{B} = \{(x, y) \in \mathcal{H} \oplus \mathcal{H} : (I - V)y = \mathrm{i}(I + V)x\}. \tag{14.7}$$

The operator V is uniquely determined by $\mathcal{B}$ and called the Cayley transform *of $\mathcal{B}$. Each unitary operator is the Cayley transform of some self-adjoint relation $\mathcal{B}$ on $\mathcal{H}$.*

Proof Suppose that $\mathcal{B}$ is self-adjoint. Let $B \in \mathcal{S}(\mathcal{H})$ be the corresponding self-adjoint operator from Lemma 14.2. By Corollary 13.6, the Cayley transform $V_B = (B - \mathrm{i}I)(B + \mathrm{i}I)^{-1}$ is a unitary operator on $\mathcal{H}_B$. Then $Vx = (I - P_B)x + V_B P_B x$, $x \in \mathcal{H}$, defines a unitary operator on $\mathcal{H}$ such that (14.7) holds. Conversely, each operator V satisfying (14.7) acts as the identity on $(I - P_B)\mathcal{H}$ and as the Cayley transform of B on $\mathcal{H}_B$. Hence, V is uniquely determined by $\mathcal{B}$.

Conversely, let V be an arbitrary unitary operator on $\mathcal{H}$. Then the closed linear subspace $\mathcal{H}_B := \mathcal{N}(I - V)^\perp$ reduces the unitary operator V, and for the restriction V_0 of V to $\mathcal{H}_B$, we have $\mathcal{N}(I - V_0) = \{0\}$. Therefore, by Corollary 13.7, V_0 is the Cayley transform (with $\lambda = \mathrm{i}$) of a self-adjoint operator B on $\mathcal{H}_B$, and we have $B = \mathrm{i}(I + V_0)(I - V_0)^{-1}$. Then the relation $\mathcal{B}$ defined by (14.5) is self-adjoint by Proposition 14.2, and we easily derive (14.7).   □

14.2 Boundary Triplets of Adjoints of Symmetric Operators

In the rest of this chapter, T denotes a *densely defined symmetric operator* on a Hilbert space $\mathcal{H}$, and we abbreviate

$$\mathcal{N}_\lambda := \mathcal{N}(T^* - \lambda I) \quad \text{for } \lambda \in \mathbb{C}.$$

Definition 14.2 A *boundary triplet* for T^* is a triplet $(\mathcal{K}, \Gamma_0, \Gamma_1)$ of a Hilbert space $(\mathcal{K}, \langle\cdot,\cdot\rangle_\mathcal{K})$ and linear mappings $\Gamma_0 : \mathcal{D}(T^*) \to \mathcal{K}$ and $\Gamma_1 : \mathcal{D}(T^*) \to \mathcal{K}$ such that:

(i) $[x, y]_{T^*} \equiv \langle T^*x, y\rangle - \langle x, T^*y\rangle = \langle \Gamma_1 x, \Gamma_0 y\rangle - \langle \Gamma_0 x, \Gamma_1 y\rangle$ for $x, y \in \mathcal{D}(T^*)$,
(ii) the mapping $\mathcal{D}(T^*) \ni x \to (\Gamma_0 x, \Gamma_1 x) \in \mathcal{K} \oplus \mathcal{K}$ is surjective.

The scalar product and norm of the "boundary" Hilbert space $\mathcal{K}$ will be denoted by $\langle\cdot,\cdot\rangle_\mathcal{K}$ and $\|\cdot\|_\mathcal{K}$, respectively, while as usual the symbols $\langle\cdot,\cdot\rangle$ and $\|\cdot\|$ refer to the scalar product and norm of the underlying Hilbert space $\mathcal{H}$.

Let $\Gamma_+ := \Gamma_1 + i\Gamma_0$ and $\Gamma_- := \Gamma_1 - i\Gamma_0$. Then condition (i) is equivalent to

$$2i[x, y]_{T^*} = \langle \Gamma_- x, \Gamma_- y\rangle_\mathcal{K} - \langle \Gamma_+ x, \Gamma_+ y\rangle_\mathcal{K} \quad \text{for } x, y \in \mathcal{D}(T^*). \tag{14.8}$$

Sometimes it is convenient to use the following reformulation of Definition 14.2:
If Γ_+ and Γ_- are linear mappings of $\mathcal{D}(T^*)$ into a Hilbert space $(\mathcal{K}, \langle\cdot,\cdot\rangle_\mathcal{K})$ such that (14.8) holds and the mapping $\mathcal{D}(T^*) \ni x \to (\Gamma_+ x, \Gamma_- x) \in \mathcal{K} \oplus \mathcal{K}$ is surjective, then $(\mathcal{K}, \Gamma_0, \Gamma_1)$ is a boundary triplet for T^*, where

$$\Gamma_1 := (\Gamma_+ + \Gamma_-)/2, \qquad \Gamma_0 := (\Gamma_+ - \Gamma_-)/2i. \tag{14.9}$$

In this case we will call the triplet $(\mathcal{K}, \Gamma_+, \Gamma_-)$ also a boundary triplet for T^*.

The vectors $\Gamma_1(x)$ and $\Gamma_2(x)$ of $\mathcal{K}$ are called *abstract boundary values* of $x \in \mathcal{D}(T^*)$, and the equation in condition (i) is called *abstract Green identity*. This terminology stems from differential operators where Γ_1 and Γ_2 are formed by means of boundary values of functions and their derivatives. In these cases integration by parts yields condition (i) resp. Eq. (14.8).

Let us first look at the simplest examples.

Example 14.1 Let $a, b \in \mathbb{R}$, $a < b$, and let $T = -i\frac{d}{dx}$ be the symmetric operator on $\mathcal{D}(T) = H_0^1(a, b)$ in $L^2(a, b)$. By integration by parts (see (1.13)) we have

$$i[f, g]_{T^*} = f(b)\overline{g(b)} - f(a)\overline{g(a)}. \tag{14.10}$$

Therefore, the triplet

$$\mathcal{K} = \mathbb{C}, \qquad \Gamma_+(f) = \sqrt{2}f(a), \qquad \Gamma_-(f) = \sqrt{2}f(b)$$

is a boundary triplet for T^*. Indeed, (14.10) implies (14.8). The surjectivity condition (ii) in Definition 14.2 is obviously satisfied, since $\mathcal{D}(T^*) = H^1(a, b)$.   ∘

Example 14.2 Let $a, b \in \mathbb{R}$, $a < b$. For the symmetric operator $T = -\frac{d^2}{dx^2}$ on $\mathcal{D}(T) = H_0^2(a, b)$ in $L^2(a, b)$, integration by parts yields (see (1.14))

$$[f, g]_{T^*} = f(b)\overline{g'(b)} - f'(b)\overline{g(b)} - f(a)\overline{g'(a)} + f'(a)\overline{g(a)}. \quad (14.11)$$

Hence, there is a boundary triplet for T^* defined by

$$\mathcal{K} = \mathbb{C}^2, \quad \Gamma_0(f) = \big(f(a), f(b)\big), \quad \Gamma_1(f) = \big(f'(a), -f'(b)\big). \quad (14.12)$$

Condition (i) holds by (14.11), while (ii) is obvious, since $\mathcal{D}(T^*) = H^2(a, b)$. $\quad\circ$

The question of when a boundary triplet for T^* exists is answered by the following:

Proposition 14.5 *There exists a boundary triplet* $(\mathcal{K}, \Gamma_0, \Gamma_1)$ *for* T^* *if and only if the symmetric operator* T *has equal deficiency indices. We then have* $d_+(T) = d_-(T) = \dim \mathcal{K}$.

Proof If $(\mathcal{K}, \Gamma_0, \Gamma_1)$ is a boundary triplet for T^*, by Lemma 14.13(ii) below Γ_0 is a topological isomorphism of $\mathcal{N}_{\pm i}$ onto $\mathcal{K}$. Thus, $d_+(T) = d_-(T) = \dim \mathcal{K}$. Conversely, if $d_+(T) = d_-(T)$, then Example 14.4 below shows that there exists a boundary triplet for T^*. $\qquad\square$

Definition 14.3 A closed operator S on $\mathcal{H}$ is a *proper extension* of T if

$$T \subseteq S \subseteq T^*.$$

Two proper extensions S_1 and S_2 of T are called *disjoint* if $\mathcal{D}(\overline{T}) = \mathcal{D}(S_1) \cap \mathcal{D}(S_2)$ and *transversal* if $\mathcal{D}(S_1) + \mathcal{D}(S_2) = \mathcal{D}(T^*)$.

Suppose that $(\mathcal{K}, \Gamma_0, \Gamma_1)$ is a boundary triple for T^*. We shall show that it gives rise to a natural parameterization of the set of proper extensions of T in terms of closed linear relations $\mathcal{B}$ on $\mathcal{K}$. In particular, the self-adjoint extensions of the symmetric operator T on $\mathcal{H}$ will be described in terms of self-adjoint relations on $\mathcal{K}$.

Let $\mathcal{B}$ be a linear relation on $\mathcal{K}$, that is, $\mathcal{B}$ is a linear subspace of $\mathcal{K} \oplus \mathcal{K}$. We denote by $T_\mathcal{B}$ the restriction of T^* to the domain

$$\mathcal{D}(T_\mathcal{B}) := \big\{ x \in \mathcal{D}(T^*) : (\Gamma_0 x, \Gamma_1 x) \in \mathcal{B} \big\}. \quad (14.13)$$

If $\mathcal{B}$ is the graph of an operator B on $\mathcal{K}$, then the requirement $(\Gamma_0 x, \Gamma_1 x) \in \mathcal{B}$ means that $\Gamma_1 x - B \Gamma_0 x = 0$, so that

$$\mathcal{D}(T_\mathcal{B}) = \mathcal{N}(\Gamma_1 - B \Gamma_0).$$

If S is a linear operator such that $T \subseteq S \subseteq T^*$, we define its *boundary space* by

$$\mathcal{B}(S) = \big\{ (\Gamma_0 x, \Gamma_1 x) : x \in \mathcal{D}(S) \big\}. \quad (14.14)$$

From Definition 14.2(ii) it follows that $\mathcal{B}(T_\mathcal{B}) = \mathcal{B}$ for any linear relation $\mathcal{B}$ on $\mathcal{K}$.

Lemma 14.6 *Let* $\mathcal{B}$ *be a linear relation on* $\mathcal{K}$, *and let* S *be a linear operator on* $\mathcal{H}$ *such that* $T \subseteq S \subseteq T^*$ *and* $\mathcal{B}(S) = \mathcal{B}$. *Then:*

(i) $S^* = T_{B^*}$.

(ii) $\overline{S} = T_{\overline{B}}$.

(iii) *If S is closed, then $B(S) = B$ is closed.*

(iv) $\overline{T} = T_{\{(0,0)\}}$, *that is*, $\mathcal{D}(\overline{T}) = \{x \in \mathcal{D}(T^*) : \Gamma_0 x = \Gamma_1 x = 0\}$.

Proof (i): First note that $T \subseteq S \subseteq T^*$ implies that $\overline{T} = T^{**} \subseteq S^* \subseteq T^*$. Hence, a vector $y \in \mathcal{D}(T^*)$ belongs to $\mathcal{D}(S^*)$ if and only if for all $x \in \mathcal{D}(S)$, we have

$$\langle T^*x, y \rangle \equiv \langle Sx, y \rangle = \langle x, T^*y \rangle,$$

or equivalently by Definition 14.2(i), if $\langle \Gamma_0 x, \Gamma_1 y \rangle_{\mathcal{K}} = \langle \Gamma_1 x, \Gamma_0 y \rangle_{\mathcal{K}}$ for $x \in \mathcal{D}(S)$, that is, if $\langle u, \Gamma_1 y \rangle = \langle v, \Gamma_0 y \rangle$ for all $(u, v) \in B(S) = B$. By the definition of the adjoint relation the latter is equivalent to $(\Gamma_0 y, \Gamma_1 y) \in B^*$, and so to $y \in \mathcal{D}(T_{B^*})$ by (14.13). Thus, we have shown that $\mathcal{D}(S^*) = \mathcal{D}(T_{B^*})$. Because both S^* and T_{B^*} are restrictions of T^*, the latter yields $S^* = T_{B^*}$.

(ii): Since $S^* = T_{B^*}$, we have $B(S^*) = B(T_{B^*}) = B^*$. Therefore, applying this to S^* and B^* and using that $B^{**} = \overline{B}$, we get $\overline{S} = S^{**} = T_{B^{**}} = T_{\overline{B}}$.

(iii): Suppose that S is closed. Then $S \subseteq T_B \subseteq T_{\overline{B}} = \overline{S} = S$ by (ii), so we have $T_B = T_{\overline{B}}$. By Definition 14.2(ii) this implies that $B = \overline{B}$.

(iv): Set $B := \mathcal{K} \oplus \mathcal{K}$. Since then $B^* = \{(0,0)\}$ and $T_B = T^*$, using (i), we obtain $\overline{T} = T^{**} = (T_B)^* = T_{B^*} = T_{\{(0,0)\}}$. $\square$

Proposition 14.7 *There is a one-to-one correspondence between all* closed *linear relations B on $\mathcal{K}$ and all* proper extensions S *of T given by $B \leftrightarrow T_B$. Furthermore, if B, B_0, and B_1 are closed relations on $\mathcal{K}$, we have:*

(i) $B_0 \subseteq B_1$ *is equivalent to* $T_{B_0} \subseteq T_{B_1}$.

(ii) T_{B_0} *and* T_{B_1} *are disjoint if and only if* $B_0 \cap B_1 = \{(0,0)\}$.

(iii) T_{B_0} *and* T_{B_1} *are transversal if and only if* $B_0 + B_1 = \mathcal{K} \oplus \mathcal{K}$.

(iv) T_B *is symmetric if and only if* B *is symmetric.*

(v) T_B *is self-adjoint if and only if* B *is self-adjoint.*

Proof All assertions are easily derived from Lemma 14.6. If B is a closed relation, then T_B is closed by Lemma 14.6(ii) applied to $S = T_B$. If S is a closed operator and $T \subseteq S \subseteq T^*$, then $B(S)$ is closed, and $S = T_{B(S)}$ by Lemma 14.6, (iii) and (ii). This gives the stated one-to-one correspondence.

(i) follows at once from (14.13) combined with the fact that $B(T_{B_j}) = B_j$, $j = 0, 1$, while (ii) and (iii) follow from (i) and the corresponding definitions.

(iv): Clearly, T_B is symmetric if and only if $T_B \subseteq (T_B)^* = T_{B^*}$ (by Lemma 14.6(i)), or equivalently, if $B \subseteq B^*$ by (i), that is, if B is symmetric.

(v): T_B is self-adjoint if and only if $T_B = T_{B^*}$, that is, if $B = B^*$ by (i). $\square$

Any boundary triplet determines two distinguished self-adjoint extensions of T.

Corollary 14.8 *If $(\mathcal{K}, \Gamma_0, \Gamma_1)$ is a boundary triplet for T^*, then there exist self-adjoint extensions T_0 and T_1 of the symmetric operator T on $\mathcal{H}$ defined by $\mathcal{D}(T_0) = \mathcal{N}(\Gamma_0)$ and $\mathcal{D}(T_1) = \mathcal{N}(\Gamma_1)$.*

Proof Clearly, $\mathcal{B}_0 = \{0\} \oplus \mathcal{K}$ and $\mathcal{B}_1 = \mathcal{K} \oplus \{0\}$ are self-adjoint relations. Hence, $T_0 = T_{\mathcal{B}_0}$ and $T_1 = T_{\mathcal{B}_1}$ are self-adjoint operators by Proposition 14.7(v).   □

Corollary 14.9 *Let $\mathcal{B}$ be a closed relation on $\mathcal{K}$. The operators $T_{\mathcal{B}}$ and T_0 are disjoint if and only if $\mathcal{B}$ is the graph of an operator on the Hilbert space $\mathcal{K}$.*

Proof By Proposition 14.7(ii), $T_{\mathcal{B}}$ and $T_0 \equiv T_{\mathcal{B}_0}$ are disjoint if and only if $\mathcal{B} \cap \mathcal{B}_0 = \{(0,0)\}$, where $\mathcal{B}_0 = \{0\} \oplus \mathcal{K}$, or equivalently, if $\mathcal{B}$ is the graph of an operator.   □

Recall from Sect. 14.1 that $\mathcal{S}(\mathcal{K})$ is the set of all self-adjoint operators B acting on a closed subspace $\mathcal{K}_B$ of $\mathcal{K}$ and P_B is the orthogonal projection onto $\mathcal{K}_B$.

For any $B \in \mathcal{S}(\mathcal{K})$, we define T_B to be the restriction of T^* to the domain

$$\mathcal{D}(T_B) := \left\{ x \in \mathcal{D}(T^*) : \Gamma_0 x \in \mathcal{D}(B) \text{ and } B\Gamma_0 x = P_B \Gamma_1 x \right\}. \qquad (14.15)$$

That is, a vector $x \in \mathcal{D}(T^*)$ is in $\mathcal{D}(T_B)$ if and only if there are vectors $u \in \mathcal{D}(B)$ and $v \in (\mathcal{K}_B)^\perp$ such that $\Gamma_0 x = u$ and $\Gamma_1 x = Bu + v$.

If $\mathcal{B}$ is the self-adjoint relation with operator part $\mathcal{B}_s := B$ and multivalued part $\mathcal{M}(\mathcal{B}) := (\mathcal{K}_B)^\perp$ (by Proposition 14.2), we conclude that $T_{\mathcal{B}} = T_B$ by comparing the definitions (14.13) and (14.15). This fact will be often used in what follows.

Proposition 14.7 gives a complete description of self-adjoint extensions of T in terms of self-adjoint relations on $\mathcal{K}$, or equivalently, of operators from $\mathcal{S}(\mathcal{K})$.

Now we develop another parameterization of self-adjoint extensions based on unitaries. If V is a unitary operator on $\mathcal{K}$, let T^V denote the restriction of T^* to

$$\mathcal{D}(T^V) := \left\{ x \in \mathcal{D}(T^*) : V\Gamma_+ x = \Gamma_- x \right\}. \qquad (14.16)$$

By Lemma 14.6(iv), the operators T^V and $T_{\mathcal{B}} = T_B$ (defined by (14.13) and (14.15)) are extensions of $\overline{T}$.

Theorem 14.10 *Suppose that $(\mathcal{K}, \Gamma_0, \Gamma_1)$ is a boundary triplet for T^*. For any operator S on $\mathcal{H}$, the following are equivalent:*

(i) *S is a self-adjoint extension of T on $\mathcal{H}$.*
(ii) *There is a self-adjoint linear relation $\mathcal{B}$ on $\mathcal{K}$ such that $S = T_{\mathcal{B}}$ (or equivalently, there is an operator $B \in \mathcal{S}(\mathcal{K})$ such that $S = T_B$).*
(iii) *There is a unitary operator V on $\mathcal{K}$ such that $S = T^V$.*

The relation $\mathcal{B}$ and the operators V and B are then uniquely determined by S.

Proof The equivalence of (i) and (ii) is already contained in Proposition 14.7. Recall that $\mathcal{B}$ is uniquely determined by the operator $T_{\mathcal{B}}$, since $\mathcal{B}(T_{\mathcal{B}}) = \mathcal{B}$.

(ii) → (iii): Let $\mathcal{B}$ be a self-adjoint relation, and let $V_{\mathcal{B}}$ be its Cayley transform (see Proposition 14.4). Using formulas (14.9), (14.7), and (14.16), one easily computes that a vector $x \in \mathcal{D}(T^*)$ satisfies $(\Gamma_0 x, \Gamma_1 x) \in \mathcal{B}$ if and only if $V_{\mathcal{B}} \Gamma_+ x = \Gamma_- x$. Therefore, $T_{\mathcal{B}} = T^{V_{\mathcal{B}}}$.

(iii) → (ii): Let V be a unitary operator on $\mathcal{K}$. By Proposition 14.4 there is a self-adjoint relation $\mathcal{B}$ with Cayley transform V. As shown in the proof of the implication

(ii) $\rightarrow$ (iii), we then have $T_B = T^V$. From (14.16) and Definition 14.2(ii) it follows that the unitary operator V is uniquely determined by S. $\qquad\square$

Example 14.3 (*Example* 14.1 *continued*) We apply Theorem 14.10 to the boundary triplet from Example 14.1. It follows that all self-adjoint extensions of the operator $T = -i\frac{d}{dx}$ on $\mathcal{D}(T) = H_0^1(a,b)$ are the operators $T^z = -i\frac{d}{dx}$ with domains $\mathcal{D}(T^z) = \{f \in H^1(a,b) : f(b) = zf(a)\}$, where $z \in \mathbb{T}$. This fact has been derived in Example 13.1 by using the Cayley transform. Note that T^z is just the operator S_z from Example 1.5. $\qquad\circ$

14.3 Operator-Theoretic Examples

In this section we assume that the densely defined symmetric operator T has *equal deficiency indices*. Our aim is to construct three examples of boundary triplets.

Example 14.4 By formula (3.10) in Proposition 3.7, each element x of $\mathcal{D}(T^*)$ can be written as $x = x_0 + x_+ + x_-$, where $x_0 \in \mathcal{D}(\overline{T})$ and $x_\pm \in \mathcal{N}_{\pm i}$ are uniquely determined by x. Define $Q_\pm x = x_\pm$. Since we assumed that $d_+(T) = d_-(T)$, there exists an isometric linear mapping W of $\mathcal{N}_i$ onto $\mathcal{N}_{-i}$. Set

$$\mathcal{K} = \mathcal{N}_{-i}, \qquad \Gamma_+ = 2WQ_+, \qquad \Gamma_- = 2Q_-.$$

Statement $(\mathcal{K}, \Gamma_+, \Gamma_-)$ *is a boundary triplet for* T^*.

Proof Let $x = x_0 + x_+ + x_-$ and $y = y_0 + y_+ + y_-$ be vectors of $\mathcal{D}(T^*)$, where $x_0, y_0 \in \mathcal{D}(\overline{T})$ and $x_\pm, y_\pm \in \mathcal{N}_{\pm i}$. A straightforward simple computation yields

$$[x_0 + x_+ + x_-, y_0 + y_+ + y_-]_{T^*} = 2i\langle x_+, y_+\rangle - 2i\langle x_-, y_-\rangle.$$

Using this equation and the fact that W is isometric, we obtain

$$2i[x,y]_{T^*} = 4\langle x_-, y_-\rangle - 4\langle x_+, y_+\rangle = 4\langle x_-, y_-\rangle - 4\langle Wx_+, Wy_+\rangle$$
$$= \langle \Gamma_- x, \Gamma_- y\rangle_{\mathcal{K}} - \langle \Gamma_+ x, \Gamma_+ y\rangle_{\mathcal{K}},$$

that is, (14.8) holds. Given $u_1, u_2 \in \mathcal{K}$ put $x = u_1 + W^{-1}u_2$. Then, $\Gamma_+ x = 2u_2$ and $\Gamma_- x = 2u_1$, so the surjectivity condition is also satisfied. $\qquad\square$

If V is a unitary operator on $\mathcal{K}$, then $U := -VW$ is an isometry of $\mathcal{N}_i$ onto $\mathcal{N}_{-i}$, and the operator T^V defined by (14.16) is just the operator T_U from Theorem 13.10 in the case $\lambda = i$. That is, for this example, the equivalence (i) $\leftrightarrow$ (iii) in Theorem 14.10 is only a restatement of von Neumann's Theorem 13.10.

Likewise, if $\tilde{W}$ denotes an isometric linear mapping of $\mathcal{N}_{-i}$ onto $\mathcal{N}_i$, then the triplet $(\mathcal{K} = \mathcal{N}_i, \Gamma_+ = 2Q_+, \Gamma_- = 2\tilde{W}Q_-)$ is a boundary triplet for T^*. $\qquad\circ$

Examples 14.5 and 14.6 below will be our guiding examples in this chapter. They are based on the following direct sum decompositions (14.17) and (14.18).

Proposition 14.11 *If A is a self-adjoint extension of T on $\mathcal{H}$ and $\mu \in \rho(A)$, then*

$$\mathcal{D}(T^*) = \mathcal{D}(\overline{T}) \dotplus (A - \mu I)^{-1}\mathcal{N}_{\overline{\mu}} \dotplus \mathcal{N}_\mu, \tag{14.17}$$

$$\mathcal{D}(T^*) = \mathcal{D}(\overline{T}) \dotplus A(A - \mu I)^{-1}\mathcal{N}_{\overline{\mu}} \dotplus (A - \mu I)^{-1}\mathcal{N}_{\overline{\mu}}, \tag{14.18}$$

$$\mathcal{D}(T^*) = \mathcal{D}(A) \dotplus \mathcal{N}_\mu, \tag{14.19}$$

$$\mathcal{D}(A) = \mathcal{D}(\overline{T}) \dotplus (A - \mu I)^{-1}\mathcal{N}_{\overline{\mu}}. \tag{14.20}$$

Proof The proof of (14.17) is similar to the proof of formula (3.10) in Proposition 3.7. Since $\mu \in \rho(A)$, we have $\mu \in \pi(T)$. Hence, Corollary 2.2 applies and yields

$$\mathcal{H} = \mathcal{R}(\overline{T} - \mu I) \oplus \mathcal{N}_{\overline{\mu}}, \quad \text{where } \mathcal{N}_{\overline{\mu}} = \mathcal{N}(T^* - \overline{\mu}I). \tag{14.21}$$

Let $x \in \mathcal{D}(T^*)$. By (14.21) there are vectors $x_0 \in \mathcal{D}(\overline{T})$ and $y_1 \in \mathcal{N}_{\overline{\mu}}$ such that $(T^* - \mu I)x = (\overline{T} - \mu I)x_0 + y_1$. Since $T \subseteq A$, we have $A = A^* \subseteq T^*$. Putting $y_2 := x - x_0 - R_\mu(A)y_1$, we compute

$$(T^* - \mu I)y_2 = (T^* - \mu I)x - (\overline{T} - \mu I)x_0 - (A - \mu I)R_\mu(A)y_1 = y_1 - y_1 = 0,$$

so that $y_2 \in \mathcal{N}_\mu$, and hence $x = x_0 + R_\mu(A)y_1 + y_2 \in \mathcal{D}(\overline{T}) + R_\mu(A)\mathcal{N}_{\overline{\mu}} + \mathcal{N}_\mu$. Since obviously $\mathcal{D}(\overline{T}) + R_\mu(A)\mathcal{N}_{\overline{\mu}} + \mathcal{N}_\mu \subseteq \mathcal{D}(T^*)$, we have proved that

$$\mathcal{D}(T^*) = \mathcal{D}(\overline{T}) + R_\mu(A)\mathcal{N}_{\overline{\mu}} + \mathcal{N}_\mu. \tag{14.22}$$

We show that the sum in (14.22) is direct. Suppose that $x_0 + R_\mu(A)y_1 + y_2 = 0$, where $x_0 \in \mathcal{D}(\overline{T})$, $y_1 \in \mathcal{N}_{\overline{\mu}}$, and $y_2 \in \mathcal{N}_\mu$. Then

$$0 = (T^* - \mu I)(x_0 + R_\mu(A)y_1 + y_2) = (\overline{T} - \mu I)x_0 + y_1. \tag{14.23}$$

By (14.21) we have $(\overline{T} - \mu I)x_0 \perp y_1$, so (14.23) implies that $(\overline{T} - \mu I)x_0 = y_1 = 0$. Thus, $R_\mu(A)(\overline{T} - \mu I)x_0 = x_0 = 0$ and $y_2 = -x_0 - R_\mu(A)y_1 = 0$. This completes the proof of (14.17).

Now we prove (14.20). Because the sum in (14.17) is direct, it suffices to show that $\mathcal{D}(A) = \mathcal{D}(\overline{T}) + R_\mu(A)\mathcal{N}_{\overline{\mu}}$. Clearly, $\mathcal{D}(\overline{T}) + R_\mu(A)\mathcal{N}_{\overline{\mu}} \subseteq \mathcal{D}(A)$. Conversely, let $x \in \mathcal{D}(A)$. By (14.17) we can write $x = x_0 + R_\mu(A)y_1 + y_2$, where $x_0 \in \mathcal{D}(\overline{T})$, $y_1 \in \mathcal{N}_{\overline{\mu}}$, and $y_2 \in \mathcal{N}_\mu$. Since $x_0 + R_\mu(A)y_1 \in \mathcal{D}(A)$, we deduce that $y_2 \in \mathcal{D}(A)$, so that $(A - \mu I)y_2 = (T^* - \mu I)y_2 = 0$, and hence $y_2 = 0$, because $\mu \in \rho(A)$. That is, $x \in \mathcal{D}(\overline{T}) + R_\mu(A)\mathcal{N}_{\overline{\mu}}$. This proves (14.20).

Equality (14.19) follows at once by comparing (14.17) and (14.20).

Finally, we verify (14.18). Replacing μ by $\overline{\mu}$ in (14.19) yields $\mathcal{D}(T^*) = \mathcal{D}(A) \dotplus \mathcal{N}_{\overline{\mu}}$. Applying once more (14.20), we get

$$\mathcal{D}(T^*) = \mathcal{D}(\overline{T}) \dotplus R_\mu(A)\mathcal{N}_{\overline{\mu}} \dotplus \mathcal{N}_{\overline{\mu}}. \tag{14.24}$$

Since $u + R_\mu(A)v = AR_\mu(A)u + R_\mu(A)(v - \mu u)$ and $AR_\mu(A)u + R_\mu(A)v = u + R_\mu(A)(v + \mu u)$, it follows that $R_\mu(A)\mathcal{N}_{\overline{\mu}} \dotplus \mathcal{N}_{\overline{\mu}} = AR_\mu(A)\mathcal{N}_{\overline{\mu}} \dotplus R_\mu(A)\mathcal{N}_{\overline{\mu}}$. Inserting the latter into (14.24), we obtain (14.18). $\square$

Example 14.5 (*First standard example*) Throughout this example we suppose that A is a *fixed* self-adjoint extension of T on $\mathcal{H}$ and μ is a *fixed* number from the resolvent set $\rho(A)$.

Let $x \in \mathcal{D}(T^*)$. By (14.18) there exist vectors $x_T \in \mathcal{D}(\overline{T})$ and $x_0, x_1 \in \mathcal{N}_{\overline{\mu}}$, all three uniquely determined by x, such that

$$x = x_T + A(A - \mu I)^{-1}x_0 + (A - \mu I)^{-1}x_1. \tag{14.25}$$

Define $\mathcal{K} = \mathcal{N}_{\overline{\mu}}$, $\Gamma_0 x = x_0$, and $\Gamma_1 x = x_1$.

Statement $(\mathcal{K}, \Gamma_0, \Gamma_1)$ *is a boundary triplet for the operator* T^*.

Proof Condition (ii) of Definition 14.2 is trivially true. We verify condition (i).

We set $R := (A - \mu I)^{-1}$ and write $x = x_T + u_x$ when x is given by (14.25), that is, $u_x := ARx_0 + Rx_1$. Let $v \in \mathcal{N}_{\overline{\mu}}$. Then $ARv = (A - \mu I)Rv + \mu Rv = (I + \mu R)v$. Using the facts that $A \subseteq T^*$ and $Rv \in \mathcal{D}(A)$, we get $T^*Rv = ARv = (I + \mu R)v$. Moreover, $T^*v = \overline{\mu}v$, since $v \in \mathcal{N}_{\overline{\mu}}$. From these relations we deduce that

$$T^*u_x = T^*ARx_0 + T^*Rx_1 = (\overline{\mu} + \mu + \mu^2 R)x_0 + (I + \mu R)x_1. \tag{14.26}$$

Now suppose that $x, y \in \mathcal{D}(T^*)$. Then

$$\begin{aligned}
\langle T^*x, y \rangle &= \langle T^*(x_T + u_x), y_T + u_y \rangle = \langle \overline{T} x_T + T^*u_x, y_T + u_y \rangle \\
&= \langle x_T, \overline{T} y_T \rangle + \langle u_x, \overline{T} y_T \rangle + \langle x_T, T^*u_y \rangle + \langle T^*u_x, u_y \rangle \\
&= \langle x, \overline{T} y_T \rangle + \langle x, T^*u_y \rangle - \langle u_x, T^*u_y \rangle + \langle T^*u_x, u_y \rangle \\
&= \langle x, T^*y \rangle - \langle u_x, T^*u_y \rangle + \langle T^*u_x, u_y \rangle.
\end{aligned}$$

Inserting the expressions for T^*u_x and T^*u_y from Eq. (14.26), we derive

$$\begin{aligned}
\langle T^*x, y \rangle - \langle x, T^*y \rangle &= \langle T^*u_x, u_y \rangle - \langle u_x, T^*u_y \rangle \\
&= \langle ((\overline{\mu} + \mu)I + \mu^2 R)x_0 + (I + \mu R)x_1, (I + \mu R)y_0 + Ry_1 \rangle \\
&\quad - \langle (I + \mu R)x_0 + Rx_1, ((\overline{\mu} + \mu)I + \mu^2 R)y_0 + (I + \mu R)y_1 \rangle \\
&= \langle [(I + \overline{\mu}R^*)((\overline{\mu} + \mu)I + \mu^2 R) - ((\overline{\mu} + \mu)I + \overline{\mu}^2 R^*)(I + \mu R)]x_0, y_0 \rangle \\
&\quad + \langle [R^*(I + \mu R) - (I + \overline{\mu}R^*)R]x_1, y_1 \rangle \\
&\quad + \langle [(I + \overline{\mu}R^*)(I + \mu R) - ((\mu + \overline{\mu})I + \overline{\mu}^2 R^*)R]x_1, y_0 \rangle \\
&\quad + \langle [R^*((\overline{\mu} + \mu)I + \mu^2 R) - (I + \overline{\mu}R^*)(I + \mu R)]x_0, y_1 \rangle.
\end{aligned}$$

Using the resolvent identity $R^* - R = (\overline{\mu} - \mu)R^*R = (\overline{\mu} - \mu)RR^*$ (by (2.5)), the first two summands after the last equality sign vanish, and the two remaining yield

$$\langle x_1, y_0 \rangle - \langle x_0, y_1 \rangle = \langle \Gamma_1 x, \Gamma_0 y \rangle_{\mathcal{K}} - \langle \Gamma_0 x, \Gamma_1 y \rangle_{\mathcal{K}}.$$

This shows that condition (i) in Definition 14.2 holds. $\qquad\square$

Since $\mathcal{D}(A) = \mathcal{D}(\overline{T}) + (A - \mu I)^{-1}\mathcal{N}_{\overline{\mu}}$ by (14.20), we have $\mathcal{D}(A) = \mathcal{N}(\Gamma_0) = \mathcal{D}(T_0)$, that is, A is the distinguished self-adjoint operator T_0 from Corollary 14.8. $\qquad\circ$

Example 14.6 (*Second standard example*) Let us assume that the symmetric operator T has a *real regular point* $\mu \in \pi(T)$. Then by Proposition 3.16 there exists a self-adjoint extension A of T on $\mathcal{H}$ such that $\mu \in \rho(A)$. Throughout this example we fix A and μ.

By Eq. (14.17), for any $x \in \mathcal{D}(T^*)$, there exist uniquely determined elements $x_T \in \mathcal{D}(\overline{T})$ and $x_0, x_1 \in \mathcal{N}_\mu$ (because μ is real!) such that

$$x = x_T + (A - \mu I)^{-1} x_1 + x_0.$$

Then we define $\mathcal{K} = \mathcal{N}_\mu$, $\Gamma_0 x = x_0$, and $\Gamma_1 x = x_1$.

Statement $(\mathcal{K}, \Gamma_0, \Gamma_1)$ *is a boundary triplet for* T^*.

Proof In this proof let $(\mathcal{K}, \Gamma_0', \Gamma_1')$ denote the triplet from Example 14.5. Since

$$A(A - \mu I)^{-1} x_0 + (A - \mu I)^{-1} x_1 = x_0 + (A - \mu I)^{-1} (\mu x_0 + x_1),$$

we have $\Gamma_0 = \Gamma_0'$ and $\Gamma_1 = \mu \Gamma_0' + \Gamma_1'$. Because μ is real and $(\mathcal{K}, \Gamma_0', \Gamma_1')$ is a boundary triplet for T^*, it follows easily that $(\mathcal{K}, \Gamma_0, \Gamma_1)$ is also a boundary triplet for T^*. $\square$

Since $\mathcal{N}(\Gamma_0) = \mathcal{N}(\Gamma_0') = \mathcal{D}(A)$, A is the operator T_0 from Corollary 14.8 for the boundary triplet $(\mathcal{K}, \Gamma_0, \Gamma_1)$.   o

The next theorem restates the equivalence (i) $\leftrightarrow$ (ii) of Theorem 14.10 for the boundary triplet $(\mathcal{K}, \Gamma_0, \Gamma_1)$ from Example 14.6. By (14.5) the boundary space of the operator T_B is $\mathcal{B} = \{(u, Bu + v) : u \in \mathcal{D}(B), v \in \mathcal{D}(B)^\perp\}$, so the description of the domain $\mathcal{D}(T_B)$ is obtained by inserting the definitions of Γ_0 and Γ_1 into (14.13). Further, $T^* R_\mu(A) = A R_\mu(A) = I + \mu R_\mu(A)$, since $A \subseteq T^*$. Thus, we obtain the following:

Theorem 14.12 *Let T be a densely defined symmetric operator on $\mathcal{H}$. Suppose that A is a fixed self-adjoint extension of T on $\mathcal{H}$ and μ is a* real *number in $\rho(A)$. Recall that $\mathcal{N}_\mu := \mathcal{N}(T^* - \mu I)$. For $B \in \mathcal{S}(\mathcal{N}_\mu)$, let*

$$\mathcal{D}(T_B) = \left\{ x + R_\mu(A)(Bu + v) + u : x \in \mathcal{D}(\overline{T}), \ u \in \mathcal{D}(B), \ v \in \mathcal{N}_\mu, \ v \perp \mathcal{D}(B) \right\},$$

$$T_B = T^* \upharpoonright \mathcal{D}(T_B), \quad \text{that is,}$$

$$T_B \big(x + R_\mu(A)(Bu + v) + u \big) = \overline{T} x + \big(I + \mu R_\mu(A) \big)(Bu + v) + \mu u.$$

Then the operator T_B is a self-adjoint extension of T on $\mathcal{H}$. Each self-adjoint extension of T on $\mathcal{H}$ is of the form T_B with uniquely determined $B \in \mathcal{S}(\mathcal{N}_\mu)$.

14.4 Examples: Differentiation Operators III

In order to derive explicit formulas for the parameterization, we assume in this section that the boundary form $[\cdot, \cdot]_{T^*}$ has the following special form:

There are linear functionals $\varphi_1, \ldots, \varphi_d, \psi_1, \ldots, \psi_d, d \in \mathbb{N}$, on $\mathcal{D}(T^*)$ such that

$$[x, y]_{T^*} = \sum_{k=1}^{d} \left(\psi_k(x)\overline{\varphi_k(y)} - \varphi_k(x)\overline{\psi_k(y)} \right), \quad x, y \in \mathcal{D}(T^*), \quad (14.27)$$

$$\left\{ (\varphi(x), \psi(x)) : x \in \mathcal{D}(T^*) \right\} = \mathbb{C}^{2d}, \quad (14.28)$$

where $\varphi(x) := (\varphi_1(x), \ldots, \varphi_d(x))$ and $\psi(x) := (\psi_1(x), \ldots, \psi_d(x))$.

Clearly, then there is a boundary triplet $(\mathcal{K}, \Gamma_0, \Gamma_1)$ for T^* defined by

$$\mathcal{K} = \mathbb{C}^d, \quad \Gamma_0(x) = \varphi(x), \quad \Gamma_1(x) = \psi(x). \quad (14.29)$$

Let $B \in \mathcal{S}(\mathcal{K})$. We choose an orthonormal basis $\{e_1, \ldots, e_n\}$ of $\mathcal{K}_B$ and a basis $\{\tilde{e}_{n+1}, \ldots, \tilde{e}_d\}$ of the vector space $(\mathcal{K}_B)^\perp$. Since B is a self-adjoint operator on $\mathcal{K}_B$, there is a hermitian $n \times n$ matrix $B = (B_{kl})$ such that $Be_l = \sum_{k=1}^{n} B_{kl}e_k$.

Obviously, $P_B \Gamma_1(x) = \sum_{k=1}^{n} \langle \Gamma_1(x), e_k \rangle e_k$, where $\langle \cdot, \cdot \rangle$ denotes the "standard" scalar product of the Hilbert space $\mathcal{K} = \mathbb{C}^d$. If $\Gamma_0(x) \in \mathcal{K}_B$, then we have

$$B\Gamma_0(x) = \sum_l \langle \Gamma_0(x), e_l \rangle Be_l = \sum_{k,l} \langle \Gamma_0(x), e_l \rangle B_{kl}e_k.$$

Recall that by (14.15) a vector $x \in \mathcal{D}(T^*)$ is in $\mathcal{D}(T_B)$ if and only if $\Gamma_0(x) \in \mathcal{D}(B)$ and $P_B \Gamma_1(x) = B\Gamma_0(x)$. Therefore, $x \in \mathcal{D}(T^*)$ belongs to $\mathcal{D}(T_B)$ if and only if

$$\langle \Gamma_0(x), \tilde{e}_j \rangle = 0, \quad j = n+1, \ldots, d,$$

$$\langle \Gamma_1(x), e_k \rangle = \sum_{l=1}^{n} B_{kl} \langle \Gamma_0(x), e_l \rangle, \quad k = 1, \ldots, n.$$

This gives an "explicit" description of the vectors of the domain $\mathcal{D}(T_B)$ by d linear equations. We discuss this in the two simplest cases $d = 1$ and $d = 2$.

Example 14.7 ($d = 1$) First, let $n = 1$. A Hermitian 1×1 matrix is just a real number B, and the operator T_B is determined by the boundary condition $B\Gamma_0(x) = \Gamma_1(x)$. If $n = 0$, then B acts on the space $\{0\}$, and hence T_B is defined by the condition $\Gamma_0(x) = 0$. If we interpret the latter as $B\Gamma_0(x) = \Gamma_1(x)$ in the case $B = \infty$, then the domains of all self-adjoint extensions T_B of T on $\mathcal{H}$ are characterized by the boundary conditions

$$B\Gamma_0(x) = \Gamma_1(x), \quad B \in \mathbb{R} \cup \{\infty\}.$$

Writing B as $B = \cot \alpha$, we obtain another convenient parameterization of self-adjoint extensions of T which will be used in Chap. 15. It is given by

$$\Gamma_0(x) \cos \alpha = \Gamma_1(x) \sin \alpha, \quad \alpha \in [0, \pi).$$

○

Example 14.8 ($d = 2$) Let $B \in \mathcal{S}(\mathcal{K})$. The operator B can act on a subspace of dimension 2, 1, or 0 of the Hilbert space $\mathcal{K} = \mathbb{C}^2$. That is, we have the following three possible cases.

Case 1 ($\mathcal{K}_B = \mathcal{K}$) Then B corresponds to a hermitian 2×2 matrix, so the relation $P_B \Gamma_1(x) = B \Gamma_0(x)$ says

$$\psi_1(x) = b_1 \varphi_1(x) + c \varphi_2(x), \qquad \psi_2(x) = \overline{c} \varphi_1(x) + b_2 \varphi_2(x), \qquad (14.30)$$

where $b_1, b_2 \in \mathbb{R}$ and $c \in \mathbb{C}$ are parameters.

Case 2 ($\dim \mathcal{K}_B = 1$) Let $P_B = e \otimes e$, where $e = (\alpha, \beta) \in \mathbb{C}^2$ is a unit vector. Then $\Gamma_0(x) \in \mathcal{D}(B) = \mathbb{C} \cdot e$ is equivalent to $\Gamma_0(x) \perp (\beta, -\alpha)$, that is, $\varphi_1(x)\beta = \varphi_2(x)\alpha$. Further, the relation $P_B \Gamma_1 x = B \Gamma_0 x$ means that

$$\psi_1(x)\overline{\alpha}\alpha + \psi_2(x)\overline{\beta}\alpha = B\varphi_1(x), \qquad \psi_1(x)\overline{\alpha}\beta + \psi_2(x)\overline{\beta}\beta = B\varphi_2(x).$$

Putting $c := \beta\alpha^{-1}$ and $b_1 := B|\alpha|^{-2}$ if $\alpha \neq 0$ and $b_1 := B$ if $\alpha = 0$, in both cases, $\alpha \neq 0$ and $\alpha = 0$, the preceding three equations are equivalent to

$$\psi_1(x) = b_1 \varphi_1(x) - \overline{c} \psi_2(x), \qquad \varphi_2(x) = c \varphi_1(x), \qquad (14.31)$$

$$\psi_2(x) = b_1 \varphi_2(x), \qquad \varphi_1(x) = 0, \qquad (14.32)$$

respectively.

Case 3 ($\mathcal{K}_B = \{0\}$) Then the domain of T_B is defined by the condition $\Gamma_0(x) = 0$, that is,

$$\varphi_1(x) = \varphi_2(x) = 0. \qquad (14.33)$$

By Theorem 14.10, the self-adjoint extensions of T are the operators T_B for $B \in \mathcal{S}(\mathcal{K})$. Since $T_B \subseteq T^*$, it suffices to describe the domains of these operators.

Summarizing, *the domains of all self-adjoint extensions of the operator T are the sets of vectors $x \in \mathcal{D}(T^*)$ satisfying one of the four sets* (14.30)–(14.33) *of boundary conditions, where $b_1, b_2 \in \mathbb{R}$ and $c \in \mathbb{C}$ are fixed.* Here different sets of parameters and boundary conditions correspond to different self-adjoint extensions.

The interesting subclass of all self-adjoint extensions of T with vanishing parameter c in the preceding equations can be described much nicer in the form

$$\psi_1(x) = \beta_1 \varphi_1(x), \qquad \psi_2(x) = \beta_2 \varphi_2(x), \qquad \text{where } \beta_1, \beta_2 \in \mathbb{R} \cup \{\infty\}.$$

As in Example 14.7, the relation $\psi_j(x) = \beta_j \varphi_j(x)$ for $\beta_j = \infty$ means that $\varphi_j(x) = 0$. This subclass can be also parameterized by $\alpha_1, \alpha_2 \in [0, \pi)$ and the equations

$$\varphi_1(x)\cos\alpha_1 = \psi_1(x)\sin\alpha_1, \qquad \varphi_2(x)\cos\alpha_2 = -\psi_2(x)\sin\alpha_2. \quad (14.34)$$

(Here the minus sign is only chosen in order to obtain more convenient formulas in applications given in Example 14.10 and Propositions 15.13 and 15.14.)   ∘

The following two examples fit nicely into the preceding setup.

Example 14.9 Let us consider the symmetric operator $T = -\frac{d^2}{dx^2}$ with domain $\mathcal{D}(T) = H_0^2(0, \infty)$ on $L^2(0, \infty)$. Integration by parts yields

$$[f, g]_{T^*} = -f(0)\overline{g'(0)} + f'(0)\overline{g(0)}.$$

Then the above assumptions are satisfied with $d = 1$, $\varphi(f) = f(0)$, $\psi(f) = f'(0)$, so there is a boundary triplet for the operator T^* given by

$$\mathcal{K} = \mathbb{C}, \qquad \Gamma_0(f) = f(0), \qquad \Gamma_1(f) = f'(0).$$

By Example 14.7 the self-adjoint extensions of T are parameterized by $\mathbb{R} \cup \{\infty\}$, and the corresponding operator T_B is defined by the boundary condition

$$Bf(0) = f'(0), \quad B \in \mathbb{R} \cup \{\infty\}. \tag{14.35}$$

The case $B = \infty$ yields the Friedrichs extension T_F of T which is the operator T_0 given by the condition $\Gamma_0(f) = f(0) = 0$.

Likewise we could have taken $\varphi(f) = -f'(0)$ and $\psi(f) = f(0)$; in this case the operator T_1 is the Friedrichs extension T_F.   ○

Example 14.10 (*Example* 14.2 *continued*) Let $T = -\frac{d^2}{dx^2}$ be the symmetric operator on $\mathcal{D}(T) = H_0^2(a, b)$, where $a, b \in \mathbb{R}$, $a < b$. Then assumptions (14.27) and (14.28) are satisfied, and the boundary triplet from Example 14.2 is of the form (14.29), where

$$d = 2, \qquad \varphi_1(f) = f(a), \qquad \varphi_2(f) = f(b),$$
$$\psi_1(f) = f'(a), \qquad \psi_2(f) = -f'(b).$$

As developed in Example 14.8, the set of all self-adjoint extensions of T is described by the following four families of boundary conditions:

$$f'(a) = b_1 f(a) + c f(b), \qquad f'(b) = -\overline{c} f(a) - b_2 f(b), \tag{14.36}$$
$$f'(a) = b_1 f(a) + \overline{c} f'(b), \qquad f(b) = c f(a), \tag{14.37}$$
$$f'(b) = -b_1 f(b), \qquad f(a) = 0, \tag{14.38}$$
$$f(a) = f(b) = 0, \tag{14.39}$$

where $c \in \mathbb{C}$ and $b_1, b_2 \in \mathbb{R}$ are arbitrary parameters.

The self-adjoint extensions T_0 and T_1 defined by Corollary 14.8 are given by *Dirichlet boundary conditions* $f(a) = f(b) = 0$ and *Neumann boundary conditions* $f'(a) = f'(b) = 0$, respectively. In particular, T_0 is the Friedrichs extension T_F of T.

In this example the subclass of self-adjoint extensions (14.34) is characterized by "*decoupled boundary conditions*"

$$f(a) \cos \alpha_1 = f'(a) \sin \alpha_1, \qquad f(b) \cos \alpha_2 = f'(b) \sin \alpha_2, \qquad \alpha_1, \alpha_2 \in [0, \pi). \quad ○$$

14.5 Gamma Fields and Weyl Functions

In this section, $(\mathcal{K}, \Gamma_0, \Gamma_1)$ is a boundary triplet for T^*, and T_0 is the self-adjoint extension of T determined by $\mathcal{D}(T_0) = \mathcal{N}(\Gamma_0)$. Recall that $\mathcal{N}_z = \mathcal{N}(T^* - zI)$.

Our aim is to define two holomorphic operator families

$$\gamma(z) = \left(\Gamma_0 \upharpoonright \mathcal{N}_z\right)^{-1} \in \mathbf{B}(\mathcal{K}, \mathcal{H}) \quad \text{and} \quad M(z) = \Gamma_1 \gamma(z) \in \mathbf{B}(\mathcal{K}), \quad z \in \rho(T_0),$$

and to derive a number of technical results. These operator fields are important tools for the study of extensions of the operator T. In particular, they will appear in the Krein–Naimark resolvent formula (14.43) and in Theorem 14.22 below.

Lemma 14.13

(i) Γ_0 and Γ_1 are continuous mappings of $(\mathcal{D}(T^*), \|\cdot\|_{T^*})$ into $\mathcal{K}$.

(ii) For each $z \in \rho(T_0)$, Γ_0 is a continuous bijective mapping of the subspace $\mathcal{N}_z$ onto $\mathcal{K}$ with bounded inverse denoted by $\gamma(z)$.

Proof (i): By the closed graph theorem it suffices to prove that the mapping (Γ_0, Γ_1) of the Hilbert space $(\mathcal{D}(T^*), \|\cdot\|_{T^*})$ into the Hilbert space $\mathcal{K} \oplus \mathcal{K}$ is closed.

Suppose that $(x_n)_{n \in \mathbb{N}}$ is a null sequence in $(\mathcal{D}(T^*), \|\cdot\|_{T^*})$ such that the sequence $((\Gamma_0 x_n, \Gamma_1 x_n))_{n \in \mathbb{N}}$ converges in $\mathcal{K} \oplus \mathcal{K}$, say $\lim_n(\Gamma_0 x_n, \Gamma_1 x_n) = (u, v)$. Then we obtain

$$\langle v, \Gamma_0 y \rangle_{\mathcal{K}} - \langle u, \Gamma_1 y \rangle_{\mathcal{K}} = \lim_{n \to \infty} \left[\langle \Gamma_1 x_n, \Gamma_0 y \rangle_{\mathcal{K}} - \langle \Gamma_0 x_n, \Gamma_1 y \rangle_{\mathcal{K}} \right]$$

$$= \lim_{n \to \infty} \left[\langle T^* x_n, y \rangle - \langle x_n, T^* y \rangle \right] = 0$$

and hence $\langle v, \Gamma_0 y \rangle_{\mathcal{K}} = \langle u, \Gamma_1 y \rangle_{\mathcal{K}}$ for all $y \in \mathcal{D}(T^*)$. By Definition 14.2(ii) there are vectors $y_0, y_1 \in \mathcal{D}(T^*)$ such that $\Gamma_0 y_0 = v$, $\Gamma_1 y_0 = 0$, $\Gamma_0 y_1 = 0$, and $\Gamma_1 y_1 = u$. Inserting these elements, we conclude that $u = v = 0$. Hence, (Γ_0, Γ_1) is closed.

(ii): Recall that $\mathcal{D}(T^*) = \mathcal{D}(T_0) \dotplus \mathcal{N}_z$ by (14.19). Since $\Gamma_0(\mathcal{D}(T^*)) = \mathcal{K}$ by Definition 14.2(ii) and $\Gamma_0(\mathcal{D}(T_0)) = \{0\}$ by the definition of T_0, Γ_0 maps $\mathcal{N}_z$ onto $\mathcal{K}$. If $x \in \mathcal{N}_z$ and $\Gamma_0 x = 0$, then $x \in \mathcal{D}(T_0) \cap \mathcal{N}_z$, and so $x = 0$, because $\mathcal{D}(T_0) \dotplus \mathcal{N}_z$ is a direct sum. The preceding proves that $\Gamma_0 : \mathcal{N}_z \to \mathcal{K}$ is bijective.

On $\mathcal{N}_z$ the graph norm $\|\cdot\|_{T^*}$ is obviously equivalent to the Hilbert space norm of $\mathcal{H}$. Hence, it follows from (i) that the bijective map $\Gamma_0 : \mathcal{N}_z \to \mathcal{K}$ is continuous. By the open mapping theorem its inverse $\gamma(z)$ is also continuous. $\square$

Let $z \in \rho(T_0)$. Recall that $\gamma(z)$ is the inverse of the mapping $\Gamma_0 : \mathcal{N}_z \to \mathcal{K}$. By Lemma 14.13(ii), $\gamma(z) \in \mathbf{B}(\mathcal{K}, \mathcal{H})$, and hence $\gamma(z)^* \in \mathbf{B}(\mathcal{H}, \mathcal{K})$. Since Γ_1 and $\gamma(z)$ are continuous (by Lemma 14.13), $M(z) := \Gamma_1 \gamma(z)$ is in $\mathbf{B}(\mathcal{K})$.

Definition 14.4 We shall call the map $\rho(T_0) \ni z \to \gamma(z) \in \mathbf{B}(\mathcal{K}, \mathcal{H})$ the *gamma field* and the map $\rho(T_0) \ni z \to M(z) \in \mathbf{B}(\mathcal{K})$ the *Weyl function* of the operator T_0 associated with the boundary triplet $(\mathcal{K}, \Gamma_0, \Gamma_1)$.

Basic properties of these operator fields are contained in the next propositions.

Proposition 14.14 *For* $z, w \in \rho(T_0)$, *we have*:

(i) $\gamma(\bar{z})^* = \Gamma_1(T_0 - zI)^{-1}$.

(ii) $\mathcal{N}(\gamma(z)^*) = \mathcal{N}_{\bar{z}}^{\perp}$ *and* $\gamma(z)^*$ *is a bijection of* $\mathcal{N}_z$ *onto* $\mathcal{K}$.

(iii) $\gamma(w) - \gamma(z) = (w - z)(T_0 - wI)^{-1}\gamma(z) = (w - z)(T_0 - zI)^{-1}\gamma(w)$.

(iv) $\gamma(w) = (T_0 - zI)(T_0 - wI)^{-1}\gamma(z)$.

(v) $\frac{d}{dz}\gamma(z) = (T_0 - zI)^{-1}\gamma(z)$.

Proof (i): Let $x \in \mathcal{H}$. Set $y = (T_0 - zI)^{-1}x$. Let $v \in \mathcal{K}$. Using the facts that $T^*\gamma(\bar{z})v = \bar{z}\gamma(\bar{z})v$, $\Gamma_0 y = 0$, and $\Gamma_0\gamma(\bar{z})v = v$ and Definition 14.2(i), we derive

$$\langle \gamma(\bar{z})^*(T_0 - zI)y, v\rangle_{\mathcal{K}} = \langle (T_0 - zI)y, \gamma(\bar{z})v\rangle$$
$$= \langle T_0 y, \gamma(\bar{z})v\rangle - \langle y, \bar{z}\,\gamma(\bar{z})v\rangle = \langle T^*y, \gamma(\bar{z})v\rangle - \langle y, T^*\gamma(\bar{z})v\rangle$$
$$= \langle \Gamma_1 y, \Gamma_0\gamma(\bar{z})v\rangle_{\mathcal{K}} - \langle \Gamma_0 y, \Gamma_1\gamma(\bar{z})v\rangle_{\mathcal{K}} = \langle \Gamma_1 y, v\rangle_{\mathcal{K}}.$$

Since $v \in \mathcal{K}$ was arbitrary, the latter yields $\gamma(\bar{z})^*(T_0 - zI)y = \Gamma_1 y$. Inserting now $y = (T_0 - zI)^{-1}x$, this gives $\gamma(\bar{z})^*x = \Gamma_1(T_0 - zI)^{-1}x$.

(ii): Since $\mathcal{R}(\gamma(z)) = \mathcal{N}_z$, we have $\mathcal{N}(\gamma(z)^*) = \mathcal{N}_z^{\perp}$, so $\gamma(z)^* \upharpoonright \mathcal{N}_z$ is injective. Let $v \in \mathcal{K}$. By Definition 14.2(ii) there exists $y \in \mathcal{D}(T^*)$ such that $\Gamma_0 y = 0$ and $\Gamma_1 y = v$. Then $y \in \mathcal{D}(T_0)$. Let y_z be the projection of $(T_0 - \bar{z}I)y$ onto $\mathcal{N}_z$. Using that $\mathcal{N}(\gamma(z)^*) = \mathcal{N}_z^{\perp}$ and (i), we obtain $\gamma(z)^*y_z = \gamma(z)^*(T_0 - \bar{z}I)y = \Gamma_1 y = v$. This shows that $\gamma(z)^* \upharpoonright \mathcal{N}_z$ is surjective.

(iii): Let $v \in \mathcal{K}$. By Lemma 14.13(ii), $v = \Gamma_0 u$ for some $u \in \mathcal{N}_z$. Putting

$$u' := u + (w - z)(T_0 - wI)^{-1}u, \tag{14.40}$$

we compute $T^*u' = zu + (w - z)T_0(T_0 - wI)^{-1}u = wu'$, that is, $u' \in \mathcal{N}_w$. Since $(T_0 - wI)^{-1}u \in \mathcal{D}(T_0)$ and hence $\Gamma_0(T_0 - wI)^{-1}u = 0$, we get $\Gamma_0 u' = \Gamma_0 u = v$. Therefore, $\gamma(z)v = \gamma(z)\Gamma_0 u = u$ and $\gamma(w)v = \gamma(w)\Gamma_0 u' = u'$. Inserting the latter into (14.40), we obtain $\gamma(w)v = \gamma(z)v + (w - z)(T_0 - wI)^{-1}\gamma(z)v$. This proves the first equality of (iii). Interchanging z and w in the first equality gives the second equality of (iii).

(iv) follows from (iii), since $(T_0 - zI)(T_0 - wI)^{-1} = I + (w - z)(T_0 - wI)^{-1}$.

(v): We divide the first equality of (iii) by $w - z$ and let $w \to z$. Using the continuity of the resolvent in the operator norm (see formula (2.7)), we obtain (v). □

Proposition 14.15 *For arbitrary* $z, w \in \rho(T_0)$, *we have*:

(i) $M(z)\Gamma_0 u = \Gamma_1 u$ *for* $u \in \mathcal{N}_z$.

(ii) $M(z)^* = M(\bar{z})$.

(iii) $M(w) - M(z) = (w - z)\gamma(\bar{z})^*\gamma(w)$.

(iv) $\frac{d}{dz}M(z) = \gamma(\bar{z})^*\gamma(z)$.

Proof (i): Since $\gamma(z) = (\Gamma_0 \upharpoonright \mathcal{N}_z)^{-1}$, we get $M(z)\Gamma_0 u = \Gamma_1\gamma(z)\Gamma_0 u = \Gamma_1 u$.

(ii): Let $u \in \mathcal{N}_z$ and $u' \in \mathcal{N}_{\bar{z}}$. Obviously, $\langle T^*u, u'\rangle = \langle u, T^*u'\rangle$. Therefore, by (i) and Definition 14.2(i) we obtain

$$\langle M(z)\Gamma_0 u, \Gamma_0 u'\rangle_{\mathcal{K}} = \langle \Gamma_1 u, \Gamma_0 u'\rangle_{\mathcal{K}} = \langle \Gamma_0 u, \Gamma_1 u'\rangle_{\mathcal{K}} = \langle \Gamma_0 u, M(\overline{z})\Gamma_0 u'\rangle_{\mathcal{K}}.$$

Since $\Gamma_0(\mathcal{N}_z) = \Gamma_0(\mathcal{N}_{\overline{z}}) = \mathcal{K}$ by Lemma 14.13, the latter shows that $M(\overline{z}) = M(z)^*$.

(iii): Using Proposition 14.14, (ii) and (i), we derive

$$M(w) - M(z) = \Gamma_1\big(\gamma(w) - \gamma(z)\big) = (w - z)\Gamma_1(T_0 - zI)^{-1}\gamma(w)$$
$$= (w - z)\gamma(\overline{z})^*\gamma(w).$$

(iv) follows by dividing (iii) by $w - z$, letting $w \to z$, and using the continuity of $\gamma(z)$ in the operator norm (by Proposition 14.14(v)). $\qquad\square$

Propositions 14.14(v) and 14.15(iv) imply that the gamma field $z \to \gamma(z)$ and the Weyl function $z \to M(z)$ are operator-valued *holomorphic* functions on the resolvent set $\rho(T_0)$. In particular, both fields are continuous in the operator norm.

Definition 14.5 An operator-valued function $F : \mathbb{C}_+ \to (\mathbf{B}(\mathcal{K}), \|\cdot\|)$ is called a *Nevanlinna function* if F is holomorphic on $\mathbb{C}_+ = \{z \in \mathbb{C} : \operatorname{Im} z > 0\}$ and

$$\operatorname{Im} F(z) = \frac{1}{2i}\big(F(z) - F(z)^*\big) \geq 0 \quad \text{for all } z \in \mathbb{C}_+.$$

Each *scalar* Nevanlinna function admits a canonical integral representation described in Theorem F.1. For general operator Nevanlinna functions and separable Hilbert spaces $\mathcal{K}$, there is a similar integral representation, where $a = a^*$ and $b = b^* \geq 0$ are in $\mathbf{B}(\mathcal{K})$, and ν is a positive operator-valued Borel measure on $\mathbb{R}$.

Corollary 14.16 *The Weyl function $M(z)$ is a Nevanlinna function on $\mathcal{K}$.*

Proof By Proposition 14.15(iv), $M(z)$ is a $\mathbf{B}(\mathcal{K})$-valued holomorphic function on $\mathbb{C}_+$. Let $z \in \mathbb{C}_+$ and $y = \operatorname{Im} z$. From Proposition 14.15, (ii) and (iii), we obtain

$$M(z) - M(z)^* = M(z) - M(\overline{z}) = (z - \overline{z})\gamma(z)^*\gamma(z) = 2iy\gamma(z)^*\gamma(z).$$

Since $y > 0$, this implies that $\operatorname{Im} M(z) \geq 0$. $\qquad\square$

The next proposition shows how eigenvalues and spectrum of an operator T_B can be detected by means of the Weyl function.

Proposition 14.17 *Suppose that B is a closed relation on $\mathcal{K}$ and $z \in \rho(T_0)$. Then the relation $B - M(z)$ (which is defined by $B - \mathcal{G}(M(z))$ is also closed, and we have:*

(i) $\gamma(z)\mathcal{N}(B - M(z)) = \mathcal{N}(T_B - zI)$.

(ii) $\dim \mathcal{N}(T_B - zI) = \dim(B - M(z))$. *In particular, z is an eigenvalue of T_B if and only if $\mathcal{N}(B - M(z)) \neq \{0\}$.*

(iii) $z \in \rho(T_B)$ *if and only if $0 \in \rho(B - M(z))$.*

Proof First, we show that the relation $\mathcal{B} - M(z)$ is closed. Let $((u_n, v_n))_{n \in \mathbb{N}}$ be a sequence from $\mathcal{B} - M(z)$ which converges to (u, v) in $\mathcal{K} \oplus \mathcal{K}$. By the definition of $\mathcal{B} - M(z)$ there are elements $(u_n, w_n) \in \mathcal{B}$ such that $v_n = w_n - M(z)u_n$. Since $u_n \to u$ and $M(z)$ is bounded, we have $M(z)u_n \to M(z)u$, and hence $w_n = v_n + M(z)u_n \to v + M(z)u$. Therefore, since $\mathcal{B}$ is closed, we conclude that $(u, v + M(z)u) \in \mathcal{B}$, which in turn implies that $(u, v) \in \mathcal{B} - M(z)$. This proves that the relation $\mathcal{B} - M(z)$ is closed.

(i): Let $v \in \mathcal{N}(\mathcal{B} - M(z))$. Then there exists $w \in \mathcal{K}$ such that $(v, w) \in \mathcal{B}$ and $w - M(z)v = 0$. Thus, $(v, M(z)v) \in \mathcal{B}$. Since $M(z) = \Gamma_1 \gamma(z)$ and $\Gamma_0 \gamma(z)v = v$, we have $(\Gamma_0 \gamma(z)v, \Gamma_1 \gamma(z)v) = (v, M(z)v) \in \mathcal{B}$, so that $\gamma(z)v \in \mathcal{D}(T_{\mathcal{B}})$ and $(T_{\mathcal{B}} - zI)\gamma(z)v = (T^* - zI)\gamma(z)v = 0$. That is, $\gamma(z)v \in \mathcal{N}(T_{\mathcal{B}} - zI)$.

Conversely, let $x \in \mathcal{N}(T_{\mathcal{B}} - zI)$. Then $(\Gamma_0 x, M(z)\Gamma_0 x) = (\Gamma_0 x, \Gamma_1 x) \in \mathcal{B}$, which yields $\Gamma_0 x \in \mathcal{N}(\mathcal{B} - M(z))$. Since $T_{\mathcal{B}} \subseteq T^*$, we have $x \in \mathcal{N}_z$, and hence $x = \gamma(z)\Gamma_0 x \in \gamma(z)\mathcal{N}(\mathcal{B} - M(z))$.

(ii) follows from (i) and the fact that $\gamma(z)$ is a bijection of $\mathcal{K}$ onto $\mathcal{N}_z$.

(iii): Suppose that $0 \in \rho(\mathcal{B} - M(z))$. Then we have $\mathcal{N}(\mathcal{B} - M(z)) = \{0\}$ and $\mathcal{R}(\mathcal{B} - M(z)) = \mathcal{K}$ by Proposition 14.1. Hence, $\mathcal{N}(T_{\mathcal{B}} - zI) = \{0\}$ by (ii). To prove that $z \in \rho(T_{\mathcal{B}})$, by Proposition 14.1 it suffices to show that $\mathcal{R}(T_{\mathcal{B}} - zI) = \mathcal{H}$.

Let $y \in \mathcal{H}$. Set $x := (T_0 - zI)^{-1}y$. Since $\mathcal{R}(\mathcal{B} - M(z)) = \mathcal{K}$, $\Gamma_1 x$ belongs to $\mathcal{R}(\mathcal{B} - M(z))$. This means that there exists $(v_1, v_2) \in \mathcal{B}$ such that $v_2 - M(z)v_1 = \Gamma_1 x$. Put $f := x + \gamma(z)v_1$. Since $x \in \mathcal{D}(T_0)$ and hence $\Gamma_0 x = 0$, it follows that $\Gamma_0 f = \Gamma_0 \gamma(z)v_1 = v_1$ and $\Gamma_1 f = \Gamma_1 x + \Gamma_1 \gamma(z)v_1 = \Gamma_1 x + M(z)v_1 = v_2$. Hence, $f \in \mathcal{D}(T_{\mathcal{B}})$, since $(v_1, v_2) \in \mathcal{B}$. Using that $\gamma(z)v_1 \in \mathcal{N}_z$ and $T_0 \subseteq T^*$, we derive

$$(T_{\mathcal{B}} - zI)f = (T^* - zI)f = (T^* - zI)x = (T_0 - zI)x = y,$$

so that $y \in \mathcal{R}(T_{\mathcal{B}} - zI)$. Thus, we have proved that $z \in \rho(T_{\mathcal{B}})$.

The reverse implication of (iii) will be shown together with the resolvent formula (14.43) in the proof of Theorem 14.18 below. $\qquad\square$

14.6 The Krein–Naimark Resolvent Formula

The *Krein–Naimark resolvent formula* in the formulation (14.43) given below expresses the difference of the resolvents of a proper extension $T_{\mathcal{B}}$ and the distinguished self-adjoint extension T_0 of T defined by $\mathcal{D}(T_0) = \mathcal{N}(\Gamma_0)$ in terms of the relation $\mathcal{B}$, the gamma field, and the Weyl function.

Theorem 14.18 *Let T be a densely defined symmetric operator on $\mathcal{H}$, and $(\mathcal{K}, \Gamma_0, \Gamma_1)$ a boundary triplet for T^*. Suppose that $\mathcal{B}$ is a closed relation on $\mathcal{K}$ and $z \in \rho(T_0)$. Then the proper extension $T_{\mathcal{B}}$ of T is given by*

$$\mathcal{D}(T_{\mathcal{B}}) = \big\{ f = (T_0 - zI)^{-1}(y + v) + \gamma(z)u :$$
$$u \in \mathcal{K},\ v \in \mathcal{N}_{\bar{z}},\ (u, \gamma(\bar{z})^* v) \in \mathcal{B} - M(z),\ y \in \mathcal{H} \ominus \mathcal{N}_{\bar{z}} \big\}, \quad (14.41)$$
$$T_{\mathcal{B}}f \equiv T_{\mathcal{B}}\big((T_0 - zI)^{-1}(y + v) + \gamma(z)u\big) = zf + y + v. \quad (14.42)$$

If $z \in \rho(T_\mathcal{B}) \cap \rho(T_0)$, *the relation* $\mathcal{B} - M(z)$ *has an inverse* $(\mathcal{B} - M(z))^{-1} \in \mathbf{B}(\mathcal{K})$, *and*

$$(T_\mathcal{B} - zI)^{-1} - (T_0 - zI)^{-1} = \gamma(z)\big(\mathcal{B} - M(z)\big)^{-1}\gamma(\overline{z})^*. \qquad (14.43)$$

Proof Let $f \in \mathcal{D}(T_\mathcal{B})$. Since $f \in \mathcal{D}(T^*)$, by (14.19) there are vectors $x_z \in \mathcal{D}(T_0)$ and $u_z \in \mathcal{N}_z$ such that $f = x_z + u_z$. Writing $y_z := (T_0 - zI)x_z = y + v$ with $v \in \mathcal{N}_{\overline{z}}$ and $y \in \mathcal{H} \ominus \mathcal{N}_{\overline{z}}$, we have $x_z = (T_0 - zI)^{-1}(y + v)$. From $f = x_z + u_z$ and $\Gamma_0 x_z = 0$ (by $x_z \in \mathcal{D}(T_0)$) we get $u := \Gamma_0 f = \Gamma_0 u_z$, and so $u_z = \gamma(z)\Gamma_0 f = \gamma(z)u$. Thus,

$$\begin{aligned}
\Gamma_1 f - M(z)\Gamma_0 f &= \Gamma_1 f - \Gamma_1 \gamma(z)\Gamma_0 f = \Gamma_1 f - \Gamma_1 u_z \\
&= \Gamma_1 x_z = \Gamma_1 (T_0 - zI)^{-1}(y + v) = \gamma(\overline{z})^*(y + v) \\
&= \gamma(\overline{z})^* v \qquad\qquad\qquad\qquad\qquad\qquad (14.44)
\end{aligned}$$

by Proposition 14.14, (i) and (ii), and therefore

$$\big(u, \gamma(\overline{z})^* v\big) = (\Gamma_0 f, \Gamma_1 f) - \big(\Gamma_0 f, M(z)\Gamma_0 f\big) = (\Gamma_0 f, \Gamma_1 x_z). \qquad (14.45)$$

Since $(\Gamma_0 f, \Gamma_1 f) \in \mathcal{B}$ by $f \in \mathcal{D}(T_\mathcal{B})$, (14.45) implies that $(u, \gamma(\overline{z})^* v) \in \mathcal{B} - M(z)$. The preceding shows that f is of the form (14.41).

From the relations $T_\mathcal{B} \subseteq T^*$, $T_0 \subseteq T^*$, and $\gamma(z)u \in \mathcal{N}_z$ we obtain

$$(T_\mathcal{B} - zI)f = \big(T^* - zI\big)\big((T_0 - zI)^{-1}(y + v) + \gamma(z)u\big) = y + v,$$

which proves (14.42).

Conversely, suppose that f is as in (14.41). Putting $x_z := (T_0 - zI)^{-1}(y + v)$ and $u_z := \gamma(z)u$, we proceed in reverse order. Since $(u, \gamma(\overline{z})^* v) \in \mathcal{B} - M(z)$ by (14.41), it follows then from (14.45) that $(\Gamma_0 f, \Gamma_1 f) \in \mathcal{B}$, that is, $f \in \mathcal{D}(T_\mathcal{B})$.

Now suppose that $z \in \rho(T_\mathcal{B}) \cap \rho(T_0)$. First, we prove that $0 \in \rho(\mathcal{B} - M(z))$. Let $w \in \mathcal{K}$. By Definition 14.2(ii) there exists $x \in \mathcal{D}(T^*)$ such that $\Gamma_0 x = 0$ and $\Gamma_1 x = w$. Then $x \in \mathcal{D}(T_0)$ and $x = (T_0 - zI)^{-1}y_z$, where $y_z := (T_0 - zI)x$. By (14.41) and (14.42) the vector $f := (T_\mathcal{B} - zI)^{-1}y_z \in \mathcal{D}(T_\mathcal{B})$ is of the form $(T_0 - zI)^{-1}y_z + \gamma(z)u$ for some $u \in \mathcal{K}$. In the above notation we then have $x = (T_0 - zI)^{-1}y_z = x_z$, and so $w = \Gamma_1 x = \Gamma_1 x_z$. Since the pair in (14.45) belongs to the relation $\mathcal{B} - M(z)$, it follows that $w = \Gamma_1 x_z \in \mathcal{R}(\mathcal{B} - M(z))$. This proves that $\mathcal{R}(\mathcal{B} - M(z)) = \mathcal{K}$. Further, by $z \in \rho(T_\mathcal{B})$ we have $\mathcal{N}(T_\mathcal{B} - zI) = \{0\}$, and hence $\mathcal{N}(\mathcal{B} - M(z)) = \{0\}$ by Proposition 14.17(ii). The relation $\mathcal{B} - M(z)$ is closed by Proposition 14.17. Therefore, Proposition 14.1, applied to $\mathcal{B} - M(z)$, yields $0 \in \rho(\mathcal{B} - M(z))$, that is, $(\mathcal{B} - M(z))^{-1} \in \mathbf{B}(\mathcal{K})$. (This also completes the proof of Proposition 14.17(iii).)

We prove the resolvent formula (14.43). Let $g \in \mathcal{H}$. Then $f := (T_\mathcal{B} - zI)^{-1}g$ is in $\mathcal{D}(T_\mathcal{B})$, so f has to be of the form (14.41), that is, $f = (T_0 - zI)^{-1}g + \gamma(z)u$. Thus, in the notation of the first paragraph, $x_z = (T_0 - zI)^{-1}g$ and $g = y + v$. Since $(\Gamma_0 f, \Gamma_1 x_z) \in \mathcal{B} - M(z)$ (by (14.45)), we conclude that $\Gamma_0 f = (\mathcal{B} - M(z))^{-1}\Gamma_1 x_z$. Further, $u = \Gamma_0 f$ and $\Gamma_1 x_z = \gamma(\overline{z})^*(y + v) = \gamma(\overline{z})^* g$ by (14.44). Therefore,

$$\gamma(z)u = \gamma(z)\Gamma_0 f = \gamma(z)\big(\mathcal{B} - M(z)\big)^{-1}\Gamma_1 x_z = \gamma(z)\big(\mathcal{B} - M(z)\big)^{-1}\gamma(\overline{z})^* g.$$

Inserting this into the equalities $f = (T_\mathcal{B} - zI)^{-1}g = (T_0 - zI)^{-1}g + \gamma(z)u$, we obtain (14.43). $\qquad\square$

We now consider the resolvent formula for our two operator-theoretic examples.

Example 14.11 (*First standard Example* 14.5 *continued*) Recall from Example 14.5 that $A = T_0$ and $\mu \in \rho(A)$. Let $v \in \mathcal{N}_{\overline{\mu}} = \mathcal{K}$. By the definition of the boundary maps the equality $v = A(A - \mu I)^{-1}v + (A - \mu I)^{-1}(-\mu v)$ yields $\Gamma_0 v = v$ and $\Gamma_1 v = -\mu v$, so $\gamma(\overline{\mu})v = v$ and $M(\overline{\mu})v = \Gamma_1\gamma(\overline{\mu})v = -\mu v$. Thus, $\gamma(\overline{\mu}) = I_\mathcal{H} \upharpoonright \mathcal{K}$ and $M(\overline{\mu}) = -\mu I_\mathcal{K}$. Therefore, by Proposition 14.14(iv),

$$\gamma(z) = (A - \overline{\mu}I)(A - zI)^{-1} \upharpoonright \mathcal{K}, \quad z \in \rho(A). \tag{14.46}$$

Let $P_\mathcal{K}$ denote the orthogonal projection of $\mathcal{H}$ onto $\mathcal{K}$. It is easily verified that

$$\gamma(z)^* = P_\mathcal{K}(A - \mu I)(A - \overline{z}I)^{-1}, \quad z \in \rho(A). \tag{14.47}$$

Applying Proposition 14.15(iii) by using the preceding formulas, we compute

$$
\begin{aligned}
M(z) &= M(\overline{\mu}) + (z - \overline{\mu})\gamma(\mu)^*\gamma(z) \\
&= -\mu I + (z - \overline{\mu})P_\mathcal{K}(A - \mu I)(A - \overline{\mu}I)^{-1}(A - \overline{\mu}I)^{-1}(A - zI)^{-1} \upharpoonright \mathcal{K} \\
&= (z - 2\operatorname{Re}\mu)I_\mathcal{K} + (z - \mu)(z - \overline{\mu})P_\mathcal{K}(A - zI)^{-1} \upharpoonright \mathcal{K}, \quad z \in \rho(A).
\end{aligned}
\tag{14.48}
$$

For the present example, Theorem 14.18 yields the following assertion:
Suppose that $\mathcal{B}$ is a closed relation on $\mathcal{K}$ and $z \in \rho(T_\mathcal{B}) \cap \rho(A)$. Then we have

$$
\begin{aligned}
&(T_\mathcal{B} - zI)^{-1} - (A - zI)^{-1} \\
&= (A - \overline{\mu}I)(A - zI)^{-1}\big(\mathcal{B} - M(z)\big)^{-1}P_\mathcal{K}(A - \mu I)(A - zI)^{-1}, \tag{14.49}
\end{aligned}
$$

where $M(z)$ is given by (14.48), *and the operator $T_\mathcal{B}$ is defined by* (14.13). $\qquad\circ$

Example 14.12 (*Second standard Example* 14.6 *continued*) Let us recall from Example 14.6 that $A = T_0$, $\mu \in \rho(A)$ is *real*, and $\mathcal{K} = \mathcal{N}_\mu$.

From the definitions of the boundary maps we get $\Gamma_0 v = v$ and $\Gamma_1 v = 0$ for $v \in \mathcal{K}$. Therefore, $\gamma(\mu)v = v$ and $M(\mu)v = \Gamma_1\gamma(\mu)v = 0$, that is, $\gamma(\mu) = I_\mathcal{H} \upharpoonright \mathcal{K}$ and $M(\mu) = 0$. Hence $\gamma(\mu)^*$ is just the orthogonal projection $P_\mathcal{K}$ of $\mathcal{H}$ onto $\mathcal{K}$. Putting these facts into Propositions 14.14(iv) and 14.15(iii) yields

$$
\begin{aligned}
\gamma(z) &= (A - \mu I)(A - zI)^{-1} \upharpoonright \mathcal{K}, \quad z \in \rho(A), \\
M(z) &= (z - \mu)\, P_\mathcal{K}(A - \mu I)(A - zI)^{-1} \upharpoonright \mathcal{K}, \quad z \in \rho(A). \tag{14.50}
\end{aligned}
$$

Inserting these expressions into (14.43), we obtain the resolvent formula for this example. We leave it to the reader to restate this formula in this example. $\qquad\circ$

In the rest of this section we give a formulation of the Krein–Naimark resolvent formula which does not depend on any boundary triplet.

For a self-adjoint operator A on $\mathcal{H}$ and a closed subspace $\mathcal{K}$ of $\mathcal{H}$, we define

$$M_{A,\mathcal{K}}(z) = zI_\mathcal{K} + \left(1 + z^2\right) P_\mathcal{K}(A - zI)^{-1} \restriction \mathcal{K}$$
$$= P_\mathcal{K}(I + zA)(A - zI)^{-1} \restriction \mathcal{K} \in \mathbf{B}(\mathcal{K}) \quad \text{for } z \in \rho(A). \quad (14.51)$$

Obviously, $M_{A,\mathcal{K}}(\pm i) = \pm iI_\mathcal{K}$. In the setup of Example 14.11, $M_{A,\mathcal{K}}(z)$ is just the Weyl function (14.48) for $\mu = -i$. In this case $M_{A,\mathcal{K}}(z)$ is a Nevanlinna function by Corollary 14.16. The next lemma shows that the latter is always true.

Lemma 14.19 $M_{A,\mathcal{K}}(z)$ *is an operator-valued Nevanlinna function.*

Proof Clearly, $M_{A,\mathcal{K}}(z) \in \mathbf{B}(\mathcal{K})$ is holomorphic on $\mathbb{C}_+$. Let $z \in \mathbb{C}_+$. Setting $x = \operatorname{Re} z$ and $y = \operatorname{Im} z$, a straightforward computation shows that

$$\operatorname{Im} M_{A,\mathcal{K}}(z) = y P_\mathcal{K}\left(I + A^2\right)\left((A - xI)^2 + y^2 I\right)^{-1} \restriction \mathcal{K} \geq 0. \qquad \square$$

The next theorem contains the Krein–Naimark resolvent formula for two *arbitrary self-adjoint extensions* of a densely defined symmetric operator. Recall that $V_C = (C - iI)(C + iI)^{-1}$ is the Cayley transform of a self-adjoint operator C.

Theorem 14.20 *Let S be a densely defined symmetric operator on $\mathcal{H}$, and let $\widetilde{A}$ and A be self-adjoint extensions of S on $\mathcal{H}$. Then there exists a closed linear subspace $\mathcal{K}$ of $\mathcal{N}(S^* - iI)$ and a self-adjoint operator B on $\mathcal{K}$ such that for all $z \in \rho(\widetilde{A}) \cap \rho(A)$, the operator $B - M_{A,\mathcal{K}}(z)$ has a bounded inverse on $\mathcal{K}$, and*

$$(\widetilde{A} - zI)^{-1} - (A - zI)^{-1}$$
$$= (A - iI)(A - zI)^{-1}\left(B - M_{A,\mathcal{K}}(z)\right)^{-1} P_\mathcal{K}(A + iI)(A - zI)^{-1}. \quad (14.52)$$

Here B and $\mathcal{K}$ are uniquely determined by $\widetilde{A}$ and A. That is, if B is a self-adjoint operator on a closed subspace $\mathcal{K}$ of $\mathcal{H}$ such that (14.52) holds for $z = -i$, then

$$\mathcal{K} = \{x \in \mathcal{H} : V_{\widetilde{A}}x = V_A x\}^\perp \quad \text{and} \quad V_B = V_A^{-1} V_{\widetilde{A}} \restriction \mathcal{K}. \quad (14.53)$$

Proof Clearly, the operator $T := A \restriction (\mathcal{D}(A) \cap \mathcal{D}(\widetilde{A}))$ is a symmetric extension of S. Hence, $T^* \subseteq S^*$ and $\mathcal{K} := \mathcal{N}(T^* - iI) \subseteq \mathcal{N}(S^* - iI)$. Let $(\mathcal{K}, \Gamma_0, \Gamma_1)$ be the boundary triplet from Example 14.5 for the operator A with $\mu := -i$. Since $\mathcal{D}(T) = \mathcal{D}(A) \cap \mathcal{D}(\widetilde{A})$, the extensions $\widetilde{A}$ and $A = T_0$ are disjoint. Therefore, by Corollary 14.9 there is a self-adjoint operator B on $\mathcal{K}$ such that $\widetilde{A} = T_B \equiv T_{\mathcal{G}(B)}$. Now formula (14.49) with $\mu = -i$ yields (14.52).

Since $M(\overline{\mu}) = -\mu I_\mathcal{K}$ as stated in Example 14.11, we have $M(i) = iI_\mathcal{K}$, and hence $M(-i) = M(i)^* = -iI_\mathcal{K}$. Therefore, setting $z = -i$ in (14.52), we obtain

$$(\widetilde{A} + iI)^{-1} - (A + iI)^{-1} = (A - iI)(A + iI)^{-1}(B + iI_\mathcal{K})^{-1} P_\mathcal{K}$$
$$= V_A(B + iI_\mathcal{K})^{-1} P_\mathcal{K}.$$

Using the equality $V_C = I - 2i(C + iI)^{-1}$ for $C = \widetilde{A}, A, B$, it follows that

$$V_{\widetilde{A}} - V_A = V_A\big(-2i(B + iI)^{-1}\big)P_{\mathcal{K}} = V_A(V_B - I)P_{\mathcal{K}}. \qquad (14.54)$$

Therefore, $V_{\widetilde{A}}x = V_A x$ if and only if $P_{\mathcal{K}}x = 0$, that is, $x \in \mathcal{K}^{\perp}$. Clearly, (14.54) implies that $V_A^{-1}V_{\widetilde{A}} \upharpoonright \mathcal{K} = V_B$. Thus, we have proved (14.53). $\square$

We rewrite the resolvent formula (14.52) in the very special case where S has *deficiency indices* $(1, 1)$. That is, we suppose that $\dim \mathcal{N}(S^* - iI) = 1$ and choose a unit vector $u \in \mathcal{N}(S^* - iI)$. Let $\widetilde{A}$ and A be self-adjoint extensions of S on $\mathcal{H}$.

First, suppose that $\widetilde{A} \neq A$. Then $\mathcal{K} \neq \{0\}$ by (14.53), so we have $\mathcal{K} = \mathbb{C} \cdot u$, and hence $M_{A,\mathcal{K}}(z)u = \langle(I + zA)(A - zI)^{-1}u, u\rangle u$ by (14.51). Therefore, by Theorem 14.20 and formula (14.52), there exists a real number B such that

$$(\widetilde{A} - zI)^{-1}x - (A - zI)^{-1}x$$
$$= \langle x, (A - iI)(A - \bar{z}I)^{-1}u\rangle\big(B - \langle(I + zA)(A - zI)^{-1}u, u\rangle\big)^{-1}(A - iI)(A - zI)^{-1}u$$

for all $x \in \mathcal{H}$ and $z \in \rho(\widetilde{A}) \cap \rho(A)$.

If $\widetilde{A} = A$, then $\mathcal{K} = \{0\}$ in (14.52). We interpret this as the case $B = \infty$ in the preceding formula.

That is, if we fix one self-adjoint extension, say A, then the other self-adjoint extension $\widetilde{A}$ is uniquely determined by $B \in \mathbb{R} \cup \{\infty\}$, and this one-to-one correspondence describes all possible self-adjoint extensions of S on $\mathcal{H}$.

14.7 Boundary Triplets and Semibounded Self-adjoint Operators

In this section we assume that T is a densely defined *lower semibounded* symmetric operator and $(\mathcal{K}, \Gamma_0, \Gamma_1)$ is a boundary triplet for T^*. Recall that T_0 is the self-adjoint operator defined in Corollary 14.8, and $\gamma(z)$ and $M(z)$ denote the gamma field and the Weyl function of T_0.

Proposition 14.21 *Suppose that T_0 is the Friedrichs extension T_F of T. Let $\mathcal{B}$ be a self-adjoint relation on $\mathcal{K}$. If the self-adjoint operator $T_{\mathcal{B}}$ is lower semibounded, so is the relation $\mathcal{B}$. More precisely, if $\lambda < m_T$ and $\lambda \leq m_{T_{\mathcal{B}}}$, then $\mathcal{B} - M(\lambda) \geq 0$.*

Proof Fix a number $\lambda' < \lambda$. Since $\lambda' < m_T = m_{T_F} = m_{T_0}$ and $\lambda' < m_{T_{\mathcal{B}}}$, we have $\lambda' \in \rho(T_{\mathcal{B}}) \cap \rho(T_0)$. Hence, the resolvent formula (14.43) applies and yields

$$\big(T_{\mathcal{B}} - \lambda'I\big)^{-1} - \big(T_0 - \lambda'I\big)^{-1} = \gamma\big(\lambda'\big)\big(\mathcal{B} - M(\lambda')\big)^{-1}\gamma\big(\lambda'\big)^*. \qquad (14.55)$$

Since T_F is the largest lower semibounded self-adjoint extension of T, we have $T_0 = T_F \geq T_{\mathcal{B}}$, and hence $(T_{\mathcal{B}} - \lambda'I)^{-1} \geq (T_0 - \lambda'I)^{-1}$ by Corollary 10.13. By Proposition 14.14(ii), $\gamma(\lambda)^*\mathcal{K} = \mathcal{K}$. Therefore, it follows from (14.55) that the operator $(\mathcal{B} - M(\lambda'))^{-1}$ is positive and self-adjoint, so $\mathcal{B} - M(\lambda')$ is a positive self-adjoint relation by Lemma 14.3. By Proposition 14.15 the Weyl function

M is a continuous $\mathbf{B}(\mathcal{K})$-valued function of self-adjoint operators on the interval $(-\infty, m_{T_0})$. Therefore, since $\lambda < m_T = m_{T_0}$, passing to the limit $\lambda' \to \lambda - 0$ yields $\mathcal{B} - M(\lambda) \geq 0$. Clearly, $\mathcal{B}$ is lower semibounded. $\square$

The converse of the assertion of Proposition 14.21 is not true in general. That is, the semiboundedness of $\mathcal{B}$ does not necessarily imply that $T_\mathcal{B}$ is semibounded.

The next theorem is the main result of this section. It expresses the form associated with a semibounded operator $T_\mathcal{B}$ in terms of the forms of the Friedrichs extension T_F and of the positive self-adjoint relation $\mathcal{B} - M(\lambda)$. Recall that the form of a relation was defined by Eq. (14.6).

Theorem 14.22 *Let T be a densely defined lower semibounded symmetric operator, and let $(\mathcal{K}, \Gamma_0, \Gamma_1)$ be a boundary triplet for T^* such that the operator T_0 is the Friedrichs extension T_F of T. Let $\lambda \in \mathbb{R}$, $\lambda < m_T$. Suppose that $\mathcal{B}$ is a self-adjoint relation on $\mathcal{K}$ such that $\mathcal{B} - M(\lambda) \geq 0$.*

Then $T_\mathcal{B} - \lambda I$ is a positive self-adjoint operator, and

$$\mathcal{D}[T_\mathcal{B}] = \mathcal{D}[T_F] \dotplus \gamma(\lambda)\mathcal{D}[\mathcal{B} - M(\lambda)], \qquad (14.56)$$

$$(T_\mathcal{B} - \lambda I)[x + \gamma(\lambda)u, x' + \gamma(\lambda)u'] = (T_F - \lambda I)[x, x'] + (\mathcal{B} - M(\lambda))[u, u'] \qquad (14.57)$$

for $x, x' \in \mathcal{D}[T_F]$ and $u, u' \in \mathcal{D}[\mathcal{B} - M(\lambda)]$.

Proof We begin by proving a simple preliminary fact. That is, we show that

$$(T_\mathcal{B} - \lambda I)[x, \gamma(\lambda)u] = 0 \qquad (14.58)$$

for $x \in \mathcal{D}[T_F]$ and $\gamma(\lambda)u \in \mathcal{D}[T_\mathcal{B}]$, $u \in \mathcal{K}$. By the definition of the Friedrichs extension there is a sequence $(x_n)_{n\in\mathbb{N}}$ from $\mathcal{D}(T)$ which converges to x in the form norm of T_F. Since $\mathfrak{t}_{T_F} \subseteq \mathfrak{t}_{T_\mathcal{B}}$ by Theorem 10.17(ii), (x_n) converges to x also in the form norm of $T_\mathcal{B}$. Using that $T \subseteq T_\mathcal{B}$ and $\gamma(\lambda)u \in \mathcal{N}_\lambda = \mathcal{N}(T^* - \lambda I)$, we derive

$$\begin{aligned}
(T_\mathcal{B} - \lambda I)[x, \gamma(\lambda)u] &= \lim_{n\to\infty} (T_\mathcal{B} - \lambda I)[x_n, \gamma(\lambda)u] \\
&= \lim_{n\to\infty} \langle (T_\mathcal{B} - \lambda I)x_n, \gamma(\lambda)u \rangle \\
&= \lim_{n\to\infty} \langle (T - \lambda I)x_n, \gamma(\lambda)u \rangle = 0.
\end{aligned}$$

Since $\mathcal{B}$ is self-adjoint, $M(\lambda) = M(\lambda)^* \in \mathbf{B}(\mathcal{K})$, and $\mathcal{B} - M(\lambda) \geq 0$ by assumption, $\mathcal{B} - M(\lambda)$ is a positive self-adjoint relation. Let C denote its operator part. Then C is a positive self-adjoint operator acting on some subspace $\mathcal{K}_C$ of $\mathcal{K}$, and the form associated with $\mathcal{B} - M(\lambda)$ is defined by $(\mathcal{B} - M(\lambda))[u, u'] = C[u, u']$ for $u, u' \in \mathcal{D}[\mathcal{B} - M(\lambda)] := \mathcal{D}[C]$.

Clearly, $\lambda < m_T = m_{T_F}$ implies that $\lambda \in \rho(T_F)$. Hence, the operator $T_\mathcal{B}$ is described by formulas (14.41) and (14.42) applied with $z = \lambda$ and $T_0 = T_F$.

Let $f \in \mathcal{D}(T_\mathcal{B})$. We write f in the form (14.41). There exist vectors $v \in \mathcal{N}_\lambda$ and $y \in \mathcal{H} \ominus \mathcal{N}_\lambda$ such that $f = x + \gamma(\lambda)u$, where $x := (T_F - \lambda I)^{-1}(y + v)$, and

$(u, \gamma(\lambda)^*v) \in \mathcal{B} - M(\lambda)$. The latter implies that $u \in \mathcal{D}(\mathcal{B} - M(\lambda)) = \mathcal{D}(C) \subseteq \mathcal{K}_C$ and $Cu = P_C\gamma(\lambda)^*v$, where P_C is the projection of $\mathcal{K}$ onto $\mathcal{K}_C$. By (14.42), $(T_\mathcal{B} - \lambda I)f = y + v$. Since $\gamma(\lambda)u \in \mathcal{N}_\lambda$, $y \perp \gamma(\lambda)u$. Using these facts, we compute

$$\langle(T_\mathcal{B} - \lambda I)f, f\rangle = \langle(T_\mathcal{B} - \lambda I)(x + \gamma(\lambda)u), x + \gamma(\lambda)u\rangle = \langle y + v, x + \gamma(\lambda)u\rangle$$
$$= \langle y + v, x\rangle + \langle v, \gamma(\lambda)u\rangle = \langle(T_F - \lambda I)x, x\rangle + \langle\gamma(\lambda)^*v, u\rangle_\mathcal{K}$$
$$= \langle(T_F - \lambda I)x, x\rangle + \langle P_C\gamma(\lambda)^*v, u\rangle_\mathcal{K}$$
$$= \langle(T_F - \lambda I)x, x\rangle + \langle Cu, u\rangle_\mathcal{K}.$$

Therefore, since $T_F - \lambda I \geq 0$ and $C \geq 0$, it follows that $T_\mathcal{B} - \lambda I \geq 0$ and

$$\left\|(T_\mathcal{B} - \lambda I)^{1/2}f\right\|^2 = \left\|(T_F - \lambda I)^{1/2}x\right\|^2 + \left\|C^{1/2}u\right\|_\mathcal{K}^2. \tag{14.59}$$

Now we prove the inclusion $\mathcal{D}[T_F] + \gamma(\lambda)\mathcal{D}[C] \subseteq \mathcal{D}[T_\mathcal{B}]$. Since $\mathcal{D}[T_F] \subseteq \mathcal{D}[T_\mathcal{B}]$ by Theorem 10.17(ii), it suffices to show that $\gamma(\lambda)\mathcal{D}[C] \subseteq \mathcal{D}[T_\mathcal{B}]$.

Let $u \in \mathcal{D}(C)$. By Proposition 14.14(ii), $\gamma(\lambda)^*$ is a bijection of $\mathcal{N}_\lambda$ onto $\mathcal{K}$, so there exists a vector $v \in \mathcal{N}_\lambda$ such that $\gamma(\lambda)^*v = Cu$. Since C is the operator part of $\mathcal{B} - M(\lambda)$, we have $(u, \gamma(\lambda)^*v) = (u, Cu) \in \mathcal{B} - M(\lambda)$. Hence, by (14.41) and (14.42), the vector $f := x + \gamma(\lambda)u$, where $x := (T_F - \lambda I)^{-1}v$, belongs to $\mathcal{D}(T_\mathcal{B})$, and $(T_\mathcal{B} - \lambda I)f = v$. Since $x \in \mathcal{D}(T_F) \subseteq \mathcal{D}[T_F] \subseteq \mathcal{D}[T_\mathcal{B}]$ and $f \in \mathcal{D}(T_\mathcal{B}) \subseteq \mathcal{D}[T_\mathcal{B}]$, we have $\gamma(\lambda)u = f - x \in \mathcal{D}[T_\mathcal{B}]$, so that $(T_\mathcal{B} - \lambda I)[x, \gamma(\lambda)u] = 0$ by (14.58). Further, $v = (T_F - \lambda I)x$. Inserting these facts, we derive

$$(T_\mathcal{B} - \lambda I)[f] = \langle(T_\mathcal{B} - \lambda I)f, f\rangle = \langle v, x + \gamma(\lambda)u\rangle$$
$$= \langle(T_F - \lambda I)x, x\rangle + \langle\gamma(\lambda)^*v, u\rangle_\mathcal{K}$$
$$= (T_F - \lambda I)[x] + \langle Cu, u\rangle_\mathcal{K},$$
$$(T_\mathcal{B} - \lambda I)[f] = (T_F - \lambda I)[x] + (T_\mathcal{B} - \lambda I)[\gamma(\lambda)u] + 2\,\mathrm{Re}(T_\mathcal{B} - \lambda I)[x, \gamma(\lambda)u]$$
$$= (T_F - \lambda I)[x] + (T_\mathcal{B} - \lambda I)[\gamma(\lambda)u].$$

Comparing both formulas yields $\langle Cu, u\rangle_\mathcal{K} = (T_\mathcal{B} - \lambda I)[\gamma(\lambda)u]$. Therefore,

$$\left\|C^{1/2}u\right\|_\mathcal{K} = \left\|(T_\mathcal{B} - \lambda I)^{1/2}\gamma(\lambda)u\right\| \tag{14.60}$$

for $u \in \mathcal{D}(C)$. Since $\mathcal{D}(C)$ is a core for the form of C, $\gamma(\lambda)$ is bounded, and the operator $(T_\mathcal{B} - \lambda I)^{1/2}$ is closed, we conclude that Eq. (14.60) extends by continuity to vectors $u \in \mathcal{D}[C] = \mathcal{D}(C^{1/2})$ and that $\gamma(\lambda)u \in \mathcal{D}[T_\mathcal{B}] = \mathcal{D}((T_\mathcal{B} - \lambda I)^{1/2})$ for such vectors. This proves the inclusion $\mathcal{D}[T_F] + \gamma(\lambda)\mathcal{D}[C] \subseteq \mathcal{D}[T_\mathcal{B}]$.

We prove the converse inclusion $\mathcal{D}[T_\mathcal{B}] \subseteq \mathcal{D}[T_F] + \gamma(\lambda)\mathcal{D}[C]$. Let $g \in \mathcal{D}[T_\mathcal{B}]$. Since $\mathcal{D}(T_\mathcal{B})$ is a form core for $T_\mathcal{B}$, there is a sequence $(f_n)_{n \in \mathbb{N}}$ of $\mathcal{D}(T_\mathcal{B})$ which converges to f in the form norm of $T_\mathcal{B}$. By (14.41), each vector $f_n \in \mathcal{D}(T_\mathcal{B})$ can be written as $f_n = x_n + \gamma(\lambda)u_n$, where $x_n = (T_F - \lambda I)^{-1}(y_n + v_n)$ and $(u_n, \gamma(\lambda)^*v_n) \in \mathcal{B} - M(\lambda)$. Since $\lambda < m_{T_F}$, the norm $\|(T_F - \lambda I)^{1/2} \cdot \|$ is equivalent to the form norm of T_F. Applying (14.59) to $f_n - f_k = x_n - x_k + \gamma(\lambda)(u_n - u_k)$, it follows therefore that (x_n) and $(C^{1/2}u_n)$ are Cauchy sequences in the Hilbert spaces

$\mathcal{D}[T_F]$ and $\mathcal{K}_C$. Because $(f_n = x_n + \gamma(\lambda)u_n)$ and (x_n) are also Cauchy sequences in $\mathcal{H}$, so is $(\gamma(\lambda)u_n)$. Since $u_n = \Gamma_0\gamma(\lambda)u_n$ and $\Gamma_0 : \mathcal{N}_\lambda \to \mathcal{K}$ is continuous by Lemma 14.13(ii), (u_n) is a Cauchy sequence in $\mathcal{K}_C$. Thus, (u_n) is a Cauchy sequence in the form norm of C. Let $x \in \mathcal{D}[T_F]$ and $u \in \mathcal{D}[C]$ be the limits of the Cauchy sequences (x_n) and (u_n) in the corresponding form Hilbert spaces. Since $\gamma(\lambda)$ is bounded, the equality $f_n = x_n + \gamma(\lambda)u_n$ implies that $f = x + \gamma(\lambda)u$. Hence, $f \in \mathcal{D}[T_F] + \gamma(\lambda)\mathcal{D}[C]$. This proves that $\mathcal{D}[T_B] \subseteq \mathcal{D}[T_F] + \gamma(\lambda)\mathcal{D}[C]$. Putting the preceding together, we have shown that $\mathcal{D}[T_B] = \mathcal{D}[T_F] + \gamma(\lambda)\mathcal{D}[C]$.

As shown in the paragraph before last, Eq. (14.59) holds for $x_n + \gamma(\lambda)u_n$. Passing to the limit, it follows that (14.59) remains valid for $x + \gamma(\lambda)u$ as well. Since we have $(\mathcal{B} - M(\lambda))[u] = C[u] = \|C^{1/2}u\|_{\mathcal{K}}^2$, this means that

$$(T_B - \lambda I)[x + \gamma(\lambda)u] = (T_F - \lambda I)[x] + (\mathcal{B} - M(\lambda))[u].$$

Equation (14.57) follows now from the latter equation by polarization.

Finally, we show that $\mathcal{D}[T_F] + \gamma(\lambda)\mathcal{D}[C]$ is a direct sum. Assume that $f = x + \gamma(\lambda)u = 0$. Since $\|(T_F - \lambda I)^{1/2} \cdot \| \geq (m_{T_F} - \lambda)\| \cdot \|$ and $m_T - \lambda > 0$, it follows from (14.59) that $x = 0$, and so $\gamma(\lambda)u = 0$.   □

Corollary 14.23 *Suppose that S is a lower semibounded self-adjoint extension of the densely defined lower semibounded symmetric operator T on $\mathcal{H}$. Let $\lambda \in \mathbb{R}$, $\lambda < m_T$, and $\lambda \leq m_S$. Then S is equal to the Friedrichs extension T_F of T if and only if $\mathcal{D}[S] \cap \mathcal{N}(T^* - \lambda I) = \{0\}$.*

Proof Since then $\lambda \in \rho(T_F)$, Example 14.6, with $A = T_F$, $\mu = \lambda$, yields a boundary triplet $(\mathcal{K}, \Gamma_0, \Gamma_1)$ for T^* such that $T_0 = T_F$. By Propositions 14.7(v) and 14.21, there is a self-adjoint relation $\mathcal{B}$ on $\mathcal{K}$ such that $S = T_B$ and $\mathcal{B} - M(\lambda) \geq 0$. Thus, all assumptions of Theorem 14.22 are fulfilled. Recall that $\mathcal{K} = \mathcal{N}(T^* - \lambda I)$. Therefore, by (14.56), $\mathcal{D}[T_B] \cap \mathcal{N}(T^* - \lambda I) = \{0\}$ if and only if $\gamma(\lambda)\mathcal{D}[\mathcal{B} - M(\lambda)] = \{0\}$, that is, $\mathcal{D}[\mathcal{B} - M(\lambda)] = \{0\}$. By (14.56) and (14.57) the latter is equivalent to $\mathfrak{t}_{T_B} = \mathfrak{t}_{T_F}$ and so to $T_B = T_F$. (Note that we even have $M(\lambda) = 0$ and $\gamma(\lambda) = I_{\mathcal{H}} \upharpoonright \mathcal{K}$ by Example 14.12.)   □

In Theorem 14.22 we assumed that the self-adjoint operator T_0 from Corollary 14.8 is equal to the Friedrichs extension T_F of T. For the boundary triplet from Example 14.6, we now characterize this property in terms of the Weyl function.

Example 14.13 (*Example* 14.12 *continued*) Suppose that the self-adjoint operator A in Example 14.6 is lower semibounded and $\mu < m_A$. Recall that A is the operator T_0 and $\mathcal{K} = \mathcal{N}(T^* - \mu I)$. Since $T \subseteq T_0 = A$, the symmetric operator T is then lower semibounded, and $\mu < m_A \leq m_T$.

Statement *The operator $A = T_0$ is equal to the Friedrichs extension T_F of T if and only if for all $u \in \mathcal{K}, u \neq 0$, we have*

$$\lim_{t \to -\infty} \langle M(t)u, u \rangle = -\infty. \tag{14.61}$$

Proof Assume without loss of generality that $\mu = 0$. Then $A \geq 0$. By Corollary 14.23, applied with $S = A$ and $\lambda = 0$, it suffices to show that any nonzero vector $u \in \mathcal{K} = \mathcal{N}(T^*)$ is not in the form domain $\mathcal{D}[A]$ if and only if (14.61) holds.

Let E denote the spectral measure of A. By formula (14.50), applied with $\mu = 0$,

$$\langle M(t)u, u \rangle = \int_0^\infty t\lambda(\lambda - t)^{-1} d\langle E(\lambda)u, u \rangle. \tag{14.62}$$

On the other hand, $u \notin \mathcal{D}[A] = \mathcal{D}(A^{1/2})$ if and only if

$$\int_0^\infty \lambda \, d\langle E(\lambda)u, u \rangle = \infty. \tag{14.63}$$

Let $\alpha > 0$. Suppose that $t \leq -\alpha$. For $\lambda \in [0, \alpha]$, we have $\lambda \leq -t$, so $\lambda - t \leq -2t$ and $1 \leq -2t(\lambda - t)^{-1}$, hence, $2t\lambda(\lambda - t)^{-1} \leq -\lambda$. Thus,

$$\int_0^\alpha 2t\lambda(\lambda - t)^{-1} d\langle E(\lambda)u, u \rangle \leq - \int_0^\alpha \lambda \, d\langle E(\lambda)u, u \rangle \quad \text{for } t \leq -\alpha. \tag{14.64}$$

If $u \notin \mathcal{D}[A]$, then (14.63) holds, and hence by (14.64) and (14.62) we obtain (14.61).

Conversely, suppose that (14.61) is satisfied. Since $-t\lambda(\lambda - t)^{-1} \leq \lambda$ for $\lambda > 0$ and $t < 0$, it follows from inequalities (14.62) and (14.61) that (14.63) holds. Therefore, $u \notin \mathcal{D}[A]$. $\qquad\qquad\qquad\qquad\qquad\qquad\qquad\qquad\qquad\square$

14.8 Positive Self-adjoint Extensions

Throughout this section we suppose that T is a densely defined symmetric operator on $\mathcal{H}$ with *positive lower bound* $m_T > 0$, that is,

$$\langle Tx, x \rangle \geq m_T \|x\|^2, \quad x \in \mathcal{D}(T), \text{ where } m_T > 0. \tag{14.65}$$

Our aim is to apply the preceding results (especially Theorem 14.22) to investigate the set of all *positive* self-adjoint extensions of T.

Since $0 < m_T = m_{T_F}$, we have $0 \in \rho(T_F)$. Hence, Theorem 14.12 applies with $\mu = 0$ and $A = T_F$. By Theorem 14.12 the self-adjoint extensions of T on $\mathcal{H}$ are precisely the operators T_B defined therein with $B \in \mathcal{S}(\mathcal{N}(T^*))$. Recall that

$$\mathcal{D}(T_B)$$
$$= \{x + (T_F)^{-1}(Bu + v) + u : x \in \mathcal{D}(\overline{T}), \ u \in \mathcal{D}(B), \ v \in \mathcal{N}(T^*) \cap \mathcal{D}(B)^\perp\},$$
$$T_B(x + (T_F)^{-1}(Bu + v) + u) = \overline{T}x + Bu + v.$$

Because of the inverse of T_F, it might be difficult to describe the domain and the action of the operator T_B explicitly. By contrast, if T_B is positive, the following theorem shows that there is an elegant and explicit formula for the associated form.

Let $\mathcal{S}(\mathcal{N}(T^*))_+$ denote the set of positive operators in $\mathcal{S}(\mathcal{N}(T^*))$.

Theorem 14.24

(i) *For $B \in \mathcal{S}(\mathcal{N}(T^*))$, we have $T_B \geq 0$ if and only if $B \geq 0$.*
   In this case the greatest lower bounds m_B and m_{T_B} satisfy the inequalities

$$m_T m_B (m_T + m_B)^{-1} \leq m_{T_B} \leq m_B.$$

(ii) *If $B \in \mathcal{S}(\mathcal{N}(T^*))_+$, then $\mathcal{D}[T_B] = \mathcal{D}[T_F] \dotplus \mathcal{D}[B]$, and*

$$T_B[y + u, y' + u'] = T_F[y, y'] + B[u, u'] \quad \text{for } y, y' \in \mathcal{D}[T_F], \ u, u' \in \mathcal{D}[B].$$

(iii) *If $B_1, B_2 \in \mathcal{S}(\mathcal{N}(T^*))_+$, then $B_1 \geq B_2$ is equivalent to $T_{B_1} \geq T_{B_2}$.*

Proof First, suppose that $B \geq 0$. Let $f \in \mathcal{D}(T_B)$. By the above formulas, f is of the form $f = y + u$ with $y = x + T_F^{-1}(Bu + v)$, where $x \in \mathcal{D}(\overline{T})$, $u \in \mathcal{D}(B)$, and $v \in \mathcal{N}(T^*) \cap \mathcal{D}(B)^\perp$, and we have $T_B f = T_F y$, since $\overline{T} \subseteq T_F$ and $y \in \mathcal{D}(T_F)$. Further, $m_T = m_{T_F}$, $T^* u = 0$, and $\langle v, u \rangle = 0$. Using these facts, we compute

$$\langle T_B f, f \rangle = \langle T_F y, y + u \rangle = \langle T_F y, y \rangle + \langle \overline{T} x + Bu + v, u \rangle = \langle T_F y, y \rangle + \langle Bu, u \rangle$$
$$\geq m_T \|y\|^2 + m_B \|u\|^2 \geq m_T m_B (m_T + m_B)^{-1} \big(\|y\| + \|u\| \big)^2$$
$$\geq m_T m_B (m_T + m_B)^{-1} \|y + u\|^2 = m_T m_B (m_T + m_B)^{-1} \|f\|^2, \quad (14.66)$$

where the second inequality follows from the elementary inequality

$$\alpha a^2 + \beta b^2 \geq \alpha \beta (\alpha + \beta)^{-1} (a + b)^2 \quad \text{for } \alpha > 0, \ \beta \geq 0, \ a \geq 0, b \geq 0.$$

Clearly, (14.66) implies that $T_B \geq 0$ and $m_{T_B} \geq m_T m_B (m_T + m_B)^{-1}$.

The other assertions of (i) and (ii) follow from Proposition 14.21 and Theorem 14.22 applied to the boundary triplet from our second standard Example 14.6, with $A = T_0 = T_F$, $\mu = 0$, by using that $\gamma(0) = I_\mathcal{H} \restriction \mathcal{K}$ and $M(0) = 0$ (see Example 14.12). Since $m_T > 0$ by assumption (14.65), we can set $\lambda = 0$ in both results. Recall that by (14.6) the form of a self-adjoint relation $\mathcal{B}$ is the form of its operator part B. Since $T_B[u] = B[u]$ for $u \in \mathcal{D}[B]$ by (14.57), it is obvious that $m_B \geq m_{T_B}$.

(iii) is an immediate consequence of (ii). $\square$

Theorem 14.24(ii) confirms an interesting phenomenon that we have seen already in Sect. 10.5 (see Examples 10.7 and 10.8): Many positive self-adjoint extensions with *different operator domains* have the *same form domains*. For instance, for all bounded operators $B \in \mathcal{S}(\mathcal{N}(T^*))_+$ acting on a fixed closed subspace $\mathcal{M}$ of $\mathcal{N}(T^*)$, the form domain $\mathcal{D}[T_B]$ is $\mathcal{D}[T_F] \dotplus \mathcal{M}$ by Theorem 14.24(ii). But, by the above formula, the operator domains $\mathcal{D}(T_{B_1})$ and $\mathcal{D}(T_{B_2})$ are different whenever $B_1 \neq B_2$. That is, form domains are more "rigid" than operator domains.

By the definition of the order relation "$\geq$" (Definition 10.5) the set $\mathcal{S}(\mathcal{N}(T^*))_+$ contains a largest operator and a smallest operator. These are the following two extreme cases:

Case 1 ($\mathcal{D}(B) = \{0\}$) This is the largest operator in $\mathcal{S}(\mathcal{N}(T^*))_+$. Then obviously $T_B \subseteq T_F$, and therefore $T_B = T_F$. That is, T_B is just the *Friedrichs extension* T_F of T. Clearly, we have $\mathcal{D}(T_F) = \mathcal{D}(\overline{T}) + T_F^{-1} \mathcal{N}(T^*)$.

Case 1 is also referred to as the case $B = \infty$, because one may think of B being infinity in order to have only the null vector in $\mathcal{D}(B)$. For this reason, the Friedrichs extension T_F is occasionally denoted by T_∞ in the literature.

Case 2 ($B = 0$, $\mathcal{D}(B) = \mathcal{N}(T^*)$) Since this is the smallest element of $\mathcal{S}(\mathcal{N}(T^*))_+$, the corresponding operator T_B is the smallest among all positive self-adjoint extensions. Therefore, by Corollary 13.15, this operator T_B coincides with the *Krein–von Neumann extension* T_N of T. Hence, we have the following formulas for the Krein–von Neumann extension:

$$\mathcal{D}(T_N) = \mathcal{D}(\overline{T}) \dotplus \mathcal{N}(T^*), \tag{14.67}$$

$$T_N(x + u) = \overline{T}x \quad \text{for } x \in \mathcal{D}(\overline{T}),\ u \in \mathcal{N}(T^*), \tag{14.68}$$

$$\mathcal{D}[T_N] = \mathcal{D}[T_F] \dotplus \mathcal{N}(T^*), \tag{14.69}$$

$$T_N[y + u, y' + u'] = T_F[y, y'] \quad \text{for } y, y' \in \mathcal{D}[T_F],\ u, u' \in \mathcal{N}(T^*). \tag{14.70}$$

Recall that we have assumed that $m_T > 0$ by (14.65).

The next theorem contains a slightly different characterization of positive self-adjoint extensions of the operator T.

Theorem 14.25 *Let T be a densely defined positive symmetric operator on $\mathcal{H}$ such that $m_T > 0$. For any positive self-adjoint operator A on $\mathcal{H}$, the following statements are equivalent:*

(i) *A is an extension of T.*
(ii) *There is an operator $B \in \mathcal{S}(\mathcal{N}(T^*))_+$ such that $A = T_B$.*
(iii) *$T_F \geq A \geq T_N$.*

Proof (i) $\rightarrow$ (ii): Combine Theorems 14.12, with $A = T_F$, $\mu = 0$, and 14.24(i).

(ii) $\rightarrow$ (iii) holds by Theorem 14.24(iii), since T_F and T_N are the two extreme cases as shown by the preceding discussion (by Corollary 13.15).

(iii) $\rightarrow$ (i): The inequalities $T_F \geq A \geq T_N$ mean that $\mathcal{D}[T_N] \supseteq \mathcal{D}[A] \supseteq \mathcal{D}[T_F]$ and $\mathfrak{t}_{T_N} \leq \mathfrak{t}_A \leq \mathfrak{t}_{T_F}$. If $y \in \mathcal{D}[T_F]$, then $T_F[y] = T_N[y]$ by (14.70), and hence $A[y] = T_F[y]$, since $\mathfrak{t}_{T_N} \leq \mathfrak{t}_A \leq \mathfrak{t}_{T_F}$. Therefore, by the polarization identity (10.2),

$$A[x, y] = T_F[x, y] \quad \text{for } x, y \in \mathcal{D}[T_F]. \tag{14.71}$$

Next, we verify that $A[x, u] = 0$ for $x \in \mathcal{D}[T_F]$ and $u \in \mathcal{N}(T^*) \cap \mathcal{D}[A]$. Let $\lambda \in \mathbb{C}$. Using (14.70) and (14.71) and the relation $\mathfrak{t}_{T_N} \leq \mathfrak{t}_A$, we compute

$$T_F[x] = T_N[x + \lambda u] \leq A[x + \lambda u] = A[x + \lambda u, x + \lambda u]$$

$$= A[x, x] + 2\operatorname{Re}\lambda A[x, u] + |\lambda|^2 A[u, u]$$

$$= T_F[x] + 2\operatorname{Re}\lambda A[x, u] + |\lambda|^2 A[u].$$

Thus, $0 \leq 2\operatorname{Re}\lambda A[x, u] + |\lambda|^2 A[u]$ for all $\lambda \in \mathbb{C}$, which implies that $A[x, u] = 0$.

Let $x \in \mathcal{D}(T)$ and $f \in \mathcal{D}[A]$. Then $f \in \mathcal{D}[T_N] = \mathcal{D}[T_F] \dotplus \mathcal{N}(T^*)$, so $f = y + u$, where $y \in \mathcal{D}[T_F]$ and $u \in \mathcal{N}(T^*)$. Since $x \in \mathcal{D}[T_F]$ and $u = f - y \in \mathcal{D}[A]$, we have $A[x, u] = 0$ as just shown. Further, $T^*u = 0$. Using (14.71), we therefore derive

$$A[x, f] = A[x, y + u] = A[x, y] = T_F[x, y] = \langle T_F x, y \rangle = \langle Tx, y \rangle = \langle Tx, f \rangle.$$

Since this holds for all $f \in \mathcal{D}[A]$, we obtain $T \subseteq A$ from Proposition 10.5(v). □

Next, we study the discreteness of spectra of positive self-adjoint extensions. To exclude trivial cases, we assume that the Hilbert space $\mathcal{H}$ is *infinite-dimensional*.

For a lower semibounded self-adjoint operator C on $\mathcal{H}$, let $(\lambda_n(C))_{n \in \mathbb{N}}$ denote the sequence defined in Sect. 12.1. Recall that C has a purely discrete spectrum if and only if $\lim_{n \to \infty} \lambda_n(C) = +\infty$.

Let A be a positive self-adjoint extension of T on $\mathcal{H}$. Since $T_N \leq A \leq T_F$ as noted above, the min–max principle (Corollary 12.3) implies that

$$\lambda_n(T_N) \leq \lambda_n(A) \leq \lambda_n(T_F) \quad \text{for } n \in \mathbb{N}.$$

Thus, if T_N has a purely discrete spectrum, so have A and T_F. Likewise, if the spectrum of A is purely discrete, so is the spectrum of T_F. Now we turn to the converse implication.

Since $m_{T_F} = m_T > 0$ by assumption (14.65), we have $\mathcal{N}(T_F) = \{0\}$. On the other hand, it follows at once from (14.68) that $\mathcal{N}(T_N) = \mathcal{N}(T^*)$.

The null space is obviously a reducing subspace for each self-adjoint operator. Hence, there exists a self-adjoint operator T'_K on the Hilbert space $\mathcal{H}' := \mathcal{N}(T_N)^\perp$ such that $T_N = T'_N \oplus 0$ on $\mathcal{H} = \mathcal{H}' \oplus \mathcal{N}(T_N)$. Clearly, $\mathcal{N}(T'_N) = \{0\}$.

It may happen that T_F has a purely discrete spectrum, but $\mathcal{N}(T_N) = \mathcal{N}(T^*)$ is infinite-dimensional (see Example 14.16 below), so the spectrum of T_N is not purely discrete. The next proposition shows that then at least the operator T'_N has a purely discrete spectrum.

Proposition 14.26 *If the Friedrichs extension T_F has a purely discrete spectrum, so has the self-adjoint operator T'_N on $\mathcal{H}'$, and $\lambda_n(T_F) \leq \lambda_n(T'_N)$ for $n \in \mathbb{N}$.*

Proof Let P denote the projection of $\mathcal{H}$ onto $\mathcal{H}'$. Let $f \in \mathcal{D}[T'_N]$. Then f is in $\mathcal{D}[T_N] = \mathcal{D}[T_F] \dotplus \mathcal{N}(T^*)$ by (14.69), so $f = y + u$ for some $y \in \mathcal{D}[T_F]$ and $u \in \mathcal{N}(T^*) = \mathcal{N}(T_N)$. Clearly, $f = P(y + u) = Py$. Using (14.70), we deduce

$$T'_N[f] = T_N[f] = T_N[y + u] = T_F[y] \geq m_T \|y\|^2 \geq m_T \|Py\|^2 = m_T \|f\|^2.$$

That is, $T'_N \geq m_T I$. Since T_N is self-adjoint and $m_T > 0$ by our assumption (14.65), this implies that $(T'_N)^{-1} \in \mathbf{B}(\mathcal{H}')$.

Let $x \in \mathcal{H}'$. Since $T_F \subseteq T^*$ and $T'_N \subseteq T^*$, we have

$$T^*\big((T_F)^{-1}x - (T'_N)^{-1}x\big) = T_F(T_F)^{-1}x - T'_N(T'_N)^{-1}x = 0,$$

so $(T_F)^{-1}x - (T'_N)^{-1}x \in \mathcal{N}(T^*)$ and $P(T_F)^{-1}x = P(T'_N)^{-1}x = (T'_N)^{-1}x$. Thus,

$$P(T_F)^{-1} \restriction \mathcal{H}' = (T'_N)^{-1}. \tag{14.72}$$

Since $0 \in \rho(T_F)$ and T_F has a purely discrete spectrum, $(T_F)^{-1}$ is compact by Proposition 2.11. Hence, its compression $P(T_F)^{-1} \upharpoonright \mathcal{H}' = (T_N')^{-1}$ to $\mathcal{H}'$ is also compact, so that T_N' has a purely discrete spectrum again by Proposition 2.11.

From (14.72) it follows that $-(T_F)^{-1} \preceq -(T_N')^{-1}$. Indeed, for $x \in \mathcal{H}'$,

$$\left\langle \left(-(T_F)^{-1}\right)x, x\right\rangle = \left\langle \left(-(T_F)^{-1}\right)x, Px\right\rangle = \left\langle \left(-P(T_F)^{-1}\right)x, x\right\rangle = \left\langle \left(-(T_N')^{-1}\right)x, x\right\rangle.$$

Hence, $\lambda_n(-(T_F)^{-1}) \le \lambda_n(-(T_N')^{-1})$ by Corollary 12.3. Since $(T_F)^{-1}$ is a positive compact operator, the eigenvalue $\lambda_n(T_F)$ is equal to $-\lambda_n(-(T_F)^{-1})^{-1}$. Likewise, $\lambda_n(T_N') = (-\lambda_n(-(T_N')^{-1}))^{-1}$. Thus, we get $\lambda_n(T_F) \le \lambda_n(T_N')$ by the preceding. (Note that the passage to the negative inverses was necessary to obtain the eigenvalues by applying the min–max principle in the formulation of Theorem 12.1. In fact, all $\lambda_n((T_F)^{-1})$ are equal to zero, which is the bottom of the essential spectrum of the compact positive operator $(T_F)^{-1}$.) $\square$

We close this section by developing three examples in detail. Note that in all three examples the operator T is closed.

Example 14.14 (*Examples* 10.4 *and* 14.10 *continued*) Let $T = -\frac{d^2}{dx^2}$ with domain $\mathcal{D}(T) = H_0^2(a, b)$ on $L^2(a, b), a, b \in \mathbb{R}$. By the Poincaré inequality (10.24) we have

$$\langle Tf, f \rangle = \left\| f' \right\|^2 \ge \pi^2 (a - b)^{-2} \|f\|^2 \quad \text{for } f \in \mathcal{D}(T).$$

That is, T has a positive lower bound, hence assumption (14.65) is satisfied.

Clearly, $T^* = -\frac{d^2}{dx^2}$ on $\mathcal{D}(T^*) = H^2(a, b)$. Hence, $\mathcal{N}(T^*) = \mathbb{C} \cdot 1 + \mathbb{C} \cdot x$. Therefore, by (14.67) we have $\mathcal{D}(T_N) = \mathcal{D}(T) \dotplus \mathcal{N}(T^*) = H_0^2(a, b) + \mathbb{C} \cdot 1 + \mathbb{C} \cdot x$. Each function of the latter set fulfills the boundary conditions

$$f'(a) = f'(b) = \left(f(a) - f(b)\right)(a - b)^{-1}. \tag{14.73}$$

This is the special case $b_1 = b_2 = -c := (a - b)^{-1}$ of Eqs. (14.36). As shown in Example 14.10, the operator T_B defined by (14.73) is a self-adjoint extension of T and of T_N by the preceding. Therefore, $T_B = T_N$. That is, the domain $\mathcal{D}(T_N)$ is the set of functions $f \in \mathcal{D}(T^*) = H^2(a, b)$ satisfying the boundary conditions (14.73). $\circ$

Example 14.15 ($T = -\frac{d^2}{dx^2} + c^2$ on $H_0^2(0, \infty), c > 0$) Then we have $\mathcal{N}(T^*) = \mathbb{C} \cdot e^{-cx}$ and $\mathcal{D}(T_N) = H_0^2(0, \infty) \dotplus \mathbb{C} \cdot e^{-cx}$ by (14.67). From the latter it follows that the Krein–von Neumann extension T_N is determined by the boundary condition $f'(0) = -cf(0)$. Thus, in particular, $(T + \lambda I)_N \ne T_N + \lambda I$ for $\lambda > 0$ in contrast to the formula $(T + \lambda I)_F = T_F + \lambda I$ by Theorem 10.17(iv).

The domain of the Friedrichs extension is $\mathcal{D}(T_F) = \{f \in H^2(0, \infty) : f(0) = 0\}$. Since $h(x) := (2c)^{-1} x e^{-cx} \in \mathcal{D}(T_F)$ and $(T_F h)(x) = -h''(x) + c^2 h(x) = e^{-cx}$ on $(0, \infty)$, we conclude that $h = (T_F)^{-1}(e^{-cx})$.

Now let $B \in \mathbb{R}$. By the description of the domain $\mathcal{D}(T_B)$ given above we have

$$
\mathcal{D}(T_B) = \mathcal{D}(T) \dotplus \mathbb{C} \cdot \left((T_F)^{-1} \left(B e^{-cx} \right) + e^{-cx} \right)
$$
$$
= \mathcal{D}(T) \dotplus \mathbb{C} \cdot \left(1 + B(2c)^{-1} x \right) e^{-cx}.
$$

This implies that the domain $\mathcal{D}(T_B)$ is characterized by the boundary condition

$$
f'(0) = \left(B(2c)^{-1} - c \right) f(0).
$$
                                                                                                                    o

Example 14.16 (*Dirichlet Laplacian on bounded domains of* $\mathbb{R}^d$) Throughout this
example, Ω is a bounded open set in $\mathbb{R}^d$ of class C^2.

We essentially use some facts on Sobolev spaces from Appendix D and about
the Dirichlet Laplacian $-\Delta_D$ from Sect. 10.6.1. Recall that $-\Delta_D$ is the Friedrichs
extension of the minimal operator $L_{\min}$ for $-\Delta$ and $L_{\min} \geq c_\Omega \cdot I$ for some constant
$c_\Omega > 0$. We fix a real number $\lambda < c_\Omega$ and set

$$
T := L_{\min} - \lambda I. \tag{14.74}
$$

Then $m_T \geq c_\Omega - \lambda > 0$, so assumption (14.65) is fulfilled.

Clearly, $T_F = (L_{\min} - \lambda I)_F = (L_{\min})_F - \lambda I = -\Delta_D - \lambda I$. Hence, we have
$\mathcal{D}(T_F) = \mathcal{D}(-\Delta_D)$ and $\mathcal{D}(T) = \mathcal{D}(L_{\min})$. Therefore, by Theorem 10.19,

$$
\mathcal{D}(T) = H_0^2(\Omega) = \left\{ f \in H^2(\Omega) : f \upharpoonright \partial\Omega = \frac{\partial f}{\partial \nu} \upharpoonright \partial\Omega = 0 \right\}, \tag{14.75}
$$

$$
\mathcal{D}(T_F) = H^2(\Omega) \cap H_0^1(\Omega) = \left\{ f \in H^1(\Omega) : f \upharpoonright \partial\Omega = 0 \right\}, \tag{14.76}
$$

where $f \upharpoonright \partial\Omega \equiv \gamma_0(f)$ and $\frac{\partial f}{\partial \nu} \upharpoonright \partial\Omega \equiv \gamma_1(f)$ denote the traces defined on $H^2(\Omega)$,
see Theorem D.7. Since $0 < c_\Omega - \lambda \leq m_T = m_{T_F}$, we have $0 \in \rho(T_F)$. Hence,
formula (14.19) with $\mu = 0$ applies and yields

$$
\mathcal{D}(T^*) = \mathcal{D}(T_F) \dotplus \mathcal{N}(T^*). \tag{14.77}
$$

Recall that $\mathcal{N}(T_N) = \mathcal{N}(T^*)$. If $\lambda = 0$ and $d \geq 2$, then $\mathcal{N}(T^*)$ is just the infinite-
dimensional vector space of all harmonic functions on Ω contained in $L^2(\Omega)$.

By Theorem D.1, the embedding of $H_0^1(\Omega)$ into $L^2(\Omega)$ is compact. This implies
that the embedding of $(\mathcal{D}(T_F), \| \cdot \|_{T_F})$ into $L^2(\Omega)$ is compact. Hence, both self-
adjoint operators T_F and T_N' (by Proposition 14.26) have purely discrete spectra.

Let $f \in H^2(\Omega)$. Then $f \in \mathcal{D}(T^*)$, so by (14.77) there exists a unique function
$\mathsf{H}(f) \in \mathcal{N}(T^*)$ such that $f - \mathsf{H}(f) \in \mathcal{D}(T_F) \subseteq H^2(\Omega)$. Therefore, by (14.76) and
$f \in H^2(\Omega)$, we have $\mathsf{H}(f) \in H^2(\Omega)$ and $f \upharpoonright \partial\Omega = \mathsf{H}(f) \upharpoonright \partial\Omega$.

Statement 1 $f \in H^2(\Omega)$ *belongs to* $\mathcal{D}(T_N)$ *if and only if* $\frac{\partial f}{\partial \nu} \upharpoonright \partial\Omega = \frac{\partial \mathsf{H}(f)}{\partial \nu} \upharpoonright \partial\Omega$.

Proof Let $f \in \mathcal{D}(T_N)$. By (14.67), f can be written as $f = g + h$, where $g \in \mathcal{D}(T)$
and $h \in \mathcal{N}(T^*)$. Then $g \in \mathcal{D}(T_F)$, and it follows from the uniqueness of the decom-
position (14.77) that $h = \mathsf{H}(f)$. Since $g \in \mathcal{D}(T)$, we have $\frac{\partial g}{\partial \nu} \upharpoonright \partial\Omega = 0$ by (14.75),
and hence $\frac{\partial f}{\partial \nu} \upharpoonright \partial\Omega = \frac{\partial h}{\partial \nu} \upharpoonright \partial\Omega = \frac{\partial \mathsf{H}(f)}{\partial \nu} \upharpoonright \partial\Omega$.

Conversely, suppose that $\frac{\partial f}{\partial \nu} \upharpoonright \partial\Omega = \frac{\partial H(f)}{\partial \nu} \upharpoonright \partial\Omega$. This means that $g := f - H(f)$ satisfies $\frac{\partial g}{\partial \nu} \upharpoonright \partial\Omega = 0$. By the definition of $H(f)$ we have $g := f - H(f) \in \mathcal{D}(T_F)$, so that $g \upharpoonright \partial\Omega = 0$ by (14.76). Therefore, $g \in H_0^2(\Omega) = \mathcal{D}(T)$ by Theorem D.6 and (14.75). Thus, we have $f = g + H(f) \in \mathcal{D}(T) + \mathcal{N}(T^*) = \mathcal{D}(T_N)$ by (14.67). □

A nice description of the domain $\mathcal{D}(T_N) \cap H^2(\Omega)$ can be given by means of the so-called Dirichlet-to-Neumann map. By formula (D.2) there is a surjective map

$$H^2(\Omega) \ni f \to \left(f \upharpoonright \partial\Omega, \frac{\partial f}{\partial \nu} \upharpoonright \partial\Omega \right) \in H^{3/2}(\partial\Omega, d\sigma) \oplus H^{1/2}(\partial\Omega, d\sigma).$$

$$(14.78)$$

Statement 2 *For any element $\varphi \in H^{3/2}(\partial\Omega)$, there exists a unique function h of $\mathcal{N}(T^*) \cap H^2(\Omega)$ such that $\varphi = h \upharpoonright \partial\Omega$.*

Proof Since the map (14.78) is surjective, $\varphi = f \upharpoonright \partial\Omega$ for some $f \in H^2(\Omega)$. By (14.77), $f \in H^2(\Omega) \subseteq \mathcal{D}(T^*)$ can be written as $f = g + h$, where $g \in \mathcal{D}(T_F)$ and $h \in \mathcal{N}(T^*)$. Since $f, g \in H^2(\Omega)$ and $g \upharpoonright \partial\Omega = 0$ by (14.76), we have $h \in H^2(\Omega)$ and $h \upharpoonright \partial\Omega = f \upharpoonright \partial\Omega = \varphi$.

Let h_1 be another such function. Then we have $g := h - h_1 \in \mathcal{N}(T^*) \cap H^2(\Omega)$ and $g \upharpoonright \partial\Omega = 0$, so that $g \in \mathcal{D}(T_F)$ by (14.76). But we also have $g \in \mathcal{N}(T^*)$. Therefore, $g = h - h_1 = 0$ by (14.77), and hence $h = h_1$. □

From Statement 2 it follows that there exists a well-defined linear map

$$\mathcal{Q}: H^{3/2}(\partial\Omega) \to H^{1/2}(\Omega), \qquad \mathcal{Q}(f \upharpoonright \partial\Omega) := \frac{\partial f}{\partial \nu} \upharpoonright \partial\Omega, \quad f \in H^2(\Omega).$$

This map $\mathcal{Q}$ is called the *Dirichlet-to-Neumann map* associated with $-\Delta$. It plays an important role in the theory of elliptic differential operators. Note that the map $\mathcal{Q}$ depends also on the parameter λ, since $\mathcal{N}(T^*) = \mathcal{N}(L_{\max} - \lambda I)$ by (14.74).

For $f \in H^2(\Omega)$, we have $\mathcal{Q}(f \upharpoonright \partial\Omega) = \mathcal{Q}(H(f) \upharpoonright \partial\Omega) = \frac{\partial H(f)}{\partial \nu} \upharpoonright \partial\Omega$. Hence, Statement 1 can be reformulated as

Statement 3 $f \in H^2(\Omega)$ *is in* $\mathcal{D}(T_N)$ *if and only if* $\frac{\partial f}{\partial \nu} \upharpoonright \partial\Omega = \mathcal{Q}(f \upharpoonright \partial\Omega)$. ○

14.9 Exercises

1. Let $\mathcal{F}$ be a linear subspace of a Hilbert space $\mathcal{H}$, and let $S_{\mathcal{F}}$ be the linear relation $S_{\mathcal{F}} = \{0\} \oplus \mathcal{F}$.
   a. Show that $(S_{\mathcal{F}})^* = \mathcal{F}^{\perp} \oplus \mathcal{H}$.
   b. Show that $S_{\mathcal{F}}$ is self-adjoint if and only if $\mathcal{F} = \mathcal{H}$.
   c. Show that $\rho(S_{\mathcal{F}}) = \emptyset$ if $\mathcal{F} \neq \mathcal{H}$ and $\rho(S_{\mathcal{F}}) = \{0\}$ if $\mathcal{F} = \mathcal{H}$.
2. Let S be a linear relation on a Hilbert space $\mathcal{H}$. Let U and V be the unitaries on $\mathcal{H} \oplus \mathcal{H}$ defined by (1.8), that is, $U(x, y) = (y, x)$ and $V(x, y) = (-y, x)$.

  a. Show that $S^{-1} = U(S)$ and $S^* = V(S^{\perp})$.
  b. Show that $(S^*)^* = \overline{S}$ and $(S^*)^{-1} = (S^{-1})^*$.
  c. Prove Eqs. (14.2) and (14.3).

*3. (*A parameterization of self-adjoint relations* [Rf])
  Let $\mathcal{K}$ be a Hilbert space, and let $A, B \in \mathbf{B}(\mathcal{K})$. Define a linear relation on $\mathcal{K}$ by
  $S := \{(x, y) \in \mathcal{K} \oplus \mathcal{K} : Ax = By\}$.
    Prove that S is self-adjoint if and only if $AB^* = B^*A$ and $\mathcal{N}(X) = \{0\}$,
  where X denotes the bounded operator acting by the block matrix on $\mathcal{K} \oplus \mathcal{K}$:

$$X := \begin{pmatrix} A & -B \\ B & A \end{pmatrix}.$$

  Sketch of proof. Define an operator $C : \mathcal{K} \oplus \mathcal{K} \to \mathcal{K}$ by $C(x, y) = Ax - By$.
  Then $S = \mathcal{N}(C)$ and $S^* = V(S^{\perp})$, where $V(x, y) := (-y, x)$. Show that
  $AB^* = B^*A$ if and only if $S^* \subseteq S$. Prove then that $S = S^*$ if and only if
  $\mathcal{N}(X) = \{0\}$.

4. Retain the notation of Exercise 3 and assume that $\mathcal{K}$ has finite dimension d.
  Show that $\mathcal{N}(X) = \{0\}$ if and only if the block matrix (A, B) has maximal
  rank d.

5. Let $(\mathcal{K}, \Gamma_0, \Gamma_1)$ be a boundary triplet for T^* with gamma field $\gamma(z)$ and Weyl
  function $M(z)$. Suppose that P is a projection of $\mathcal{K}$ and set $\widetilde{\mathcal{K}} := P\mathcal{K}$. Define an
  operator $S := T^* \restriction \mathcal{D}(S)$ on $\mathcal{D}(S) := \{x \in \mathcal{D}(T^*) : \Gamma_0 x = 0, \Gamma_1 x \in (I - P)\mathcal{K}\}$.
  a. Show that S is symmetric and $\mathcal{D}(S^*) = \{x \in \mathcal{D}(T^*) : \Gamma_0 x \in \widetilde{\mathcal{K}}\}$.
  b. Show that $(\widetilde{\mathcal{K}}, \widetilde{\Gamma}_0 := P\Gamma_0, \widetilde{\Gamma}_1 := P\Gamma_1)$ is a boundary triplet for S^*.
  c. Show that the boundary triplet $(\widetilde{\mathcal{K}}, \widetilde{\Gamma}_0, \widetilde{\Gamma}_1)$ has the gamma field $\widetilde{\gamma}(z) = \gamma(z)P$ and the Weyl function $\widetilde{M}(z) = PM(z)P$.

6. Let $(\mathcal{K}, \Gamma_0, \Gamma_1)$ be a boundary triplet for T^*. Let $\widetilde{\mathcal{K}}$ be another Hilbert space,
  $B \in \mathbf{B}(\widetilde{\mathcal{K}}, \mathcal{K})$ an operator with bounded inverse $B^{-1} \in \mathbf{B}(\mathcal{K}, \widetilde{\mathcal{K}})$, and $A = A^* \in \mathbf{B}(\mathcal{K})$. Define $\widetilde{\Gamma}_0 := B^{-1}\Gamma_0$ and $\widetilde{\Gamma}_1 := B^*(\Gamma_1 + A\Gamma_0)$.
  a. Show that $(\widetilde{\mathcal{K}}, \widetilde{\Gamma}_0, \widetilde{\Gamma}_1)$ is also a boundary triplet for T^* and $\mathcal{N}(\widetilde{\Gamma}_0) = \Gamma_0$.
  b. Show that the Weyl function of $(\widetilde{\mathcal{K}}, \widetilde{\Gamma}_0, \widetilde{\Gamma}_1)$ is $\widetilde{M}(z) = B^*(M(z) + A)B$,
  $z \in \mathbb{C} \setminus \mathbb{R}$, where $M(z)$ is the Weyl function of $(\mathcal{K}, \Gamma_0, \Gamma_1)$.

*7. Assume that $(\mathcal{K}, \Gamma_0, \Gamma_1)$ and $(\widetilde{\mathcal{K}}, \widetilde{\Gamma}_0, \widetilde{\Gamma}_1)$ are boundary triplets for T^* such that
  $\mathcal{N}(\widetilde{\Gamma}_0) = \Gamma_0$. Prove that $(\widetilde{\mathcal{K}}, \widetilde{\Gamma}_0, \widetilde{\Gamma}_1)$ is of the form described in Exercise 6, that
  is, $\widetilde{\Gamma}_0 = B^{-1}\Gamma_0$ and $\widetilde{\Gamma}_1 = B^*(\Gamma_1 + A\Gamma_0)$, where $A = A^* \in \mathbf{B}(\mathcal{K})$, $B \in \mathbf{B}(\widetilde{\mathcal{K}}, \mathcal{K})$,
  and $B^{-1} \in \mathbf{B}(\mathcal{K}, \widetilde{\mathcal{K}})$.

8. Let $(\mathcal{K}, \Gamma_0, \Gamma_1)$ be a boundary triplet for T^*, and let $\mathcal{G}_0$ and $\mathcal{G}_1$ be closed linear subspaces of $\mathcal{K}$. Define a proper extension $S_{\mathcal{G}_1, \mathcal{G}_2}$ of the operator T by
  $\mathcal{D}(S_{\mathcal{G}_0, \mathcal{G}_1}) = \{x \in \mathcal{D}(T^*) : \Gamma_0 x \in \mathcal{G}_0, \Gamma_1 x \in \mathcal{G}_1\}$. Determine $(S_{\mathcal{G}_1, \mathcal{G}_2})^*$.

9. Consider the relations $\mathcal{B}_j$, $j = 1, 2, 3$, on the Hilbert space $\mathcal{H} = \mathbb{C}^2$ defined
  by the sets of pairs $((x_1, x_2), (y_1, y_2)) \in \mathbb{C}^2 \oplus \mathbb{C}^2$ satisfying the equations:
  $\mathcal{B}_1$: $x_1 + x_2 = 0$, $y_1 = y_2$, $\mathcal{B}_2$: $y_1 = x_1 + x_2$, $y_2 = 0$,
  $\mathcal{B}_3$: $y_1 = x_1 + zx_2$, $y_2 = wx_1 + x_2$, where z, w are fixed complex numbers.
  a. Decide which relations are graphs of operators and decompose the others
    into operator parts and multivalued parts.
  b. Describe the adjoint relations.

10. Consider the operators $S_j = -\frac{d^2}{dx^2}$, $j = 1, 2, 3$, on $L^2(0, 1)$, where the domains consists of all $f \in H^2(0, 1)$ satisfying the following boundary conditions:
S_1: $f(0) + f(1) = f'(0) + f'(1) = 0$, S_2: $f'(0) = f(0) + f(1)$, $f'(1) = 0$,
S_3: $f'(0) = f(0) + zf(1)$, $-f'(1) = wf(0) + f(1)$, where $z, w \in \mathbb{C}$.
Determine the domains of the adjoint operators $(S_j)^*$, $j = 1, 2, 3$.
Hint: Use the boundary triplet (14.12), Exercise 9 and Lemma 14.6.

11. Let $T = -\frac{d^2}{dx^2}$ on $\mathcal{D}(T) = H_0^2(0, 1)$. Show that the self-adjoint operator S_1 from Exercise 10 and the Friedrichs extension T_F of T have the same lower bound $m_S = m_{T_F} = \pi^2$.

12. Let $T = -\frac{d^2}{dx^2} + c^2 I$ on $\mathcal{D}(T) = H_0^2(0, 1)$, where $c > 0$. Find the boundary conditions for the Krein–von Neumann extension T_N.

13. Let $T = -\frac{d^2}{dx^2}$ on $\mathcal{D}(T) = \{f \in H^2(a, b) : f(a) = f(b) = f'(b) = 0\}$, $a, b \in \mathbb{R}$.
    a. Show that T is a densely defined closed positive symmetric operator on $L^2(a, b)$.
    b. Show that $\mathcal{D}(T^*) = \{f \in H^2(a, b) : f(a) = 0\}$.
    c. Use Exercise 5 to construct a boundary triplet for T^*.
    d. Determine the Friedrichs extension T_F.
    e. Show that $m_T > 0$ and determine the Krein–von Neumann extension T_N.
    f. Determine the Krein–von Neumann extension of $T + c^2 I$, where $c > 0$.

14. Let T be a densely defined lower semibounded symmetric operator. Let $\mathcal{A}$ be a subset of the commutant $\{T\}'$ (see (1.23)) such that $B^* \in \mathcal{A}$ when $B \in \mathcal{A}$.
    a. Suppose that $m_T > 0$. Prove that $\mathcal{A} \subseteq \{T_N\}'$.
    Hint: Combine Exercises 1.8 and 1.19 with formula (14.67).
    b. Prove that there exists a self-adjoint extension A of T such that $\mathcal{A} \subseteq \{A\}'$.
    Hint: Obviously, $(T + \lambda I)_N - \lambda I$ is a self-adjoint extension of T.

Chapter 15
Sturm–Liouville Operators

Fundamental ideas on self-adjoint extensions appeared in H. Weyl's classical work on Sturm–Liouville operators long before the abstract Hilbert space theory was developed. This chapter is devoted to a short treatment of the Hilbert space theory of Sturm–Liouville operators. Section 15.1 is concerned with the case of regular end points. In Sect. 15.2 we develop the basics of H. Weyl's limit point–limit circle theory. In Sect. 15.3 we define boundary triplets in the various cases and determine the Weyl functions and the self-adjoint extensions. The final Sect. 15.4 deals with formulas for the resolvents of some self-adjoint extensions.

15.1 Sturm–Liouville Operators with Regular End Points

In this chapter we will be concerned with the second-order differential expression

$$\mathcal{L} = -\frac{d^2}{dx^2} + q(x)$$

treated as an operator on the Hilbert space $\mathcal{H} = L^2(a, b)$, where $-\infty \leq a < b \leq \infty$.

Throughout this chapter we assume that q is a *real-valued continuous* function on (a, b). The behavior of the function q in the neighborhoods of the end points a and b will be crucial for our considerations.

Remark The theory remains valid almost verbatim for the differential operator $\mathcal{L}f = -(pf')' + qf$, where $p \in C^1(a, b)$ and $p(x) > 0$ on (a, b).

As shown in the proof of the following lemma, the operator L_0 given by $L_0 f = -f'' + qf$ for $f \in \mathcal{D}(L_0) := C_0^\infty(a, b)$ is closable. We denote the closure of L_0 by $L_{\min}$. The maximal operator $L_{\max}$ is defined by $L_{\max} f = -f'' + qf$ for f in

$$\mathcal{D}(L_{\max}) := \big\{ f \in L^2(a, b) : f, f' \in AC[\alpha, \beta] \text{ for } [\alpha, \beta] \subset (a, b),$$
$$-f'' + qf \in L^2(a, b) \big\}.$$

K. Schmüdgen, *Unbounded Self-adjoint Operators on Hilbert Space*,
Graduate Texts in Mathematics 265,
DOI 10.1007/978-94-007-4753-1_15, © Springer Science+Business Media Dordrecht 2012

Lemma 15.1 $L_{\min}$ *is a densely defined symmetric operator, and* $(L_{\min})^* = L_{\max}$.

Proof Let $\varphi \in C_0^\infty(a, b)$ and $f \in \mathcal{D}(L_{\max})$. Then $f \in L^1_{\mathrm{loc}}(a, b)$. Hence, applying integration by parts (see (E.2)) twice and using that the boundary terms vanish, we obtain $\langle L_0\varphi, f \rangle = \langle \varphi, L_{\max}f \rangle$. Therefore, $L_{\max} \subseteq (L_0)^*$, so $(L_0)^*$ is densely defined. Hence L_0 is closable, and for its closure $L_{\min}$ we have $L_{\max} \subseteq (L_{\min})^*$. Since obviously $L_{\min} \subseteq L_{\max}$, this implies that $L_{\min}$ is symmetric.

To prove the converse inclusion $(L_{\min})^* \subseteq L_{\max}$, let $f \in \mathcal{D}((L_{\min})^*)$ and set $g := (L_{\min})^*f$. Fix $c \in (a, b)$. Since $f, g \in L^2(a, b)$ and $q \in C(a, b)$, we have $qf - g \in L^1_{\mathrm{loc}}(a, b)$. Hence, the function

$$h(x) := \int_c^x \int_c^s \big(q(t)f(t) - g(t)\big)\, dt\, ds$$

and its derivative h' are in $AC[\alpha, \beta]$ for each compact interval $[\alpha, \beta] \subseteq (a, b)$, and we have $h'' = qf - g$ on (a, b). If $\varphi \in C_0^\infty(a, b)$, then $\overline{\varphi} \in \mathcal{D}(L_{\min})$, and we obtain

$$\big\langle f, -\overline{\varphi}'' + q\overline{\varphi} \big\rangle = \langle f, L_{\min}\overline{\varphi} \rangle = \big\langle (L_{\min})^*f, \overline{\varphi} \big\rangle = \langle g, \overline{\varphi} \rangle = \big\langle qf - h'', \overline{\varphi} \big\rangle,$$

so that $\langle f, \overline{\varphi}'' \rangle = \langle h'', \overline{\varphi} \rangle$. Using integration by parts, the latter yields

$$\int_a^b f(x)\varphi''(x)\, dx = \int_a^b h''(x)\varphi(x)\, dx = \int_a^b h(x)\varphi''(x)\, dx.$$

This implies that the second distributional derivative of $f - h$ is zero on (a, b). Hence, there exist constants $c_0, c_1 \in \mathbb{C}$ such that $f(x) = h(x) + c_0 + c_1 x$ on (a, b). Therefore, f and f' are in $AC[\alpha, \beta]$ for $[\alpha, \beta] \subseteq (a, b)$ and $f'' = h'' = qf - g$, so that $-f'' + qf = g \in L^2(a, b)$. That is, $f \in \mathcal{D}(L_{\max})$ and $g = L_{\max}f$. $\square$

In what follows we denote the symmetric operator $L_{\min}$ by T. Then $T^* = L_{\max}$ by Lemma 15.1. Suppose that $f, g \in \mathcal{D}(T^*)$. For $c \in (a, b)$, we abbreviate

$$[f, g]_c := f(c)\overline{g'(c)} - f'(c)\overline{g(c)}. \tag{15.1}$$

For each compact interval $[\alpha, \beta]$ contained in (a, b), integration by parts yields the *Green's formula*

$$\int_\alpha^\beta \big[(\mathcal{L}f)(x)\overline{g(x)} - f(x)\overline{(\mathcal{L}g)(x)}\big]\, dx = [f, g]_\beta - [f, g]_\alpha. \tag{15.2}$$

Since $f, g \in \mathcal{D}(T^*)$, the integral of $\mathcal{L}f\overline{g} - f\overline{\mathcal{L}g}$ over the whole interval (a, b) is finite. Hence, it follows from (15.2) that the limits

$$[f, g]_a := \lim_{\alpha \to a+0} [f, g]_\alpha \quad \text{and} \quad [f, g]_b := \lim_{\beta \to b-0} [f, g]_\beta \tag{15.3}$$

exist, and we obtain

$$[f, g]_{T^*} \equiv \big\langle T^*f, g \big\rangle - \big\langle f, T^*g \big\rangle = [f, g]_b - [f, g]_a \quad \text{for } f, g \in \mathcal{D}(T^*). \tag{15.4}$$

Lemma 15.2 *If* $f \in \mathcal{D}(\overline{T})$ *and* $g \in \mathcal{D}(T^*)$, *then* $[f, g]_a = [f, g]_b = 0$.

Proof We carry out the proof of the equality $[f, g]_a = 0$. Let us choose $g_0 \in \mathcal{D}(T^*)$ such that $g = g_0$ in some neighborhood of a and $g_0 = 0$ in some neighborhood of b. Then we have $[f, g]_a = [f, g_0]_a$ and $[f, g_0]_b = 0$. Therefore, by (15.4),

$$0 = \langle \overline{T} f, g_0 \rangle - \langle f, T^* g_0 \rangle = -[f, g_0]_a = -[f, g]_a. \qquad \square$$

Definition 15.1 The expression $\mathcal{L}$ is *regular* at a if $a \in \mathbb{R}$ and $\int_a^c |q(x)| \, dx < \infty$ for some (hence for all) $c \in (a, b)$; otherwise $\mathcal{L}$ is said to be *singular* at a.

Likewise, $\mathcal{L}$ is called *regular* at b if $b \in \mathbb{R}$ and $\int_c^b |q(x)| \, dx < \infty$ for some (hence for all) $c \in (a, b)$; otherwise $\mathcal{L}$ is *singular* at b.

Thus, $\mathcal{L}$ is regular at both end points a and b if and only if the interval (a, b) is bounded and q is integrable on (a, b).

Let $\lambda \in \mathbb{C}$ and $g \in L^1_{\text{loc}}(a, b)$. By a solution of the equation $\mathcal{L}(f) - \lambda f = g$ on (a, b) we mean a function f on (a, b) such that $f, f' \in AC[\alpha, \beta]$ for any compact interval $[\alpha, \beta] \subset (a, b)$ and $\mathcal{L}(f)(x) - \lambda f(x) = g(x)$ a.e. on (a, b).

We shall use the following results on ordinary differential equations. Proofs can be found, e.g., in [Na2, 16.2] and in most text books on ordinary differential equations.

Proposition 15.3 *Let $\lambda \in \mathbb{C}$ and $g \in L^1_{\text{loc}}(a, b)$.*

(i) *If $\mathcal{L}$ is regular at a, then for any solution of $\mathcal{L}(f) - \lambda f = g$, the (finite) limits*

$$f(a) := \lim_{x \to a+0} f(x) \quad \text{and} \quad f'(a) := \lim_{x \to a+0} f'(x)$$

exist, so f and f' can be extended to continuous functions on $[a, b)$. A similar statement holds if $\mathcal{L}$ is regular at b.

(ii) *Let c be an element of (a, b) or a regular end point. Given $c_1, c_2 \in \mathbb{C}$, there is a unique solution of the differential equation $\mathcal{L}(f) - \lambda f = g$ on (a, b) such that $f(c) = c_1$ and $f'(c) = c_2$.*

For arbitrary functions $f, g \in C^1(a, b)$, we define their *Wronskian* $W(f, g)$ by

$$W(f, g)_x = \begin{vmatrix} f(x) & g(x) \\ f'(x) & g'(x) \end{vmatrix} = f(x) g'(x) - f'(x) g(x), \quad x \in (a, b).$$

Note that $W(f, \overline{g})_x = [f, g]_x$ for $x \in (a, b)$.

From Proposition 15.3(ii), applied with $g = 0$, it follows in particular that for any $\lambda \in \mathbb{C}$, the solutions of the homogeneous differential equation

$$\mathcal{L}(f) \equiv -f'' + qf = \lambda f \quad \text{on } (a, b) \tag{15.5}$$

form a two-dimensional complex vector space. A basis of this vector space is called a *fundamental system*. It is easily verified that for any fundamental system $\{u_1, u_2\}$, the Wronskian $W(u_1, u_2)_x$ is a nonzero constant on (a, b).

Since $T^* = L_{\max}$ by Lemma 15.1, the solution of (15.5) belonging to the Hilbert space $L^2(a, b)$ are precisely the elements of $\mathcal{N}(T^* - \lambda I)$.

Proposition 15.4 *The operator T has deficiency indices* $(0, 0)$ *or* $(1, 1)$ *or* $(2, 2)$.

Proof Since q is real-valued, a function f belongs to $\mathcal{N}(T^* - \lambda I)$ if and only if its complex conjugate $\overline{f}$ is in $\mathcal{N}(T^* - \overline{\lambda} I)$. Hence, T has equal deficiency indices. As just noted, the dimension of $\mathcal{N}(T^* - \lambda I)$ is less than or equal to 2. $\square$

One of our main aims is to analyze these three possible cases. The outcome will be summarized in Theorem 15.10 below. In order to study the deficiency indices of T, the following definition is convenient.

Definition 15.2 Let f be a function on (a, b). We say that f *is in* L^2 *near a* (resp. b) if there is a number $c \in (a, b)$ such that $f \in L^2(a, c)$ (resp. $f \in L^2(c, b)$).

Let $\lambda \in \mathbb{C}$, and let f be a solution of (15.5) on (a, b). Since then $f, f' \in AC[\alpha, \beta]$ for any interval $[\alpha, \beta] \subset (a, b)$, it is clear that $f \in \mathcal{N}(T^* - \lambda I)$, or equivalently $f \in L^2(a, b)$, if and only if f is in L^2 near a *and* f is in L^2 near b. This simple fact will be often used without mention when we study the deficiency indices of T.

Some results concerning regular end points are collected in the next proposition.

Proposition 15.5 *Suppose that the end point a is regular for $\mathcal{L}$.*

 (i) *If $f \in \mathcal{D}(T^*)$, then f and f' can be extended to continuous functions on $[a, b)$.*
 (ii) *Let f be a solution of (15.5) on (a, b), where $\lambda \in \mathbb{C}$. Then f and f' extend to continuous functions on $[a, b)$. In particular, f is in L^2 near a.*
 (iii) *The vector space $\{(f(a), f'(a)) : f \in \mathcal{D}(T^*)\}$ is equal to $\mathbb{C}^2$.*
 (iv) *If $f \in \mathcal{D}(\overline{T})$, then $f(a) = f'(a) = 0$.*
 (v) *If $\mathcal{L}$ is regular at a and b, then $f(a) = f'(a) = f(b) = f'(b) = 0$ for any $f \in \mathcal{D}(\overline{T})$, and the vector space $\{(f(a), f'(a), f(b), f'(b)) : f \in \mathcal{D}(T^*)\}$ is equal to $\mathbb{C}^4$.*

Proof Since $T^* f \in L^2(a, b) \subseteq L^1_{\mathrm{loc}}(a, b)$, Proposition 15.3(i) applies to $g := T^* f$ and $\lambda = 0$ and yields the assertion of (i). (ii) follows also from Proposition 15.3(i) now applied with $g = 0$.

(iii): Let $\{u_1, u_2\}$ be a fundamental system of solutions of $\mathcal{L}(f) = 0$ on (a, b). By Proposition 15.3(i), u_j and u'_j extend to continuous functions on $[a, b)$. Then the Wronskian $W = u_1 u'_2 - u'_1 u_2$ is a nonzero constant on $[a, b)$. For $(\alpha_1, \alpha_2) \in \mathbb{C}^2$, set $g := \alpha_1 u_1 + \alpha_2 u_2$. Given $(c_1, c_2) \in \mathbb{C}^2$, the requirements $g(a) = c_1, g'(a) = c_2$ lead to a system of two linear equations for α_1 and α_2. It has a unique solution, because its determinant is the nonzero Wronskian W. Let g be the corresponding function. Choosing a function $f \in \mathcal{D}(T^*)$ such that $f = g$ in some neighborhood of a and $f = 0$ in some neighborhood of b, we have $f(a) = c_1, f'(a) = c_2$.

(iv): Let $f \in \mathcal{D}(\overline{T})$ and $g \in \mathcal{D}(T^*)$. By (i), f, f', g, and g' extend by continuity to a. From Lemma 15.2 we obtain $[f, g]_a = f(a)g'(a) - f'(a)g(a) = 0$. Because the values $g(a), g'(a)$ are arbitrary by (iv), the latter implies that $f(a) = f'(a) = 0$.

(v): The first assertion follows at once from (iii) and the corresponding result for b. Let $c = (c_1, c_2, c_3, c_4) \in \mathbb{C}^4$ be given. By (iv) there are $g, h \in \mathcal{D}(T^*)$ satisfying

$g(a) = c_1$, $g'(a) = c_2$, $h(b) = c_3$, $h'(b) = c_4$. We choose functions $g_0, h_0 \in \mathcal{D}(T^*)$ such that $g = g_0$ and $h_0 = 0$ in some neighborhood of a and $g_0 = 0$ and $h = h_0$ in some neighborhood of b. Setting $f := g_0 + h_0$, we then have $f(a) = c_1$, $f'(a) = c_2$, $f(b) = c_3$, $f'(b) = c_4$. $\qquad\square$

Proposition 15.6 *Let $\lambda \in \mathbb{C}\backslash\mathbb{R}$. Then for each end point of the interval (a, b), there exists a nonzero solution of (15.5) which is in L^2 near to it.*

Proof We carry out the proof for the end point b; the case of a is similar.

Fix $c \in (a, b)$ and let T_c denote the corresponding operator on the interval (c, b) in $L^2(c, b)$. Choosing functions $f, g \in C_0^\infty(a, b)$ such that $f(c) = g'(c) = 0$ and $f'(c) = g(c) = 1$ and applying (15.4) to the operator T_c^* we obtain

$$\langle T_c^* f, g \rangle - \langle f, T_c^* f \rangle = -[f, g]_c = 1.$$

Hence, T_c^* is not symmetric. Therefore, by Proposition 15.4, the symmetric operator T_c has nonzero equal deficiency indices. Hence, there exists a nonzero f_0 in $\mathcal{N}(T_c^* - \lambda I)$. Then $\mathcal{L}(f_0)(x) = \lambda f_0(x)$ on (c, b). We fix a number $d \in (c, b)$. By Proposition 15.3(ii) there exists a solution f of the equation $\mathcal{L}(f) = \lambda f$ on (a, b) such that $f(d) = f_0(d)$ and $f'(d) = f_0'(d)$. From the uniqueness assertion of Proposition 15.3(ii) it follows that $f(x) = f_0(x)$ on (c, b). Then f is a nonzero solution of (15.5), and f is in L^2 near b, because $f_0 \in L^2(c, b)$. $\qquad\square$

Corollary 15.7 *If at least one end point is regular, then the deficiency indices of T are $(1, 1)$ or $(2, 2)$. If both end points are regular, then T has deficiency indices $(2, 2)$.*

Proof Let $\lambda \in \mathbb{C}\backslash\mathbb{R}$. If both end points are regular, by Proposition 15.5(ii) all solutions f of (15.5) are in L^2 near a and in L^2 near b, so $f \in \mathcal{N}(T^* - \lambda I)$. Thus, $\dim \mathcal{N}(T^* - \lambda I) = 2$, and T has deficiency indices $(2, 2)$.

Suppose that $\mathcal{L}$ is regular at one end point, say a. By Proposition 15.6 there is a nonzero solution f of (15.5) which is in L^2 near b. Since $\mathcal{L}$ is regular at a, f is also in L^2 near a by Proposition 15.5(ii). Therefore, $f \in L^2(a, b)$, and hence $f \in \mathcal{N}(T^* - \lambda I)$. Hence, by Proposition 15.4, the deficiency indices of T are $(1, 1)$ or $(2, 2)$. $\qquad\square$

15.2 Limit Circle Case and Limit Point Case

The following classical result of H. Weyl is crucial for the study of deficiency indices and of self-adjointness questions for Sturm–Liouville operators.

Theorem 15.8 (*Weyl's alternative*) *Let d denote an end point of the interval (a, b). Then precisely one of the following two possibilities is valid:*

(i) *For each $\lambda \in \mathbb{C}$, all solutions of (15.5) are in L^2 near d.*
(ii) *For each $\lambda \in \mathbb{C}$, there exists one solution of (15.5) which is not in L^2 near d.*

In case (ii), *for any $\lambda \in \mathbb{C}\backslash\mathbb{R}$, there is a unique (up to a constant factor) nonzero solution of (15.5) which is in L^2 near d.*

Proof We shall carry out the proof for $d = b$. Since the solution space of (15.5) is two-dimensional, the last assertion only restates Proposition 15.6.

To prove the alternative, it suffices to show that if for *some* $\lambda_0 \in \mathbb{C}$, all solutions of (15.5) are in L^2 near b, this is also true for *each* $\lambda \in \mathbb{C}$. Let u be a solution of $\mathcal{L}(f) = \lambda f$. We have to prove that $u \in L^2(c, b)$ for some $c \in (a, b)$.

We fix a basis $\{u_1, u_2\}$ of the solution space of $\mathcal{L}(f) = \lambda_0 f$. Then the Wronskian $W(u_1, u_2)$ is a nonzero constant on (a, b), so we can assume that $W(u_1, u_2) = 1$ by normalization. For $g \in L^1_{\mathrm{loc}}(a, b)$, let f_g denote the function given by

$$f_g(x) := u_1(x) \int_c^x u_2(t)g(t)\,dt - u_2(x) \int_c^x u_1(t)g(t)\,dt, \quad x \in (a, b). \quad (15.6)$$

From the variation-of-constants method it is known (and easily verified) that f_g satisfies the equation $\mathcal{L}(f_g) - \lambda_0 f_g = g$ on (a, b). Setting $v = u + (\lambda_0 - \lambda)f_u$, we get

$$\mathcal{L}(v) = \mathcal{L}(u) + (\lambda_0 - \lambda)\mathcal{L}(f_u) = \lambda u + (\lambda_0 - \lambda)(\lambda_0 f_u + u) = \lambda_0 v.$$

Hence, there exist complex numbers α_1, α_2 such that $v = \alpha_1 u_1 + \alpha_2 u_2$. That is,

$$u = \alpha_1 u_1 + \alpha_2 u_2 + (\lambda - \lambda_0)f_u. \quad (15.7)$$

Put $\varphi := \max(|u_1|, |u_2|)$ and $\alpha := \max(|\alpha_1|, |\alpha_2|)$. By assumption all solutions of $\mathcal{L}(f) = \lambda_0 f$ are in L^2 near b, so there exists $e \in (a, b)$ such that $u_1, u_2 \in L^2(e, b)$. Hence, $\varphi \in L^2(e, b)$, so that $C := 8|\lambda - \lambda_0|^2 \int_e^b \varphi(x)^2\,dx < \infty$. Using first equations (15.7) and (15.6), applied to f_u, and then the Hölder inequality, we derive

$$|u(x)|^2 \le 8\alpha^2\varphi(x)^2 + 8|\lambda - \lambda_0|^2\varphi(x)^2 \left(\int_e^x \varphi(t)|u(t)|\,dt \right)^2$$

$$\le 8\alpha^2\varphi(x)^2 + \varphi(x)^2 C \int_e^x |u(t)|^2\,dt, \quad x \in (e, b). \quad (15.8)$$

Since $\varphi \in L^2(e, b)$, there is $c \in (e, b)$ such that $\int_c^b \varphi(x)^2\,dx \le (2C)^{-1}$.

Fix $y \in (c, b)$. Integrating both sides of (15.8) over (c, y), we obtain

$$2\int_c^y |u(x)|^2\,dx \le 16\alpha^2 \int_c^y \varphi(x)^2\,dx + 2C \int_c^y \varphi(x)^2 \left(\int_e^y |u(t)|^2\,dt \right) dx$$

$$\le 16\alpha^2 \int_c^y \varphi(x)^2\,dx + \int_e^y |u(t)|^2\,dt,$$

which in turn implies that

$$\int_c^y |u(x)|^2\,dx \le 16\alpha^2 \int_c^b \varphi(x)^2\,dx + \int_e^c |u(x)|^2\,dx. \quad (15.9)$$

Because u is continuous on (a, b) and $e, c \in (a, b)$, we have $\int_e^c |u(x)|^2 \, dx < \infty$. Since $\varphi \in L^2(e, b)$ as noted above, $\varphi \in L^2(c, b)$. Hence, the right-hand side of (15.9) is a finite constant. Therefore, since $y \in (c, b)$ is arbitrary, $u \in L^2(c, b)$. □

The following terminology was introduced by H. Weyl.

Definition 15.3 Case (i) in Theorem 15.8 is called the *limit circle case* at d and case (ii) the *limit point case* at d.

By Proposition 15.5(ii), T is in the limit circle case at d if the end point d is regular.

Let us illustrate the preceding for the simplest possible example.

Example 15.1 ($q(x) \equiv 0$ *on* (a, b)) For $\lambda \in \mathbb{C}$, $\lambda \neq 0$, a fundamental system of Eq. (15.5) is given by

$$\{u_1(x; \lambda) := \sin \sqrt{\lambda} x, \ u_2(x; \lambda) := \cos \sqrt{\lambda} x\}.$$

If a or b is in $\mathbb{R}$, it is a regular end point, and T is in the limit circle case at a resp. b. If $a = -\infty$ or $b = \infty$, then $u_1(\cdot; \lambda) \notin L^2(a, b)$, so T is in the limit point case. Let $\lambda \in \mathbb{C} \backslash \mathbb{R}$, and let $\sqrt{\lambda}$ denote the square root satisfying $\operatorname{Im} \sqrt{\lambda} > 0$. For $a = -\infty$ resp. $b = \infty$, by Theorem 15.8(ii), there is a unique (up to a constant factor) solution of (15.5) which is in L^2 near a resp. b. These solutions are

$$e^{-i\sqrt{\lambda} x} = u_2(x; \lambda) - iu_1(x; \lambda) \in L^2(-\infty, 0),$$
$$e^{i\sqrt{\lambda} x} = u_2(x; \lambda) + iu_1(x; \lambda) \in L^2(0, +\infty).$$

○

In the proof of our next theorem we shall use the following lemma.

Lemma 15.9 *If T is in the limit point case at a (resp. b), then $[f, g]_a = 0$ (resp. $[f, g]_b = 0$) for all $f, g \in \mathcal{D}(T^*)$.*

Proof We prove this for the end point b. As in the proof of Proposition 15.6, we fix $c \in (a, b)$ and consider the operator T_c on the interval (c, b) in $L^2(c, b)$. Let $\lambda \in \mathbb{C} \backslash \mathbb{R}$. Since $\mathcal{L}$ is regular at c by the assumption $q \in C(a, b)$, all solutions of $\mathcal{L}(f) = \lambda f$ on (c, b) are in L^2 near c by Proposition 15.5(ii). Because T is in the limit point case at b, up to a constant factor only one solution is in L^2 near b. Hence, T_c has deficiency indices $(1, 1)$, so $\dim \mathcal{D}(T_c^*)/\mathcal{D}(\overline{T_c}) = 2$ by Proposition 3.7.

Take $f_1, f_2 \in C_0^\infty(a, b)$ satisfying $f_1(c) = f_2'(c) = 0$ and $f_1'(c) = f_2(c) = 1$. Then $f_1, f_2 \in \mathcal{D}(T_c^*)$. Since $h(c) = h'(c) = 0$ for $h \in \mathcal{D}(\overline{T_c})$ by Proposition 15.5(iv), f_1 and f_2 are linearly independent modulo $\mathcal{D}(\overline{T_c})$. Therefore,

$$\mathcal{D}(T_c^*) = \mathcal{D}(\overline{T_c}) + \operatorname{Lin}\{f_1, f_2\}. \tag{15.10}$$

Since $T_c \subseteq T$, the restrictions of $f, g \in \mathcal{D}(T^*)$ to (c, b) are in $\mathcal{D}(T_c^*)$. Therefore, by (15.10), there are functions $f_0, g_0 \in \mathcal{D}(\overline{T_c})$ such that $f - f_0$ and $g - g_0$

vanish in some neighborhood of b, because f_1 and f_2 do so. Clearly, $[f, g]_b = [f_0, g_0]_b$. Since $f_0, g_0 \in \mathcal{D}(\overline{T_c})$, we have $[f_0, g_0]_b = 0$ by Lemma 15.2. Hence, $[f, g]_b = 0$. $\square$

Theorem 15.10 *The symmetric operator T has deficiency indices*

$(2, 2)$ *if T is in the limit circle case at both end points,*

$(1, 1)$ *if T is in the limit circle case at one end point and in the limit point case at the other, and*

$(0, 0)$ *if T is in the limit point case at both end points.*

Proof If T is in the limit circle case at both end points, all solution of (15.5) are in L^2 near a and near b; hence, they are in $L^2(a, b)$, and T has deficiency indices $(2, 2)$.

Now suppose that T is the limit circle case at one end point, say a, and in the limit circle case at the other end point b. Then, by the last assertion in Theorem 15.8, for any $\lambda \in \mathbb{C} \setminus \mathbb{R}$, there is a unique (up to a factor) nonzero solution of (15.5) which is in L^2 near b. Since T is in the limit circle case at a, this solution is in L^2 near a and hence in $L^2(a, b)$. Therefore, the deficiency indices of T are $(1, 1)$.

Assume that T is in the limit point case at both end points. Then, by Lemma 15.9, $[f, g]_a = [f, g]_b = 0$ for $f, g \in \mathcal{D}(T^*)$. Hence, (15.4) implies that T^* is symmetric. Therefore, T is essentially self-adjoint and has deficiency indices $(0, 0)$. $\square$

The following propositions contain two useful criteria that allow us to decide which of the two cases happens.

Proposition 15.11 *Let $b = +\infty$ and suppose that q is bounded from below near to $+\infty$ (that is, there are numbers $c > a$ and $\gamma \in \mathbb{R}$ such that $q(x) \geq \gamma$ for all $x \geq c$). Then T is in the limit point case at $b = +\infty$.*

Proof By Proposition 15.3(ii), there is a unique solution u of the differential equation $\mathcal{L}(f) \equiv -f'' + qf = \gamma f$ on $(a, +\infty)$ satisfying $u(c) = 1$, $u'(c) = 0$. Since $\overline{u}$ is a solution with the same initial data, u is real-valued by the uniqueness. Then

$$\left(u^2\right)'' = \left(2u'u\right)' = 2\left(u'\right)^2 + 2u''u = 2\left(u'\right)^2 + 2(q - \gamma)u \cdot u \geq 0 \quad \text{on } (c, +\infty).$$

Therefore, since $(u^2)'(c) = 2u'(c)u(c) = 0$, we obtain

$$\left(u^2\right)'(x) = \int_c^x \left(u^2\right)''(t) \, dt \geq 0 \quad \text{for } x \in (c, +\infty).$$

Hence, the function u^2 is increasing on $(c, +\infty)$. Since $u(c) = 1$, it follows that u is not in L^2 near $b = +\infty$, so T is in the limit point case at $b = +\infty$. $\square$

Proposition 15.12 *Suppose that $a = 0$ and $b \in (0, +\infty]$.*

(i) *If there exists a positive number c such that $c < b$ and $q(x) \geq \frac{3}{4}x^{-2}$ for all $x \in (0, c)$, then T is in the limit point case at $a = 0$.*

(ii) *If there are positive numbers ε and c such that $c < b$ and $|q(x)| \leq (\frac{3}{4} - \varepsilon)x^{-2}$ for all $x \in (0, c)$, then T is in the limit circle case at $a = 0$.*

Proof (i): Put $u_0(x) := x^{-1/2}$. Clearly, the function u_0 is not in L^2 near $a = 0$, and it satisfies the equation $-u_0''(x) + \frac{3}{4}x^{-2}u_0(x) = 0$ on $(0, b)$.

By Proposition 15.3(ii) there is a unique solution u of the equation $-u'' + qu = 0$ on $(0, b)$ satisfying $u(c) = u_0(c) + 1$ and $u'(c) = u_0'(c)$. Then u is real-valued on $(0, b)$. Since $u(c) > u_0(c)$, $u(x) > u_0(x)$ in some neighborhood of c. Set

$$\alpha := \inf\{t \in (0, c) : u(x) > u_0(x) \text{ for all } x \in (t, c)\}.$$

We prove that $\alpha = 0$. Assume to the contrary that $\alpha > 0$. Then $u(\alpha) = u_0(\alpha)$ by the definition of α. Using the assumption on q, we derive

$$u''(x) = q(x)u(x) \geq \frac{3}{4}x^{-2}u(x) > \frac{3}{4}x^{-2}u_0(x) = u_0''(x) \quad \text{for } x \in (\alpha, c).$$

From this inequality it follows that the function $u' - u_0'$ is strictly increasing on (α, c), so $u'(x) - u_0'(x) < u'(t) - u_0'(t)$ for $\alpha < x < t < c$. Letting $t \to c - 0$ and using that $u'(c) = u_0'(c)$, we obtain $u'(x) - u_0'(x) < 0$ for $x \in (\alpha, c)$. Therefore, $u - u_0$ is strictly decreasing on (α, c), that is, $u(x) - u_0(x) > u(t) - u_0(t)$ for $\alpha < x < t < c$. Letting first $t \to c - 0$ and then $x \to \alpha + 0$, we get $u(x) - u_0(x) \geq 1$ and $u(\alpha) - u_0(\alpha) \geq 1$, which contradicts $u(\alpha) = u_0(\alpha)$. Thus, $\alpha = 0$.

The equality $\alpha = 0$ means that $u(x) > u_0(x) = x^{-1/2}$ on $(0, c)$. Therefore, u is not in L^2 near $a = 0$, so T is in the limit point case at $a = 0$.

(ii): To prove that T is in the limit circle case at $a = 0$, by Theorem 15.8 it suffices to show that any solution u of the equation $-u'' + qu = 0$ is in L^2 near $a = 0$. We fix such a solution u and assume without loss of generality that $\varepsilon < 1$.

By Proposition 15.3(ii) there exists a unique solution v of the equation

$$-v'' + (3/4 - \varepsilon)x^{-2}v = 0 \quad \text{on } (0, b) \tag{15.11}$$

satisfying the boundary conditions

$$v(c) = |u(c)| + 1 \quad \text{and} \quad v'(c) = -|u'(c)| - 1. \tag{15.12}$$

A fundamental system $\{v_+, v_-\}$ of Eq. (15.11) is given by

$$v_+(x) = x^{n_+}, \quad v_-(x) = x^{n_-}, \quad \text{where } n_\pm := 1/2 \pm \sqrt{1 - \varepsilon}.$$

Since $n_\pm > -\frac{1}{2}$ and v is a linear combination of v_+ and v_-, we have $v \in L^2(0, c)$.

The boundary conditions of v imply that $v(x) > |u(x)|$ and $v'(x) < -|u'(x)|$ in some neighborhood of c. Similarly to the proof of (i), we define

$$\alpha := \inf\{t \in (0, c) : v(x) > |u(x)|, v'(x) < -|u'(x)| \text{ for } x \in (t, c)\} \tag{15.13}$$

and prove that $\alpha = 0$. Assume to the contrary that $\alpha > 0$.

Using (15.12), (15.11), and the inequalities $|q(x)| \leq (\frac{3}{4} - \varepsilon)x^{-2}$, $v(x) > |u(x)|$
and $v'(x) < -|u'(x)|$ on the interval (α, c), we estimate

$$v'(\alpha) = -|u'(c)| - 1 - \int_{\alpha}^{c} v''(x)\,dx = -|u'(c)| - 1 - \int_{\alpha}^{c} (3/4 - \varepsilon)x^{-2}v(x)\,dx$$

$$< -|u'(c)| - 1 - \int_{\alpha}^{c} |q(x)u(x)|\,dx \leq -|u'(c)| - 1 - \left|\int_{\alpha}^{c} u''(x)\,dx\right|$$

$$= -|u'(c)| - 1 - |u'(c) - u'(\alpha)| < -|u'(\alpha)|,$$

$$v(\alpha) = |u(c)| + 1 - \int_{\alpha}^{c} v'(x)\,dx > |u(c)| + 1 + \int_{\alpha}^{c} |u'(x)|\,dx$$

$$\geq |u(c)| + 1 + \left|\int_{\alpha}^{c} u'(x)\,dx\right| = |u(c)| + 1 + |u(c) - u(\alpha)| > |u(\alpha)|.$$

These two inequalities imply that the set in (15.13) contains a smaller positive number than α. This is a contradiction which proves that $\alpha = 0$.

Since $\alpha = 0$, $v(x) > |u(x)|$ on $(0, c)$. But v is in L^2 near 0, and so is u.   □

There is an extensive literature concerning limit point and limit circle criteria,
see, e.g., [DS, XIII.6]. Without proof we state one additional result [DS, p. 1407]
which can be considered as a counterpart of Proposition 15.12(i) at $b = +\infty$:

*Suppose that $b = +\infty$. If there are numbers $c \in (a, b)$ and $C > 0$ such that
$q(x) \geq -Cx^2$ for all $x \in (c, +\infty)$, then T is in the limit point case at $b = +\infty$.*

Example 15.2 ($q(x) = \gamma x^{-2}$, $\gamma \in \mathbb{R}$, *on* $(0, +\infty)$) Then $\mathcal{L}(f) = -f'' + qf = 0$ is
one form of Bessel's equation. The preceding criterion implies that T is in the limit
point case at $b = +\infty$. By Proposition 15.12, T is in the limit point case at $a = 0$
if and only if $\gamma \geq 3/4$. Therefore, by Theorem 15.10, T is essentially self-adjoint
if and only if $\gamma \geq 3/4$. Otherwise, the symmetric operator T has deficiency indices
$(1, 1)$.   ○

15.3 Boundary Triplets for Sturm–Liouville Operators

In this section we develop some boundary triplets for the adjoint of the symmetric
operator $T = L_{\min}$ depending on the various cases.

Example 15.3 (*Regular end points a and b*) Throughout this example we assume
that *both end points a and b are regular*. Recall that by Definition 15.1 this holds if
and only if a and b are in $\mathbb{R}$ and $q \in L^1(a, b)$.

Then the boundary triplet in Example 14.2 for $q = 0$ generalizes verbatim to
the present case. Indeed, if $f \in \mathcal{D}(T^*)$, then f and f' can be extended to continuous functions on $[a, b]$ by Proposition 15.5(i). Therefore, by combining (15.1) and
(15.4), or by integration by parts, the boundary form $[\cdot, \cdot]_{T^*}$ can be written as

$$[f, g]_{T^*} = f(b)\overline{g'(b)} - f'(b)\overline{g(b)} - f(a)\overline{g'(b)} + f'(a)\overline{g(a)}. \quad (15.14)$$

There is a boundary triplet $(\mathcal{K}, \Gamma_0, \Gamma_1)$ for the operator T^* given by

$$\mathcal{K} = \mathbb{C}^2, \quad \Gamma_0(f) = \big(f(a), f(b)\big), \quad \Gamma_1(f) = \big(f'(a), -f'(b)\big). \quad (15.15)$$

Indeed, Eq. (15.14) shows that condition (i) in Definition 14.2 holds. The surjectivity condition (ii) is satisfied by Proposition 15.5(v). Note that (15.14) and (15.15) are just the formulas (14.11) and (14.12) appearing in Example 14.2.

Thus, we are in the setup of Sect. 14.4 and Example 14.8 therein. Therefore, all self-adjoint extensions of T are parameterized by Eqs. (14.36), (14.37), (14.38), (14.39).

Next we determine the gamma field and Weyl function for the boundary triplet (15.15). Let $z \in \mathbb{C}$. Since a is a regular end point, by Proposition 15.3(ii) there are unique solutions $\mathsf{s} = \mathsf{s}(\cdot; z)$ and $\mathsf{c} = \mathsf{c}(\cdot; z)$ of the equation $-f'' + qf = zf$ on (a, b) satisfying the boundary conditions

$$\mathsf{s}(a; z) = 0, \quad \mathsf{s}'(a; z) = 1, \quad \mathsf{c}(a; z) = 1, \quad \mathsf{c}'(a; z) = 0. \quad (15.16)$$

Since both end points are regular, s and c are in $L^2(a, b)$ by Proposition 15.5(ii), and hence $\mathsf{s}, \mathsf{c} \in \mathcal{N}_z := \mathcal{N}(T^* - zI)$. Clearly, $W(\mathsf{c}, \mathsf{s})_a = 1$, so that $W(\mathsf{c}, \mathsf{s})_x = 1$ on $[a, b]$, because the Wronskian is constant. Thus, $\{\mathsf{s}(\cdot; z), \mathsf{c}(\cdot; z)\}$ is a fundamental system of the equation $-f'' + qf = zf$ and a basis of the vector space $\mathcal{N}_z$.

Recall from Corollary 14.8 that T_0 denotes the self-adjoint extension of T defined by $\mathcal{D}(T_0) = \mathcal{N}(\Gamma_0)$. Suppose that $z \in \rho(T_0)$. Then $\mathsf{s}(b; z) \neq 0$. (Indeed, otherwise we have $\mathsf{s}(a; z) = \mathsf{s}(b; z) = 0$, and hence $\mathsf{s} \in \mathcal{D}(T_0)$; since $\mathsf{s} \in \mathcal{N}_z$, we obtain $T_0\mathsf{s} = z\mathsf{s}$, which contradicts $z \in \rho(T_0)$.)

Any element $u \in \mathcal{N}_z$ is a linear combination of the basis $\{\mathsf{s}(\cdot; z), \mathsf{c}(\cdot; z)\}$. Considering the boundary values $u(a)$ and $u(b)$ and using (15.16), we verify for $x \in [a, b]$,

$$u(x) = \big(u(b) - u(a)\mathsf{c}(b; z)\big)\mathsf{s}(b; z)^{-1}\mathsf{s}(x; z) + u(a)\mathsf{c}(x; z). \quad (15.17)$$

For the *gamma field* $\gamma(z) = (\Gamma_0 {\upharpoonright} \mathcal{N}_z)^{-1}$, we therefore obtain for $(c_1, c_2) \in \mathcal{K}$,

$$\gamma(z)(c_1, c_2) = \big(c_2 - c_1\mathsf{c}(b; z)\big)\mathsf{s}(b; z)^{-1}\mathsf{s}(\cdot; z) + c_1\mathsf{c}(\cdot; z). \quad (15.18)$$

The equation $M(z)\Gamma_0 u = \Gamma_1 u$ means that $M(z)(u(a), u(b)) = (u'(a), -u'(b))$ for $u \in \mathcal{N}_z$. Inserting the values $u'(a)$ and $u'(b)$ obtained from (15.17) a simple computation shows that the *Weyl function* $M(z)$ is given by

$$M(z) = -\mathsf{s}(b; z)^{-1} \begin{pmatrix} \mathsf{c}(b; z) & -1 \\ -1 & \mathsf{s}'(b; z) \end{pmatrix}. \qquad \circ$$

Example 15.4 (*Limit circle case at a and b*) In this example we suppose that T is in the *limit circle case at both end points a and b*. In order to construct a boundary triplet for T^*, we assume that u_1 and u_2 are real-valued functions of $\mathcal{D}(T^*)$ satisfying

$$[u_1, u_2]_a = [u_1, u_2]_b = 1. \quad (15.19)$$

First, let us show that such functions always exist. Let $c \in (a, b)$ and $\lambda \in \mathbb{R}$. By Proposition 15.3(ii) there exist unique solutions u_1, u_2 of the differential equation

$-u'' + qu = \lambda u$ on (a, b) satisfying $u_1(c) = u_2'(c) = 1$ and $u_1'(c) = u_2(c) = 0$. Since the complex conjugates $\overline{u}_1, \overline{u}_2$ are solutions with the same initial data, u_1 and u_2 are real-valued on (a, b), so that $W(u_1, u_2)_x = [u_1, u_2]_x$ on (a, b). Because T is in the limit circle case at a and b (see Theorem 15.8), u_1 and u_2 are in $L^2(a, b)$ and hence in $\mathcal{D}(T^*)$. The Wronskian $W(u_1, u_2)_x$ is constant on (a, b), so we have $[u_1, u_2]_x = [u_1, u_2]_c = 1$ for $x \in (a, b)$. Since $u_1, u_2 \in \mathcal{D}(T^*)$, the limits $x \to a + 0$ and $x \to b - 0$ exist (see (15.3)) and yield (15.19).

Let $f, g \in \mathcal{D}(T^*)$. Next, we prove the equation

$$[f, g]_d = [f, u_1]_d \overline{[g, u_2]}_d - [f, u_2]_d \overline{[g, u_1]}_d \quad \text{for } d = a, b. \tag{15.20}$$

Equation (15.20) will be derived from the *Plücker identity* for the Wronskian

$$W(f_1, f_2)_x W(f_3, f_4)_x - W(f_1, f_3)_x W(f_2, f_4)_x + W(f_1, f_4)_x W(f_2, f_3)_x = 0,$$

which holds for arbitrary functions $f_1, f_2, f_3, f_4 \in C^1(a, b)$ and $x \in (a, b)$. It follows by a direct computation or from the Laplace expansion of the determinant

$$\begin{vmatrix} f_1(x) & f_2(x) & f_3(x) & f_4(x) \\ f_1'(x) & f_2'(x) & f_3'(x) & f_4'(x) \\ f_1(x) & f_2(x) & f_3(x) & f_4(x) \\ f_1'(x) & f_2'(x) & f_3'(x) & f_4'(x) \end{vmatrix} = 0.$$

Inserting $f_1 = f$, $f_2 = \overline{g}$, $f_3 = u_1$, $f_4 = u_2$ into the Plücker identity and using the relation $W(u, \overline{v})_x = [u, v]_x$ and the fact that u_1, u_2 are real-valued, we get

$$[f, g]_x [u_1, u_2]_x - [f, u_1]_x [\overline{g}, u_2]_x + [f, u_2]_x [\overline{g}, u_1]_x = 0.$$

Since $[\overline{g}, u_j]_x = \overline{[g, u_j]}_x$ and $[u_1, u_2]_a = [u_1, u_2]_b = 1$ by assumption (15.19), Eq. (15.20) follows by letting $x \to a + 0$ and $x \to b - 0$.

Combining Eqs. (15.4) and (15.20), we finally obtain

$$[f, g]_{T^*} = [f, g]_b - [f, g]_a$$
$$= [f, u_1]_b \overline{[g, u_2]}_b - [f, u_2]_b \overline{[g, u_1]}_b - [f, u_1]_a \overline{[g, u_2]}_a + [f, u_2]_a \overline{[g, u_1]}_a. \tag{15.21}$$

Then there is a boundary triplet $(\mathcal{K}, \Gamma_0, \Gamma_1)$ for the operator T^* defined by

$$\mathcal{K} = \mathbb{C}^2, \quad \Gamma_0(f) = \big([f, u_1]_a, [f, u_1]_b\big), \quad \Gamma_1(f) = \big([f, u_2]_a, -[f, u_2]_b\big). \tag{15.22}$$

Indeed, condition (i) in Definition 14.2 follows at once from (15.21). To verify condition (ii), we take functions $u_{1a}, u_{1b}, u_{2a}, u_{2b} \in \mathcal{D}(T^*)$ such that $u_{ka} = u_k$ (resp. $u_{kb} = u_k$) in some neighborhood of a (resp. b) and zero in some neighborhood of b (resp. a). Since $u_j \in \mathcal{D}(T^*)$ is real-valued, we have $[u_j, u_j]_a = [u_j, u_j]_b = 0$ for $j = 1, 2$. Therefore, choosing f as a linear combination of $u_{1a}, u_{1b}, u_{2a}, u_{2b}$ and using (15.19), we easily see that any vector of $\mathbb{C}^4$ arises as $(\Gamma_0(f), \Gamma_1(f))$.

Since the boundary triplet (15.22) is of the form treated in Sect. 14.4 by setting

$$\varphi_1(f) = [f, u_1]_a, \qquad \varphi_2(f) = [f, u_1]_b,$$
$$\psi_1(f) = [f, u_2]_a, \qquad \psi_2(f) = -[f, u_2]_b,$$

all self-adjoint extensions of T are described by Eqs. (14.30), (14.31), (14.32), (14.33).   ○

Example 15.5 (*Regular end point a and limit point case at b*) Assume now that the *end point a is regular* and T is in the *limit point case at b*.

Then, by Proposition 15.5(i) and Lemma 15.9, f and f' extend to continuous functions on $[a, b)$, and we have $[f, g]_b = 0$ for $f, g \in \mathcal{D}(T^*)$. Therefore, by (15.4),

$$[f, g]_{T^*} = f'(a)\overline{g(a)} - f(a)\overline{g'(a)}, \qquad f, g \in \mathcal{D}(T^*).$$

Hence, it follows that there is a boundary triplet $(\mathcal{K}, \Gamma_0, \Gamma_1)$ for T^* defined by

$$\mathcal{K} = \mathbb{C}, \qquad \Gamma_0(f) = f(a), \qquad \Gamma_1(f) = f'(a).$$

By Example 14.7, all self-adjoint extensions of T on $L^2(a, b)$ are parameterized by the boundary conditions

$$f(a)\cos\alpha = f'(a)\sin\alpha, \quad \alpha \in [0, \pi). \tag{15.23}$$

As in Example 15.3, we consider the fundamental system $\{s(\cdot; z), c(\cdot; z)\}$ of the equation $-f'' + qf = zf$ determined by the boundary conditions (15.16).

Since T is in the limit circle case at a (because a is a regular end point) and in the limit point case at b, T has deficiency indices $(1, 1)$ by Theorem 15.10. Hence, for any $z \in \mathbb{C}\backslash\mathbb{R}$, the space of solutions of $-f'' + qf = zf$ belonging to $L^2(a, b)$ has dimension 1. But $s(\cdot; z)$ is not in $L^2(a, b)$. (Otherwise it would be in $\mathcal{D}(T_0)$ and so an eigenfunction of the self-adjoint operator T_0 with eigenvalue $z \in \mathbb{C}\backslash\mathbb{R}$, which is impossible.) Hence, there exists a unique complex number $m(z)$ such that

$$u_b(t; z) := c(t; z) + m(z)s(t; z) \in L^2(a, b), \quad z \in \mathbb{C}\backslash\mathbb{R}. \tag{15.24}$$

The function $m(z)$ is called the *Weyl–Titchmarsh function*. It plays a crucial role for the spectral analysis of Sturm–Liouville operators.

Since $\mathcal{N}_z = \mathbb{C} \cdot u_b(\cdot; z)$, we have $\gamma(z) = u_b(\cdot; z)$. Combining (15.24) and (15.16), we conclude that $u_b'(a; z) = m(z)$. Hence, $M(z) = \Gamma_1\gamma(z) = u_b'(a; z) = m(z)$. That is, the *Weyl function* $M(z)$ of the boundary triplet $(\mathcal{K}, \Gamma_0, \Gamma_1)$ defined above is just the Weyl–Titchmarsh function $m(z)$.   ○

In the next example we compute the function $m(z)$ in the simplest case.

Example 15.6 (*Example* 15.1 *continued*: $q(x) \equiv 0$ *on* $(0, +\infty)$) Then $a = 0$ is a regular end point, and T is in the limit point case at $b = +\infty$. The fundamental system from (15.16) is given by $s(x; 0) = x$, $c(x; 0) = 1$ and

$$s(x; z) = \frac{1}{\sqrt{z}}\sin\sqrt{z}x, \qquad c(x; z) = \cos\sqrt{z}x, \quad z \in \mathbb{C}, \ z \neq 0.$$

For $z \in \mathbb{C}\backslash[0, \infty)$, let $\sqrt{z}$ denote the square root of z satisfying $\operatorname{Im}\sqrt{z} > 0$. Then

$$u_b(x; z) = e^{i\sqrt{z}x} = \mathsf{c}(x; z) + i\sqrt{z}\mathsf{s}(x; z) \in L^2(0, \infty). \tag{15.25}$$

Hence, the Weyl function is $M(z) = \mathsf{m}(z) = i\sqrt{z}$.   ∘

Example 15.7 (*Limit circle case at a and limit point case at b*) Suppose that T is in the *limit circle case at a* and in the *limit point case at b*.

Let us assume that u_1 and u_2 are real-valued functions from $\mathcal{D}(T^*)$ such that

$$[u_1, u_2]_a = 1. \tag{15.26}$$

To show that such functions exist, we modify the reasoning from Example 15.4. Let us denote the functions u_1, u_2 constructed therein by $\tilde{u}_1, \tilde{u}_2$, respectively. Since T is in the limit point case at b, we cannot conclude that $\tilde{u}_1, \tilde{u}_2 \in L^2(a, b)$. However, if we choose real-valued functions $u_1, u_2 \in \mathcal{D}(T^*)$ such that $u_j = \tilde{u}_j$ in a neighborhood of a and $u_j = 0$ in a neighborhood of b, then $[u_1, u_2]_a = [\tilde{u}_1, \tilde{u}_2]_a = 1$.

Now we proceed as in Example 15.4. Since $[f, g]_b = 0$ for $f, g \in \mathcal{D}(T^*)$ by Lemma 15.9, we obtain

$$[f, g]_{T^*} = [f, u_2]_a \overline{[g, u_1]}_a - [f, u_1]_a \overline{[g, u_2]}_a, \quad f, g \in \mathcal{D}(T^*).$$

Hence, there is a boundary triplet $(\mathcal{K}, \Gamma_0, \Gamma_1)$ for T^* defined by

$$\mathcal{K} = \mathbb{C}, \qquad \Gamma_0(f) = [f, u_1]_a, \qquad \Gamma_1(f) = [f, u_2]_a.$$

Note that $u_1 \in \mathcal{D}(T_0)$ and $u_2 \in \mathcal{D}(T_1)$ by (15.26), where T_0 and T_1 are the two distinguished self-adjoint extensions of T from Corollary 14.8.   ∘

15.4 Resolvents of Self-adjoint Extensions

First we recall the *variation-of-constants formula* for the inhomogeneous equation. Let $z \in \mathbb{C}$ and $g \in L^1_{\text{loc}}(a, b)$. Let $\{u_1, u_2\}$ be a basis of the vector space of solution of $\mathcal{L}(u) - zu = 0$ on (a, b), and let c be a point of (a, b) or a regular end point. Then a function f on (a, b) is a solution of the equation $\mathcal{L}(f) - zf = g$ on (a, b) if and only if there are numbers $\alpha_1, \alpha_2 \in \mathbb{C}$ such that

$$\begin{aligned} f(x) &= \alpha_1 u_1(x) + \alpha_2 u_2(x) \\ &\quad + W(u_1, u_2)_x^{-1}\left(u_1(x)\int_c^x u_2(t)g(t)\, dt - u_2(x)\int_c^x u_1(t)g(t)\, dt\right), \end{aligned} \tag{15.27}$$

where we set $\int_c^x := -\int_x^c$ if $x \le c$. To prove this fact, it suffices to check that the function on the second line is a solution of the inhomogeneous equation. This follows from a simple computation using that the Wronskian is constant on (a, b).

Let us turn to the resolvent formulas. Throughout $\{u_a, u_b\}$ will be a basis of the solution space of $\mathcal{L}(u) - zu = 0$ on (a, b). The Wronskian $W(u_b, u_a)_x$ is a nonzero constant $W(u_b, u_a)$ on (a, b). We define a kernel $K_z(\cdot, \cdot)$ on $(a, b) \times (a, b)$ by

$$K_z(x, t) = \begin{cases} W(u_b, u_a)^{-1} u_a(x) u_b(t) & \text{for } x \le t \\ W(u_b, u_a)^{-1} u_b(x) u_a(t) & \text{for } x > t \end{cases} \tag{15.28}$$

and an integral operator R_z on $L^2(a, b)$ by

$$(\mathsf{R}_z g)(x) = \int_a^b K_z(x, t) g(t) \, dt, \quad g \in L^2(a, b). \tag{15.29}$$

In the following two propositions both functions u_a and u_b are in $L^2(a, b)$. Since therefore $K_z \in L^2((a, b) \times (a, b))$, by Theorem A.6 the integral operator R_z is a Hilbert–Schmidt operator on $L^2(a, b)$. In particular, R_z is then compact.

Proposition 15.13 *Suppose that both end points a and b are regular and $\alpha = (\alpha_1, \alpha_2)$, where $\alpha_1, \alpha_2 \in [0, \pi)$. Let A_α denote the restriction of T^* to the domain*

$$\mathcal{D}(A_\alpha) = \{ f \in \mathcal{D}(T^*) : f(a) \cos \alpha_1 = f'(a) \sin \alpha_1, \ f(b) \cos \alpha_2 = f'(b) \sin \alpha_2 \}.$$

For $z \in \mathbb{C}$, we denote by u_a and u_b the unique solutions (by Proposition 15.3(ii)) of the differential equation $\mathcal{L}(u) \equiv -u'' + qu = zu$ satisfying the boundary conditions

$$u_a(a) = \sin \alpha_1, \quad u_a'(a) = \cos \alpha_1, \quad u_b(b) = \sin \alpha_2, \quad u_b'(b) = \cos \alpha_2. \tag{15.30}$$

Then A_α is a self-adjoint operator on $L^2(a, b)$. For any $z \in \rho(A_\alpha)$, the resolvent $R_z(A_\alpha)$ is equal to the operator R_z defined by (15.29) with kernel (15.28), that is,

$$\left(R_z(A_\alpha) g \right)(x) = W(u_b, u_a)^{-1} \left(u_b(x) \int_a^x u_a(t) g(t) \, dt + u_a(x) \int_x^b u_b(t) g(t) \, dt \right). \tag{15.31}$$

It should be emphasized that u_a (resp. u_b) satisfies the boundary condition for $\mathcal{D}(A_\alpha)$ at the end point a (resp. b), but not at b (resp. a) when $z \in \rho(A_\alpha)$. Also, note that the functions u_a and u_b depend on the number z and they are in $L^2(a, b)$ for all $z \in \mathbb{C}$ by Proposition 15.5(ii), because both end points are regular.

Proof The operator A_α is self-adjoint, because it is defined by Eqs. (14.34) in Example 14.8 applied to the boundary triplet (15.15) of Example 15.3.

Suppose that $z \in \rho(A_\alpha)$. Then u_a and u_b are linearly independent, and hence the Wronskian is a nonzero constant. (Indeed, if u_a were a multiple of u_b, then u_a would satisfy both boundary conditions of A_α, so $u_a \in \mathcal{D}(A_\alpha)$. But then u_a would be an eigenvector of A_α with eigenvalue z, which contradicts $z \in \rho(A_\alpha)$.)

Let $L_0^2(a, b)$ denote the set of $g \in L^2(a, b)$ which vanish in neighborhoods of a and b. Since R_z and $R_z(A_\alpha)$ are bounded operators and $L_0^2(a, b)$ is obviously dense in $L^2(a, b)$, it suffices to prove the equality $\mathsf{R}_z g = R_z(A_\alpha) g$ for $g \in L_0^2(a, b)$.

By (15.28) and (15.29) the function $f := \mathsf{R}_z g$ is given by the right-hand side of formula (15.31). Writing the integral $\int_x^b$ in (15.31) as $\int_a^b - \int_a^x$, the function f is of the form (15.27), where $c = a$, $\alpha_1 = 0$, $\alpha_2 = W(u_b, u_a)^{-1} \int_b^a u_b(t) g(t) \, dt$, $u_1 = u_b$, $u_2 = u_a$. Hence, f is a solution of the inhomogeneous equation $\mathcal{L}(f) - zf = g$ on (a, b). That is, we have $T^* f - zf = g$. Since $g \in L_0^2(a, b)$, it follows from (15.31) that f is a constant multiple of u_a in some neighborhood of a and a constant multiple of u_b in some neighborhood of b. Therefore, $f \in \mathcal{D}(A_\alpha)$. These facts imply that $f \equiv \mathsf{R}_z g = R_z(A_\alpha) g$. $\qquad \square$

Proposition 15.14 *Suppose that T is in the limit circle case at a and b. Assume that u_1 and u_2 are as in Example 15.4 (that is, u_1 and u_2 are real-valued functions of $\mathcal{D}(T^*)$ satisfying (15.19)). Let $\alpha_1, \alpha_2 \in [0, \pi)$ and set $\alpha = (\alpha_1, \alpha_2)$. We denote by $\mathcal{D}(A_\alpha)$ the set of all functions $f \in \mathcal{D}(T^*)$ such that*

$$[f, u_1]_a \cos\alpha_1 = [f, u_2]_a \sin\alpha_1, \qquad [f, u_1]_b \cos\alpha_2 = [f, u_2]_b \sin\alpha_2.$$
$$(15.32)$$

Then $A_\alpha := T^* {\restriction} \mathcal{D}(A_\alpha)$ is a self-adjoint operator on $L^2(a, b)$. For any $z \in \rho(A_\alpha)$ there exist nonzero solutions u_a, u_b of the equation $\mathcal{L}(u) = zu$ on (a, b) such that

$$[u_a, u_1]_a \cos\alpha_1 = [u_a, u_2]_a \sin\alpha_1, \qquad [u_b, u_1]_b \cos\alpha_2 = [u_b, u_2]_b \sin\alpha_2$$
$$(15.33)$$

and the resolvent $R_z(A_\alpha)$ acts by formula (15.31). The functions u_a and u_b are in $L^2(a, b)$, and they are uniquely determined up to some constant factors by (15.33).

Note that the kernel (15.28) remains unchanged if u_a or u_b are replaced by nonzero constant multiples. As in Proposition 15.13, for any $z \in \rho(A_\alpha)$, the functions u_a and u_b satisfy the boundary conditions for A_α at one end point, but not at the other.

Proof Since conditions (15.32) are just Eqs. (14.34) in Example 14.8 applied to the boundary triplet (15.22) of Example 15.4, the operator A_α is self-adjoint.

Fix $z \in \rho(A_\alpha)$. Let $\{v_1, v_2\}$ be a fundamental system of the equation $\mathcal{L}(u) = zu$ on (a, b). Put $v = c_1 v_1 + c_2 v_2$ for $c_1, c_2 \in \mathbb{C}$. Since the condition

$$[v, u_1]_a \cos\alpha_1 = [v, u_2]_a \sin\alpha_1 \qquad (15.34)$$

is a homogeneous linear equation in c_1, c_2, it has a nonzero solution v. Set $u_a := v$. To prove that this solution is uniquely determined up to a factor, we assume the contrary. Then (15.34) holds for all such v. We choose a nonzero v satisfying $[v, u_1]_b \cos\alpha_2 = [v, u_2]_b \sin\alpha_2$. Then $v \in \mathcal{D}(A_\alpha)$. Hence, v is an eigenvector of A_α with eigenvalue z, which contradicts the assumption $z \in \rho(A_\alpha)$.

The assertions concerning u_b are proved by a similar reasoning.

Since T is in the limit circle case at a and b, we have $u_a, u_b \in L^2(a, b)$ by Theorem 15.8(i). Now the proof of the resolvent formula (15.31) is verbatim the same as the corresponding proof of Proposition 15.13. $\qquad \square$

Corollary 15.15 *Suppose that T is in the limit circle case at a and b. Then each self-adjoint extension A of T on $L^2(a, b)$ has a purely discrete spectrum. If $v_n(A)$, $n \in \mathbb{N}$, is an enumeration of all nonzero eigenvalues of A counted with multiplicities, then*

$$\sum_{n=1}^{\infty} v_n(A)^{-2} < \infty. \tag{15.35}$$

Proof Recall that the resolvent $R_i(A_\alpha)$ of the operator A_α from Proposition 15.14 is an integral operator with kernel $K_i \in L^2((a, b) \times (a, b))$, since $u_a, u_b \in L^2(a, b)$. Hence $R_i(A_\alpha)$ is a Hilbert–Schmidt operator by Theorem A.6.

Since A and A_α are extensions of T, the operator $C := R_i(A) - R_i(A_\alpha)$ vanishes on $\mathcal{R}(T - iI)$. By Theorem 15.10, T has deficiency indices $(2, 2)$. Hence, $\dim \mathcal{R}(T - iI)^\perp = 2$, so C is a bounded operator of rank not more than 2. Since C and $R_i(A_\alpha)$ are Hilbert–Schmidt operators, so is $R_i(A)$. By Corollary A.5 the sequence $(\mu_n; n \in \mathbb{N})$ of eigenvalues of $R_i(A)$ is in $l^2(\mathbb{N})$. By Propositions 2.11 and 2.10(ii), A has a purely discrete spectrum, and the eigenvalues of A are $\mu_n^{-1} + i$. It is easily verified that $(\mu_n) \in l^2(\mathbb{N})$ implies (15.35). $\square$

For the rest of this section, we assume that the *end point a is regular* and T is in the *limit point case at b*. Then, by Theorem 15.10, T has deficiency indices $(1, 1)$. As noted in Example 15.5, the self-adjoint extensions of T on $L^2(a, b)$ are the operators $A_\alpha := T^* \restriction \mathcal{D}(A_\alpha)$, where

$$\mathcal{D}(A_\alpha) = \left\{ f \in \mathcal{D}(T^*) : f(a) \cos \alpha = f'(a) \sin \alpha \right\}, \quad \alpha \in [0, \pi). \tag{15.36}$$

Fix $\alpha \in [0, \pi)$ and $z \in \rho(A_\alpha)$. We show that there exists a fundamental system $\{u_a, u_b\}$ for the equation $\mathcal{L}(u) = zu$ such that $u_b \in L^2(a, b)$ and u_a satisfies the boundary condition in (15.36). Indeed, since a is a regular end point, by Proposition 15.3(ii) such a solution u_a exists. By Proposition 15.6 there is a nonzero solution u_b which is in L^2 near b. Then $u_b \in L^2(a, b)$, because a is a regular end point. But $u_a \notin L^2(a, b)$. (Otherwise, $u_a \in \mathcal{D}(A_\alpha)$, so $z \in \rho(A_\alpha)$ would be an eigenvalue of the operator A_α, which is impossible.) Clearly, u_a and u_b are linearly independent, since u_b is in $L^2(a, b)$, but u_a is not.

Since $u_a \notin L^2(a, b)$, we cannot conclude that the kernel K_z defined by (15.28) is in $L^2((a, b) \times (a, b))$. But since $u_b \in L^2(a, b)$ and $u_a \in L^2(a, c)$ for $a < c < b$, it follows easily from (15.28) that $K_z(x, \cdot) \in L^2(a, b)$ for any fixed $x \in (a, b)$.

In particular, $K_z(x, \cdot) \in L^2(a, b)$ a.e. on (a, b). An integral operator (15.29) with kernel K_z satisfying the latter condition is called a *Carleman integral operator*.

Proposition 15.16 *Suppose that a is a regular end point and T is in the limit point case at b. Let $\alpha \in [0, \pi)$ and $z \in \rho(A_\alpha)$. Then there exists a fundamental system $\{u_a, u_b\}$ for the equation $\mathcal{L}(u) = zu$ such that $u_a(a) \cos \alpha = u_a'(a) \sin \alpha$ and $u_b \in L^2(a, b)$. The resolvent $R_z(A_\alpha)$ of the self-adjoint operator A_α defined by (15.36) is the integral operator R_z with kernel (15.28).*

Proof By the preceding discussion only the assertion concerning the resolvent formula remains to be proved.

First, suppose that $g \in L_0^2(a, b)$. Set $f := R_z(A_\alpha)g$. Then $\mathcal{L}(f) - zf = g$ on (a, b) and $g \in L_{\text{loc}}^1(a, b)$. Therefore, by formula (15.27) applied to the fundamental system $\{u_b, u_a\}$, f has to be of the form

$$f(x) = u_b(x)\left(\alpha_1 + W^{-1} \int_a^x u_a(t)g(t)\, dt\right)$$

$$+ u_a(x)\left(\alpha_2 - W^{-1} \int_a^x u_b(t)g(t)\, dt\right), \tag{15.37}$$

where $W := W(u_b, u_a)$ and $\alpha_1, \alpha_2 \in \mathbb{C}$. Since $g \in L_0^2(a, b)$, there are numbers $c, d \in (a, b)$ such that $g(x) = 0$ on $(a, c) \cup (d, b)$. By (15.37) we have $f(x) = u_b(x)\alpha_1 + u_a(x)\alpha_2$ on (a, c). Both $f \in \mathcal{D}(A_\alpha)$ and u_a satisfy the boundary condition (15.36) at the end point a, but u_b does not (otherwise u_b would be a multiple of u_a by the uniqueness assertion of Proposition 15.3(ii)). Hence $\alpha_1 = 0$.

Recall that $g(t) = 0$ on (d, b). Therefore, by (15.37) we have for $x \in (d, b)$,

$$f(x) = u_b(x)W^{-1} \int_a^b u_a(t)g(t)\, dt$$

$$+ u_a(x)\left(\alpha_2 - W^{-1} \int_a^b u_b(t)g(t)\, dt\right).$$

Since $u_a \notin L^2(d, b)$ and $f, u_b \in L^2(d, b)$, it follows from the preceding formula that $\alpha_2 = W^{-1} \int_a^b u_b(t)g(t)\, dt$. By inserting this into (15.37) the right-hand side of (15.37) can be rewritten as $\mathsf{R}_z g$. This proves that $R_z(A_\alpha)g = f = \mathsf{R}_z g$.

Now let $g \in L^2(a, b)$. Since $L_0^2(a, b)$ is dense in $L^2(a, b)$, there is a sequence (g_n) from $L_0^2(a, b)$ such that $g_n \to g$. Then $R_z(A_\alpha)g_n \to R_z(A_\alpha)g$, because $R_z(A_\alpha)$ is bounded. Since $K_z(x, \cdot) \in L^2(a, b)$ for fixed $x \in (a, b)$ as noted above, we obtain

$$\left|(\mathsf{R}_z g_n)(x) - (\mathsf{R}_z g)(x)\right| = \left|\int_a^b K_z(x, t)\big(g_n(t) - g(t)\big)\, dt\right|$$

$$\leq \int_a^b \left|K_z(x, t)\big(g_n(t) - g(t)\big)\right|\, dt$$

$$\leq \|g_n - g\| \left(\int_a^b \left|K_z(x, t)\right|^2 dt\right)^{1/2} \to 0$$

as $n \to \infty$ for all fixed $x \in (a, b)$. But $\mathsf{R}_z g_n = R_z(A_\alpha)g_n \to R_z(A_\alpha)g$ as $n \to \infty$. Therefore, the preceding implies that $\mathsf{R}_z g = R_z(A_\alpha)g$. $\square$

Example 15.8 (*Example 15.6 continued*: $q(x) \equiv 0$ *on* $(0, +\infty)$) Recall that 0 is a regular end point and T is in the limit point case at $+\infty$. The function u_b was

already given by formula (15.25). A solution u_a of the equation $\mathcal{L}(u) \equiv -u'' = zu$
satisfying the boundary condition $u(0) \cos \alpha = u'(0) \sin \alpha$ is

$$u_a(x; z) = \sin \alpha \cos \sqrt{z}x + \frac{\cos \alpha}{\sqrt{z}} \sin \sqrt{z}x. \tag{15.38}$$

Hence, by Proposition 15.16, the resolvent $R_z(A_\alpha)$ is the integral operator R_z with
kernel (15.28) obtained by inserting the functions u_a and u_b defined by (15.38) and
(15.25), respectively.   ◦

15.5 Exercises

1. Decide whether or not the end points are regular and whether or not T is in the
   limit circle or in the limit point case. Determine the deficiency indices of the
   corresponding Sturm–Liouville operator.
   a. $q(x) = -x^2$ on $(0, +\infty)$,
   b. $q(x) = x^2 + x^{-1/2}$ on $(0, +\infty)$,
   c. $q(x) = -x^2 + x^{-2}$ on $(0, 1)$,
   d. $q(x) = x^2 + \frac{1}{2}x^{-2}$ on $(0, 1)$,
   e. $q(x) = \sqrt{x(1 - x)}$ on $(0, 1)$,
   f. $q(x) = 2x^{-1}\sqrt{1 - x}$ on $(0, 1)$.
2. Let J be the conjugation on $L^2(a, b)$ defined by $(Jf)(x) = \overline{f(x)}$.
   a. Describe all self-adjoint extensions of T in Example 15.3 which are J-real
      (see Definition 13.1).
   b. Suppose that a and b are regular end points. Show that there exists a self-
      adjoint extension of T which has eigenvalues of multiplicity at least 2.
      Hint: Use an extension which is not J-real.
   c. Suppose that a is a regular end point and T is in the limit point case at b.
      Show that all self-adjoint extensions of T are J-real.
3. Let A_α be the self-adjoint operator with decoupled boundary conditions from
   Proposition 15.13. Show that each eigenvalue of A_α has multiplicity one.
4. Suppose that T is in the limit circle case at a and in the limit point case at b. Let
   $u \in \mathcal{D}(T^*)$ be real-valued and suppose that $[v, u]_a \neq 0$ for some $v \in \mathcal{D}(T^*)$.
   Let A_u denote the restriction of T^* to $\mathcal{D}(A_u) := \{f \in \mathcal{D}(T^*) : [f, u]_a = 0\}$.
   a. Use the Plücker identity to show that $[f_1, f_2]_a = 0$ for $f_1, f_2 \in \mathcal{D}(A_u)$.
   b. Show that A_u is self-adjoint.
*5. Let $R_z(A_\alpha)$ be the resolvent operator from Proposition 15.16. Let B be the
   multiplication operator by the characteristic function of a compact subinterval
   $[c, d]$ of (a, b). Show that the operator $BR_z(A_\alpha)B$ is compact.
6. Suppose that T is in the limit circle case at a and in the limit point case at b.
   a. Describe all self-adjoint extensions of T.
   *b. Find and prove the resolvent formulas for these self-adjoint extensions.
      Hint: Compare with Proposition 15.14 and modify the proof of Proposi-
      tion 15.16.

*7. Suppose that T is in the limit point case at a and b. Find and prove the resolvent formula for the self-adjoint operator $\overline{T}$.

8. Consider the operators $T_j = -\frac{d^2}{dx^2}$, $j = 1, 2, 3$, on $L^2(0, 1)$, where the domains consists of all functions of $H^1(0, 1)$ satisfying the boundary conditions: $T_1 : f'(0) = f'(1) = 0$, $T_2 : f'(0) = f(1) = 0$, $T_3 : f(0) = f(1) + f'(1) = 0$. Find the eigenvalues, eigenfunctions, and the resolvents of the self-adjoint operators T_j, $j = 1, 2, 3$.

Chapter 16
The One-Dimensional Hamburger Moment Problem

The close relationship between the moment problem and operator theory dates back to the early days of functional analysis. In this chapter we develop the one-dimensional Hamburger moment problem with the emphasis on self-adjoint extension theory. Section 16.1 deals with orthogonal polynomials and the Jacobi operator associated with a moment sequence. Basic results on the existence, the set of solutions, and the uniqueness are given in terms of self-adjoint extensions. In Sect. 16.2, polynomials of second kind are studied. Section 16.3 is devoted to the advanced theory in the indeterminate case. Fundamental notions such as Nevanlinna functions and Weyl circles are treated, and von Neumann solutions are described therein. In Sect. 16.4, Nevanlinna's parameterization of the set of all solutions is developed.

16.1 Moment Problem and Jacobi Operators

Let us begin with some notation. We denote by $\mathcal{M}(\mathbb{R})$ the set of all positive Borel measures μ on $\mathbb{R}$ for which all polynomials are μ-integrable and by $\langle \cdot, \cdot \rangle_\mu$ the scalar product of $L^2(\mathbb{R}, \mu)$. For a polynomial $p(x) = \sum_n \gamma_n x^n \in \mathbb{C}[x]$, we set $\overline{p}(x) := \sum_n \overline{\gamma}_n x^n$. If $s = (s_n)_{n \in \mathbb{N}_0}$ is a real sequence, let L_s be the linear functional on $\mathbb{C}[x]$ defined by $L_s(x^n) = s_n$ for $n \in \mathbb{N}_0$. The vector space of all finite complex sequences $\gamma = (\gamma_0, \dots, \gamma_k, 0, 0, \dots)$ is denoted by d.

Definition 16.1 For a measure $\mu \in \mathcal{M}(\mathbb{R})$, we call the number

$$s_n \equiv s_n(\mu) = \int_{\mathbb{R}} x^n \, d\mu(x), \quad n \in \mathbb{N}_0, \tag{16.1}$$

the nth *moment* of μ, and the sequence $s = (s_n)_{n \in \mathbb{N}_0}$ the *moment sequence* of μ.

The *Hamburger moment problem*, briefly the *moment problem*, asks when a given real sequence s is the moment sequence of some measure $\mu \in \mathcal{M}(\mathbb{R})$. In this case we say the moment problem for s is *solvable* and μ is a *solution* or μ is a *representing measure* for s. The set of all representing measures for s is denoted by $\mathcal{M}_s$.

K. Schmüdgen, *Unbounded Self-adjoint Operators on Hilbert Space*,
Graduate Texts in Mathematics 265,
DOI 10.1007/978-94-007-4753-1_16, © Springer Science+Business Media Dordrecht 2012

Definition 16.2 A moment sequence s is *determinate* if it has only one representing measure. A measure $\mu \in \mathcal{M}(\mathbb{R})$ is called *determinate* if its moment sequence is determinate. Otherwise, we say that s and μ are *indeterminate*.

The aim of this section is to study the *existence problem* and the *uniqueness problem* by means of self-adjoint operators on Hilbert space.

Example 16.1 (*Examples of indeterminate moment sequences*) Explicit examples of indeterminate moment sequences were already constructed by T. Stieltjes (1894). He observed that the measures $\mu, \mu_k \in \mathcal{M}(\mathbb{R})$, $k \in \mathbb{Z}$, given by

$$d\mu(x) = \chi_{(0,+\infty)}(x) e^{-x^{1/4}} dx \quad \text{and} \quad d\mu_k(x) = \chi_{(0,+\infty)}(x) x^{k-\ln x} dx, \quad k \in \mathbb{Z},$$

are indeterminate. We will prove this fact for the measures μ_k.

Substituting $t = \ln x$, the moments of μ_k, $k \in \mathbb{Z}$, are computed by

$$s_n(\mu_k) = \int_{\mathbb{R}} x^n \, d\mu_k(x) = \int_0^\infty x^n x^{k-\ln x} \, dx = \int_{\mathbb{R}} e^{tn} e^{t(k-t)} e^t \, dt$$

$$= \int_{\mathbb{R}} e^{-(t-(n+k+1)/2)^2} e^{(n+k+1)^2/4} \, dt = \sqrt{\pi} e^{(n+k+1)^2/4}, \quad n \in \mathbb{N}_0.$$

Also, this proves that all powers x^n are μ_k-integrable and hence $\mu_k \in \mathcal{M}(\mathbb{R})$.

For arbitrary $c \in [-1, 1]$, we define the measure $\rho_{c,k} \in \mathcal{M}(\mathbb{R})$ by

$$d\rho_{c,k}(x) = \left[1 + c \sin(2\pi \ln x)\right] d\mu_k(x).$$

Each measure $\rho_{c,k}$ has the same moments as μ_k, and hence μ_k is indeterminate, since

$$\int_{\mathbb{R}} x^n \left(\sin(2\pi \ln x)\right) d\mu_k(x)$$

$$= \int_{\mathbb{R}} e^{tn} (\sin 2\pi t) e^{t(k-t)} e^t \, dt$$

$$= \int_{\mathbb{R}} e^{-(t-(n+k+1)/2)^2} e^{(n+k+1)^2/4} (\sin 2\pi t) \, dt$$

$$= e^{(n+k+1)^2/4} \int_{\mathbb{R}} e^{-s^2} \left(\sin\left(2\pi s + (n+k+1)\pi\right)\right) ds$$

$$= (-1)^{n+k+1} e^{(n+k+1)^2/4} \int_{\mathbb{R}} e^{-s^2} (\sin 2\pi s) \, ds = 0, \quad n \in \mathbb{N}_0.$$

Let $a > 0$ and $b > 0$ and define two measures $\nu_1, \nu_2 \in \mathcal{M}(\mathbb{R})$ by

$$d\nu_1(x) = e^{-a|x|^b} dx \quad \text{and} \quad d\nu_2(x) = \chi_{(0,+\infty)}(x) e^{-ax^b} dx.$$

It can be shown that ν_1 is determinate if and only if $b \geq 1$ and that ν_2 is determinate if and only if $b \geq 1/2$ (see, e.g., [ST], p. 22).   ∘

Now let $\mu \in \mathcal{M}(\mathbb{R})$, and let s be the moment sequence of μ. Then

$$L_s(p\overline{p}) = \sum_{k,l} \gamma_k \overline{\gamma}_l s_{k+l} = \sum_{k,l} \gamma_k \overline{\gamma}_l \int x^{k+l} d\mu(x) = \int |p(x)|^2 d\mu(x) \geq 0 \quad (16.2)$$

for all polynomials $p(x) = \sum_k \gamma_k x^k \in \mathbb{C}[x]$ or all sequences $\gamma = (\gamma_k) \in d$. That is, the sequence s is *positive semi-definite*, or equivalently, $L_s(p\overline{p}) \geq 0$ for all $p \in \mathbb{C}[x]$.

Further, if $L_s(p\overline{p}) = 0$ for some $p \in \mathbb{C}[x]$, $p \neq 0$, it follows from (16.2) that μ is supported by the finite set of zeros of p. We want to exclude this simple case.

Let $s = (s_n)_{n \in \mathbb{N}_0}$ be a real sequence. From linear algebra it is well known that the sequence s is positive definite if and only if

$$
\begin{vmatrix}
s_0 & s_1 & s_2 & \dots & s_n \\
s_1 & s_2 & s_3 & \dots & s_{n+1} \\
s_2 & s_3 & s_4 & \dots & s_{n+2} \\
\dots & & & & \\
s_n & s_{n+1} & s_{n+2} & \dots & s_{2n}
\end{vmatrix} > 0 \quad \text{for all } n \in \mathbb{N}_0.
$$

From now on we assume that $s = (s_n)_{n \in \mathbb{N}_0}$ is a positive definite real sequence and $s_0 = 1$.

The requirement $s_0 = 1$ simplifies various formulas. For instance, it implies that $P_0(x) = 1$, where P_0 is the first of the polynomials from Proposition 16.3 below.

That s is positive definite means that $L_s(p\overline{p}) > 0$ for all $p \in \mathbb{C}[x]$, $p \neq 0$. Then

$$\langle p, q \rangle_s := L_s(p\,\overline{q}), \quad p, q \in \mathbb{C}[x], \quad (16.3)$$

defines a scalar product $\langle \cdot, \cdot \rangle_s$ on the vector space $\mathbb{C}[x]$. We denote the Hilbert space completion of the unitary space $(\mathbb{C}[x], \langle \cdot, \cdot \rangle_s)$ by $\mathcal{H}_s$. Note that $\langle p, q \rangle_s$ is real for $p, q \in \mathbb{R}[x]$, because the sequence $s = (s_n)$ is real.

Let M_x denote the multiplication operator by the variable x, that is,

$$M_x p(x) := x p(x) \quad \text{for } p \in \mathcal{D}(M_x) := \mathbb{C}[x].$$

For $p, q \in \mathbb{C}[x]$, we have $\langle M_x p, q \rangle_s = L_s(xp\overline{q}) = L_s(p\overline{x}\overline{q}) = \langle p, M_x q \rangle_s$. That is, M_x is a *densely defined symmetric operator* on the Hilbert space $\mathcal{H}_s$.

The self-adjoint extension theory of this symmetric operator M_x is the thread of the operator-theoretic approach to the moment problem.

The *existence of a solution* of the moment problem is clarified by the following:

Theorem 16.1 *Let s be a positive definite real sequence. Then the moment problem for s has a solution.*

If A is a self-adjoint operator on a Hilbert space $\mathcal{G}$ such that $\mathcal{H}_s$ is a subspace of $\mathcal{G}$ and $M_x \subseteq A$ and E_A denotes the spectral measure of A, then $\mu_A(\cdot) = \langle E_A(\cdot)1, 1 \rangle$ is a representing measure for s. Every representing measure for s is of this form.

Proof Let A be a self-adjoint extension of M_x on $\mathcal{G}$. Since $M_x \subseteq A$ and hence $(M_x)^n \subseteq A^n$, the polynomial 1 is in the domain $\mathcal{D}(A^n)$, and we have

$$\int_{\mathbb{R}} x^n \, d\langle E_A(x)1, 1\rangle = \langle A^n 1, 1\rangle = \langle (M_x)^n 1, 1\rangle_s = \langle x^n, 1\rangle_s = L_s(x^n) = s_n$$

for $n \in \mathbb{N}_0$. This shows that μ_A is a solution of the moment problem for s.

That each solution is of this form follows from Lemma 16.2 below.

Since the symmetric operator M_x has a self-adjoint extension on a larger Hilbert space (by Proposition 3.17), the moment problem for s is always solvable. $\square$

Next, we will give a more precise description of the relevant self-adjoint extensions of the symmetric operator M_x. A self-adjoint extension A of M_x on a possibly larger Hilbert space $\mathcal{G}$ (that is, $\mathcal{H}_s \subseteq \mathcal{G}$ and $M_x \subseteq A$) is called *minimal* if the polynomial 1 is a cyclic vector for A (see Definition 5.1), or equivalently, if $\{f(A)1 : f \in \mathcal{F}_1\}$ is dense in $\mathcal{G}$. Here $\mathcal{F}_1$ denotes the set of all E_A-a.e. finite measurable functions $f : \mathbb{R} \to \mathbb{C} \cup \{\infty\}$ for which $1 \in \mathcal{D}(f(A))$.

Lemma 16.2 *Let $\mu \in \mathcal{M}_s$. Then we have*

$$L_s(p) = \int_{\mathbb{R}} p(x) \, d\mu(x) \quad \text{and} \quad \langle p, q\rangle_s = \langle p, q\rangle_\mu \quad \text{for } p, q \in \mathbb{C}[x]. \quad (16.4)$$

The inclusion $\mathbb{C}[x] \subseteq L^2(\mathbb{R}, \mu)$ extends to a unitary operator of $\mathcal{H}_s$ onto a closed subspace of $L^2(\mathbb{R}, \mu)$. The self-adjoint operator A_x on $L^2(\mathbb{R}, \mu)$ defined by

$$(A_x f)(x) = x f(x) \quad \text{for } f \in \mathcal{D}(A_x) := \{ f \in L^2(\mathbb{R}, \mu) : x f(x) \in L^2(\mathbb{R}, \mu) \}$$

is a minimal extension of the operator M_x, and $\mu(\cdot) = \mu_{A_x}(\cdot) \equiv \langle E_{A_x}(\cdot)1, 1\rangle_\mu$.

Proof From $\mu \in \mathcal{M}(\mathbb{R})$ it follows that $\mathbb{C}[x] \subseteq \mathcal{D}(A_x)$. Clearly, $M_x \subseteq A_x$. Since μ is a representing measure for s, we have

$$L_s(p) = \sum_k \gamma_k s_k = \sum_k \gamma_k \int_{\mathbb{R}} x^k \, d\mu(x) = \int_{\mathbb{R}} p(x) \, d\mu(x)$$

for $p(x) = \sum_k \gamma_k x^k \in \mathbb{C}[x]$, which proves the first equality of (16.4). Obviously, this implies the second equality. Hence, the inclusion $\mathbb{C}[x] \subseteq L^2(\mathbb{R}, \mu)$ can be extended by continuity to a unitary embedding of $\mathcal{H}_s$ into $L^2(\mathbb{R}, \mu)$.

By Examples 2.1, 4.6, and 5.2, A_x is self-adjoint, $\mu(\cdot) = \langle E_{A_x}(\cdot)1, 1\rangle$, and the operators $f(A_x)$ act as multiplication operators by $f(x)$ on $L^2(\mathbb{R}, \mu)$. Clearly, $\mathcal{F}_1$ contains the dense subset $L^\infty(\mathbb{R}, \mu)$ of $L^2(\mathbb{R}, \mu)$, so A_x is a minimal extension. $\square$

Let A be an arbitrary self-adjoint extension of M_x on some Hilbert space $\mathcal{G}$. Let P be the projection onto the closure $\mathcal{G}_1$ of $\{f(A)1 : f \in \mathcal{F}_1\}$ in $\mathcal{G}$. Since P commutes with the spectral projections of A, by Proposition 1.15 it reduces A to a self-adjoint operator A_1 on the Hilbert space $P\mathcal{G} = \mathcal{G}_1$. Clearly, A_1 is a minimal self-adjoint extension of M_x, and $\langle E_{A_1}(\cdot)1, 1\rangle = \langle E_A(\cdot)1, 1\rangle$. Therefore, by Theorem 16.1, all representing measures for s are of the form $\mu_{A_1}(\cdot) = \langle E_{A_1}(\cdot)1, 1\rangle$.

We now introduce a notion for those solutions coming from self-adjoint extensions acting on the *same* Hilbert space $\mathcal{H}_s$.

Definition 16.3 A measure $\mu \in \mathcal{M}_s$ is called a *von Neumann solution* of the moment problem for s if $\mathbb{C}[x]$ is dense in $L^2(\mathbb{R}, \mu)$, or equivalently, if the embedding of $\mathbb{C}[x]$ into $L^2(\mathbb{R}, \mu)$ (by Lemma 16.2) extends to a unitary operator of $\mathcal{H}_s$ onto $L^2(\mathbb{R}, \mu)$.

Our next aim is to represent the multiplication operator M_x as a Jacobi operator. For this, we need a convenient orthonormal basis.

Proposition 16.3 *There exists an orthonormal basis* $\{P_n(x) : n \in \mathbb{N}_0\}$ *of the unitary space* $(\mathbb{C}[x], \langle \cdot, \cdot \rangle_s)$ *such that* degree $P_n = n$ *and the leading coefficient of* P_n *is positive for all* $n \in \mathbb{N}_0$. *The basis* $\{P_n : n \in \mathbb{N}_0\}$ *is uniquely determined by these properties. Moreover,* $P_n \in \mathbb{R}[x]$ *for all* $n \in \mathbb{N}_0$.

Proof For the existence, it suffices to apply the Gram–Schmidt procedure to the vector space basis $\{1, x, x^2, \ldots\}$ of the unitary space $(\mathbb{C}[x], \langle \cdot, \cdot \rangle_s)$. Since the scalar product is real for polynomials of $\mathbb{R}[x]$, we obtain an orthonormal sequence $P_n \in \mathbb{R}[x]$. Upon multiplying by -1 if necessary, the leading coefficient of P_n becomes positive. The uniqueness assertion follows by a simple induction argument. $\square$

Definition 16.4 The polynomials $P_n, n \in \mathbb{N}_0$, are called *orthogonal polynomials of first kind*, or simply *orthogonal polynomials*, associated with the positive definite sequence s.

The next proposition is a basic fact about zeros of orthogonal polynomials.

Proposition 16.4 *For* $n \in \mathbb{N}$, *the polynomial* $P_n(x)$ *has* n *distinct real zeros.*

Proof Let $n \in \mathbb{N}_0$. Let $\lambda_1, \ldots, \lambda_k$ be the real points, where P_n changes sign, and put $r(x) := (x - \lambda_1) \ldots (x - \lambda_k)$. If there is no such point, we set $r = 1$. Then $r P_n$ does not change sign on $\mathbb{R}$, so $q(x) := \tau r(x) P_n(x) \geq 0$ on $\mathbb{R}$ for $\tau = 1$ or $\tau = -1$.

We prove that $k = n$. Assume to the contrary that $k < n$. By Theorem 16.1 there exists a solution, say μ, of the moment problem for s. Because r is in the linear span of $P_0, \ldots, P_k$ and P_n is orthogonal to these polynomials, we obtain

$$\int_{\mathbb{R}} q(x) \, d\mu(x) = \tau \int_{\mathbb{R}} r(x) P_n(x) \, d\mu(x) = \tau \langle r, P_n \rangle_s = 0.$$

Since $q(x) \geq 0$ on $\mathbb{R}$, this implies that μ has a finite support. But then s is not positive definite, which contradicts our standing assumption. Thus, $k = n$.

Since P_n has degree n and changes sign as $\lambda_1, \ldots, \lambda_n$, P_n has n distinct real zeros $\lambda_1, \ldots, \lambda_n$. $\square$

Now we represent the operator M_x in terms of the orthonormal basis $\{P_n\}$.

Proposition 16.5 *There are numbers $a_n > 0$ and $b_n \in \mathbb{R}$ for $n \in \mathbb{N}_0$ such that*

$$M_x P_n \equiv x P_n(x) = a_n P_{n+1}(x) + b_n P_n(x) + a_{n-1} P_{n-1}(x), \quad n \in \mathbb{N}_0, \quad (16.5)$$

where we set $a_{-1} = 1$ and $P_{-1}(x) = 0$. In particular,

$$P_0(x) = 1, \qquad P_1(x) = a_0^{-1}(x - b_0),$$
$$P_2(x) = a_0^{-1} a_1^{-1}(x - b_0)(x - b_1) - a_0 a_1^{-1}.$$

Proof Let $n \in \mathbb{N}_0$. By Proposition 16.3, $x P_n(x)$ has degree $n + 1$, and $\{P_0, \ldots, P_{n+1}\}$ is a basis of the vector space of real polynomials of degree less than or equal to $n + 1$. Hence, there are real numbers c_{nk} such that $x P_n(x) = \sum_{k=0}^{n+1} c_{nk} P_k(x)$. Since the basis $\{P_k\}$ is orthonormal and M_x is symmetric, $c_{nk} = \langle M_x P_n, P_k \rangle_s = \langle P_n, M_x P_k \rangle_s$. Because $M_x P_k$ is in the span of $P_0, \ldots, P_{k+1}$, it follows from the latter equality that $c_{nk} = 0$ when $k + 1 < n$. Using that $c_{n-1,n}$ is real, we derive

$$c_{n,n-1} = \langle P_n, M_x P_{n-1} \rangle_s = \left\langle P_n, \sum_{k=0}^{n} c_{n-1,k} P_k \right\rangle_s = c_{n-1,n}. \quad (16.6)$$

Set $a_n = c_{n,n+1}$ and $b_n = c_{n,n}$. Then $a_{n-1} = c_{n,n-1}$ by (16.6). The preceding proves (16.5). By (16.5), a_n is obviously the quotient of the leading coefficients of P_n and P_{n+1}. Since these coefficients are positive, we have $a_n > 0$.

Clearly, $P_0(x) = 1$, since $s_0 = 1$ by assumption. The formulas for P_1 and P_2 are derived from (16.5) by simple computations. $\square$

Let $\{e_n : n \in \mathbb{N}_0\}$ be the standard orthonormal basis of the Hilbert space $l^2(\mathbb{N}_0)$ given by $e_n := (\delta_{kn})$. We define a unitary isomorphism U of $\mathcal{H}_s$ onto $l^2(\mathbb{N}_0)$ by $U P_n = e_n$. Then $T := U M_x U^{-1}$ is a symmetric operator on $l^2(\mathbb{N}_0)$ which acts by

$$T e_n = a_n e_{n+1} + b_n e_n + a_{n-1} e_{n-1}, \quad n \in \mathbb{N}_0, \quad (16.7)$$

where $e_{-1} := 0$. The domain $\mathcal{D}(T) = U(\mathbb{C}[x])$ is the linear span of vectors e_n, that is, $\mathcal{D}(T) = \mathsf{d}$. For any finite sequence $\gamma = (\gamma_n) \in \mathsf{d}$, we obtain

$$T\left(\sum_n \gamma_n e_n\right) = \sum_n \gamma_n (a_n e_{n+1} + b_n e_n + a_{n-1} e_{n-1})$$

$$= \sum_n (\gamma_{n-1} a_{n-1} + \gamma_n b_n + \gamma_{n+1} a_n) e_n$$

$$= \sum_n (a_n \gamma_{n+1} + b_n \gamma_n + a_{n-1} \gamma_{n-1}) e_n,$$

where we have set $\gamma_{-1} := 0$. That is,

$$(T\gamma)_0 = a_0 \gamma_1 + b_0 \gamma_0, \qquad (T\gamma)_n = a_n \gamma_{n+1} + b_n \gamma_n + a_{n-1} \gamma_{n-1}, \quad n \in \mathbb{N}. \quad (16.8)$$

Equation (16.8) means that the operator T acts on a sequence $\gamma \in \mathsf{d}$ by multiplication with the infinite matrix

$$\mathfrak{A} = \begin{pmatrix} b_0 & a_0 & 0 & 0 & 0 & \cdots \\ a_0 & b_1 & a_1 & 0 & 0 & \cdots \\ 0 & a_1 & b_2 & a_2 & 0 & \cdots \\ 0 & 0 & a_2 & b_3 & a_3 & \cdots \\ \cdots & \cdots & \ddots & \ddots & \ddots & \end{pmatrix}. \tag{16.9}$$

Such a matrix $\mathfrak{A}$ is called a (semi-finite) *Jacobi matrix*, and the corresponding operator T (likewise its closure) is called a *Jacobi operator*. Thus, we have shown that the multiplication operator M_x is unitarily equivalent to the Jacobi operator T. The moments s_n are recovered from the Jacobi operator T by

$$s_n = L_s\left(x^n\right) = \langle x^n 1, 1 \rangle_s = \langle (M_x)^n 1, 1 \rangle_s = \langle T^n e_0, e_0 \rangle.$$

Conversely, let a Jacobi matrix (16.9) be given, where $a_n, b_n \in \mathbb{R}$ and $a_n > 0$. We define the linear operator T with domain $\mathcal{D}(T) = \mathsf{d}$ by (16.8). Then $T\mathcal{D}(T) \subseteq \mathcal{D}(T)$. It is easily verified that $s = (s_n := \langle T^n e_0, e_0 \rangle)_{n \in \mathbb{N}_0}$ is a positive definite real sequence and that T is the Jacobi operator associated with s.

Summarizing, we have obtained a *one-to-one correspondence between positive definite sequences s and Jacobi matrices of the form* (16.9), where $b_n \in \mathbb{R}$ and $a_n > 0$ for $n \in \mathbb{N}_0$, and a *unitary equivalence between the multiplication operator M_x and the corresponding Jacobi operator T*.

Remarks Jacobi operators are discrete analogues of Sturm–Liouville operators. Indeed, we define left and right difference operators Δ_l and Δ_r by $(\Delta_l \gamma)_n = \gamma_n - \gamma_{n-1}$ and $(\Delta_r \gamma)_n = \gamma_{n+1} - \gamma_n$ and set $c_n = b_n + a_n + a_{n-1}$, where $\gamma_{-1} = a_{-1} := 0$. Then we compute

$$\left(\Delta_l(a\Delta_r \gamma)\right)_n + c_n \gamma_n = a_n \gamma_{n+1} + b_n \gamma_n + a_{n-1}\gamma_{n-1} = (T\gamma)_n.$$

This expression can be viewed as a discrete analogue of the differential operator $\mathcal{L}(f)(x) = \frac{d}{dx} a(x) \frac{d}{dx} f + c(x) f(x)$. Most notions and results of this chapter have their counterparts for Sturm–Liouville operators. We leave it to the reader to discover these and mention only a sample: The relation $\mathfrak{q}_z + I_\mu(z)\mathfrak{p}_z \in l^2(\mathbb{N}_0)$ in Proposition 16.18 below corresponds to the assertion of Proposition 15.6.

Let $\gamma = (\gamma_n)_{n \in \mathbb{N}_0}$ and $\beta = (\beta_n)_{n \in \mathbb{N}_0}$ be complex sequences. We define their *Wronskian* as the sequence $W(\gamma, \beta) = (W(\gamma, \beta)_n)_{n \in \mathbb{N}_0}$ with terms

$$W(\gamma, \beta)_n := a_n(\gamma_{n+1}\beta_n - \gamma_n \beta_{n+1}), \quad n \in \mathbb{N}_0, \tag{16.10}$$

and a linear mapping $\mathcal{T}$ of the vector space of all complex sequences by

$$(\mathcal{T}\gamma)_n = a_n \gamma_{n+1} + b_n \gamma_n + a_{n-1}\gamma_{n-1} \quad \text{for } n \in \mathbb{N}_0, \text{ where } \gamma_{-1} := 0.$$

Lemma 16.6 *Let $\gamma = (\gamma_n)$, $\beta = (\beta_n)$ be complex sequences, and $z, w \in \mathbb{C}$. Then*

$$\sum_{k=0}^{n}\left[(\mathcal{T}\gamma)_k \beta_k - \gamma_k(\mathcal{T}\beta)_k\right] = W(\gamma, \beta)_n \quad \text{for } n \in \mathbb{N}_0. \tag{16.11}$$

If $(\mathcal{T}\gamma)_k = z\gamma_k$ and $(\mathcal{T}\beta)_k = w\beta_k$ for all $k = m+1, \ldots, n$, then

$$(z - w) \sum_{k=m+1}^{n} \gamma_k \beta_k = W(\gamma, \beta)_n - W(\gamma, \beta)_m \quad \text{for } m, n \in \mathbb{N}_0, \ m < n.$$

$$(16.12)$$

Proof We prove the first identity (16.11) by computing

$$\sum_{k=0}^{n} \left[(\mathcal{T}\gamma)_k \beta_k - \gamma_k (\mathcal{T}\beta)_k \right]$$

$$= (a_0\gamma_1 + b_0\gamma_0)\beta_0 - \gamma_0(a_0\beta_1 + b_0\beta_0)$$

$$\quad + \sum_{k=1}^{n} \left[(a_k\gamma_{k+1} + b_k\gamma_k + a_{k-1}\gamma_{k-1})\beta_k - \gamma_k(a_k\beta_{k+1} + b_k\beta_k + a_{k-1}\beta_{k-1}) \right]$$

$$= a_0(\gamma_1\beta_0 - \gamma_0\beta_1) + \sum_{k=1}^{n} \left[(a_k(\gamma_{k+1}\beta_k - \gamma_k\beta_{k+1}) - a_{k-1}(\gamma_k\beta_{k-1} - \gamma_{k-1}\beta_k) \right]$$

$$= W(\gamma, \beta)_0 + \sum_{k=1}^{n} \left[W(\gamma, \beta)_k - W(\gamma, \beta)_{k-1} \right] = W(\gamma, \beta)_n.$$

Equation (16.12) is obtained by applying (16.11) to n and m and taking the difference of these sums.   □

Just as in the case of differential operators, the adjoint T^* of the symmetric operator T is the "maximal operator" acting on $l^2(\mathbb{N}_0)$ by the same formula.

Proposition 16.7 *The adjoint operator T^* is given by*

$$T^*\gamma = \mathcal{T}\gamma \quad \text{for } \gamma \in \mathcal{D}(T^*) = \left\{ \gamma \in l^2(\mathbb{N}_0) : \mathcal{T}\gamma \in l^2(\mathbb{N}_0) \right\}.$$

For $\gamma, \beta \in \mathcal{D}(T^*)$, the limit $W(\gamma, \overline{\beta})_\infty := \lim_{n \to \infty} W(\gamma, \overline{\beta})_n$ exists, and

$$\langle T^*\gamma, \beta \rangle - \langle \gamma, T^*\beta \rangle = W(\gamma, \overline{\beta})_\infty.$$

$$(16.13)$$

Proof Let $\gamma \in l^2(\mathbb{N}_0)$ be such that $\mathcal{T}\gamma \in l^2(\mathbb{N}_0)$. A straightforward computation shows that $\langle T\beta, \gamma \rangle = \langle \beta, \mathcal{T}\gamma \rangle$ for $\beta \in$ d. Therefore, $\gamma \in \mathcal{D}(T^*)$ and $T^*\gamma = \mathcal{T}\gamma$.

Conversely, let $\gamma \in \mathcal{D}(T^*)$ and $n \in \mathbb{N}_0$. Using (16.7), we derive

$$\langle e_n, T^*\gamma \rangle = \langle Te_n, \gamma \rangle = a_n\gamma_{n+1} + b_n\gamma_n + a_{n-1}\gamma_{n-1} = (T^*\gamma)_n,$$

so that $T^*\gamma = \mathcal{T}\gamma$. This proves the first assertion concerning T^*.

Further, by (16.11) we have

$$\sum_{k=0}^{n} \left[(\mathcal{T}\gamma)_k \overline{\beta}_k - \gamma_k \overline{(\mathcal{T}\beta)}_k \right] = W(\gamma, \overline{\beta})_n.$$

Since $\gamma, \beta, \mathcal{T}\gamma, \mathcal{T}\beta \in l^2(\mathbb{N}_0)$, the limit $n \to \infty$ in the preceding equality exists and we obtain Eq. (16.13).   □

Let $z \in \mathbb{C}$. Let us consider the *three-term recurrence relation*

$$(\mathcal{T}\gamma)_n \equiv a_n\gamma_{n+1} + b_n\gamma_n + a_{n-1}\gamma_{n-1} = z\gamma_n, \tag{16.14}$$

where $\gamma_{-1} := 0$, for an arbitrary complex sequence $\gamma = (\gamma_n)_{n \in \mathbb{N}_0}$. Clearly, since $a_n > 0$, if we fix two initial data γ_{k-1} and γ_k and assume that relation (16.14) is satisfied for all $n \geq k$, all terms γ_n, where $n \geq k+1$, are uniquely determined.

We set $\gamma_{-1} = 0$ and $\gamma_0 = 1$ and assume that (16.14) holds for all $n \in \mathbb{N}_0$. Comparing (16.14) with (16.5) by using that $P_{-1}(z) = 0$ and $P_0(z) = 1$, we conclude that γ_n is just the value $P_n(z)$ of the polynomial P_n from Proposition 16.3.

Let us introduce the notation

$$\mathfrak{p}_z = \big(P_0(z), P_1(z), P_2(z), \dots\big), \quad \text{where } z \in \mathbb{C}. \tag{16.15}$$

Lemma 16.8 *Let* $z \in \mathbb{C}$.

(i) $\mathcal{N}(T^* - zI) = \{0\}$ *if* $\mathfrak{p}_z \notin l^2(\mathbb{N}_0)$ *and* $\mathcal{N}(T^* - zI) = \mathbb{C}\cdot\mathfrak{p}_z$ *if* $\mathfrak{p}_z \in l^2(\mathbb{N}_0)$.
(ii) *If* $h \in \mathcal{N}(M_x^* - zI)$ *and* $\langle h, 1\rangle_s = 0$, *then* $h = 0$.

Proof (i): From Proposition 16.7 it follows that a sequence γ is in $\mathcal{N}(T^* - zI)$ if and only if $\gamma \in l^2(\mathbb{N}_0)$, $\mathcal{T}\gamma \in l^2(\mathbb{N}_0)$ and (16.14) holds for all $n \in \mathbb{N}_0$, where $\gamma_{-1} := 0$. Since $\gamma_{-1} = 0$, any solution γ of (16.14) is uniquely determined by the number γ_0, so we have $\gamma = \gamma_0 \mathfrak{p}_z$. This implies the assertions.

(ii): Passing to the unitarily equivalent operator T, the assertion says that $\gamma_0 = 0$ and $\gamma \in \mathcal{N}(T^* - zI)$ imply $\gamma = 0$. Since $\gamma = \gamma_0 \mathfrak{p}_z$, this is indeed true. $\qquad\square$

Corollary 16.9 *The symmetric operator* T *(or* M_x*) has deficiency indices* $(0,0)$ *or* $(1,1)$. *The operator* T *(or* M_x*) is essentially self-adjoint if and only if* $\mathfrak{p}_z$ *is not in* $l^2(\mathbb{N}_0)$, *or equivalently* $\sum_{n=0}^{\infty} |P_n(z)|^2 = \infty$, *for one (hence for all)* $z \in \mathbb{C}\backslash\mathbb{R}$.

Proof Since $P_n(x) \in \mathbb{R}[x]$ and hence $\overline{P_n(z)} = P_n(\overline{z})$, we have $\mathfrak{p}_z \in l^2(\mathbb{N}_0)$ if and only if $\mathfrak{p}_{\overline{z}} \in l^2(\mathbb{N}_0)$ for $z \in \mathbb{C}$. Therefore, by Lemma 16.8(i), T has deficiency indices $(0,0)$ or $(1,1)$, and T has deficiency indices $(0,0)$ if and only if $\mathfrak{p}_z \notin l^2(\mathbb{N}_0)$ for some (then for all) $z \in \mathbb{C}\backslash\mathbb{R}$. $\qquad\square$

Lemma 16.10 *If* A *and* B *are different self-adjoint extensions of the multiplication operator* M_x *on* $\mathcal{H}_s$, *then* $\langle (A - zI)^{-1}1, 1\rangle_s \neq \langle (B - zI)^{-1}1, 1\rangle_s$ *for all* $z \in \mathbb{C}\backslash\mathbb{R}$.

Proof Fix $z \in \mathbb{C}\backslash\mathbb{R}$ and assume to the contrary that

$$\big\langle (A - zI)^{-1}1, 1\big\rangle_s = \big\langle (B - zI)^{-1}1, 1\big\rangle_s. \tag{16.16}$$

Put $f := (A - zI)^{-1}1 - (B - zI)^{-1}1$. Since $A \subseteq M_x^*$ and $B \subseteq M_x^*$, we have $f \in \mathcal{D}(M_x^*)$ and

$$\big(M_x^* - zI\big)f = \big(M_x^* - z\big)(A - zI)^{-1}1 - \big(M_x^* - zI\big)(B - zI)^{-1}1 = 1 - 1 = 0,$$

so $f \in \mathcal{N}(M_x^* - zI)$. Since $\langle f, 1\rangle_s = 0$ by (16.16), Lemma 16.8(ii) yields $f = 0$.

Set $g := (A - zI)^{-1}1$. If g were in $\mathcal{D}(\overline{M_x})$, then for $h \in \mathcal{N}(M_x^* - \bar{z}I)$, we would get

$$0 = \langle h, (\overline{M_x} - zI)g \rangle_s = \langle h, (A - zI)(A - zI)^{-1}1 \rangle_s = \langle h, 1 \rangle_s,$$

so $h = 0$ again by Lemma 16.8(ii). Thus, $\mathcal{N}(M_x^* - \bar{z}I) = \{0\}$. This is a contradiction, since M_x has two different self-adjoint extensions, and hence its deficiency indices are $(1, 1)$ by Corollary 16.9. This proves that g is not in $\mathcal{D}(\overline{M_x})$.

Let S denote the restriction of A to $\mathcal{D}(\overline{M_x}) + \mathbb{C} \cdot g$. Then S is symmetric, because A is self-adjoint. Since M_x, hence $\overline{M_x}$, has deficiency indices $(1, 1)$ by Corollary 16.9 and $g \notin \mathcal{D}(\overline{M_x})$, S has deficiency indices $(0, 0)$ by Proposition 3.6. Therefore, $\overline{S}$ is self-adjoint, and hence $\overline{S} = A$. (In fact, S is closed, but this is not needed.)

Since $f = 0$ and hence $g = (B - zI)^{-1}1$, the same reasoning with B in place of A shows that $\overline{S} = B$. Thus, $A = B$, which contradicts our assumption and shows that Eq. (16.16) cannot hold. $\qquad \square$

The following theorem is the second main result of this section. It gives an operator-theoretic answer to the *uniqueness problem*.

Theorem 16.11 *The moment problem for s is determinate if and only if the multiplication operator M_x (or equivalently, the corresponding Jacobi operator T) is essentially self-adjoint.*

Proof Assume that M_x is not essentially self-adjoint. Then, by Corollary 16.9, M_x has deficiency indices $(1, 1)$, so it has at least two different self-adjoint extensions A and B on $\mathcal{H}_s$. By Theorem 16.1, $\mu_A(\cdot) = \langle E_A(\cdot)1, 1 \rangle$ and $\mu_B(\cdot) = \langle E_B(\cdot)1, 1 \rangle$ are representing measures for s. If μ_A were equal to μ_B, then $\langle (A - zI)^{-1}1, 1 \rangle_s = \langle (B - zI)^{-1}1, 1 \rangle_s$ for $z \in \mathbb{C} \backslash \mathbb{R}$ by the functional calculus. This contradicts Lemma 16.10. Thus, $\mu_A \neq \mu_B$ and s is indeterminate.

Suppose now that M_x is essentially self-adjoint. Fix $z \in \mathbb{C} \backslash \mathbb{R}$. Since M_x is essentially self-adjoint, $\mathcal{R}(M_x - zI)$ is dense in $\mathcal{H}_s$. Hence there exists a sequence $(r_n(x))$ of polynomials such that $1 = \lim_n (x - z)r_n$ in $\mathcal{H}_s$. (The sequence (r_n) may depend on the number z, but it is crucial that it is independent on any representing measure.) Let μ be an arbitrary representing measure for s. Obviously, the bounded function $\frac{1}{x-z}$ is in $L^2(\mathbb{R}, \mu) \cap L^1(\mathbb{R}, \mu)$. Using the equations $L_s(p) = \int p \, d\mu$ for $p \in \mathbb{C}[x]$ and $\|1\|_{L^2(\mathbb{R}, \mu)}^2 = s_0 = 1$ and the Hölder inequality, we derive

$$\left| I_\mu(z) - L_s(r_n) \right|^2 = \left| \int_\mathbb{R} (x - z)^{-1} \, d\mu(x) - \int_\mathbb{R} r_n(x) \, d\mu(x) \right|^2$$

$$\leq \left(\int_\mathbb{R} \left| (x - z)^{-1} - r_n(x) \right| d\mu(x) \right)^2$$

$$\leq \|1\|_{L^2(\mathbb{R}, \mu)}^2 \int_\mathbb{R} \left| (x - z)^{-1} - r_n(x) \right|^2 d\mu(x)$$

$$= \int_{\mathbb{R}} |x - z|^{-2} \big|1 - (x - z)r_n(x)\big|^2 \, d\mu(x)$$

$$\leq |\operatorname{Im} z|^{-2} \int_{\mathbb{R}} \big|1 - (x - z)r_n(x)\big|^2 \, d\mu(x)$$

$$= |\operatorname{Im} z|^{-2} \big\|1 - (x - z)r_n(x)\big\|_s^2 \to 0.$$

Therefore, $I_\mu(z) = \lim_n L_s(r_n)$ is independent on the representing measure μ. By Theorem F.2 the values of the Stieltjes transform I_μ determine the measure μ. Hence, μ is uniquely determined by s, that is, the moment problem for s is determinate. $\qquad\square$

By Proposition 3.8, the multiplication operator M_x is essentially self-adjoint if and only if the vector space $(x - z)\mathbb{C}[x]$ is dense in $\mathcal{H}_s$ for all $z \in \mathbb{C}\backslash\mathbb{R}$. However, in this particular situation the following stronger criterion is valid.

Proposition 16.12 *If there exist a number $z_0 \in \mathbb{C}\backslash\mathbb{R}$ and a sequence $(r_n(x))_{n\in\mathbb{N}}$ of polynomials $r_n \in \mathbb{C}[x]$ such that*

$$1 = \lim_{n \to \infty} (x - z_0)r_n(x) \quad \text{in } \mathcal{H}_s,$$

then M_x is essentially self-adjoint, and the moment problem for s is determinate.

Proof Fix $p \in \mathbb{C}[x]$. Since z_0 is a zero of the polynomial $p(x) - p(z_0)$, there is a polynomial $q \in \mathbb{C}[x]$ such that $p(x) - p(z_0) = (x - z_0)q(x)$. Then

$$(x - z_0)\big(q + p(z_0)r_n\big) = p - p(z_0) + p(z_0)(x - z_0)r_n$$

$$\to p - p(z_0) + p(z_0)1 = p.$$

Therefore, because $\mathbb{C}[x]$ is dense in $\mathcal{H}_s$, $\mathcal{R}(M_x - z_0 I)$ is dense in $\mathcal{H}_s$. Since M_x has equal deficiency indices by Corollary 16.9, M_x is essentially self-adjoint. $\qquad\square$

Proposition 16.13 *For a measure $\mu \in \mathcal{M}_s$ the following statements are equivalent:*

(i) *μ is von Neumann solution.*
(ii) *$f_z(x) := \frac{1}{x-z}$ is in the closure of $\mathbb{C}[x]$ in $L^2(\mathbb{R}, \mu)$ for all $z \in \mathbb{C}\backslash\mathbb{R}$.*
(iii) *$f_z(x) := \frac{1}{x-z}$ is in the closure of $\mathbb{C}[x]$ in $L^2(\mathbb{R}, \mu)$ for one $z \in \mathbb{C}\backslash\mathbb{R}$.*

Proof (i) $\to$ (ii): That μ is a von Neumann solution means that $\mathbb{C}[x]$ is dense in $L^2(\mathbb{R}, \mu)$. Hence, the function $f_z \in L^2(\mathbb{R}, \mu)$ is in the closure of $\mathbb{C}[x]$.

(ii) $\to$ (iii) is trivial.

(iii) $\to$ (i): Set $b := \operatorname{Im} z$. Let $\mathcal{G}$ denote the closure of $\mathbb{C}[x]$ in $L^2(\mathbb{R}, \mu)$. We first prove by induction on k that $f_z^k \in \mathcal{G}$ for all $k \in \mathbb{N}$. For $k = 1$, this is true by assumption. Suppose that $f_z^k \in \mathcal{G}$. Then there is sequence $(p_n)_{n\in\mathbb{N}}$ of polynomials such that $p_n \to f_z^k$ in $L^2(\mathbb{R}, \mu)$. We can write $p_n(x) = p_n(z) + (x - z)q_n(x)$ with $q_n \in \mathbb{C}[x]$. Using that $|f_z(x)| \leq |b|^{-1}$ on $\mathbb{R}$, we derive

$$\left\| f_z^{k+1} - p_n(z)(x-z)^{-1} - q_n(x) \right\|_{L^2(\mu)}$$
$$= \left\| f_z \big(f_z^k - p_n(z) - (x-z)q_n(x) \big) \right\|_{L^2(\mu)}$$
$$= \left\| f_z \big(f_z^k - p_n(x) \big) \right\|_{L^2(\mu)} \le |b|^{-1} \left\| f_z^k - p_n(x) \right\|_{L^2(\mu)} \to 0 \quad \text{as } n \to \infty.$$

Thus, $f_z^{k+1} \in \mathcal{G}$, which completes the induction proof.

Clearly, $f_z^k \in \mathcal{G}$ implies that $f_{\bar{z}}^k \in \mathcal{G}$. Hence the linear span of elements f_z^n and $f_{\bar{z}}^k$, $n, k \in \mathbb{N}_0$, is contained in $\mathcal{G}$. Using that $2bi f_z^n f_{\bar{z}}^k = f_z^n f_{\bar{z}}^{k-1} - f_z^{n-1} f_{\bar{z}}^k$ it follows by induction on $n + k$ that $f_z^n f_{\bar{z}}^k \in \mathcal{G}$ for all $n, k \in \mathbb{N}_0$. Therefore the unital $*$-algebra $\mathcal{F}$ generated by f_z is contained in $\mathcal{G}$.

We prove that $\mathbb{C}[x]$ is dense in $L^2(\mathbb{R}, \mu)$. For notational simplicity, assume that $\operatorname{Re} z = 0$. Let X denote the unit circle in $\mathbb{R}^2$. The map $\phi(x) = (u, v)$, where

$$u = \frac{2bx}{x^2 + b^2} \quad \text{and} \quad v = \frac{x^2 - b^2}{x^2 + b^2},$$

is a homeomorphism of $\mathbb{R}$ onto $X \backslash \{(0, 1)\}$ such that $2b f_z(x) - \mathrm{i} = u - \mathrm{i}v$ and $2b f_{\bar{z}}(x) + \mathrm{i} = u + \mathrm{i}v$. The latter implies that $\mathcal{F}$ is equal to the $*$-algebra $\mathcal{C}$ of all complex polynomials in real variables u, v satisfying $u^2 + v^2 = 1$. Let $C_c(\mathbb{R})$ denote the compactly supported functions of $C(\mathbb{R})$. Since $\mathcal{C}$ is dense in $C(X)$, $C_c(\mathbb{R})$ is in the closure of $\mathcal{F}$ in the uniform convergence on $\mathbb{R}$ and therefore in $L^2(\mathbb{R}, \mu)$, since $\mu(\mathbb{R}) < \infty$. Because $C_c(\mathbb{R})$ is dense in $L^2(\mathbb{R}, \mu)$, so are $\mathcal{F}$ and hence $\mathcal{G}$ and $\mathbb{C}[x]$.

Since $\mathbb{C}[x]$ is dense in $L^2(\mathbb{R}, \mu)$, μ is a von Neumann solution. $\qquad\square$

16.2 Polynomials of Second Kind

In this section we introduce another family $\{Q_n : n \in \mathbb{N}_0\}$ of polynomials which are associated with the positive definite sequence s.

We fix $z \in \mathbb{C}$ and return to the recurrence relation (16.14). Now we set $\gamma_0 = 0$ and $\gamma_1 = a_0^{-1}$ and suppose that (16.14) holds for all $n \in \mathbb{N}$. (Note that we do *not* assume that (16.14) is valid for $n = 0$.) The numbers γ_n are then uniquely determined, and we denote γ_n by $Q_n(z)$, $n \in \mathbb{N}_0$. (The same solution is obtained if we start with the initial data $\gamma_{-1} = -1$, $\gamma_0 = 0$ and require (16.14) for all $n \in \mathbb{N}_0$.)

Using relation (16.14), it follows easily by induction on n that $Q_n(z)$, $n \in \mathbb{N}$, is a polynomial in z of degree $n - 1$. We denote the corresponding sequence by

$$\mathfrak{q}_z := \big(Q_0(z), Q_1(z), Q_2(z), \ldots \big), \quad \text{where } z \in \mathbb{C}. \tag{16.17}$$

By definition, $Q_0(z) = 0$. We compute $Q_1(z) = a_0^{-1}$ and $Q_2(z) = (z - b_1)(a_0 a_1)^{-1}$.

Definition 16.5 The polynomials $Q_n(z)$, $n \in \mathbb{N}_0$, are the *polynomials of second kind* associated with the sequence s.

Lemma 16.14 $T\mathfrak{q}_z = e_0 + z\mathfrak{q}_z$ for all $z \in \mathbb{C}$.

Proof By the recurrence relation for $Q_n(z)$, $(\mathcal{T}\mathfrak{q}_z)_n = zQ_n(z) \equiv z(\mathfrak{q}_z)_n$ for $n \in \mathbb{N}$. Using that $\gamma_0 = 0$ and $\gamma_1 = a_0^{-1}$, we compute the zero component $(\mathcal{T}\mathfrak{q}_z)_0$ by

$$(\mathcal{T}\mathfrak{q}_z)_0 = a_0\gamma_1 + b_0\gamma_0 = a_0a_0^{-1} + 0 = 1 + z\gamma_0 = 1 + zQ_0(z). \qquad \square$$

From the defining relations for $Q_n(z)$ it follows that the polynomials $\widetilde{P}_n(z) := a_0 Q_{n+1}(z)$ satisfy the recurrence relation (16.14) with a_n replaced by $\widetilde{a}_n := a_{n+1}$ and b_n by $\widetilde{b}_n := b_{n+1}$ and the initial data $\widetilde{P}_{-1} = 0$ and $\widetilde{P}_0 = 1$. Therefore, the polynomials $\widetilde{P}_n(z) = a_0 Q_{n+1}(z)$, $n \in \mathbb{N}_0$, are the orthogonal polynomials of first kind with respect to the Jacobi matrix

$$\widetilde{\mathfrak{A}} = \begin{pmatrix} b_1 & a_1 & 0 & 0 & 0 & \dots \\ a_1 & b_2 & a_2 & 0 & 0 & \dots \\ 0 & a_2 & b_3 & a_3 & 0 & \dots \\ 0 & 0 & a_3 & b_4 & a_4 & \dots \\ \dots & \dots & & \ddots & \ddots & \ddots \end{pmatrix}. \qquad (16.18)$$

That is, $\widetilde{P}_n(z)$ are orthogonal polynomials for the moment sequence $\widetilde{s} = (\widetilde{s}_n)$ which is defined by $\widetilde{s}_n = \langle \widetilde{T}^n e_0, e_0 \rangle$, where $\widetilde{T}$ is the Jacobi operator corresponding to the Jacobi matrix $\widetilde{\mathfrak{A}}$. Therefore, Proposition 16.4 applies to $\widetilde{P}_n$ and yields the following:

Corollary 16.15 *For $n \in \mathbb{N}$, the polynomial $Q_{n+1}(z)$ has n distinct real zeros.*

Corollary 16.16 *Let $n \in \mathbb{N}$ and $z, z' \in \mathbb{C}$. If $|\operatorname{Im} z| \leq |\operatorname{Im} z'|$, then*

$$\left| P_n(z) \right| \leq \left| P_n(z') \right| \quad and \quad \left| Q_n(z) \right| \leq \left| Q_n(z') \right|. \qquad (16.19)$$

Proof Let r be a polynomial of degree n which has n distinct real zeros, say $x_1, \dots, x_n$. Then $r(z) = c(z - x_1) \dots (z - x_n)$ for some $c \in \mathbb{C}$. Hence, it is easily seen that $|r(z)| \leq |r(z')|$ when $|\operatorname{Im} z| \leq |\operatorname{Im} z'|$. By Proposition 16.4 and Corollary 16.15 this applies to P_n and Q_{n+1} and gives the assertion. $\qquad \square$

The next proposition contains another description of these polynomials Q_n. Let $r(x) = \sum_{k=0}^{n} \gamma_k x^k \in \mathbb{C}[x]$, $n \in \mathbb{N}$, be a polynomial. For any fixed $z \in \mathbb{C}$,

$$\frac{r(x) - r(z)}{x - z} = \sum_{k=0}^{n} \gamma_k \frac{x^k - z^k}{x - z} = \sum_{k=1}^{n} \sum_{l=0}^{k-1} \gamma_k z^{k-l} x^l$$

is also a polynomial in x, so we can apply the functional L_s to it. We shall write $L_{s,x}$ to indicate that x is the corresponding variable.

Proposition 16.17 $Q_n(z) = L_{s,x}\left(\frac{P_n(x) - P_n(z)}{x - z}\right)$ *for $n \in \mathbb{N}_0$ and $z \in \mathbb{C}$.*

Proof Let us denote the polynomial at the right-hand side by $r_n(z)$. From the recurrence relation (16.5) we obtain for $n \in \mathbb{N}$,

$$a_n \frac{P_{n+1}(x) - P_{n+1}(z)}{x - z} + b_n \frac{P_n(x) - P_n(z)}{x - z} + a_{n-1} \frac{P_{n-1}(x) - P_{n-1}(z)}{x - z}$$

$$= \frac{x P_n(x) - z P_n(z)}{x - z} = z \frac{P_n(x) - P_n(z)}{x - z} + P_n(x).$$

Applying the functional $L_{s,x}$ to this identity and using the orthogonality relation $0 = \langle P_n, 1 \rangle_s = L_{s,x}(P_n)$ for $n \geq 1$, we get

$$a_n r_{n+1}(z) + b_n r_n(z) + a_{n-1} r_{n-1}(z) = z r_n(z) \quad \text{for } n \in \mathbb{N}.$$

Since $P_1(x) = a_0^{-1}(x - b_0)$ and $P_0 = 1$, we have $r_0(z) = L_{s,x}(0) = 0$ and

$$r_1(z) = L_{s,x}\left(\frac{P_1(x) - P_1(z)}{x - z} \right) = L_{s,x}\left(\frac{x - z}{a_0(x - z)} \right) = a_0^{-1} L_{s,x}(1) = a_0^{-1}.$$

This shows that the sequence $(r_n(z))$ satisfies the same recurrence relation and initial data as $(Q_n(z))$. Thus, $r_n(z) = Q_n(z)$ for $n \in \mathbb{N}_0$. □

In what follows we will often use the function f_z and the Stieltjes transform I_μ (see Appendix F) of a finite Borel measure μ on $\mathbb{R}$. Recall that they are defined by

$$f_z(x) := \frac{1}{x - z} \quad \text{and} \quad I_\mu(z) = \int_{\mathbb{R}} \frac{1}{x - z} d\mu(x), \quad z \in \mathbb{C} \backslash \mathbb{R}. \quad (16.20)$$

Proposition 16.18 *Let $\mu \in \mathcal{M}_s$. For $z \in \mathbb{C} \backslash \mathbb{R}$ and $n \in \mathbb{N}_0$,*

$$\langle f_z, P_n \rangle_{L^2(\mathbb{R}, \mu)} = Q_n(z) + I_\mu(z) P_n(z), \quad (16.21)$$

$$\left\| q_z + I_\mu(z) p_z \right\|^2 = \sum_{n=0}^{\infty} \left| Q_n(z) + I_\mu(z) P_n(z) \right|^2 \leq \frac{\operatorname{Im} I_\mu(z)}{\operatorname{Im} z}. \quad (16.22)$$

In particular, $q_z + I_\mu(z) p_z \in l^2(\mathbb{N}_0)$. Moreover, we have equality in the inequality of (16.22) if and only if the function f_z is in $\mathcal{H}_s$.

Proof Clearly, the bounded function $f_z(x) = \frac{1}{x - z}$ is in $L^2(\mathbb{R}, \mu)$. We compute

$$\langle f_z, P_n \rangle_{L^2(\mathbb{R}, \mu)} = \int_{\mathbb{R}} \frac{P_n(x) - P_n(z)}{x - z} d\mu(x) + \int_{\mathbb{R}} \frac{P_n(z)}{x - z} d\mu(x)$$

$$= L_{s,x}\left(\frac{P_n(x) - P_n(z)}{x - z} \right) + P_n(z) I_\mu(z) = Q_n(z) + I_\mu(z) P_n(z)$$

which proves (16.21). Here the equality before last holds, since μ is a representing measure for s, and the last equality follows from Proposition 16.17.

The equality in (16.22) is merely the definition of the norm in $l^2(\mathbb{N}_0)$. Since $(\mathbb{C}[x], \langle \cdot, \cdot \rangle_s)$ is a subspace of $L^2(\mathbb{R}, \mu)$, $\{P_n : n \in \mathbb{N}_0\}$ is an orthonormal subset of $L^2(\mathbb{R}, \mu)$. The inequality in (16.22) is just Bessel's inequality for the Fourier coefficients of f_z with respect to this orthonormal set, because

$$\| f_z \|_{L^2(\mathbb{R}, \mu)}^2 = \int \frac{1}{|x - z|^2} d\mu(x) = \int \frac{1}{z - \bar{z}} \left(\frac{1}{x - z} - \frac{1}{x - \bar{z}} \right) d\mu(x) = \frac{\operatorname{Im} I_\mu(z)}{\operatorname{Im} z}.$$

Clearly, equality in Bessel's inequality holds if and only if f_z belongs to the closed subspace generated by the polynomials P_n, that is, $f_z \in \mathcal{H}_s$. $\square$

The following polynomials and facts will be used in the next section.

Lemma 16.19 *For $z, w \in \mathbb{C}$ and $k \in \mathbb{N}_0$, we have*

$$A_k(z, w) := (z - w) \sum_{n=0}^{k} Q_n(z) Q_n(w) = a_k \big(Q_{k+1}(z) Q_k(w) - Q_k(z) Q_{k+1}(w) \big),$$

$$B_k(z, w) := -1 + (z - w) \sum_{n=0}^{k} P_n(z) Q_n(w)$$
$$= a_k \big(P_{k+1}(z) Q_k(w) - P_k(z) Q_{k+1}(w) \big),$$

$$C_k(z, w) := 1 + (z - w) \sum_{n=0}^{k} Q_n(z) P_n(w) = a_k \big(Q_{k+1}(z) P_k(w) - Q_k(z) P_{k+1}(w) \big),$$

$$D_k(z, w) := (z - w) \sum_{n=0}^{k} P_n(z) P_n(w) = a_k \big(P_{k+1}(z) P_k(w) - P_k(z) P_{k+1}(w) \big).$$

Proof All four identities are easily derived from Eq. (16.12). As a sample, we verify the identity for $B_k(z, w)$. Recall that the sequences $\mathfrak{p}_z = (P_n(z))$ and $\mathfrak{q}_w = (Q_n(w))$ satisfy the relations

$$(\mathcal{T} \mathfrak{p}_z)_n = z P_n(z) = z(\mathfrak{p}_z)_n, \qquad (\mathcal{T} \mathfrak{q}_w)_n = w Q_n(w) = w(\mathfrak{q}_w)_n$$

for $n \in \mathbb{N}$. Hence, (16.12) applies with $m = 0$. We have $Q_0(w) = 0$ and

$$W(\mathfrak{p}_z, \mathfrak{q}_w)_0 = a_0 \big(P_1(z) Q_0(w) - P_0(z) Q_1(w) \big) = a_0 \big(0 - a_0^{-1} \big) = -1$$

by (16.10). Using these facts and Eqs. (16.12) and (16.10), we derive

$$(z - w) \sum_{n=0}^{k} P_n(z) Q_n(w) = (z - w) \sum_{n=1}^{k} P_n(z) Q_n(w)$$
$$= W(\mathfrak{p}_z, \mathfrak{q}_w)_k - W(\mathfrak{p}_z, \mathfrak{q}_w)_0$$
$$= a_k \big(P_{k+1}(z) Q_k(w) - P_k(z) Q_{k+1}(w) \big) + 1. \qquad \square$$

Lemma 16.20 *For any $z, w \in \mathbb{C}$, we have*

$$A_k(z, w) D_k(z, w) - B_k(z, w) C_k(z, w) = 1, \tag{16.23}$$

$$D_k(z, 0) B_k(w, 0) - B_k(z, 0) D_k(w, 0) = -D_k(z, w). \tag{16.24}$$

Proof Inserting the four identities from Lemma 16.19, we compute

$$A_k(z, w) D_k(z, w) - B_k(z, w) C_k(z, w)$$
$$= a_k^2 \big(P_{k+1}(z) Q_k(z) - P_k(z) Q_{k+1}(z) \big) \big(P_{k+1}(w) Q_k(w) - Q_{k+1}(w) P_k(w) \big)$$
$$= B_k(z, z) B_k(w, w) = (-1)(-1) = 1.$$

Likewise, we derive

$$D_k(z, 0)B_k(w, 0) - B_k(z, 0)D_k(w, 0)$$
$$= a_k^2\big(P_{k+1}(z)P_k(w) - P_k(z)P_{k+1}(w)\big)\big(P_{k+1}(0)Q_k(0) - P_k(0)Q_{k+1}(0)\big)$$
$$= D_k(z, w)B_k(0, 0) = -D_k(z, w). \qquad \square$$

16.3 The Indeterminate Hamburger Moment Problem

Throughout this section we assume that s is an *indeterminate* moment sequence.

Lemma 16.21 *For any $z \in \mathbb{C}$, the series $\sum_{n=0}^{\infty} |P_n(z)|^2$ and $\sum_{n=0}^{\infty} |Q_n(z)|^2$ converge. The sums are uniformly bounded on compact subsets of the complex plane.*

Proof Because the moment problem for s is indeterminate, it has at least two different solutions μ and ν. The corresponding Stieltjes transforms I_μ and I_ν are holomorphic functions on $\mathbb{C}\backslash\mathbb{R}$. Since $\mu \neq \nu$, they do not coincide by Theorem F.2. Hence, the set $\mathcal{Z} := \{z \in \mathbb{C}\backslash\mathbb{R} : I_\mu(z) = I_\nu(z)\}$ has no accumulation point in $\mathbb{C}\backslash\mathbb{R}$.

Let M be a compact subset of $\mathbb{C}$. We choose $b > 0$ such that $|z| \leq b$ for all $z \in M$ and the line segment $L := \{z \in \mathbb{C} : \operatorname{Im} z = b, |\operatorname{Re} z| \leq b\}$ does not intersect $\mathcal{Z}$. By the first condition and Corollary 16.16, the suprema of $|P_n(z)|$ and $|Q_n(z)|$ over M are less than or equal to the corresponding suprema over L. Hence, it suffices to prove the uniform boundedness of both sums on the set L.

Suppose that $z \in L$. Applying (16.22), we obtain

$$\big|I_\mu(z) - I_\nu(z)\big|^2 \sum_n \big|P_n(z)\big|^2$$

$$= \sum_n \big|(Q_n(z) + I_\mu(z)P_n(z)) - (Q_n(z) + I_\nu(z)P_n(z))\big|^2$$

$$\leq 2\sum_n \big|Q_n(z) + I_\mu(z)P_n(z)\big|^2 + 2\sum_n \big|Q_n(z) + I_\nu(z)P_n(z)\big|^2$$

$$\leq 2b^{-1}\big(\operatorname{Im} I_\mu(z) + \operatorname{Im} I_\nu(z)\big) \leq 2b^{-1}\big(\big|I_\mu(z)\big| + \big|I_\nu(z)\big|\big).$$

Since the function $|I_\mu(z) - I_\nu(z)|$ has a positive infimum on L (because L has a positive distance from the set $\mathcal{Z}$) and $I_\mu(z)$ and $I_\nu(z)$ are bounded on L, the preceding inequality implies that the sum $\sum_n |P_n(z)|^2$ is finite and uniformly bounded on L.

Using once more (16.22) and proceeding in a similar manner, we derive

$$\sum_n \big|Q_n(z)\big|^2 \leq 2\sum_n \big|Q_n(z) + I_\mu(z)P_n(z)\big|^2 + 2\big|I_\mu(z)\big|^2 \sum_n \big|P_n(z)\big|^2$$

$$\leq 2b^{-1}\big|I_\mu(z)\big| + 2\big|I_\mu(z)\big|^2 \sum_n \big|P_n(z)\big|^2$$

for $z \in L$. Hence, the boundedness of the sum $\sum_n |P_n(z)|^2$ on L implies the boundedness of $\sum_n |Q_n(z)|^2$ on L. $\square$

Combining Lemma 16.21, Proposition 16.7, and Lemma 16.14, we therefore obtain the following:

Corollary 16.22 *If s is an indeterminate moment sequence, then for all (!) numbers $z \in \mathbb{C}$, the sequences $\mathfrak{p}_z = (P_n(z))_{n \in \mathbb{N}_0}$ and $\mathfrak{q}_z = (Q_n(z))_{n \in \mathbb{N}_0}$ are in $\mathcal{D}(T^*)$,*

$$T^* \mathfrak{p}_z = z \mathfrak{p}_z \quad and \quad T^* \mathfrak{q}_z = e_0 + z \mathfrak{q}_z. \tag{16.25}$$

Lemma 16.23 *For any sequence $c = (c_n) \in l^2(\mathbb{N}_0)$, the equations*

$$f(z) = \sum_{n=0}^{\infty} c_n P_n(z) \quad and \quad g(z) = \sum_{n=0}^{\infty} c_n Q_n(z) \tag{16.26}$$

define entire functions $f(z)$ and $g(z)$ on the complex plane.

Proof We carry out the proof for $f(z)$. Since $(P_n(z)) \in l^2(\mathbb{N}_0)$ by Lemma 16.21 and $(c_n) \in l^2(\mathbb{N}_0)$, the series $f(z)$ converges for all $z \in \mathbb{C}$. We have

$$\left| f(z) - \sum_{n=0}^{k} c_n P_n(z) \right|^2 = \left| \sum_{n=k+1}^{\infty} c_n P_n(z) \right|^2 \leq \left(\sum_{n=k+1}^{\infty} |c_n|^2 \right) \left(\sum_{n=0}^{\infty} |P_n(z)|^2 \right)$$

for $k \in \mathbb{N}$ and $z \in \mathbb{C}$. Therefore, since $(c_n) \in l^2(\mathbb{N}_0)$ and the sum $\sum_n |P_n(z)|^2$ is bounded on compact sets (by Lemma 16.21), it follows that $\sum_{n=0}^{k} c_n P_n(z) \to f(z)$ as $k \to \infty$ uniformly on compact subsets of $\mathbb{C}$. Hence, $f(z)$ is holomorphic on the whole complex plane. $\square$

From Lemmas 16.21 and 16.23 we conclude that

$$A(z, w) := (z - w) \sum_{n=0}^{\infty} Q_n(z) Q_n(w),$$

$$B(z, w) := -1 + (z - w) \sum_{n=0}^{\infty} P_n(z) Q_n(w),$$

$$C(z, w) := 1 + (z - w) \sum_{n=0}^{\infty} Q_n(z) P_n(w),$$

$$D(z, w) := (z - w) \sum_{n=0}^{\infty} P_n(z) P_n(w)$$

are entire functions in each of the complex variables z and w.

Definition 16.6 The four functions $A(z, w), B(z, w), C(z, w), D(z, w)$ are called the *Nevanlinna functions* associated with the indeterminate moment sequence s.

These four functions are a fundamental tool for the study of the indeterminate moment problem. It should be emphasized that they depend only on the indeterminate moment sequence s, but not on any representing measure for s.

We shall see by Theorems 16.28 and 16.34 below that the entire functions

$$A(z) := A(z, 0), \qquad B(z) := B(z, 0), \qquad C(z) := C(z, 0), \qquad D(z) := D(z, 0)$$

will enter in the parameterization of solutions. Often these four entire functions $A(z)$, $B(z)$, $C(z)$, $D(z)$ are called the *Nevanlinna functions* associated with s.

Some technical facts on these functions are collected in the next proposition.

Proposition 16.24 *Let $z, w \in \mathbb{C}$. Then we have:*

(i) $A(z, w)D(z, w) - B(z, w)C(z, w) = 1$.

(ii) $D(z, 0)B(w, 0) - B(z, 0)D(w, 0) = -D(z, w)$.

(iii) $A(z, w) = (z - w)\langle \mathfrak{q}_z, \mathfrak{q}_{\overline{w}} \rangle_s$, $D(z, w) = (z - w)\langle \mathfrak{p}_z, \mathfrak{p}_{\overline{w}} \rangle_s$,
   $B(z, w) + 1 = (z - w)\langle \mathfrak{p}_z, \mathfrak{q}_{\overline{w}} \rangle_s$, $C(z, w) - 1 = (z - w)\langle \mathfrak{q}_z, \mathfrak{p}_{\overline{w}} \rangle_s$.

(iv) $\mathrm{Im}(B(z)\overline{D(z)}) = \mathrm{Im}\, z \|\mathfrak{p}_z\|_s^2$.

(v) $D(z) \neq 0$ and $D(z)t + B(z) \neq 0$ *for* $z \in \mathbb{C}\backslash\mathbb{R}$ *and* $t \in \mathbb{R}$.

(vi) $D(z)\zeta + B(z) \neq 0$ *for all* $z, \zeta \in \mathbb{C}$, $\mathrm{Im}\, z > 0$ *and* $\mathrm{Im}\, \zeta \geq 0$.

Proof (i) and (ii) follow from Lemma 16.20 by passing to the limit $k \to \infty$, while (iii) is easily obtained from the definitions of A, B, C, D and $\mathfrak{p}_z, \mathfrak{q}_w$.

(iv): Using (ii) and the second equality from (iii), we compute

$$B(z)\overline{D(z)} - \overline{B(z)}D(z) = D(z, \overline{z}) = (z - \overline{z})\|\mathfrak{p}_z\|_s^2.$$

(v): Suppose that $z \in \mathbb{C}\backslash\mathbb{R}$. Then $\mathrm{Im}\, z\|\mathfrak{p}_z\|_s^2 \neq 0$, and hence $D(z) \neq 0$ by (iv). Assume to the contrary that $D(z)t + B(z) = 0$ for some $z \in \mathbb{C}\backslash\mathbb{R}$ and $t \in \mathbb{R}$. Then we have $-t = B(z)D(z)^{-1}$, and hence

$$0 = \mathrm{Im}\big(B(z)D(z)^{-1}|D(z)|^2\big) = \mathrm{Im}\big(B(z)\overline{D(z)}\big) = \mathrm{Im}\, z\|\mathfrak{p}_z\|_s^2$$

by (iv), which is a contradiction.

(vi) follows in a similar manner as the last assertion of (v). $\square$

Proposition 16.25 *If $\mu \in \mathcal{M}_s$ is a von Neumann solution, then*

$$I_\mu(z) = -\frac{A(z, w) + I_\mu(w)C(z, w)}{B(z, w) + I_\mu(w)D(z, w)} \quad \text{for } z, w \in \mathbb{C}\backslash\mathbb{R}. \tag{16.27}$$

In particular, formula (16.27) determines *all* values of $I_\mu(z)$ on $\mathbb{C}\backslash\mathbb{R}$, provided that *one* fixed value $I_\mu(w)$ is given.

Proof Since μ is a von Neumann solution, $\mathcal{H}_s \cong L^2(\mathbb{R}, \mu)$. Hence, $\{P_n : n \in \mathbb{N}_0\}$ is an orthonormal basis of $L^2(\mathbb{R}, \mu)$, so by (16.21) for all $z, w \in \mathbb{C}\backslash\mathbb{R}$, we have

$$f_z = \sum_{n=0}^{\infty} \big(Q_n(z) + I_\mu(z)P_n(z)\big) P_n, \qquad f_{\overline{w}} = \sum_{n=0}^{\infty} \big(Q_n(\overline{w}) + I_\mu(\overline{w})P_n(\overline{w})\big) P_n.$$

Using these formulas, the Parseval identity, and Lemma 16.19, we derive

$$I_\mu(z) - I_\mu(w) = (z - w) \int \frac{1}{x-z} \frac{1}{x-w} \, d\mu(x) = (z-w)\langle f_z, f_{\overline{w}}\rangle_\mu$$

$$= (z-w) \sum_{n=0}^{\infty} \big(Q_n(z) + I_\mu(z)P_n(z)\big)\big(\overline{Q_n(\overline{w}) + I_\mu(\overline{w})P_n(\overline{w})}\big)$$

$$= (z-w) \sum_{n=0}^{\infty} \big(Q_n(z) + I_\mu(z)P_n(z)\big)\big(Q_n(w) + I_\mu(w)P_n(w)\big)$$

$$= A(z,w) + I_\mu(z)\big(B(z,w) + 1\big) + I_\mu(w)\big(C(z,w) - 1\big)$$
$$+ I_\mu(z)I_\mu(w)D(z,w),$$

from which we obtain (16.27).   □

Next, we construct a *boundary triplet* for the adjoint operator T^*. For this, it is crucial to know that the sequences $\mathfrak{p}_0$ and $\mathfrak{q}_0$ are in $l^2(\mathbb{N}_0)$ and so in $\mathcal{D}(T^*)$.

Note that $P_0(z) = 1$, $Q_0(z) = 0$, $T^*\mathfrak{p}_0 = 0$, and $T^*\mathfrak{q}_0 = e_0$ by Corollary 16.22. Using these relations and the fact that the operator $\overline{T}$ is symmetric, we compute

$$\langle T^*(\gamma + c_0\mathfrak{q}_0 + c_1\mathfrak{p}_0), \beta' + c_0'\mathfrak{q}_0 + c_1'\mathfrak{p}_0\rangle$$
$$- \langle \gamma + c_0\mathfrak{q}_0 + c_1\mathfrak{p}_0, T^*\big(\beta' + c_0'\mathfrak{q}_0 + c_1'\mathfrak{p}_0\big)\rangle$$
$$= \langle \overline{T}\gamma + c_0 e_0, \beta' + c_0'\mathfrak{q}_0 + c_1'\mathfrak{p}_0\rangle - \langle \gamma + c_0\mathfrak{q}_0 + c_1\mathfrak{p}_0, \overline{T}\beta' + c_0'e_0\rangle$$
$$= \langle \gamma, c_0'e_0\rangle + \langle c_0 e_0, \beta' + c_1'\mathfrak{p}_0\rangle - \langle c_0 e_0, \beta'\rangle - \langle \gamma + c_1\mathfrak{p}_0, c_0'e_0\rangle$$
$$= \gamma_0\overline{c_0'} + c_0\big(\overline{\beta_0'} + \overline{c_1'}\big) - c_0\overline{\beta_0'} - (\gamma_0 + c_1)\overline{c_0'} = c_0\overline{c_1'} - c_1\overline{c_0'} \qquad (16.28)$$

for $\gamma, \beta, \gamma', \beta' \in \mathcal{D}(\overline{T})$ and $c_0, c_1, c_0', c_1' \in \mathbb{C}$. From this equality we conclude that

$$\mathcal{D}(T^*) = \mathcal{D}(\overline{T}) \dotplus \mathbb{C} \cdot \mathfrak{q}_0 \dotplus \mathbb{C} \cdot \mathfrak{p}_0. \qquad (16.29)$$

Indeed, since T has deficiency indices $(1,1)$, we have $\dim \mathcal{D}(T^*)/\mathcal{D}(\overline{T}) = 2$ by Proposition 3.7. Therefore it suffices to show $\mathfrak{q}_0$ and $\mathfrak{p}_0$ are linearly independent modulo $\mathcal{D}(\overline{T})$. If $c_0\mathfrak{q}_0 + c_1\mathfrak{p}_0 \in \mathcal{D}(\overline{T})$, then

$$\langle T^*(c_0\mathfrak{q}_0 + c_1\mathfrak{p}_0), c_0'\mathfrak{q}_0 + c_1'\mathfrak{p}_0\rangle = \langle c_0\mathfrak{q}_0 + c_1\mathfrak{p}_0, T^*\big(c_0'\mathfrak{q}_0 + c_1'\mathfrak{p}_0\big)\rangle$$

for arbitrary (!) numbers $c_0', c_1' \in \mathbb{C}$. By (16.28) this implies that $c_0 = c_1 = 0$, whence (16.29) follows.

By (16.29) each vector $\varphi \in \mathcal{D}(T^*)$ is of the form $\varphi = \gamma + c_0\mathfrak{q}_0 + c_1\mathfrak{p}_0$, where $\gamma \in \mathcal{D}(\overline{T})$ and $c_0, c_1 \in \mathbb{C}$ are uniquely determined by φ. Then we define

$$\mathcal{K} = \mathbb{C}, \qquad \Gamma_0(\gamma + c_0\mathfrak{q}_0 + c_1\mathfrak{p}_0) = c_0, \qquad \Gamma_1(\gamma + c_0\mathfrak{q}_0 + c_1\mathfrak{p}_0) = -c_1. \quad (16.30)$$

From Definition 14.2 and formulas (16.28) and (16.30) we immediately obtain the following:

Proposition 16.26 $(\mathcal{K}, \Gamma_0, \Gamma_1)$ *is a boundary triplet for* T^*.

For the description of von Neumann solutions (Theorem 16.28 below), we need

Corollary 16.27 *The self-adjoint extensions of the operator T on the Hilbert space $\mathcal{H}_s \cong l^2(\mathbb{N}_0)$ are the operators $T_{(t)} = T^* \upharpoonright \mathcal{D}(T_{(t)})$, $t \in \mathbb{R} \cup \{\infty\}$, where*

$$\mathcal{D}(T_{(t)}) = \mathcal{D}(\overline{T}) \dotplus \mathbb{C} \cdot (\mathfrak{q}_0 + t\mathfrak{p}_0) \quad \text{for } t \in \mathbb{R}, \ \mathcal{D}(T_{(\infty)}) = \mathcal{D}(\overline{T}) \dotplus \mathbb{C} \cdot \mathfrak{p}_0. \quad (16.31)$$

Proof By Example 14.7 the self-adjoint extensions of T on $l^2(\mathbb{N}_0)$ are the operators T_t, $t \in \mathbb{R} \cup \{\infty\}$, where $\mathcal{D}(T_t) = \{\varphi \in \mathcal{D}(T^*) : t\Gamma_0\varphi = \Gamma_1\varphi\}$. Set $T_{(t)} := T_{-t}$. Then, by (16.30), the domain $\mathcal{D}(T_{(t)})$ has the form (16.31). $\qquad\square$

For $t \in \mathbb{R} \cup \{\infty\}$, we set $\mu_t(\cdot) := \langle E_t(\cdot)e_0, e_0 \rangle$, where E_t denotes the spectral measure of the self-adjoint operator $T_{(t)}$.

Theorem 16.28 *The measures μ_t, where $t \in \mathbb{R} \cup \{\infty\}$, are precisely the von Neumann solutions of the moment problem for the indeterminate moment sequence s. For $z \in \mathbb{C}\backslash\mathbb{R}$, we have*

$$I_{\mu_t}(z) \equiv \int_{\mathbb{R}} \frac{1}{x - z} d\mu_t(x) = \langle (T_{(t)} - zI)^{-1}e_0, e_0 \rangle = -\frac{A(z) + tC(z)}{B(z) + tD(z)},$$
$$(16.32)$$

where for $t = \infty$, the fraction at the right-hand side has to be set equal to $-\frac{C(z)}{D(z)}$.

The following lemma is used in the proof of Theorem 16.28.

Lemma 16.29 $\lim_{y \in \mathbb{R}, y \to 0} I_{\mu_t}(yi) = \lim_{y \in \mathbb{R}, y \to 0} \langle (T_{(t)} - yiI)^{-1}e_0, e_0 \rangle = t$ *for $t \in \mathbb{R}$.*

Proof Let $t \in \mathbb{R}$. Since $\mathcal{N}(T_{(t)}) \subseteq \mathcal{N}(T^*) = \mathbb{C} \cdot \mathfrak{p}_0$ by Lemma 16.8(i) and $\mathfrak{p}_0 \notin \mathcal{D}(T_{(t)})$ by (16.31) and (16.29), we have $\mathcal{N}(T_{(t)}) = \{0\}$, so the operator $T_{(t)}$ is invertible. From (16.25) we obtain $T_{(t)}(\mathfrak{q}_0 + t\mathfrak{p}_0) = T^*(\mathfrak{q}_0 + t\mathfrak{p}_0) = e_0$. Hence, $e_0 \in \mathcal{D}(T_{(t)}^{-1})$ and $T_{(t)}^{-1}e_0 = \mathfrak{q}_0 + t\mathfrak{p}_0$. Since $e_0 \in \mathcal{D}(T_{(t)}^{-1})$, the function $h(\lambda) = \lambda^{-2}$ is μ_t-integrable (by (5.11)), and hence $\mu_t(\{0\}) = 0$. Set

$$h_y(\lambda) = \left| (\lambda - yi)^{-1} - \lambda^{-1} \right|^2 \quad \text{for } y \in (-1, 1).$$

Then $h_y(\lambda) \to 0$ μ_t-a.e. on $\mathbb{R}$ as $y \to 0$ and $|h_y(\lambda)| \leq h(\lambda)$ for $y \in (-1, 1)$. Hence Lebesgue's dominated convergence theorem (Theorem B.1) applies and yields

$$\left\| (T_{(t)} - yiI)^{-1}e_0 - T_{(t)}^{-1}e_0 \right\|^2 = \int_{\mathbb{R}} \left| (\lambda - yi)^{-1} - \lambda^{-1} \right|^2 d\mu_t(\lambda) \to 0 \quad \text{as } y \to 0.$$

This proves that $\lim_{y \to 0}(T_{(t)} - yiI)^{-1}e_0 = T_{(t)}^{-1}e_0 = \mathfrak{q}_0 + t\mathfrak{p}_0$. Therefore,

$$\lim_{y \to 0} I_{\mu_t}(yi) = \lim_{y \to 0} \langle (T_{(t)} - yiI)^{-1}e_0, e_0 \rangle = \langle \mathfrak{q}_0 + t\mathfrak{p}_0, e_0 \rangle = t. \qquad\square$$

Proof of Theorem 16.28 By Corollary 16.27 the operators $T_{(t)}$, $t \in \mathbb{R} \cup \{\infty\}$, are the self-adjoint extensions of T on $\mathcal{H}_s$. Hence, the measures μ_t are the von Neumann solutions.

The first equality of (16.32) follows at once from the definition of μ_t. We now prove the second equality. For this, we essentially use formula (16.27). Note that $D(z) \neq 0$ and $D(z)t + B(z) \neq 0$ for $z \in \mathbb{C}\backslash\mathbb{R}$ and $t \in \mathbb{R}$ by Proposition 16.24(v).

First, let $t \in \mathbb{R}$. Set $w = yi$ with $y \in \mathbb{R}$ in (16.27). Letting $y \to 0$, we obtain the second equality of (16.32) by Lemma 16.29.

Now let $t = \infty$. We apply (16.27) to $\mu = \mu_\infty$. Fixing w, we see that $I_{\mu_\infty}(z)$ is a quotient of two entire functions and hence meromorphic. Put $w = yi$ with $y \in \mathbb{R}$, $y \neq 0$, in (16.27) and let $y \to 0$. Since each meromorphic function is a quotient of two entire functions which have no common zeros, $\lim_{y \to 0} I_{\mu_\infty}(yi)$ is either ∞ or a number $s \in \mathbb{C}$ depending on whether or not the denominator function vanishes at the origin. In the first case the right-hand side of (16.27) becomes $-C(z)D(z)^{-1}$.

To complete the proof, it suffices to show that the second case is impossible. Since $\overline{I_{\mu_\infty}(yi)} = I_{\mu_\infty}(-yi)$, the number s is real. As $y \to 0$, the right-hand side of (16.27) tends to the right-hand side of (16.32) for $t = s$. Comparing the corresponding left-hand sides yields $I_{\mu_\infty}(z) = I_{\mu_s}(z)$. Hence,

$$\langle (T_{(\infty)} - zI)^{-1} e_0, e_0 \rangle = I_{\mu_\infty}(z) = I_{\mu_s}(z) = \langle (T_{(s)} - zI)^{-1} e_0, e_0 \rangle, \quad z \in \mathbb{C}\backslash\mathbb{R}.$$

Since $T_{(\infty)}$ and $T_{(s)}$ are different self-adjoint extensions of T on $\mathcal{H}_s$, the latter contradicts Lemma 16.10. (Here we freely used the fact that M_x and T are unitarily equivalent by a unitary which maps 1 onto e_0.)   □

Corollary 16.30 *Let $t \in \mathbb{R} \cup \{\infty\}$. Then $I_{\mu_t}(z)$ is a meromorphic function, and the self-adjoint operator $T_{(t)}$ has a purely discrete spectrum consisting of eigenvalues of multiplicity one. The eigenvalues of $T_{(t)}$ are precisely the zeros of the entire function $B(z) + tD(z)$ for $t \in \mathbb{R}$ resp. $D(z)$ for $t = \infty$. They are distinct for different values of $t \in \mathbb{R} \cup \{\infty\}$. Each real number is an eigenvalue of some $T_{(t)}$.*

Proof Since $A(z), B(z), C(z), D(z)$ are entire functions, (16.32) implies that $I_{\mu_t}(z)$ is meromorphic. By Theorem F.5 the support of μ_t is the set of poles of the meromorphic function I_{μ_t}. Therefore, since $\text{supp}\,\mu_t = \sigma(T_{(t)})$ by (5.25), $T_{(t)}$ has a purely discrete spectrum. By Corollary 5.19 all eigenvalues of $T_{(t)}$ have multiplicity one.

Recall that by Proposition 16.24(i) we have

$$A(z)D(z) - B(z)C(z) = 1 \quad \text{for } z \in \mathbb{C}. \tag{16.33}$$

The functions $A + tC$ and $B + tD$ have no common zero z, since otherwise

$$A(z)D(z) - B(z)C(z) = \big(-tC(z)\big)D(z) - \big(-tD(z)\big)C(z) = 0.$$

By (16.33), C and D have no common zero. The eigenvalues of $T_{(t)}$ are the poles of $I_{\mu_t}(z)$. Since numerator and denominator in (16.32) have no common zero as just shown, the poles of $I_{\mu_t}(z)$ are precisely the zeros of the denominator.

Let $z \in \mathbb{R}$. If z were an eigenvalue of $T_{(t_1)}$ *and* $T_{(t_2)}$ with $t_1 \neq t_2$, then we would have $B(z) = D(z) = 0$, which contradicts (16.33). If $D(z) = 0$, then z is an eigenvalue of $T_{(\infty)}$. If $D(z) \neq 0$, then $t := -B(z)D(z)^{-1}$ is real (because B and D have real coefficients and z is real), and z is an eigenvalue of $T_{(t)}$. □

Another important tool of the indeterminate moment theory are the Weyl circles.

Definition 16.7 For $z \in \mathbb{C}\backslash\mathbb{R}$, the *Weyl circle* K_z is the closed circle in the complex plane with radius ρ_z and center C_z given by

$$
\rho_z := \frac{1}{|z - \overline{z}|\,\|p_z\|_s^2} = \frac{z - \overline{z}}{|z - \overline{z}|} \frac{1}{D(z, \overline{z})},
$$

$$
C_z := -\frac{(z - \overline{z})^{-1} + \langle q_z, p_z \rangle_s}{\|p_z\|_s^2} = -\frac{C(z, \overline{z})}{D(z, \overline{z})}.
$$

The two equalities in this definition follow easily from Proposition 16.24(iii).

The proof of the next proposition shows that in the indeterminate case inequality (16.22) means that the number $I_\mu(z)$ belongs to the Weyl circle K_z.

Proposition 16.31 *Let* $\mu \in \mathcal{M}_s$. *Then the number* $I_\mu(z)$ *lies in the Weyl circle* K_z *for each* $z \in \mathbb{C}\backslash\mathbb{R}$. *The measure* μ *is a von Neumann solution if and only if* $I_\mu(z)$ *belongs to the boundary* ∂K_z *for one (hence for all)* $z \in \mathbb{C}\backslash\mathbb{R}$.

Proof We fix $z \in \mathbb{C}\backslash\mathbb{R}$ and abbreviate $\zeta = I_\mu(z)$. Then (16.22) says that

$$
\|q_z\|_s^2 + \zeta\overline{\langle q_z, p_z \rangle_s} + \overline{\zeta}\langle q_z, p_z \rangle_s + |\zeta|^2\|p_z\|_s^2 \equiv \|q_z + \zeta p_z\|_s^2 \leq \frac{\zeta - \overline{\zeta}}{z - \overline{z}}.
$$

$$(16.34)$$

The inequality in (16.34) can be rewritten as

$$
|\zeta|^2\|p_z\|_s^2 + \zeta\left(\overline{\langle q_z, p_z \rangle_s + (z - \overline{z})^{-1}}\right) + \overline{\zeta}\left(\langle q_z, p_z \rangle_s + (z - \overline{z})^{-1}\right) + \|q_z\|_s^2 \leq 0.
$$

The latter inequality is equivalent to

$$
\|p_z\|_s^2\left|\zeta + \|p_z\|_s^{-2}\left(\langle q_z, p_z \rangle_s + (z - \overline{z})^{-1}\right)\right|^2
$$
$$
\leq \|p_z\|_s^{-2}\left|\langle q_z, p_z \rangle_s + (z - \overline{z})^{-1}\right|^2 - \|q_z\|_s^2
$$

and hence to

$$
\left|\zeta + \|p_z\|_s^{-2}\left(\langle q_z, p_z \rangle_s + (z - \overline{z})^{-1}\right)\right|^2
$$
$$
\leq \|p_z\|_s^{-4}\left(\left|\langle q_z, p_z \rangle_s + (z - \overline{z})^{-1}\right|^2 - \|p_z\|_s^2\|q_z\|_s^2\right).
$$

This shows that ζ lies in a circle with center C_z given above and radius

$$
\tilde{\rho}_z = \|p_z\|_s^{-2}\sqrt{\left|\langle q_z, p_z \rangle_s + (z - \overline{z})^{-1}\right|^2 - \|p_z\|_s^2\|q_z\|_s^2}, \qquad (16.35)
$$

provided that the expression under the square root is nonnegative. Using Proposition 16.24(iii), the relation $\overline{C(z, \overline{z})} = -B(z, \overline{z})$, and Proposition 16.24(i), we get

$$\tilde{\rho}_z = \frac{z - \overline{z}}{D(z, \overline{z})} \sqrt{\frac{|C(z, \overline{z})|^2}{|z - \overline{z}|^2} - \frac{A(z, \overline{z})D(z, \overline{z})}{(z - \overline{z})^2}}$$

$$= \frac{z - \overline{z}}{D(z, \overline{z})} \sqrt{\frac{-C(z, \overline{z})B(z, \overline{z})}{|z - \overline{z}|^2} + \frac{A(z, \overline{z})D(z, \overline{z})}{|z - \overline{z}|^2}}$$

$$= \frac{z - \overline{z}}{D(z, \overline{z})} \frac{1}{|z - \overline{z}|} = \frac{1}{\|p_z\|_s^2 |z - \overline{z}|}.$$

Thus, $\tilde{\rho}_z$ is equal to the radius ρ_z of the Weyl circle, and we have proved that $\zeta \in K_z$.

The preceding proof shows that $I_\mu(z) = \zeta \in \partial K_z$ if and only if we have equality in the inequality (16.34) and hence in (16.22). The latter is equivalent to the relation $f_z \in \mathcal{H}_s$ by Proposition 16.18 and so to the fact that μ is a von Neumann solution by Proposition 16.13. This holds for fixed and hence for all $z \in \mathbb{C} \backslash \mathbb{R}$. □

Let $z, w \in \mathbb{C}$. Since

$$A(z, w)D(z, w) - B(z, w)C(z, w) = 1$$

by Proposition 16.24(i), the fractional linear transformation $H_{z,w}$ defined by

$$\xi = H_{z,w}(\zeta) := -\frac{A(z, w) + \zeta C(z, w)}{B(z, w) + \zeta D(z, w)} \tag{16.36}$$

is a bijection of the extended complex plane $\overline{\mathbb{C}} = \mathbb{C} \cup \{\infty\}$ with inverse given by

$$\zeta = H_{z,w}^{-1}(\xi) = -\frac{A(z, w) + \xi B(z, w)}{C(z, w) + \xi D(z, w)}. \tag{16.37}$$

We will use only the transformations $H_z := H_{z,0}$. Some properties of the general transformations $H_{z,w}$ can be found in Exercise 13. Set $\overline{\mathbb{R}} := \mathbb{R} \cup \{\infty\}$ and recall that $\mathbb{C}_+ = \{z \in \mathbb{C} : \operatorname{Im} z > 0\}$.

Lemma 16.32

(i) H_z is a bijection of $\overline{\mathbb{R}}$ onto the boundary $\partial K_z = \{I_{\mu_t}(z) : t \in \overline{\mathbb{R}}\}$ of the Weyl circle K_z for $z \in \mathbb{C} \backslash \mathbb{R}$.

(ii) H_z is a bijection of $\mathbb{C}_+$ onto the interior $\mathring{K}_z$ of the Weyl circle K_z for $z \in \mathbb{C}_+$.

(iii) $K_z \subseteq \mathbb{C}_+$ for $z \in \mathbb{C}_+$.

Proof (i): By Theorem 16.28, H_z maps $\overline{\mathbb{R}}$ on the set $\{I_{\mu_t}(z) : t \in \overline{\mathbb{R}}\}$. Since $I_{\mu_t}(z)$ is in ∂K_z by Proposition 16.31, H_z maps $\overline{\mathbb{R}}$ into ∂K_z. But fractional linear transformations map generalized circles bijectively on generalized circles. Hence, H_z maps $\overline{\mathbb{R}}$ onto ∂K_z.

(ii): From (i) it follows that H_z is a bijection of either the upper half-plane or the lower half-plane on the interior of K_z. It therefore suffices to find one point $\xi \in \overset{\circ}{K}_z$ for which $H_z^{-1}(\xi) \in \mathbb{C}_+$. Since $I_{\mu_0}(z), I_{\mu_\infty}(z) \in \partial K_z$ by (i),

$$\xi := \left(I_{\mu_0}(z) + I_{\mu_\infty}(z)\right)/2 = \left(-A(z)B(z)^{-1} - C(z)D(z)^{-1}\right)/2 \in \overset{\circ}{K}_z.$$

Here the second equality follows from Theorem 16.28. Inserting this expression into (16.37), we easily compute

$$H_z^{-1}(\xi) = B(z)D(z)^{-1} = |D(z)|^{-2} B(z)\overline{D(z)}.$$

Hence, $H_z^{-1}(\xi) \in \mathbb{C}_+$ by Proposition 16.24(iv), since $z \in \mathbb{C}_+$.

(iii): Since $z \in \mathbb{C}_+$, we have $I_{\mu_t}(z) \in \mathbb{C}_+ \cap \partial K_z$ by (i). Hence, $K_z \subseteq \mathbb{C}_+$.   □

16.4 Nevanlinna Parameterization of Solutions

First we prove a classical result due to Hamburger and Nevanlinna that characterizes solutions of the moment problem in terms of their Stieltjes transforms.

Proposition 16.33

(i) *If $\mu \in \mathcal{M}_s$ and $n \in \mathbb{N}_0$, then*

$$\lim_{y \in \mathbb{R}, y \to \infty} y^{n+1} \left(I_\mu(\mathrm{i}y) + \sum_{k=0}^{n} \frac{s_k}{(\mathrm{i}y)^{k+1}}\right) = 0, \qquad (16.38)$$

where for fixed n, the convergence is uniform on the set $\mathcal{M}_s$.

(ii) *If $\mathcal{I}(z)$ is a Nevanlinna function such that (16.38) is satisfied (with I_μ replaced by $\mathcal{I}$) for all $n \in \mathbb{N}_0$, then there exists a measure $\mu \in \mathcal{M}_s$ such that $\mathcal{I}(z) = I_\mu(z)$ for $z \in \mathbb{C}\backslash\mathbb{R}$.*

Proof (i): First, we rewrite the sum in Eq. (16.38) by

$$\sum_{k=0}^{n} \frac{s_k}{(\mathrm{i}y)^{k+1}} = \sum_{k=0}^{n} \frac{(-\mathrm{i})^{k+1}}{y^{k+1}} \int_\mathbb{R} x^k \, d\mu(x) = -\mathrm{i}\frac{1}{y^{n+1}} \int_\mathbb{R} \sum_{k=0}^{n} (-\mathrm{i}x)^k y^{n-k} \, d\mu(x)$$

$$= -\mathrm{i}\frac{1}{y^{n+1}} \int_\mathbb{R} \frac{(-\mathrm{i}x)^{n+1} - y^{n+1}}{-\mathrm{i}x - y} \, d\mu(x)$$

$$= \frac{1}{y^{n+1}} \int_\mathbb{R} \frac{(-\mathrm{i}x)^{n+1} - y^{n+1}}{x - \mathrm{i}y} \, d\mu(x). \qquad (16.39)$$

Therefore, since $|x - \mathrm{i}y|^{-1} \le |y|^{-1}$ for $x, y \in \mathbb{R}$, $y \neq 0$, we obtain

$$\left| y^{n+1} \left(I_\mu(iy) + \sum_{k=0}^{n} \frac{s_k}{(iy)^{k+1}} \right) \right|$$

$$= \left| \int_\mathbb{R} \frac{y^{n+1}}{x - iy} d\mu(x) + \int_\mathbb{R} \frac{(-ix)^{n+1} - y^{n+1}}{x - iy} d\mu(x) \right|$$

$$= \left| \int_\mathbb{R} \frac{(-ix)^{n+1}}{x - iy} d\mu(x) \right| \le |y|^{-1} \int_\mathbb{R} |x|^{n+1} d\mu(x) \le |y|^{-1} c_n,$$

where $c_n := s_{n+1}$ if n is odd and $c_n := s_n + s_{n+2}$ if n is even. Since c_n does not depend on the measure μ, we conclude that (16.38) holds uniformly on the set $\mathcal{M}_s$.

(ii): Condition (16.38) for $n = 0$ yields $\lim_{y \to \infty} y \mathcal{I}(iy) = i s_0$. Therefore, by Theorem F.4, the Nevanlinna function $\mathcal{I}$ is the Stieltjes transform I_μ of a finite positive Borel measure μ on $\mathbb{R}$. Since $\mu(\mathbb{R}) < \infty$, Lebesgue's dominated convergence theorem implies that

$$s_0 = \lim_{y \to \infty} -iy \mathcal{I}(iy) = \lim_{y \to \infty} -iy I_\mu(iy) = \lim_{y \to \infty} \int_\mathbb{R} \frac{-iy}{x - iy} d\mu(x) = \int_\mathbb{R} d\mu(x).$$

The main part of the proof is to show that $\mu \in \mathcal{M}_s$, that is, the nth moment of μ exists and is equal to s_n for all $n \in \mathbb{N}_0$. We proceed by induction on n. For $n = 0$, this was just proved. Let $n \in \mathbb{N}$ and assume that μ has the moments $s_0, \ldots, s_{2n-2}$. Then, by the preceding proof of (i), formula (16.39) is valid with n replaced by $2n - 2$. We use this formula in the case $2n - 2$ to derive the second equality below and compute

$$(iy)^{2n+1} \left(\sum_{k=0}^{2n} \frac{s_k}{(iy)^{k+1}} + I_\mu(iy) \right)$$

$$= s_{2n} + iy s_{2n-1} + i^{2n+1} y^2 \left(y^{2n-1} \sum_{k=0}^{2n-2} \frac{s_k}{(iy)^{k+1}} + y^{2n-1} I_\mu(iy) \right)$$

$$= s_{2n} + iy s_{2n-1} + i^{2n+1} y^2 \left(\int_\mathbb{R} \frac{(-ix)^{2n-1} - y^{2n-1}}{x - iy} d\mu(x) + \int_\mathbb{R} \frac{y^{2n-1}}{x - iy} d\mu(x) \right)$$

$$= s_{2n} + iy s_{2n-1} - y^2 \int_\mathbb{R} \frac{x^{2n-1}}{x - iy} d\mu(x)$$

$$= s_{2n} - \int_\mathbb{R} \frac{x^{2n}}{(x/y)^2 + 1} d\mu(x) + iy \left(s_{2n-1} - \int_\mathbb{R} \frac{x^{2n-1}}{(x/y)^2 + 1} d\mu(x) \right).$$

By assumption (16.38) the expression in the first line converges to zero as $y \to \infty$. Considering the real part and using Lebesgue's monotone convergence theorem, we conclude that

$$s_{2n} = \lim_{y \to \infty} \int_\mathbb{R} \frac{x^{2n}}{(x/y)^2 + 1} d\mu(x) = \int_\mathbb{R} x^{2n} d\mu(x) < \infty. \tag{16.40}$$

The imaginary part, hence the imaginary part divided by y, converges also to zero as $y \to \infty$. Since

$$\frac{|x|^{2n-1}}{(x/y)^2 + 1} \le |x|^{2n-1} \le 1 + x^{2n}$$

and $1 + x^{2n}$ is μ-integrable by (16.40), the dominated convergence theorem yields

$$s_{2n-1} = \lim_{y \to \infty} \int_{\mathbb{R}} \frac{x^{2n-1}}{(x/y)^2 + 1} \, d\mu(x) = \int_{\mathbb{R}} x^{2n-1} \, d\mu(x). \tag{16.41}$$

By (16.40) and (16.41) the induction proof is complete. $\qquad\square$

Let $\mathfrak{N}$ denote the set of Nevanlinna functions (see Appendix F). If we identify $t \in \mathbb{R}$ with the constant function equal to t, then $\mathbb{R}$ becomes a subset of $\mathfrak{N}$. Set $\overline{\mathfrak{N}} := \mathfrak{N} \cup \{\infty\}$.

The main result in this section is the following theorem of *R. Nevanlinna*.

Theorem 16.34 *Let s be an indeterminate moment sequence. There is a one-to-one correspondence between functions $\phi \in \overline{\mathfrak{N}}$ and measures $\mu \in \mathcal{M}_s$ given by*

$$I_{\mu_\phi}(z) \equiv \int_{\mathbb{R}} \frac{1}{x - z} \, d\mu_\phi(x) = -\frac{A(z) + \phi(z)C(z)}{B(z) + \phi(z)D(z)} \equiv H_z\big(\phi(z)\big), \quad z \in \mathbb{C}_+. \tag{16.42}$$

Proof Let $\mu \in \mathcal{M}_s$. If μ is a von Neumann solution, then by Theorem 16.28 there exists a $t \in \overline{\mathbb{R}}$ such that $I_\mu(z) = H_z(t)$ for all $z \in \mathbb{C}_+$.

Assume that μ is not a von Neumann solution and define $\phi(z) := H_z^{-1}(I_\mu(z))$. Let $z \in \mathbb{C}_+$. Then $I_\mu(z) \in \mathring{K}_z$ by Proposition 16.31, and hence $\phi(z) = H_z^{-1}(I_\mu(z)) \in \mathbb{C}_+$ by Lemma 16.32(ii). That is, $\phi(\mathbb{C}_+) \subseteq \mathbb{C}_+$. We show that $C(z) + I_\mu(z)D(z) \neq 0$. Indeed, otherwise $I_\mu(z) = -C(z)D(z)^{-1} = H_z(\infty) \in \partial K_z$ by Lemma 16.32(i), which contradicts the fact that $I_\mu(z) \in \mathring{K}_z$. Thus, $\phi(z)$ is the quotient of two holomorphic functions on $\mathbb{C}_+$ with nonvanishing denominator function. Therefore, ϕ is holomorphic on $\mathbb{C}_+$. This proves that $\phi \in \mathfrak{N}$. By the definition of ϕ we have $H_z(\phi(z)) = I_\mu(z)$ on $\mathbb{C}_+$, that is, (16.42) holds.

Conversely, suppose that $\phi \in \overline{\mathfrak{N}}$. If $\phi = t \in \overline{\mathbb{R}}$, then by Theorem 16.28 there is a von Neumann solution $\mu_t \in \mathcal{M}_s$ such that $I_{\mu_t}(z) = H_z(t)$.

Suppose now that ϕ is not in $\overline{\mathbb{R}}$. Let $z \in \mathbb{C}_+$ and define $\mathcal{I}(z) = H_z(\phi(z))$. Then $\phi(z) \in \mathbb{C}_+$ and hence $\mathcal{I}(z) = H_z(\phi(z)) \in \mathring{K}_z \subseteq \mathbb{C}_+$ by Lemma 16.32, (ii) and (iii). From Proposition 16.24(vi) it follows that $B(z) + \phi(z)D(z) \neq 0$. Therefore, $\mathcal{I}$ is a holomorphic function on $\mathbb{C}_+$ with values in $\mathbb{C}_+$, that is, $\mathcal{I} \in \mathfrak{N}$.

To prove that $\mathcal{I} = I_\mu$ for some $\mu \in \mathcal{M}_s$, we want to apply Proposition 16.33(ii). For this, we have to check that condition (16.38) is fulfilled. Indeed, by Proposition 16.33(i), given $\varepsilon > 0$, there exists $Y_\varepsilon > 0$ such that

$$\left| y^{n+1} \left(I_\mu(iy) + \sum_{k=0}^{n} \frac{s_k}{(iy)^{k+1}} \right) \right| < \varepsilon \quad \text{for all } y \geq Y_\varepsilon \tag{16.43}$$

and for all $\mu \in \mathcal{M}_s$. (Here it is crucial that Y_ε does not depend on μ and that (16.43) is valid for all measures $\mu \in \mathcal{M}_s$!) Fix a $y \geq Y_\varepsilon$. Since $\mathcal{I}(iy) = H_{iy}(\phi(iy))$ is in the interior of the Weyl circle K_{iy} by Lemma 16.32(ii), $\mathcal{I}(iy)$ is a convex combination of two points from the boundary ∂K_{iy}. By Lemma 16.32(i), all points of ∂K_{iy} are of the form $I_{\mu_t}(iy)$ for some $t \in \overline{\mathbb{R}}$. Since (16.43) holds for all $I_{\mu_t}(iy)$ and $\mathcal{I}(iy)$ is a convex combination of values $I_{\mu_t}(iy)$, (16.43) remains valid if $I_\mu(iy)$ is replaced by $\mathcal{I}(iy)$. This shows that $\mathcal{I}(z)$ fulfills the assumptions of Proposition 16.33(ii), so that $\mathcal{I} = I_\mu$ for some measure $\mu \in \mathcal{M}_s$.

By Theorem F.2 the positive measure μ_ϕ is uniquely determined by the values of its Stieltjes transform I_{μ_ϕ} on $\mathbb{C}_+$. Therefore, since I_{μ_ϕ} and $\phi \in \overline{\mathfrak{N}}$ correspond to each other uniquely by the relation $I_\mu(z) = H_z(\phi(z))$ on $\mathbb{C}_+$, (16.42) gives a one-to-one correspondence between $\mu \in \mathcal{M}_s$ and $\phi \in \overline{\mathfrak{N}}$. □

Theorem 16.34 contains a complete parameterization of the solution set $\mathcal{M}_s$ in the indeterminate case in terms of the set $\overline{\mathfrak{N}}$. Here the subset $\mathbb{R} \cup \{\infty\}$ of $\overline{\mathfrak{N}}$ corresponds to the von Neumann solutions, or equivalently, to the self-adjoint extensions of T on the Hilbert space $\mathcal{H}_s \cong l^2(\mathbb{N}_0)$, while the nonconstant Nevanlinna functions correspond to self-adjoint extension on a strictly larger Hilbert space.

Fix $z \in \mathbb{C}_+$. Then the values $I_\mu(z)$ for all von Neumann solutions $\mu \in \mathcal{M}_s$ fill the boundary ∂K_z of the Weyl circle, while the numbers $I_\mu(z)$ for all other solutions $\mu \in \mathcal{M}_s$ lie in the interior $\mathring{K}_z$. By taking convex combinations of von Neumann solutions it follows that each number of $\mathring{K}_z$ is of the form $I_\mu(z)$ for some representing measure $\mu \in \mathcal{M}_s$.

16.5 Exercises

1. Let $s = (s_n)_{n \in \mathbb{N}_0}$ be a positive semi-definite real sequence.
   a. Suppose that s is *not* positive definite. Show that s is a determinate moment sequence and that the representing measure is supported by a finite set.
   b. Show that $s_0 = 0$ implies that $s_n = 0$ for all $n \in \mathbb{N}$.
   c. Suppose that $s_0 > 0$. Show that $\tilde{s} := (\frac{s_n}{s_0})_{n \in \mathbb{N}_0}$ is a moment sequence.
2. Let s be a moment sequence that has a representing measure supported by a compact set. Prove that s is determinate.
3. Let s be a positive definite real sequence.
   a. Show that s has a representing measure supported by a compact set if and only if $\liminf_{n \to \infty} s_{2n}^{\frac{1}{2n}} < \infty$.
   b. Let $s_0 = 1$. Show that s has a representing measure supported by the interval $[-1, 1]$ if and only if $\liminf_{n \to \infty} s_{2n} < \infty$.
4. Let $s = (s_n)_{n \in \mathbb{N}_0}$ be a positive definite real sequence. If $K \subseteq \mathbb{R}$ is a closed set, we say that s is a K-*moment sequence* if there exists a measure $\mu \in \mathcal{M}_s$ such that supp $\mu \subseteq K$. For $k \in N$, we define a sequence $E^k s$ by $(E^k s)_n := s_{n+k}$.
   a. (*Stieltjes moment problem*)
   Show that s is a $[0, +\infty)$-moment sequence if and only if Es is positive definite.
   Hint: Show that the symmetric operator M_x is positive if Es is positive definite and consider a positive self-adjoint extension of M_x.
   b. (*Hausdorff moment problem*)
   Show that s is a $[0, 1]$-moment sequence if and only if $Es - E^2 s$ is positive definite.

   c. Let $a, b \in \mathbb{R}$, $a < b$. Show that s is an $[a, b]$-moment sequence if and only if the sequence $bEs + aEs - E^2s - abs$ is positive definite.

   d. Show that the preceding assertions remain if "positive definite" is replaced by "positive semi-definite."

5. Let s be a positive definite real sequence. Prove that the following statements are equivalent:

   (i) The moment problem for s is determinate.

   (ii) There exists a number $z \in \mathbb{C} \backslash \mathbb{R}$ such that $\mathfrak{p}_z \notin l^2(\mathbb{N}_0)$ and $\mathfrak{q}_z \notin l^2(\mathbb{N}_0)$.

   (iii) There is a representing measure μ for s such that $\mathbb{C}[x]$ is dense in $L^2(\mathbb{R}, \nu)$, where $d\nu(x) = (x^2 + 1) \, d\mu(x)$.

6. Prove the "Plücker identity" for the Wronskian defined by (16.10):

$$W(\alpha, \beta)_n W(\gamma, \delta)_n - W(\alpha, \gamma)_n W(\beta, \delta)_n + W(\alpha, \delta)_n W(\beta, \gamma)_n = 0, \quad n \in \mathbb{N}_0,$$

where $\alpha, \beta, \gamma, \delta$ are arbitrary complex sequences.

Note: Compare this with the Plücker identity occurring in Example 15.4.

7. Prove that each moment sequence s satisfying the *Carleman condition*

$$\sum_{n=0}^{\infty} s_{2n}^{-1/2n} = +\infty$$

is determinate.

Hint: Show that 1 is a quasi-analytic vector for $M_x \cong T$. Apply Theorem 7.18, with $A = M_x$ and $\mathcal{Q} = \mathbb{C}[M_x]$, and Theorem 16.11.

8. Let μ be a positive Borel measure on $\mathbb{R}$ such that $\int e^{\varepsilon x^2} d\mu(x) < \infty$ for some $\varepsilon > 0$. Show that $\mu \in \mathcal{M}(\mathbb{R})$ and that μ is determinate.

Hint: Prove that 1 is an analytic vector for M_x.

9. Let $z_1, z_2, z_3, z_4 \in \mathbb{C}$. Prove the following identities:

$$A(z_1, z_2)D(z_3, z_4) - B(z_3, z_2)C(z_1, z_4) + B(z_3, z_1)C(z_2, z_4) = 0,$$

$$A(z_1, z_2)C(z_3, z_4) + A(z_3, z_1)C(z_2, z_4) + A(z_2, z_3)C(z_1, z_4) = 0,$$

$$D(z_1, z_2)B(z_3, z_4) + D(z_3, z_1)B(z_2, z_4) + D(z_2, z_3)B(z_1, z_4) = 0.$$

Hint: Verify the corresponding identities for A_k, B_k, C_k, D_k. Use Lemma 16.19.

10. Let $(\mathcal{K}, \Gamma_0, \Gamma_1)$ be the boundary triplet from Proposition 16.26.

   a. Show that $\Gamma_0 \mathfrak{p}_z = D(z)$ and $\Gamma_1 \mathfrak{p}_z = B(z)$ for $z \in \mathbb{C}$.

   Hint: Verify that $\mathfrak{p}_z + B(z)\mathfrak{p}_0 - D(z)\mathfrak{q}_0 \in \mathcal{D}(\overline{T})$ for any $z \in \mathbb{C}$.

   b. Show that $\gamma(z)1 = D(z)^{-1}\mathfrak{p}_z$ and $M(z) = B(z)D(z)^{-1}$ for $z \in \mathbb{C} \backslash \mathbb{R}$.

   c. What does the Krein–Naimark resolvent formula say in this case?

11. Prove the assertion of Corollary 16.27 without using boundary triplets.

   Hint: Verify (16.29). Use the Cayley transform or symmetric subspaces of $\mathcal{D}(T^*)$.

12. Let $T_{(t)}$, $t \in \mathbb{R} \cup \{\infty\}$, be the self-adjoint operator from Corollary 16.27, and let λ be an eigenvalue of $T_{(t)}$. Prove that $\mathfrak{p}_\lambda$ is an eigenvector for λ and $\mu_t(\{\lambda\}) = \|\mathfrak{p}_\lambda\|^{-2}$.

13. Prove that the fractional linear transformation $H_{z,w}$ defined by (16.36) maps
   a. ∂K_w onto ∂K_z if $\mathrm{Im}\, z \neq 0$ and $\mathrm{Im}\, w \neq 0$,
   b. ∂K_w onto $\overline{\mathbb{R}}$ if $\mathrm{Im}\, z = 0$ and $\mathrm{Im}\, w \neq 0$,
   c. $\overline{\mathbb{R}}$ onto ∂K_z if $\mathrm{Im}\, z \neq 0$ and $\mathrm{Im}\, w = 0$,
   d. $\overline{\mathbb{R}}$ onto $\overline{\mathbb{R}}$ if $\mathrm{Im}\, z = \mathrm{Im}\, w = 0$.

16.6 Notes to Part VI

Chapter 13:
The self-adjoint extension theory based on the Cayley transform is essentially due to von Neumann [vN1]. The theory of positive self-adjoint extensions using the Krein transform was developed by Krein [Kr3]. For strictly positive symmetric operators, the existence of a smallest positive extension was already shown by von Neumann [vN1]. The Ando–Nishio theorem was proved in [AN]. The factorization construction of the Krein–von Neumann extension and the Friedrichs extension was found by Prokaj, Sebestyen, and Stochel [SeS, PS].

There is an extensive literature on self-adjoint extension theory and on positive self-adjoint extensions of symmetric operators and relations; see, for instance, [Br, Cd2, CdS, DdS, AN, AHS, AS, AT, HMS].

Chapter 14:
Linear relations as multivalued linear operators have been first studied by Arens [As]; see, e.g., [Cd2] and [HS].

The notion of an abstract boundary triplet was introduced by Kochubei [Ko] and Bruk [Bk] (see, e.g., [GG]), but the idea can be traced back to Calkin [Ck]. Gamma fields and Weyl functions associated with boundary triplets have been invented and investigated by Derkach and Malamud [DM]. The celebrated Krein–Naimark resolvent formula was obtained by Naimark [Na1] and Krein [Kr1, Kr2] for finite deficiency indices and by Saakyan [Sa] for infinite deficiency indices. The Krein–Birman–Vishik theory has its origin in the work of the three authors [Kr3, Bi, Vi]. Theorem 14.25 goes back to Krein [Kr3]. Theorem 14.24 is essentially due to Birman [Bi], while Theorem 14.22 was obtained by Malamud [Ma].

Boundary triplets and their applications have been studied in many papers such as [DM, Ma, MN, BMN, Po, BGW, BGP].

Chapter 15:
The limit point–limit circle theory of Sturm–Liouville operators is due Weyl [W2]. It was developed many years before the self-adjoint extension theory appeared. A complete spectral theory of Sturm–Liouville operators was obtained by Titchmarsh [Tt] and Kodaira [Kd1, Kd2]. This theory can be found in [DS, JR], and [Na2]. An interesting collection of surveys on the Sturm–Liouville theory and its historical development is the volume [AHP], see especially the article [BE] therein.

Chapter 16:

The moment problem was first formulated and solved by T. Stieltjes in his memoir [Stj] in the case where the measure is supported by the half-axis $[0, \infty)$. The moment problem on the whole real line was first studied by Hamburger [Hm], who obtained the existence assertion of Theorem 16.1. Nevanlinna's Theorem 16.34 was proved in [Nv]; another approach can be found in [BC].

The standard monograph on the one-dimensional moment problem is Akhiezer's book [Ac]; other treatment are given in [Be, Sm4], and [GrW].

Appendix A
Bounded Operators and Classes of Compact Operators on Hilbert Spaces

Most of the results on bounded or compact operators we use can be found in standard texts on bounded Hilbert space operators such as [RS1, AG, GK, We].

Let $\mathcal{H}$, $\mathcal{H}_1$, $\mathcal{H}_2$ be complex Hilbert spaces with scalar products and norms denoted by $\langle\cdot,\cdot\rangle$, $\langle\cdot,\cdot\rangle_1$, $\langle\cdot,\cdot\rangle_2$ and $\|\cdot\|$, $\|\cdot\|_1$, $\|\cdot\|_2$, respectively.

A linear operator $T : \mathcal{H}_1 \to \mathcal{H}_2$ is called *bounded* if there is a constant $C > 0$ such that $\|Tx\|_2 \le C\|x\|_1$ for all $x \in \mathcal{D}(T)$; the *norm* of T is then defined by

$$\|T\| = \sup\{\|Tx\|_2 : \|x\|_1 \le 1,\ x \in \mathcal{D}(T)\}.$$

Let $\mathbf{B}(\mathcal{H}_1, \mathcal{H}_2)$ denote the set of bounded linear operators $T : \mathcal{H}_1 \to \mathcal{H}_2$ such that $\mathcal{D}(T) = \mathcal{H}_1$. Then $(\mathbf{B}(\mathcal{H}_1, \mathcal{H}_2), \|\cdot\|)$ is a Banach space. For each $T \in \mathbf{B}(\mathcal{H}_1, \mathcal{H}_2)$, there exists a unique *adjoint operator* $T^* \in \mathbf{B}(\mathcal{H}_2, \mathcal{H}_1)$ such that

$$\langle Tx, y\rangle_2 = \langle x, T^*y\rangle_1 \quad \text{for all } x \in \mathcal{H}_1,\ y \in \mathcal{H}_2.$$

The set $\mathbf{B}(\mathcal{H}) := \mathbf{B}(\mathcal{H}, \mathcal{H})$ is a $*$-algebra with composition of operators as product, adjoint operation as involution, and the identity operator $I = I_{\mathcal{H}}$ on $\mathcal{H}$ as unit.

An operator $T \in \mathbf{B}(\mathcal{H})$ is said to be *normal* if $T^*T = TT^*$.

An operator $T \in \mathbf{B}(\mathcal{H}_1, \mathcal{H}_2)$ is called an *isometry* if $T^*T = I_{\mathcal{H}_1}$ and *unitary* if $T^*T = I_{\mathcal{H}_1}$ and $TT^* = I_{\mathcal{H}_2}$. Clearly, $T \in \mathbf{B}(\mathcal{H}_1, \mathcal{H}_2)$ is an isometry if and only if $\langle Tx, Ty\rangle_2 = \langle x, y\rangle_1$ for all $x, y \in \mathcal{H}_1$. Further, $T \in \mathbf{B}(\mathcal{H}_1, \mathcal{H}_2)$ is called a *finite-rank operator* if $\dim \mathcal{R}(T) < \infty$. If $T \in \mathbf{B}(\mathcal{H}_1, \mathcal{H}_2)$ is a finite-rank operator, so is T^*.

A sequence $(T_n)_{n\in\mathbb{N}}$ from $\mathbf{B}(\mathcal{H})$ is said to *converge strongly* to $T \in \mathbf{B}(\mathcal{H})$ if $\lim_{n\to\infty} T_n x = Tx$ for all $x \in \mathcal{H}$; in this case we write $T = \text{s-}\lim_{n\to\infty} T_n$.

A sequence $(x_n)_{n\in\mathbb{N}}$ of vectors $x_n \in \mathcal{H}$ *converges weakly* to a vector $x \in \mathcal{H}$ if $\lim_{n\to\infty}\langle x_n, y\rangle = \langle x, y\rangle$ for all $y \in \mathcal{H}$; then we write $x = \text{w-}\lim_{n\to\infty} x_n$.

Now we collect some facts on *projections*. Let $\mathcal{K}$ be a closed linear subspace of $\mathcal{H}$. The operator $P_{\mathcal{K}} \in \mathbf{B}(\mathcal{H})$ defined by $P_{\mathcal{K}}(x + y) = x$, where $x \in \mathcal{K}$ and $y \in \mathcal{K}^{\perp}$, is called the (orthogonal) *projection* of $\mathcal{H}$ on $\mathcal{K}$. An operator $P \in \mathbf{B}(\mathcal{H})$ is a projection if and only if $P = P^2$ and $P = P^*$; in this case, P projects on $P\mathcal{H}$.

Let P_1 and P_2 be projections of $\mathcal{H}$ onto subspaces $\mathcal{K}_1$ and $\mathcal{K}_2$, respectively. The sum $Q := P_1 + P_2$ is a projection if and only if $P_1 P_2 = 0$, or equivalently, $\mathcal{K}_1 \perp \mathcal{K}_2$; then Q projects on $\mathcal{K}_1 \oplus \mathcal{K}_2$. The product $P := P_1 P_2$ is a projection if

K. Schmüdgen, *Unbounded Self-adjoint Operators on Hilbert Space*,
Graduate Texts in Mathematics 265,
DOI 10.1007/978-94-007-4753-1, © Springer Science+Business Media Dordrecht 2012

and only if $P_1 P_2 = P_2 P_1$; in this case, P projects on $\mathcal{K}_1 \cap \mathcal{K}_2$. Further, $P_1 \leq P_2$, or equivalently $\mathcal{K}_1 \subseteq \mathcal{K}_2$, holds if and only if $P_1 P_2 = P_1$, or equivalently $P_2 P_1 = P_1$; if this is fulfilled, then $P_2 - P_1$ is the projection on $\mathcal{K}_2 \ominus \mathcal{K}_1$.

Let $P_n, n \in \mathbb{N}$, be pairwise orthogonal projections (that is, $P_k P_n = 0$ for $k \neq n$) on $\mathcal{H}$. Then the infinite sum $\sum_{n=1}^{\infty} P_n$ converges strongly to a projection P, and P projects on the closed subspace $\bigoplus_{n=1}^{\infty} P_n \mathcal{H}$ of $\mathcal{H}$.

If $T \in \mathbf{B}(\mathcal{H})$ and P is a projection of $\mathcal{H}$ on $\mathcal{K}$, then the operator $PT \restriction \mathcal{K}$ is in $\mathbf{B}(\mathcal{K})$; it is called the *compression* of T to $\mathcal{K}$.

In the remaining part of this appendix we are dealing with *compact operators*.

Definition A.1 A linear operator $T : \mathcal{H} \to \mathcal{H}$ with domain $\mathcal{D}(T) = \mathcal{H}$ is called *compact* if the image $T(M)$ of each bounded subset M is relatively compact in $\mathcal{H}$, or equivalently, if for each bounded sequence $(x_n)_{n \in \mathbb{N}}$, the sequence $(T x_n)_{n \in \mathbb{N}}$ has a convergent subsequence in $\mathcal{H}$.

The compact operators on $\mathcal{H}$ are denoted by $\mathbf{B}_\infty(\mathcal{H})$. Each compact operator is bounded. Various characterizations of compact operators are given in the following:

Proposition A.1 *For any $T \in \mathbf{B}(\mathcal{H})$, the following statements are equivalent:*

(i) *T is compact.*
(ii) *$|T| := (T^* T)^{1/2}$ is compact.*
(iii) *$T^* T$ is compact.*
(iv) *T maps each weakly convergent sequence $(x_n)_{n \in \mathbb{N}}$ into a convergent sequence in $\mathcal{H}$, that is, if $\langle x_n, y \rangle \to \langle x, y \rangle$ for all $y \in \mathcal{H}$, then $T x_n \to T x$ in $\mathcal{H}$.*
(v) *There exists a sequence $(T_n)_{n \in \mathbb{N}}$ of finite-rank operators $T_n \in \mathbf{B}(\mathcal{H})$ such that $\lim_{n \to \infty} \|T - T_n\| = 0$.*

Proof [RS1, Theorems VI.11 and VI.13] or [We, Theorems 6.3, 6.4, 6.5].   □

Proposition A.2 *Let $T, T_n, S_1, S_2 \in \mathbf{B}(\mathcal{H})$ for $n \in \mathbb{N}$. If $\lim_{n \to \infty} \|T - T_n\| = 0$ and each operator T_n is compact, so is T. If T is compact, so are $S_1 T S_2$ and T^*.*

Proof [RS1, Theorem VI.12] or [We, Theorem 6.4].   □

Basic results about the spectrum of compact operators are collected in the following two theorems.

Theorem A.3 *Let $T \in \mathbf{B}_\infty(\mathcal{H})$. The spectrum $\sigma(T)$ is an at most countable set which has no nonzero accumulation point. If $\mathcal{H}$ is infinite-dimensional, then $0 \in \sigma(T)$. If λ is a nonzero number of $\sigma(T)$, then λ is an eigenvalue of T of finite multiplicity, and $\bar{\lambda}$ is an eigenvalue of T^* which has the same multiplicity as λ.*

Proof [RS1, Theorem VI.15] or [We, Theorems 6.7, 6.8] or [AG, Nr. 57].   □

From now on we assume that $\mathcal{H}$ is an *infinite-dimensional separable* Hilbert space.

Theorem A.4 *Suppose that* $T \in \mathbf{B}_\infty(\mathcal{H})$ *is normal. Then there exist a complex sequence* $(\lambda_n)_{n \in \mathbb{N}}$ *and an orthonormal basis* $\{e_n : n \in \mathbb{N}\}$ *of* $\mathcal{H}$ *such that*

$$\lim_{n \to \infty} \lambda_n = 0 \quad and \quad T e_n = \lambda_n e_n, \qquad T^* e_n = \overline{\lambda_n} e_n \quad for \, n \in \mathbb{N}.$$

Proof [AG, Nr. 62] or [We, Theorem 7.1], see, e.g., [RS1, Theorem VI.16].   □

Conversely, let $\{e_n : n \in \mathbb{N}\}$ be an orthonormal basis of $\mathcal{H}$, and $(\lambda_n)_{n \in \mathbb{N}}$ a bounded complex sequence. Then the normal operator $T \in \mathbf{B}(\mathcal{H})$ defined by $T e_n = \lambda_n e_n$, $n \in \mathbb{N}$, is compact if and only if $\lim_{n \to \infty} \lambda_n = 0$.

Next we turn to *Hilbert–Schmidt operators*.

Definition A.2 An operator $T \in \mathbf{B}(\mathcal{H})$ is called a *Hilbert–Schmidt operator* if there exists an orthonormal basis $\{e_n : n \in \mathbb{N}\}$ of $\mathcal{H}$ such that

$$\|T\|_2 := \left(\sum_{n=1}^{\infty} \|T e_n\|^2 \right)^{1/2} < \infty. \tag{A.1}$$

The set of Hilbert–Schmidt operators on $\mathcal{H}$ is denoted by $\mathbf{B}_2(\mathcal{H})$. The number $\|T\|_2$ in (A.1) does not depend on the particular orthonormal basis, and $\mathbf{B}_2(\mathcal{H})$ is a Banach space equipped with the norm $\| \cdot \|_2$. Each Hilbert–Schmidt operator is compact.

An immediate consequence of Theorem A.4 and the preceding definition is the following:

Corollary A.5 *If* $T \in \mathbf{B}_2(\mathcal{H})$ *is normal, then the sequence* $(\lambda_n)_{n \in \mathbb{N}}$ *of eigenvalues of* T *counted with multiplicities belongs to* $l^2(\mathbb{N})$.

Theorem A.6 *Let* M *be an open subset of* $\mathbb{R}^d$, *and* $K \in L^2(M \times M)$. *Then the integral operator* T_K *defined by*

$$(T_K f)(x) = \int_M K(x, y) f(y) \, dy, \quad f \in L^2(M),$$

is a Hilbert–Schmidt operator on $\mathcal{H} = L^2(M)$, *and* $\|T_K\|_2 = \|K\|_{L^2(M \times M)}$. *In particular,* T_K *is a compact operator on* $L^2(M)$.

Proof [RS1, Theorem VI.23] or [AG, Nr. 32] or [We, Theorem 6.11].   □

Finally, we review some basics on *trace class operators*.

Let $T \in \mathbf{B}_\infty(\mathcal{H})$. Then $|T| = (T^* T)^{1/2}$ is a positive self-adjoint compact operator on $\mathcal{H}$. Let $(s_n(T))_{n \in \mathbb{N}}$ denote the sequence of eigenvalues of $|T|$ counted with multiplicities and arranged in decreasing order. The numbers $s_n(T)$ are called *singular numbers* of the operator T. Note that $s_1(T) = \|T\|$.

Definition A.3 An operator $T \in \mathbf{B}_\infty(\mathcal{H})$ is said to be of *trace class* if

$$\|T\|_1 := \sum_{n=1}^{\infty} s_n(T) < \infty.$$

The set $\mathbf{B}_1(\mathcal{H})$ of all trace class operators on $\mathcal{H}$ is a Banach space equipped with the trace norm $\| \cdot \|_1$. Trace class operators are Hilbert–Schmidt operators.

If $T \in \mathbf{B}_1(\mathcal{H})$ and $S_1, S_2 \in \mathbf{B}(\mathcal{H})$, then $T^* \in \mathbf{B}_1(\mathcal{H})$, $S_1 T S_2 \in \mathbf{B}_1(\mathcal{H})$, and

$$\|T\| \le \|T\|_1 = \left\| T^* \right\|_1 \quad \text{and} \quad \|S_1 T S_2\|_1 \le \|T\|_1 \|S_1\| \|S_2\|. \tag{A.2}$$

Proposition A.7 *The set of finite-rank operators of* $\mathbf{B}(\mathcal{H})$ *is a dense subset of the Banach space* $(\mathbf{B}_1(\mathcal{H}), \| \cdot \|_1)$.

Proof [RS1, Corollary, p. 209]. □

Theorem A.8 (*Trace of trace class operators*) *Let* $T \in \mathbf{B}_1(\mathcal{H})$. *If* $\{x_n : n \in \mathbb{N}\}$ *is an orthonormal basis of* $\mathcal{H}$, *then*

$$\operatorname{Tr} T := \sum_{n=1}^{\infty} \langle T x_n, x_n \rangle \tag{A.3}$$

is finite and independent of the orthonormal basis. It is called the trace *of* T.

Proof [RS1, Theorem VI.24] or [GK, Chap. III, Theorem 8.1] or [AG, Nr. 66]. □

Combining Theorems A.4 and A.8, we easily obtain the following:

Corollary A.9 *For* $T = T^* \in \mathbf{B}_1(\mathcal{H})$, *there are a real sequence* $(\alpha_n)_{n \in \mathbb{N}} \in l^1(\mathbb{N})$ *and an orthonormal basis* $\{x_n : n \in \mathbb{N}\}$ *of* $\mathcal{H}$ *such that*

$$T = \sum_{n=1}^{\infty} \alpha_n \langle \cdot, x_n \rangle x_n, \qquad \|T\|_1 = \sum_{n=1}^{\infty} |\alpha_n|, \qquad \operatorname{Tr} T = \sum_{n=1}^{\infty} \alpha_n. \tag{A.4}$$

(A.4) implies that $T \to \operatorname{Tr} T$ is a continuous linear functional on $(\mathbf{B}_1(\mathcal{H}), \| \cdot \|_1)$.

Theorem A.10 (*Lidskii's theorem*) *Let* $T \in \mathbf{B}_1(\mathcal{H})$. *Let* $\lambda_n(T)$, $n \in \mathbb{N}$, *denote the eigenvalues of* T *counted with multiplicities; if* T *has only finitely many eigenvalues* $\lambda_1(T), \ldots, \lambda_k(T)$, *set* $\lambda_n(T) = 0$ *for* $n > k$; *if* T *has no eigenvalue, put* $\lambda_n(T) = 0$ *for all* $n \in \mathbb{N}$. *Then*

$$\sum_{n=1}^{\infty} |\lambda_n(T)| \le \|T\|_1 \quad \text{and} \quad \operatorname{Tr} T = \sum_{n=1}^{\infty} \lambda_n(T). \tag{A.5}$$

Proof [Ld] or [RS4, Corollary, p. 328] or [GK, Chap. III, Theorem 8.4]. □

Appendix B
Measure Theory

The material collected in this appendix can be found in advanced standard texts on measure theory such as [Ru2] or [Cn].

Let Ω be a nonempty set, and let $\mathfrak{A}$ be a σ-algebra on Ω. A *positive measure* (briefly, a *measure* on $(\Omega, \mathfrak{A})$) is a mapping $\mu : \mathfrak{A} \to [0, +\infty]$ which is σ-additive, that is,

$$\mu\left(\bigcup_{n=1}^{\infty} M_n\right) = \sum_{n=1}^{\infty} \mu(M_n) \tag{B.1}$$

for any sequence $(M_n)_{n \in \mathbb{N}}$ of disjoints sets $M_n \in \mathfrak{A}$. A positive measure μ is said to be *finite* if $\mu(\Omega) < \infty$ and *σ-finite* if Ω is a countable union of sets $M_n \in \mathfrak{A}$ with $\mu(M_n) < \infty$. A *measure space* is a triple $(\Omega, \mathfrak{A}, \mu)$ of a set Ω, a σ-algebra $\mathfrak{A}$ on Ω, and a positive measure μ on $\mathfrak{A}$. A property is said to hold μ-*a.e.* on Ω if it holds except for a μ-null set. A function f on Ω with values in $(-\infty, +\infty]$ is called $\mathfrak{A}$-*measurable* if for all $a \in \mathbb{R}$, the set $\{t \in \Omega : f(t) \leq a\}$ is in $\mathfrak{A}$. A function f with values in $\mathbb{C} \cup \{+\infty\}$ is said to be μ-*integrable* if the integral $\int_{\Omega} f d\mu$ exists and is finite.

A *complex measure* on $(\Omega, \mathfrak{A})$ is a σ-additive map $\mu : \mathfrak{A} \to \mathbb{C}$. Note that complex measures have only finite values, while positive measures may take the value $+\infty$.

Let μ be a complex measure. Then μ can written as $\mu = \mu_1 - \mu_2 + i(\mu_3 - \mu_4)$ where $\mu_1, \mu_2, \mu_3, \mu_4$ are finite positive measures. We say that μ is *supported* by a set $K \in \mathfrak{A}$ if $\mu(M) = \mu(M \cap K)$ for all $M \in \mathfrak{A}$. Given $M \in \mathfrak{A}$, let

$$|\mu|(M) = \sup \sum_{n=1}^{k} \mu(M_n), \tag{B.2}$$

where the supremum is taken over all finite partitions $M = \bigcup_{n=1}^{k} M_n$ of disjoint sets $M_n \in \mathfrak{A}$. Then $|\mu|$ is a positive measure on $(\Omega, \mathfrak{A})$ called the *total variation* of μ.

There are three basic limit theorems. Let $(\Omega, \mathfrak{A}, \mu)$ be a measure space.

K. Schmüdgen, *Unbounded Self-adjoint Operators on Hilbert Space*,
Graduate Texts in Mathematics 265,
DOI 10.1007/978-94-007-4753-1, © Springer Science+Business Media Dordrecht 2012

Theorem B.1 (*Lebesgue's dominated convergence theorem*)   Let f_n, $n \in \mathbb{N}$, and f be $\mathfrak{A}$-measurable complex functions on Ω, and let $g : \Omega \to [0, +\infty]$ be a μ-integrable function. Suppose that

$$\lim_{n \to \infty} f_n(t) = f(t) \quad and \quad |f_n(t)| \leq g(t), \quad n \in \mathbb{N}, \ \mu\text{-a.e. on } \Omega. \qquad (B.3)$$

Then the functions f and f_n are μ-integrable,

$$\lim_{n \to \infty} \int_\Omega |f_n - f| \, d\mu = 0, \quad and \quad \lim_{n \to \infty} \int_\Omega f_n \, d\mu = \int_\Omega f \, d\mu.$$

Proof [Ru2, Theorem 1.34].   □

Theorem B.2 (*Lebesgue's monotone convergence theorem*)   Let f_n, $n \in \mathbb{N}$, and f be $[0, +\infty]$-valued $\mathfrak{A}$-measurable functions on Ω such that

$$\lim_{n \to \infty} f_n(t) = f(t) \quad and \quad f_n(t) \leq f_{n+1}(t), \quad n \in \mathbb{N}, \ \mu\text{-a.e. on } \Omega. \qquad (B.4)$$

Then

$$\lim_{n \to \infty} \int_\Omega f_n \, d\mu = \int_\Omega f \, d\mu.$$

Proof [Ru2, Theorem 1.26].   □

Theorem B.3 (*Fatou's lemma*)   If f_n, $n \in \mathbb{N}$, are $[0, +\infty]$-valued $\mathfrak{A}$-measurable functions on Ω, then

$$\int_\Omega \left(\liminf_{n \to \infty} f_n \right) d\mu \leq \liminf_{n \to \infty} \int_\Omega f_n \, d\mu.$$

Proof [Ru2, Lemma 1.28].   □

Let $p, q, r \in (1, \infty)$ and $p^{-1} + q^{-1} = r^{-1}$. If $f \in L^p(\Omega, \mu)$ and $g \in L^q(\Omega, \mu)$, then $fg \in L^r(\Omega, \mu)$, and the *Hölder inequality* holds:

$$\|fg\|_{L^r(\Omega, \mu)} \leq \|f\|_{L^p(\Omega, \mu)} \|g\|_{L^q(\Omega, \mu)}. \qquad (B.5)$$

Let φ be a mapping of Ω onto another set Ω_0. Then the family $\mathfrak{A}_0$ of all subsets M of Ω_0 such that $\varphi^{-1}(M) \in \mathfrak{A}$ is a σ-algebra on Ω_0 and $\mu_0(M) := \mu(\varphi^{-1}(M))$ defines an measure on $\mathfrak{A}_0$. Thus, $(\Omega_0, \mathfrak{A}_0, \mu_0)$ is also a measure space.

Proposition B.4 (*Transformations of measures*)   If f is a μ_0-a.e. finite $\mathfrak{A}_0$-measurable function on Ω_0, then $f \circ \varphi$ is a μ-a.e. finite $\mathfrak{A}$-measurable function on Ω, and

$$\int_{\Omega_0} f(s) \, d\mu_0(s) = \int_\Omega f(\varphi(t)) \, d\mu(t), \qquad (B.6)$$

where if either integral exists, so does the other.

Proof [Ha, §39, Theorem C].   □

Let $(\Omega, \mathfrak{A}, \mu)$ and $(\Phi, \mathfrak{B}, \nu)$ be σ-finite measure spaces. Let $\mathfrak{A} \times \mathfrak{B}$ denote the σ-algebra on $\Omega \times \Phi$ generated by the sets $M \times N$, where $M \in \mathfrak{A}$ and $N \in \mathfrak{B}$. Then there exists a unique measure $\mu \times \nu$, called the *product* of μ and ν, on $\mathfrak{A} \times \mathfrak{B}$ such that $(\mu \times \nu)(M \times N) = \mu(M)\nu(N)$ for all $M \in \mathfrak{A}$ and $N \in \mathfrak{B}$.

Theorem B.5 (*Fubini's theorem*) *Let f be an $\mathfrak{A} \times \mathfrak{B}$-measurable complex function on $\Omega \times \Phi$. Then*

$$\int_{\Omega} \left(\int_{\Phi} |f(x, y)|\, d\nu(y) \right) d\mu(x) = \int_{\Phi} \left(\int_{\Omega} |f(x, y)|\, d\mu(x) \right) d\nu(y). \quad (B.7)$$

If the double integral in (B.7) is finite, then we have

$$\int_{\Omega} \left(\int_{\Phi} f(x, y)\, d\nu(y) \right) d\mu(x) = \int_{\Omega} \left(\int_{\Phi} f(x, y)\, d\mu(x) \right) d\nu(y)$$
$$= \int_{\Omega \times \Phi} f(x, y)\, d(\mu \times \nu)(x, y).$$

Proof [Ru2, Theorem 8.8].   □

Thus, if f is $[0, +\infty]$-valued, one can always interchange the order of integrations. If f is a complex function, it suffices to check that the integral in (B.7) is finite.

Let μ and ν be complex or positive measures on $\mathfrak{A}$. We say that μ and ν are *mutually singular* and write $\mu \perp \nu$ if μ and ν are supported by disjoint sets.

Suppose that ν is positive. We say that μ is *absolutely continuous* with respect to ν and write $\mu \ll \nu$ if $\nu(N) = 0$ for $N \in \mathfrak{A}$ implies that $\mu(N) = 0$.

For any $f \in L^1(\Omega, \nu)$, there is a complex measure μ defined by

$$\mu(M) = \int_M f\, d\nu, \quad M \in \mathfrak{A}.$$

Obviously, $\mu \ll \nu$. We then write $d\mu = f\, d\nu$.

Theorem B.6 (*Lebesgue–Radon–Nikodym theorem*) *Let ν be a σ-finite positive measure, and μ a complex measure on $(\Omega, \mathfrak{A})$.*

(i) *There is a unique pair of complex measures μ_a and μ_s such that*

$$\mu = \mu_a + \mu_s, \quad \mu_s \perp \nu, \quad \mu_a \ll \nu. \quad (B.8)$$

(ii) *There exists a unique function $f \in L^1(\Omega, \nu)$ such that $d\mu_a = f\, d\nu$.*

Proof [Ru2, Theorem 6.10].   □

Assertion (ii) is the *Radon–Nikodym theorem*. It says that each complex measure μ which is absolutely continuous w.r.t. ν is of the form $d\mu = f d\nu$ with

$f \in L^1(\Omega, \nu)$ uniquely determined by ν and μ. The function f is called the *Radon–Nikodym derivative* of ν w.r.t. μ and denoted by $\frac{d\nu}{d\mu}$.

Assertion (i) is the *Lebesgue decomposition theorem*. We refer to the relation $d\mu = f \, d\nu + d\mu_s$ as the *Lebesgue decomposition* of μ relative to ν.

Let μ be a σ-finite positive measure. Then there is also a decomposition (B.8) into a unique pair of positive measures μ_a and μ_s, and there exists a $[0, +\infty]$-valued $\mathfrak{A}$-measurable function f, uniquely determined ν-a.e., such that $d\mu_a = f \, d\nu$.

Next we turn to Borel measures. Let Ω be a locally compact Hausdorff space that has a countable basis. The *Borel algebra* $\mathfrak{B}(\Omega)$ is the σ-algebra on Ω generated by the open subsets of Ω. A $\mathfrak{B}(\Omega)$-measurable function on Ω is called a *Borel function*. A *regular Borel measure* is a measure μ on $\mathfrak{B}(\Omega)$ such that for all M in $\mathfrak{B}(\Omega)$,

$$\mu(M) = \sup\{\mu(K) : K \subseteq M, K \text{ compact}\} = \inf\{\mu(U) : M \subseteq U, \ U \text{ open}\}.$$

The *support* of μ is the smallest closed subset M of Ω such that $\mu(\Omega \backslash M) = 0$; it is denoted by $\text{supp}\,\mu$. A *regular complex Borel measure* is a complex measure of the form $\mu = \mu_1 - \mu_2 + i(\mu_3 - \mu_4)$, where μ_1, μ_2, μ_3, μ_4 are finite regular positive Borel measures.

Theorem B.7 (*Riesz representation theorem*) *Let $C(\Omega)$ be the Banach space of continuous functions on a compact Hausdorff space Ω equipped with the supremum norm. For each continuous linear functional F on $C(\Omega)$, there exists a unique regular complex Borel measure μ on Ω such that*

$$F(f) = \int_\Omega f \, d\mu, \quad f \in C(\Omega).$$

Moreover, $\|F\| = |\mu|(\Omega)$.

Proof [Ru2, Theorem 6.19].   □

The Lebesgue measure on $\mathbb{R}^d$ is denoted by m. We write simply dx instead of $dm(x)$ and *a.e.* when we mean m-a.e. If Ω is an open or closed subset of $\mathbb{R}^d$, then $L^p(\Omega)$ always denotes the space $L^p(\Omega, m)$ with respect to the Lebesgue measure.

In the case where ν is the Lebesgue measure on $\Omega = \mathbb{R}$ we specify our terminology concerning Theorem B.6(i). In this case we write μ_{ac} for μ_a and μ_{sing} for μ_s and call μ_{ac} and μ_{sing} the *absolutely continuous part* and *singular part* of μ, respectively.

Let μ be a positive regular Borel measure on $\mathbb{R}$. A point $t \in \mathbb{R}$ is called an *atom* of μ if $\mu(\{t\}) \neq 0$. The set P of atoms of μ is countable, since μ is finite for compact sets. There is a measure μ_p on $\mathfrak{B}(\mathbb{R})$ defined by $\mu_p(M) = \mu(M \cap P)$. Then μ_p is a *pure point measure*, that is, $\mu_p(M) = \sum_{t \in M} \mu(\{t\})$ for $M \in \mathfrak{B}(\mathbb{R})$. Clearly, $\mu_{sc} := \mu_{sign} - \mu_p$ is a positive measure. Further, μ_{sc} and m are mutually singular, and μ_{sc} is a *continuous measure*, that is, μ_{sc} has no atoms. Such a measure is called *singularly continuous*. We restate these considerations.

Proposition B.8 *For any regular positive Borel measure μ on $\mathbb{R}$, there is a unique decomposition $\mu = \mu_{\mathrm{p}} + \mu_{\mathrm{sc}} + \mu_{\mathrm{ac}}$ into a pure point part μ_{p}, a singularly continuous part μ_{sc}, and an absolutely continuous part μ_{ac}. These parts are mutually singular.*

Let μ be a complex Borel measure on $\mathbb{R}$. The *symmetric derivative* $D\mu$ of μ with respect to the Lebesgue measure is defined by

$$(D\mu)(t) = \lim_{\varepsilon \to +0} (2\varepsilon)^{-1}\mu\big([t - \varepsilon, t + \varepsilon]\big), \quad t \in \mathbb{R}. \tag{B.9}$$

The following theorem relates singular and absolutely continuous parts of a measure and its symmetric derivative. The description of the singular part in (ii) is the classical *de la Vallée Poussin theorem*.

Theorem B.9 *Suppose that μ is a complex regular Borel measure on $\mathbb{R}$. Let $d\mu = f\,dm + d\mu_{\mathrm{sign}}$ be its Lebesgue decomposition with respect to the Lebesgue measure.*

(i) *$(D\mu)(t)$ exists a.e. on $\mathbb{R}$, and $D\mu = f$ a.e. on $\mathbb{R}$, that is, $d\mu_{\mathrm{ac}}(t) = (D\mu)(t)\,dt$.*

(ii) *Suppose that the measure μ is positive. Then its singular part μ_{sign} is supported by the set $S(\mu) = \{t \in \mathbb{R} : (D\mu)(t) = +\infty\}$, and its absolutely continuous part μ_{ac} is supported by $L(\mu) = \{t \in \mathbb{R} : 0 < (D\mu)(t) < \infty\}$.*

Proof (i): [Ru2, Theorem 7.8].

(ii): The result about μ_{sign} is proved in [Ru2, Theorem 7.15].

The assertion concerning μ_{ac} is derived from (i). Set $M_0 := \{t : (D\mu)(t) = 0\}$ and $M_\infty := \{t : (D\mu)(t) = +\infty\}$. Because μ is a positive measure, $(D\mu)(t) \geq 0$ a.e. on $\mathbb{R}$. Since $d\mu_{\mathrm{ac}} = (D\mu)\,dt$ by (i) and $D\mu = f \in L^1(\mathbb{R})$, we have $\mu_{\mathrm{ac}}(M_0) = 0$ and $m(M_\infty) = 0$, so $\mu_{\mathrm{ac}}(M_\infty) = 0$. Hence, μ_{ac} is supported by $L(\mu)$. $\square$

Finally, we discuss the relations between complex regular Borel measures and functions of bounded variation. Let f be a function on $\mathbb{R}$. The *variation* of f is

$$V_f := \sup \sum_{k=1}^{n} \big| f(t_k) - f(t_{k-1}) \big|,$$

the supremum being taken over all collections of numbers $t_0 < \cdots < t_n$ and $n \in \mathbb{N}$. We say that f is of *bounded variation* if $V_f < \infty$. Clearly, each bounded nondecreasing real function is of bounded variation. Let *NBV* denote the set of all right-continuous functions f of bounded variation satisfying $\lim_{t \to -\infty} f(t) = 0$.

Theorem B.10 (i) *There is a one-to-one correspondence between complex regular Borel measures μ on $\mathbb{R}$ and functions f in NBV. It is given by*

$$f_\mu(t) = \mu\big((-\infty, t]\big), \quad t \in \mathbb{R}. \tag{B.10}$$

(ii) *Each function $f \in NBV$ is differentiable a.e. on $\mathbb{R}$, and $f' \in L^1(\mathbb{R})$. Further, if $f = f_\mu$ is given by (B.10), then $f'(t) = (D\mu)(t)$ a.e. on $\mathbb{R}$.*

Proof (i): [Cn, Proposition 4.4.3] or [Ru1, Theorem 8.14].
   (ii): [Ru1, Theorem 8.18].   □

Obviously, if μ is a positive measure, then the function f_μ is nondecreasing.

By a slight abuse of notation one usually writes simply $\mu(t)$ for the function $f_\mu(t)$ defined by (B.10).

Appendix C
The Fourier Transform

In this appendix we collect basic definitions and results on the Fourier transform. Proofs and more details can be found, e.g., in [RS2, Chap. IX], [GW], and [VI].

Definition C.1 The *Schwartz space* $\mathcal{S}(\mathbb{R}^d)$ is the set of all functions $\varphi \in C^\infty(\mathbb{R}^d)$ satisfying

$$p_{\alpha,\beta}(\varphi) := \sup\left\{\left|\left(x^\alpha D^\beta \varphi\right)(x)\right| : x \in \mathbb{R}^d\right\} < \infty \quad \text{for all } \alpha, \beta \in \mathbb{N}_0^d.$$

Equipped with the countable family of seminorms $\{p_{\alpha,\beta} : \alpha, \beta \in \mathbb{N}_0^d\}$, $\mathcal{S}(\mathbb{R}^d)$ is a locally convex Fréchet space. Its dual $\mathcal{S}'(\mathbb{R}^d)$ is the vector space of continuous linear functionals on $\mathcal{S}(\mathbb{R}^d)$. The elements of $\mathcal{S}'(\mathbb{R}^d)$ are called *tempered distributions*.

Each function $f \in L^p(\mathbb{R}^d)$, $p \in [1, \infty]$, gives rise to an element $F_f \in \mathcal{S}'(\mathbb{R}^d)$ by defining $F_f(\varphi) = \int f \varphi \, dx$, $\varphi \in \mathcal{S}(\mathbb{R}^d)$. We shall consider $L^p(\mathbb{R}^d)$ as a linear subspace of $\mathcal{S}'(\mathbb{R}^d)$ by identifying f with F_f.

For $f \in L^1(\mathbb{R}^d)$, we define two functions $\hat{f}, \check{f} \in L^\infty(\mathbb{R}^d)$ by

$$\hat{f}(x) = (2\pi)^{-d/2} \int_{\mathbb{R}^d} e^{-ix\cdot y} f(y) \, dy, \tag{C.1}$$

$$\check{f}(y) = (2\pi)^{-d/2} \int_{\mathbb{R}^d} e^{ix\cdot y} f(x) \, dx, \tag{C.2}$$

where $x \cdot y := x_1 y_1 + \cdots + x_d y_d$ denotes the scalar product of $x, y \in \mathbb{R}^d$.

Lemma C.1 (*Riemann–Lebesgue lemma*) If $f \in L^1(\mathbb{R}^d)$, then $\lim_{\|x\| \to \infty} \hat{f}(x) = 0$.

Proof [RS2, Theorem IX.7] or [GW, Theorem 17.1.3]. $\qquad\square$

Theorem C.2 *The map $\mathcal{F} : f \to \mathcal{F}(f) := \hat{f}$ is a continuous bijection of $\mathcal{S}(\mathbb{R}^d)$ onto $\mathcal{S}(\mathbb{R}^d)$ with inverse $\mathcal{F}^{-1}(f) = \check{f}$, $f \in \mathcal{S}(\mathbb{R}^d)$. For $f \in \mathcal{S}(\mathbb{R}^d)$ and $\alpha \in \mathbb{N}_0^d$,*

$$D^\alpha f = \mathcal{F}^{-1} x^\alpha \mathcal{F} f. \tag{C.3}$$

K. Schmüdgen, *Unbounded Self-adjoint Operators on Hilbert Space*, Graduate Texts in Mathematics 265, DOI 10.1007/978-94-007-4753-1, © Springer Science+Business Media Dordrecht 2012

Proof [RS2, Formula (IX.1) and Theorem IX.1] or [GW, Theorem 19.3.1].   □

For $F \in \mathcal{S}'(\mathbb{R}^d)$, define $\mathcal{F}(F)(\varphi) := F(\hat{\varphi}) \equiv F(\mathcal{F}(\varphi))$, $\varphi \in \mathcal{S}(\mathbb{R}^d)$. By Theorem C.2, we have $\mathcal{F}(F) \in \mathcal{S}'(\mathbb{R}^d)$. The tempered distribution $\mathcal{F}(F)$ is called the *Fourier transform* of F. An immediate consequence (see, e.g., [RS2, Theorem IX.2]) of Theorem C.2 is the following:

Corollary C.3 $\mathcal{F}$ *is a bijection of* $\mathcal{S}'(\mathbb{R}^d)$ *onto* $\mathcal{S}'(\mathbb{R}^d)$. *The inverse of* $\mathcal{F}$ *is given by* $\mathcal{F}^{-1}(F)(\varphi) = F(\check{\varphi}) \equiv F(\mathcal{F}^{-1}(\varphi))$, $\varphi \in \mathcal{S}(\mathbb{R}^d)$.

Since $L^p(\mathbb{R}^d)$ is contained in $\mathcal{S}'(\mathbb{R}^d)$, we know from Corollary C.3 that $\mathcal{F}(f)$ and $\mathcal{F}^{-1}(f)$ are well-defined elements of $\mathcal{S}'(\mathbb{R}^d)$ for any $f \in L^p(\mathbb{R}^d)$, $p \in [1, \infty]$. The following two fundamental theorems describe some of these elements.

Theorem C.4 (*Plancherel theorem*) $\mathcal{F}$ *is a unitary linear map of* $L^2(\mathbb{R}^d)$ *onto* $L^2(\mathbb{R}^d)$. *For* $f \in L^2(\mathbb{R}^d)$, *the corresponding functions* $\mathcal{F}(f)$ *and* $\mathcal{F}^{-1}(f)$ *of* $L^2(\mathbb{R}^d)$ *are given by*

$$\hat{f}(x) := \mathcal{F}(f)(x) = \lim_{R \to \infty} (2\pi)^{-d/2} \int_{\|y\| \leq R} e^{-ix \cdot y} f(y) \, dy, \qquad (C.4)$$

$$\check{f}(y) := \mathcal{F}^{-1}(f)(y) = \lim_{R \to \infty} (2\pi)^{-d/2} \int_{\|x\| \leq R} e^{ix \cdot y} f(x) \, dx. \qquad (C.5)$$

Proof [RS2, Theorem IX.6] or [GW, Theorem 22.1.4].   □

Theorem C.5 (*Hausdorff–Young theorem*) *Suppose that* $q^{-1} + p^{-1} = 1$ *and* $1 \leq p \leq 2$. *Then the Fourier transform* $\mathcal{F}$ *maps* $L^p(\mathbb{R}^d)$ *continuously into* $L^q(\mathbb{R}^d)$.

Proof [RS2, Theorem IX.8] or [Tl, Theorem 1.18.8] or [LL, Theorem 5.7].   □

Definition C.2 Let $f, g \in L^2(\mathbb{R}^d)$. The *convolution* of f and g is defined by

$$(f * g)(x) = \int_{\mathbb{R}^d} f(x - y) g(y) \, dy, \quad x \in \mathbb{R}^d.$$

Note that the function $y \to f(x - y)g(y)$ is in $L^1(\mathbb{R}^d)$, because $f, g \in L^2(\mathbb{R}^d)$.

Theorem C.6 $\mathcal{F}^{-1}(\mathcal{F}(f) \cdot \mathcal{F}(g)) = (2\pi)^{-d/2} f * g$ *for* $f, g \in L^2(\mathbb{R}^d)$.

Proof [GW, Proposition 33.3.1] or [LL, Theorem 5.8]; see also [RS2, Theorem IX.3].   □

Appendix D
Distributions and Sobolev Spaces

Let Ω be an open subset of $\mathbb{R}^d$. In this appendix we collect basic notions and results about distributions on Ω and the Sobolev spaces $H^n(\Omega)$ and $H_0^n(\Omega)$.

Distributions are treated in [GW, Gr, LL, RS2], and [Vl]. Sobolev spaces are developed in many books such as [At, Br, EE, Gr, LL, LM], and [Tl].

We denote by $\mathcal{D}(\Omega) = C_0^\infty(\Omega)$ the set of all $f \in C^\infty(\Omega)$ for which its support

$$\operatorname{supp} f := \overline{\left\{ x \in \mathbb{R}^d : f(x) \neq 0 \right\}}$$

is a compact subset of Ω. Further, recall the notations $\partial^\alpha := (\frac{\partial}{\partial x_1})^{\alpha_1} \cdots (\frac{\partial}{\partial x_d})^{\alpha_d}$, $\partial_k := \frac{\partial}{\partial x_k}$, and $|\alpha| = \alpha_1 + \cdots + \alpha_d$ for $\alpha = (\alpha_1, \dots, \alpha_d) \in \mathbb{N}_0^d$, $k = 1, \dots, d$.

Definition D.1 A linear functional F on $\mathcal{D}(\Omega)$ is called a *distribution* on Ω if for every compact subset K of Ω, there exist numbers $n_K \in \mathbb{N}_0$ and $C_K > 0$ such that

$$\left| F(\varphi) \right| \leq C_K \sup_{x \in K, |\alpha| \leq n_K} \left| \partial^\alpha \varphi(x) \right| \quad \text{for all } \varphi \in \mathcal{D}(\Omega), \operatorname{supp} \varphi \subseteq K.$$

The vector space of distributions on Ω is denoted by $\mathcal{D}'(\Omega)$.

Let $L_{\mathrm{loc}}^1(\Omega)$ denote the vector space of measurable functions f on Ω such that $\int_K |f| \, dx < \infty$ for each compact set $K \subseteq \Omega$. Each $f \in L_{\mathrm{loc}}^1(\Omega)$ yields a distribution $F_f \in \mathcal{D}'(\Omega)$ defined by $F_f(\varphi) = \int f\varphi \, dx$, $\varphi \in \mathcal{D}(\Omega)$. By some abuse of notation, we identify the function $f \in L_{\mathrm{loc}}^1(\Omega)$ with the distribution $F_f \in \mathcal{D}'(\Omega)$.

Suppose that $F \in \mathcal{D}'(\Omega)$. If $a \in C^\infty(\Omega)$, then $aF(\cdot) := F(a \cdot)$ defines a distribution $aF \in \mathcal{D}'(\Omega)$. For $\alpha \in \mathbb{N}_0^d$, there is a distribution $\partial^\alpha F$ defined by

$$\partial^\alpha F(\varphi) := (-1)^{|\alpha|} F\left(\partial^\alpha \varphi\right), \quad \varphi \in \mathcal{D}(\Omega).$$

Let $u \in L_{\mathrm{loc}}^1(\Omega)$. A function $v \in L_{\mathrm{loc}}^1(\Omega)$ is called the αth *weak derivative* of u if the distributional derivative $\partial^\alpha u \in \mathcal{D}'(\Omega)$ is given by the function v, that is, if

$$(-1)^{|\alpha|} \int_\Omega u \partial^\alpha(\varphi) \, dx = \int_\Omega v\varphi \, dx \quad \text{for all } \varphi \in \mathcal{D}(\Omega).$$

In this case, v is uniquely determined a.e. on Ω by u and denoted also by $\partial^\alpha u$.

K. Schmüdgen, *Unbounded Self-adjoint Operators on Hilbert Space*,
Graduate Texts in Mathematics 265,
DOI 10.1007/978-94-007-4753-1, © Springer Science+Business Media Dordrecht 2012

If $u \in C^n(\Omega)$, integration by parts implies that the αth weak derivative $\partial^\alpha u$ exists when $|\alpha| \leq n$ and is given by the "usual" derivative $\partial^\alpha u(x)$ on Ω.

Next, we introduce the Sobolev spaces $H^n(\Omega)$ and $H_0^n(\Omega)$, $n \in \mathbb{N}_0$.

Definition D.2 $H^n(\Omega)$ is the vector space of functions $f \in L^2(\Omega)$ for which the weak derivative $\partial^\alpha f$ exists and belongs to $L^2(\Omega)$ for all $\alpha \in N_0^d$, $|\alpha| \leq n$, endowed with the scalar product

$$\langle f, g \rangle_{H^n(\Omega)} := \sum_{|\alpha| \leq n} \langle \partial^\alpha f, \partial^\alpha g \rangle_{L^2(\Omega)},$$

and $H_0^n(\Omega)$ is the closure of $C_0^\infty(\Omega)$ in $(H^n(\Omega), \| \cdot \|_{H^n(\Omega)})$.

Then $H^n(\Omega)$ and $H_0^n(\Omega)$ are Hilbert spaces.
The two compact embedding Theorems D.1 and D.4 are due to *F. Rellich*.

Theorem D.1 *If Ω is bounded, the embedding $H_0^1(\Omega) \rightarrow L^2(\Omega)$ is compact.*

Proof [At, A 6.1] or [EE, Theorem 3.6].   □

Theorem D.2 (*Poincaré inequality*) *If Ω is bounded, there is a constant $c_\Omega > 0$ such that*

$$c_\Omega \|u\|_{L^2(\Omega)}^2 \leq \sum_{k=1}^{d} \int_\Omega |\partial_k u|^2 \, dx \equiv \|\nabla u\|_{L^2(\Omega)}^2 \quad \textit{for all } u \in H_0^1(\Omega). \quad (\text{D.1})$$

Proof [At, 4.7] or [Br, Corollary 9.19].   □

The Poincaré inequality implies that $\|\nabla \cdot \|_{L^2(\Omega)}$ defines an equivalent norm on the Hilbert space $(H_0^1(\Omega), \| \cdot \|_{H_0^1(\Omega)})$.

For $\Omega = \mathbb{R}^d$, the Sobolev spaces can be described by the Fourier transform.

Theorem D.3 $H^n(\mathbb{R}^d) = H_0^n(\mathbb{R}^d) = \{f \in L^2(\mathbb{R}^d) : \|x\|^{n/2} \hat{f}(x) \in L^2(\mathbb{R}^d)\}$.
Further, if $k \in \mathbb{N}_0$ and $n > k + d/2$, then $H^n(\mathbb{R}^d) \subseteq C^k(\mathbb{R}^d)$.

Proof [LM, p. 30] or [RS2, p. 50 and Theorem IX.24].   □

The next results require some "smoothness" assumption on the boundary $\partial\Omega$. Let $m \in \mathbb{N} \cup \{\infty\}$. An open set $\Omega \subseteq \mathbb{R}^d$ is said to be of *class* C^m if for each point $x \in \partial\Omega$, there are an open neighborhood $\mathcal{U}$ of x in $\mathbb{R}^d$ and a bijection φ of $\mathcal{U}$ on some open ball $\mathcal{B} = \{y \in \mathbb{R}^d : \|y\| < r\}$ such that φ and φ^{-1} are C^m-mappings,

$$\varphi(\mathcal{U} \cap \Omega) = \{(y', y_d) \in \mathcal{B} : y_d > 0\}, \quad \text{and} \quad \varphi(\mathcal{U} \cap \partial\Omega) = \{(y', 0) \in \mathcal{B}\}.$$

Here $y \in \mathbb{R}^d$ was written as $y = (y', y_d)$, where $y' \in \mathbb{R}^{d-1}$ and $y_d \in \mathbb{R}$.

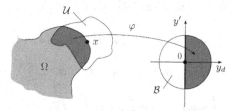

It should be noted that, according to this definition, $\Omega \cap \mathcal{U}$ lies "at one side of $\partial\Omega$." It may happen that $\partial\Omega$ is a C^m-manifold, but Ω is not of class C^m; for instance, $\{(y_1, y_2) \in \mathbb{R}^2 : y_1 \neq 0\}$ is not of class C^1. But $\{(y_1, y_2) \in \mathbb{R}^2 : y_1 > 0\}$ is C^∞.

From now on we assume that Ω is a *bounded open set of class* C^1. We denote by $\nu(x) = (\nu_1, \ldots, \nu_d)$ the *outward unit normal vector* at $x \in \partial\Omega$ and by $d\sigma$ the *surface measure* of $\partial\Omega$.

Theorem D.4 *The embedding map* $H^1(\Omega) \to L^2(\Omega)$ *is compact.*

Proof [At, A 6.4] or [Br, Theorem 9.19] or [LM, Theorem 16.1].   □

Theorem D.5 $C^\infty(\mathbb{R}^d){\upharpoonright}\Omega$ *is a dense subset of* $H^n(\Omega)$ *for* $n = 1, 2$.

Proof [At, A 6.7] or [EE, Chap. V, Theorem 4.7].   □

The following two results are called *trace theorems*. Theorem D.6 gives the possibility to assign "boundary values" $\gamma_0(f)$ along $\partial\Omega$ to any function $f \in H^1(\Omega)$.

Theorem D.6 *There exists a uniquely determined continuous linear mapping* $\gamma_0 : H^1(\Omega) \to L^2(\partial\Omega, d\sigma)$ *such that* $\gamma_0(f) = f{\upharpoonright}\partial\Omega$ *for* $f \in C^\infty(\mathbb{R}^d)$. *Moreover,* $H_0^1(\Omega) = \{f \in H^1(\Omega) : \gamma_0(f) = 0\}$.

Proof [At, A 6.6] or [LM, Theorem 8.3].   □

Theorem D.7 *Let* Ω *be of class* C^2 *and bounded. There is a continuous linear map* $(\gamma_0, \gamma_1) : H^2(\Omega) \to L^2(\partial\Omega, d\sigma) \oplus L^2(\partial\Omega, d\sigma)$ *such that* $\gamma_0(g) = g{\upharpoonright}\partial\Omega$ *and* $\gamma_1(g) = \frac{\partial g}{\partial\nu}{\upharpoonright}\partial\Omega$ *for* $g \in C^\infty(\mathbb{R}^d)$. *The kernel of this map is* $H_0^2(\Omega)$.

Proof [LM, Theorem 8.3].   □

The maps γ_0 and (γ_0, γ_1) are called *trace maps*. Taking fractional Sobolev spaces on the boundary $\partial\Omega$ for granted, their ranges can be nicely described (see [LM, p. 39]) by

$$\gamma_0\big(H^1(\Omega)\big) = H^{1/2}(\partial\Omega), \ (\gamma_0, \gamma_1)\big(H^2(\Omega)\big) = H^{3/2}(\partial\Omega) \oplus H^{1/2}(\partial\Omega).$$

$$(D.2)$$

For $f \in H^1(\Omega)$ and $g \in H^2(\Omega)$, we consider $\gamma_0(f)$ and $\gamma_1(g)$ as *boundary values*, and we also write $f \lceil \partial\Omega$ for $\gamma_0(f)$ and $\frac{\partial g}{\partial v} \lceil \partial\Omega$ for $\gamma_1(g)$.

Theorem D.8 (*Gauss' formula*) *For $f, g \in H^1(\Omega)$ and $k = 1, \ldots, d$, we have*

$$\int_\Omega \partial_k f \, dx = \int_{\partial\Omega} f v_k \, d\sigma, \tag{D.3}$$

$$\int_\Omega (\partial_k f \overline{g} + f \overline{\partial_k g}) \, dx = \int_{\partial\Omega} f \overline{g} v_k \, d\sigma. \tag{D.4}$$

Proof [At, A 6.8] or [Br, p. 316]. ☐

Theorem D.9 (*Green's formulas*) *Let Ω be of class C^2. For all $h \in H^1(\Omega)$ and $f, g \in H^2(\Omega)$, we have*

$$\int_\Omega (-\Delta f) \overline{h} \, dx = \sum_{k=1}^d \int_\Omega \partial_k f \overline{\partial_k h} \, dx - \int_{\partial\Omega} \frac{\partial f}{\partial v} \overline{h} \, d\sigma, \tag{D.5}$$

$$\int_\Omega (-\Delta f) \overline{g} \, dx - \int_\Omega f \overline{(-\Delta g)} \, dx = \int_{\partial\Omega} \left(f \frac{\overline{\partial g}}{\partial v} - \frac{\partial f}{\partial v} \overline{g} \right) d\sigma. \tag{D.6}$$

Proof [Br, p. 316]. ☐

Strictly speaking, the corresponding traces $\gamma_0(f)$, $\gamma_0(\overline{g})$, $\gamma_1(f)$, $\gamma_0(\overline{h})$, $\gamma_1(f)$, $\gamma_1(\overline{g})$ are meant in the integrals along the boundary $\partial\Omega$ in (D.3)–(D.6). Clearly, (D.3) implies (D.4), while (D.5) implies (D.6).

In the literature these formulas are often stated only for functions from $C^\infty(\overline{\Omega})$. The general case is easily derived by a limit procedure. Let us verify (D.5). By Theorem D.5, $h \in H^1(\Omega)$ and $f \in H^2(\Omega)$ are limits of sequences (h_n) and (f_n) from $C^\infty(\mathbb{R}^d) \lceil \Omega$ in $H^1(\Omega)$ resp. $H^2(\Omega)$. Then (D.5) holds for h_n and f_n. Passing to the limit by using the continuity of the trace maps, we obtain (D.5) for h and f.

The next theorem deals with the *regularity of weak solutions* of the Dirichlet and Neumann problems. To formulate these results, we use the following equation:

$$\int_\Omega \nabla f \cdot \nabla h \, dx + \int_\Omega f h \, dx = \int_\Omega g h \, dx. \tag{D.7}$$

Theorem D.10 *Let $\Omega \subseteq \mathbb{R}^d$ be an open bounded set of class C^2, and $g \in L^2(\Omega)$.*

(i) (*Regularity for the Dirichlet problem*)
   Let $f \in H_0^1(\Omega)$. If (D.7) holds for all $h \in H_0^1(\Omega)$, then $f \in H^2(\Omega)$.
(ii) (*Regularity for the Neumann problem*)
   Let $f \in H^1(\Omega)$. If (D.7) holds for all $h \in H^1(\Omega)$, then $f \in H^2(\Omega)$.

Proof [Br, Theorems 9.5 and 9.6]. ☐

Appendix E
Absolutely Continuous Functions

Basics on absolutely continuous functions can be found, e.g., in [Ru2, Cn], or [HS]. Throughout this appendix we suppose that $a, b \in \mathbb{R}$, $a < b$.

Definition E.1 A function f on the interval $[a, b]$ is called *absolutely continuous* if for each $\varepsilon > 0$, there exists a $\delta > 0$ such that

$$\sum_{k=1}^{n} |f(b_k) - f(a_k)| < \varepsilon$$

for every finite family of pairwise disjoint subintervals (a_k, b_k) of $[a, b]$ satisfying

$$\sum_{k=1}^{n} (b_k - a_k) < \delta.$$

The set of all absolutely continuous functions on $[a, b]$ is denoted by $AC[a, b]$. Each $f \in AC[a, b]$ is continuous on $[a, b]$ and a function of bounded variation.

Theorem E.1 *A function f on $[a, b]$ is absolutely continuous if and only if there is a function $h \in L^1(a, b)$ such that*

$$f(x) - f(a) = \int_a^x h(t)\, dt \quad \text{for } x \in [a, b]. \tag{E.1}$$

In this case, f is differentiable a.e., and we have $f'(x) = h(x)$ a.e. on $[a, b]$. The function $h \in L^1(a, b)$ is uniquely determined by f.

Theorem E.2 *For $f, g \in AC[a, b]$, the integration-by-parts formula holds:*

$$\int_a^b f'(t)g(t)\, dt + \int_a^b f(t)g'(t)\, dt = f(b)g(b) - f(a)g(a). \tag{E.2}$$

Theorems E.1 and E.2 are proved, e.g., in [Cn, Corollaries 6.3.7 and 6.3.8].

K. Schmüdgen, *Unbounded Self-adjoint Operators on Hilbert Space*,
Graduate Texts in Mathematics 265,
DOI 10.1007/978-94-007-4753-1, © Springer Science+Business Media Dordrecht 2012

The Sobolev spaces $H^n(a, b)$ and $H_0^n(a, b)$, $n \in \mathbb{N}$, can be expressed in terms of absolutely continuous functions. That is,

$$H^1(a, b) = \{ f \in AC[a, b] : f' \in L^2(a, b) \},$$
$$H^n(a, b) = \{ f \in C^{n-1}([a, b]) : f^{(n-1)} \in H^1(a, b) \},$$
$$H_0^n(a, b) = \{ f \in H^n(a, b) : f(a) = f'(a) = \cdots = f^{(n-1)}(a) = 0,$$
$$f(b) = f'(b) = \cdots = f^{(n-1)}(b) = 0 \}.$$

In particular, $C^n([a, b]) \subseteq H^n(a, b) \subseteq C^{n-1}([a, b])$ for $n \in \mathbb{N}$.

Now let $\mathcal{J}$ be an unbounded open interval of $\mathbb{R}$, and let $\overline{\mathcal{J}}$ denote the closure of $\mathcal{J}$ in $\mathbb{R}$, that is, $\overline{\mathcal{J}} = [a, +\infty)$ if $\mathcal{J} = (a, +\infty)$ and $\overline{\mathcal{J}} = \mathbb{R}$ if $\mathcal{J} = \mathbb{R}$. Then

$$H^1(\mathcal{J}) = \{ f \in L^2(\mathcal{J}) : f \in AC[a, b] \text{ for } [a, b] \subseteq \overline{\mathcal{J}} \text{ and } f' \in L^2(\mathcal{J}) \},$$
$$H^n(\mathcal{J}) = \{ f \in L^2(\mathcal{J}) : f \in C^{n-1}(\overline{\mathcal{J}}) \text{ and } f^{(n-1)} \in H^1(\mathcal{J}) \},$$
$$H_0^n(a, \infty) = \{ f \in H^n(a, \infty) : f(a) = f'(a) = \cdots = f^{(n-1)}(a) = 0 \},$$
$$H_0^n(\mathbb{R}) = H^n(\mathbb{R}).$$

For $f \in H^n(\mathcal{J})$, we have $f \in L^2(\mathcal{J})$ and $f^{(n)} \in L^2(\mathcal{J})$ by the preceding formula. It can be shown that $f \in H^n(\mathcal{J})$ implies that $f^{(j)} \in L^2(\mathcal{J})$ for *all* $j = 1, \ldots, n$.

The preceding formulas are proved in [Br, Chap. 8] and [Gr, Sect. 4.3], where Sobolev spaces on intervals are treated in detail.

Appendix F
Nevanlinna Functions and Stieltjes Transforms

In this appendix we collect and discuss basic results on Nevanlinna functions and on Stieltjes transforms and their boundary values. For some results, it is difficult to localize easily accessible proofs in the literature, and we have included complete proofs of those results.

Let $\mathbb{C}_+ = \{z \in \mathbb{C} : \operatorname{Im} z > 0\}$ denote the upper half-plane.

Definition F.1 A holomorphic complex function f on $\mathbb{C}_+$ is called a *Nevanlinna function* (or likewise, a *Pick function* or a *Herglotz function*) if

$$\operatorname{Im} f(z) \geq 0 \quad \text{for all } z \in \mathbb{C}_+.$$

We denote the set of all Nevanlinna functions by $\mathfrak{N}$.

If $f \in \mathfrak{N}$ and $f(z_0)$ is real for some point $z_0 \in \mathbb{C}_+$, then f is not open and hence a constant. That is, all nonconstant Nevanlinna functions map $\mathbb{C}_+$ into $\mathbb{C}_+$.

Each Nevanlinna function f can be extended to a holomorphic function on $\mathbb{C}\backslash\mathbb{R}$ by setting $f(\bar{z}) := \overline{f(z)}$ for $z \in \mathbb{C}_+$.

Theorem F.1 (*Canonical integral representation of Nevanlinna functions*) *For each Nevanlinna function f, there exist numbers $a, b \in \mathbb{R}$, $b \geq 0$, and a finite positive regular Borel measure μ on the real line such that*

$$f(z) = a + bz + \int_{\mathbb{R}} \frac{1+zt}{t-z}\, d\mu(t), \quad z \in \mathbb{C}/\mathbb{R}, \tag{F.1}$$

where the numbers a, b and the measure μ are uniquely determined by f.
Conversely, any function f of this form is a Nevanlinna function.

Proof [AG, Nr. 69, Theorem 2] or [Dn1, p. 20, Theorem 1]. □

Often the canonical representation (F.1) is written in the form

$$f(z) = a + bz + \int_{\mathbb{R}} \left(\frac{1}{t-z} - \frac{t}{1+t^2} \right) d\nu(t), \quad z \in \mathbb{C}/\mathbb{R}, \tag{F.2}$$

K. Schmüdgen, *Unbounded Self-adjoint Operators on Hilbert Space*,
Graduate Texts in Mathematics 265,
DOI 10.1007/978-94-007-4753-1, © Springer Science+Business Media Dordrecht 2012

where ν is a regular positive Borel measure on $\mathbb{R}$ satisfying $\int (1+t^2)^{-1} \, d\nu(t) < \infty$. The two forms (F.1) and (F.2) are related by the formula $d\mu(t) = (1+t^2)^{-1} \, d\nu(t)$. Let now μ be a *regular complex* Borel measure on $\mathbb{R}$.

Definition F.2 The *Stieltjes transform* of μ is defined by

$$I_\mu(z) = \int_{\mathbb{R}} \frac{1}{t-z} \, d\mu(t), \quad z \in \mathbb{C}\backslash\mathbb{R}. \tag{F.3}$$

Stieltjes transforms are also called *Cauchy transforms* or *Borel transforms*.

Theorem F.2 (*Stieltjes–Perron inversion formula*) *The measure μ is uniquely determined by the values of its Stieltjes transform on $\mathbb{C}\backslash\mathbb{R}$. In fact, for $a, b \in \mathbb{R}, a < b$, we have*

$$\mu\big((a,b)\big) + \frac{1}{2}\mu(\{a\}) + \frac{1}{2}\mu(\{b\}) = \lim_{\varepsilon \to +0} \frac{1}{2\pi\mathrm{i}} \int_a^b \big[I_\mu(t+\mathrm{i}\varepsilon) - I_\mu(t-\mathrm{i}\varepsilon)\big] dt,$$

$$\tag{F.4}$$

$$\mu(\{a\}) = \lim_{\varepsilon \to +0} -\mathrm{i}\varepsilon I_\mu(a+\mathrm{i}\varepsilon). \tag{F.5}$$

Proof Since μ is a linear combination of finite positive Borel measures, one can assume without loss of generality that μ is positive. In this case proofs are given in [AG, Nr. 69] and [We, Appendix B].

(F.4) can be proved by similar arguments as used in the proof of Proposition 5.14. Since $\frac{-\mathrm{i}\varepsilon}{t-(a+\mathrm{i}\varepsilon)} \to \chi_{\{a\}}(t)$ as $\varepsilon \to +0$, (F.5) follows from (F.3) by interchanging limit and integration using the dominated convergence Theorem B.1. □

From formula (F.4) it follows at once that $I_\mu \equiv 0$ on $\mathbb{C}\backslash\mathbb{R}$ implies that $\mu = 0$. However, there are complex measures $\mu \neq 0$ for which $I_\mu = 0$ on $\mathbb{C}_+$.

The following fundamental theorem, due to I.I. Privalov, is about *boundary values* of Stieltjes transforms. Again, we assume that μ is a regular complex Borel measure on $\mathbb{R}$. Recall that the function $\mu \equiv f_\mu$ defined by (B.10) is of bounded variation and differentiable a.e. on $\mathbb{R}$ by Theorem B.10.

Theorem F.3 (*Sokhotski–Plemelj formula*) *The limits $I_\mu(t \pm \mathrm{i}0) :=$* $\lim_{\varepsilon\to\pm0} I_\mu(t+\mathrm{i}\varepsilon)$ *exist, are finite, and*

$$I_\mu(t \pm \mathrm{i}0) = \pm\mathrm{i}\pi \frac{d\mu}{dt}(t) + (\mathrm{PV}) \int_{\mathbb{R}} \frac{1}{s-t} \, d\mu(s) \quad a.e. \ on \ \mathbb{R}, \tag{F.6}$$

where (PV) $\int$ *denotes the principal value of the integral.*

Proof [Pv] or [Mi]. □

From now on we assume that μ is a *finite positive* regular Borel measure on $\mathbb{R}$. Then $\overline{I_\mu(z)} = I_\mu(\bar{z})$ on $\mathbb{C} \backslash \mathbb{R}$. Hence, formulas (F.4) and (F.5) can be written as

$$\mu\big((a,b)\big) + \frac{1}{2}\mu\big(\{a\}\big) + \frac{1}{2}\mu\big(\{b\}\big) = \lim_{\varepsilon \to +0} \pi^{-1} \int_a^b \operatorname{Im} I_\mu(t + i\varepsilon)\,dt, \qquad \text{(F.7)}$$

$$\mu\big(\{a\}\big) = \lim_{\varepsilon \to +0} \varepsilon \operatorname{Im} I_\mu(a + i\varepsilon). \qquad \text{(F.8)}$$

Thus, the *positive* measure μ is uniquely determined by the values of I_μ on $\mathbb{C}_+$.

The Stieltjes transform I_μ of the finite *positive* (!) Borel measure μ is obviously a Nevanlinna function, since

$$\operatorname{Im} I_\mu(z) = \operatorname{Im} z \int_\mathbb{R} \frac{1}{|t - z|^2}\,d\mu(t), \qquad z \in \mathbb{C}\backslash\mathbb{R}.$$

The next result characterizes these Stieltjes transforms among Nevanlinna functions.

Theorem F.4 *A Nevanlinna function f is the Stieltjes transform I_μ of a finite positive Borel measure μ on $\mathbb{R}$ if and only if*

$$\sup\big\{\,|yf(iy)| : y \in \mathbb{R},\ y \ge 1\big\} < \infty. \qquad \text{(F.9)}$$

Proof [AG, Nr. 69, Theorem 3].   □

The necessity of condition (F.9) is easily seen:

Since $\mu(\mathbb{R}) < \infty$, Lebesgue's convergence Theorem B.1 applies and implies that $\mu(\mathbb{R}) = \lim_{y \in \mathbb{R}, y \to \infty} -iy I_\mu(iy)$, which yields (F.9).   □

Theorem F.5 *Let K be a closed subset of $\mathbb{R}$. The Stieltjes transform $I_\mu(z)$ has a holomorphic extension to $\mathbb{C}\backslash K$ if and only if $\operatorname{supp}\mu \subseteq K$.*

Proof [Dn1, Lemma 2, p. 26].   □

Let us sketch the proof of Theorem F.5:

If $\operatorname{supp}\mu \subseteq K$, then (F.3) with $z \in \mathbb{C}\backslash K$ defines a holomorphic extension of I_μ to $\mathbb{C}\backslash K$. Conversely, suppose that I_μ has a holomorphic extension, say f, to $\mathbb{C}\backslash K$. Then $\lim_{\varepsilon \to +0} I_\mu(t \pm i\varepsilon) = f(t)$, and hence $\lim_{\varepsilon \to +0} \operatorname{Im} I_\mu(t + i\varepsilon) = 0$ for $t \in \mathbb{R}\backslash K$. Therefore, by (F.7) and (F.8), μ has no mass in $\mathbb{R}\backslash K$, that is, $\mu(\mathbb{R}\backslash K) = 0$.   □

Since μ is a positive measure, the integral $(PV)\int$ in (F.6) is real, and we obtain

$$\frac{d\mu}{dt}(t) = \pi^{-1} \operatorname{Im} I_\mu(t + i0) \quad \text{a.e. on } \mathbb{R}. \qquad \text{(F.10)}$$

Obviously, if for some $t \in \mathbb{R}$, the boundary value $I_\mu(t + i0)$ of I_μ exists, so does the boundary value $(\operatorname{Im} I_\mu)(t + i0)$ of $\operatorname{Im} I_\mu$ and $\operatorname{Im}(I_\mu(t + i0)) = (\operatorname{Im} I_\mu)(t + i0)$. However, $(\operatorname{Im} I_\mu)(t + i0)$ may exist, but $I_\mu(t + i0)$ does not. But we always have $\operatorname{Im}(I_\mu(t + i0)) = (\operatorname{Im} I_\mu)(t + i0)$ a.e. on $\mathbb{R}$.

The next result describes the parts of the Lebesgue decomposition (B.8) of the measure μ with respect to the Lebesgue measure $\nu = m$ in terms of boundary values of the imaginary part of its Stieltjes transform I_μ.

Theorem F.6 (i) *The singular part* μ_{sign} *of* μ *is supported by the set*

$$S_\mu := \big\{ t \in \mathbb{R} : (\operatorname{Im} I_\mu)(t + i0) = +\infty \big\}.$$

(ii) *The absolutely continuous part* μ_{ac} *is given by* $d\mu_{\text{ac}}(t) = \pi^{-1}(\operatorname{Im} I_\mu)(t + i0)\,dt$ *and supported by the set*

$$L_\mu := \big\{ t \in \mathbb{R} : 0 < (\operatorname{Im} I_\mu)(t + i0) < +\infty \big\}.$$

Proof (i): The de la Vallée Poussin theorem (Theorem B.9(ii)) states that μ_{sign} is supported by the set $S(\mu) = \{t \in \mathbb{R} : (D\mu)(t) = +\infty\}$. From the inequalities

$$(\operatorname{Im} I_\mu)(t + i\varepsilon) = \int_{\mathbb{R}} \frac{\varepsilon}{(s-t)^2 + \varepsilon^2}\, d\mu(s) \geq \varepsilon^{-1} \int_{t-\varepsilon}^{t+\varepsilon} \frac{\varepsilon^2}{(s-t)^2 + \varepsilon^2}\, d\mu(s)$$

$$\geq \varepsilon^{-1} \int_{t-\varepsilon}^{t+\varepsilon} \frac{\varepsilon^2}{\varepsilon^2 + \varepsilon^2}\, d\mu(s) = (2\varepsilon)^{-1} \mu\big([t+\varepsilon, t-\varepsilon] \big)$$

we conclude that $(D\mu)(t) = +\infty$ implies $(\operatorname{Im} I_\mu)(t + i0) = +\infty$, so $S(\mu) \subseteq S_\mu$. This proves (i).

(ii): From formula (F.10) it follows that

$$(D\mu)(t) = \frac{d\mu}{dt}(t) = \pi^{-1} \operatorname{Im} I_\mu(t + i0) = \pi^{-1}(\operatorname{Im} I_\mu)(t + i0) \quad \text{a.e. on } \mathbb{R},$$

$$(\text{F.11})$$

so that $d\mu_{\text{ac}}(t) = \pi^{-1}(\operatorname{Im} I_\mu)(t + i0)\,dt$ by Theorem B.9(i). Hence, μ_{ac} is supported by the set $\tilde{L}_\mu := \{t : 0 < (\operatorname{Im} I_\mu)(t + i0)\}$. By (F.11) and Theorem B.9(i) we have $\pi^{-1} \operatorname{Im} I_\mu(\cdot + i0) = D\mu = f \in L^1(\mathbb{R})$. Therefore, $\mu_{\text{ac}}(S_\mu) = 0$. (This follows also from (i).) Hence, μ_{ac} is also supported by the set $L_\mu = \tilde{L}_\mu \backslash S_\mu$. $\qquad\square$

Theorem F.7 *Suppose that* f *is a Nevanlinna function such that* (F.9) *holds and*

$$\sup\big\{ |\operatorname{Im} f(z)| : z \in \mathbb{C}_+ \big\} < \infty. \tag{F.12}$$

Then the function $\operatorname{Im} f(t + i0)$ *is in* $L^1(\mathbb{R})$, *and we have*

$$f(z) = \pi^{-1} \int_{\mathbb{R}} \frac{\operatorname{Im} f(t + i0)}{t - z}\, dt, \quad z \in \mathbb{C}_+. \tag{F.13}$$

Proof Since we assumed that condition (F.9) holds, Theorem F.4 applies, and hence $f = I_\mu$ for some finite positive Borel measure μ.

We prove that (F.12) implies that μ is absolutely continuous and (F.13) holds. From (F.8) and (F.12) it follows that $\mu(\{c\}) = 0$ for all $c \in \mathbb{R}$. Therefore, by (F.7),

$$\mu\big((a,b) \big) = \lim_{\varepsilon \to +0} \pi^{-1} \int_a^b \operatorname{Im} I_\mu(t + i\varepsilon)\, dt = \pi^{-1} \int_a^b \operatorname{Im} I_\mu(t + i0)\, dt, \tag{F.14}$$

where assumption (F.12) ensures that limit and integration can be interchanged by Lebesgue's convergence Theorem B.1. From (F.14) we easily conclude that $d\mu(t) = \pi^{-1} \operatorname{Im} f(t + i0)\, dt$. Inserting this into (F.3), we obtain (F.13). $\qquad\square$

We close this appendix by stating another related but finer result. It was not used in this text. Let v be a positive regular Borel measure on $\mathbb{R}$ satisfying

$$\int_{\mathbb{R}} \frac{1}{1+t^2}\, dv(t) < \infty.$$

Then the *Poisson–Stieltjes transform* P_v is defined by

$$P_v(x, y) = \int_{\mathbb{R}} \frac{y}{(t-x)^2 + y^2}\, dv(t), \quad x \in \mathbb{R},\ y > 0.$$

Note that $P_v(x, y)$ is a nonnegative harmonic function on $\mathbb{C}_+$. If the measure v is finite, the Stieltjes transform I_v is defined, and obviously $P_v(x, y) = \operatorname{Im} I_v(x + iy)$. If v is the Lebesgue measure, one computes that $P_v(x, y) = \pi$ for $x + iy \in \mathbb{C}_+$.

Then there is the following *Fatou-type theorem*; a proof is given in [Dn3].

Theorem F.8 *Let $t \in \mathbb{R}$.*

(i) *The limit $P_v(t, +0) := \lim_{\varepsilon \to +0} P_v(t, \varepsilon)$ exists and is finite if and only if the symmetric derivative $(Dv)(t)$ (defined by (B.9)) does. In this case we have $P_v(t, +0) = \pi(Dv)(t)$.*

(ii) *If $(Dv)(t) = +\infty$, then $P_v(t, +0) \equiv \lim_{\varepsilon \to +0} P_v(t, \varepsilon) = +\infty$; the converse is not true in general.*

Suppose that the positive measure v is finite. Since $P_v(x, y) = \operatorname{Im} I_v(x + iy)$, it is obvious that $P_v(t, +0) = (\operatorname{Im} I_v)(t + i0)$. Hence, the assertion of Theorem F.6(ii) can be derived from Theorem F.8(i), thereby avoiding the use of the Sokhotski–Plemelj formula (F.10). Indeed, since $\pi(Dv)(t) = (\operatorname{Im} I_v)(t+i0)$ by Theorem F.8(i), we have $dv_{ac}(t) = \pi^{-1}(\operatorname{Im} I_v)(t + i0)\, dt$ by Theorem B.9(i). This implies the assertion of Theorem F.6(ii).

References

Classical Articles

[Ck] Calkin, J.W.: Abstract symmetric boundary conditions. Trans. Am. Math. Soc. **45**, 369–442 (1939)

[Cl] Carleman, T.: Sur les équations intégrales singulières à noyau réel symétrique. Almquist and Wiksells, Uppsala (1923)

[Cr] Courant, R.: Über die Eigenwerte bei den Differentialgleichungen der Mathematischen Physik. Math. Z. **7**, 1–57 (1920)

[Fi] Fischer, E.: Über quadratische Formen mit reellen Koefficienten. Monatshefte Math. Phys. **16**, 234–249 (1905)

[Fl] Friedrichs, K.O.: Spektraltheorie halbbeschränkter Operatoren und Anwendung auf die Spektralzerlegung von Differentialoperatoren. Math. Ann. **109**, 465–487 (1934)

[Hm] Hamburger, H.L.: Über eine Erweiterung des Stieltjesschen Momentenproblems. Math. Ann. **81**, 235–319 (1920) and **82**, 120–164, 168–187 (1920)

[HT] Hellinger, E., Toeplitz, O.: Integralgleichungen und Gleichungen mit unendlich vielen Unbekannten. Enzyklopädie d. Math. Wiss. II.C **13**, 1335–1616 (1928)

[Hi] Hilbert, D.: Grundzüge einer allgemeinen Theorie der linearen Integralgleichungen. Teubner-Verlag, Leipzig (1912)

[Hl] Hille, E.: Functional Analysis and Semigroups. Am. Math. Soc. Coll. Publ., vol. 38. Am. Math. Soc., New York (1948)

[K1] Kato, T.: Fundamental properties of Hamiltonians of Schrödinger type. Trans. Am. Math. Soc. **70**, 195–211 (1951)

[Kr1] Krein, M.G.: On Hermitian operators with defect numbers one. Dokl. Akad. Nauk SSSR **43**, 339–342 (1944)

[Kr2] Krein, M.G.: On resolvents of an Hermitian operator with defect index (m, m). Dokl. Akad. Nauk SSSR **52**, 657–660 (1946)

[Kr3] Krein, M.G.: The theory of self-adjoint extensions of semibounded Hermitian operators and its applications. Mat. Sb. **20**, 365–404 (1947)

[Lö] Löwner, K.: Über monotone Matrixfunktionen. Math. Z. **38**, 117–216 (1934)

[Na1] Naimark, M.A.: Spectral functions of a symmetric operator. Izv. Akad. Nauk SSSR, Ser. Mat. **7**, 285–296 (1943)

[Nv] Nevanlinna, R.: Asymptotische Entwicklungen beschränkter Funktionen und das Stieltjessche Momentenproblem. Ann. Acad. Sci. Fenn. A **18**, 1–53 (1922)

[Re] Rellich, F.: Störungstheorie der Spektralzerlegung II. Math. Ann. **116**, 555–570 (1939)

[Ri1] Riesz, F.: Les systèmes d'équations linéaires à une infinité d'inconnues. Gauthiers–Villars, Paris (1913)

K. Schmüdgen, *Unbounded Self-adjoint Operators on Hilbert Space*,
Graduate Texts in Mathematics 265,
DOI 10.1007/978-94-007-4753-1, © Springer Science+Business Media Dordrecht 2012

[Ri2] Riesz, F.: Über die linearen Transformationen des komplexen Hilbertschen Raumes. Acta Sci. Math. Szeged **5**, 23–54 (1930)

[Stj] Stieltjes, T.J.: Recherches sur les fractions continues. Ann. Fac. Sci. Toulouse **8**, 1–122 (1894)

[St1] Stone, M.H.: Linear transformations in Hilbert space. Proc. Natl. Acad. Sci. USA **16**, 172–175 (1930)

[St2] Stone, M.H.: Linear Transformations in Hilbert Space. Am. Math. Soc., New York (1932)

[St3] Stone, M.H.: On one-parameter unitary groups in Hilbert space. Ann. Math. **33**, 643–648 (1932)

[vN1] Von Neumann, J.: Allgemeine Eigenwerttheorie Hermitescher Funktionaloperatoren. Math. Ann. **102**, 49–131 (1929)

[vN2] Von Neumann, J.: Zur Theorie der unbeschränkten Matrizen. J. Reine Angew. Math. **161**, 208–236 (1929)

[vN3] Von Neumann, J.: Mathematische Grundlagen der Quantenmechanik. Berlin (1932)

[vN4] Von Neumann, J.: Über adjungierte Funktionaloperatoren. Ann. Math. **33**, 294–310 (1932)

[W1] Weyl, H.: Über beschränkte quadratische Formen, deren Differenz vollstetig ist. Rend. Circ. Mat. Palermo **27**, 373–392 (1909)

[W2] Weyl, H.: Über gewöhnliche Differentialgleichungen mit Singularitäten und die zugehörigen Entwicklungen willkürlicher Funktionen. Math. Ann. **68**, 220–269 (1910)

[W3] Weyl, H.: Das asymptotische Verteilungsgesetz der Eigenwerte linearer partieller Differentialgleichungen. Math. Ann. **71**, 441–469 (1912)

[Yo] Yosida, K.: On the differentiability and the representation of one-parameter semigroups of linear operators. J. Math. Soc. Jpn. **1**, 15–21 (1949)

Books

[Ag] Agmon, S.: Lectures on Elliptic Boundary Problems. Van Nostrand, Princeton (1965)

[Ac] Akhiezer, N.I.: The Classical Moment Problem. Oliver and Boyd, Edinburgh and London (1965)

[AG] Akhiezer, N.I., Glazman, I.M.: Theory of Linear Operators in Hilbert Space. Ungar, New York (1961)

[AGH] Albeverio, S., Gesztesy, S., Hoegh-Krohn, R., Holden, H.: Solvable Models in Quantum Mechanics. Am. Math. Soc., Providence (2005)

[At] Alt, H.W.: Lineare Funktionalanalysis, 5. Auflage. Springer-Verlag, Berlin (2006)

[AHP] Amrein, W.O., Hinz, A.M., Pearson, D.B. (eds.): Sturm–Liouville Theory. Past and Present. Birkhäuser-Verlag, Basel (2005)

[Ao] Ando, T.: Topics on Operator Inequalities, Sapporo (1978)

[Ap] Apostol, T.: Mathematical Analysis. Addison-Wesley, Reading (1960)

[Bd] Berard, P.: Lectures on Spectral Geometry. Lecture Notes Math., vol. 1207. Springer-Verlag, New York (1986)

[Bn] Berberian, S.K.: Notes on Spectral Theory. van Nostrand, Princeton (1966)

[BG] Berenstein, C.A., Gay, R.: Complex Variables. Springer-Verlag, New York (1991)

[Be] Berezansky, Y.M.: Expansions in Eigenfunctions of Selfadjoint Operators. Am. Math. Soc., Providence (1968)

[BEH] Blank, J., Exner, P., Havlicek, M.: Hilbert Space Operators in Quantum Physics. Springer-Verlag, Berlin (2010)

[BSU] Berezansky, Y.M., Sheftel, Z.G., Us, G.F.: Functional Analysis, vol. II. Birkhäuser-Verlag, Basel (1996)

[BS] Birman, M.S., Solomyak, M.Z.: Spectral Theory of Selfadjoint Operators in Hilbert Space. Kluwer, Dordrecht (1987)

[Br] Brezis, H.: Functional Analysis, Sobolev Spaces and Partial Differential Equations. Springer-Verlag, Berlin (2011)

[Cn] Cohn, D.L.: Measure Theory. Birkhäuser-Verlag, Boston (1980)

[Cw] Conway, J.B.: A Course in Functional Analysis. Springer-Verlag, New York (1990)

[CH] Courant, R., Hilbert, D.: Methods of Mathematical Physics. Interscience Publ., New York (1990)

[D1] Davies, E.B.: One-Parameter Semigroups. Academic Press, London (1980)

[D2] Davies, E.B.: Spectral Theory and Differential Operators. Cambridge University Press, Cambridge (1994)

[Dn1] Donoghue Jr., W.F.: Monotone Matrix Functions and Analytic Continuation. Springer-Verlag, Berlin (1974)

[DS] Dunford, N., Schwartz, J.T.: Linear Operators, Part II. Spectral Theory. Interscience Publ., New York (1963)

[EN] Engel, K.-J., Nagel, R.: One-Parameter Semigroups for Linear Evolution Equations. Springer-Verlag, Berlin (2000)

[EE] Edmunds, D.E., Evans, W.D.: Spectral Theory and Differential Operators. Clarendon Press, Oxford (1987)

[Fe] Federer, H.: Geometric Measure Theory. Springer-Verlag, New York (1969)

[GW] Gasquet, C., Witomski, P.: Fourier Analysis and Applications. Springer-Verlag, New York (1999)

[GGK] Gokhberg, I.C., Goldberg, S., Krupnik, N.: Traces and Determinants of Linear Operators. Birkhäuser-Verlag, Basel (2000)

[GK] Gokhberg, I.C., Krein, M.G.: Introduction to the Theory of Linear Nonselfadjoint Operators. Am. Math. Soc., Providence (1969)

[GG] Gorbachuk, V.I., Gorbachuk, M.L.: Boundary Value Problems for Operator Differential Equations. Kluwer, Dordrecht (1991)

[Gr] Grubb, G.: Distributions and Operators. Springer-Verlag, Berlin (2009)

[Ha] Halmos, P.R.: Measure Theory. Springer-Verlag, Berlin (1974)

[HS] Hewitt, E., Stromberg, K.: Real and Abstract Analysis. Springer-Verlag, Berlin (1965)

[HP] Hille, E., Phillips, R.S.: Functional Analysis and Semigroups. Am. Math. Soc. Coll. Publ., New York (1957)

[Hr] Hörmander, L.: The Analysis of Linear Partial Differential Operators I. Springer-Verlag, Berlin (1973)

[JR] Jörgens, K., Rellich, F.: Eigenwerttheorie Gewöhnlicher Differentialgleichungen. Springer-Verlag, Berlin (1976)

[K2] Kato, T.: Perturbation Theory for Linear Operators. Springer-Verlag, Berlin (1966)

[Kz] Katznelson, Y.: An Introduction to Harmonic Analysis. Dover, New York (1968)

[Kr6] Krein, M.G.: Topics in Differential and Integral Equations and Operator Theory, pp. 107–172. Birkhäuser-Verlag, Basel (1983)

[LU] Ladyzhenskaya, O.A., Uraltseva, N.N.: Linear and Quasilinear Elliptic Equations. Academic Press, New York (1968)

[LL] Lieb, E.H., Loss, M.: Analysis. Am. Math. Soc., Providence (1997)

[Ls] Lions, J.L.: Equations différentielles opérationnelles et problèmes aux limites. Springer-Verlag, Berlin (1961)

[LM] Lions, J.L., Magenes, E.: Non-Homogeneous Boundary Value Problems and Applications, vol. I. Springer-Verlag, Berlin (1972)

[Mi] Muskhelishvili, N.I.: Singular Integral Equations. Noordhoff, Groningen (1972)

[Na2] Naimark, M.A.: Linear Differential Operators. Ungar, New York (1968)

[Pa] Pazy, A.: Semigroups of Linear Operators and Applications to Partial Differential Equations. Springer-Verlag, Berlin (1983)

[Pv] Privalov, I.I.: Boundary Properties of Analytic Functions. GITTL, Moscow (1950). German translation, DVW, Berlin (1956)

[RA] Rade, L., Westergren, B.: Springers Mathematische Formeln. Springer-Verlag, Berlin (1991)

[RS1] Reed, M., Simon, B.: Methods of Modern Mathematical Physics I. Functional Analysis. Academic Press, New York (1972)

[RS2] Reed, M., Simon, B.: Methods of Modern Mathematical Physics II. Fourier Analysis and Self-Adjointness. Academic Press, New York (1975)

[RS3] Reed, M., Simon, B.: Methods of Modern Mathematical Physics III. Scattering Theory. Academic Press, New York (1979)

[RS4] Reed, M., Simon, B.: Methods of Modern Mathematical Physics IV. Analysis of Operators. Academic Press, New York (1978)

[RN]   Riesz, F., Nagy, Sz.-B.: Functional Analysis. Dover, New York (1990)

[Ru1] Rudin, W.: Real and Complex Analysis, 2nd edn. McGraw-Hill Inc., New York (1974)

[Ru2] Rudin, W.: Real and Complex Analysis, 3rd edn. McGraw-Hill Inc., New York (1987)

[Ru3] Rudin, W.: Functional Analysis. McGraw-Hill, New York (1973)

[Sch1] Schmüdgen, K.: Unbounded Operator Algebras and Representation Theory. Birkhäuser-Verlag, Basel (1990)

[ST]   Shohat, J.A., Tamarkin, J.D.: The Problem of Moments. Am. Math. Soc., Providence (1943)

[Sm1] Simon, B.: Trace Ideals and Their Applications. Cambridge University Press, Cambridge (1979)

[Tn]   Tanabe, H.: Equations of Evolution. Pitman, London (1979)

[Th]   Thirring, W.: A Course in Mathematical Physics, vol. 3. Springer-Verlag, New York (1981)

[Tt]   Titchmarsh, E.C.: Eigenfunction Expansions Associated with Second-Order Differential Equations. Part I. Clarendon Press, Oxford (1962). Part II (1970)

[Tl]   Triebel, H.: Interpolation Theory, Function Spaces, Differential Operators. North-Holland, Dordrecht (1978)

[Vl]   Vladimirov, V.S.: Equations of Mathematical Physics. Marcel Dekker, Inc., New York (1971)

[Wa]   Warner, G.: Harmonic Analysis on Semi-Simple Lie Groups. Springer-Verlag, Berlin (1972)

[We]   Weidmann, J.: Linear Operators in Hilbert Spaces. Springer-Verlag, Berlin (1987)

[Wr]   Wiener, N.: The Fourier Integral and Certain of Its Applications. Cambridge Univ. Press, New York (1933)

[Yf]   Yafaev, D.R.: Mathematical Scattering Theory: General Theory. Am. Math. Soc., Providence, RI (1992)

Articles

[AS]   Alonso, A., Simon, B.: The Birman–Krein–Vishik theory of selfadjoint extensions of semibounded operators. J. Oper. Theory **4**, 51–270 (1980)

[AN]   Ando, T., Nishio, K.: Positive selfadjoint extensions of positive symmetric operators. Tohoku Math. J. **22**, 65–75 (1970)

[ALM] Ando, T., Li, C.-K., Mathias, R.: Geometric means. Linear Algebra Appl. **385**, 305–334 (2004)

[As]   Arens, R.: Operational calculus of linear relations. Pac. J. Math. **11**, 9–23 (1961)

[AHS] Arlinski, Y.M., Hassi, S., Sebestyen, Z., de Snoo, H.S.V.: On the class of extremal extensions of a nonnegative operator. Oper. Theory Adv. Appl. **127**, 41–81 (2001)

[AT]   Arlinski, Y., Tsekanovskii, E.: The von Neumann problem for nonnegative symmetric operators. Integral Equ. Oper. Theory **51**, 319–356 (2005)

[Ar]   Aronszajn, N.: On a problem of Weyl in the theory of Sturm–Liouville equations. Am. J. Math. **79**, 597–610 (1957)

[AD]   Aronzajn, N., Donoghue, W.F.: On exponential representations of analytic functions in the upper half-plane with positive imaginary part. J. Anal. Math. **5**, 321–388 (1957)

[BE]   Bennewitz, C., Everitt, W.N.: The Weyl–Titchmarsh eigenfunction expansion theorem for Sturm–Liouville operators. In [AHP], pp. 137–172

[Bi]   Birman, M.S.: On the theory of selfadjoint extensions of positive operators. Mat. Sb. **38**, 431–450 (1956)

[BVS]   Birman, M.Sh., Vershik, A.M., Solomjak, M.Z.: The product of commuting spectral measures may fail to be countably additive. Funkc. Anal. Prilozh. **13**, 61–62 (1979)

[BK]   Birman, M.Sh., Krein, M.G.: On the theory of wave operators and scattering operators. Dokl. Akad. Nauk SSSR **144**, 475–478 (1962)

[BP]   Birman, M.Sh., Pushnitski, A.B.: Spectral shift function, amazing and multifaceted. Integral Equ. Oper. Theory **30**, 191–199 (1998)

[BY]   Birman, M.Sh., Yafaev, D.R.: The spectral shift function. The work of M.G. Krein and its further development. St. Petersburg Math. J. **4**, 833–870 (1993)

[BMN]   Brasche, J.F., Malamud, M.M., Neidhardt, N.: Weyl function and spectral properties of selfadjoint extensions. Integral Equ. Oper. Theory **43**, 264–289 (2002)

[BGW]   Brown, B.M., Grubb, G., Wood, I.: M-functions for closed extensions of adjoint pairs of operators with applications to elliptic boundary problems. Math. Nachr. **282**, 314–347 (2009)

[Bk]   Bruk, V.M.: On a class of boundary value problems with a spectral parameter in the boundary condition. Mat. Sb. **100**, 210–216 (1976)

[BGP]   Brüning, J., Geyler, V., Pankrashkin, K.: Spectra of selfadjoint extensions and applications to solvable Schrödinger operators. Rev. Math. Phys. **20**, 1–70 (2008)

[BC]   Buchwalther, H., Casier, G.: La paramétrisation de Nevanlinna dans le problème des moments de Hamburger. Expo. Math. **2**, 155–178 (1984)

[CSS]   Cimprič, J., Savchuk, Y., Schmüdgen, K.: On q-normal operators and the quantum complex plane. Trans. Am. Math. Soc. (to appear)

[Cd1]   Coddington, E.A.: Formally normal operators having no normal extensions. Can. J. Math. **17**, 1030–1040 (1965)

[Cd2]   Coddington, E.A.: Extension theory of formally normal and symmetric subspaces. Mem. Am. Math. Soc. **134** (1973)

[CdS]   Coddington, E.A., de Snoo, H.S.V.: Positive selfadjoint extensions of positive subspaces. Math. Z. **159**, 203–214 (1978)

[DM]   Derkach, V.A., Malamud, M.M.: Generalized resolvents and the boundary value problem for Hermitian operators with gaps. J. Funct. Anal. **95**, 1–95 (1991)

[DdS]   Dijksma, A., de Snoo, H.S.V.: Selfadjoint extensions of symmetric subspaces. Pac. J. Math. **54**, 71–100 (1974)

[Dn2]   Donoghue, W.F.: On the perturbation of spectra. Commun. Pure Appl. Math. **18**, 559–579 (1965)

[Dn3]   Donoghue, W.F.: A theorem of the Fatou type. Monatsh. Math. **67**, 225–228 (1963)

[Dr]   Dritschel, M.: On factorization of trigonometric polynomials. Integral Equ. Oper. Theory **49**, 11–42 (2004)

[F2]   Friedrichs, K.O.: Symmetric positive linear differential equations. Commun. Pure Appl. Math. **11**, 333–418 (1955)

[Fu]   Fuglede, B.: A commutativity theorem for normal operators. Proc. Nat. Acad. Sci. **36**, 35–40 (1950)

[GT]   Gesztesy, F., Tsekanosvkii, E.: On matrix-valued Herglotz functions. Math. Nachr. **218**, 61–138 (2000)

[GMN]   Gesztesy, F., Makarov, K.A., Naboko, S.N.: The spectral shift operator. Oper. Theory Adv. Appl. **108**, 59–99 (1999)

[GrW]   Grampp, U., Wähnert, Ph.: Das Hamburger'sche Momentenproblem und die Nevanlinna-Parametrisierung. Diplomarbeit, Leipzig (2009)

[Hs]   Harish-Chandra: Representations of a semi-simple Lie group on a Banach space I. Trans. Am. Math. Soc. **75**, 185–243 (1953)

[HMS] Hassi, S., Malamud, M., de Snoo, H.S.V.: On Krein's extension theory of nonnegative operators. Math. Nachr. **274**, 40–73 (2004)

[Hz]   Heinz, E.: Beiträge zur Störungstheorie der Spektralzerlegung. Math. Ann. **123**, 415–438 (1951)

[Kc]   Kac, M.: Can you hear the shape of a drum? Am. Math. Monthly **73**, 1–23 (1966)

[K3]   Kato, T.: Quadratic forms in Hilbert spaces and asymptotic perturbations series. Techn. Report No. 9, University of California (1955)

[Ka]   Kaufman, W.F.: Representing a closed operator as a quotient of continuous operators. Proc. Am. Math. Soc. **72**, 531–534 (1978)

[Ko]   Kochubei, A.N.: Extensions of symmetric operators. Math. Notes **17**, 25–28 (1975)

[Kd1] Kodaira, K.: Eigenvalue problems for ordinary differential equations of the second order and Heisenberg's theory of S-matrices. Am. J. Math. **71**, 921–945 (1950)

[Kd2] Kodaira, K.: On ordinary differential equations of any even order and the corresponding eigenfunction expansions. Am. J. Math. **72**, 502–544 (1950)

[Kp]   Koplienko, L.S.: On the trace formula for perturbations of non-trace class type. Sib. Mat. Zh. **25**, 72–77 (1984)

[Kr4] Krein, M.G.: On the trace formula in perturbation theory. Mat. Sb. **33**, 597–626 (1953)

[Kr5] Krein, M.G.: On perturbation determinants and the trace formula for unitary and self-adjoint operators. Dokl. Akad. Nauk SSSR **144**, 268–273 (1962)

[LaM] Lax, P.D., Milgram, A.N.: Parabolic equations. In: Contributions to the Theory of Partial Differential Equations. Ann. Math. Stud., vol. 33, pp. 167–190. Princeton (1954)

[Ld]   Lidskii, V.B.: Non-selfadjoint operators with a trace. Dokl. Akad. Nauk **125**, 485–487 (1959)

[Lf]   Lifshits, I.M.: On a problem in perturbation theory. Usp. Mat. Nauk **7**, 171–180 (1952)

[Lv]   Livsic, M.S.: On the spectral resolution of linear non-selfadjoint operators. Mat. Sb. **34**, 145–179 (1954)

[Ma]   Malamud, M.M.: Certain classes of extensions of a lacunary Hermitian operator. Ukr. Mat. Zh. **44**, 215–233 (1992)

[MN]   Malamud, M.M., Neidhardt, H.: On the unitary equivalence of absolutely continuous parts of self-adjoint extensions. J. Funct. Anal. **260**, 613–638 (2011)

[Mc]   McCarthy, C.A.: c_p. Isr. J. Math. **5**, 249–271 (1967)

[Ne1] Nelson, E.: Analytic vectors. Ann. Math. **70**, 572–614 (1959)

[Ne2] Nelson, E.: Interactions of nonrelativistic particles with a quantized scalar field. J. Math. Phys. **5**, 1190–1197 (1964)

[Nu]   Nussbaum, A.E.: Quasi-analytic vectors. Ark. Math. **6**, 179–191 (1965)

[Ph]   Phillips, R.S.: Dissipative hyperbolic systems. Trans. Am. Math. Soc. **86**, 109–173 (1957)

[Po]   Posilicano, A.: A Krein-type formula for singular perturbations of self-adjoint operators and applications. J. Funct. Anal. **183**, 109–147 (2001)

[PS]   Prokaj, V., Sebestyen, Z.: On extremal positive extensions. Acta Sci. Math. Szeged **62**, 485–491 (1996)

[Rf]   Rofe–Beketov, F.S.: Self-adjoint extensions of differential operators in a space of vector-valued functions. Dokl. Akad. Nauk **184**, 1034–1037 (1959)

[Ro]   Rosenblum, M.: On a theorem of Fuglede and Putnam. J. Lond. Math. Soc. **33**, 376–377 (1958)

[Sa]   Saakyan, S.N.: Theory of resolvents of symmetric operators with infinite deficiency indices. Dokl. Akad. Nauk Arm. SSR **41**, 193–198 (1965)

[Sch2] Schmüdgen, K.: On domains of powers of closed symmetric operators. J. Oper. Theory **9**, 53–75 (1983)

[Sch3] Schmüdgen, K.: On commuting unbounded selfadjoint operators. Acta Sci. Math. (Szeged) **47**, 131–146 (1984)

[Sch4] Schmüdgen, K.: A formally normal operator having no normal extension. Proc. Am. Math. Soc. **98**, 503–504 (1985)

[SeS] Sebestyen, Z., Stochel, J.: Restrictions of positive selfadjoint operators. Acta Sci. Math. (Szeged) **55**, 149–154 (1991)

[Sm2] Simon, B.: Notes on infinite determinants. Adv. Math. **24**, 244–273 (1977)

[Sm3] Simon, B.: Spectral analysis of rank one perturbations and applications. CRM Lecture Notes, vol. 8, pp. 109–149. Am. Math. Soc., Providence (1995).

[Sm4] Simon, B.: The classical moment problem as a self-adjoint finite difference operator. Adv. Math. **137**, 82–203 (1998)

[SW] Simon, B., Wolff, T.: Singular continuous spectrum under rank one perturbations and localization for random Hamiltonians. Commun. Pure Appl. Math. **39**, 75–90 (1986)

[Sn] Steen, L.A.: Highlights in the history of spectral theory. Am. Math. Monthly **80**, 359–381 (1973)

[SS1] Stochel, J., Szafraniec, F.H.: On normal extensions of unbounded operators. I. J. Oper. Theory **14**, 31–45 (1985)

[SS2] Stochel, J., Szafraniec, F.H.: On normal extensions of unbounded operators. II. Acta Sci. Math. (Szeged) **53**, 153–177 (1989)

[Str] Strichartz, R.: Multipliers on fractional Sobolev spaces. J. Math. Mech. **16**, 1031–1060 (1967)

[Ti] Tillmann, H.G.: Zur Eindeutigkeit der Lösungen der quantenmechanischen Vertauschungsrelationen. Acta Sci. Math. (Szeged) **24**, 258–270 (1963)

[Tr] Trotter, H.: On the product of semigroups of operators. Proc. Am. Math. Soc. **10**, 545–551 (1959)

[Vi] Vishik, M.: On general boundary conditions for elliptic differential equations. Tr. Moskv. Mat. Obc. **1**, 187–246 (1952)

[W4] Weyl, H.: Ramifications, old and new, of the eigenvalue problem. Bull. Am. Math. Soc. **56**, 115–139 (1950)

[Wo] Woronowicz, S.L.: Unbounded elements affiliated with C^*-algebras and non-compact quantum groups. Commun. Math. Phys. **136**, 399–432 (1991)

[WN] Woronowicz, S.L., Napiorkowski, K.: Operator theory in the C^*-algebra framework. Rep. Math. Phys. **31**, 353–371 (1992)

[Wt] Wüst, R.: Generalizations of Rellich's theorem on perturbations of (essentially) self-adjoint operators. Math. Z. **119**, 276–280 (1971)

Subject Index

K. Schmüdgen, *Unbounded Self-adjoint Operators on Hilbert Space*,
Graduate Texts in Mathematics 265,
DOI 10.1007/978-94-007-4753-1, © Springer Science+Business Media Dordrecht 2012

Symbol Index

K. Schmüdgen, *Unbounded Self-adjoint Operators on Hilbert Space*,
Graduate Texts in Mathematics 265,
DOI 10.1007/978-94-007-4753-1, © Springer Science+Business Media Dordrecht 2012